AF784623

DISCARDED MATERIAL
FROM UVU LIBRARY

Geophysical Monograph Series

Including

IUGG Volumes
Maurice Ewing Volumes
Mineral Physics Volumes

92 **Mauna Loa Revealed: Structure, Composition, History, and Hazards** *J. M. Rhodes and John P. Lockwood (Eds.)*

93 **Cross-Scale Coupling in Space Plasmas** *James L. Horowitz, Nagendra Singh, and James L. Burch (Eds.)*

94 **Double-Diffusive Convection** *Alan Brandt and H. J. S. Fernando (Eds.)*

95 **Earth Processes: Reading the Isotopic Code** *Asish Basu and Stan Hart (Eds.)*

96 **Subduction Top to Bottom** *Gray E. Bebout, David Scholl, Stephen Kirby, and John Platt (Eds.)*

97 **Radiation Belts: Models and Standards** *J. F. Lemaire, D. Heynderickx, and D. N. Baker (Eds.)*

98 **Magnetic Storms** Bruce *T. Tsurutani, Walter D. Gonzalez, Yohsuke Kamide, and John K. Arballo (Eds.)*

99 **Coronal Mass Ejections** *Nancy Crooker, Jo Ann Joselyn, and Joan Feynman (Eds.)*

100 **Large Igneous Provinces** *John J. Mahoney and Millard F. Coffin (Eds.)*

101 **Properties of Earth and Planetary Materials at High Pressure and Temperature** *Murli Manghnani and Takehiki Yagi (Eds.)*

102 **Measurement Techniques in Space Plasmas: Particles** *Robert F. Pfaff, Joseph E. Borovsky, and David T. Young (Eds.)*

103 **Measurement Techniques in Space Plasmas: Fields** *Robert F. Pfaff, Joseph E. Borovsky, and David T. Young (Eds.)*

104 **Geospace Mass and Energy Flow: Results From the International Solar-Terrestrial Physics Program** *James L. Horwitz, Dennis L. Gallagher, and William K. Peterson (Eds.)*

105 **New Perspectives on the Earth's Magnetotail** *A. Nishida, D. N. Baker, and S. W. H. Cowley (Eds.)*

106 **Faulting and Magmatism at Mid-Ocean Ridges** *W. Roger Buck, Paul T. Delaney, Jeffrey A. Karson, and Yves Lagabrielle (Eds.)*

107 **Rivers Over Rock: Fluvial Processes in Bedrock Channels** *Keith J. Tinkler and Ellen E. Wohl (Eds.)*

108 **Assessment of Non-Point Source Pollution in the Vadose Zone** *Dennis L. Corwin, Keith Loague, and Timothy R. Ellsworth (Eds.)*

109 **Sun-Earth Plasma Interactions** *J. L. Burch, R. L. Carovillano, and S. K. Antiochos (Eds.)*

110 **The Controlled Flood in Grand Canyon** *Robert H. Webb, John C. Schmidt, G. Richard Marzolf, and Richard A. Valdez (Eds.)*

111 **Magnetic Helicity in Space and Laboratory Plasmas** *Michael R. Brown, Richard C. Canfield, and Alexei A. Pevtsov (Eds.)*

112 **Mechanisms of Global Climate Change at Millennial Time Scales** *Peter U. Clark, Robert S. Webb, and Lloyd D. Keigwin (Eds.)*

113 **Faults and Subsurface Fluid Flow in the Shallow Crust** *William C. Haneberg, Peter S. Mozley, J. Casey Moore, and Laurel B. Goodwin (Eds.)*

114 **Inverse Methods in Global Biogeochemical Cycles** *Prasad Kasibhatla, Martin Heimann, Peter Rayner, Natalie Mahowald, Ronald G. Prinn, and Dana E. Hartley (Eds.)*

115 **Atlantic Rifts and Continental Margins** *Webster Mohriak and Manik Talwani (Eds.)*

116 **Remote Sensing of Active Volcanism** *Peter J. Mouginis-Mark, Joy A. Crisp, and Jonathan H. Fink (Eds.)*

117 **Earth's Deep Interior: Mineral Physics and Tomography From the Atomic to the Global Scale** *Shun-ichiro Karato, Alessandro Forte, Robert Liebermann, Guy Masters, Lars Stixrude (Eds.)*

118 **Magnetospheric Current Systems** *Shin-ichi Ohtani, Ryoichi Fujii, Michael Hesse, and Robert L. Lysak (Eds.)*

119 **Radio Astronomy at Long Wavelengths** *Robert G. Stone, Kurt W. Weiler, Melvyn L. Goldstein, and Jean-Louis Bougeret (Eds.)*

120 **GeoComplexity and the Physics of Earthquakes** *John B. Rundle, Donald L. Turcotte, and William Klein (Eds.)*

121 **The History and Dynamics of Global Plate Motions** *Mark A. Richards, Richard G. Gordon, Rob D. van der Hilst (Eds.)*

122 **Dynamics of Fluids in Fractured Rock** *Boris Faybishenko, Paul A. Witherspoon, and Sally M. Benson (Eds.)*

123 **Atmospheric Science Across the Stratopause** *David E. Siskind, Stephen D. Eckerman, and Michael E. Summers (Eds.)*

124 **Natural Gas Hydrates: Occurrence, Distribution, and Detection** *Charles K. Paull and Willam P. Dillon (Eds.)*

125 **Space Weather** *Paul Song, Howard J. Singer, George L. Siscoe (Eds.)*

126 **The Oceans and Rapid Climate Change: Past, Present, and Future** *Dan Seidov, Bernd J. Haupt, and Mark Maslin (Eds.)*

127 **Gas Transfer at Water Surfaces** *M. A. Donelan, W. M. Drennan, E. S. Saltzman, and R. Wanninkhof (Eds.)*

Geophysical Monograph 128

Hawaiian Volcanoes
Deep Underwater Perspectives

Eiichi Takahashi
Peter W. Lipman
Michael O. Garcia
Jiro Naka
Shigeo Aramaki
Editors

American Geophysical Union
Washington, DC

Published under the aegis of the AGU Books Board

John E. Costa, Chair; David Bercovici, Andrew Dessler, Jeffrey M. Forbes, W. Rockwell Geyer, Rebecca Lange, Douglas S. Luther, Darrell Strobel, and R. Eugene Turner, members.

Library of Congress Cataloging-in-Publication Data
Hawaiian Volcanoes: deep underwater perspectives / Eiichi Takahashi...[et al.], editors.
p. cm.--(Geophysical monograph ; 128)
Includes bibliographical references.
ISBN 0-87590-987-6
1. Volcanoes--Hawaii. 2. Volcanism--Hawaii. I. Takahashi, E. (Eiichi), 1951-. II. Series.
QE524.2.U6 H39 2002
551.21'909969--dc21

2001053840

ISBN 0-87590-987-6
ISSN 0065-8448

Copyright 2002 by the American Geophysical Union
2000 Florida Avenue, N.W.
Washington, DC 20009

Cover: Kilauea coast. Photograph by Peter Lipman. Back cover 3D ocean floor image by J. Smith.

Figures, tables, and short excerpts may be reprinted in scientific books and journals if the source is properly cited.

Authorization to photocopy items for internal or personal use, or the internal or personal use of specific clients, is granted by the American Geophysical Union for libraries and other users registered with the Copyright Clearance Center (CCC) Transactional Reporting Service, provided that the base fee of $1.50 per copy plus $0.35 per page is paid directly to CCC, 222 Rosewood Dr., Danvers, MA 01923. 0065-8448/02/$01.50+0.35.

This consent does not extend to other kinds of copying, such as copying for creating new collective works or for resale. The reproduction of multiple copies and the use of full articles or the use of extracts, including figures and tables, for commercial purposes requires permission from the American Geophysical Union.

Printed in the United States of America.

CONTENTS

CONTENTS

PREFACE

In the summer of 1963, when a group of Japanese scientists arrived at the aged building of the Hawaiian Volcano Observatory, run by the U. S. Geological Survey, there began a program of cooperation and friendship between American and Japanese volcanologists that continues to the present. The late Professor Takeshi Minakami, a top volcano-physicist long involved in research at various volcanoes, including Asama, then the most active volcano in Japan, led the Japanese group. The time coincided with a changeover in Hawaii, from the pioneering volcanologic studies of Harold Stearns, Gordon Macdonald, and Jerry Eaton to more comprehensive research by younger volcanologists. The Japanese team was also fortunate enough to witness a rift zone eruption at Kilauea volcano (Alae eruption, July 1963), a direct, eye-opening encounter with a volcano plumbing system in action.

The time was also ripe for the field application of modern volcanological tools. We were especially fortunate to encounter a period of frequent eruption at Kilauea, which has since had continuing activity both at the summit and on the rift zones. During the last 40 years, Japanese volcanologists have thus become well acquainted with the research results of our American friends and their concepts of how Hawaiian volcanoes work despite the challenge posed by the great differences in structure and eruptive activity between Japanese and Hawaiian volcanoes. Nonetheless, from geophysical and paleomagnetic points of view, the Hawaii islands and the surrounding region is most likely underlain by a thick layer of electrically high conductive material, which in turn bears evidence that the volcanic activity in this region may be a manifestation of deep seated upwelling of hot mantle material, i.e., a plume.

The great story of how the Hawaiian shields grew from the ocean floor to >4 km above sea level now appears in every textbook of volcanology, a story based mainly on painstaking research on the subaerial part of the shield. Because of the difficulty of observation, the subaqueous part of the story is somewhat weaker. Up to that time, submarine research, involving a combination of bathymetric mapping, dredging, and submersible dives, had been done mainly by various American research groups, with most dives restricted to depths shallower than 2000 m. As a result, when Shigeo Aramaki met with Jim Moore of USGS in 1985, we concluded that the newly developed Japanese submersible, Shinkai 6500, should be used to further promote the study of the Hawaiian volcanoes at greater depths. Naturally, many other scientists, including Pete Lipman and Alex Malahoff, reached the same conclusion. A basic concept for implementing the present program was actually formed when Mike Garcia and Jim Kinoshita met in Hawaii in 1997. Backed by the enthusiasm of these scientists, and thanks to the Japan Marine Science and Technology Center (JAMSTEC) for operating research submersibles outside Japanese waters, we implemented our vision for further deep-sea exploration around Hawaii.

During a four week cruise in 1998, the ROV Kaiko supported by its mother ship RV Kairei, made 10 dives at depths to 5,200 m for direct sampling and video observations, supplemented by dredging, piston cores, and SeaBeam bathymetric surveys. During a seven-week cruise in August-September 1999, the Shinkai 6500 submersible made 29 dives from the RV Yokosuka for sampling and direct observations as deep as 5,560 m, and further SeaBeam surveys were obtained. Participating scientists were largely from JAMSTEC, the University of Hawaii, several universities in Japan, the U.S. Geological Survey, and the Geological Survey of Japan.

The overall coverage of the scientific results gained from this two-year program is presented in this volume as initial documentation and discussion. The most important results include four aspects: 1) A series of detailed submarine bathymetric maps were made around the Hawaiian Islands. 2) Petrological characteristics of the Hilina slump off Kilauea volcano were clarified, and the existence of abundant alkaline basalt that is inferred to correspond to the "Loihi stage" of Kilauea became evident. 3) Topography and geology of the giant Nuuanu and Wailau slides were documented, the geometry of the original volcanic edifices consisting largely of volcaniclastic materials (hyaloclastite) was reconstructed, and the tsunamis which would have been triggered by the collapses were modeled. 4) The growth history of the restored Koolau volcano of Oahu island was found to include a multistage evolution of magmas, changing from the Kea to Loa types, then to distinctive tholeiitic rocks with high SiO_2 contents at the last stage.

Readers of this volume will immediately recognize that these initial scientific results will modify ideas about Hawaiian volcanoes now and in the future. We anticipate additional studies and reports will follow this publication. As participants in this bilateral cooperative program, we thank all our colleagues, American and Japanese, for their friendly partnership and mutual stimulation. We are especially gratified to be able to introduce younger Japanese researchers into this international cooperative program and expose them to new ideas and experiences. We are also certain that, among the older generation, our long-standing warm friendships will continue and deepen. Both 98 and 99 JAMSTEC Hawaii

Cruises were carried out under the Chief Scientist Jiro Naka. The strenuous work of scientifically managing the whole program and editing this publication with much zeal was done by Eiichi Takahashi. Special appreciation by all the members of this project is due to both of them.

This book also contains several papers highlighting scientific results obtained by previous cruises around Hawaii (not by JAMSTEC) and some geophysical observations related to the Hawaiian hot spot. We thank those contributors who have made this book a truly outstanding research summary on the evolution of Hawaiian volcanoes.

Shigeo Aramaki
Professor Emeritus, University of Tokyo

Hajimu Kinoshita
Executive Director, JAMSTEC)

Introduction to Section 1
Progress in Deep Underwater Geology around Hawaii

Jiro Naka

Japan Marine Science and Technology Center

As many as 2000 seamounts in the ocean basins have elevations of 1000 m or more. The largest of these seamounts rise above sea level and become oceanic islands. Most oceanic island rocks are products of volcanism. From base to surface, such islands generally consist of submarine pillow lava in deep water, volcaniclastic rocks in shallow water, and lava flows on land. The major portions of these materials exist under the sea level, and research on submarine portions of volcanic islands and seamounts is important.

The Hawaii Islands, as type examples of intraoceanic volcanic islands, illustrate every stage of their growth cycle, and many of them are active. Moreover, the subaerial volcanoes of these islands are well-studied and record over 70 million years of geological history. As a result, they are advantageous for submarine study when compared with other Pacific volcanic islands. In the 1980's, the U.S. Geological Survey carried out GLORIA side-scan sonar surveys of the entire U.S. Exclusive Economic Zone (EEZ) around the Hawaiian Islands. These surveys documented significant geological features, such as widespread submarine landslide and/or debris-avalanche deposits on and around the flanking the islands, and extensive lava fields on the Hawaiian arch. Detailed bathymetry compiled for the southern flank of Hawaii Island also provided new interpretations for the submarine slopes.

A series of Japan-U.S. cooperative research cruises, using research vessels of the Japan Marine Science and Technology Center (JAMSTEC), were carried out in 1998-1999 around the islands of Hawaii and Oahu. As reported by *Smith and Satake* [this section, maps in CD], the detailed bathymetry obtained from these cruises covers a wider area then any preexisting maps. These maps provide a framework not only for siting submersible dives but also for interpretive studies that are discussed in this volume, especially in section 3. *Smith et al.* [this section] describes bathymetric features on the northeast flank of the Hawaii Island and the east rift zone of Haleakala volcano, Maui. Based on the new bathymetry, they compare the slope structures on the northeast flank of Hawaii to those of the Hilina slump on the island's southeast flank, as interpreted in previous studies and new results included in this volume.

In contrast, the North Arch lava field, also initially identified during the USGS GLORIA surveys and studied petrologically with limited dredge samples, was examined in detail and sampled by submersible dives for the first time during the 1998-99 cruises. These cruises also made bathymetric surveys for about one-third of the field. *Clague et al.* [this section] used these data to describe the topographic features of some representative North Arch volcanic edifices and discuss their eruption style.

Umino et al. [this section] examined submarine lava flow morphology and estimated deep-water supply rates for basaltic lava using direct observations from KAIKO and SHINKAI 6500 dives of flows on Loihi and north of Oahu. Their study also makes detailed comparisons with high-resolution studies of subaerial lava flows.

The smooth sea-floor slopes surrounding large oceanic islands for several hundred kilometers have been described as archpelagic aprons. These aprons result from deposition of volcaniclastic materials derived from the islands. Previous DSDP/ODP drilling confirmed that thick volcaniclastic turbidities extend as much as 400 km from the source islands. Turbidites collected during ODP 136 are interpreted to have been deposited from Hawaiian landslides. The maximum volume of archpelagic clastic materials derived from a volcanic island can almost equal the volume of the island. Thus, volcaniclastic materials in the archipelagic

Hawaiian Volcanoes: Deep Underwater Perspectives
Geophysical Monograph 128
Copyright 2002 by the American Geophysical Union

apron contain a significant stratigraphic record of the growth of such islands.

Kanamatsu et al. [this section] used magneto-stratigraphy of archpelagic sediments collected south of Hawaii and northeast of Oahu to interpret ages of turbidities that may be related to the giant landslides such as Nuuanu and Wailau. *Naka et al.* [this section] shows that the volcaniclastic sediments in piston cores from the archipelagic apron south of Hawaii record turbidite events for at least the last 1 Ma. The shallow sediments are inferred to record eruptive events and landslides from Hawaii; deeper and older turbidites must have come from volcanoes on older Hawaiian islands prior to the growth of the island of Hawaii.

Deepwater Multibeam Sonar Surveys Along the Southeastern Hawaiian Ridge: Guide to the CD-ROM

John R. Smith

School of Ocean and Earth Science and Technology, University of Hawaii, Honolulu, Hawaii

Kenji Satake

Active Fault Research Center, AIST-GSJ, Tsukuba, Japan

Kiyoshi Suyehiro

Deep Sea Research Department, JAMSTEC, Yokosuka, Japan

Approximately 115,000-km^2 was surveyed with a multibeam seafloor mapping sonar system in the southeastern Hawaiian Islands during a collaborative research program between scientists from Japan and U.S. institutions that was funded by the Japan Marine Science and Technology Center (JAMSTEC). The primary focus of the cruises was to carry out manned and robotic submersible diving programs on the deep underwater flanks and rift zones of several Hawaiian Islands, the giant submarine landslide deposits along the Hawaiian Ridge, and the expansive lava flows located on the Hawaiian Arch. Details of the mapping program are provided here to accompany the digital data on the CD-ROM that is included with this monograph. In addition, other authors of manuscripts in this monograph have contributed their extended datasets, which can also be found on the archive.

1. INTRODUCTION

The two cruises to Hawaii sponsored by the Japan Marine Science and Technology Center (JAMSTEC) in 1998 and 1999 included significant seafloor mapping [*Naka et al.*, 2000], because only a few deepwater bathymetric surveys around the island of Hawaii had adequate resolution. High quality bathymetric data were essential for program objectives including basic submersible vehicle operations, dive and sample location, and geologic mapping.

Hawaiian Volcanoes: Deep Underwater Perspectives
Geophysical Monograph 128
Copyright 2002 by the American Geophysical Union

Because identical *SeaBeam 2112* sonar mapping systems are mounted on duplicate hulls of the sister ships *Kairei* and *Yokosuka*, merging their data is relatively simple and seamless. The *SeaBeam* system was run nearly every night for 12-14 hours during all cruise legs. Several full days were dedicated to *SeaBeam* surveying during the 1998 site survey cruise and on submersible or remotely operated vehicle (ROV) maintenance days.

Products from the seafloor mapping include standard contour maps, shaded relief bathymetry, beam amplitude, and acoustic backscatter imagery. Bathymetry data portrays seafloor depth, while the backscatter data provides textural information that distinguishes bare rock, sedimented areas, small blocks, and structural lineations. Acoustic backscatter is analogous to sidescan imagery acquired with towed deepwater systems such as GLORIA [*Laughton*, 1981], SeaMARC II [*Blackinton et al.*, 1983], and HAWAII MR1 [*Rognstad,* 1992].

2. PREVIOUS MULTIBEAM WORK

The only large areas of multibeam coverage prior to the JAMSTEC surveys were west, south, and east of the island of Hawaii. U.S. governmental agencies collected these data using early *SeaBeam* systems with a narrower swath width [*Chadwick et al.*, 1993a,b, 1994a,b; *Clague et al.*, 1994; *Smith*, 1994; *Moore and Chadwick*, 1995]. The data, at reduced resolution to match the surrounding data, are included in a summary map and digital database of the southeastern Hawaiian Islands that is upgraded as additional data become available [*Duennebier et al.*, 1994]. A Krupp Atlas *Hydrosweep* multibeam system was used to partially map the Nuuanu debris avalanche deposit off northeastern Oahu in 1998 during a deep seismic reflection survey [*Moore et al.*, 1998]. Another *Hydrosweep* survey covers the Hilo Ridge on the northeastern flank of Hawaii island [*Holcomb et al.*, 2000]. High-resolution bathymetry and acoustic backscatter surveys were acquired in 1998, using a Kongsberg *Simrad EM 300* multibeam system [*Dartnell and Gardner*, 1999; *MBARI*, 2000], in intermediate water depths (roughly from 70 to 3000 m) for select areas from Niihau to Hawaii [*Gardner and Hughes-Clarke*, 1998; *Torresan and Gardner,* 2000; *Clague et al.* 2000].

3. *SEABEAM 2112* SYSTEM DESCRIPTION

SeaBeam 2112 is a multibeam survey system built by L-3 Communications Sea Beam Instruments, Inc. for producing wide-swath contour maps and acoustic backscatter images of the seafloor. Two main subsystems compose the hardware — a 12 kHz transmitter and a receiver. The sonar beams, with a narrow 2° beam angle fore/aft, are projected as a swath and travel through the water column to the sea floor and are reflected off the bottom.

An array of hydrophones mounted across the bottom of the ship receives the reflected sonar signals. This receiver array detects and processes the returning echoes through multiple stabilized narrow athwartship beams (2° x 2°) in a fan shape. The swath width varies from 120° to 90°, decreasing with increasing water depth. The system electronics process the signals, and based on the travel time of the received signals and signal intensity, calculates the bottom depth and other characteristics for echoes received across the swath. The bathymetry data represents a maximum of 120 data points per sonar ping, while the acoustic backscatter data contains up to 2000 pixels per ping. Swath location is based on navigation from the Global Positioning System satellites, using a differential GPS system, and ship motion input from an onboard vertical reference unit.

The horizontal resolution of the bathymetry data depends on depth and ship speed. The accuracy of the depth measurement is specified as 0.5% of the depth or better. One important parameter for accuracy of depth and position of the bathymetric survey is the sound velocity profile. Data quality also depends greatly on the sea state. Favorable sea conditions during the typical reduced trade wind period of August and September, during which time the cruises were scheduled, allowed for surveying at speeds from 10 to 16 knots, though at times ship speed was reduced on some northeast transects headed into the seas. The typical swath width for this system around the Hawaiian Ridge (3000-5000 m water depth) is 10 km, or 2-3 times the depth.

Post-processing includes editing of the cross track and navigation data (deletion of bad data, correction of position, etc.), making grid files, and various maps. The software used for post-processing includes MB-System [*Caress and Chayes*, 1996] and the GMT-System [*Wessel and Smith*, 1995]. Bathymetry processing for all survey lines was performed onboard using the MB-System interactive program *mbedit*. Amplitude and acoustic backscatter data processing were carried out in batch mode shoreside after the cruise using other MB-System programs and the recommended processing sequence outlined in the MB-System manual and summarized on the CD-ROM. After all processing is completed, the remaining points are gridded or mosaicked and various maps are produced. The grid cell interval is 150 m for the bathymetric grids and 75 m for backscatter grids.

4. SURVEY AREAS

4.1 North Arch Lava Flows

In 1999, the R/V *Yokosuka* mapped ~15,000 km^2 of the North Arch lava flows for the first time synoptically with a multibeam system (Fig. 1). Approximately two-thirds of the flows were surveyed, based on previous wide-swath GLORIA sidescan sonar surveys covering the entire region [*Clague et al.*, 1990]. The high reflectivity measured with the acoustic backscatter component of the *SeaBeam 2112* revealed the low-relief lava flows much better than the bathymetry. However, the bathymetric component was key in identifying the detail of numerous lava vents distributed throughout the flow field [*Clague et al.,* this volume].

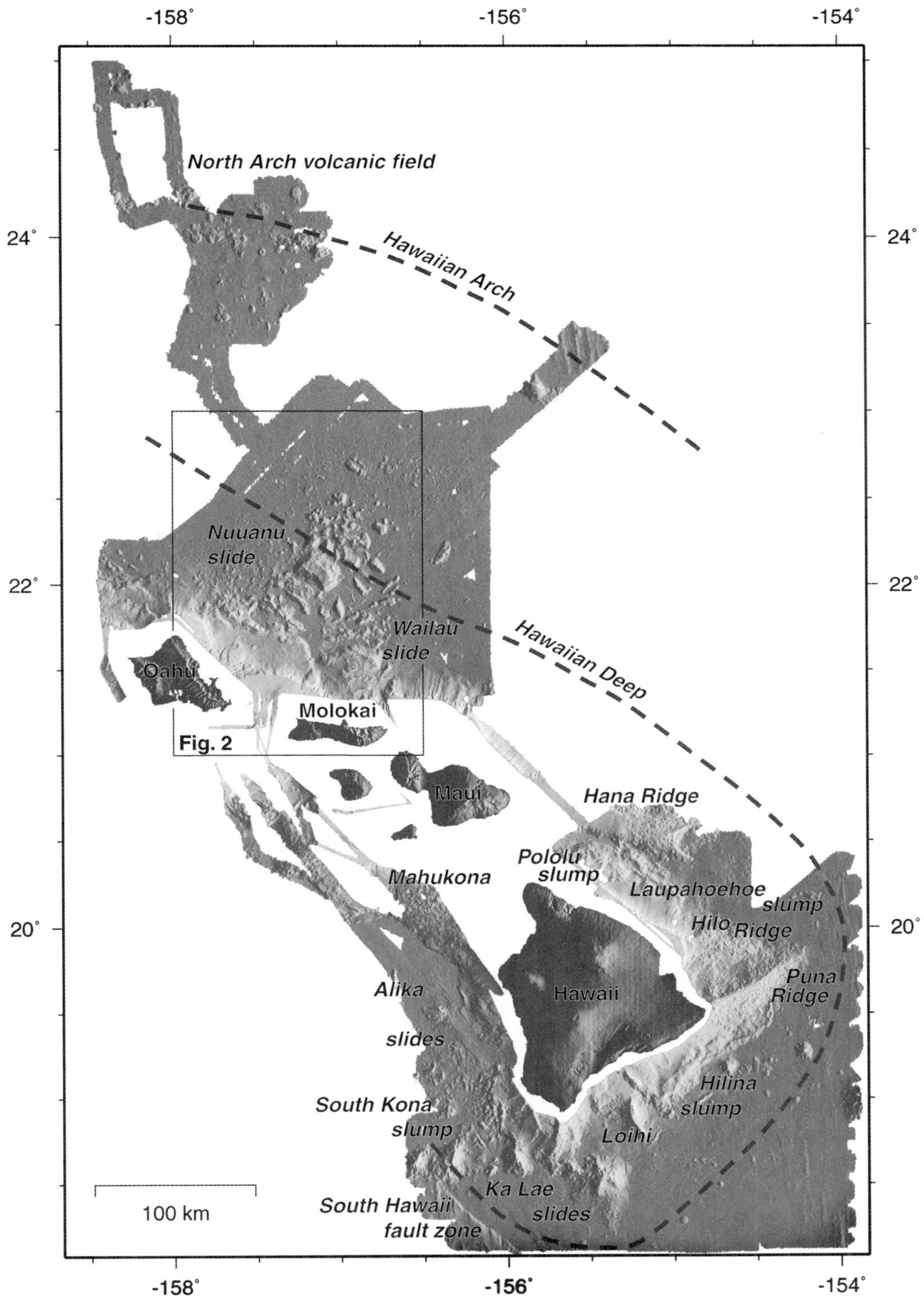

Figure 1. Main Hawaii Islands (USGS topography) with *SeaBeam 2112* shaded relief bathymetry showing coverage during the JAMSTEC surveys with main offshore features mapped in this study labeled. Illumination from the east. Box shows location of Figure 2.

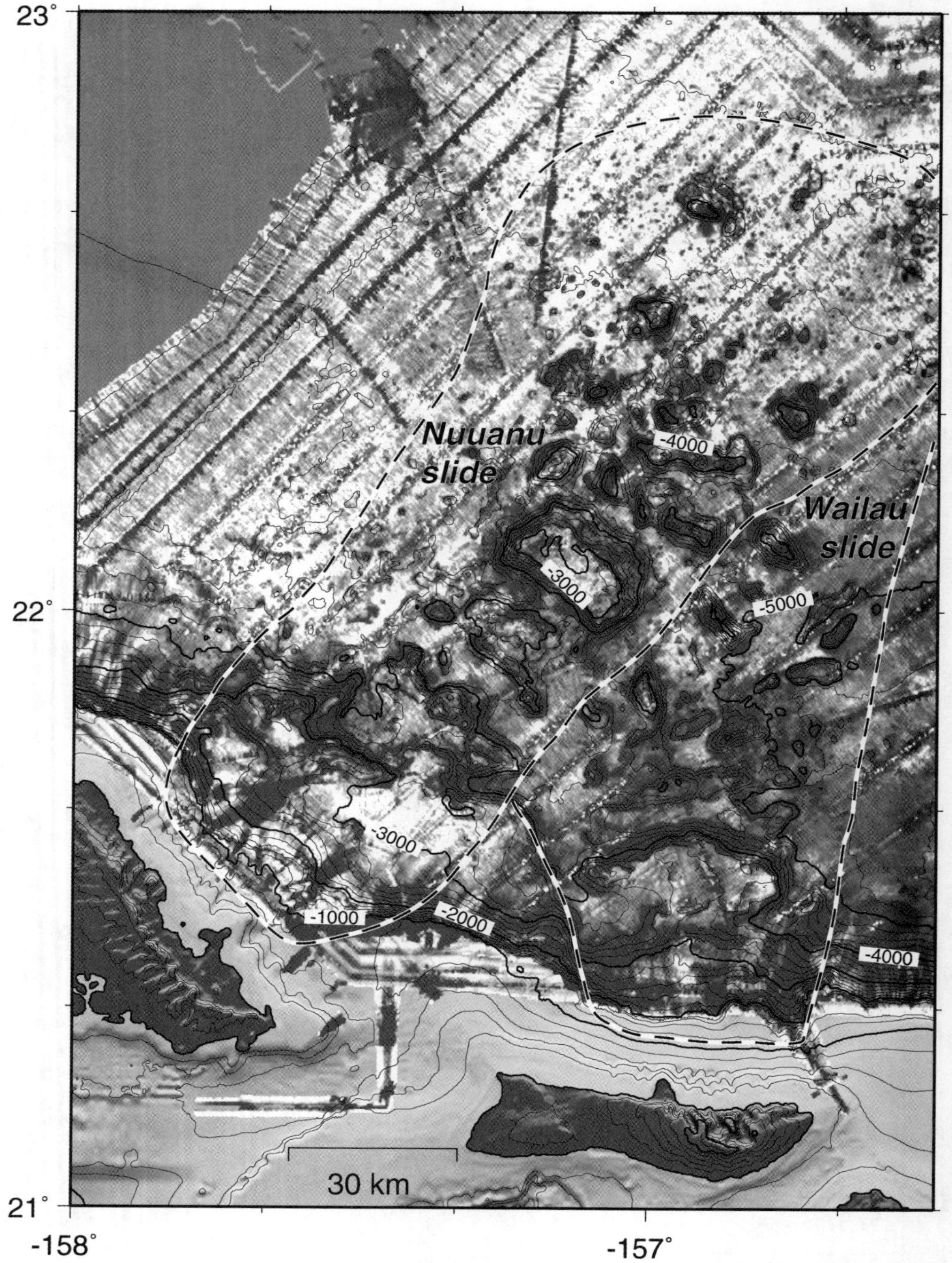

Figure 2. One application of the JAMSTEC *SeaBeam 2112* multibeam data available on the CD-ROM. (A) *SeaBeam* bathymetry merged with SOEST compilation that includes USGS topographic and various bathymetric datasets. Continuous light gray shaded relief bathymetry and topography with illumination from the north, contours every 200 m, bold every 1000 m. (B) *SeaBeam* backscatter, with darker gray tones as stronger reflectors. Boundaries of Nuuanu and Wailau landslide deposits from *Moore and Clague* and *Satake et al.* [both this volume]. For additional examples of multibeam data applications and research using it, see *Clague et al., Moore and Clague, Satake et al.,* and *Smith et al.* [all this volume].

4.2 Offshore Oahu and Molokai

Roughly 50,000 km^2 of seafloor was surveyed off the northern coasts of Oahu and Molokai islands, covering the entire Nuuanu and Wailau landslide deposits (Figs. 1, 2). The main survey area extended to 230 km from the Oahu shore, with exploratory lines out to 300 km on top of the northern Hawaiian Arch (Fig. 1). The distal limit of the slides appears to be 230 km, which matches the interpretation from the GLORIA sidescan sonar surveys [*Moore et al.*, 1989].

The improved resolution over previous survey data allows detailed mapping of the block distribution, orientation, and structure of the landslide. Several attempts have been made to delineate the boundary between the two slide deposits, determine the relative timing of the events, and reconstruct the pre-slide volcanoes from the overlapping slide deposits mapped in this study [*Moore and Clague, Satake et al.,* and *Yokose*; all this volume]. Fifty to sixty large blocks lie within the boundaries of the Nuuanu and Wailau slide deposits. These blocks are 5-20 km long, 2-20 km wide, and 300-1500 m in relief above the surrounding seabed. Numerous smaller blocks are located within the major block facies while others are dispersed farther afield.

The volcanic platform of Koolau volcano near the northern tip of Oahu, undisturbed by the Nuuanu landslide, was surveyed west to nearly 158°30'W. Dredge and dive samples confirm that submarine lava sequences of mostly pillow lava and some sheet flows compose the outcrop in this area.

4.3 Insular flanks of Hawaii and Maui

The 1998 *Kairei* and 1999 *Yokosuka* cruises mapped ~49,000 km^2 on all sides of the island of Hawaii (except some areas to the north and northwest), from nearshore to abyssal depths (Fig. 1). The surveys included exploratory lines out to 260 km on the crest of the southeastern portion of the Hawaiian Arch. Papers in chapters 1 and 2 of this volume present results from around the island of Hawaii.

The mobile south flank of Kilauea (the Hilina slump) was thoroughly mapped by the addition of survey lines between the initial tracks to increase the data quality by producing a greater percentage of overlapping swath data on the southeast flank of the island. The mostly flat region southeast of the island has now been mapped south to 18°10'N and east to 154°W (Fig. 1), extending earlier multibeam surveys [*Chadwick et al.*, 1993a,b; *Smith*, 1994]. Only a few more blocks or small seamounts were located and mapped on the abyssal plain southeast of Hawaii island in more detail than presented in the GLORIA images [*Groome et al.*, 1997].

The Puna Ridge (submarine continuation of the Kilauea east rift zone) and Hilo Ridge (offshore extension of either a Kohala or a Mauna Kea east rift zone) were mapped in their entirety. The eastern submarine flanks of Mauna Kea and Kohala volcanoes were surveyed north to nearly 20°30'N covering the distal portion of the Hana Ridge (Haleakala east rift zone) extending from Maui (Fig. 1). A newly identified landslide, the Laupahoehoe slump, was mapped, and the Pololu slump better defined [*Smith et al.*, this volume]. Additionally, ~4100 km^2 of the Alika and Ka Lae debris avalanche chutes and the South Kona slump off the Kona (west) coast were re-mapped (Fig. 1) in order to gain the acoustic backscatter component that was absent in previous multibeam surveys of the slide complex [*Chadwick et al.*, 1994a,b; *Smith*, 1994]. The northern portion of the South Hawaii fault zone (far south of Ka Lae, or South Point) was surveyed, but its southeastern extent could not be completed because of transit time constraints. Finally, some improvements were made to bathymetric coverage of the submerged Mahukona volcano (west of Kohala) during Hawaii to Oahu transits (Fig. 1).

5. GUIDE TO THE CD-ROM

The *SeaBeam* data and images produced from it are available on the CD-ROM included with this monograph. An HTML web style interface is included on the CD-ROM that guides the user to the products. Other authors from this volume have contributed extended datasets related to their papers and these are found using the same index. Simply open the index.htm file in the top directory of the CD-ROM with a web-browsing program.

Plot files of the *SeaBeam* data are included for those who do not wish to manipulate the data. Metadata files describing parameters of the geospatial data are also located on the CD-ROM.

SeaBeam data products are presented in different formats for importing into several common software packages. For GMT-System and MB-System users, binary netCDF grid files are available. For Earth Systems Research Institute (ESRI) ArcInfo and ArcView Geographic Information Software (GIS) users, *.e00 export files are provided that can be imported as grids using the *import* command in ArcInfo or the *import71* utility in ArcView. Image files are furnished for both bathymetry and backscatter, which can be opened in many image processing software packages.

Acknowledgments. The authors and all onboard scientists wish to thank the Japan Marine Science and Technology Center for sponsoring this work. We are also indebted to Captains O. Yukawa of *Kairei* and H. Tanaka of *Yokosuka*, their crews, and the technicians who assisted with *SeaBeam* data collection and processing. The Hawaii Undersea Research Lab and the Geological Survey of Japan provided salary, travel (also JAMSTEC), and computer support for JRS and KS to attend the cruises, post-process the data shoreside, and present the results at meetings. Reviews by editors M.O. Garcia, E. Takahashi, and P.W. Lipman improved the manuscript. SOEST contribution #5814.

REFERENCES

Blackinton, J.G., D.M. Hussong, and J.G. Kosalos, First results from a combination sides-scan sonar and sea floor mapping system (SeaMARC II), in *Proc. Offshore Techn. Conf.*, pp. 307-314, Dallas, TX, 1983.

Caress, D.W., and D.N. Chayes, Improved processing of Hydrosweep DS multibeam data on the R/V Maurice Ewing, *Mar. Geophys. Res.*, *18*, 631-650, 1996.

Chadwick, W.W., Jr., J.G. Moore, M.O. Garcia, and C.G. Fox, Bathymetry of southern Mauna Loa Volcano, Hawaii, *U.S. Geol. Surv. Misc. Field Studies Map MF-2233*, 1993a.

Chadwick, W.W., Jr., J.G. Moore, and C.G. Fox, Bathymetry of the southwest flank of Mauna Loa Volcano, Hawaii, *U.S. Geol. Surv. Misc. Field Studies Map MF-2255*, 1994a.

Chadwick, W.W., Jr., J.G. Moore, and C.G. Fox, Bathymetry of the west-central slope of the island of Hawaii, *U.S. Geol. Survey Miscellaneous Field Studies Map MF-2269*, 1994b.

Chadwick, W.W., Jr., J.R. Smith, Jr., J.G. Moore, D.A. Clague, M.O. Garcia, and C.G. Fox, Bathymetry of south flank of Kilauea Volcano, Hawaii, *U.S. Geol. Surv. Misc. Field Studies Map MF-2231*, 1993b.

Clague, D.A., R.T. Holcomb, J.M. Sinton, R.S. Detrick, and M.E. Torresan, Pliocene and Pleistocene alkalic flood basalts on the seafloor north of the Hawaiian Islands, *Earth Planet. Sci. Let.*, *98*, 175-191, 1990.

Clague, D.A., K.A. Hon, J.L. Anderson, W.W. Chadwick, Jr., and C.G. Fox, Bathymetry of Puna Ridge, Kilauea Volcano, Hawaii, *U.S. Geol. Surv. Misc. Field Studies Map MF-2237*, 1994.

Clague, D.A., J.G. Moore, and J.R. Reynolds, Formation of submarine flat-topped volcanic cones in Hawaii, *Bull. Volcanol.*, *62*, 214-233, 2000.

Clague, D.A., K. Uto, and A.S. Davis, Eruption style and flow emplacement in the submarine North Arch Volcanic Field, Hawaii, this volume.

Dartnell, P., and J.V. Gardner, Sea-floor images and data from multibeam surveys in San Francisco Bay, Southern California, Hawaii, the Gulf of Mexico, and Lake Tahoe, California-Nevada, *Digital Data Series DDS-55,* U.S. Geological Survey, 1999.

Duennebier, T., T. Reed, and H. Staff, Northwestern Hawaiian Islands: merged bathymetry and topography, in *Hawaii Seafloor Atlas, Sheet #2*, Hawaii Institute of Geophysics and Planetology, Honolulu, 1994.

Gardner, J.V., and J. Hughes-Clarke, Cruise report: RV ocean alert cruise A1-98-HV: mapping Hawaii insular slopes (January 30 through February 23, 1998, Honolulu to Honolulu, Hawai`i), *U.S. Geol. Surv. Open-File Rep. 98-212*, 26 p, 1998.

Groome, M.G., C.E. Gutmacher, and A.J. Stevenson, Atlas of GLORIA sidescan-sonar imagery of the Exclusive Economic Zone of the United States: EEZ-View, *U.S. Geol. Surv. Open-File Rep. 97-540*, 1997.

Holcomb, R.T., B.K. Nelson, P.W. Reiners, and N.-L. Sawyer, Overlapping volcanoes: the origin of Hilo Ridge, Hawaii, *Geology*, *28*, 547-550, 2000.

Laughton, A.S., The first decade of GLORIA, *J. Geophys. Res. 86* (B12), 11,511-534, 1981.

MBARI, MBARI Multibeam Survey, Version 1, *Digital Data Series No. 2*, Monterey Bay Aquarium Research Institute, Monterey, 2000.

Moore, G.F., J.K. Morgan, S. Leslie, B. Taylor, and C. Berndt, Morphology and structure of the Nu`uanu debris avalanche, north of O`ahu, Hawai`i, *Eos Trans. AGU,*, *79*, F1008, 1998.

Moore, J.G., and D.A. Clague, Mapping of the Nuuanu landslides in Hawaii, this volume.

Moore, J.G., D.A. Clague, R.T. Holcomb, P.W. Lipman, W.R. Normark, and M.E. Torresan, Prodigious submarine landslides on the Hawaiian Ridge, *J. Geophys. Res.*, *94*, 17,465-17,484, 1989.

Moore, J.G., and W.W.J. Chadwick, Offshore geology of Mauna Loa and adjacent areas, Hawaii, in *Am. Geophys. Union Monogr. 92*, edited by J.M. Rhodes, and J.P. Lockwood, pp. 21-44, American Geophysical Union, Washington, DC, 1995.

Naka, J., and Scientific Team, Tectono-magmatic processes investigated at deep-water flanks of Hawaiian volcanoes, *Eos Trans. AGU*, *81*, 221, 226-227, 2000.

Rognstad, M., HAWAII MR1: A new underwater mapping tool, in *Proc. Internatl. Conf. on Signal Processing and Techn.*, pp. 900-905, 1992.

Satake, K., J.R. Smith, and K. Shinozaki, Three-dimensional reconstruction and tsunami model of the Nuuanu and Wailau giant landslides, Hawaii, this volume.

Smith, J.R., Island of Hawaii and Loihi submarine volcano, high resolution multibeam bathymetry around the island of Hawaii, in *Hawaii Seafloor Atlas, Sheet #6*, Hawaii Institute of Geophysics and Planetology, Honolulu, 1994.

Smith, J.R., K. Satake, J.K. Morgan, and P.W. Lipman, Submarine landslides and volcanic features on Kohala and Mauna Kea volcanoes and the Hana Ridge, Hawaii, this volume.

Torresan, M.E., and J.V. Gardner, Acoustic mapping of the regional seafloor geology in and around Hawaiian ocean dredged-material disposal sites, *U.S. Geol. Surv. Open-File Rep. 00-124*, 63 p., 8 map sheets, 2000.

Wessel, P., and W.H.F. Smith, New version of the Generic Mapping Tools released, *Eos Trans. AGU*, *76*, 329, 1995.

Yokose, H., Landslides on the windward flanks of Oahu and Molokai, Hawaii: SHINKAI 6500 submersible investigation, this volume.

K. Satake, Active Fault Research Center, AIST-GSJ, Tsukuba, Ibaraki 305-8567, Japan (kenji.satake@aist.go.jp).

J. R. Smith, Hawaii Undersea Research Lab, University of Hawaii, 1000 Pope Rd., MSB 303, Honolulu, HI 96822 (jrsmith@soest.hawaii.edu).

K. Suyehiro, Deep Sea Research Department, JAMSTEC, 2-15 Natsushima-cho, Yokosuka, 237-0061 Japan. (suyehiro@jamstec.go.jp).

Submarine Landslides and Volcanic Features on Kohala and Mauna Kea Volcanoes and the Hana Ridge, Hawaii

John R. Smith[1], Kenji Satake[2], Julia K. Morgan[3], and Peter W. Lipman[4]

The deep submarine eastern flanks of Mauna Kea, Kohala, and Haleakala volcanoes were mapped for the first time with a multibeam bathymetric and sidescan sonar system during joint Japan-US cruises aboard the JAMSTEC vessel R/V *Yokosuka* in 1999. The Pololu slump off northeast Kohala is overlain by a carbonate platform in the shallow region and the deeper areas are incised by downslope oriented channels. It is cut by several faults and slump scars and appears to override an older slide located farther east, here named the Laupahoehoe slump. The structures characteristic of the Laupahoehoe slump are NW-SE oriented scarp-and-bench topographic features analogous to the Hilina slump on the mobile SE flank of Kilauea. Enclosed basins lie at 3000-5000 m, fronted by ridges on their seaward sides. The basins may result from local rotational slumps or from uplift above discontinuous thrust faults. The Laupahoehoe slump appears to be overlapped by shield margins of both Kohala and Mauna Kea and thus may have been derived from an elongate Kohala edifice. A large debris apron continues from the base of the two-slide complex, abutting the distal Hana Ridge (Haleakala east rift zone). The tip of the Hana Ridge displays a curious steep-sided arcuate rift tip that resembles the classic amphitheater scarp of a landslide. Numerous flat-topped volcanic domes are distributed along the broad crest of the lower rift zone. Similar cones, though fewer in number, are also present on the Hilo Ridge, which may be the continuation of a Kohala rift zone.

1. INTRODUCTION

Mapping of the Hawaiian Exclusive Economic Zone (HEEZ) by the U.S. Geological Survey in the late 1980s, using the GLORIA sidescan sonar system, resulted in an improved understanding of the regional geology of the sea floor to a distance of 200 n.mi. from the coastline [*Groome et al.*, 1997]. This synoptic picture provided a framework within which more thorough study of the Hawaiian ridge, deep, and arch could occur. The GLORIA surveys showed the existence of many large submarine landslides [*Lipman et al.*, 1988; *Moore et al.*, 1989, 1994]. *Moore et al.* [1989] interpreted the landslides as evidence for major processes involved in island and seamount morphogenesis. The surveys confirmed that this process is pervasive along the Hawaiian chain, as far as Midway Island and beyond [*Moore et al.*, 1994].

Prior to the present study, deepwater bathymetric data along the windward (northeastern) coasts of Maui and Hawaii were poor in coverage, resolution, and navigation. These coasts are exposed to nearly constant trade winds, making hydrographic work difficult, and the lack of local

[1]School of Ocean and Earth Science and Technology, University of Hawaii, Honolulu
[2]Active Fault Research Center, AIST-GSJ, Tsukuba, Japan
[3]Department of Earth Science, Rice University, Houston, Texas
[4]Volcano Hazards Program, U.S. Geological Survey, Menlo Park, California

Hawaiian Volcanoes: Deep Underwater Perspectives
Geophysical Monograph 128
Copyright 2002 by the American Geophysical Union

ports reduces the volume of bathymetric data acquired by passing vessels of opportunity. The MBARI surveys in 1998 [*MBARI,* 2000] collected the first high quality data using modern methods in these areas. However, because of their limited survey time and the use of a higher-resolution system, the full extent and deepest portions of the volcanic flanks were not mapped.

JAMSTEC sponsored deep-diving and over-the-side sampling cruises in 1998 and 1999 that also acquired multibeam bathymetric and backscatter data from the flanks and deep sea floor surrounding the islands of Oahu, Molokai, Maui, and Hawaii. Some surveying was required as site location for the ROV and submersible dives, but most was aimed at better defining the large landslide units, as well as some exploratory mapping to locate future dive targets.

This paper describes the structure, volcanic morphology, and geological processes which have modified the submarine northeast flank of Hawaii and the tip of the Hana Ridge (Haleakala east rift zone, HERZ) off the island of Maui, based on detailed bathymetric and acoustic backscatter (sidescan) maps (Fig. 1). Our study uses the SeaBeam multibeam bathymetry and backscatter data to provide improved bathymetric information, locate underwater faults, define landslide boundaries (including scarps and deposits), and locate submarine volcanic centers on the insular slopes. These data provide evidence for the style, and possibly the timing, of deformation. An understanding of the processes leading to large-scale collapse requires information on the foundation and internal structure of the island itself. While those problems could not be addressed directly in this study, analysis of the morphology of the slopes and structures provides clues to the deeper properties of the volcanic pile and underlying crust.

2. METHODS

2.1 SeaBeam 2112 System

Bathymetric data were collected by the SeaBeam 2112 built by SeaBeam Instruments, Inc. The SeaBeam 2112 is a 12 kHz, deep-water multibeam sonar survey system that allows the production of wide-swath, contiguous contour maps, and backscatter (acoustic reflective intensity) images of the seafloor. Positioning of depth soundings on the seafloor is based on GPS and ship motion input. The system, survey statistics, and the digital data are described in *Smith and Satake* [this volume].

2.2 Post-Processing

The UNIX freeware software packages GMT-System [*Wessel and Smith*, 1995] and MB-System [*Caress and Chayes*, 1996] were used to post-process, grid, merge, and display the data. JAMSTEC SeaBeam bathymetry data were merged with various lower and higher resolution data sets including National Ocean Survey nearshore hydrographic soundings, offshore trackline data from vessels of opportunity archived at the National Geophysical Data Center, and U.S. Geological Survey (USGS) topographic data [*Smith*, 1994]. We also incorporated the recently released higher-resolution medium-depth Simrad EM-300 data collected by the Monterey Bay Aquarium Research Institute (MBARI) in 1998 [*MBARI*, 2000]. The MBARI data cover the shallower portions of the Pololu slide and the Hana, Hilo, and Puna ridges (Fig. 1).

The grids used to generate the bathymetric and topographic digital elevation models were built using a continuous curvature surface gridding algorithm [*Smith and Wessel*, 1990], while backscatter data were mosaicked using MB-System programs [*Caress and Chayes*, 1996]. The bathymetry and topography figures were produced from 100-meter grids, and backscatter images are based on 50-meter grids. Coverage of the JAMSTEC SeaBeam 2112 bathymetry is shown in the index map of Figure 1, and the main figure as the higher-resolution texture superimposed on the smooth, relatively featureless, background surface. Additionally, since the backscatter images show only the JAMSTEC data, which is coincident with the bathymetry, one can recognize the boundaries of the SeaBeam 2112 surveys by looking at the backscatter figures. Artifacts often appear in merged bathymetric and topographic data sets along the join lines as a result of differences in resolution and navigation. These artifacts are difficult to remove fully without degrading the resolution of the better data.

2.3 Survey Coverage

The area mapped with SeaBeam was extended in all directions from the southeast flank of the island of Hawaii, which was the focus of the submersible and ROV diving programs. Vehicle service, weather days, and overnight transects allowed additional surveying to the south, east, and northeast of Hawaii Island with a total of ~49,000 km^2 of seafloor mapped. Approximately 12,000 km^2 of new area was surveyed between Hilo Ridge and Hana Ridge. Some of these areas, such as those discussed in this paper, were exploratory and without dives, dredges, or cores.

3. NORTHEAST FLANKS OF KOHALA AND MAUNA KEA

The JAMSTEC data over the submarine flanks of northeastern Kohala and Mauna Kea volcanoes reveal flank

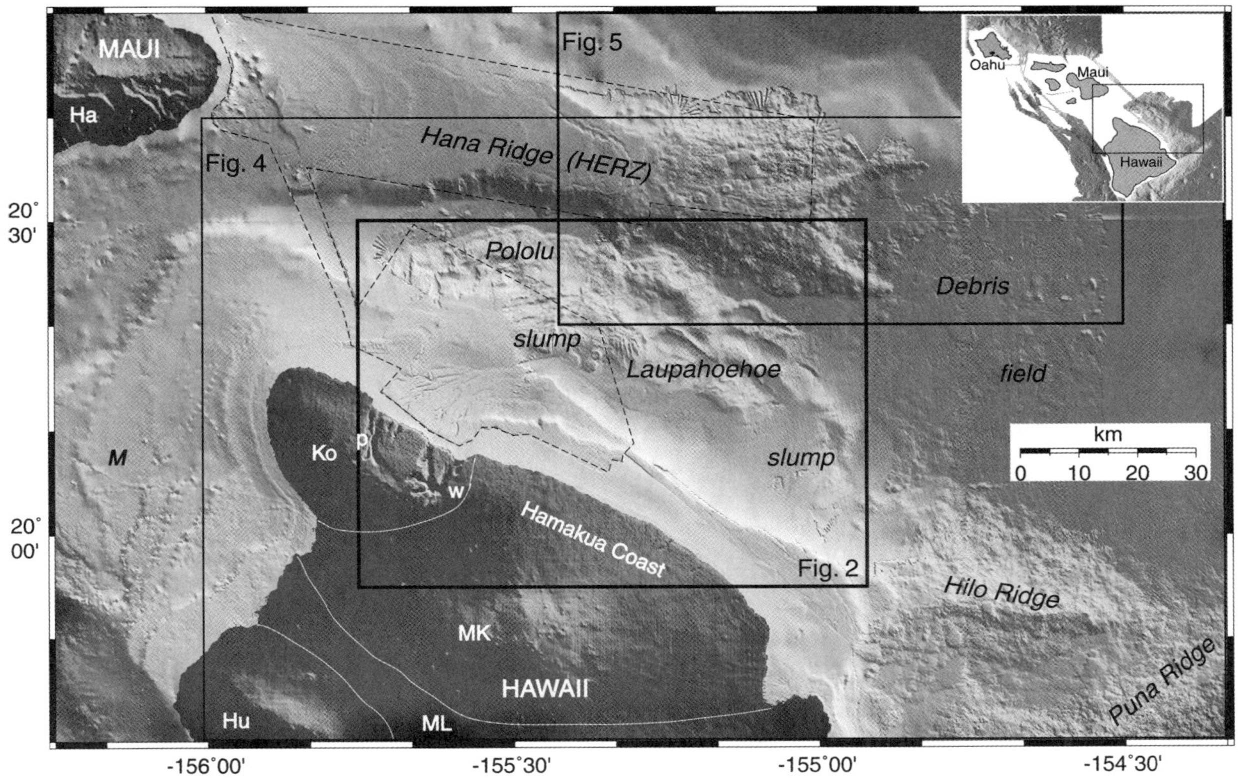

Figure 1. Shaded relief map showing the east tip of Maui and the Hana Ridge, the northern half of the island of Hawaii and its submarine northeast flank. Obliquely illuminated from the northeast. Numbered boxes correspond to following figures. Volcano boundaries are solid white lines. *MBARI* [2000] multibeam coverage shown by dashed lines. Boundaries of JAMSTEC multibeam data are evident on index map and Figs. 2b, 5c. Abbreviations as follows: Ha, Haleakala; MK, Mauna Kea; Hu, Hualalai; Ko, Kohala; ML, Mauna Loa; *M*, Mahukona; p, Pololu Valley; w, Waipio Valley. Index map shows main Hawaiian Islands, JAMSTEC SeaBeam coverage, and location of figure.

morphology in unprecedented detail. Contrasts in surface texture distinguish several regions; rough, apparently eroded terrain with regionally high relief and steep slopes characterizes the northwestern flank (from –155° 40' to –155° 20'); to the southeast, broad, smooth slopes, quite steep in the upslope regions, give way to bench and basin morphology at the toe (from –155° 20' to –155° 00') (Fig. 2a). The Hilo Ridge to the east represents the submarine extension of a volcanic rift zone associated either with Mauna Kea [*Fiske and Jackson*, 1972] or Kohala volcano [*Holcomb et al.*, 2000; *Kauahikaua et al*., 2000] and is similar to the Puna Ridge offshore extension of Kilauea's east rift zone (Fig. 1). It is distinguished by much smaller-scale surface roughness defined by flat-topped cones and terraces formed by submarine volcanism and by a systematic narrowing and deepening of the rift axis.

Two terraces overlap the flanks, at 1100 m and 400 m, associated with cessation of subaerial eruptive shield building activity from Kohala and Mauna Kea volcanoes, respectively. The terraces are identified as breaks in slope at the boundary between gently sloping subaerial lava flows and steeply dipping layers of subaqueously chilled and fragmented material deposited below sea level [*Moore and Clague*, 1992]. Once high volumes of lava flows cease at the end of shield building, coral reefs can grow and are gradually drowned as the volcano(es) and preserved slope breaks subside. Dating the fossil reefs allows one to gain subsidence rates, as well as approximate dates of the terraces and end of shield building on each volcano. A detailed explanation and further examples are presented in *Moore and Clague* [1992].

Moore et al. [1989] were the first to identify the seaward bulging bathymetry [*National Ocean Survey*, 1983; *Campbell*, 1987] and hummocky seafloor deposits north of Kohala seen in the GLORIA sidescan images as the Pololu landslide, which they described as a debris avalanche cov-

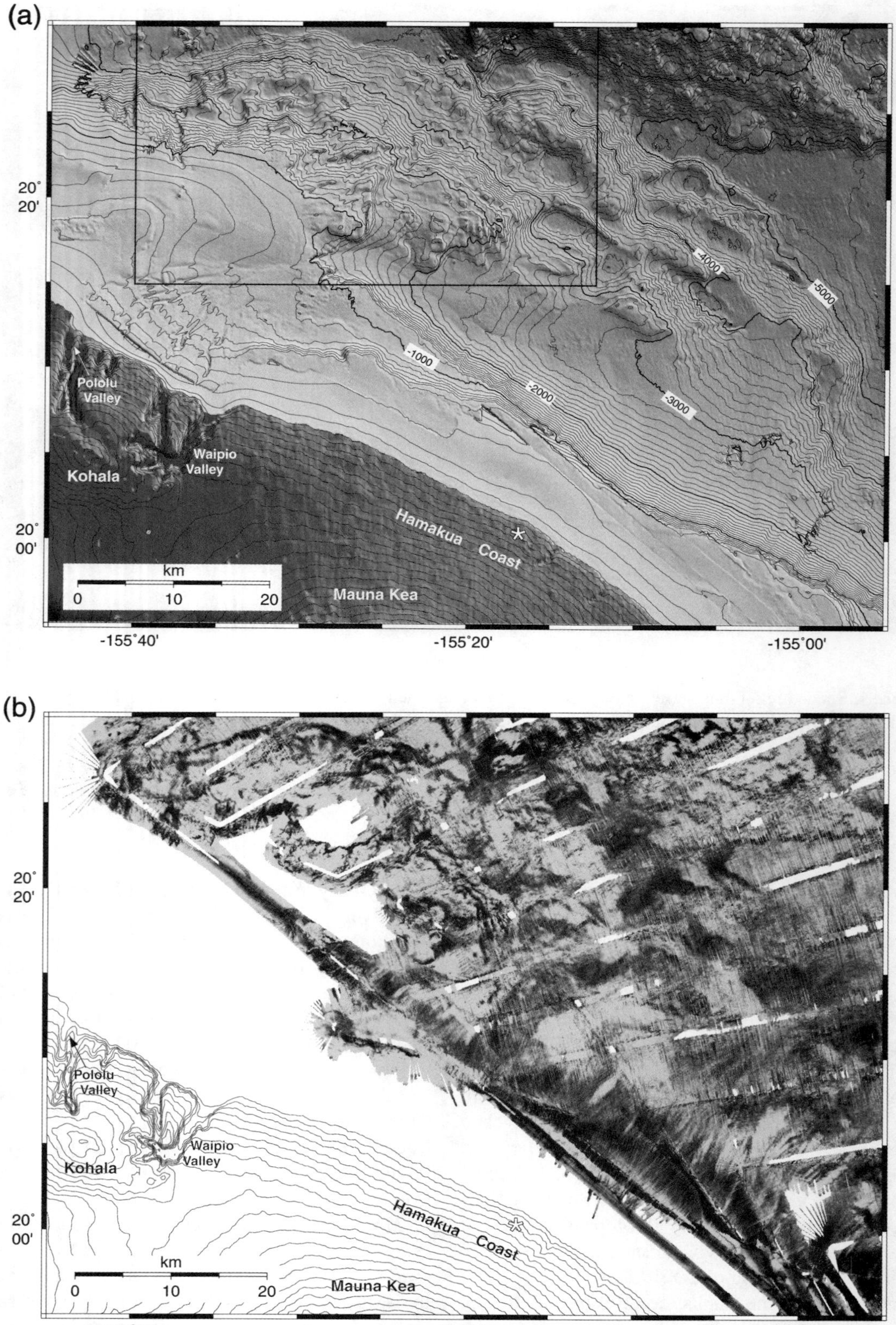
(a)
20° 20'
20° 00'
-1000
-2000
-3000
-4000
-5000
Pololu Valley
Waipio Valley
Kohala
Hamakua Coast
Mauna Kea
km
0
10
20
-155°40'
-155°20'
-155°00'
(b)
20° 20'
20° 00'
Pololu Valley
Waipio Valley
Kohala
Hamakua Coast
Mauna Kea
km
0
10
20
-155°40'
-155°20'
-155°00'

Figure 2

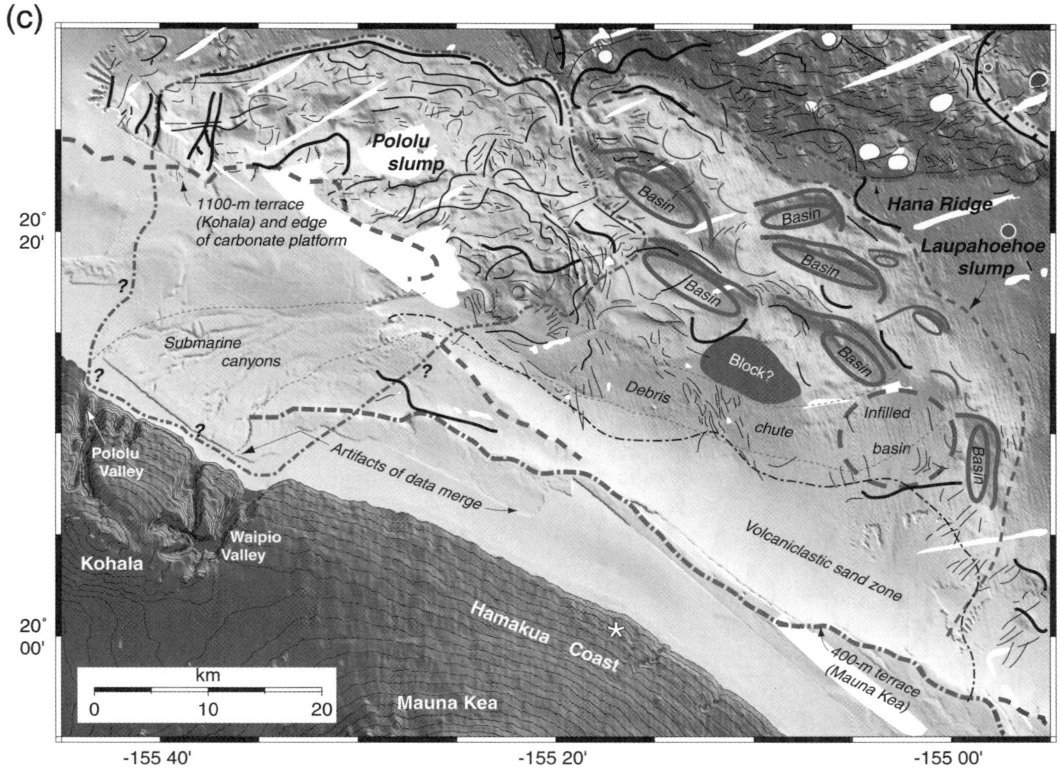

Figure 2. Northeast portions of Kohala and Mauna Kea volcanoes and their offshore slopes that include the Pololu and Laupahoehoe slumps. (a) Bathymetry and topography at 100-m contours, shaded and obliquely illuminated from the northeast. Town of Laupahoehoe shown by asterisk (*). Artifacts of data set merging shown on (c). Box shows location of Fig. 3. (b) Backscatter image, with high acoustic reflectivity as darker gray tones. Topography at 100-m contours. Limit of JAMSTEC data equals the backscatter coverage in this figure. (c) Simplified interpretation of (a) and (b) appropriate for this scale. Variously weighted dashed and dotted lines outline different geologic features. Thick solid gray lines mark the frontal ridges and surround the basins discussed in the text. Three thickness levels of solid black lines are reflective lineations, structures (faults, fractures), and erosional gullies that represent the regional topographic texture (from bathymetry) and acoustic character (from backscatter). Some may be debris lobes or a series of lava flow fronts. Thickness of the lines increases for wider reflectors, including higher and more consistent reflectivity. A detailed morphological and structural interpretation of the chaotic lineation patterns on the Pololu slump is presented in Figure 3. Solid white circular or elliptical features are interpreted to be volcanic constructions. Solid gray circular features are interpreted as debris blocks. Thick gray dash-dot line is former Mauna Kea coastline at the end of shield building, corresponding to the 400-m-deep terrace. Thick gray dashed line is former Kohala coastline at the end of shield building, corresponding to the 1100-m-deep terrace. Areas of no backscatter data in white, most correspond to ship nadir.

ering an area of 3500 km^2. We maintain the use of their terminology to represent the highly eroded northwestern terrain, although its origin is ambiguous, and we introduce here the Laupahoehoe slump that encompasses the previously unrecognized bench and scarp terrain found to the east. The name is after the prominent Laupahoehoe Point on the Hamakua Coast that lies approximately in the middle of the upper edge of the slump and overlooks this submarine region (Fig. 2). The entire Pololu and Laupahoehoe complex is 90 km wide and extends seaward at least 45 km, abutting the distal Hana Ridge.

3.1 Pololu Slump

The Pololu landslide was reported to head above the deeply carved subaerial canyons (~600 m from original surface to valley floor) on the northeast side of Kohala volcano (Fig. 2a) in a fault-bound graben at the summit that is oriented generally perpendicular to the proposed direction of landsliding [*Moore et al.*, 1989; *Moore and Clague*, 1992]. The landslide amphitheater extends to a water depth of 900 m and is marked by a 20-km wide shoreline reentrant that reaches 2 km landward. The upper portion of the landslide amphitheater is cut by four submarine canyons [*Moore*, 1987] traceable to water depths >1000 m. The bulk of the landslide deposit was interpreted to turn eastward between the Hana Ridge to the north and the steep submarine upper flanks of Kohala and Mauna Kea [*Moore et al.*, 1989]. *Moore et al.* [1989] inferred that the landslide reached 60 km beyond the steep-sided trough separating the Hana Ridge from Kohala, and it produced a broad depositional lobe traceable to 130 km from the shoreline based on a hummocky character in the GLORIA images.

The JAMSTEC data show that the Pololu slump is characterized by an east-northeast directed lobe marked by ridges and basins and hummocky terrain containing blocks or ridges 2-5 km in length with 50-200 m relief (Figs. 2a,b). Eight to ten bathymetric benches are evident on the feature, some at mid-depth and others near the foot of the slump on the north side (Figs. 2a,c, 3). For most, a linear ridge ~100 m high fronts a basin that is 10-40 m deep and appears well sedimented in the backscatter data (Fig. 2b). This morphology is similar to the scarp-and-bench topography recognized on other clearly defined Hawaii slumps [*Moore et al.*, 1989] such as the Hilina slump on the south flank of Kilauea [*Smith et al.*, 1999]. However, at Pololu, the slopes leading to the basins and the scarps below the benches are more eroded, showing shallow downslope channels. Along the south flank of Kilauea, such benches form perpendicular to the direction of transport; the east-west trend of the northern benches on Pololu suggests north-directed displacement over this part of the slump. The basins here have not filled with sediment and are not enclosed; each has a downslope outflow, implying on this region of the slump that the topography rather than sediment supply controls fill level. In addition, this may, in part result from most of the present day runoff being channeled to the east through the submarine canyons (Fig. 2a) and might explain the difference in texture between the Pololu and Laupahoehoe slumps described in section 3.2.

The landslide interpreted by *Moore et al.* [1989] is thought to have formed late during the shield stage of Kohala, before formation of the 370-ka terrace (now at ~1100 m water depth) that marks the end of shield building at this volcano [*Moore and Campbell*, 1987]. A 400-m terrace stretching 70 km along the northeast flank of Mauna Kea and marking the end of shield building on this volcano [*Moore et al.*, 1989] at about 128 ka [*Moore and Campbell*, 1987] also grew across the southern portion of the Pololu amphitheater, which is now cut by channels (Figs. 2a,c).

Pololu is likely one of the oldest landslides on the island of Hawaii [*Moore and Clague*, 1992; their fig. 5], and a carbonate platform and associated reef terraces have grown across at least the shallower portions of the slide feature [*Clague et al.*, 1998]. Their data extend closer to the Kohala shoreline than the JAMSTEC SeaBeam, and more submarine canyons and/or branches can be seen, some not visible farther down the slope. The canyons become less pronounced as they merge, across a gently sloping carbonate covered platform, into a large debris channel at the edge of the 1100-m terrace (Figs. 2a,c). This carbonate platform is cut by canyons on all sides, possibly by dissolution from submarine freshwater seepage [*Clague et al.*, 1998].

In addition to incision of the edge of the carbonate platform [*Clague et al.*, 1998], erosional gullies of various dimensions weave through the bulk of the landslide welt, obscuring the original structure and mode of failure (Fig. 3). If these gullies do extend throughout the deposit, they seem unlikely to be totally due to freshwater seepage since freshwater rises upon effusing from the seafloor. The gullies and intervening ridges are oriented radially away from the edge of the carbonate platform and in line with the flow directions expected from runoff or turbidites. They are evident in the bathymetric (Fig. 3) and backscatter data (Fig. 2b) as sinuous patterns of steep slope and high reflectivity, respectively. It is possible that the heads of these conduits were started by freshwater effusion, then their downslope continuation was maintained by sediment-laden flows when the platform was shallower and closer to sea level (*i.e.*, shore) prior to subsidence.

Several of the channel features evident on the Pololu slide are atypical in form. One near the center of the slide is dammed on its downslope side by a prominent east-west

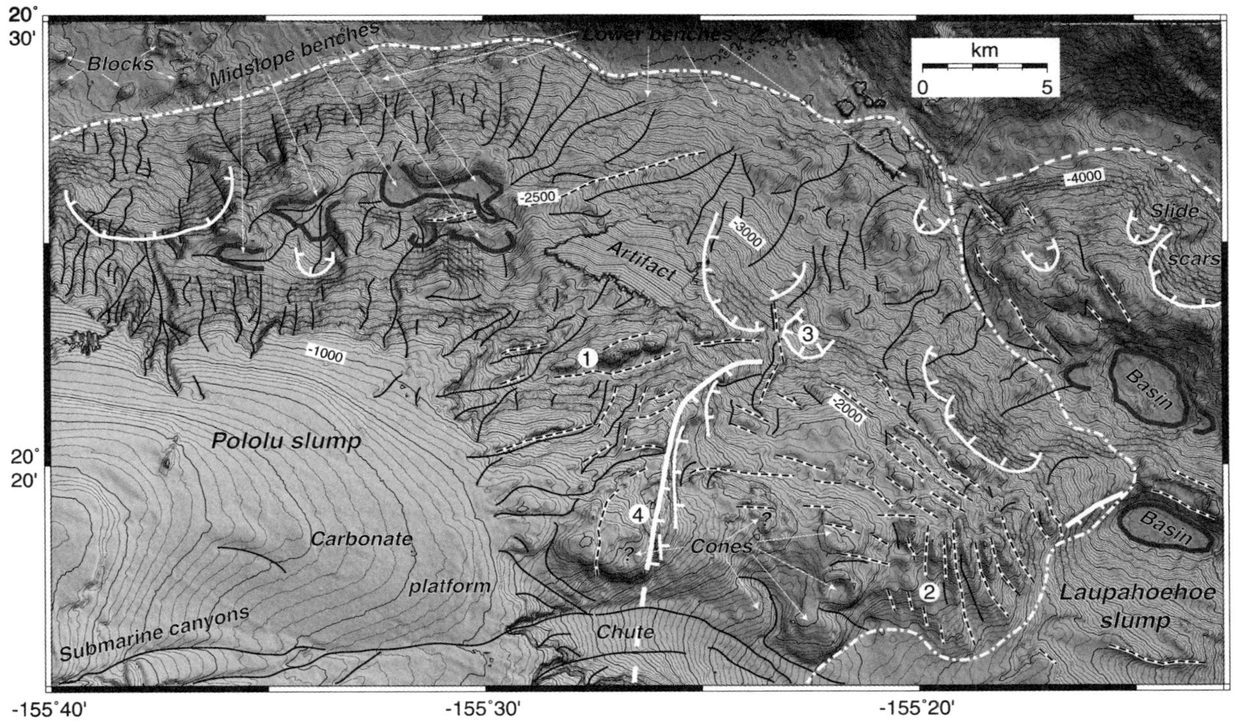

Figure 3. Detailed structure and morphology map of Pololu and Laupahoehoe slumps overlain on bathymetry at 20-m contours, bold every 100-m, shaded and obliquely illuminated from the northeast. Slump boundaries as in Fig. 2c (but white here). Basins/benches outlined in thick solid gray. Thick, solid white, ticked lines are landslide scarps. Thin black lines are erosive channels. Black and white dashed lines are fractures. Features adjacent to -- described in text.

trending ridge (Fig. 3,). Several adjacent highs define linear trends parallel to this ridge, suggesting these define fractures or faults introduced during emplacement of the slide. Similar parallel lineations recognized on a high east of the tilted cones (Fig. 3,) and slope-parallel basins fronted by ridges in the north also demonstrate the highly segmented nature of this slide. Most of the lineations are concentrated in the southeastern region above about 2500-m water depth. Elsewhere such features may be absent or have been buried by thick slope cover. Detailed mapping and analysis of these deformation structures is complicated by the highly eroded nature of the Pololu landslide.

Although a few large arcuate reentrants (1.5-7 km wide) are present at ~2000 m water depth, the deeper reaches of the Pololu landslide deposit (*i.e.*, below ~2500 m), particularly to the northeast, exhibit more numerous and less distinct reentrants, usually characterized by an arcuate form (Fig. 3). We interpret all of these to reflect local slope collapse, probably enhanced by the broken nature of the slide deposit. In some cases, the local slump scars are located immediately downslope of a major channel (Fig. 3,) and collapse may have been triggered by erosional events. The presence of minor channels within the scars themselves indicates they are of great age.

A distinct scarp, oriented almost north-south, crosses the southeastern part of the lobe (Fig. 3,) and can be followed northward to where it curves off to the northeast. Maximum relief is about 500 m and decreases to the northeast. We hypothesize that this north trending scarp may denote the headwall of a large slump that displaced the eastern block down and to the northeast. The scarp may once have continued to the south, but is now truncated by an east-west trending cliff line that overlooks a broad debris channel.

3.2. *Laupahoehoe Slump*

In comparison to other large Hawaiian volcanoes, Mauna Kea appears to be a geometrically simple cone without distinct rift zones or fault systems, either on geologic maps [*Wolfe and Morris*, 1996] or in the magnetic and gravity data imaging its internal structure [*Hildenbrand et al.*, 1993; *Kauahikaua et al.*, 2000]. Bathymetric data available prior to the JAMSTEC surveys showed no major disruption of the northeast submarine Mauna Kea flank that might indicate landslide features [*Smith*, 1994].

In contrast, the new JAMSTEC data show that the Pololu slump is juxtaposed against structures with a different morphology and texture at ~3000 m water depth (Figs. 2a, 3). These features are northwest-to-southeast oriented bench and basin structures, analogous to those of the Hilina slump on the southeast flank of Kilauea [*Smith et al.*, 1999]. Six enclosed basins (100-400 m deep, 4-10 km long, and 1-5 km wide) lie at 3000-5000 m water depth, fronted by arcuate ridges 50-200 m high on their seaward sides. A large solitary block (Fig. 2c) that may have a similar origin lies just upslope. The basins and ridges show smooth, sedimented surfaces in the backscatter images (Fig. 2b). It is possible that several small benches near the foot of the Pololu slump on its northern side also belong to the Laupahoehoe slump and were covered by the later landslide. However, ridge trends in the two regions are very different, suggesting that they represent a distal facies of the Pololu slump (Fig. 3). Although the low reflectivity areas of the Pololu and Laupahoehoe slumps are similar in acoustic intensity and shape, the most highly reflective features of the Pololu slump are sinuous in geometry and more acoustically intense than the broad, diffuse areas of highest reflectivity of the Laupahoehoe slump (Fig. 2b).

Numerous smaller concave landslide scars 1-4 km in diameter are evident in the steep scarps both upslope and downslope from the bathymetric benches (Figs. 2a-c). Unlike those on the Pololu deposit, these do not appear to connect to upslope erosional gullies and do not contain similar small channels. Instead, they resemble the scars on the outer scarp below the midslope bench of the Hilina slump [*Smith et al.*, 1999]. The contrast in morphology with Pololu scars may reflect the inability of erosional sediment laden flows on Laupahoehoe to reach beyond the slope basins.

The Laupahoehoe slump contains at least one large debris pathway and may be bounded to the east by another such channel or slump. A long debris chute directed east-southeast across the back of the landslide lower zone begins at the (present day) 1000-m-deep contour near the subaerial valley reentrant and continues to 3000-m depth where it disperses onto a broad sloping bathymetric bench (possibly an infilled basin) that extends to 4000-m depth (Figs. 2a,c). The head of the debris chute may have originally been close to sea level, prior to subsidence of Kohala. The 5-km-wide head of the chute marks the coalescence of the submarine canyons downslope from the subaerial valley reentrant. The channel cuts through the 1100-m-deep terrace, flattening and broadening to approximately 10 km at its distal point.

Several origins can by hypothesized for the sediment chute. It may reflect the cumulative erosion of the flank due to focussed delivery of sediments from the Kohala reentrant and submarine canyons. If this is the case, it is not presently a highly active conduit as alternating high and low acoustic intensity sonar imagery indicate a downslope transport direction for recent slope sediments that is oblique to the east-southeast channel trend (Fig. 2b). These sediments are likely fragmental basalt from subaerial lava flows as on Kilauea [*Lee et al.*, 1994; *Moore and Chadwick*, 1995; *Smith et al.*, 1999]. Such a channel may have been localized along a preexisting fracture or depression induced by block faulting or slumping. Alternatively, the chute may define a debris avalanche track that incised the deformed flank, similar to the Alika debris avalanche which cut through the South Kona slump along the southwest flank of Mauna Loa [*Lipman et al.*, 1988]. This hypothesis would explain the absence of basin and ridge structures where it widens and levels out at 3000-4000 m water depth, invoking a catastrophic erosive event that planed off preexisting basin and ridge structures on the foot of the Laupahoehoe slump. Such a cut into the flank would redirect the bulk of future sediments from Kohala into this channel and leave the Pololu slump sediment-starved.

The second, even wider chute (11-19 km in width) is bounded by our inferred southeastern lateral edge of the Laupahoehoe slump delineated by bulging bathymetry and the visible northwest extent of the Hilo Ridge (Fig. 4). This chute feeds directly onto the abyssal seafloor. The slope defining the chute may represent original volcano flank continuous with the Hilo Ridge but blanketed with volcaniclastic debris derived from shoreline processes. A similar downrift transition from sediment-smoothed slope to irregular lobate lava downslope from submarine vents is evident on the southeast flank of the Puna Ridge [*Moore and Chadwick,* 1995]. However, distinct slope breaks on both margins of the chute suggest instead that it may represent a separate debris-avalanche scar that cut across the Laupahoehoe slump or an earlier configuration of the flank. A large debris field, covering an area of approximately 2000 km^2, extends another 50 km away from this chute and the submerged base of the island and probably beyond the margin of the well-surveyed region (Fig. 4).

4. HANA RIDGE

Prior to the past few years, little work had been done on the submarine Haleakala east rift zone (HERZ), also referred to as the Hana Ridge [*Clague et al.*, 1998]. The ridge extends from the tip of subaerial Maui, 140 km to the east-southeast where it reaches a water depth of nearly 5500 m at the edge of the Hawaiian Deep (Figs. 1, 5). It is twice as

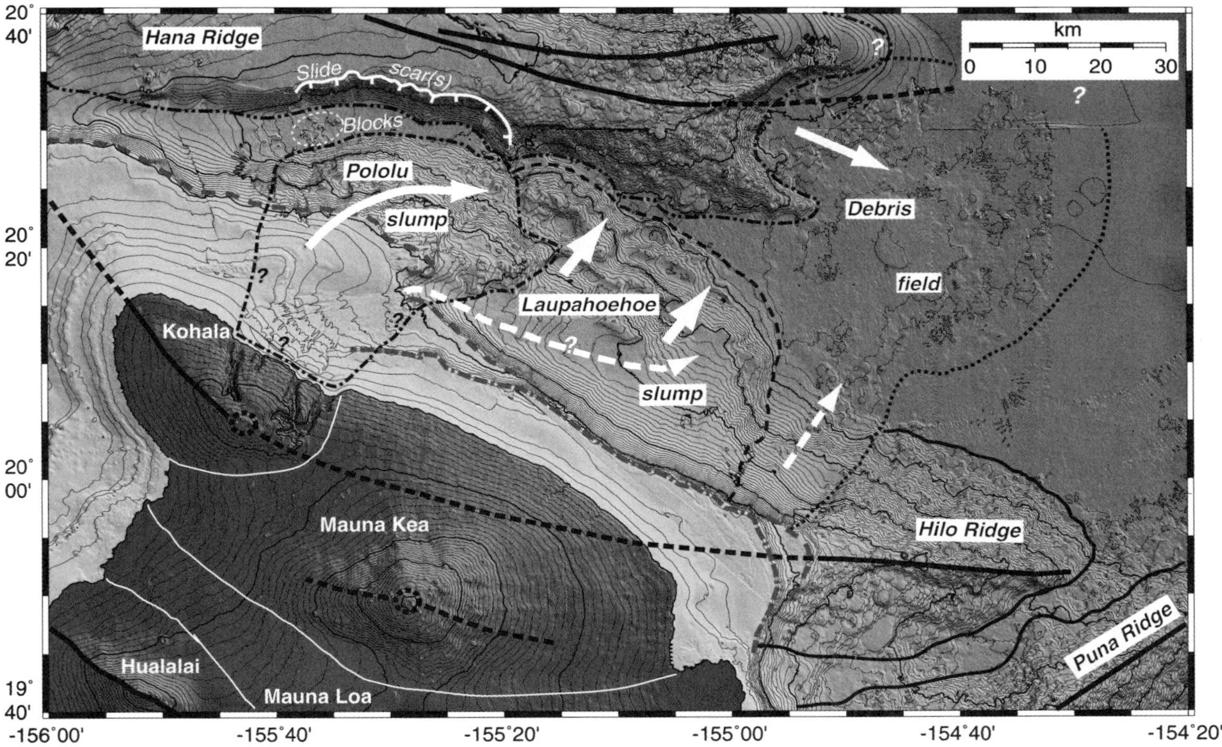

Figure 4. Summary of interpretations, including inferred volcano affinities of submarine ridges (thick dashed line where buried, solid black where exposed), modified from *Holcomb et al.* [2000] and *Kauahikaua et al.* [2000]. Location of proposed buried calderas (dashed circles) drawn near the summits of Kohala and Mauna Kea after same authors. Thick dashed line on Hana Ridge is projected HERZ tip prior to collapse. Volcano boundaries are thinner solid black lines. Solid white arrows indicate direction of slide failure and flow of material to abyssal floor into the debris field outlined by the black dotted line. White dashed arrows are secondary debris pathways. Thick gray dash-dot line is former Mauna Kea coastline, at the end of shield building, corresponding to the 400-m-deep terrace. Thick gray dashed line is former Kohala coastline, at the end of shield building, now at 1100-m water depth. Bathymetry shaded and obliquely illuminated from the northeast; 100 m contours; bold every 1000 m.

long and nearly twice as wide as Puna Ridge, the submarine east rift zone of Kilauea volcano, making it the largest identified rift zone in Hawaii [*Moore et al.*, 1990]. The Hana Ridge has a complex geometry, including a distinct steep-sided arcuate tip, at 3000-5000 m water depth with a diameter of 22 km (Fig. 5a). It is quite unlike the prototypical Hawaiian submarine rift zone (*e.g.*, Puna Ridge) that is wedge-shaped and narrows to a point (Figs. 1, 4).

Separate from the arcuate tip, two subparallel ridge crests are evident on the Hana Ridge, particularly in the backscatter data (Fig. 5b). The northern axis is an east-northeast directed zone of high reflectivity offset from the deep ridge tip, narrowing from a width of 9 km near the fossil reefs to 3 km wide near its distal eastern end (Fig. 5c). A reentrant in the bathymetry (Figs. 5a,c,) topographically separates two axes of volcanism, also corresponding to a change from high acoustic reflectivity of the northern axis of vents (Figs. 5b,c, *Primary axis*) to lower reflectivity along the southern axis. Analysis of aeromagnetic anomaly data for the Hana Ridge has identified a normally magnetized, narrower, shorter rift zone extending from subaerial Maui that is similar in size to the Puna Ridge of Kilauea [*Malahoff and Woollard*, 1968]. A separate set of wider, reversely magnetized source bodies (*i.e.*, dikes) lies in line with the Hana Ridge, extending to the east from the previous anomalies, but are offset slightly to the south. Dual rift structures similar to the Hana Ridge occur on many older and younger volcanoes in the Hawaiian-Emperor chain. These include seamounts at the Hawaii-Emperor bend [*Smoot*, 1982], probably the Kaena rift zone offshore of

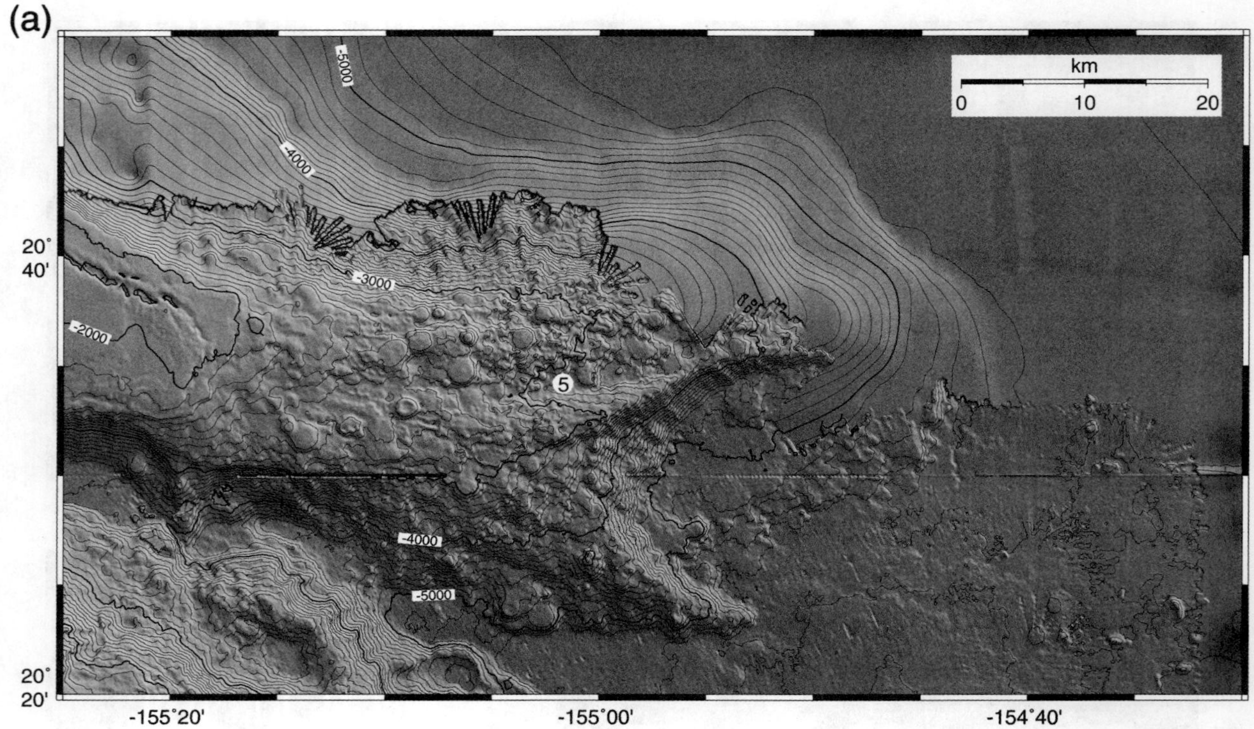
(a)
km
0
10
20
-5000
-4000
-3000
-2000
5
-4000
-5000
20°
40'
20°
20'
-155°20'
-155°00'
-154°40'

(b)
km
0
10
20
20°
40'
20°
20'
-155°20'
-155°00'
-154°40'

Figure 5

Figure 5. The Hana Ridge tip east of the island of Maui. (a) Bathymetry at 100 m contours, shaded and obliquely illuminated from the northeast. Artifacts of data set merging shown on (c). Feature labeled as described in text. (b) Backscatter image, with high acoustic reflectivity as darker gray tones. Limit of JAMSTEC data equals the backscatter coverage in this figure. (c) Interpretation of (a) and (b). Variously weighted dashed and dotted lines outline geologic features as described in the caption for Fig. 2c. *PS*, Pololu Slump.

Oahu [*Duennebier et al.*, 1994] and the north rift zone of the active Loihi submarine volcano southeast of Hawaii [*Chadwick et al.*, 1993; *Garcia et al.*, 1995].

Moore et al. [1989] mapped a landslide, the Hana slump, as covering an area of 4900 km^2 and extending 85 km to the northeast from the Hana Ridge. The landslide begins on the northeast flank of Haleakala and continues 110 km along the Hana Ridge. A 30-km long landslide scarp in the middle of the Hana Ridge, but on its south side, was mapped in both the JAMSTEC and MBARI surveys, but much better detail on the shallow portion was acquired by MBARI (Figs. 1, 4). *Reynolds et al.* [1998] state that this landslide, and slides like it on other Hawaii rift zones, occurred while the rifts were volcanically active. They contend that such a catastrophic episode could lead to a change in the location and propagation direction of the rift axis and associated zone of volcanism.

A series of subsided and tilted reefs are evident along a 55-km long by 20-km wide, smooth section of the shallower portion of the rift zone that was once above sea level [*Moore et al.*, 1990; *Clague et al.*, 1998]. The reefs have now subsided under the weight of both islands and are tilted toward Hawaii [*Moore et al.*, 1990]. A series of pointy-top cones interpreted to have formed by shallow-water or subaerial eruption of postshield-stage alkalic lavas lies close to shore (1000-1400 m water depth) along the western end of the Hana Ridge [*Reynolds et al.*, 1998; *Clague et al.*, 2000]. Farther down rift, these researchers found numerous flat-topped pancake cones composed of tholeiitic lavas, probably erupted during the shield stage [*Clague et al.*, 2000].

Analysis of the bathymetry and backscatter images (Figs. 5a,b) has allowed us to identify 73 volcanic constructs, many of them flat-topped volcanic cones (1-2.5 km dia; 50-150 m relief), along the broad 8-13 km wide crest of the distal rift zone (Fig. 5c). On the northern axis of relatively recent volcanism relative to the remainder of the more highly sedimented Hana Ridge, we mapped 37 flat-topped cones and/or series of lava flow fronts, but few structural lineations (Fig. 5c). As previously noted, *Clague et al.* [2000] mapped such cones on the shallower portion of the Hana Ridge adjacent to our data set and along other Hawaiian rift zones. The cones mapped here correspond in dimension and orientation with those mapped in the *Clague et al.* study.

5. HILO RIDGE

The Hilo Ridge (Fig. 1) has been interpreted as the submarine continuation of an approximately eastern rift zone of either Mauna Kea [*Fiske and Jackson*; 1972; *Moore and Clague*, 1992] or Kohala [*Holcomb et al.*, 2000; *Kauahikaua et al.*, 2000]. *Holcomb et al.* [2000] show multibeam bathymetric data for the Hilo Ridge collected prior to the expanded JAMSTEC surveys. They use those data plus isotopic geochemical analyses of dredged rocks from key locations to extrapolate a new rift zone orientation for the volcanoes composing the northern half of the island of Hawaii. They hypothesize that the Hilo Ridge is a submarine rift zone for Kohala. This idea is reinforced by recent three-dimensional gravity data modeling showing a gravity high extending from Kohala to the east through the present day northeast slope of Mauna Kea to the Hilo Ridge [*Kauahikaua et al.*, 2000], also mimicked in early aeromagnetic data [*Malahoff and Woollard*, 1968]. A reversely polarized anomaly over Hilo Ridge, from analysis of aeromagnetic surveys [*Malahoff and Woollard*, 1968] and marine magnetic anomaly data collected during the JAMSTEC cruises [*Kanamatsu and Tsuboyama*, 2000; *Naka et al.*, this volume], indicates the ridge is much older (>780 ka) than subaerial exposures of any Hawaii Island volcano. From analysis of our new mapping data, we can say that the general size and morphology of the Hilo Ridge (Fig. 4) appear similar to that of Puna Ridge, although flat-topped volcanic cones are less obvious on Puna Ridge [compare with *Clague et al.*, 2000; *Smith and Satake*; *D. Smith et al.*, both this volume]. A more detailed morphometric analysis of the SeaBeam bathymetry and backscatter data for volcanic features on the Hilo Ridge is in progress by S. Umino.

6. DISCUSSION

6.1 Structure and Origin of Pololu and Laupahoehoe Slides

In order to fully understand the volcano flanks and related slumps mapped in this study, it is useful to compare them to other volcanic slopes and associated landslides that have been investigated in more detail. The closest modern analogs to the Pololu and Laupahoehoe slumps may be the mobile south flank of Kilauea volcano (also known as the Hilina slump) and the South Kona and Alika slide complex of southwest Mauna Loa. Historical deformation and seismicity of the mobile south flank of Kilauea are well documented on land [*Lipman et al.*, 1985; *Delaney et al.*, 1993; *Owen et al.*, 1995, 2000; *Denlinger and Okubo*, 1995], and the underwater slopes have been the subject of numerous studies including sonar surveys, seismic reflection work, and petrologic analysis of recovered samples from submersible dives [*Moore and Chadwick*, 1995; *Smith et al.*, 1999; *Morgan et al.*, 2000; *Hills et al.*, this volume; *Lipman et al.*, 2000, this volume; *Naka et al.*, this volume; *Sisson et al.*, this volume]. Studies of the Kona and Alika landslides, primarily using sonar mapping, were carried out by *Lipman et al.* [1988], *Normark et al.* [1993], and *Moore et al.* [1995].

The south flank of Kilauea is located seaward of its summit caldera, east rift zone and southwest rift zone [*Swanson et al.*, 1976; *Holcomb*, 1987]. Two major fault systems are evident on the south flank: the Koae and Hilina fault systems [*Wolfe and Morris,* 1996]. The Koae faults are an enigmatic pattern of antithetic normal faults having a throw opposite to the general slope and direction of seaward motion of the flank, though also accompanied by numerous grabens with downslope facing faults [*Duffield*, 1975]. A second normal fault system, closer to the coast and downthrown seaward , is the Hilina fault system.

The headwall structure of the Pololu slump was previously identified as the graben along the subaerial axial rift zone of Kohala [*Moore et al.*, 1989; *Moore and Clague*, 1992]. It is unlikely that the axial graben-like structure is the headwall because it faces upslope toward the caldera and could not accommodate the necessary northeast directed downslope displacement to explain the slump complex. Kohala's axial faults appear to be more analogous to the Koae faults of Kilauea, which are interpreted to link up the radial rift arms across the summit region and accommodate normal displacement as the rift zones extend [*e.g.*, *Swanson et al.*, 1976; *Denlinger and Okubo*, 1995]. Instead, we propose that the headwall, if one exists, must lie farther downslope, possibly within the previously described reentrant or in the offshore region (Fig. 2c) consistent with the current thinking for Nuuanu and Wailau landslides north of Oahu and Molokai respectively [e.g., *Moore and Clague*, this volume]. In support of this idea is the fact that the original shield surface preserved between the Pololu and Waipio Valleys is continuous with that surrounding it (Figs. 2a, 4), indicating that the subaerial valleys were carved into the original volcanic surface. Headwall erosion to form the high sea cliffs and valleys would have been enhanced by truncation and exposure of deeper strata. Any headwall faults responsible for displacement of the Pololu slump, *i.e.*, similar to the Hilina fault system on Kilauea volcano, are probably now covered by the carbonate platform (Fig. 2c). Similarly, the headwall of the Laupahoehoe slump, if it is a distinct feature, has been long buried by volcanic activity from Mauna Kea.

Interpreting the internal structure and origin of the Pololu slump is problematic, as we still lack important data sets such as subsurface images and direct sampling. *Clague et al.* [2000] suggest that the Pololu slide may be a rotational slump, analogous to the Hilina slump on the southeast flank of Kilauea, because several cones apparently slid intact and came to rest tilted 4-5° toward the island (Fig. 3). In the new bathymetric compilation over this area, we recognize possible slump detachments and fault scarps that may have accommodated east or northeast directed displacement of the eastern volcanic block. The bulk of the Pololu slump is cut by faults and fractures, suggesting it defines a highly broken unit of rock subject to large-scale collapse. The broad frontal slopes of the northern Pololu unit, however, are not obviously transected by fractures but instead show many narrow, closely spaced, downslope channels (Fig. 3). The eroded but coherent character of this zone suggests long exposure without significant tectonic or gravitational deformation, although the low relief east-west trending ridges and basins found at ~2000 m water depth on the northern slopes of Pololu do suggest a past history of slumping or thrust faulting at the base of the slope.

The highly eroded nature and irregular morphology of the Pololu slump mask the materials that compose unit. The flat-topped lava cones noted along the eastern edge by *Clague et al.* [2000] demonstrate late stage volcanic activity took place, at least in that area. However, the broad slopes and presence of sediment filled basins fronted by narrow ridges along the northern edge point to a mantle of sediments (Fig. 3), probably fragmental basalts as found at Kilauea [*Lee et al.*, 1994] or uplifted debris deposited in an older volcanic apron [*e.g.*, *Morgan et al.*, 2000]. Without direct observation, it is difficult to infer more about the composition of this region.

Morphologically, the offshore slope associated with the newly imaged Laupahoehoe slump resembles the submarine south flank of Kilauea volcano where four distinct zones have been recognized: the upper slope, midslope bench, transition zone, and outer scarp/foot [*Smith et al.*, 1999]. The Kilauea upper slope equivalent is the apparently undeformed and gently dipping slope of Mauna Kea at a current water depth of 1200-3000 m, upslope of the Laupahoehoe ridge and basin terrain (Fig 2a). This slope is proposed to be a zone of volcanic sand covering an area of 375 km^2 downslope from the northeast coastline of Mauna Kea (Figs. 2b,c). Ongoing eruptions at Kilauea suggest that downslope directed reflective streaks in the sonar imagery represent a mantle of fragmental basalt [*Smith et al.*, 1999], formed by shattering of subaerially erupted lavas entering the sea, as well as debris derived from larger collapsed lava deltas [*Lee et al.*, 1994; *Moore and Chadwick*, 1995]. The presence of similar streaks offshore Mauna Kea (Fig. 2b), which has ceased erupting, suggests that reflective sediments are still exposed and derive in part from erosion of previously emplaced flows, evident in the ubiquitous sea cliffs along the Hamakua Coast. The upper slopes of both Kilauea and Mauna Kea may be underlain by primary volcanics [*Morgan et al.*, 2000].

The central region of Kilauea's south flank is characterized by a single, extensive midslope bench, basin, and seaward transition zone (bench outer high of *Hills et al.*, this volume) [*Smith et al.*, 1999; *Morgan et al.*, 2000], in contrast to the multiple small basins and frontal ridges observed on the Pololu and Laupahoehoe slumps. The basins and ridges are well expressed in the bathymetry; they begin at a water depth greater than 3000 m and occur as deep as 5000 m at the base of the volcano (Fig 2a). The basins are consistently underfilled, suggesting recent formation or limited clastic sediment input. However, the low contrast, smooth character of the frontal ridges in the backscatter data (Fig. 2b) indicates well sedimented ridges with few structural lineations apparent at the surface (Fig. 2c) suggesting the toe of the Laupahoehoe slump has been inactive for some time. This is in contrast to the rough surface defining the transition zone at Kilauea, which appears highly fractured in a coherent, along-strike pattern [*Smith et al.*, 1999].

The origin of the distal basins and frontal ridges on the Laupahoehoe slump, and by analogy those on the northern Pololu slump, is uncertain; they may result from local rotational slumps or from uplift above discontinuous thrust faults in the bench. They are similar in scale and structural position to the series of broad basal benches and linear ridges of the proximal Puna Ridge, consisting of pillow lavas of transitional composition that may represent structural terraces related to gravitational spreading and slumping of the rift flanks [*Clague et al.*, 1994; *Moore and Chadwick*, 1995; *Lipman et al.*, 2000, this volume]. Analogous lobate-terraced debris slide deposits also surround the bases of Cretaceous seamounts southwest of Hawaii island [*Normark et al.*, 1987; *Lipman et al.*, 1988]. Those authors interpreted the deposits as an older assemblage of gravitational debris slide material.

The gravitational debris origin is difficult to reconcile with the morphology of these bench and basin constructs. Specifically, the frontal ridges extend beyond the confines of the basins, essentially enclosing them. Although debris avalanche and flow deposits may be lobate in form, they tend to be thickest near their centers and taper toward their edges [*e.g.*, *Varnes*, 1978]. However, basins can form above discrete thrust faults proposed at a larger scale for Kilauea's mobile south flank, as shown by multi-channel seismic reflection data [*Morgan et al.*, 2000; *Hills et al.*, this volume]. Compression near the base of the slope, whether due to volcanic spreading [*e.g.*, *Borgia,* 1994; *Denlinger and Okubo*, 1995] or slumping [*Varnes,* 1978], will cause overthrusting, uplifting, and tilting to transform accumulated debris into ridges and ponded basins. The small size and close spacing of the basins suggest that they formed upon relatively thin thrust sheets, which have experienced only moderate seaward displacement, again in contrast to Kilauea's south flank [*Morgan et al.,* 2000; *Hills et al.,* this volume]. Interestingly, several of the elongate basins, although morphologically distinct, do line up into rows (Figs. 2a,c), suggesting that they may have formed above laterally continuous thrust faults that have since been offset and segmented along a lateral fault similar to that interpreted for Kilauea by *Hills et al.* [this volume].

On Kilauea, an outer scarp (lower scarp of *Lipman et al.*, 2000, this volume) extends from the edge of the midslope bench and transition zone to the abyssal seafloor, revealing chaotic patterns of high and low amplitude reflections in sidescan images [*Smith et al.*, 1999]. JAMSTEC-sponsored submersible surveys revealed that this outer scarp is cut into a thick accumulation of massive volcaniclastic breccias and highly jointed indurated sandstones [*Naka et al.*, 2000], probably derived from progressive mass wasting of the upper volcanic slopes. Although much smaller in scale, the Laupahoehoe and Pololu slumps reveal similar features. The segmented nature of Laupahoehoe results in multiple small-scale outer scarps with well developed scalloped scars 1.5-4 km wide at ~4000 m (Figs. 2a-c). Their shape and acoustic reflectivity are similar to the concave failure scars that are 5-7 km wide on the Hilina outer scarp [*Smith et al.,* 1999]. As with Kilauea, continuing deformation, perhaps in the form of thrust toes emerging in the lower parts of the flank, might steepen the outer scarps and thus intermittently initiate these localized failures. Once the entire slump body is emplaced and stalled, as is likely the case at Laupahoehoe, these concave scars may still form or enlarge, possibly triggered by island earthquakes or local gravitational collapse on the leading edge of the slump.

The debris field directed to the northeast of Hawaii Island off the Hamakua Coast, represented by a welt in the bathymetry data with a relief of 200-300 m (Fig. 4), is not distinctive in the backscatter data for the entire flank [*Smith and Satake*, this volume]. An oddity is that the debris field, though well imaged by SeaBeam at least proximally, contains few large blocks, in contrast to the debris fields at the base of the Kilauea flank [*Moore and Chadwick*, 1995; *Smith et al.*, 1999] and South Kona slump [*Lipman et al.,* 1988; *Moore et al.,* 1995]. It appears the material could derive from several surrounding source areas (*i.e.*, Pololu, Laupahoehoe, and Hana Ridge) and may represent turbidite sequences, debris avalanche material, and blocks from the arcuate landslide scars on the lower slopes. We interpret this to define a volcanic apron, probably continuous with offscraped strata in the lobate terraces and benches composing the Laupahoehoe slump.

The juxtaposition of the Pololu and Laupahoehoe slumps, which show such distinctly different morphologies, leads us to speculate on the relationship between the two and the relative timing of their emplacement. A first order interpretation is that the two slides share a common history, possibly resulting from compression and uplift, forming the ridge and basin features at the toes of the slopes. Displacements on basal thrusts resulting in uplift and deformation may have been greatest in the current vicinity of the Pololu slide, accounting for greater uplift in this region. In this case, we might argue that both features derive from Kohala volcano, which spread laterally under gravitational load as hypothesized for Kilauea volcano [*Borgia and Treves*, 1992; *Denlinger and Okubo*, 1995; *Morgan et al.*, 2000].

However, the different trends of the bench and basin domains, nearly east to west on the northern Pololu unit (Fig. 3) and southeast to northwest on the Laupahoehoe slump (Figs. 2a,c), suggest that the two features are separate. This argument is strengthened by their division by the volcanic high at the south edge of Pololu and the broad sediment chute directed outward from the Pololu amphitheater (Figs. 2c, 3). The latter feature, in combination with waning eruptive activity on Kohala, reduced the sediment supply to the Pololu slump; the subsequent evolution of Pololu was one of erosional degradation, while the Laupahoehoe slump continued to accumulate a smoothing blanket of sediment.

Despite the more eroded appearance of the Pololu slump, it appears to override Laupahoehoe features to the southeast suggesting that Pololu may in fact be younger, at least in part. By its location, Pololu cannot originate from Mauna Kea volcano, therefore Kohala is the likely candidate. This suggests Laupahoehoe is also related to Kohala, implying that the more ancient volcano extends much farther to the southeast than previously proposed [*Fiske and Jackson*; 1972], in agreement with recent models for the origin of Hilo Ridge as an eastern rift zone of Kohala volcano, presented in section 5.0 [*Holcomb et al.*, 2000; *Kauahikaua et al.*, 2000]. Both the Pololu and Laupahoehoe slumps are overlain by the two slope breaks at 1100 and 400 m water depth, the deeper of which is attributed to the end of shield growth on Kohala volcano (Figs. 2c, 4). In further defense of this model, we call upon the magnetic anomaly data over Hilo Ridge described earlier [*Malahoff and Woollard*, 1968; *Kanamatsu and Tsuboyama*, 2000; *Naka et al.*, this volume] to suggest that the Hilo Ridge is old enough to be a former southeast rift zone of Kohala instead of Mauna Kea, as previously accepted.

The structural picture is complicated by the possibility that the eastern part of Pololu may have partially slumped, overriding the distal Laupahoehoe ridges; the boundary between the two domains is difficult to identify precisely in this area (Fig. 3). This would allow for Pololu to be of greater age than Laupahoehoe, consistent with its more broken and eroded appearance. In addition, the east to west trend of the ridge and basin structures to the north does not parallel the flanks of either Kohala or Mauna Kea volcanoes (Fig. 2a), as one might expect if the compressional structures formed perpendicular to the axial rift zones as observed elsewhere [*e.g.*, *Smith et al.*, 1999; *Hills et al.*, this volume]. This leads us to speculate that the Pololu slump may actually relate to an even older volcano. One possibility is the completely submerged Mahukona volcano found to the west of Kohala (Fig. 1). Although the summit area for Mahukona has been previously interpreted to lie in the western waters [*Garcia et al.*, 1990; *Clague and Moore*, 1991], we recognize an eastward continuation of the deep contours that outline Mahukona on the west side of Kohala [*Smith*, 1994] and appear to define the Pololu lobe. If Mahukona outcrops to the east of Kohala, then the volcanic features observed on eastern Pololu may actually define a submarine rift zone of the ancient Mahukona volcano. The mantling sediments over the northern slopes of Pololu slump may also have derived from Mahukona, although lava flows and fragmental basalts from Kohala have subsequently blanketed the nearshore portions of the submarine edifice, masking its origin. These intriguing possibilities remain to be tested by future surveys over northeastern flanks of Hawaii.

6.2 Evolution of Hana Ridge

The Hana Ridge tip displays an unusual (for a rift zone) gaping concave geometry that we have interpreted as a collapse scar, even though the limited bathymetric and backscatter sonar data immediately downslope from this feature do not show a large debris field (Figs. 5a-c). The striking divergence at the arcuate tip may either be due to a landslide failure of the central portion, or a primary constructional feature. The tip exhibits the classic amphitheater shape of a landslide, here suprisingly circular, but the volume of the debris lobe embraced by the branching arms appears rather small (Fig. 5a). However, the northern tip and base level upon which the debris lies were not fully imaged in this study. The original HERZ axis of volcanism may have been projected along this trend (Fig. 4). At a later point in the development of the Hana Ridge, perhaps prior to 400 ka in keeping with the *Moore and Clague* [1992] model for the peak of shield building on Haleakala, this tip collapsed, possibly by oversteepening and/or magma pressure with no buttressing from adjacent features. Such a catastrophic modification to a volcanically active conduit may have had extreme results. The neovolcanic zone may have shifted to the most recently active axis with a more east-northeast trend just north of the original morphologically dominant and geometrically central axis (Fig. 4).

McGuire and Pullen [1989] showed that the gravitational stress regime generated by the gross morphology of the volcanic edifice is modified by the local topography of the vertical walls of the Valle del Bove caldera on Etna, which served to steer dikes around the caldera rather than projecting through its walls. This observation was further reinforced by their qualitative physical models using gelatin molds in the same paper, following the work of Fiske and Jackson [1972]. The northern splay of the Hana Ridge neovolcanic zone (Figs. 5b,c) may indicate such a redirection of plumbing around the failure scar. However, in the case of subaerial stratovolcanoes where collapses of volcanic edifices have been observed, many of the volcanoes completely recover the lost portion with little, if any, trace of failure (*e.g.*, Fuji). Following this model, the collapse event(s) would not have affected the shallow level magma plumbing system. Other scenarios may include (1) that the Haleakala rift shifted northward on its own prior to the collapse or (2) that both rift branches were active a long time prior to failure and then the southern rift was abandoned following the collapse.

Alternatively, could a classic "Y"-shaped seamount with three subequal rift zones have been overridden and partly buried, along its northern flank, by the younger trend of the HERZ that is so impressive on the backscatter image (Fig. 5b)? The magnetic anomaly data show source bodies of reverse polarity discontinuous from the proximal Hana Ridge, larger in size than typical offshore Hawaiian rift zones and centered directly over this proposed seamount [Malahoff and Woollard, 1968]. While it would be regarded as a large submarine edifice, this "Y"-shaped seamount would have failed to reach sea level by about 500 m, perhaps because it lay well north of the hotspot axis. As such, it should contain a high proportion of alkalic rocks, in comparison to the distal Hana Ridge, a relation that could readily be tested by submersible or even dredge sampling. Based on the aeromagnetic data over Hana Ridge, it is possible that either hypothesis is true, but because of the reversed anomaly, the distal end with the branching arms must be older than the proximal section of the Haleakala east rift zone.

7. CONCLUSIONS

A newly identified landslide unit off the Hamakua Coast is named the Laupahoehoe slump. It is morphologically and perhaps structurally similar to the mobile south flank of Kilauea (Hilina slump) and the Kona/Alika landslide complex. The Laupahoehoe slump appears to be older than the active Hilina slump in the backscatter imagery and may have halted in an earlier stage of development during off-scraping of debris at the distal edge of the slump. The feature likely originated from the Kohala edifice, thus predating Mauna Kea volcanics which overlapped the upper flank, buried pre-existing headwall faults, and delivered mantling volcaniclastic debris to the slopes below. This study supports results of other recent investigations [*Holcomb et al.,* 2000; *Kauahikaua et al.*, 2000] that suggest the Hilo rift zone belongs to Kohala volcano rather than Mauna Kea, as previously accepted.

The Pololu slide has been mapped in more detail than before. The backscatter data show a chaotic surface morphology and structure that is overlain by a carbonate platform and reef terrace deposits, subsequently cut throughout by submarine canyons formed by erosion from submarine turbidites and/or dissolution from freshwater seeps. The eastern edge of the Pololu slide appears to override the northwestern portion of the Laupahoehoe slump, possibly due to slumping of a large volcanic block along a north trending scarp. The northern flank of Pololu contains several slope parallel ridges and basins similar to those on the newly recognized Laupahoehoe slump. The east-to-west trend of these structures suggests that they may have formed along the much older Mahukona volcano, previously recognized only off the west coast of Kohala.

Mechanisms of emplacement for the mobile south flank of Kilauea recently published by other researchers (including this volume) may be applicable to explain generation of the similar morphologies and structures of the Pololu and Laupahoehoe slide complex. For example, the Laupahoehoe slump may record seaward displacement and overthrusting at the toe of a spreading volcano, as interpreted for Kilauea's mobile south flank by *Morgan et al.* [2000] and *Hills et al.* [this volume]. Secondly, the Pololu slide may have detached from the Kohala (or Mahukona) edifice forming a slump block/debris avalanche complex similar to that interpreted for the South Kona/Alika slide southeast of Mauna Loa [*Lipman et al.,* 1988; *Moore et al.,* 1995]. Future studies employing seismic reflection methods and/or sampling of the slide components by coring, dredging, or submersible will be necessary to build more refined interpretations of the subsurface structure of this region. Analysis of this two-slide overlapping complex may shed new light on the results acquired from the Hawaii Scientific Drill Project, especially if both slides originated from Kohala prior to being subaerially resurfaced by Mauna Kea lavas.

The nearby Hana Ridge is approximately twice as long and wide as either Hilo or Puna Ridges. It has a complicated history that may include collapse of the rift zone tip prior to, or following, a major redirection of volcanic plumbing to a new volcanic zone adjacent to the original rift but splayed to the north. A more intriguing explanation, requiring additional study, is that the distal Hana Ridge, as mapped here, is a separate seamount juxtaposed against the proximal Haleakala east rift zone.

Acknowledgments. We are indebted to the captains, marine crews, and operations teams of the R/V's *Kairei* and *Yokosuka* of the Japan Marine Science and Technology Center (JAMSTEC). JAMSTEC also generously provided the funding for these expeditions. The Hawaii Undersea Research Laboratory at the University of Hawaii furnished computer and salary support for JRS to carry out shoreside post-processing and analysis of the data used in this study. Rapid release of the MBARI multibeam data sets (PI, D.A. Clague) significantly enhanced the interpretations we could make in the intermediate water depths. The manuscript was greatly improved from reviews by J. Sinton, S. Umino, and S. Day. SOEST contribution #5575.

REFERENCES

Borgia, A., Dynamic basis of volcanic spreading, *J. Geophys. Res.*, *99*, 17,791-17,804, 1994.

Borgia, A., and B. Treves, Volcanic plates overriding the oceanic crust: structure and dynamics of Hawaiian volcanoes, in *Geol. Soc. Spec. Pub.*, *60*, edited by L.M. Parson, B.J. Murton, and P. Browning, pp. 277-299, The Geological Society, London, 1992.

Campbell, J.F., Bathymetric atlas of the southeast Hawaiian Islands, Hawaii Institute of Geophysics, Honolulu, 1987.

Caress, D.W., and D.N. Chayes, Improved processing of Hydrosweep DS multibeam data on the R/V Maurice Ewing, *Mar. Geophys. Res.*, *18*, 631-650, 1996.

Chadwick, W.W., Jr., J.R. Smith, Jr., J.G. Moore, D.A. Clague, M.O. Garcia, and C.G. Fox, Bathymetry of south flank of Kilauea Volcano, Hawaii, *U.S. Geol. Surv. Misc. Field Studies Map MF-2231*, 1993.

Clague, D.A., K.A. Hon, J.L. Anderson, W.W. Chadwick, Jr., and C.G. Fox, Bathymetry of Puna Ridge, Kilauea Volcano, Hawaii, *U.S. Geol. Surv. Misc. Field Studies Map MF-2237*, 1994.

Clague, D.A., and J.G. Moore, Geology and petrology of Mahukona Volcano, Hawaii, *Bull. Volcanol.*, *53*, 159-172, 1991.

Clague, D.A., J.G. Moore, and J.R. Reynolds, Formation of submarine flat-topped volcanic cones in Hawaii, *Bull. Volcanol.*, *62*, 214-233, 2000.

Clague, D.A., J.R. Reynolds, N. Maher, G. Hatcher, W. Danforth, and J.V. Gardner, High-resolution Simrad EM300 multibeam surveys near the Hawaiian Islands: canyons, reefs, and landslides, (abstract), *Eos Trans. AGU*, *79*, F826, 1998.

Delaney, P.T., A. Miklius, T. Arnadottir, A.T. Okamura, and M.K. Sako, Motion of Kilauea Volcano during sustained eruption from the Puu Oo and Kupaianaha vents, 1983-1991, *J. Geophys. Res.*, *98*, 17,801-17,820, 1993.

Denlinger, R.P., and P. Okubo, Structure of the mobile south flank of Kilauea Volcano, Hawaii, *J. Geophys. Res.*, *100*, 24,499-24,507, 1995.

Duennebier, T., T. Reed, and H. Staff, Northwestern Hawaiian Islands: merged bathymetry and topography, Sheet #2, *Hawaii Seafloor Atlas,* Hawaii Institute of Geophysics and Planetology, Honolulu, 1994.

Duffield, W.A., Structure and origin of the Koae fault system, Kilauea Volcano, Hawaii, *U.S. Geol. Surv. Prof. Pap. 856*, 12 p, 1975.

Garcia, M.O., D.J.P. Foss, H.B. West, and J.J. Mahoney, Geochemical and isotopic evolution of Loihi Volcano, Hawaii, *J. Petrology*, *36*, 1647-1674, 1995.

Garcia, M.O., M.D. Kurz, and D.W. Muenow, Mahukona: the missing Hawaiian volcano, *Geology*, *18*, 1111-1114, 1990.

Fiske, R.S., and E.D. Jackson, Orientation and growth of Hawaiian volcanic rifts: the effect of regional structure and gravitational stresses, *Royal Society of London Proceedings*, *Series A. Mathematical and Physical Sciences* (329), 299-326, 1972.

Groome, M.G., C.E. Gutmacher, and A.J. Stevenson, Atlas of GLORIA sidescan-sonar imagery of the Exclusive Economic Zone of the United States: EEZ-View, *U.S. Geol. Surv. Open-File Rep. 97-540*, 1997.

Hildenbrand, T.G., J.G. Rosenbaum, and J.P. Kauahikaua, Aeromagnetic study of the island of Hawaii, *J. Geophys. Res.*, *98*, 4099-4119, 1993.

Hills, D.J., J.K. Morgan, G.F. Moore, and S.C. Leslie, Structural variability along the submarine south flank of Kilauea volcano, Hawaii, from a multichannel seismic reflection survey, this volume.

Holcomb, R.T., Eruptive history and long-term behavior of Kilauea Volcano, in *U.S. Geol. Surv. Prof. Pap. 1350*, edited by R.W. Decker, T.L. Wright, and P.H. Stauffer, pp. 261-350, 1987.

Holcomb, R.T., B.K. Nelson, P.W. Reiners, and N.-L. Sawyer, Overlapping volcanoes: the origin of Hilo Ridge, Hawaii, *Geology*, *28*, 547-550, 2000.

Kanamatsu, T., and N. Tsuboyama, Results of underway geophysical surveys, gravity, magnetic, and single-channel seismic surveys, during cruises KR98-08, 09 and YK99-07, 08, (abstract), *Eos Trans. AGU*, *81*, 238, 2000.

Kauahikaua, J., T. Hildenbrand, and M. Webring, Deep magmatic structures of Hawaiian volcanoes, imaged by three-dimensional gravity models, *Geology*, 28, 883-886, 2000.

Lee, H.J., M.E. Torresan, and W. McArthur, Stability of submerged slopes on the flanks of the Hawaiian Islands, a simplified approach, *U.S. Geol. Surv. Open-File Rep. 94-638*, 54 p, 1994.

Lipman, P.W., J.P. Lockwood, R.T. Okamura, D.A. Swanson, and K.M. Yamashita, Ground deformation associated with the 1975 magnitude-7.2 earthquake and resulting changes in activity of Kilauea Volcano, Hawaii, *U.S. Geol. Surv. Prof. Pap. 1276*, 45 p, 1985.

Lipman, P.W., W.R. Normark, J.G. Moore, J.B. Wilson, and C.E. Gutmacher, The giant submarine Alika debris slide, Mauna Loa, Hawaii, *J. Geophys. Res.*, *93*, 4279-4299, 1988.

Lipman, P.W., T.W. Sisson, T. Ui, and J. Naka, In search of ancestral Kilauea volcano, *Geology*, *28*, 1079-1082, 2000.

Lipman, P.W., T.W. Sisson, T. Ui, J. Naka, and J.R. Smith, Ancestral submarine growth of Kilauea Volcano and instability of its south flank, this volume.

Malahoff, A., and G.P. Woollard, Magnetic and tectonic trends over the Hawaiian Ridge, in *Am. Geophys. Union Monogr. 15*, edited by L. Knopoff, C.L. Drake, and P.J. Hart, pp. 241-276, 1968.

MBARI, MBARI Multibeam Survey, Version 1, CD-ROM, Monterey Bay Aquarium Research Institute, Monterey, 2000.

McGuire, W.J., and A.D. Pullen, Location and orientation of eruptive fissures and feeder-dykes at Mount Etna; influence of gravitational and regional tectonic stress regimes, *J. Volcanol. Geotherm Res.*, *38*, 325-344, 1989.

Moore, J.G., Subsidence of the Hawaiian Ridge, in *U.S. Geol. Surv. Prof. Pap. 1350*, edited by R.W. Decker, T.L. Wright, and P.H. Stauffer, pp. 85-100, 1987.

Moore, J.G., W.B. Bryan, M.H. Beeson, and W.R. Normark, Giant blocks in the South Kona landslide, Hawaii, *Geology*, *23*, 125-128, 1995.

Moore, J.G., and J.F. Campbell, Age of tilted reefs, Hawaii, *J. Geophys. Res.*, *92*, 2641-2646, 1987.

Moore, J.G., and W.W.J. Chadwick, Offshore geology of Mauna Loa and adjacent areas, Hawaii, in *Am. Geophys. Union Monogr. 92*, edited by J.M. Rhodes, and J.P. Lockwood, pp. 21-44, American Geophysical Union, Washington, DC, 1995.

Moore, J.G., and D.A. Clague, Volcano growth and evolution of the island of Hawaii, *Geol. Soc. Amer. Bull.*, *104*, 1471-1484, 1992.

Moore, J.G. and D.A. Clague, Mapping the Nuuanu and Wailau landslides in Hawaii, this volume.

Moore, J.G., D.A. Clague, R.T. Holcomb, P.W. Lipman, W.R. Normark, and M.E. Torresan, Prodigious submarine landslides on the Hawaiian Ridge, *J. Geophys. Res.*, *94*, 17,465-17,484, 1989.

Moore, J.G., D.A. Clague, K.R. Ludwig, and R.K. Mark, Subsidence and volcanism of the Haleakala Ridge, Hawaii, *J. Volcanol. Geotherm. Res.*, *42*, 273-284, 1990.

Moore, J.G., W.R. Normark, and R.T. Holcomb, Giant Hawaiian landslides, *Ann. Rev. Earth Planet. Sci.*, *22*, 119-144, 1994.

Morgan, J.K., G.F. Moore, D.J. Hills, and S. Leslie, Overthrusting and sediment accretion along Kilauea's mobile south flank, Hawaii: evidence for volcanic spreading from marine seismic reflection data, *Geology*, *28*, 667-670, 2000.

Naka, J., T. Kanamatsu, P.W. Lipman, T.W. Sisson, N. Tsuboyama, J.K. Morgan, J.R. Smith, and T. Ui, Deep-sea volcaniclastic sedimentation around the southern flank of Hawaii, this volume.

Naka, J., and Scientific Team, Tectono-magmatic processes investigated at deep-water flanks of Hawaiian volcanoes, *Eos Trans. AGU, 81*, 221, 226-227, 2000.

National Ocean Survey, Hydrographic survey sheet H-9974 (Offshore Upolu Pt. to Laupahoehoe Pt.), Rockville, MD, 1983.

Normark, W.R., J.G. Moore, and M.E. Torresan, Giant volcano-related landslides and the development of the Hawaiian Islands, in *U.S. Geol. Surv. Bull. 2002*, edited by W.C. Schwab, H.J. Lee, and D.C. Twichell, pp. 184-196, 1993.

Normark, W.R., P.W. Lipman, J.B. Wilson, C.J. Jacobs, D.P. Johnson, and C.E. Gutmacher, Preliminary cruise report, Hawaiian GLORIA Leg 2, F6-86-HW, November 1986, *U.S. Geol. Surv. Open-File Rep. 87-298*, 197 p., 1987.

Owen, S., P. Segall, J. Freymueller, A. Miklius, R. Denlinger, T. Arnadottir, M. Sako, and R. Burgmann, Rapid deformation of the south flank of Kilauea Volcano, Hawaii, *Science, 267*, 1328-1332, 1995.

Owen, S., P. Segall, M. Lisowski, A. Miklius, R. Denlinger, and M. Sako, Rapid deformation of Kilauea Volcano: global positioning system measurements between 1990 and 1996, *J. Geophys. Res., 105*, 18,983-18,998, 2000.

Reynolds, J.R., D.A. Clague, N. Maher, and G. Hatcher, Evolutionary sequence of submarine volcanic rift zones in Hawaii, (abstract), *Eos Trans. AGU, 79*, F825, 1998.

Sisson, T.W., P.W. Lipman, and J. Naka, Submarine alkalic to tholeiitic shield-stage development of Kilauea volcano, Hawaii, this volume.

Smith, D.K., L. S. Kong, K.T.M. Johnson, J. Reynolds, Detailed volcanic morphology of the submarine Puna ridge, Kilauea volcano, this volume.

Smith, J.R., Island of Hawaii and Loihi submarine volcano, high resolution multibeam bathymetry around the Island of Hawaii, Sheet #6, *Hawaii Seafloor Atlas,* Hawaii Institute of Geophysics and Planetology, Honolulu, 1994.

Smith, J.R., A. Malahoff, and A.N. Shor, Submarine geology of the Hilina slump and morpho-structural evolution of Kilauea Volcano, Hawaii, *J. Volcanol. Geotherm Res., 94*, 59-88, 1999.

Smith, J.R., and K. Satake, Deepwater multibeam sonar surveys along the southeastern Hawaiian Ridge, this volume.

Smith, W.H.F., and P. Wessel, Gridding with continuous curvature splines in tension, *Geophysics, 55*, 293-305, 1990.

Smoot, N.C., Guyots of the mid-Emperor Chain mapped with multi-beam sonar, *Mar. Geol., 47*, 153-163, 1982.

Swanson, D.A., W.A. Duffield, and R.S. Fiske, Displacement of the south flank of Kilauea Volcano: the result of forceful intrusion of magma into the rift zones, *U.S. Geol. Surv. Prof. Pap. 963*, 39 p, 1976.

Varnes, D.J., Slope movement types and processes, in *Landslides, Analysis and Control, Special Report 176*, edited by R.L. Schuster, and R.J. Krizek, pp. 11-33, National Academy of Sciences, Washington, D.C., 1978.

Wessel, P., and W.H.F. Smith, New version of the Generic Mapping Tools released, *Eos Trans. AGU, 76*, 329, 1995.

Wolfe, E.W., and J. Morris, Geologic map of the island of Hawaii, *U.S. Geol. Surv. Misc. Invest. Series Map I-2524-A*, 1996.

P. W. Lipman, Volcano Hazards Program, U.S. Geological Survey, 345 Middlefield Rd., MS 910, Menlo Park, CA 94025 (plipman@usgs.gov).

J. K. Morgan, Department of Earth Science, Rice University, 6100 Main St., MS-126, Houston, TX 77005 (morganj@esci.rice.edu).

K. Satake, Active Fault Research Center, AIST-GSJ, Tsukuba, Ibaraki 305-8567, Japan (kenji.satake@aist.go.jp).

J. R. Smith, Hawaii Undersea Research Lab, University of Hawaii, 1000 Pope Rd., MSB 303, Honolulu, HI 96822 (jrsmith@soest.hawaii.edu).

Deep-Sea Volcaniclastic Sedimentation Around the Southern Flank of Hawaii

Jiro Naka[1], Toshiya Kanamatsu[1], Peter W. Lipman[2], Thomas W. Sisson[2], Nohiro Tsuboyama[1], Julia K. Morgan[3], John R. Smith[4] and Tadahide Ui[5]

Most slopes of the Hilina slump are steep, but local small benches, mantled by volcaniclastic sand and fine sediments, were sampled in 1998-1999 with ROV KAIKO and DSRV SHINKAI 6500. Most surficial glass sands on the Hilina slump have compositions of subaerially erupted Kilauea lava, which fragmented and quenched as they entered the sea. Samples from the base of the Puna Ridge contain both subaerially and submarine-erupted fragments of Kilauea composition. Some glass sands from the base of Loihi contain both subaerial Kilauea and submarine Loihi compositions in the same bed. Two piston cores, collected 120 km (P6) and 250 km (P5) southeast of Hawaii, are composed of fine sediments interbedded with volcaniclastic turbidite layers. Although some volcaniclastic fragments in the fine sediments intervals are disturbed by bioturbation, nearly continuous volcanostratigraphic sequences are preserved. P6 and the upper 2.3 m of P5 are normally magnetized: the lower part of P5 is reversed and is therefore older than 0.78 Ma. At depths of 4 m below sea floor (mbsf) in P6 and 1 mbsf in P5, the dominant glass compositions change down core from Kilauea to Mauna Loa type. In the P6, an interval of abundant submarine-erupted alkalic glasses lies between 3.3 and 1.75 mbsf and may record the ancestral alkalic phase of Kilauea volcano. Magnetic susceptibility trends and glass compositions suggest that the entire P6 core correlates with only the uppermost 1.2 m of the P5, and that the average sedimentation rate at the P6 is about 5 times greater than that at P5.

[1]Deep Sea Research Department, JAMSTEC, Yokosuka, Japan
[2]Volcano Hazards Program, U.S. Geological Survey, Menlo Park, California
[3]Department of Geology and Geophysics, Rice Univ. Houston, Texas
[4]School of Ocean & Earth Science & Technology, University of Hawaii, Honolulu, Hawaii
[5]Graduate School of Science, Hokkaido University, Sapporo, Japan

Hawaiian Volcanoes: Deep Underwater Perspectives
Geophysical Monograph 128
Copyright 2002 by the American Geophysical Union

1. INTRODUCTION

The southern flank of Hawaii is unstable in response to the growth of Kilauea and Mauna Loa and to magmatic activity within these shield volcanoes [*Swanson et al.,* 1976; *Lipman et al.,* 1985; *Owens et al.,* 2000]. Such instability has triggered giant submarine landslides and volcaniclastic turbidites that surround the Hawaiian Islands [*Moore,* 1964; *Lipman et al.,* 1988; *Moore et al.,* 1989; *Garcia,* 1996]. The resulting volcaniclastic sequences, which likely form during the main shield-building phase of oceanic volcanoes [*Kelts and Arthur,* 1981], can provide a long-term volcano-stratigraphic record of hotspot volcanism.

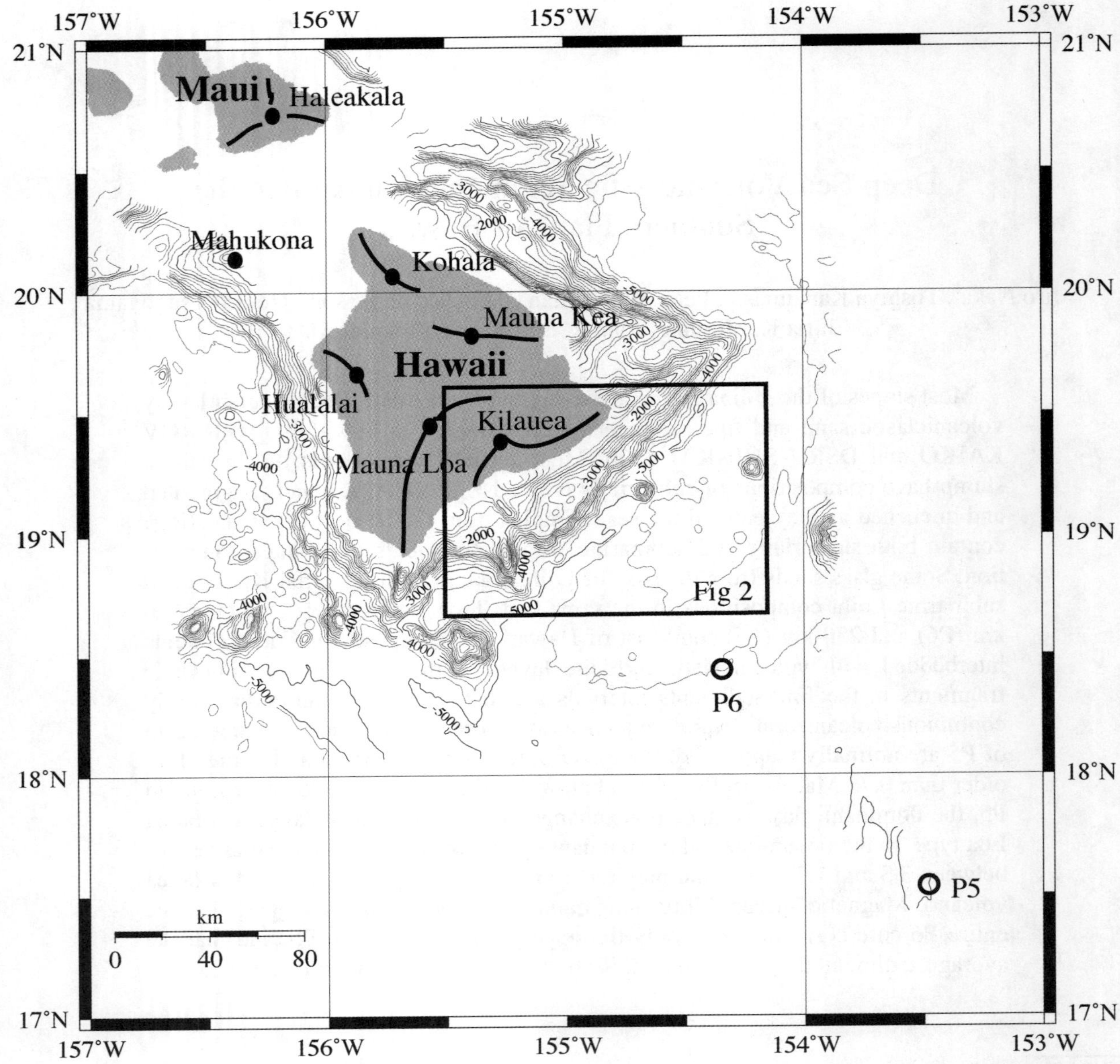

Figure 1. Index map of Hawaiian Islands, showing study area. Bathymetry and new SeaBeam data were acquired during 1998-1999 JAMSTEC cruises. Rectangle, area of Fig.2; circles, piston core sites.

Deep-water slopes south of Hawaii were investigated by the 11,000-m-capable unmanned Remotely Operated Vehicle (ROV) KAIKO in 1998 and the manned Deep-Sea Research Vehicle (DSRV) SHINKAI 6500 in 1999 (Fig. 1 and 2A). Geophysical surveys (bathymetry, gravity, magnetics) and sampling (piston coring) were also carried out, using the mother ships R/V KAIREI and R/V YOKOSUKA [*Naka et al.,* 2000]. In this paper we relate volcanic activity to volcaniclastic sedimentation on the undersea southern flank of Hawaii and the adjacent Hawaiian Deep and Arch to the south.

2. GEOLOGICAL SETTING AND SAMPLING SITES

The south flank of Hawaii Island has long been recognized as dominated by the active Hilina slump, related to

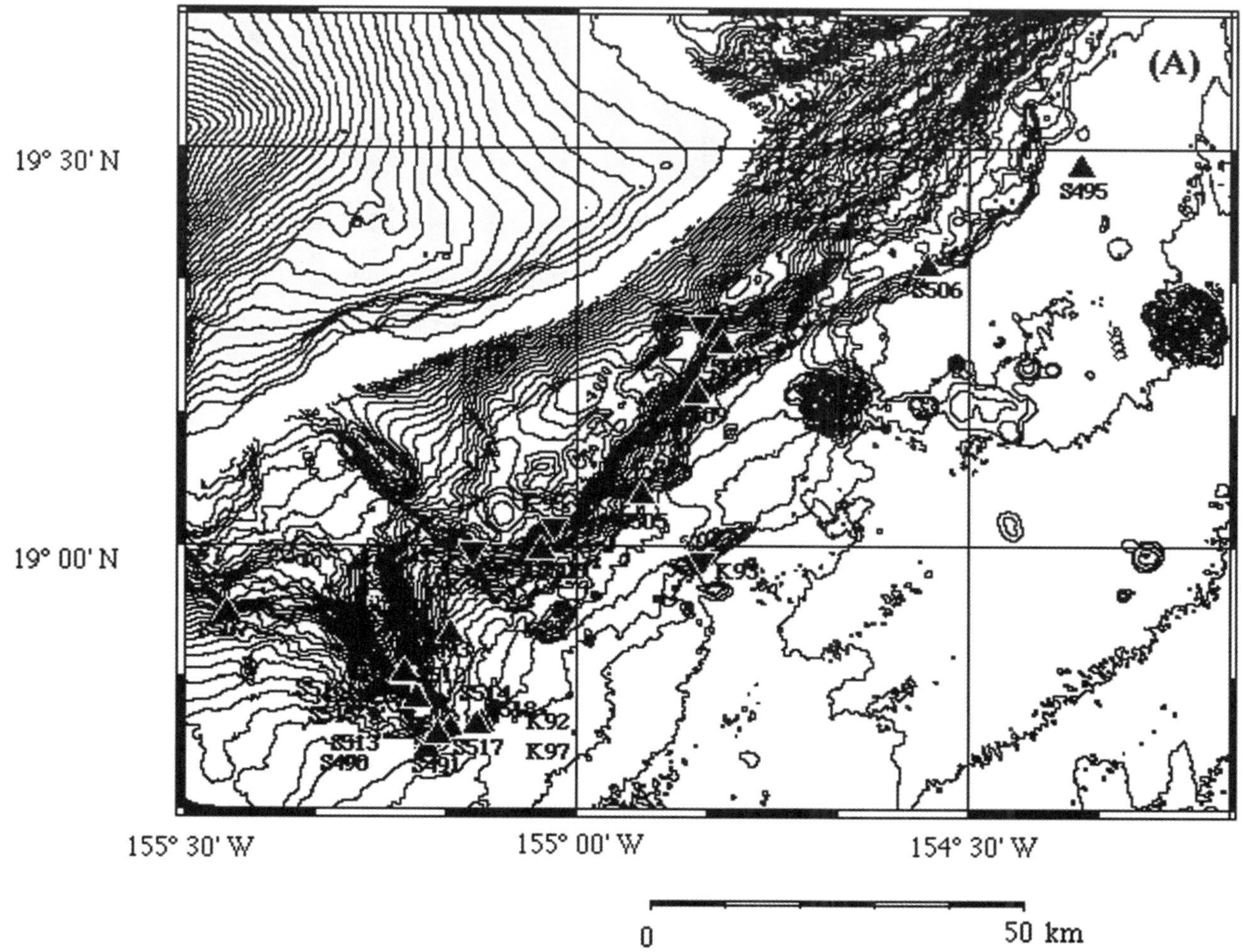

Figure 2A. Bathymetric map of south flank of Hawaii. Upright triangles, SHINKAI 6500 dive sites; inverted triangles, KAIKO dive sites. These positions are also shown on table 1 (appendix).

growth of Kilauea volcano [*Swanson et al.*, 1976; *Lipman et al.*, 1985]. Detailed SeaBeam bathymetry [*Chadwick et al.*, 1993; *Smith and Satake*, this volume] has provided a framework for interpreting the morphology of the underwater flank [*Moore and Chadwick*, 1995; *Smith et al.*, 1999; *Lipman* et al., this volume]. The linear western margin of the Hilina slump is bounded by Papa'u Seamount [*Fornari et al.*, 1979a] and a steep, southwestward-facing scarp (Fig.2B); the eastern side merges with the Puna Ridge. Following *Smith et al.* [1999] and *Lipman et al.* [this volume], we divide the Hilina slump into: 1) Upper slope, 2) Mid-slope bench, 3) Transitional zone, 4) Lower scarp, 5) Lobate terraces (or Remnant blocks), and 6) Outlying blocks (K93 dive site). For adjacent areas, we follow *Moore and Chadwick* [1995] in designating: 7) Puna Ridge (including its base) 8) Loihi (south rift), 9) Loihi base, and 10) Punaluu slump (Fig. 2B).

During the 1998 and 1999 cruises, 8 KAIKO dives and 18 SHINKAI 6500 dives provided direct observations and samples from southern flank of Hawaii and nearby Loihi (Table 1; see accompanying CD-ROM). Existing bathymetry [*Chadwick et al.*, 1993] and seismic studies [*Smith et al.*, 1999; *Morgan et al.*, 2000] indicated that extensive bedrock exposures were likely to be found along the steep lower scarp of the Hilina slump. Concurrent with bedrock studies using 4 KAIKO and 7 SHINKAI dives [*Lipman et al*, 2000, this volume], bottom-sediment samples were obtained adjacent to steep areas of the Hilina slump region (including the Punaluu slump and the base of the Puna Ridge). During 4 KAIKO and 11 SHINKAI 6500 dives to investigate volcanic

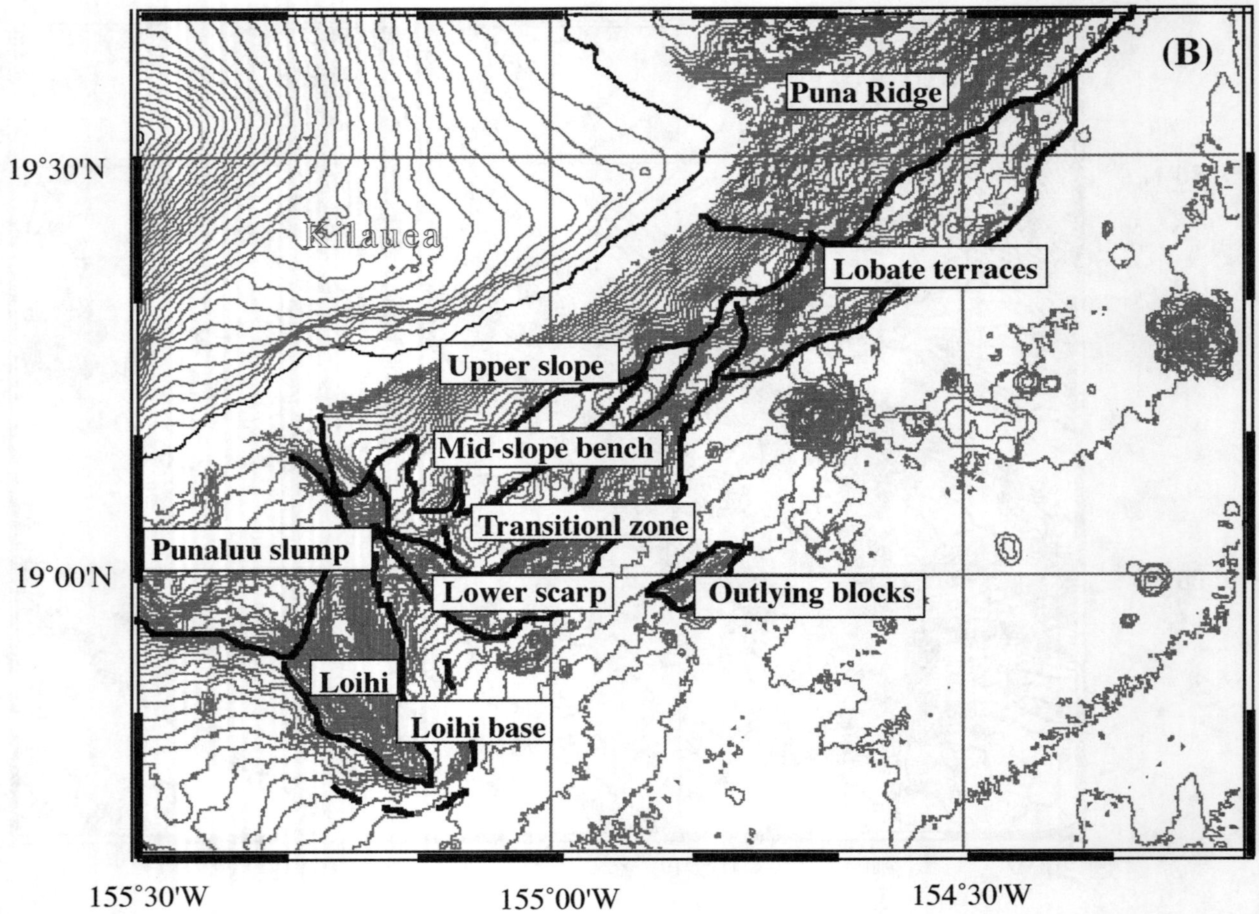

Figure 2B. Geologic domains, southern flank of Hawaii . Modified from *Moore and Chadwick* [1995], *Smith et al.* [1999] and *Lipman et al.* [this volume].

and hydrothermal activity at Loihi [*Naka et al., 2000*], surficial sediments were collected along its south rift zone and at its eastern base. Two piston cores were obtained, one (P6) in the Hawaiian Deep about 120 km southeast and the other (P5) on the gentle slope of the Hawaiian Arch 250 km southeast of Hawaii. Descriptions and interpretation of bottom features are based on visual observations (representative photographs are shown in Plate 1; see accompanying CD-ROM), , modal and grain-size analyses (Tables, 2-3; see accompanying CD-ROM), chemical analyses of basaltic glass fragments (Table 4; see accompanying CD-ROM), and magneto-stratigraphic studies.

Nine of the 22 surficial sediments samples collected by push- and grab-core samplers, and 9 of them preserve continuous sedimentary sections. Many surficial beds are composed of coarse sand. However, the topography at some sample sites was too steep to recover original sedimentary structures in small push cores; such cores are described as bulk samples. The cores were visually described from halved surfaces (Figs. 3 and 4); microscope observations were made on smear slides and thin sections (Table 2; see accompanying CD-ROM). The samples were sieved by 1/2 ϕ scale intervals between $-1\ \phi$ (2 mm) and $4\ \phi$ (0.063 mm) and then weighed. Fractions smaller than fine silt (< 0.015 mm, or 6ϕ) are described as fine sediment. Material $<4\ \phi$ was analyzed by pipette: grains suspended in a dispersion of sodium hexametaphosphate were weighed at every 1/2 ϕ time interval. Water in the suspension cylinder (1 liter) was kept at the same temperature as that in the water tank to eliminate thermal effects on settling velocity. Grain-size analyses (Table 3; see accompanying CD-ROM) were calculated as weight % for each 1/2 ϕ scale.

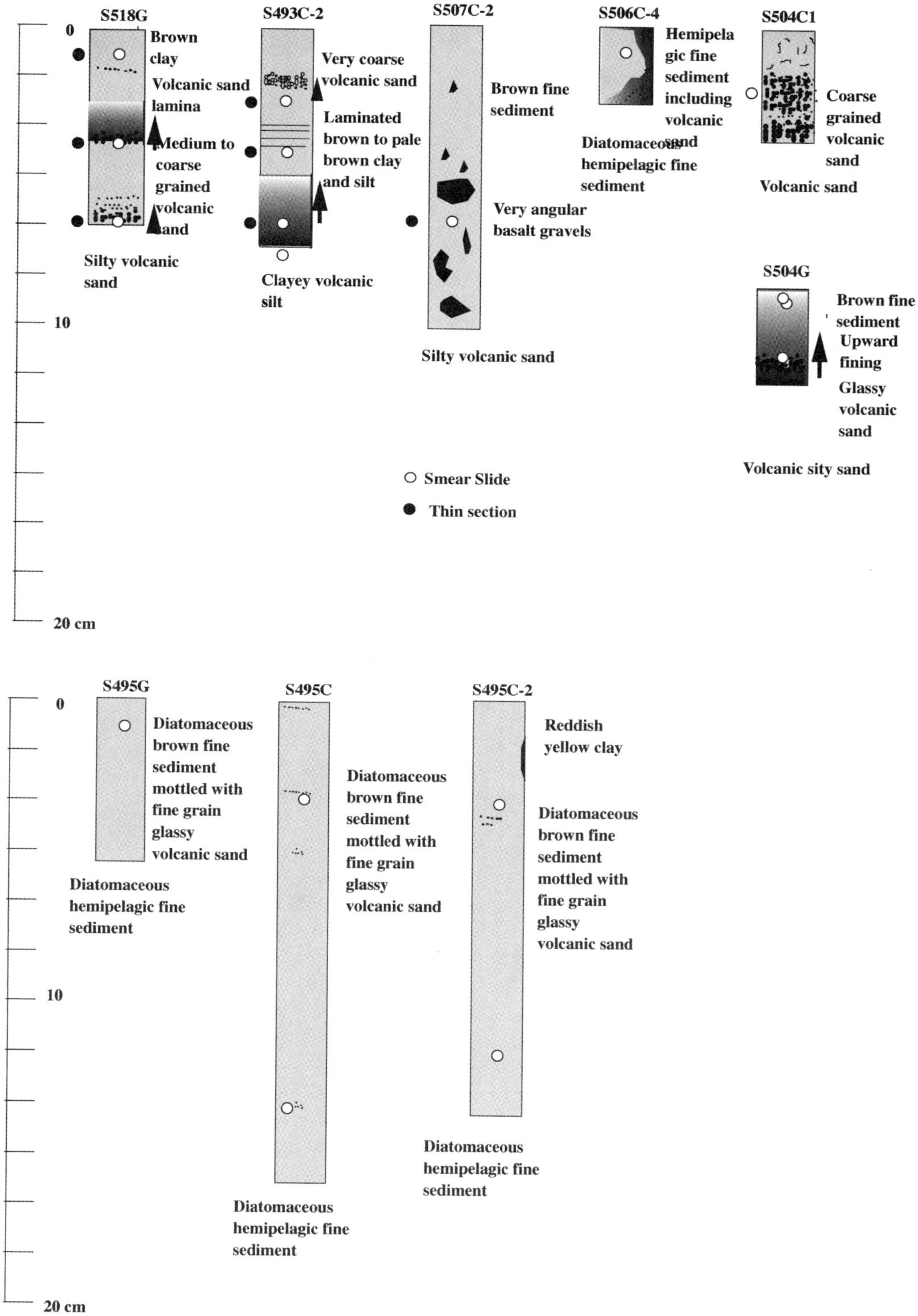

Figure 3. Visually logged push-core lithologies. Black parts indicate basaltic materials; light-colored intervals are pelagic and hemipelagic fine sediments. Open circles, smear-slide samples; solid circles, thin section samples. Scale on left side indicates depth in centimeters below sea floor.

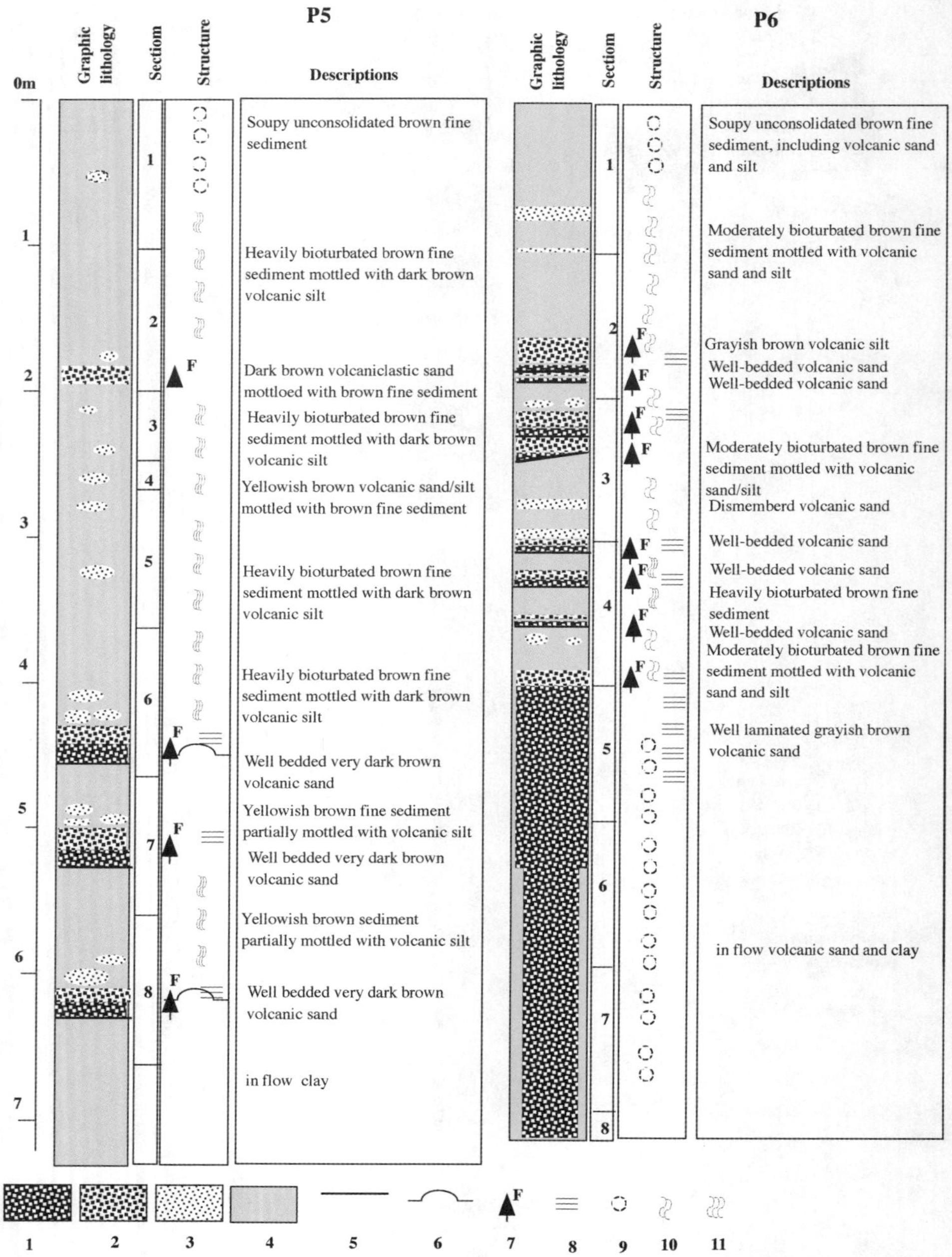

Figure 4. Piston-core lithologies, visually logged from halved surfaces: 1, volcanic sand; 2, volcanic silt; 3, disseminated volcanic silt/sand; 4, pelagic/hemipelagic fine sediment; 5, sharp boundary; 6, fold; 7, normal grading; 8, parallel lamination; 9, soupy sediments; 10, moderately bioturbated sediment; 11, heavily bioturbated sediment. Scales on left side indicate depth in meters below sea floor. In-flow material was the stuff sucked in by the piston action during piston corer retrieval.

2.1. Push Core and Grab Samples

The southwest margin and lower scarp of the Hilina slump have steep slopes that expose well-indurated, massive and bedded volcaniclastic breccias and sandstones in cliffs as much as 300 m high (Plate 1-2, Photo 4; see accompanying CD-ROM). Some of these outcrops are strongly jointed (Plate 1-2, Photo 3; see accompanying CD-ROM), and white zeolite veins fill many fractures. Intact pillow lava and broken pillow breccia of transitional basalt [*Sisson et al.*, this volume] crop out in lower parts of the upper slope (K95, S504). Talus breccias are common below cliffs (Plate 1-2, Photo 2; see accompanying CD-ROM), and semiconsolidated volcanic silt and sand locally mantle rock outcrops. Pelagic and hemipelagic fine sediments (Plate 1-1, Photo 4; see accompanying CD-ROM), some with ripple marks, cover gentle slopes on small steps and benches between the cliffs. Near the transition from lobate terraces to mid-slope bench, much dark volcanic sand is mixed with lighter-colored clay (Plate 1-1, Photo 1 and 2).

Three continuous cores (S504: C1, G, C4) from the Hilina slump are near the boundary between the transitional area, upper slope, and lobate terraces. The poor recovery in the C1 and G cores, only about 4 and 5 cm, respectively, may reflect the coarse grain size of the sediment encountered at these sites. In the C1 and G cores, beds are normally graded, and dominant materials are volcaniclastic sand and silt. Pelagic and hemipelagic fine sediments become dominant upward in graded layers. Bulk samples (including K93) are composed of volcanic sand and minor pelagic and hemipelagic fine sediment. Bulk sample S509-C contains a larger percentage of altered volcanic glass than the other samples.

Fractured lava is exposed along the south base of the Puna Ridge, in lobate terraces at about 5100 m depth (S506). Bottom materials at the southeast base of Puna Ridge (S495) consist chiefly of pelagic and hemipelagic fine sediments, but a young alkalic lava flow, first identified by USGS GLORIA surveys [*Holcomb et al.*, 1988], was also studied. The topography is very flat, even on the surface of the lava flow, which is only discontinuously covered with thin sediment. Benthic animal traces and burrows are common in the sediments, and dark volcanic sand is present locally (Plate 1-2, Photo 1; see accompanying CD-ROM). Two push cores (>10 cm) and one grab sample recovered brown pelagic and hemipelagic fine sediments. The upper part of core S495-C2 includes reddish fine-grained yellow material, probably of hydrothermal origin. These sediments contain 10 to 20 % siliceous fossils, mainly diatoms and radiolarians, and minor sponge spicules and silicoflagellata. These cores are variably bioturbated and include only minor amounts of volcaniclastic grains. Most of the volcaniclastic material is mottled with pelagic and hemipelagic fine sediments, but some disrupted volcaniclastic layers are present (i.e., S495-C1, 6 cm bsf).

Steep slopes and outcrops of fractured massive basalt characterize the middle Punaluu slump (S507). Talus and sand are present below most cliffs, and pelagic and hemipelagic fine sediments cover benches. The single continuous core from the Punaluu slump is poorly sorted pelagic and hemipelagic fine sediment, mixed with discrete angular volcaniclastic sand and silt. The bulk samples are mainly volcanic sand and silt, along with small amounts of pelagic and hemipelagic fine sediment.

Dominant bottom materials on the south rift of Loihi are young intact pillow basalt and sheet flows, some thinly covered by fine sediment. Black volcanic sand is locally trapped in depressions on the flows. Altered gray volcaniclastic rock crops out on the steep south slope of a small ridge at the base of the south rift (Fig. 2B), near the low-temperature hydrothermal system found in 1991 by the Russian submersible MIR [*A. Malahoff, et al.*, 1991]. The flat north side is covered with thick pelagic and hemipelagic fine sediment, with some scattered dark volcanic sand. On the eastern flank of Loihi (S493), pillow lavas crop out in steep areas, but the basal slope is mantled by fine sediment (Plate 1-1, Photo 3; see accompanying CD-ROM). Most surficial sediments on the Loihi south rift are thin, and no continuous cores could be collected. The dominant lithology is basaltic silt and sand. In one continuous push core from the eastern base, and in four others at the southern base of Loihi, thin graded-bedded volcaniclastic sand and silt layers are interlayered with pelagic and hemipelagic fine sediment.

Most surficial push-core samples on the Hilina lower scarp and adjacent areas contain sparse planktonic fossils, including siliceous diatoms and radiolarians, calcareous planktonic and benthic foraminifer, and nannofossils. Planktonic calcareous fossils are present only in samples shallower than about 4200 m, indicating that the carbonate compensation depth (CCD) around the south flank of Kilauea is about 4000 m at present. This depth agrees with the general CCD around the Hawaii Islands [*Berger and Winterer*, 1974]. A few calcareous planktonic fossils are present in thin turbidite layers in a push-core sample at the base of Loihi (S493-C) at a water depth of 4513 m, somewhat deeper than the expected CCD.

2.2. Piston-Core Samples

Piston-core site P6, 120 km south of Hawaii, is located on the gentle northwest-dipping slope between the Hawaiian Deep and Arch. Site P5, 250 km south of Hawaii, is on the Hawaiian Arch. Total core recoveries were about 7.4 m in

P5, and 7.1 m in P6 (Fig. 4), but the lithology and other data show that the material deeper than 6.3 m for P5 and 5.3 m for P6 is in-flow material that stuff sucked in by the piston action during corer retrieval.

The distal P5 core consists dominantly of pelagic and hemipelagic fine sediment and lesser amounts of bedded volcaniclastic sand and silt (Fig. 4). These sediments are brown to yellowish brown clay and fine silt; most horizons contain volcanic materials (4-10%) and are heavily bioturbated. Poorly preserved or fragmental planktonic siliceous bioclasts (mostly diatoms) are common (to 20%), especially at depths of 0 to 0.4 mbsf and 2.4 to 3.3 mbsf. Four bedded volcaniclastic turbidite layers are conspicuous in P5. One bed near 1.8 mbsf (5-10 cm thick) is somewhat bioturbated; the other three are greater than 4.5 mbsf and preserve primary depositional features. These three beds are graded, having sharp basal contacts with the underlying pelagic and hemipelagic fine sediments. Toward the top, the volcaniclastic beds are mottled, becoming yellowish, and are locally discontinuous or interbedded with pelagic and hemipelagic fine sediment. These beds contain parallel-laminated fine sand and silt in the upper part of each layer. Two horizons (4.45-4.49 mbsf, 6.05-6.12 mbsf) contain folded (slump?) layers. Poorly preserved diatoms and radiolarians of Early to Middle Pleistocene age are contained in P5 in diatomaceous horizons [*Kanamatsu et al.*, this volume], but magnetic data indicate that these siliceous fossils were redeposited by bottom currents and are not indicative of P5 sediment ages.

The proximal P6 core contains two major sediment types (Fig. 4): 1) pelagic and hemipelagic brown clay and fine silt, and 2) eight dark-brown layers of volcaniclastic sand and silt. The proportion of volcanic material is greater than in P5 and makes up about 20% of the core. Most of the pelagic and hemipelagic clay is brown, and most beds are heavily bioturbated. Most of these fine sediment layers contain about 10 percent dispersed volcanic detritus (see Table 2; see accompanying CD-ROM). All eight of the P6 volcaniclastic layers are graded, with sharp basal contacts against pelagic and hemipelagic fine sediments. Some layers consist of two or three graded beds that have merged together, with no intervening pelagic and hemipelagic fine sediment. The base of the lowest volcaniclastic layer is below the effective bottom of the P6 core (5.3 mbsf), so this layer is >1.2 m thick (Fig. 4). Upward, the volcaniclastic layers become mottled with pelagic and hemipelagic fine sediments; these layers become yellowish, and some are discontinuous or interbedded with pelagic and hemipelagic fine sediments. Siliceous fossils are common, except in the interval 2.7-3.2 mbsf. At 0.7-1.8 m, planktonic siliceous fossils, mostly diatoms, and form up to 40 percent of the deposit. Generally these siliceous fossils are poorly preserved or fragmental. At 2.3-2.5 mbsf, a few calcareous shell fragments are present. Most of the calcareous fossils occur within or just below volcaniclastic turbidite layers. They may have been transported and buried quickly by deep water turbidity currents, explaining their presence below the CCD. The P6 core is slightly coarser than P5, and it contains a higher percentage of volcaniclastic debris and biogenic grains—features that probably reflect proximity to the Island of Hawaii.

3. TEXTURES AND COMPOSITIONS OF VOLCANICLASTIC FRAGMENTS

Volcaniclastic fragments are composed of fresh basaltic glass (sideromelane), altered glass (palagonite), finely crystallized volcanic fragments (mostly opaque tachylite), plagioclase, olivine, pyroxene, opaque minerals, and rare evolved micropumice. Basaltic glass is dominant, typically making up 50-80% of the volcanic sand and silt fractions and 30-50% of the pelagic and hemipelagic fine sediments. In piston cores, glass is less abundant in thick well-bedded sand layers than in the thin sand/silt layers. For example, the thick sand bed at the base of the proximal P6 core contains 30-40 % glass, but a thin layer at 2.42 mbsf contains 77%. Opaque materials show an opposite tendency. Glass is less abundant as discrete volcaniclastic fragments in pelagic and hemipelagic fine sediments than in the volcaniclastic sand. Fresh glass is nearly absent in some P5 beds (i.e. 5.36 mbsf), but plagioclase and opaque materials are common. Small spheres of dark brownish opaque materials are sparsely present in the pelagic and hemipelagic fine sediment; these appear similar to the "red brown semi-opaque oxide" described by *Yeats et al.* [1976]. Trace micropumice grains were recovered from all of the volcaniclastic layers in both piston cores and may be present, but overlooked, in the intervening pelagic and hemipelagic intervals.

3.1. Textures

About 20 representative SEM (Scanning Electron Microscopy) photographs (Plate 2; see accompanying CD-ROM) of the coarse grained fractions (>125 or >250 microns) and 120 thin sections were used to characterize the textures of volcaniclastic fragments. Hilina samples generally include abundant angular to subangular blocky fragments and minor bubble-wall or scoria glass fragments (Plate 2-1, Photos 1 and 2; see accompanying CD-ROM). A sample (K93-P1B) from the southern base of a northeast-southwest-trending block, shows the same character as the Hilina samples. One sample (S509-C1B) includes rounded fragments, perhaps reflecting a large amount of altered volcanic glass. Samples from the southern base of

Puna Ridge (S495-C1, 2, G) contain glass fragments similar to the Hilina samples, but they also contain much scoria and ash with convex bubble-wall shapes (Plate 2-1, Photo 3; see accompanying CD-ROM). Glass fragments in Loihi rift samples S515 (Plate 2-1, Photo 3; see accompanying CD-ROM) and S516 have angular, flaky, and blocky shapes. The S516 fragments are more vesicular, probably due to the shallower depth. Generally, these glass fragments have higher aspect ratios than the Hilina glass fragments. Loihi base samples (S493, S518 K97) show a mix of blocky Hilina style and flaky Loihi rift type shapes (Plate 2-2, Photos 1 and 2; see accompanying CD-ROM).

Volcanic fragments in bedded volcaniclastic layers of cores P5 and P6 have essentially the same shapes. Angular to subangular blocky volcanic fragments are abundant; bubble-wall, scoria, or pumice fragments are rare (Plate 2-2, Photos 3-4; see accompanying CD-ROM). No significant changes are evident from the coarse grained base to the fine-grained top in single graded beds. Particle shapes in the disseminated volcaniclastic samples between turbidite layers are generally similar to those in the bedded volcaniclastic layers, but some disseminated samples include more bubble-wall (concave surfaces) or scoria fragments than in the bedded volcaniclastic layers.

3.2. Glass Chemistry

Glass fragments were analyzed with electron microprobes (5-spectrometer JEOL 8900) at JAMSTEC and at the USGS, yielding similar compositional variations with depth. USGS analyses examined only volcaniclastic layers in piston cores P5 and P6, whereas JAMSTEC examined both volcaniclastic layers and fine sediments. Only JAMSTEC analyses are discussed here, except for evolved micropumice glasses that were identified only at the USGS. Beam conditions for the JAMSTEC analyses were 15 kV, 20 nA sample current, and scan mode under magnifications of 5000 to 20000. Peak counting times were 10 sec for every element. To examine uniformity, three points were measured for most grains and averaged. If the measurements differed significantly, the data were omitted. ZAF oxide corrections were made. Standards were wollastonite for SiO_2 and CaO, pyrophanite for TiO_2 and MnO, corundum for Al_2O_3, fayalite for FeO*, forsterite for MgO, jadeite for Na_2O, K-feldspar for K_2O, barite for SO_3 and potassium titanium phosphate for P_2O_5. The same thin sections were used for modal and glass microprobe analyses. For each sample, 10 to 12 fresh glass fragments were measured, except for a few samples of P5 that contain only sparse fresh glass (Table 4; see accompanying CD-ROM). To classify the glass compositions, all data (JAMSTEC) are plotted on alkali-silica diagrams (Fig. 5).

Glass fragments from the Hilina slump area vary widely in composition. They include major amounts of tholeiitic basalt and basaltic andesite, and minor alkali basalt and basanite. Tholeiite and basaltic andesite glasses range from 47 to 53 wt % SiO_2, although most are 50.5 to 52 wt %. Most of these glasses have about 3 wt % Na_2O+K_2O. Glass grains in some samples (K95-C2, S504-C and G) show only small compositional variations (only 0.16% STD for SiO_2 in S504-G3.5). Fragments of alkali basalt glass are present in S505-C2, S506-C4, S506-G S507-C2B, and S508-C2; basanite glass in S505-C2 and K93-C1B. Only 1-2 alkali basalt grains are present per thin section, except for S505-C2-4.

Loihi rift samples S494, S515, and S516 contain abundant fragments of alkali basalt and two of these (S515 and S516) also contain some tholeiite. They vary widely in composition, clustering near the alkali-tholeiite boundary. SiO_2 ranges from 46.5-49.5 wt%, and Na_2O+K_2O from 3.3-4.7 wt%. Loihi base samples vary more widely than samples from its rift zone. The east base samples (S493-C2-3) include alkali and tholeiitic glass showing the same variations as Loihi rift samples, but much glass contains high silica and low alkali contents (alkali-tholeiite transitional basalt, SiO_2, 49-50 wt%; Na_2O+K_2O <3 wt%). Glasses from the south base are similar to those from the east base and rift. They include alkali basalt, tholeiite basalt, and basaltic andesite, but some fragments contain higher silica and lower alkali contents (SiO_2 >50.5 wt%, Na_2O+K_2O about 3 wt%). Differences can be large within a single core; S518-G7 contains only fragments of low-alkali tholeiite and basaltic andesite, but S518 G4 includes abundant clasts intermediate between alkali and tholeiite basalts, as well as sparse fragments of low-alkali tholeiite.

Glass grains from the P5 and P6 cores vary widely in composition. Throughout the P5 core, the common glass is tholeiite and low-alkali basaltic andesite ($Na_2O + K_2O$ <3.0 wt%) that is high in silica (>50.5 wt% SiO_2). Rare glass analyses have apparent total alkalis <2 wt% due to low Na_2O. This feature has not been replicated, does not correlate with other geochemical features, and probably is a product of incipient alteration; those data are not included in Figs. 6, 9 and 10. Alkali basalt glass is confined to two comparatively thick beds at 4.35 to 5.25 mbsf. A single basanite grain was found at 0.8 mbsf. Low-alkali tholeiite basalt in an upper horizon of bedded basaltic glass (1.8 to 1.9 mbsf) has relatively low SiO_2 (50-51.5 wt %), whereas a lower horizon (below 4.35 mbsf) containing scattered grains of tholeiite has lower SiO_2 (49.5 to 51.5 wt %) and low alkalis. Only sparse grains of fresh glass are present in some pelagic and hemipelagic fine sediment horizons of P5; in these, the compositional variations are larger than in the conspicuous volcaniclastic layers.

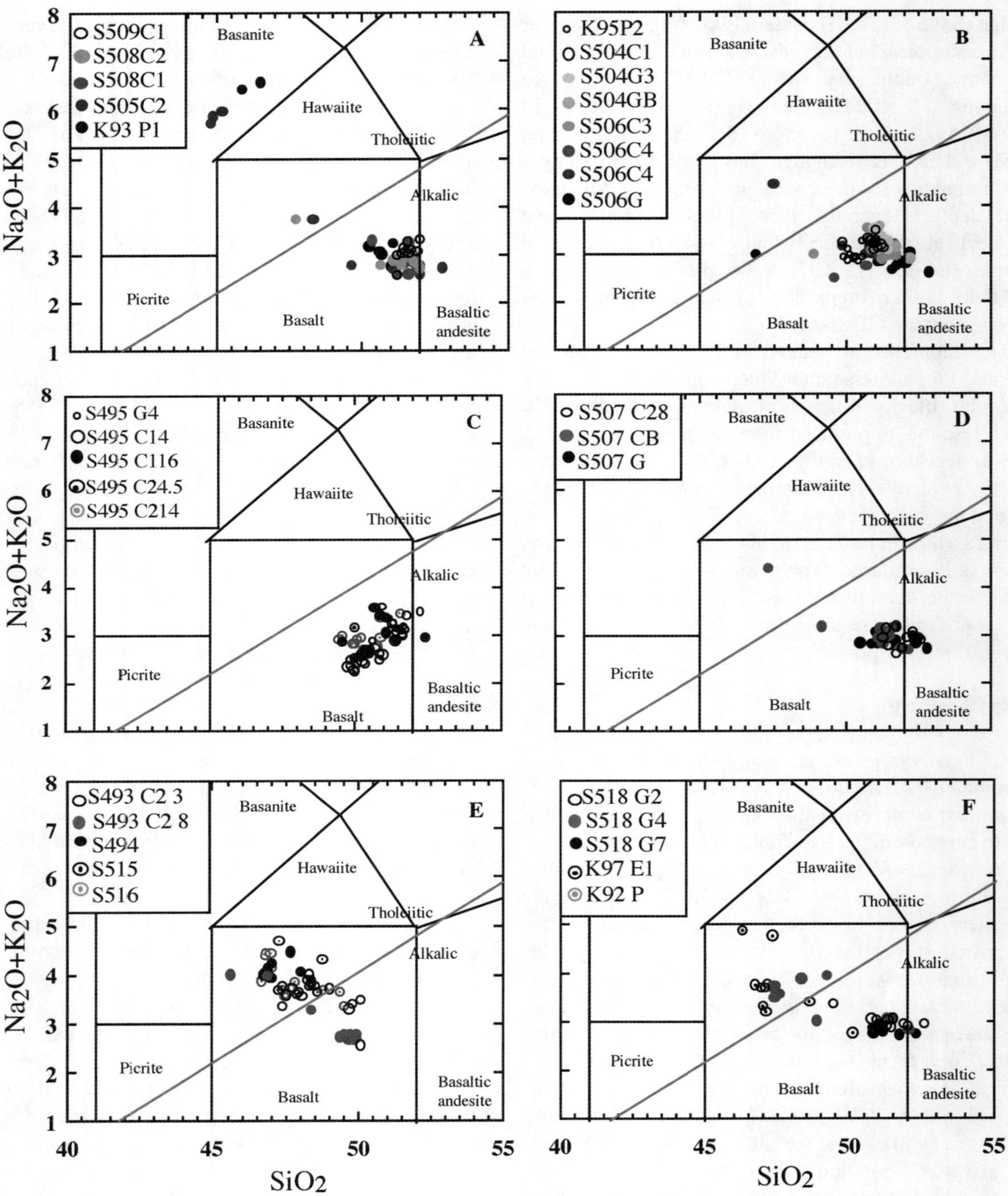

Figure 5. Alkali-silica variation diagrams for glass fragments (JAMSTEC data only), showing IUGG compositional classification [*LeBas et al.*, 1986], and the *Macdonald and Katsura* [1964] boundary line (dashed line) to distinguish alkalic basalt vs. tholeiite. 6A, Lower Hilina scarp and outlying block; 6B, Upper slope and lobate terraces/remnant blocks; 6C, Puna Ridge base; 6D, Punaluu slump; 6E, Loihi east base; 6F, Loihi south base; 6G, P5 0 – 1 mbsf; 6H, P5 1 – 2 mbsf; 6I, P5 2 – 4 mbsg; 6J, P5 4 – 4.8 mbsf; 6K, P5 4.8 – 5.3 mbsf; 6L, P5 5.3 mbsf –; 6M, P6 0 – 1 mbsf; 6N, P6 1 – 2 mbsf; 6O, P6 2 – 3 mbsf; 6P, P6 3-4 mbsf; 6Q, P6 4-4.4 mbsf; 6R, P6 4.4-5.3 mbsf .

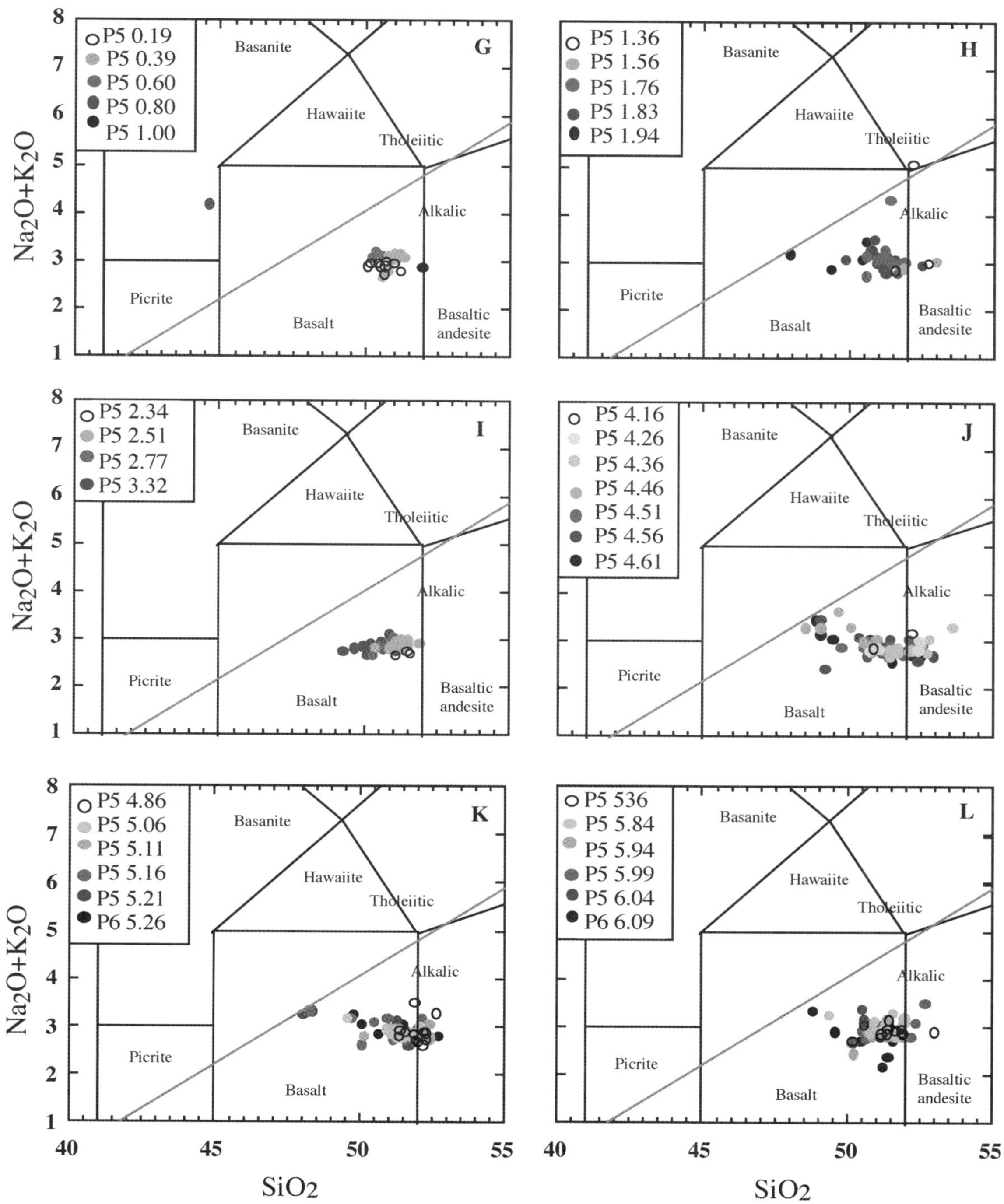

Figure 5 (continued)

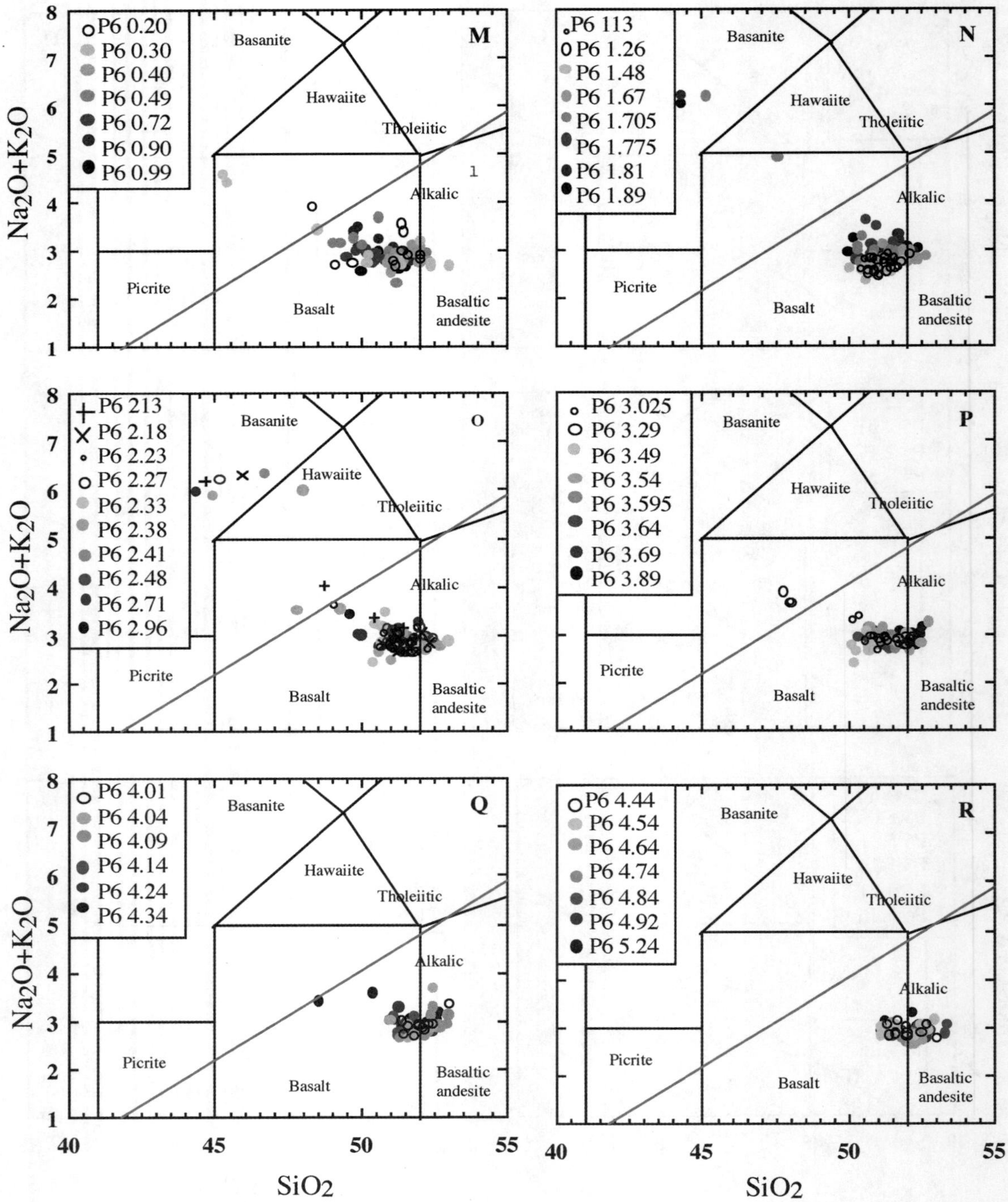

Figure 5 (continued)

Throughout core P6, glass fragments are dominantly low-alkali tholeiite and basaltic andesite. The thick basal volcaniclastic layer in P6 contains abundant tholeiitic basaltic andesite, and minor alkali basalt grains. An interval at 1.75-3.3 mbsf contains basanite and alkali basalt fragments. Below 3.3 mbsf, only fragments of low-alkali tholeiite and basaltic andesite glass were found. Below 3.9 m, SiO_2 of the glass increases by about 0.5%. Basal coarse parts of volcaniclastic glass layers in the P6 core tend to have relatively small compositional variations, and compositions do not vary between well-bedded layers and intervals containing scattered glass fragments.

Light brown-to-white, finely vesicular grains of micropumice were recovered from all volcaniclastic layers in the P6 core. Glasses in these micropumice grains are low-MgO (1.4 - 2 wt.%) mugearite and benmoreite composition (USGS analyses), similar to the minor evolved magmas of the post-caldera alkalic stage of Hawaiian shield volcanoes [*Clague,* 1987]. Similar appearing micropumice grains were subsequently found in all of the volcaniclastic layers in the distal P5 core, but have not been analyzed.

The sulfur content of basaltic glass is a well-established indicator of eruption depth [*Moore and Schilling* 1973; *Dixon et al* 1990]. Basaltic glass erupted in deep water (> ~1000 m) generally contains more than 1000 ppm S. Sulfur degasses efficiently at ocean depths less than about 200 m, and the sulfur content of subaerial basalt is typically less than 200 ppm. Here, we define three groups, based on S content (Fig. 6):

1) Subaerially erupted: < 200ppm S
2) Shallow water erupted: 200 to 1200 ppm S
3) Deep water erupted: > 1200 ppm S

Most samples from the lower Hilina scarp are mainly composed of low-S glass (subaerially erupted), along with small amount of deep-water fragments. Fragments from the base of the Puna Ridge vary widely in S content and have values typical of both deep-sea and subaerial basalts. Loihi south rift samples include shallow and deep water fragments, depending largely upon sample depth. In contrast, samples from the southwest extension of the Loihi south rift contain a mixture of subaerial, shallow water, and deep water fragments.

4. ORIGIN OF THE VOLCANICLASTIC LAYERS

Graded bedding is common in volcaniclastic layers worldwide, and deposits in deep-water settings have been interpreted as being the product of volcanic air falls (into the sea) or turbidity currents [*Hung*, 1980]. The volcaniclastic beds on the south flank are here interpreted as turbidity-current deposits. The dominant shape of volcaniclastic fragments on the south flank is blocky, not bubble wall, and some volcaniclastic layers in cores P6 and P5 have typical parallel turbidite laminations in their upper parts. Volcaniclastic layers at nearby ODP136 sites 842 and 843 display similar characteristics and have been interpreted as turbidite deposits [*Naka et al.*, 1993; *Garcia*, 1993, 1996].

4.1. Bioturbation

Moderate to heavy bioturbation is common in pelagic and hemipelagic fine sediment layers collected in push cores and piston cores. Bioturbation is also the likely cause of the disseminated volcaniclastic material in pelagic sediment recovered at ODP 136 [*Naka et al.*, 1993]. For push-core samples, the minimum thickness of visible volcaniclastic turbidite layers having sharp basal contacts is about 0.5 cm. In comparison, the minimum bedded turbidite thickness in cores P6, and P5 is about 2 and 5 cm, respectively (including amalgamated layers). This contrast may reflect different sedimentation rates, as suggested by *Sparks et al.* [1984]. If the sedimentation rate is much higher than the disturbance by bioturbation, original sediment layers can be preserved. Such processes, suggested by the sharp basal boundaries between volcaniclastic turbidites and the underlying pelagic and hemipelagic fine sediments, could explain the preservation of rapidly deposited thick turbidites, as well as the thin turbidite layers in high sedimentation rate environments.

According to *Ruddiman and Glover* [1972], bioturbation in pelagic environments disturbs sediments mainly in a vertical direction, especially upwards. The degree of disturbance from a lower horizon decreases exponentially upwards, and downward mixing is generally limited to 4 - 5 cm, which is the burrowing limit of benthic animals. Thus, the discrete volcanic fragments in bioturbated layers may have been derived from lower levels, or they could have remained at almost the same horizon, even if affected by bioturbation.

4.2. Glass Abundance and Composition.

Fresh glass makes up 50 to 70% of most volcaniclastic layers. In contrast, glassy rims of both pillow lava and subaerial pahoehoe are only 1 to 2 cm thick (Plate 3; see accompanying CD-ROM), and this glassy material makes up at most 10-20% of the lava. If the deep volcaniclastic fragments were solely the products of erosion of previously cooled lava flows in the surf zone, glass should make up less than 20 % of the debris downslope. These relationships

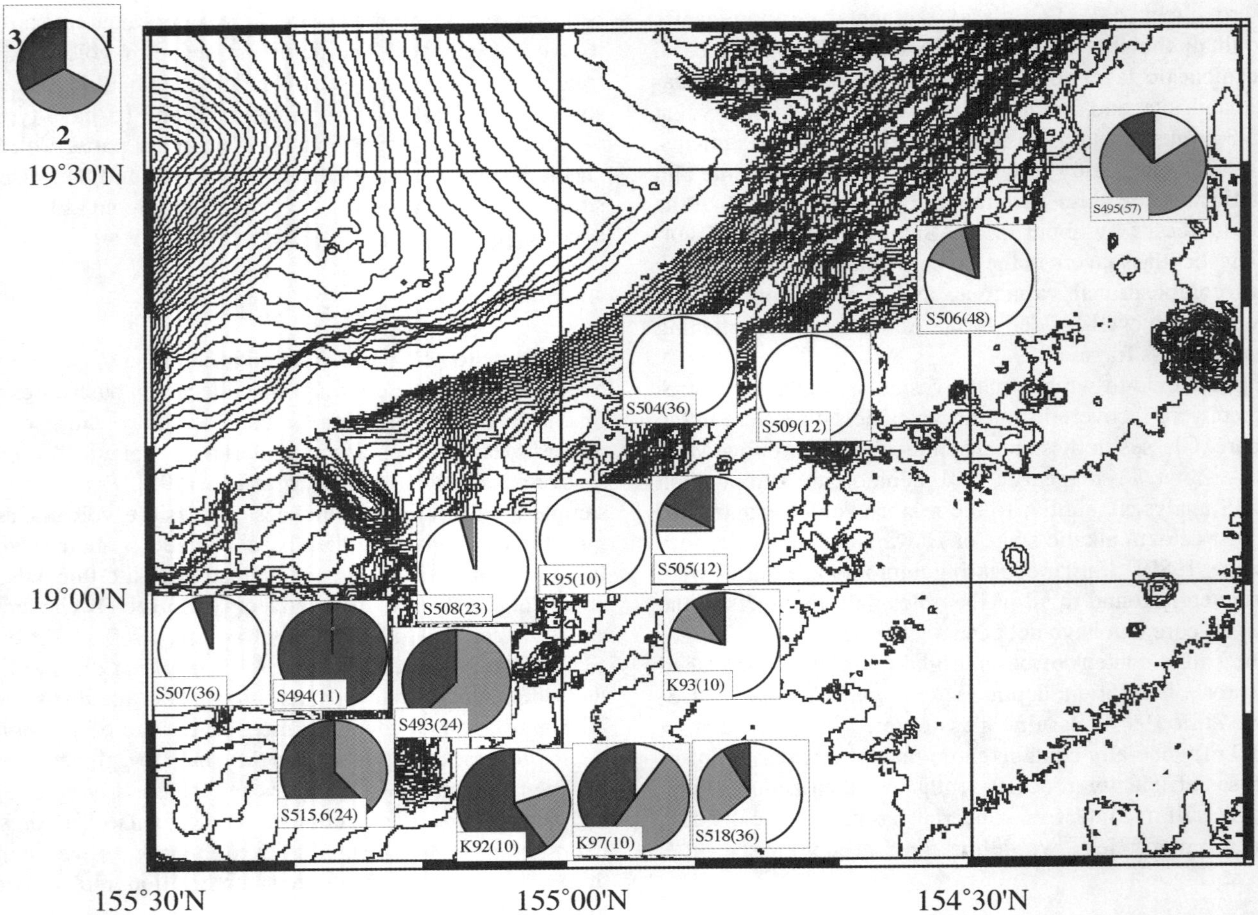

Figure 6. Summary of sulfur contents in south flank glass fragments.1) < 200 ppm S, 2) 200 – 1200 ppm S, 3) >1200 ppm S. Each pie-diagram indicates the different sulfur contents of basaltic glasses. Each number indicates the dive number and number of the measured glass grains.

suggest that most of the volcaniclastic fragments originated by surf-zone hydrovolcanic fragmentation of actively flowing lava entering the sea, plus fragmentation by erosion of both subaerial and submarine lava flows [*i.e., Jones and Nelson*, 1970; *Furnes and Fredleifsson*, 1974; *Moore et al.*, 1973; *Furnes and Sturt*, 1976]. The mixtures of glass and crystallized lava fragments suggest that most south flank volcaniclastics were produced by collapse of such mixed materials that formed at and just below sea level.

The composition of fresh volcaniclastic glass provides some information about its provenance. For purposes of calibration, we analyzed recently erupted tholeiitic glass sand from Kilauea's shoreline and found it to be remarkably homogeneous, with the standard deviation of SiO_2 in 11 grains being only 0.124 wt%. The compositional ranges of basaltic glass in south flank and piston-core samples are much wider than the Kilauea coastal sand. This relationship was observed both for discrete volcaniclastic fragments in pelagic and hemipelagic fine sediment layers and in well-bedded volcaniclastic turbidites. The same relationships at ODP 136 site 842 were attributed to redeposition and mixing during transportation by turbidity currents [*Garcia*, 1993, 1996]. However, on the Kilauea submarine south flank at the boundary between the western edges of the lobate terraces, outer scarp, and upper slope (S504), glass sands are fairly homogeneous. This contrast probably reflects the shorter transport distance and simpler topography as compared with the other sites, and indicates that transportation distance affects the degree of mixing. The poor sorting of volcaniclastic sands on the south flank and adja-

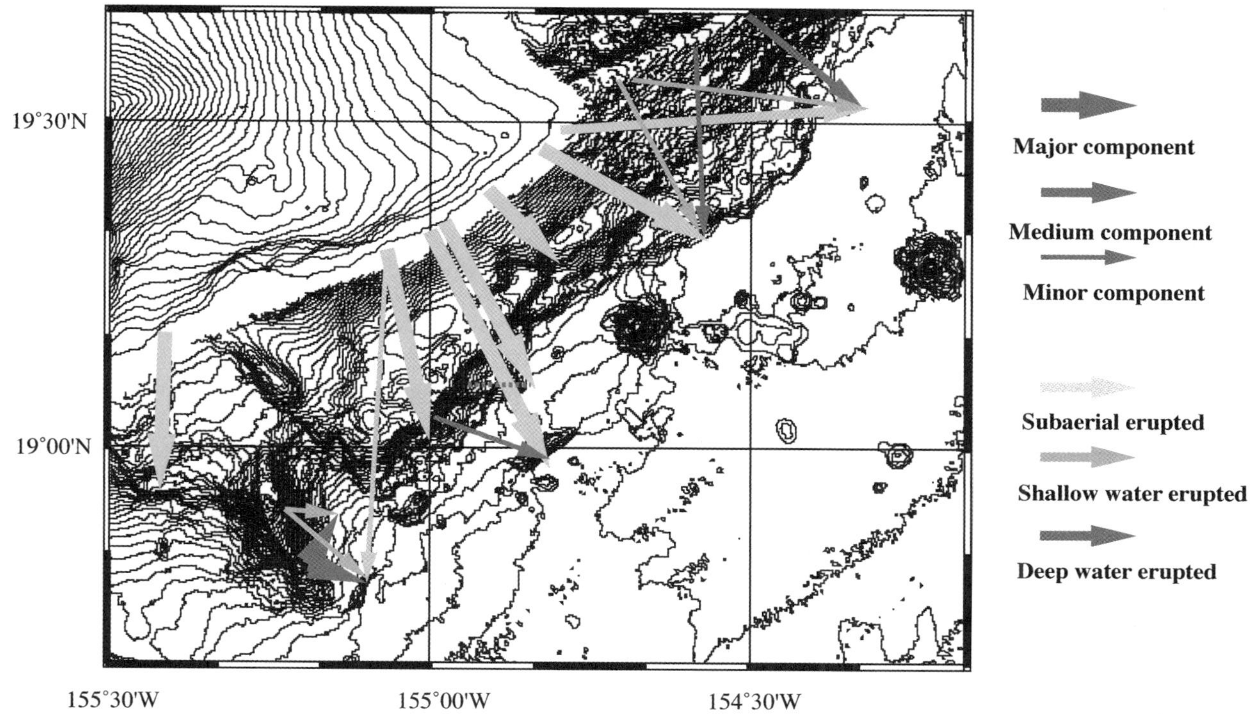

Figure 7. Inferred source areas for the surficial volcaniclastic fragments. Arrows show the deposition sites and possible source areas. The thickness of the arrows indicates the relative abundance, and the darkness indicates the inferred eruption depth of the glass fragments.

cent areas supports such an interpretation (1 to 3, in (ϕ84-ϕ16)/2) compared with the sandy fraction of the turbidites in cores P5 and P6 (< 1, in (ϕ84-ϕ16)/2: see table 3).

4.3. Low-S Blocky Glass Fragments

Most south flank volcaniclastic glass fragments have blocky shapes and low sulfur contents that indicate derivation from actively flowing subaerial lava that had degassed prior to fragmentation. *Swanson* [1973] and *Swanson and Fabbi* [1973] showed that far-traveled basaltic lava erupted from Kilauea volcano degassed during subaerial flowage (vesicularity decreased from 50-70 % to 18 % while traveling about 12 km). Thus, we infer that the source of the degassed, blocky glass fragments was a large subaerial shield.

5. SEDIMENT SOURCE AND DEPOSITIONAL PROCESSES ON THE SOUTH FLANK

Textures and compositions of the volcaniclastic fragments collected from the south flank of Hawaii provide information on source and water depth, transport pathways, and depositional processes (Fig. 7). Many glass fragments on the Hilina slump have low S contents and blocky shapes, indicating fragmentation as actively flowing lava crossed the surf zone. However, the basal Puna Ridge samples contain bubble-wall or scoria fragments. Some fragments have moderate to high S contents, indicating eruption in shallow to deep water. Glass fragments on the Hilina slump area have tholeiitic compositions similar to present-day Kilauea and were derived from this volcano.

Most of the debris mantling the upper scarp of the Hilina slump is sandy and muddy, and is interpreted to be the products of phreatic processes in shallow water or the surf zone [*Moore and Fiske*, 1969]. The upper Hilina slope fragments seem to have had a comparatively simple transportation history, as reflected by the homogeneous glass compositions at the eastern end of lobate terraces. Recent seismic studies of the Hilina slump. [*Smith et al.*, 1999 and *Morgan et al.*, 2000] suggest that a large sediment body (500-2000 m or more thick) accumulated on the mid slope bench. The lower scarp samples show much wider compositional variations, reflecting a multi-depositional and com-

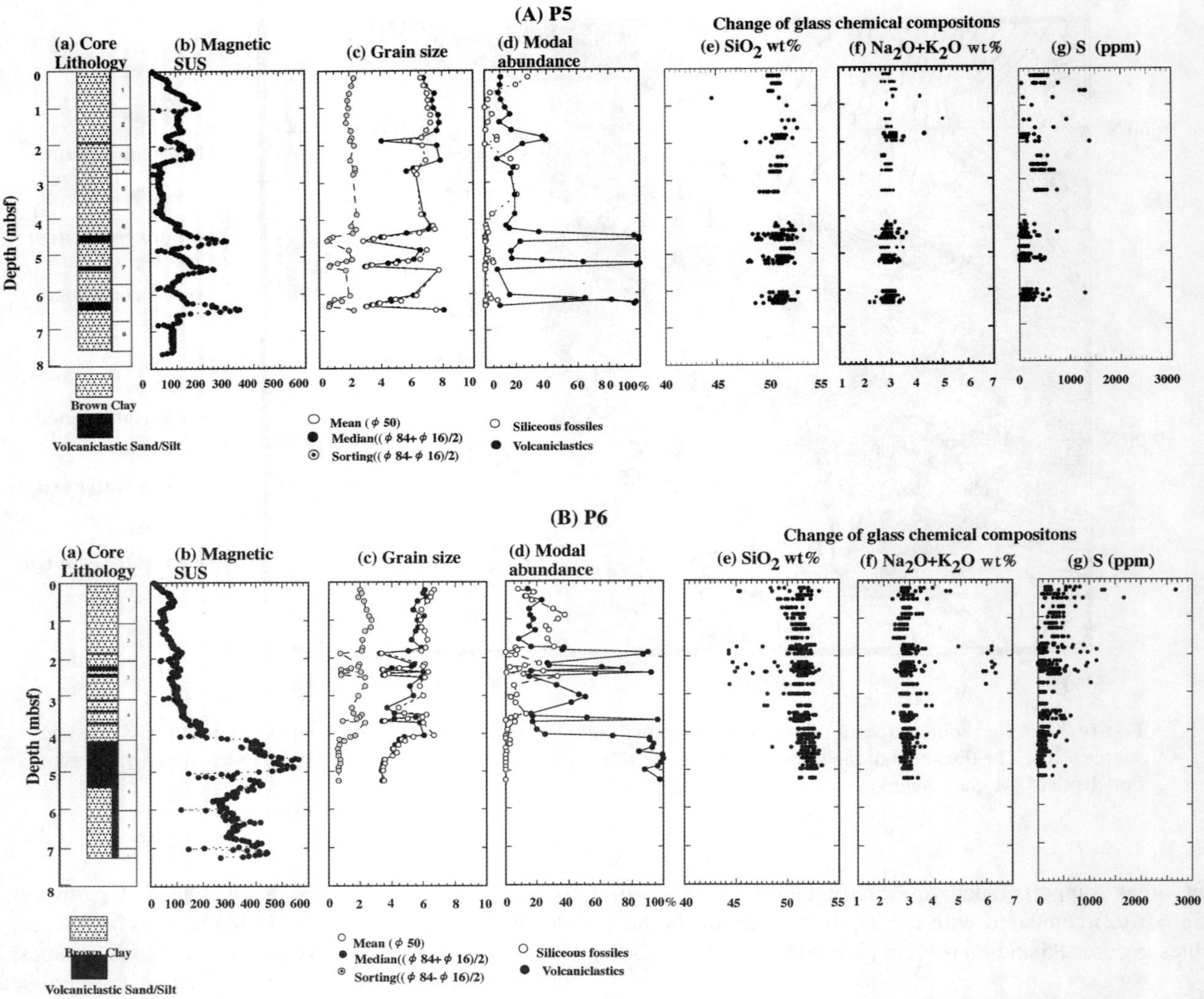

Figure 8. Vertical changes in piston cores P5 (A) and P6 (B). (a) Simplified visual core lithology; (b) Magnetic susceptibility (SUS, n 10^{-5} SI unit); (c) Grain size (mean, medium and sorting, after *Inman* [1954)]; (d). Modal abundance of volcaniclastic fragments (glass, opaque materials, feldspar and mafic minerals, etc.), and siliceous fossils (diatom, radiolarians, sponge spicules etc.); (e), (f), and (g) are SiO_2, Na_2O+K_2O, TiO_2 and S contents range of glass fragments, respectively. All sections are plotted against the depth in meters below the sea floor.

plex topographic environment separating the source area and depositional site. Ripple marks in the Hilina slump area suggest that some volcaniclastics may also have been reworked by bottom currents. The sediments in the boundary area between the mid-slope bench and lobate terraces are rich in rippled black basaltic sand, as also observed on the top of Papa‘u Seamount [*Fornari et al.*, 1979b]. These large amounts of volcaniclastic sand reflect the simple pathways and short distance from the surf zone (S504, K93).

Previous studies have used basaltic glass chemistry, especially alkali-silica diagrams, to evaluate the sources of volcaniclastic turbidites [i.e. *Garica*, 1996; *McMurtry et al.*, 1999]. Based on present-day topography, three volcanoes are possible sources for the south flank sediment samples: 1) Loihi, 2) Kilauea, and 3) Mauna Loa. Most glass fragments of tholeiitic basalt and basaltic andesite are within the reported compositional ranges of these volcanoes and samples that contain alkali basalt and basanite. One possible source for the alkali basalt and basanite sands on the Hilina lower scarp is the recently discovered early alkali part of ancestral Kilauea within the Hilina slump [*Lipman et al.*, 2000].

The Loihi south rift samples contain tholeiitic and alkali

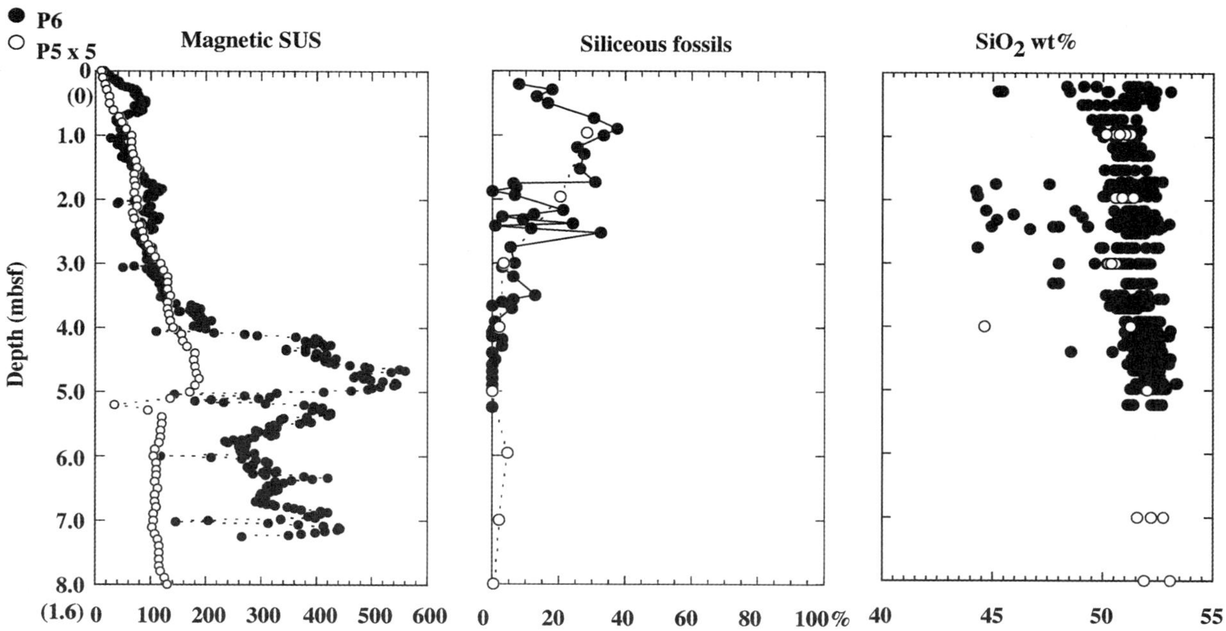

Figure 9. Vertical changes in piston core P6 and topmost 1.2 m of core P5: magnetic susceptibility (SUS), siliceous fossils (diatom, radiolarians, sponge spicules etc.), and SiO_2 contents of glass fragments. Solid circles, P6; open circles, P5 (5x times actual depth).

basalt fragments derived from Loihi. They have deep-sea hyaloclastite and bubble-wall shapes, and shallow to deep-sea S contents. Such textures show that they erupted along the rift and fragmented almost in place, or were transported from upper part of the rift. Two continuous push cores from the base of Loihi recovered thin turbidite layers 1-3 cm thick as well as discrete volcaniclastic fragments. Two volcaniclastic turbidite layers in S493-C2 contain Loihi-type tholeiite and alkali basalt fragments of probable Loihi origin. Most S518-G7 glass fragments are subaerially erupted Kilauea tholeiite, but the other two layers (G2 and G4) include both subaerially erupted Kilauea tholeiite and Loihi alkali basalt fragments in the same layers. Thin, evenly bedded volcaniclastic turbidites at the base of Loihi contain volcaniclastics of both deep-sea Loihi and shallow water Kilauea origin. These relationships suggest that reworking and mixing of volcanic sands are common processes on the south flank.

In the pelagic environment, volcaniclastic turbidites can mix with preexisting ambient pelagic sediment during transport [i.e., *Sparks and Wilson*, 1983]. Volcaniclastic turbidites, representing distal faces of giant landslides from Lanai and Molokai, were also redeposited in ODP 136 site 842 [*Garcia*, 1996]. The Loihi base samples indicate that this kind of redeposition or mixing occurred even in small turbidites. Push core sample S493-C contains two well-bedded volcaniclastic turbidite layers, 2-4 cm thick, in which deep-water-erupted volcaniclastics may have been derived from the eastern flank of Loihi by mass wasting [*Fornari et al.*, 1988; *Malahoff* 1987]. The mixture of both tholeiite and alkali basalt clasts indicates that this stage was similar to the upper stratigraphic sequence of Loihi, where both types of eruptions took place at about the same time [*Garcia et al.*, 1993]. Generally, the south flank surficial volcaniclastic sediments were mixed during downslope turbidite emplacement, more than by surface currents as recorded in ripple marks on the seafloor.

6. EVENTS RECORDED IN PISTON CORES

The volcaniclastic material recovered in piston cores from sites P5 and P6 record a long history of past volcanic events in Hawaii. The P6 core contains eight well-bedded volcaniclastic turbidite layers, all below about 1.75 mbsf (Fig. 8). At shallower levels, sparse glass fragments, probably dispersed by bioturbation, provide an almost continuous volcano-stratigraphic record to the present. Most tholeiite and basaltic andesite glass fragments have low S contents and blocky shapes, reflecting a shoreline origin. At a depth of about 4 m the average SiO_2 content increases, changing from

Kilauea to Mauna Loa tholeiite type, suggesting that the Kilauea shield became well developed only at after an age corresponding to 4 mbsf. Many of the glass grains in layers shallower than 4 mbsf also have slightly elevated S contents (200 – 600 ppm, in comparison to ~100 ppm for the recently erupted coastal glass sand from Kilauea), perhaps indicating that the subaerial Kilauea shield at that stage was too small for effective low-pressure degassing during subaerial flowage from vent to the coast. Between 1.75 and 3.3 mbsf, glass fragments of alkali basalt or basanite are more common than at other levels (Fig. 8). This interval may correlate with the initial alkalic stage of Kilauea volcanism recognized on the Hilina lower scarp [*Lipman et al.*, 2000].

The P6 core has normal magnetic polarity throughout, whereas all P5 layers below about 2.3 m depth are reversed and must be older than 0.78 Ma [*Kanamatsu et al.*, this volume]. Magnetic susceptibility (SUS) was measured for P5 and P6 cores at 2 cm intervals (Fig.8). The SUS generally reflects the average abundance of magnetic materials for the interval measured. Generally, volcaniclastic horizons rich in glass and mafic minerals have high magnetic SUS, but if a magnetic-rich layer is thin, its intensity is low. The vertical SUS variation in the uppermost 1.2 m of P5 has a pattern similar to all of P6 (Fig. 9), suggesting that the average sedimentation rate at the P6 site was about 5 time faster than that in the upper 1.2 m at P5. This interpretation is supported by SiO_2 contents in glass fragments, which tend to vary similarly (Fig. 9). Abundances of siliceous bioclasts also show similar variations between P6 and topmost part of P5. The intensity of magnetic susceptibility and therefore the amounts of volcanic glass, generally decreases upward in the P6 core, especially above the boundary between Mauna Loa- and Kilauea-type glasses. This suggests that the importance of gravitational flowage (turbidity currents, debris flows) diminished as the Kilauea shield grew. Such an interpretation is supported by the increased S content in glass fragments above the Kilauea-Mauna Loa boundary in the core.

No identifiable volcaniclastic layer in P5 correlates with the thick basal volcaniclastic layer in P6, suggesting that the P5 turbidites record only exceptionally large events. Mean grain sizes (6-8ϕ) in the pelagic and hemipelagic fine sediment intervals are smaller than in P6 (5-6ϕ), probably reflecting the greater distance from Hawaiian volcanic sources. Some samples contain only sparse volcaniclastic material (especially glass). A transition from Kilauea- to Mauna Loa-type tholeiite at 1.0 mbsf (even for glass fragments in the pelagic and hemipelagic fine sediment intervals) is nearly at the expected depth based on the estimated average sedimentation rates. Mauna Loa-type glass is dominant from 1 to 1.7 mbsf, but Kilauea-type glass reappears from 1.8 to 4 mbsf. Because most of this interval is older than 0.78 Ma (reversely magnetized), the glass must have come from a pre-Kilauea volcano. Three thick turbidite layers below 4 m contain mixed fragments of high SiO_2 (50 to 52 wt%) tholeiite and basaltic andesite and low S basalt (mostly shallow or subaerially erupted). These turbidite layers likely record large collapse events on old well-developed shields.

Sediments on the outer Hawaiian archipelagic apron have a total thickness of about 200 m [*Winterer*, 1989]. The age of oceanic crust at the ODP 136 site 843 is about 110 Ma [*Waggoner*, 1993], indicating an average long-term sedimentation rate of about 2 m/my. Because the more deeply buried sediments have been compacted, the 2 m/my rate is a lower limit. Several recent studies summarized below indicate slightly higher Quaternary rates, especially following large landslide events. The magneto-stratigraphy of two piston cores on the Hawaiian Arch north of Oahu (P2 and P3) indicates a pelagic sedimentation rate of about 4 or 2.5 m/my since 1.8 or 2.8 Ma [*Kanamatsu et al.*, this volume]. The Arch south of Oahu at ODP 842 and 843 has an average sedimentation rate of about 4 m/my since 3.8 Ma [*Dziewonski et al.*, 1992]. Post-collapse sedimentation after the Alika landslide from the west flank of Mauna Loa (~110 ka) is about 2.5 m/my [*Moore et al.*, 1989; *McMurtry et al*, 1999]. The average sedimentation rate since 0.78 Ma (Brunhes epoch) is 2.9 m/my in the P5 core [*Kanamatsu et al.*, this volume], similar to the previous determinations and appropriate for pelagic conditions with the introduction of only small amounts of volcaniclastic material.

The P5 core contains no recognizable well-bedded turbidites from Kilauea, suggesting that, (1) coastal activity at this volcano has been insufficiently large to generate turbidity currents that traveled as far as the P5 site, and (2) that no large landslides have occurred on the southern flank of Hawaii since the late stage of Mauna Loa. Some turbidites low in the P5 core may record large slides from Kohala or older volcanoes, such as those in the reversed magnetized interval that must be older than 0.78 Ma. Our magnetic surveys [*Shipboard Scientists,* 2000] demonstrate that Hilo Ridge is largely composed of reversely magnetized rocks, and thus is older than 0.78 Ma. If Hilo Ridge is a rift zone of Mauna Kea, as commonly inferred, it is much older than the 200-300 ka age proposed by *Moore and Clague* [1992]. Even if Hilo Ridge is a rift zone of Kohala volcano [*Holcomb et al.*, 2000], the oldest known subaerial lavas from this volcano [0.43 Ma: *Clague and Dalrymple*, 1987] are younger than the reversely magnetized bulk of Hilo Ridge. The dated lavas, however, represent just the younger part of the Kohala shield, and older parts, now deeply buried, may well have a genetic relationship with the Hilo Ridge.

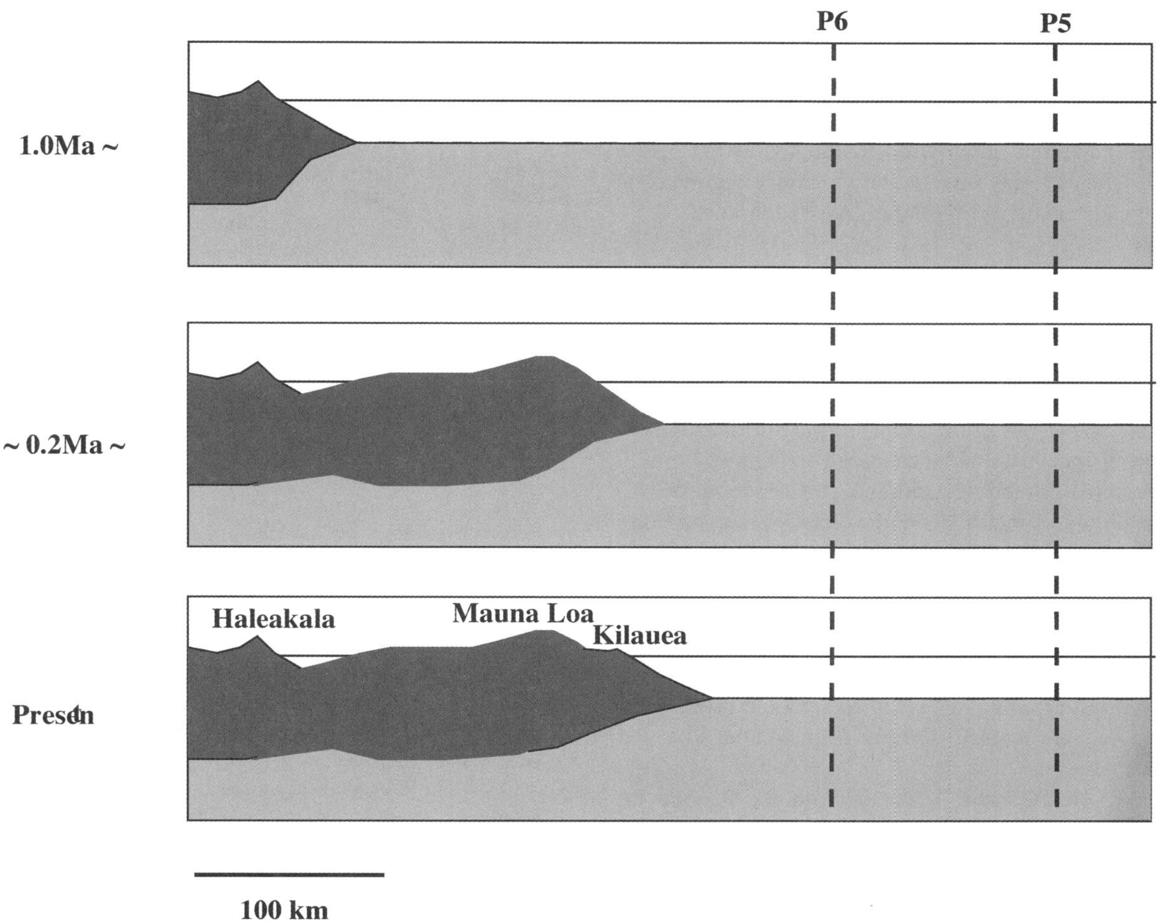

Fig.ure 10. Schematic illustration of the possible past relations between the piston core sites (P6 and P5) and volcanoes to the north.

If a simple propagation model [*Clague and Dalrymple*, 1987] is used, the estimated distance between the P5 site and expected location of the Hawaiian hotspot at the top of the Jaramillo Subchron (0.9 Ma) is about 300 km. Possible active volcanoes according to this model are Kohala, Mahukona, and Haleakala. Three thick turbidite layers in lower portions of P5 may document large collapse events that should have had a source in a large subaerial shield, because the dominant constituents are degassed. Therefore, Kohala or Haleakala, the subaerial volcanoes among these three possibilities, are the most likely sources (Fig. 10). Volcanic ash layers more than 500 km south of Hawaii (14° to 15° N), with ages of 1-2 Ma, have been interpreted as turbidites from West Maui or Haleakala [*Rehm and Halbach*, 1982; *Garcia*, 1996]. If the lower volcaniclastic turbidites in the P5 core correlate with those reported by *Rehm and Halbach* [1982], the P5 layers may record large landslides that have been buried by younger Hawaiian volcanoes.

Small volumes of benmoreite and mugearite have erupted widely in the post-caldera alkalic stage of Hawaiian shield volcanism, and there are many potential sources for the micropumice in the P5 and P6 cores. In the last ~1 Ma, volcanoes known to have erupted benmoreite or mugearite are West Maui, Haleakala, Kohala, and Mauna Kea [*Clague*, 1987; *Wolfe, et al.*, 1997]. Hualalai has erupted trachyte and phonolite [*Clague*, 1987] and it is likely that mugearites and benmoreites are also present, but have not been recognized. Unlike the common poorly vesicular glass grains, which result from seawater fragmentation of basaltic lavas, the micropumice are magmatically inflated pyroclasts that were probably dispersed by subaerial eruption columns before being redeposited by turbidity currents. Multiple sources and possible redeposition prevent use of the sparse micropumice for stratigraphic correlations in the P5 and P6 cores. However, their ubiquity suggests that small, pumice-producing eruptions are relatively common and that piston cores collected down-wind to the southwest of the Hawaiian Islands may encounter primary pumiceous ash-fall deposits useful as time-stratigraphic markers.

7. SUMMARY

1) Recent volcaniclastic materials, deposited around the south flank of Hawaii, were mostly derived from coastal areas of Kilauea volcano.

2) Around the base of Loihi, volcaniclastic turbidites have sources both from coastal Kilauea and deep parts of Loihi.

3) Piston cores P5 and P6, 250 and 120 km south of Hawaii, respectively, contain bedded volcaniclastic turbidite layers rich in basaltic glass. Volcanic fragments are also scattered through pelagic clay zones. The average sedimentation rate at the P6 locality is about 5 times faster than at the more distal P5.

4) The volcaniclastic fragments in the P6 core originated largely or entirely from Kilauea and Mauna Loa; lower parts of the P5 core may include detritus from Mauna Loa and older volcanoes.

5) Bedded volcaniclastic layers in both the P5 and P6 cores may contain turbidites related to large collapse events. The P5 core may contain detritus from events originating at volcanoes that are older than Kohala.

Acknowledgements. This study was carried out during US-Japan cooperative research cruises using RV KAIREI and YOKOSUKA in 1998 and 1999. We thank Captains Osamu Yukawa and Hitoshi Tanaka, the ship crews, and commanders of Tustomu Fukui and Yoshij Imai and the operation teams of KAIKO and SHINKAI 6500. We also acknowledge Dr. Richard S. Fiske of Smithsonian Institute, Dr. Ralph Moberly, and Dr. Michael O. Garcia of the University of Hawaii for their critical reviews and editorial help with this manuscript.

REFERENCES

Berger, W. H., and E. L. Winterer, Plate stratigraphy and the fluctuating carbonate line, *in Pelagic Sediments on Land and Under the Sea*, edited by K. J. Hsu and H. C. Jenkins, Spec. Publ. 1, 11-98, International Association of Sedimentologists, Blackwell, Oxford, 1974.

Chadwick, W. W., J. R. Smith, Jr., J. G. Moore, D. A. Clague, M. O. Garcia, and C. G. Fox, Bathymetry of south flank of Kilauea volcano, Hawaii. *U.S. Geol. Survey Map MF-2231*, 1993.

Clague, D. A., Hawaiian alkaline volcanism, *in Alkaline Igneous Rocks*, edited by J. G. Fitton and B. G. J. Upton, Geol. Soc. Spec. Publ. 30, 227-252, 1987.

Clague, D. A., and G. B. Dalrymple, The Hawaiian-Emperor volcanic chain. Part 1. Geologic evolution, *U. S. Geol. Survey Prof. Paper 1350*, 5-54, 1987.

Dixon, J.E., D.A. Clague, and E. M.Stolper, Degassing history of water, sulfur, and carbon in submarine lavas from Kilauea volcano, Hawaii, *J. Geol., 99,* 371-394, 1991.

Dziewonski, A., Wilkens, R., Firth, J., et al., *Proc. ODP, Init. Rept. 136,* College Station, TX, 156 p, 1992,

Foranri, D. J., M. O. Garcia, R. C. Tyce, and D. G. Gallo, Morphology and structure of Loihi Seamount based on Seabeam sonar mapping, *J. Geophys. Res., 93*, 15227-15238, 1988.

Foranri, D. J., J. G. Moore, and L. Calk, A large submarine sand-rubble flow on Kilauea volcano, Hawaii, *J. Volc. Geotherm Res., 5*, 239-256, 1979a.

Fornari, D. J., A. Malahoff, and B. C. Heezen, Submarine slope micromorphology and volcanic substructure of the island of Hawaii inferred from visual observations made from U. S. Navy deep-submergence vehicle (DSV) "Sea Cliff", *Marine Geol., 32*, 1-20, 1979b.

Furnes, H., and I. B. Fridleifsson, Tidal effects on the formation of pillow lava/hyaloclastite deltas, *Geology*, 4, 381-384, 1974.

Furnes, H., and B. A. Sturt, Beach/shallow marine hyaloclastite deposits and their geological significance – an example from Gran Canaria, *J. Geology, 84,* 439-453, 1976.

Garcia, M. O., Pliocene-Pleistocene volcanic sands from site 842: Products of giant landslides, *in* Wilkens, R. H, Firth, J., Bender, J., et al., eds. *Proc. Ocean Drill., Sci. Res., 136*, 53-64, 1993.

Garcia, M.O., Turbidite from slope failure on Hawaiian volcanoes, *in* McGuire,W. J. et al. eds. Volcano instability on the Earth and other planets, *Geol. Soc. Spec. Pub., 110,* 281-29, 1996.

Garcia, M. O., B. A. Jorgenson, J. J. Mahorney, E. Ito, and A. J. Irving, An evaluation of temporal geochemical evolution of Loihi summit lavas; Results from Alvin submersible dives, *J. Geophys. Res.*, 98, 537-550, 1993Holcomb, R. T., B. K. Nelson, P.W. Reiners, and N-L. Sawyer, Overlapping volcanoes: The origin of Hilo Ridge, Hawaii, *Geology, 28,* 547-550, 2000.

Holcomb, R. T., J. G. Moore, P. W. Lipman, and R. H. Belderson, Voluminous submarine lava flows from Hawaiian volcanoes, *Geology, 16*, 400-404, 1988.

Huang, T. C., A volcanic sedimentation model: implications of processes and responses of deep-sea ashes, *Marine Geol., 38,* 103-122, 1980.

Inman, D. I, Measures for describing the size distribution of sediments, *Jour. Sediment. Petrol., 22, 125-145,* 1952.

Jones, J. G., and P. H. H. Nelson, The flow of basalt from air into water: its structural expression and stratigraphic significance, *Geol. Mag., 107*, 13-21, 1970.

Kanamatsu, T., E. Herrero-Bervera, and G. M. McMurtry. Magnetostratigraphy of deep-sea sediments from areas adjacent to the Hawaiian Islands [this volume].

Kelts, K., and M. A. Arthur, Turbidites after the years of deep-sea drilling - wringing out the mop, *SEPM Spec. Pub., 32,* 91-127, 1981.

LeBas, M. J., R. W. Maitre, A. Streckeisen, and B. Zanettin, A chemical classification of volcanic rocks based on the total alkali-silica diagram, *J. Petrol., 27,* 745-750, 1986.

Lipman, P. W., W. R. Normark, J. G. Moore, J. B. Wilson, and C. E. Gutmacher, The giant submarine Alika debris slide, Mauna Loa, Hawaii, *J., Geophys. Res., 93,* 4279-4299, 1988.

Lipman, P. W., J. P.Lockwood, R. T. Okamura, D. A. Swanson, and K. M. Yamashita, Ground deformation associated with the 1975 magnitude 7.2 earthquake and resulting changes in activity of Kilauea volcano 1975-1977, Hawaii, *U.S. Geol. Surv. Prof. Paper 1276*, 44 p., 1985.

Lipman, P. W., T. W. Sisson, T. Ui, and J. Naka, Ancestral growth of Kilauea volcano: submarine alkalic through tholeiitic shield-stage development and landsliding on the offshore south flank of Hawaii Island, *Geology, 28*, 1079-1082, 2000.

Lipman, P. W., T. W. Sisson, T. Ui, J. Naka, and J, R. Smith, Ancestral submarine growth of Kilauea Volcano and instability of its south flank, this volume.

Macdonald, G. A., and T. Katsura, Chemical compositions of Hawaiian lavas, *J. Petrol., 5*, 82-133, 1964.

Malahoff, A., Geology of the summit of Loihi submarine volcano, Geologic evolution, *U. S. Geol. Survey Prof. Paper 1350*, 133-148, 1987.

Malahoff, A., A.M. Sagalevitch, O.G. Sorokhtin, and J.R Smith, 1991, Volcanic processes, hydrothermal fields and mass wasting on Loihi submarine volcano, Hawaii, *Geol. Soc. Amer. Abstracts with Programs*, vol. 23, no. 2, p. 75.

McMurtry, G. M., E. Herrero-Bervera, M. D. Cremer, J. R. Smith, J. Resig, V. Sherman, and M. E. Torresan, Stratigraphic constraints on the timing and emplacement of the Alika 2 giant Hawaiian submarine landslide, *J. Volcanol, Geotherm. Res., 94,* 35-58, 1999.

Menard, H. W., Archipelagic apron, *Bull. Amer. Assc. Petrol. Geol., 40,* 2195-2210, 1956.

Moore, J. G., Giant submarine landslides on the Hawaiian Ridge, *U.S.Geol. Surv. Prof. Paper, 501D*, 95-98, 1964.

Moore, J. G., and D. A. Clague, Volcano growth and evolution of the island of Hawaii, *Geol. Soc. Amer. Bull., 104*, 1471-1484, 1992.

Moore, J. G., and R. S. Fiske, Volcanic substructure inferred from dredge samples and ocean-bottom photographs, Hawaii, *Geol. Soc. Amer. Bull., 80,* 1191-1202, 1969.

Moore, J. G., D. A. Clague, R. T. Holcomb, P. W. Lipman, W. R. Normark, and M. E. Torresan, Prodigious submarine landslides of the Hawaiian ridge. *J. Geophys. Res., 94*, 17465-17484, 1989.

Moore, J. G., W. R. Normark, and R. T. Holcomb, Giant Hawaiian underwater landlide , *Science*, 264, 46-47, 1994.

Moore, J.G., and W. W. Chadwick,Offshore geology of Mauna Loa and adjacent area, Hawaii, *Am. Geophys. U Monograph, 92,* 21-44, 1995.

Moore, J. G., R. L.Phillips, R. W.Grigg, D. W.Peterson. and D. A Swanson, Flow of lava into the sea, 1969-1971, Kilauea volcano, Hawaii, *Geol. Soc. Amer. Bull., 84,* 537-546, 1973.

Moore, G. J., and J. G. Schilling, Vesicles, water, and sulfur in Reykjanes ridge basalt, *Contrib. Mineral. Petrol., 41,* 105-118, 1973.

Morgan, J. K., G. F. Moore, D. J. Hills, and S. Leslie, Overthrusting and sediment accretion along Kilauea's mobile south flank, Hawaii, from marine seismic reflection data: test of volcanic spreading model, *Geology, 28,* 667-670, 2000.

Naka, J., et al., Tectono-magmatic processes investigated at deep-water flanks of Hawaiian volcanoes, *EOS, 81,* 211,226-227, 2000.

Naka, J., R. Tsugaru, T. Danhara, T. Tanaka, and K. Fujioka, Sedimentary processes of volcaniclastic sediments. Leg 136, in Wilkens, R. H, Firth, J., Bender, J., et al., eds. *Proc. Ocean Drill., Scie. Res.*, 136,85-98, 1993.

Owen, S., P. Segall, M. Lisowski, A. Miklius, R. Denlinger. and M. Saklo, The rapid deformation of Kilauea Volcano: GPS measurement between 1990 and 1996, *J. Gepphys. Res.* in press.

Rehm, E., and P. Halbach, Hawaiian-derived volcanic ash layers in equatorial northeastern Pacific sediments, *Marine Geology, 50,* 25-40, 1982.

Ruddiman, W. F., and L. K. Glover, Vertical mixing of ice-rafted volcanic ash in north Atlantic sediments, *Geol. Soc. Amer. Bull.*, 83, 2817-2836, 1972.

Shipboard scientists, Results of underway geophysical surveys, gravity, magnetic, and single channel seismic surveys, during cruises KR98-08, 09 and YK99-07, 08, *WPGM abs.*, 2000.

Sisson, T. W., P. W. Lipman, and J. Naka, Submarine alkalic through tholeiitic shied-stage development of Kilauea Volcano, Hawaii, *this volume.*

Smith, J. R., A. Malahoff, and A. N. Shor, Submarine geology of the Hilina slump and morph-structural evolution of Kilauea volcano, Hawaii, *J. Geotherm. Res., 94,* 59-88, 1999.

Smith, J. R. and K. Satake, Deepwater multibeam sonar surveys along the southeastern Hawaiian Ridge, *this volume.*

Sparks. R. S. J., S. Brazier, T. C. Huang, and D. Muerdter, Sedimentology of the Minoan deep-sea tephra layer in the Aegean and eastern Mediterranean, *Marine Geol., 54*, 131-167, 1984.

Swanson, D. A., Pahoehoe flows from the 1969-1971 Mauna Ulu eruption, Kilauea volcano, Hawaii, *Geol. Soc. Amer. Bull., 84,* 6156-626, 1973.

Swanson, D. A., and B. P. Fabbi, Loss of volatiles during fountaining and flowage of basalt lava at Kilauea volcano, Hawaii, *J. Res. U.S.Geol. Survey, 1,* 649-658, 1973.

Swanson, D. A., W. A., Duffield, and R. S. Fiske, Displacements of the south flank of Kilauea volcano: The result of forceful intrusion of magma into the rift zone, *U. S. Geol. Surv. Prof. Pap 963*, 30 pp., 1976.

Waggoner, D.G., The age and alteration of basalts from Leg 136, central Pacific oceanic crust near, Site 843, *in* Wilkens, R. H, Firth, J., Bender, J., et al., eds. *Proc. Ocean Drill., Sci. Res., 136,* 119-132, 1993.

Winterer, E.L., Sediment thickness map of the northeast Pacific, in Winterer, E. L., et al., eds, *The Geology of North America*, N., 307-310, 1989.

Wolfe, E.W., Wise, S. W., and Dalrymple, B., The geology and petrology of Mauna Kea Volcano, Hawaii -- A study of post-shield volcansim, *U.S. Geol. Surv. Prof. Paper 1557*, 129 pp., 4 plates, 1997.

Yeats, R.S., Hart, S.R., et al., *Init. Repts. DSDP 34*, Washington (US Govt. Print. Office), 1976.

Toshiya Kanamatsu, Deep Sea Research Department, JAMSTEC, 2-15 Natsushima-cho, Yokosuka 237, Japan. (e-mail; toshiyak@jamstec.go.jp)

Peter W. Lipman, Volcano Hazards Program, U.S. Geological Survey, MS 910, 345 Middlefield Road, Menlo Park, California 94025. (plipman@usgs.gov)

Julia K. Morgan, Department of Geology and Geophysics, Ms-126, Rice University, 6100, 6100 Main Street, Houston, TX 77005 (morganj@tienshan.rice.edu)

Jiro Naka, Deep Sea Research Department, JAMSTEC, 2-15 Natsushima-cho, Yokosuka 237, Japan. (nakaj@jamstec.go.jp)

Thomas W. Sisson, Volcano Hazards Program, U.S. Geological Survey, MS 910, U.S. Geological Survey, 345 Middlefield Road, Menlo Park, CA 94025. (tsisson@usgs.gov)

John R. Smith, School of Ocean & Earth Science & Technology, University of Hawaii, 1000 Pope Road, MSB 205, Honolulu, Hawaii 96822-2336. (jrsmith@iniki.soest.hawaii.edu)

Nohiro Tsuboyama, 1-33-411, kita 21 nishi 7, kitaku, Sapporo, Japan (unagi@orchid.plala.or.jp).

Tadahide Ui, Graduate School of Science, Hokkaido University, N10 W8, Kita-ku, Sapporo 060-0810, Japan.
(ui@ep.sci.hokudai.ac.jp)

Plates (in Appendix, on CD):

Plate 1-1. Representative seafloor photographs
Photo 1: Rippled dark volcanic sand-rich sediment, basal part of upper slope (S504).
Photo 2: Subsurface outcrop of black volcanic sand, steep slope at eastern end of lower scarp (S509).
Photo 3: Thick sediment cover, eastern base of Loihi (S493).
Photo 4: Fine-grained sediment, poor in volcanic sand, lower scarp (S508).

Plate 1-2. Representative seafloor photographs (cont.)
Photo 1: Seafloor covered with fine sediment, base of Puna Ridge. Benthic animal tracks are present (S495).
Photo 2: Talus breccia at cliff base, lower scarp (S508).
Photo 3: Jointed indurated volcaniclastic sandstone, lower scarp (S508)
Photo 4: Moderately indurated breccia, steep slope at eastern end of lower scarp (S509).

Plate 2-1. Representative SEM photographs of coarse sediment fractions
Photo 1: S504 G3-5, Mostly blocky fragments, Hilina upper slope.
Photo 2: S505 C2, Mostly blocky fragments, Hilina lower scarp
Photo 3: S495 C1 2-5, Blocky and bubble-wall fragments, basal Puna Ridge
Photo 4: S515 S1, Blocky and flaky fragments, Loihi rift.

Plate 2-2. Representative SEM photographs of coarse sediment fractions (cont.)
Photo 1: S493 C2cc, Mostly blocky fragments, Loihi east base
Photo 2: S518 G4, Mostly blocky fragments, Loihi south base
Photo 3: P5 S7 58-62, Mostly blocky fragments, bedded volcaniclastic layer in P5 core.
Photo 4: P6 S5 88-92, Mostly blocky fragments, bedded volcaniclastic layer in P6 core.

Plate 3. Photomicrographs of subaerial Kilauea pahoehoe and submarine Puna Ridge pillow lava. .
Photo 1: Glassy rim of subaerial pahoehoe, 0-1 cm from outer surface
Photo 2: Devitrified interior of subaerial pahoehoe, 1-2 cm from outer surface
Photo 3: Glassy rim of submarine pillow lava, 0-1 cm from outer surface
Photo 4: Devitrified interior of submarine pillow lava, 1-2 cm from outer surface

TABLES (In Appendix CD)

Table 1. Sample locations.

Table 2. Smear slide observations (a) and modal abundances data (b). All data of table 2a are approximate relative abundances from microscope observation. Modal abundances were measured by 200-point counts of thin sections under the microscope. FS: Fine sediment, FD: Feldspar, RF: Rock fragment, AG: Altered glass, FG: Fresh glass, AC: Accessory minerals, OP: Opaque materials, PO: Pore space, MF: Mafic minerals, CF: Calcareous fossils (FO:Foraminifera, NA: Nannofossil,), SF: Siliceous fossils (DI: Diatom, RD: Radiolarian, SS: Sponge spicule, SI: Silicoflagellate), Ot: Others

Table 3. Summary of grain size analyses (every 1/2 ϕ interval).

Table 4. Summary of glass analyses. Averages of multiple analyses per grain. *, indicates low-Na_2O values that were omitted in figs 6, 9 and 10

Magnetostratigraphy of Deep-Sea Sediments From Piston Cores Adjacent to the Hawaiian Islands: Implication for Ages of Turbidites Derived From Submarine Landslides

Toshiya Kanamatsu

Deep Sea Research Department, Japan Marine Science and Technology Center, Yokosuka, Japan

Emilio Herrero-Bervera

Paleomagnetic and Petrofabrics Laboratory, School of Ocean and Earth Science and Technology, Hawaii Institute of Geophysics and Planetology, University of Hawaii, Honolulu, Hawaii

Gary M. McMurtry

Department of Oceanography, University of Hawaii, Honolulu, Hawaii

Volcaniclastic turbidite beds in ocean-floor piston cores provide a distal record of large Hawaiian submarine landslides adjacent to the Hawaiian Islands. Cores were obtained north of Oahu, where the Nuuanu and Wailau debris avalanches were derived from Koolau and East Molokai volcanoes, and in the front of the active Hilina slump south of the Island of Hawaii. Magnetic measurements were carried out on the cores to determine turbidite ages. Two magnetostratigraphic models are possible for the cores north of Oahu, using paleontological age markers. Both age models yield turbidite ages that are close to published ages for the source volcanoes. A near-constant sedimentation-rate model indicates ages of about 2 Ma and older than 2.58 Ma for turbidites inferred to record the Wailau and Nuuanu debris avalanches, respectively. These ages are close to the initial-shield building stages of each source volcano. The second model, involving variable sedimentation rates, indicates ages of about 1 Ma and older than 1.77 Ma for these debris avalanches, ages that are close to post-shield stages. The proximal core in the front of the Hilina slump contains only turbidites younger than 0.78 Ma. Turbidites recovered in the distal core are 0.78 Ma or older, predating any published volcano ages for the island of Hawaii, and suggesting sources from volcanoes on other islands.

1. INTRODUCTION

The existence of giant submarine landslides from Hawaiian volcanoes has been recognized since the 1960's [*Moore*, 1964]. GLORIA side-scan sonar surveys of the seafloor in

Hawaiian Volcanoes: Deep Underwater Perspectives
Geophysical Monograph 128
Copyright 2002 by the American Geophysical Union

the Hawaiian Exclusive Economic Zone imaged 15 large submarine landslides adjacent to the Hawaii Islands, mapping their extent and the distribution of debris blocks [*Moore et al.*, 1989]. In 1998 and 1999, joint U.S.-Japan surveys using research vessels and submersibles of the Japan Marine Science and Technology Center, further studied these massive landslides. Landslide areas north of Oahu and the south of the island of Hawaii were the main targets of this project. The northern area includes the submarine flank of the Koolau volcano, the Nuuanu and Wailau debris avalanches, and the North Arch lava field. The area south of Hawaii includes Loihi Seamount, the Hilina slump, and the Puna Ridge.

The Hawaiian submarine landslides have been classified into two types by their morphology and transport distance: debris avalanches and slumps [*Moore et al.*, 1989]. The debris avalanches are associated with large-mass and long-distance transport to the archipelagic apron. The slumps are considered to be rooted in the volcanic edifice, as the flank of the volcano creeps seaward. Understanding the mechanisms and timing of submarine landsliding is important for hazard mitigation, because the slides are closely associated with earthquakes, coastline submergence, and tsunamis.

The mechanisms of large submarine landslide are poorly understood, because little is known about structure of the volcano flank when slope failure takes place. The timing of landsliding is also important to determine. Because Hawaiian shield volcanoes grow rapidly, to several thousand meters above the ocean floor within about 1 m.y. [*Clague and Dalrymple*, 1987], slope instability and slides can be expected any time from the initial growth stage to the post-shield stage of volcanic activity.

A distal turbidite record of the landslides should be preserved in archipelagic sediment in the Hawaiian Deep (Figure 1), which results from isostatic subsidence of the Hawaiian Ridge. Turbidite and fine-grained debris-avalanche deposits should preferentially accumulate in the axis of the Deep, where high sedimentation rates are enhanced by windblown dust, volcanic ash, and turbidity currents from the volcanoes [*Moore and Chadwick.*, 1995]. The sedimentation history of the archipelagic environment has been interpreted from seismic profiling. A 0.9-km thick landslide unit in the Hawaiian Deep was recognized between a 0.2-km layer of sediment above and a 1.5-km layer below [*Brocher and Brink*, 1987]. *Rees et al.*, [1993] used single-channel seismic reflection data to classify the sequence into four units: 1) the old pelagic unit, 2) the landslide unit, 3) an off-lapping unit, and 4) the youngest unit. The off-lapping unit was considered to be a product of large-scale mass wasting, either the distal facies of the landslide or turbidites associated with smaller-scale slope failures.

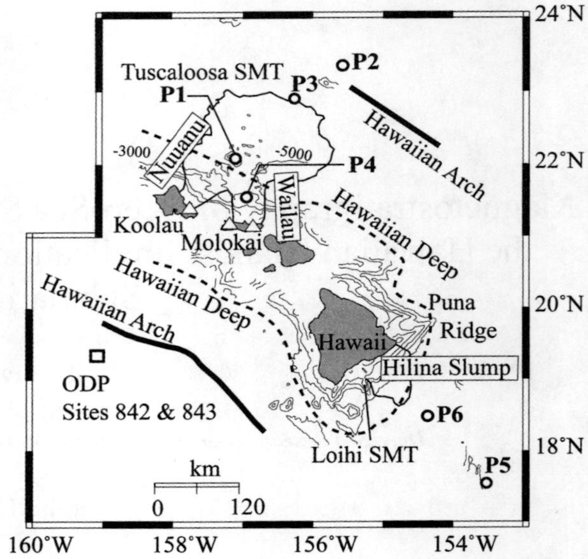

Figure 1. Locations of piston-core sites during 1998 cruise north of Oahu and south of Hawaii Island (circles), and distribution of Hawaiian submarine landslides: Nuuanu, Wailau debris avalanches, and Hilina slump [Moore *et al.*, 1989]. Square indicates location of Ocean Drilling Program Leg 136, sites 842 and 843. Dashed line, axis of Hawaiian Deep: heavy solid line, crest of the Hawaiian Arch.

In this study, we provide constraints for the timing of submarine landsliding using piston cores from north of Oahu and south of Hawaii. Distal portions of the landslides were chosen to recover cores, because we expect that sediment accumulations near islands would include turbidite deposits from small slope failures and sediment eroded from the shoreline, not just from the large landslides.

2. SUBMARINE LANDSLIDES

2.1. The Nuuanu and Wailau Debris Avalanches

The Nuuanu debris avalanche, which occupies the broadest area of any Hawaiian landslide, extends from the flank of the Koolau volcano northeast across the Hawaiian Deep. The avalanche volume is estimated to be $5x10^{12}$ m^3, and the maximum run-out distance is about 230 km [*Moore et al.*, 1989]. Distribution of avalanche blocks starts at a depth of 4500 m at the foot of the island and ends at a depth of 4300 m on the flank of the Hawaiian Arch (Figure 1), indicating that the distal debris avalanche moved upslope. The debris avalanche contains many huge blocks, the largest of which (Tuscaloosa Seamount) is 30 km long, 17 km wide, and 1.8 km thick, with a flat sloping surface. Large blocks in the debris avalanche are oriented in a direction perpendicular to the travel direction. Smaller blocks (<1 km in size) are scattered in distal areas, seaward of the Tuscaloosa Seamount [*Moore et al.*, 1989].

The Wailau debris avalanche, with a volume of 1000 km^3, contains many large blocks seaward from the north flank of East Molokai volcano. It merges with the southeast side of the Nuuanu debris avalanche. Distributions and orientations of the hummocky structures have suggested that the Wailau debris avalanche is younger than Nuuanu [*Moore et al.*, 1989]. The age of the Wailau slide, based on the subsidence rate of a terrace on the north side of Molokai, was estimated as 1.4 ± 0.2 Ma [*Moore et al.*, 1989].

2.2. *Hilina Slump, South of the Island of Hawaii*

The active Hilina slump on the south flank of Kilauea is bounded by the Hilina fault system on land. The west and east ends terminate at Loihi Seamount and the Puna Ridge, respectively [*Moore and Chadwick*, 1995; *Smith et al.*, 1999]. A large, 3000-m deep, 20-km wide submarine bench extends parallel to the fault system. The on-land part is moving seaward at 5-10 cm/yr [*Swanson et al.*, 1976; *Owens et al.*, 2000]. This motion was punctuated by a magnitude 7.2 earthquake in 1975, accompanied by 3.5-m subsidence of the lower slope of the volcano in the Hilina fault system [*Lipman et al.*, 1985]. Motion of the flank has been interpreted to be strongly related to volcanic activity, as dike injection in the rift zone pushes the flank seaward [*Swanson et al.*, 1976]. The kinematic model for seaward sliding of the upper flank has been proposed as rotational slump blocks [*Lipman et al.*, 1985; *Moore et al.*, 1994]. Compressional thrust structures and a decollment have been inferred from seismic-reflection data for lower portions of the submarine slope, suggesting an accretionary prism-like model for the Hilina slump [*Morgan et al.*, 2000].

2.3. *Deep-sea Sedimentation Record of Submarine Landslides*

Long-distance sediment transport of turbidites generated by Hawaiian submarine landslides has been suggested by several previous studies. Consolidated turbidite-like sediments recovered 270 km north of Oahu, covering part of the North Arch lava field close to Hawaiian Arch, are regarded as derived from the Nuuanu debris avalanche [*Moore et al.*, 1989]. Cores sampled by the Ocean Drilling Program (ODP) [*Dziewonski, et al.*, 1992] at ODP site 842, 265 km south of the islands on the outer side of the Hawaiian Arch [*Garcia*, 1993], contain turbidites with compositions of volcanic glass fragments that indicate origin from the Hawaii Islands, and sulfur contents suggest derivation from shallow to subaerial eruptions [*Garcia and Hull*, 1994]. Volcaniclastic sediment layers in these cores yielded ages of 100 ±20 ka, 0.78-0.9 Ma, and 1.4-1.6 Ma [*Garcia*, 1993; *Naka et al.*, 1993]. *Garcia* [1996] further interpreted material in a core [*Rehm and Halbach*, 1982] obtained 930 km south of Hawaii (14°N, 153°W) as an even farther-traveled turbidite. Mixed assemblages of Pleistocene and Eocene microfossils in the volcaniclastic layers suggests these were fed by turbidity currents and not by airborne ash.

Table 1. List of coring locations during cruises KR9808 and 09

Site	Location	Latitude	Longitude	depth
P1	North of Oahu	22-04.78'N	157-07.68'W	3012m
P2	North of Oahu	23-20.21'N	155-36.78'W	4199m
P3	North of Oahu	22-53.06'N	156-17.49'W	4323m
P4	North of Oahu	21-32.32'N	156-56.21'W	3282m
P5	South of Big Is.	17-33.04'N	153-32.03'W	4932m
P6	South of Big Is.	18-27.13'N	154-22.32'W	5204m

Thus, long-traveled turbidite sediment can provide a record of large debris avalanches adjacent to the Hawaiian Islands. The timing of landslides can be inferred by the stratigraphy of turbidite-bearing cores. For example, U-series dating, detailed magnetostratigraphy, and oxygen-isotope analysis on cores containing turbidite layers from the vicinity of the Alika landslide west of the Island of Hawaii yielded information on the age of this slide [*McMurtry et al.* 1999].

3. CORE LITHOLOGY

During the 1998 survey, six piston cores were obtained, four from northeast of Oahu, and two from southeast of Hawaii (Figure 1, Table 1). Site surveys were conducted with a multi-narrow-beam sounding system with an attached sub-bottom profiling system on the *R/V KAIREI* [*Naka et al.*, this volume]. Thick volcanic turbidite layers appear several meters from the top of some piston cores. In the P1, P2, P3, and P6 cores, sediments below the thick turbidites were sucked into the barrels, and provide no stratigraphic control for lower parts of these cores.

The cores were visually described from halved surfaces (Figure 2); microscope observations were made on smear slides and thin sections (Table 2, in appendix). Fractions smaller than fine silt (< 0.015 mm, or 6 ϕ) are described as fine sediments. The samples were sieved by 1/2 ϕ scale intervals between –1 ϕ (2 mm) and 4 ϕ (0.063 mm) and

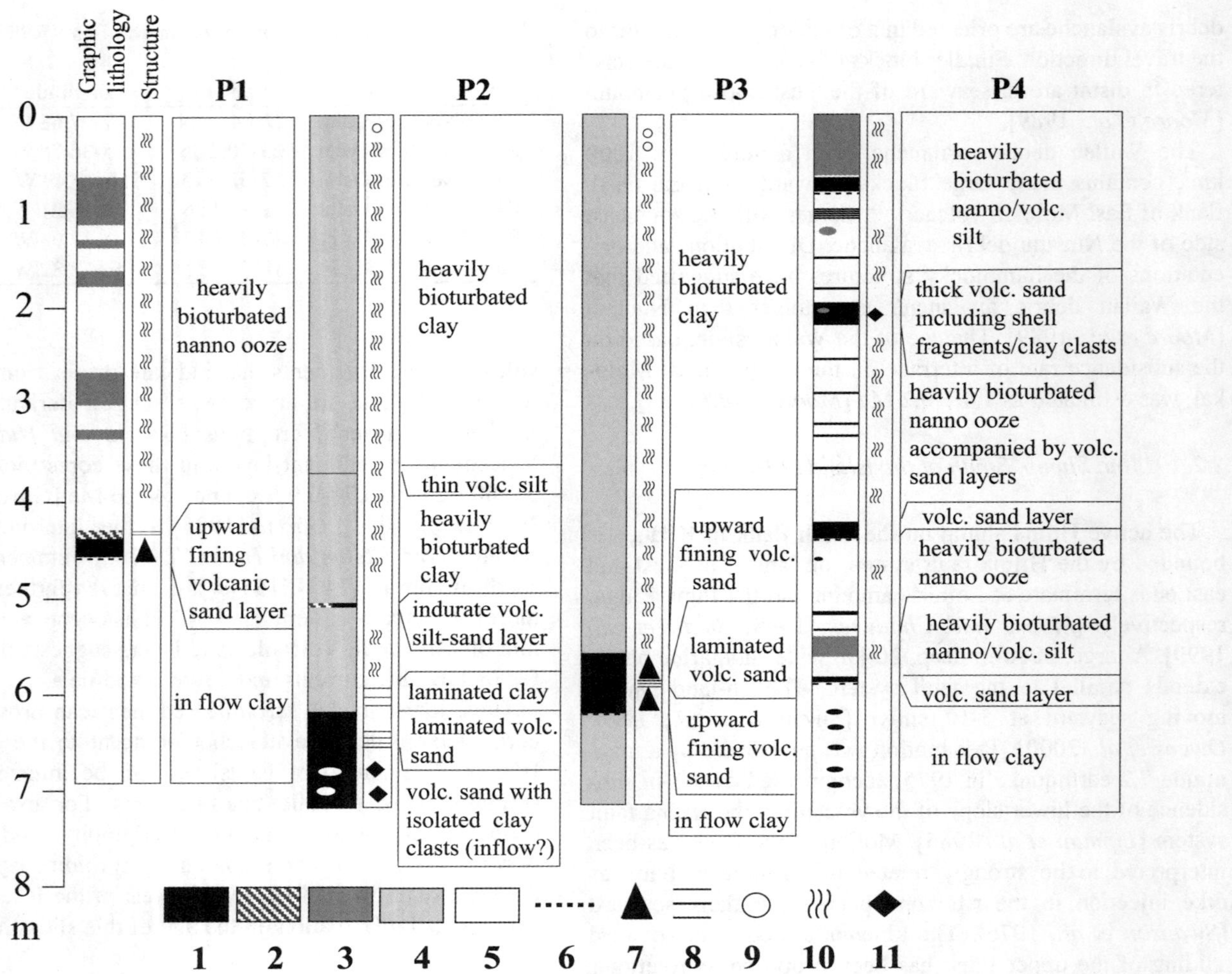

Figure 2. Simplified lithology of sediment cores P1 through 4. 1, volcanic sand; 2, volcanic silt; 3, discrete volcanic sand/silt or mix of volcanic sand/silt and pelagic/hemipelagic fine sediment; 4, pelagic/hemipelagic fine sediment; 5, calcareous ooze; 6, sharp contact; 7, normal grading; 8, parallel lamination; 9, soupy sediments; 10, bioturbated sediment; 11, isolated fine sediment.

then weighed. Fractions smaller than fine silt (< 0.015 mm, or 6 ϕ) are described as fine sediment. Material <4 ϕ was analyzed by pipette: grains suspended in a dispersion of sodium hexametaphosphete were weighed at every 1/2 ϕ time interval. Water in the suspension cylinder (1 liter) was kept the same temperature as that in the water tank to eliminate thermal effects on settling velocity. Grain-size analyses (Table 3, in appendix) were calculated as weight % for each 1/2 ϕ scale.

3.1. Core Lithology North of Oahu and Molokai

The P1 site is on the top of Tuscaloosa Seamount, the largest avalanche block of the Nuuanu landslide. A strong sub-bottom reflection indicative of a sedimentary layer was detected about 3.8 m (5 ms: two-way travel time) below the sea floor. Therefore, the sediment sequence on the flat surface of the seamount may include material younger than the Nuuanu landslide. Core P1 consists of two major lithologies and one minor lithology (Figure 2). Of the two major lithologies, one is pale yellowish brown nannofossil ooze, and the other is dark yellowish brown bioturbated volcanic silt/sand mottled with calcareous fossils. The nannofossil ooze is heavily bioturbated; it includes small amounts of foraminifera, radiolarians, sponge spicules, and volcanic sand. The minor lithology is thick-bedded volcanic sand, showing normal graded and laminated structures in the interval 4.34-4.63 m below the seafloor (mbsf). The volcanic

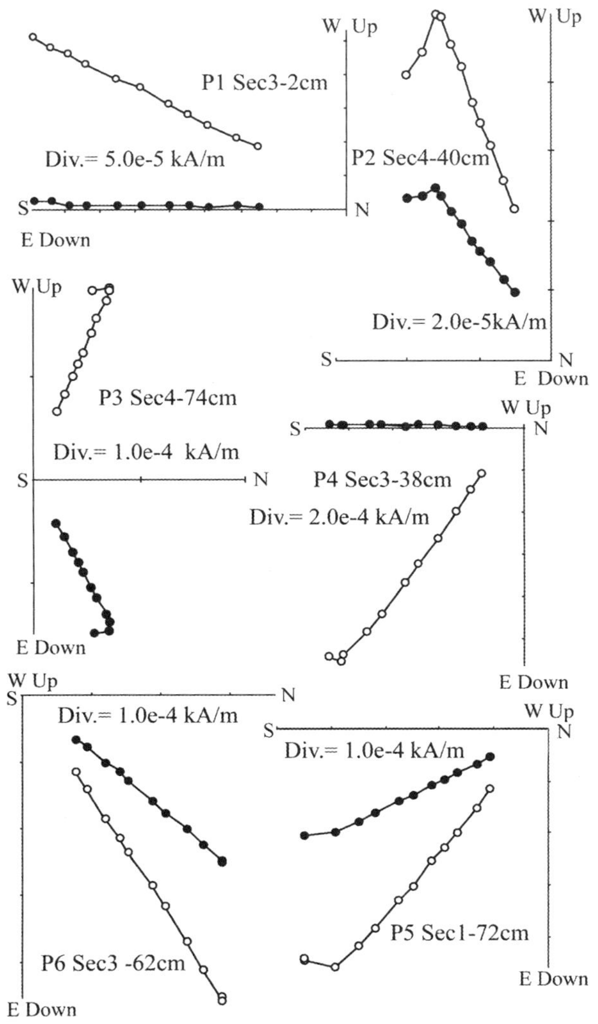

Figure 3. Representative orthogonal projections of alternating field demagnetization experiments. The demagnetization steps are 5, 10, 15, 20, 25, 30, 35, 40, 50, and 60 mT. Open and closed symbols indicate the vertical and horizontal, respectively.

sand and silt are of finely crystallized opaque material (tachylite), fresh volcanic glass, and rock fragments. Clay below the sand bed (> 4.63 mbsf) is considered as core inflow.

The P2 site, located near the axis of the Hawaiian Arch approximately 300 km from the flank of Oahu, showed strong sub-bottom reflections, at about 3.8 and 8.3 m (5 and 11 ms) below the sea floor. Core P2 consists dominantly of brown fine sediment and lesser amounts of bedded volcanic sand in the lower part of core. The fine sediment beds are heavily bioturbated, and small amounts of patchy yellowish clay are included at some horizons. The fine sediment is mostly composed of clay and fine silt. At 3.52 mbsf, a thin bed (5 mm) of volcanic clayey silt is present. An indurated volcanic silt–sand bed occurs at 5.02-5.10 mbsf. Thin beds of dark gray volcanic silt occur at 5.92-6.20 mbsf. A thick volcanic laminated silt-sand bed appears at 6.20-6.57 mbsf; below 6.57 mbsf, this bed contains isolated brown clay clasts of uncertain origin. The volcanic silt consists mainly of feldspars, opaque materials, basaltic glass, and mafic minerals.

The P3 site, at the margin of the Nuuanu debris blocks (200 km from Oahu), has sub-bottom reflections at about 5.3 and 9 m (7 and 12 ms) below the sea floor. Core P3 consists dominantly of brown fine sediment and a lesser amount of thick-bedded volcanic sand. The fine sediment is heavily bioturbated and is composed of clay and fine silt, along with minor volcanic material. Sparse radiolarians and diatoms are locally present. A thick-bedded volcanic sand layer is present at 5.60-6.22 mbsf. This volcanic sand bed shows normal grading and parallel laminations. The volcanic sand is composed of opaque materials, basaltic glass, rock fragments, mafic minerals, and feldspars. Fine sediments below 6.22 mbsf seem to be inflow.

The P4 site is located on the distal Wailau slide (Figure 1). Strong reflections at P4 were detected at about 1.5 and 3.8 m (2 and 5 ms), although the signal depth was unstable in the record. Core P4 contains two major lithologies. One is calcareous ooze, including abundant nannofossils and moderate to minor amounts of foraminifera and volcanic materials. The second is thin-bedded volcanic sand. The sand consists of opaque materials, rock fragments, volcanic glass (mostly altered), feldspars, and mafic minerals. The sand beds include shell fragments, especially at 1.97-2.23 m, and some isolated mud clasts. Most sand beds lack sedimentary structures. Volcanic sand beds are distorted in the clay below the 5.92 mbsf, with abnormal structures coming from coring disturbance.

3.2. Core Lithology South of Hawaii

The P6 and P5 sites are approximately 110 and 250 km, respectively, south of the island of Hawaii. Strong sub-bottom reflections are present at both sites at depths of several meters that probably relate to the turbidite layers. The reflections are discontinuous between P5 and P6 sites [*Naka et al.*, this volume]. Core P5 is composed of brown clay in the upper portion; the lower portion contains three volcaniclastic turbidite layers. Core P6 has eight or more turbidite sand layers in the middle and lower horizons of yellow-brown clay. Detail descriptions are in *Naka et al.* [this volume].

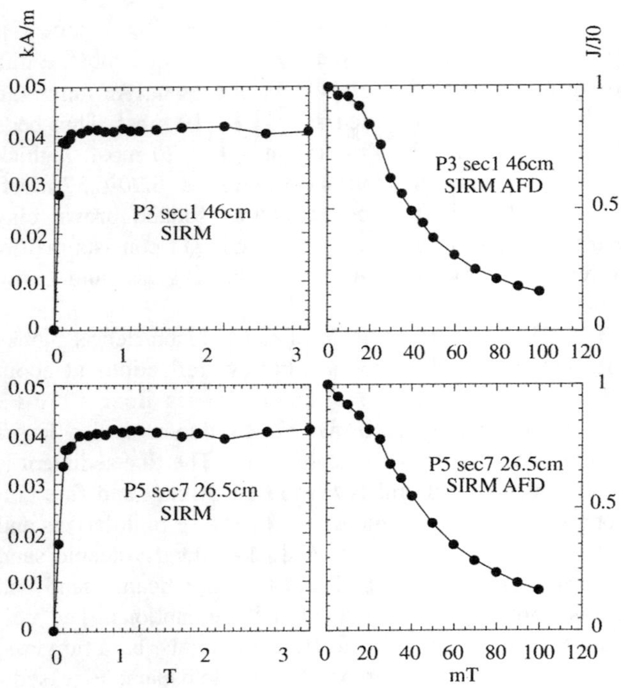

Figure 4. Representative progressive IRM acquisition and associated demagnetization curves.

4. PALEOMAGNETISM OF THE PISTON CORES

We made continuous magnetic measurements on U-channel samples from the cores in order to determine the magnetostratigraphy. The U-channel samples are 2 x 2-cm square cross-section samples that were collected from one half of the split cores. Magnetic measurements of the U-channels were made at 2-cm intervals using a horizontal 2G cryogenic magnetometer (2G Enterprises). Natural remanent magnetization (NRM) and demagnetizations were routinely measured at levels of 5, 10, 15, 20, 25, 30, 35, 40, 50, and 60 mT. Additional magnetic mineralogical experiments were carried out to confirm the nature of remanent magnetization. Results of the NRM demagnetization procedures show a stable single component between 20 mT and 60 mT (Figure 3).

5. MAGNETIC CHARACTERISTICS OF THE SEDIMENTS

Additional experiments were performed in order to identify the magnetic carriers of NRM in the sediment and to confirm the stability of magnetic directions. Saturation isothermal remanent magnetizations (SIRM) experiments were applied to selected core samples. The samples were magnetized in steps up to 2T, using an impulse magnetizer, and subsequently subjected to stepwise alternating field (AF) demagnetization (Figure 4). Low-field susceptibility versus temperature (k-T) measurements were made for a heating-cooling cycle to a maximum temperature of 700 °C, in an argon atmosphere to avoid oxidation of the magnetic minerals. A CS-3 unit coupled with KLY-3 susceptibility-bridge apparatus was used for this procedure. Samples from P1 (section 4, 35-cm level) and P4 (section 5, 80-cm level) were taken from calcareous ooze; other samples were from the yellow-brown pelagic clays.

All samples were saturated in the range of 0.2-0.4 T; the value of the median destructive field of saturation IRM was 20-40 mT. Saturation IRM suggests the presence of magnetite. Typical susceptibility versus temperature (k-T) curves (Figure 5) show that in most cases the magnetic susceptibility disappears at temperatures above 580 °C, both in the heating and in the cooling sequences. Cooling curves commonly have larger susceptibility values than the heating curves; this behavior probably results from the reduction of

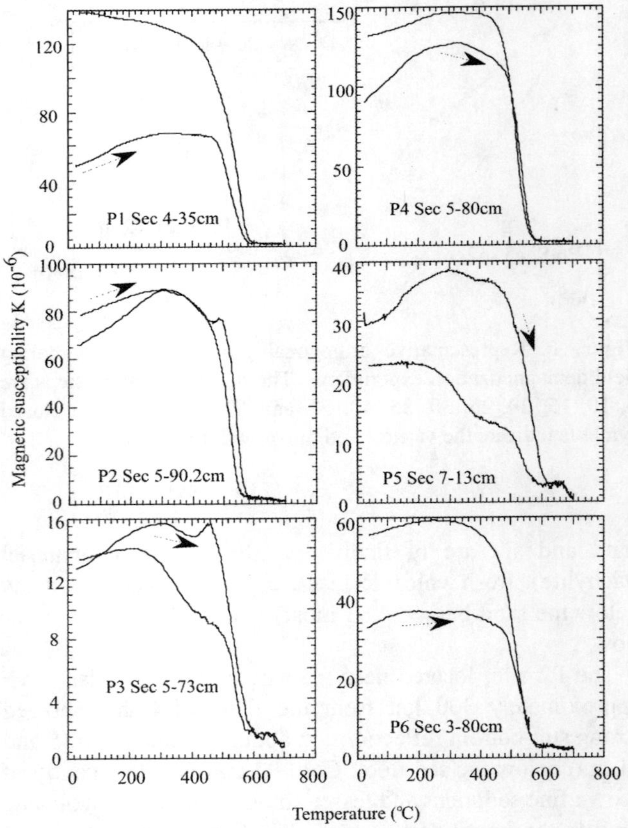

Figure 5. Representative susceptibility curves versus temperature.

maghemite or hematite in the Ar atmosphere. These rock magnetic experiments indicate that the magnetic carriers are primarily magnetite.

6. MAGNETOSTRATIGRAPHY

Several polarity changes were found in cores P1, P2, P3 and P5 (Figure 6), but none in cores P4 and P6 . The polarity transitions are generally distinct, occurring within a few centimeters of sediment. Mean inclinations are 38, 40, 35, 34, 26, and 30 degrees for cores P1 through P6, respectively, close to those predicted by the geocentric axial-dipole field model (35-39 degrees).

Declination and inclination changes are synchronous in cores, except in the interval 4.45-5.15 mbsf in P5. After demagnetization to 20 mT, this interval shows shallow inclinations. Otherwise, declinations in this interval are close to the overlying and underlying sections, but with a systematic difference between them of about 30 degrees. We suspect that the AF demagnetization was not effective in removing spurious components of magnetization for this interval. Additional sub-samples were demagnetized by thermal demagnetization (Figure 7). The magnetic mineralogy tests show no significant differences between this interval and the others. Thermal demagnetization experiments show that, for that the interval in question, the inclination values are still shallow with respect to the expected inclination values for latitudes of the sampling site. On the other hand, declinations in this interval show systematic shifts from both the upper and lower polarities. Possibly, there was some disturbance during coring, although we observed no evidence for this during the ship operations.

Another characteristic of the polarity profiles is that swings of declinations and inclinations occurred synchronously within several cm in P2 (~1.49 mbsf) and P3 (~1.97 mbsf). Interpretation of these results is discussed in a later section.

7. PALEONTOLOGY

Paleontological examinations were performed by Palynosurvey Co., Ltd. Japan on selected spot samples: P1 (4.01 mbsf), P3 (3.94 mbsf), P4 (4.41 mbsf), P4 (5.47 mbsf), P5 (2.54 mbsf), P5 (3.51 mbsf), P6 (1.76 mbsf), and P6 (3.81mbsf). Four microfossil groups examined for stratigraphic purpose: calcareous nannofossils, planktonic foraminifers, diatoms, and radiolarians. The microfossil zonations were refered from *Okada and Bukry* [1980]; calcareous nannoplankton zone, *Berggren* et al., [1995]; planktonic foraminifera zone, *Barron*, [1985]; diatom zone, and *Sanfilippo and Nigrini* [1998]; radiolarian Zone. The results of examination are summarized in Table 4.

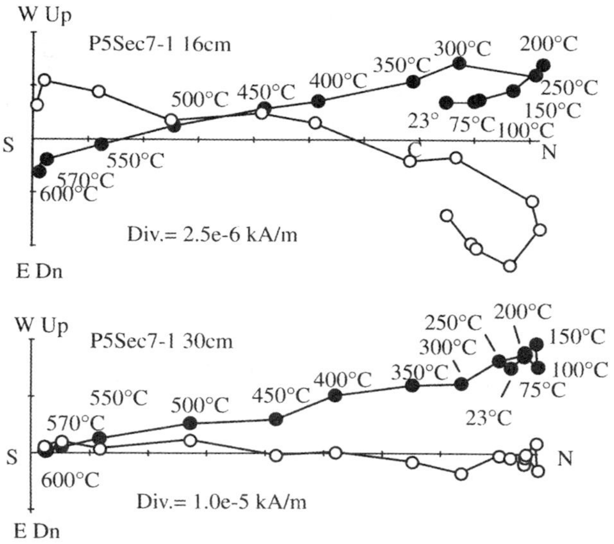

Figure 6. Declination, inclination, and intensity for cores P1 through 6, obtained from the U-channel samples after 20 mT AF demagnetization, and simplified lithologic columns.

Reworking of microfossils has been reported in the archipelagic sediment of Hawaiian Arch region [c.f. *Wintner* et al., 1971: *Hull*, 1993]. Our examinations also suggest microfossil reworking in P4. P4 (5.47 mbsf) yields a younger apparent age than a higher horizon of P4 (4.41 mbsf). Because reworking is difficult to evaluate from spot samples, we consider that the obtained ages are oldest limits for each sample, as follows: 2 Ma for P1 (4.01 mbsf), 1.87 Ma for P3 (394 cm), 0.95 Ma for P4 (4.41 mbsf), 0.51 Ma for P4 5.47 mbsf, 1.12 Ma for P5 2.54 mbsf and P5 3.51 mbsf, 1.12 Ma for P6 (1.76 mbsf) and P6 (3.81 mbsf).

8. MAGNETOSTRATIGRAPHIC MODELS

The polarity boundaries are defined at sharp transitions of synchronous changes in declination and inclination. Two magnetostratigraphic models can be constructed using alternative correlations of magnetic polarities in the cores with the polarity time scale [*Cande and Kent*, 1995], supplemented by the paleontologic determinations for the possible oldest ages in each core. It is generally assumed that sedimentation rate is more or less constant for fitting of the observed polarity pattern to the time scale, but archipelagic sequences around the Hawaii Ridge may show variable sedimentation rates caused by volacaniclastic contributions

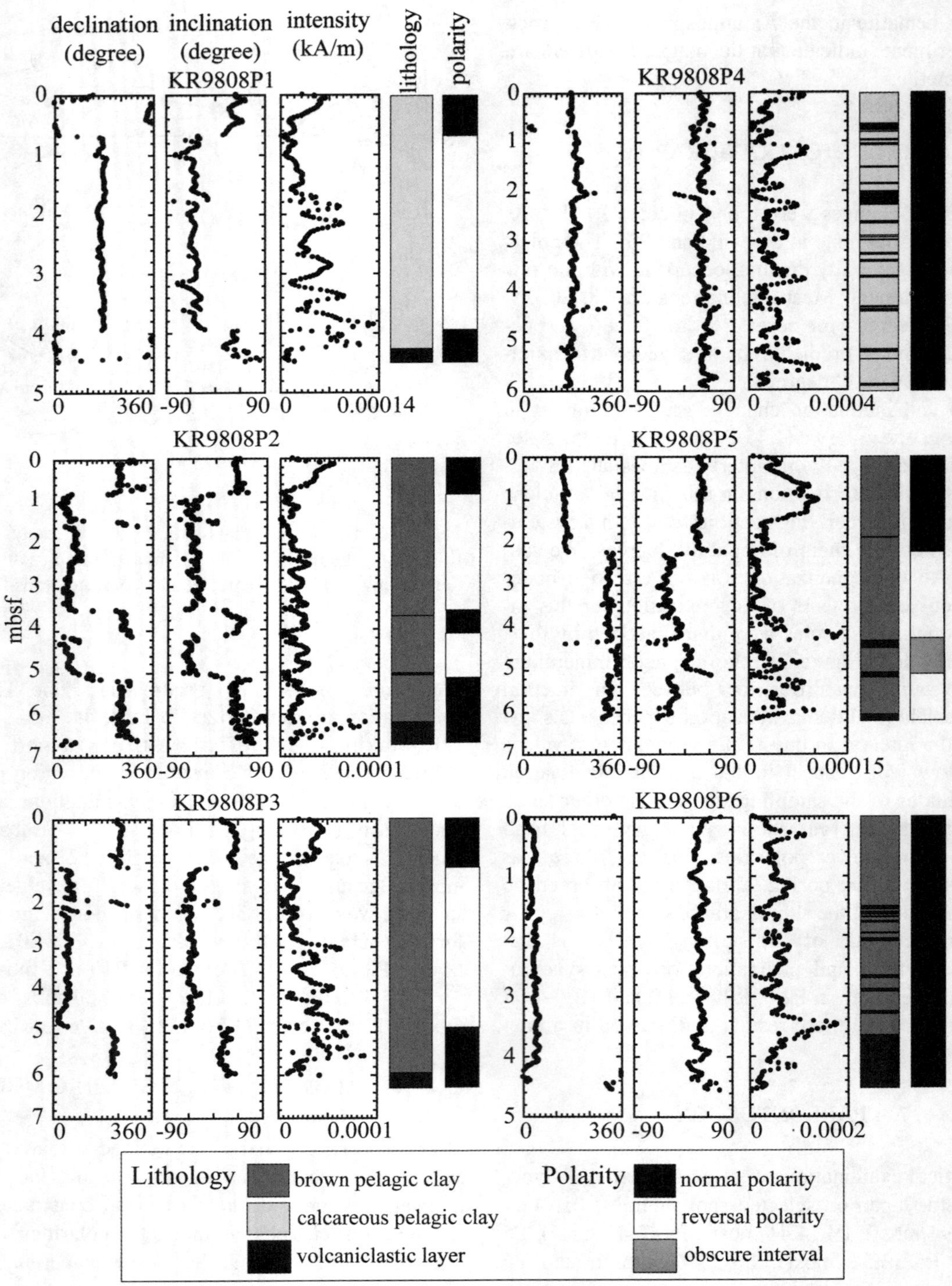

Figure 7. Additional thermal demagnetization experiments of sub-samples in the interval of anomalous inclination of P5. The demagnetization steps are 23, 75, 100, 150, 200, 250, 300, 350, 400, 450, 500, 550, 570, and 600 ˚C

Table 4. Summary of paleontological examination of P1, P3, P4, P5, and P6

Sample	Microfossil zones	Age of zone	Age range
P1 (4.01 mbsf)	CN13a	2-1.6 Ma	2 to 1.77 Ma
	upper zone of PL6	2-1.77 Ma	
P3 (3.94 mbsf)	NTD16	1.87-1 Ma	1.87 tol Ma
P4 (4.41 mbsf)	CN14	0.95-0.41 Ma	0.95 to 0.65 Ma
	Pt1a	1.77-0.65 Ma	
P4 (5.47mbsf)	CN14a	0.51-0.41 Ma	0.51 to 0.41 Ma
	Pt1a	1.77-0.65 Ma	
P5 (2.54 mbsf)	younger than NTD16	<1.87-1 Ma	1.10-1.12 Ma to
	RN14	1.10-1.12 Ma to 0.61-0.47 Ma	0.61-0.47 Ma
P5 (3.51 mbsf)	middle zone of NTD16	1.87 to 1 Ma	1.10-1.12 to 1 Ma
	RN14	1.10-1.12 Ma to 0.61-0.47 Ma	
P6 (1.76 mbsf)	younger than NTD16	1.87 to 1 Ma	1.10-1.12 Ma to
	RN14	1.10-1.12 Ma to 0.61-0.47 Ma	0.61-0.47 Ma
P6 (3.81 mbsf)	middle zone of NTD16	1.87 to 1 Ma	1.10-1.12 to 1 Ma
	RN14	1.10-1.12 Ma to 0.61-0.47 Ma	

Microfossile zones: Diatom, Radiolarian, Calcareous Nanno, Planktonic foraminifera

from the islands (discussed in more detail in section 9.1). One model yields a rather constant slow sedimentation rate under pelagic conditions (2-3 mm/ky); the second model yields more variable rates. The results of correlation are summarized in Table 5.

8.1. Model 1: Near-constant Sedimentation Rate

The Brunhes-Matuyama (B-M) boundary (0.78 Ma) is identified in P1 (0.69 mbsf), P2 (0.85 mbsf), P3 (1.16 mbsf), and P5 (2.28 mbsf) (Figure 6). The top of the second normal chron at 3.97 mbsf in P1, 3.53 mbsf in P2, and 4.89 mbsf in P3 are correlated with the top of Olduvai Subchron (1.77 Ma). Then, the top of the third normal chron in P2 at 5.08 mbsf can be correlated to the Matuyama-Gauss boundary (2.58 Ma). The sedimentation rates for all interval ranges 0.8-3.3 mm/ky. The horizons that show short swings in directions, at 1.49 mbsf in P2 and 1.97 mbsf in P3, are calculated, by the ages of horizons of the B-M boundary and the top of the Olduvai Subchron of P2 and P3, to have ages of 1.01 Ma, and 0.99 Ma, respectively. These ages are close to the Jaramillo Subchron of short duration; these cores may have incomplete records of this brief subchron (80 ky) caused by slow sedimentation.

8.2. Model 2: Variable Sedimentation Rate

The B-M boundary is interpreted at the same horizons as well as model 1. The second normal chron from the top is correlated with the Jaramillo Subchron, at 3.97 mbsf in P1, 3.53 mbsf in P2, and 4.89 mbsf in P3 (Figure 6). The top of the third normal chron of P2 is correlated with the top of Olduvai Subchron. This model suggests very high sedimentation rate of 12.8-17.7 mm/ky. between the B-M boundary and the top of Jaramillo Subchron as deep sea sediment. However, the other intervals are smaller than 3 mm/ky. The interval of high sedimentation rate include excursion-like short swings of declinations and inclinations at 1.49 mbsf in P2 and 1.97 in P3 mbsf. The most likely correlation is to the Kamikatsura at 0.85-0.88 Ma [*Maenaka*, 1983; *Champion et al.*, 1988; *Singer et al.*, 1999], because the record in the P2 and P3 near the top of the Matuyama Chron is estimated to be 0.83-0.84 Ma, based on assumption of a constant sedimentation rate within the upper Matuyama Chron. The successful recording of rapid changes of the geomagnetic field coincides with the high sedimentation rate in the upper Matuyama Chron.

9. DISCUSSION

9.1. Sedimentation Rates

The striking contrast between the two age models is the different sedimentation rates between the B-M boundary and the top of Jaramillo Subchron. The sedimentation rate for this interval by model 1 is 2.7-3.8 mm/ky. in P1, P2, and P3. For model 2, the sedimentation rates are 12.8-17.7

Table 5. List of polarity chron boundaries by Model 1 and Model 2

Model 1 Polarity chron	Ma	Model 2 Polarity chron	Ma	P1 cm	P2 cm	P3 cm	P4 cm	P5 cm	P6 cm
B-M boundary	0.78	B-M boundary	0.78	69	85	116	-	228	-
Jaramillo	0.99-1.07	Kamikatsura	0.85	-	149	197	-	-	-
Olduvai Top	1.77	Jaramillo Top	0.99	397	353	489	-	-	-
Olduvai Base	1.95	Jaramillo Base	1.07	-	412	-	-	-	-
M-G boundary	2.58	Olduvai Top	1.77	-	508	-	-	-	-

B-M boundary: Brunhes-Matuyama boundary, M-G boundary: Matuyama-Gauss boundary

mm/ky. The sedimentation rate for model 2 is very high for the deep ocean, but sedimentation rates near Hawaii may have been enhanced by volcaniclastic contributions. Magnetic susceptibility reflects the quantity of volcanic material in the sediment. The turbidite layers and other volcanic-rich horizons in fine sediments have high magnetic susceptibility (Figure 8), suggesting a significant volcaniclastic supply to the archipelagic basin.

Magnetostratigraphy at ODP sites 842 and 843 south of Oahu shows a sedimentation pattern similar to model 2. The B-M boundary is about 2 mbsf, and the top and base of Jaramillo Subchron is at 4 to 4.75 mbsf. [*Dziewonski et al.*, 1992]. The sedimentation rate is about 2.5 mm/ky in the Brunhes section, 11.6 mm/ ky. in the interval between the B-M boundary and the top of Jaramillo Subchron in hole 842B. *Dziewonski et al.*, [1992] considered that there must be a hiatus in the core above the B-M boundary, because the apparent sedimentation rate is slower than for the older intervals.

Clague et al., [1990] estimated the sedimentation rate for the pelagic sediments on the North Arch lava, using magnetostratigraphy and palagonitized glass rinds on the lavas for age control. The sedimentation rates are 5.8 mm/ky for the period 1.67 to 0.98 Ma, and 1.8 mm/ky for the period 0.90 to 0.73 Ma at a distance of 140 km from the Island. These non-uniform sedimentation rates of archipelagic sequences around the Hawaii Ridge suggest varying sedimentation processes, as proposed previously by *Theyer et al.*, [1985] from studies of cores close to the Hawaiian Islands.

These sedimentation patterns favor Model 2, rather than Model 1, but we could not identify any anomalous lithologies as evidence of high sedimentation rates. If the rate change were simply due to varying volcaniclastic input, ~80% of the pre-B-M section should be volcaniclastic, which would be obvious visually and petrographically. For model 2, it also remains unclear why the volcaniclastic component would have decreased since the B-M boundary, during a time interval when volcanic production rates along the Hawaiian chain have been increasing [*Shaw*, 1973; *Clague and Dalrymple*, 1987]. Additional age information is needed in order to resolve this problem.

9.2. Ages of Turbidites North of Oahu

Using the age models constructed by magnetostratigraphy, we recognize two prominent turbidite layers northeast of Oahu: one 20-40 cm below the top of the second normal polarity interval for P1 and P3, and the other about 100 cm below the top of third normal polarity interval in P2. Magnetic susceptibility profiles in P1-P3 correspond well with one another (Figure 8). The profiles show decreases at the B-M boundary, and peaks below the top of second polarity interval. At these susceptibility peaks, thick sand layers are present in P1 and P3, whereas no sand layers were observed in P2. The thin fine-grained volcaniclastic layer at 3.52 mbsf may be equivalent to this peak. Other smaller peaks between the B-M and the top of second normal polarity could also correlate with one another. Therefore, changes in the volcaniclastic supply appear to have occurred synchronously in P1, P2, and P3.

If the second normal polarity is the Jaramillo Subchron (model 2), the ages of the two turbidites are about 1 Ma (P1 and P3) and >1.77 Ma (P2). If second normal polarity interval is the Olduvai Subchron (model 2), the ages of the turbidites are about 2 Ma (P1 and P3) and older than 2.58 Ma (P2) (Figure 9).

The Nuuanu debris avalanche originated from Koolau volcano on Oahu, and the Wailau debris avalanche from the East Molokai volcano [e.g. *Moore et al.,* 1989]. Compositions of volcaniclastic glass in the turbidites show that the P1 and P3 turbidites are from East Molokai, and the P2 turbidite is from Koolau [*Sherman et al.*, this volume].

Clague and Dalrymple, [1987] summarized available K-Ar ages for Hawaiian volcanoes. The best ages for Koolau range from 1.8 to 2.7 Ma [*Doell and Dalrymple*, 1973]. Ages of the East Molokai volcano range from 1.76 to 1.24 Ma [*Naughton*, 1980]. For model 1, the ages of both turbidites match the initial shield-building stage of the source

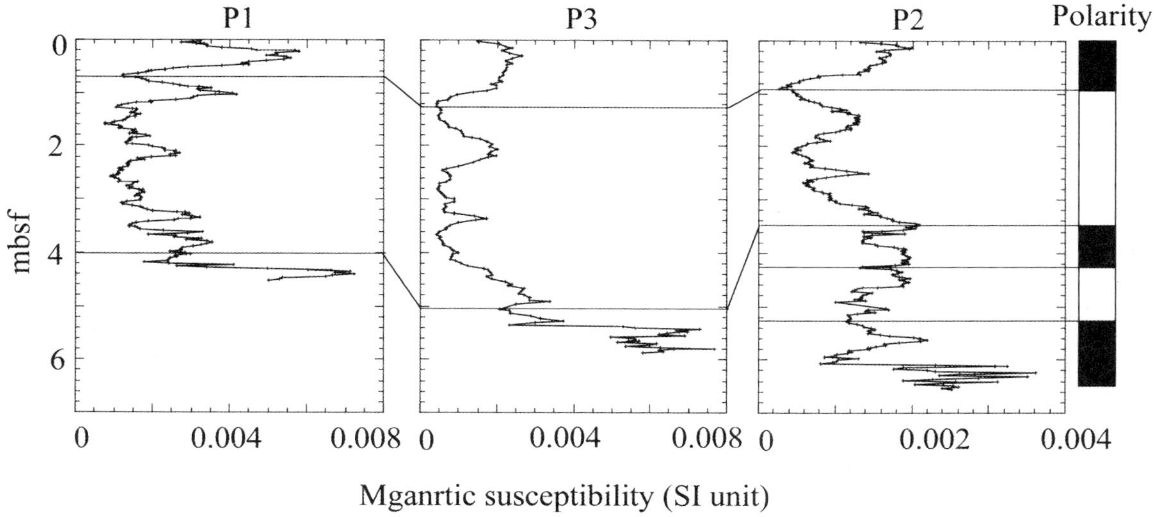

Figure 8 Magnetic susceptibility profile of P1, 2, and 3 versus depth and polarity boundaries.

volcanoes. The younger turbidite in P1 and P3 is about 2 Ma, close to older ages for East Molokai volcano, and the P2 turbidite is >2.58 Ma, close to the older ages for Koolau volcano. The ages inferred by model 2, about 1 Ma for P1 and P3 turbidites and >1.77 Ma for the P2 turbidite, are close to the younger post-shield building ages for these volcanoes [e.g. *Naka et al.*, 2000: E. Herrero-Bervera, collapse of the Koolau Volcano, submitted to *Physics of the Earth and Planetary Interiors*, 2001].

Because core P4 was obtained only about 40 km from Molokai, the sand layers in it probably are related to volcanism on that island and also to small slope failures.

9.3. Ages of Turbidites South of Hawaii

Multiple thin sand layers in the middle to bottom of P6 are included within the Brunhes Chron (0-0.78 Ma) (Figure 9). *Naka et al.* [this volume] shows that the volcanic fragments in P6 originated largely or entirely from Kilauea and Mauna Loa, Hawaii Island. In contrast, three turbidite layers in P5 are older than B-M boundary (0.78 Ma), yet document collapse events of subaerial shields, because the fragment contain low S basalt (mostly shallow or subaerially erupted). It seems likely that the volcaniclastic materials of P5 were supplied from older islands, rather than from the Island of Hawaii, because all volcanoes on Hawaii are considered to be younger than 0.7 Ma [*Clague and Dalrymple*, 1987].

10. SUMMARY

We have examined the timing of turbidites as proxies for two types of Hawaiian submarine landslides (debris avalanches and slumps), and have provided alternate age models for emplacement of the Nuuanu and Wailau debris avalanches. A constant-rate sedimentation model indicates >2.58 Ma and ~2 Ma for Nuuanu and Wailau debris avalanches. The variable-rate model indicates >1.77 Ma and ~1 Ma for the same avalanches. A proximal piston core in the front of the Hilina slump contains only turbidites younger than 0.78 Ma, but turbidites at the distal site have ages older than 0.78 Ma.

Acknowledgments: We thank Captain O. Yukawa and his crew for the complex operation of the ship, T. Sugawara, N. Togashi, and M. Matsuhashi for on-board scientific support, J. Lau and K. Matsuo for their laboratory support work at SOEST-HIGP, University of Hawaii and JAMSTEC, respectively, and the entire scientific party and W. Soh for the fruitful discussions throughout all of the cruise legs and subsequent meetings. We greatly appreciated useful reviews by D. Champion and W. Normark. Especially, Champion encouraged us to evaluate the near-constant sedimentation model. We thank also P. Lipman for critical and useful comments to improve this manuscript. This is SOEST contribution 5563 and HIGP contribution 1147.

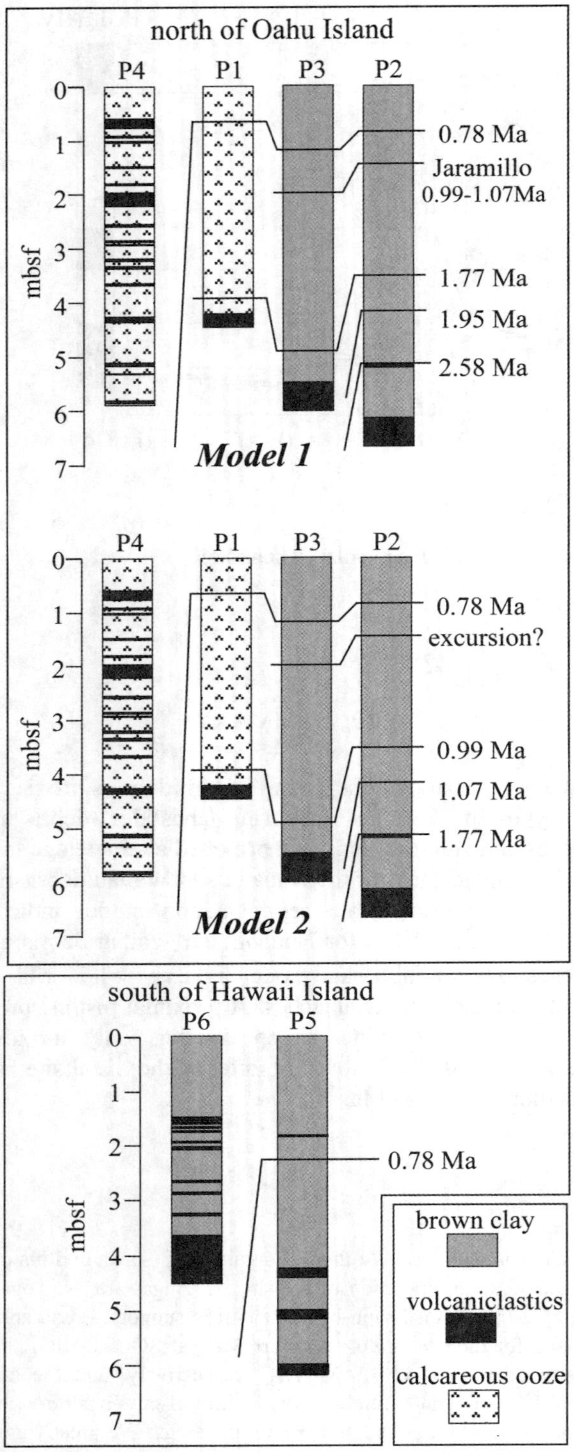

Figure 9. Stratigraphic positions of volcaniclastic layers of P1 through 6.

REFERENCES

Barron, J. A., Miocene to Holocene planktonic diatoms, *in Plankton Stratigraphy*, edited by H. M.Boli, and J. B. Saunders, pp. 763-809, Cambridge University Press, New York, 1985.

Berggren, W. A., D. V. Kent, C. C. Swisher, and M-P. Aubry, Revised Cenozoic geochronology and chronostratigraphy, *in Geochronology time Scales and Global Stratigraphic Correlation, SEPM Special Pub, 54*, 129-212, 1995.

Brocher, T. M., and U.S. Brink, Variation in oceanic layer 2 elastic velocities near Hawaii and their correlation to lithospheric flexure, *J. Geophys. Res., 92*, 2647-2661, 1987.

Cande S.C., and D. V. Kent, Revised calibration of the geomagnetic polarity time scale for Cretaceous and Cenozoic. *J. Geophys. Res., 100*, 6093-6095, 1995.

Champion, D. E., M. A. Lanphere, and M. A. Kuntz, Evidence for a new geomagnetic reversal from lava flows in Idaho: Discussion of short polarity reversals in the Brunhes and late Matuyama polarity chrons, *J. Geophys. Res., 93*, 1167-1168, 1988.

Clague, D., and G. Dalrymple, The Hawaiian-Emperor volcanic chain, part 1, *U.S. Geol. Surv. Prof. Pap., 1350*, 5-54, 1987.

Clague, D., R. Holcomb, J. Sinton, R. Detrick, and M. Torresan, Pliocene and Pliocene alkalic flood basalts on the seafloor north of the Hawaiian islands, *Earth Planet. Sci. Lett., 98*, 175-191, 1990.

Doell, R., and B. Dalrymple, Potassium-argon age and paleomagnetism of Waianae and Koolau Volcanic series, Oahu, Hawaii: *Geol. Soc. Am. Bull., 84*, 89-121, 1973.

Dziewonski, A., Wilkens, R., Firth, J, *et al., Proc. ODP, Init. Repts., 136*, Ocean Drilling Program, College Station, TX, 1992.

Garcia, M., Pliocene-Pleistocene volcanic sands, in *Proc. ODP, Sci. Results, 136*, Wilkerns, R., J. Firth, and J. Bender, et al., Ocean Drilling Program, College Station, TX, 53-63, 1993.

Garcia, M., and D. Hull, Turbidite from giant Hawaiian landslides: results from Ocean Drilling Program Site 842, *Geology, 22*, 159-162, 1994.

Garcia, M., Turbidite from slope failure on Hawaii volcanoes, *in Volcano Instability on the Earth and Other Planets,* W. McGuire, A. P. Jones, and J. Neuberg, eds., *Geol. Soc. Spec. Publ., 110*, 281-294, 1996.

Herrero-Bervera, E., G.P.L. Walker, J.C. Guerrero-Garcia, and E. Canon-Tapia, The Nuuanu and Wailau giant landslides: Insights from paleomagnetic and anisotropy of magnetic susceptibility (AMS) studies, *Phys. Earth Planet. Int.*, IN REVIEW. [NOT A VALID CITATION, UNLESS "IN PRESS"]

Hull, D. M., Quaternary, Eocene and Cretaceous radiolarians from the Hawaiian Arch, northern equatorial Pacific, *Proc. ODP, Sci. Results,* 136, Wilkerns, R., J. Firth, and J. Bender, et al., Ocean Drilling Program, College Station, TX, 3-25, 1993.

Lipman, P., Declining growth of Mauna Loa during the last 100,000 years: rates of lava accumulation vs. gravitational subsidence, in J.M. Rhodes and J.P. Lockwood, eds., Mauna Loa

Revealed: structure, composition, history, and Hazards, *Am. Geophys Union Geophys. Monogr. Ser., 92*, 45-80, 1995.

Maenaka, K., Magnetostratigraphic study on the Osaka Group, with special reference to the existence of pre- and post Jaramillo episodes in the late Matuyama Polarity Epoch, *Mem. Hanazono Univ*., 14, 1-65, 1983.

McMurtry, G., E. Herrero-Bervera, M. D. Cremer, J. R. Smith, J. Resig, and C. Sherman, and M. E. Torresan, Stratigraphic constraints on the timing and emplacement of the Alika 2 giant Hawaiian submarine landslide, *J. Volcan. Geoth. Res*., *94*, 35-58, 1999.

Moore, J. G., 1964, Giant submarine landslide on the Hawaiian ridge, *U.S. Geol. Survey Prof. Pap., 501-D,* D95-D98

Moore, J. G., D. A. Clague, R. T. Holcomb, P. W. Lipman, W. R. Normark and M. E. Torresan, Progressive submarine landslides on the Hawaiian Ridge, *J. Geophys. Res*., 94, 17465-17484, 1989.

Moore, J. G., W. R. Normark and R. T. Holcomb, Giant Hawaiian landslides, *Annu. Rev. Earth Planet. Sci*., *22*, 119-144, 1994.

Moore, J. G., W. Chadwick., Offshore geology of Mauna Loa and adjucent areas, Hawaii, in J.M Rhodes and J.P. Lockwood, eds., Mauna Loa Revealed: structure, composition, history, and Hazards, *Am. Geophys. Union Geophys. Monogr. Ser.,. 92*, 21-44, 1995.

Morgan, J., G. Moore, and D. Hills, and S. Leslie, Overthrusting and sediment accretion along Kilauea's mobile south flank, Hawaii, from marine seismic reflection data: tests of volcanic spreading models, *Geology, 28*, 667-670, 2000.

Naka, J., R. Tsugaru, T. Danhara, T. Tanaka, and K. Fujioka, Sedimentary processes of volcaniclastic sediments, in *Proc. ODP, Sci. Results, 136*, Wilkerns, R., J. Firth, and J. Bender, et al., pp. 85-95, Ocean Drilling Program, College Station, TX, 1993.

Naka et al., 2000. Tectono-magmatic processes investigated at deep-water flanks of Hawaiian volcanoes, *EOS, 81*, 221-227.

Naka, J., T. Kanamatsu, P. Lipman, T. Sisson, N. Tsuboyama, J. Morgan, J. Smith, and T. Ui, Deep-Sea Volcaniclastic sedimentation around the southern flank of Hawaii Island, *this volume*

Naughton, J.J., G.A.Macdonald, and V.A.Greenberg, Some additional potasium-argon ages of Hawaiian rocks: the Maui volcanic complex of Molkai, Maui, Lanai, and Kahoolawe, *J. Volcan. Geoth. Res*., *7*, 339-355, 1980.

Okada, H., and D. Bukry, Supplementary modification and introduction of code numbers to the low-latitude cocolith biostratigraphic zonation., *Mar. Micropaleontol., 5*, 321-325, 1980.

Owen, S., P. Segall, M. Lisowski, A. Miklius, R. Denlinger, and M. Sako, The rapid deformation of Kilauea Volcano: GPS measurements between 1990 and 1996, *J. Geophys. Res., 105,* 18,983-18998, 2000.

Rees, B., R. Detrick, and B. Coakley, Seismic stratigraphy of the Hawaiian flexural moat, *Geol. Soc. Am. Bull., 105*, 189-205, 1993

Rehm, E., and P. Halbach, Hawaiian-derived volcanic ash layers in equatorial northeastern Pacific sediments, *Mar. Geol., 50*, 25-40, 1982

Sanfilippo, A. and C. Nigrini, Code numbers for Cenozoic low latitude radiolarian biostratigraphic zones and GPTS conversion tables, *Mar. Micropaleontol., 33*, 109-156, 1998.

Shaw, H. R., Mantle convection and volcanic periodicity in the Pacific: evidence from Hawaii, *Geol. Soc. Am. Bull., 84*, 1505-1526, 1973.

Sherman, S., M. Garcia and E. Takahashi, Geochemistry of volcanic glasses from piston cores taken north of Oahu and Molokai Islands, *this volume*.

Singer B.S., K.A. Hoffman, A. Chauvin, R. S. Coe, and M. S. Pringle, Dating transitionally magnetized lavas of Matuyama Chron: Toward a new $^{40}Ar/^{39}Ar$ timescale of reversals and events, *J. Geophys. Res*., *104*, 679-693, 1999.

Smith, J, Malahoff, A., and Shor, Submarine geology of the Hilina slump and morpho-structural evolution of Kilauea volcano, Hawaii, *J. Volcan. Geoth. Res*., *94*, 59-88, 1999.

Swanson, D.A., W. A. Duffield, and R. S. Fiske, Displacement of the south flank of Kilauea volcano: The result of forceful intrusion of magma into the rift zone, *U. S. Geol. Surv. Prof. Pap., 963*, 30 pp., 1976

Theyer, F., E. Herrero-Bervera, V. Hsu and S.R. Hammond, The zonal harmonic model of polarity transitions: A test using successive reversals, *J. Geophys. Res*., 90, 1963-1982, 1985.

Wintner, E. L., W. R. Riedel, et al., *Initial Rep. Deep Sea Drilling Project 7*, 1757pp, 1971.

E. Herrero-Bervera, Paleomagnetic and Petrofabrics Laboratory, School of Ocean and Earth Science and Technology, Hawaii Institute of Geophysics and Planetology, University of Hawaii, Honolulu, Hawaii, 96822, USA.

T. Kanamatsu, Deep Sea Research Department, Japan Marine Science and Technology Center, Yokosuka, 237-0061, Japan.

G. M. McMurtry, Department of Oceanography, University of Hawaii, Honolulu, Hawaii, 96822, USA.

Eruption Style and Flow Emplacement in the Submarine North Arch Volcanic Field, Hawaii

David A. Clague

Research and Development Division, Monterey Bay Aquarium Research Institute, Moss Landing, California

Kozo Uto, Kenji Satake

Geological Survey of Japan, Higashi, Japan

Alicé S. Davis

Research and Development Division, Monterey Bay Aquarium Research Institute, Moss Landing, California

The North Arch Volcanic Field covers about 24,000 km^2 of seafloor north of Oahu and has an estimated volume between 1,000 and 1,250 km^3. The field straddles the Hawaiian flexural arch about 250 km north of the axis of the island chain and surrounds numerous Cretaceous volcanic ridges, circular flat-topped volcanoes, and low-relief regions of sediment-covered seafloor. New SeaBeam bathymetric maps that cover about 1/3 of the flow field reveal nearly 100 volcanic structures ranging from low shields to steep cones. One shield is modified by a pit crater, approximately 1.1x1.25 km and 300 m deep. A lava flow in the SE part of the volcanic field covers about 3,600 km^2, has an estimated volume of 36-72 km^3, and apparently erupted from a 75-km-long NNW-trending fissure system. A 108-km-long flow advanced north in a graben parallel to the Cretaceous mid-ocean ridge that formed the crust; its surface gradient is 1.9 m/km (slope of 0.1°). *Shinkai 6500* submersible dive 502 explored one of the composite volcanoes and observed and collected dense alkalic basalt sheet flows erupted after vesicular basanite pillow basalts and fragmental hyaloclastite that make up the steep-sided cone. Dive 503 collected alkalic basalt sheet flows and pillow basalt from the top 122 m of the southern wall of a pit crater that formed by collapse caused by a decrease in magma volume from a shallow storage chamber located 1-2 km below the surface. The volume change may have been caused by loss of gas bubbles from the stored magma when replenishment ceased at the end of the eruption. The surficial drapery-folded sheet flow is covered by only a few cm of

Hawaiian Volcanoes: Deep Underwater Perspectives
Geophysical Monograph 128
Copyright 2002 by the American Geophysical Union

sediment, indicating that it is younger than the 0.5-1.5 Ma ages previously estimated for North Arch flows and vents. The near-vent constructs and flow characteristics indicate that vigorous eruption of highly vesicular lava constructed steep-sided cones of pillow basalt and hyaloclastite whereas steady eruption of dense lava that had lost its bubbles prior to eruption formed low-relief lava shields. Vesiculation, and even fragmentation, of these alkalic magmas at water depths as great as 4300 m is caused by exsolution of much of their high initial volatile content, mainly CO_2 with less H_2O. One hyaloclastite sample contains thin glass bubble-wall fragments similar to those formed by mild explosions as seawater boils following mixing of magma and seawater. Low-viscosity dense lava is emplaced as sheet and channelized tube-fed flows that travel down gradients as small as 1.5 m/km (0.08°) and flow as far as 108 km from their vents.

1. INTRODUCTION

The Plio-Pleistocene North Arch Volcanic Field is located on the deep seafloor extending from 100 to 390 km north of Oahu [Figure 1; *Clague et al.*, 1990, 2000d]. The field covers about 24,000 km^2 and is the largest known submarine lava field, exclusive of the globe-girdling mid-ocean ridge system. Several flows in the lava field appear to be areally extensive and to have traveled long distances down gentle slopes at water depths in excess of 4 km. The types of near-vent structures, eruption styles, and emplacement mechanisms of these flows were largely unknown before detailed bathymetry was collected over part of the flow field in 1999.

The eruption and emplacement of lava flows in the submarine environment is still poorly understood and documented, primarily because there have been no observations of submarine eruptions and only limited observations of lava flowing under water, always near shore after lava entered the sea [e.g., *Moore et al.*, 1973; *Moore*, 1975]. These and subsequent observations have documented the formation of pillow lava and shown that, at least on steep slopes, narrow lava channels can form and flows can advance in a manner analogous to subaerial ‘a‘a flows [e.g., *Lipman and Banks*, 1987]. Most of our inferences about the eruption and emplacement of submarine lava flows has been derived from observations of inactive flows and vents [e.g. *Ballard et al.*, 1979; *Fornari et al.*, 1984; *Batiza et al.*, 1989; *Perfit and Chadwick*, 1998; *Gregg et al.*, 1996, 2000; *Clague et al.*, 2000b] coupled with laboratory wax experiments and numerical modeling [e.g., *Fink and Griffiths*, 1990; *Griffiths and Fink*, 1992, 1993; *Gregg and Fink*, 1995; *Gregg and Fornari*, 1998]. Most of the direct observations have taken place along the mid-ocean ridge spreading ridges where lava occurs as pillow lava, lobate flows, sheet flows, and a range of intermediate flow types [summary in *Batiza and White*, 2000] that form a variety of vent constructs including pillow ridges and mounds, small steep-sided cones of pillow lava, and flat-topped cones [e.g., *Smith*, 1988; *Smith and Jordan*, 1988; *Smith and Cann*, 1992; *Perfit and Chadwick*, 1998; *Batiza and White*, 2000].

Recent high-resolution mapping on the flanks of a number of Hawaiian Islands [*Dartnell and Gardner*, 1999; *MBARI Mapping Team*, 2000] has revealed a variety of cone morphologies on and near the submarine rift zones of Loihi Seamount, Kilauea, Kohala, Hualalai, Mahukona, and Haleakala volcanoes, and on the eastern and southern flanks of Oahu, the eastern and southern flanks of Kauai, and on the northwest flank of Niihau. *Clague et al.* [2000b] described many of these near-vent structures and suggested that the pancake-like, flat-topped volcanic cones form as continuously overflowing submarine lava lakes during effusive eruptions of long duration. These cones are similar in form to many described from mid-ocean ridges [e.g., *Lonsdale*, 1983; *Batiza et al.*, 1989; *Rappaport et al.*, 1997] which probably formed under similar conditions. *Clague et al.* [2000b] reviewed and summarized the conditions that led to development of submarine flat-topped volcanic cones and more steep-sided conical near-vent structures offshore Hawaii.

Extensive submarine lava flows around Hawaii have been identified at the base of the Kilauea East rift zone [*Holcomb et al.*, 1988a] and in the South Arch flow field [*Lipman et al.*, 1989] based on their high backscatter on GLORIA images. However, in neither case have the eruptive vents been identified. The North Arch Volcanic Field provides an unprecedented opportunity to further examine the near-vent structures formed during the submarine eruption of basaltic lava, particularly deep eruption of highly alkalic composition lava, and to evaluate mode of emplacement of the flows. In addition, on the steep submarine island flanks and along mid-ocean ridges, where lava flows are stacked one on top of another, it is difficult to map the distribution of any individual flow. However, in the North Arch, several of the flows on the edges of the flow field can be mapped in detail since they were emplaced on top of thick sediment, which provides a significant difference in sonar backscatter.

2. PREVIOUS WORK

Menard [1964] first hypothesized that young lava flows might occur on the Hawaiian flexural Arch, and the presence of young flows on the Arch was documented during a near-bottom geophysical survey (Figure 1) only a few years later by *Normark and Shor* [1968] and by *Spiess et al.* [1969]. However, the region surveyed was small—only about 400 km^2—and the extent of these young flows remained unknown for more than a decade. *Wallin* [1982], in an unpublished thesis, mapped the distribution of hyperbolic reflections observed in seismic reflection profiles to conclude that a substantial region north of Oahu was covered by young lava flows (Figure 1). He estimated that the flows were between 2 and 6 Ma, based on their thin sediment cover.

Wallin's [1982] interpretations lay fallow until 1988 when the U.S. Geological Survey's GLORIA sidescan survey of the United States Exclusive Economic Zone around Hawaii revealed the full extent of the North Arch Volcanic Field [*Holcomb et al.*, 1988b; *Normark et al.*, 1989]. The flows could be readily identified by their high backscatter in the GLORIA images (Figure 1), and their extent and distribution could be mapped in detail. However, because the GLORIA data did not include any bathymetric component, the number and locations of the near-vent constructs remained largely unknown. These GLORIA surveys further established that young submarine lava flows and cones occur along the entire Hawaiian volcanic chain as far west as Midway Island [*Holcomb et al.*, 1988b; *Moore et al.*, 1994], but that nowhere do they cover an area as large as in the North Arch Volcanic Field.

A SeaBeam survey of a small part of the flow field and a few preliminary dredges were collected soon after the GLORIA surveys were completed [*Detrick et al.*,1988] and were followed by a more extensive sampling cruise in late 1988 [*Clague et al.*, 1989]. Several small areas that appeared to have even higher backscatter than the surrounding flows were explored and found to be small hills interpreted to be volcanic vents. The samples collected during these cruises, together with a few additional samples gleaned from the core lockers at the University of Hawaii and the Lamont-Doherty Geological Observatory (Figure 1), formed the basis for a series of articles on the general chemistry, volatile chemistry, and isotopic and trace-element chemistry of the lavas from the North Arch Volcanic Field [*Clague et al.*, 1990; *Dixon et al.*, 1997; *Frey et al.*, 2000].

These studies established that the lavas of the North Arch Volcanic Field consist of alkalic basalt, basanite, and nephelinite lavas that are compositionally similar to rejuvenated stage lavas on the Hawaiian Islands such as the Honolulu [*Clague and Frey*, 1982] and Koloa Volcanics [*Clague and Dalrymple*, 1988] on Oahu and Kauai, respectively. They also established that some steeper hills are composed of highly vesicular pillow lava and hyaloclastite whereas the high-backscatter flows are dominantly sheet flows. The pillow lavas and hyaloclastites form from degassing of abundant magmatic volatiles leading to vesiculation and even fragmentation of the lava. However, due to the low viscosity of these lavas, the bubbles are quickly lost as the flows progress away from their eruptive vents resulting in dense, fluid sheet flows that appear to comprise most of the flow field. The ages of the lavas were estimated from the thickness of palagonite on the glassy margins of the flows, calibrated to a few glass samples collected in sediment cores whose age was determined paleomagnetically. The sampled flows were estimated to range from about 0.5 to 1.6 Ma.

No additional sampling or surveys specific to studying the North Arch Volcanic Field took place for a decade until 1999 when JAMSTEC's *R/V Yokosuka* collected SeaBeam bathymetric survey data and conducted the two *Shinkai 6500* submersible dives described here. In the intervening period, several transit SeaBeam lines crossed the area (data stored at the National Geophysical Data Center) and are incorporated in our analysis.

3. NEW DATA

3.1 SeaBeam 2100 Mapping

In 1999, the *R/V Yokosuka* used a SeaBeam 2100 mapping system to collect bathymetric and sidescan data over approximately 12,000 km^2 of seafloor including many kipukas consisting of Cretaceous seamounts and volcanic ridges and roughly 1/3 of the known North Arch Volcanic Field (Plate 1). The survey was navigated using continuous GPS and was designed to collect overlapping swaths with 100% coverage. A single swath of data was also collected in the northwestern part of the field that tracks a narrow long lava flow. The surveyed region includes the locations where 15 of 27 previous dredges and cores recovered lava samples [*Clague et al.*, 1990]. We have processed the bathymetric data with 140-m resolution. The associated sidescan data were not as high resolution as the pre-

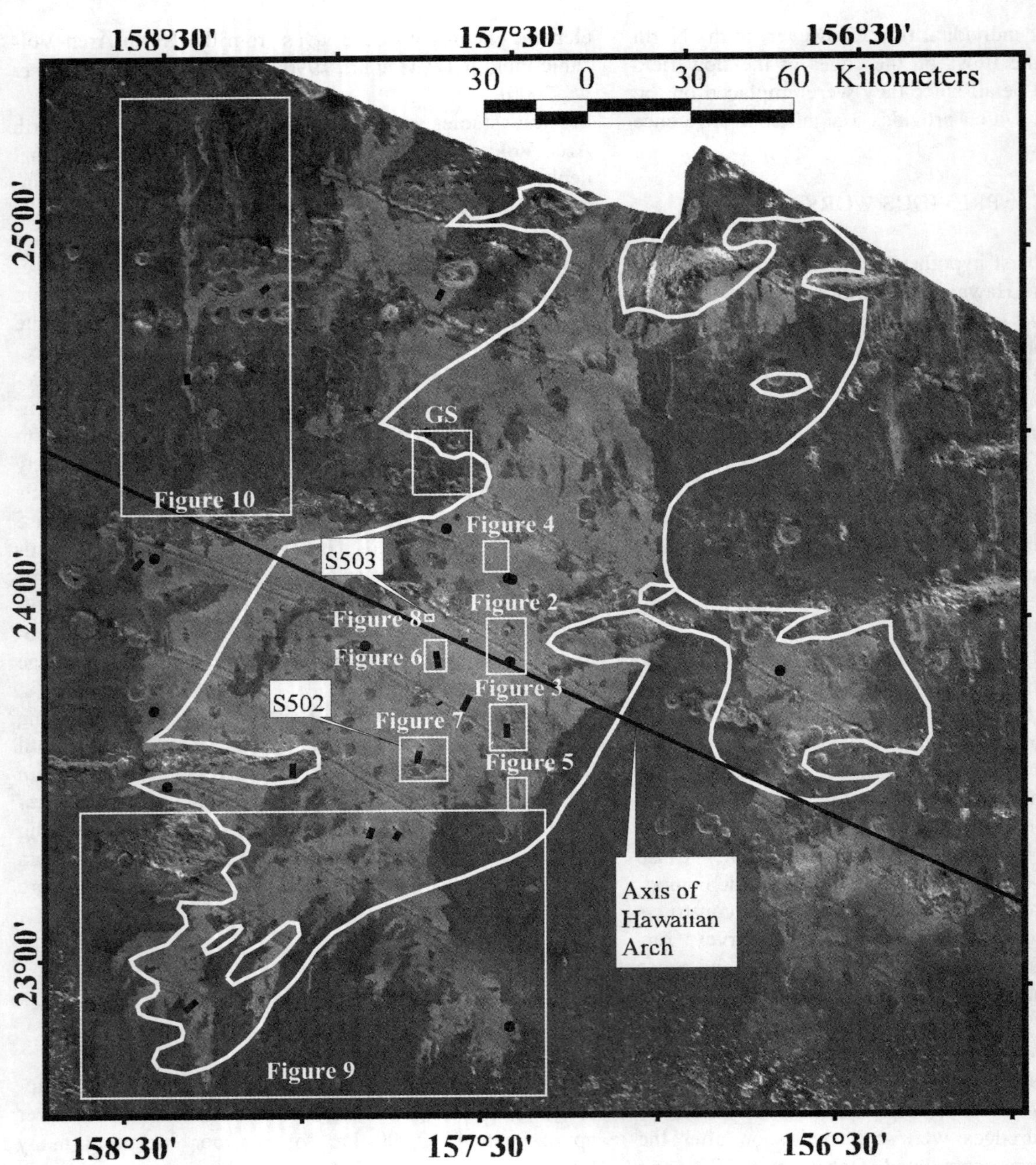

Figure 1. GLORIA side-scan image of the North Arch Volcanic Field. The light regions have high backscatter and are interpreted to be Plio-Pleistocene lava flows and vents with thin sediment cover. Steep slopes on some older volcanic cones and ridges also have high backscatter. The dark areas are sediment covered Cretaceous seafloor. Locations of dredges (black bars) and cores (black dots) that recovered young lava are shown, and *Shinkai 6500* submersible dives are labeled S502 and S503. The white irregular line outlines the region inferred on the basis of seismic reflection profiles by *Wallin* [1982] to be young lava flows and the white box labeled GS shows the area of a detailed geophysical survey [*Normark and Shor*, 1968; *Spiess et al.*, 1969]. The axis of the Hawaiian Arch is shown, based on the new compilation of bathymetry. The approximate locations of Figures 2 through 10 are shown as labeled boxes. The speckled pattern in the southeastern part of the image is the combined distal blocky ends of the giant Wailau and Nuuanu landslides [*Moore et al.*, 1989]. The landslide blocks at 22°55'N, 156°40'W rest on top of the North Arch flow field.

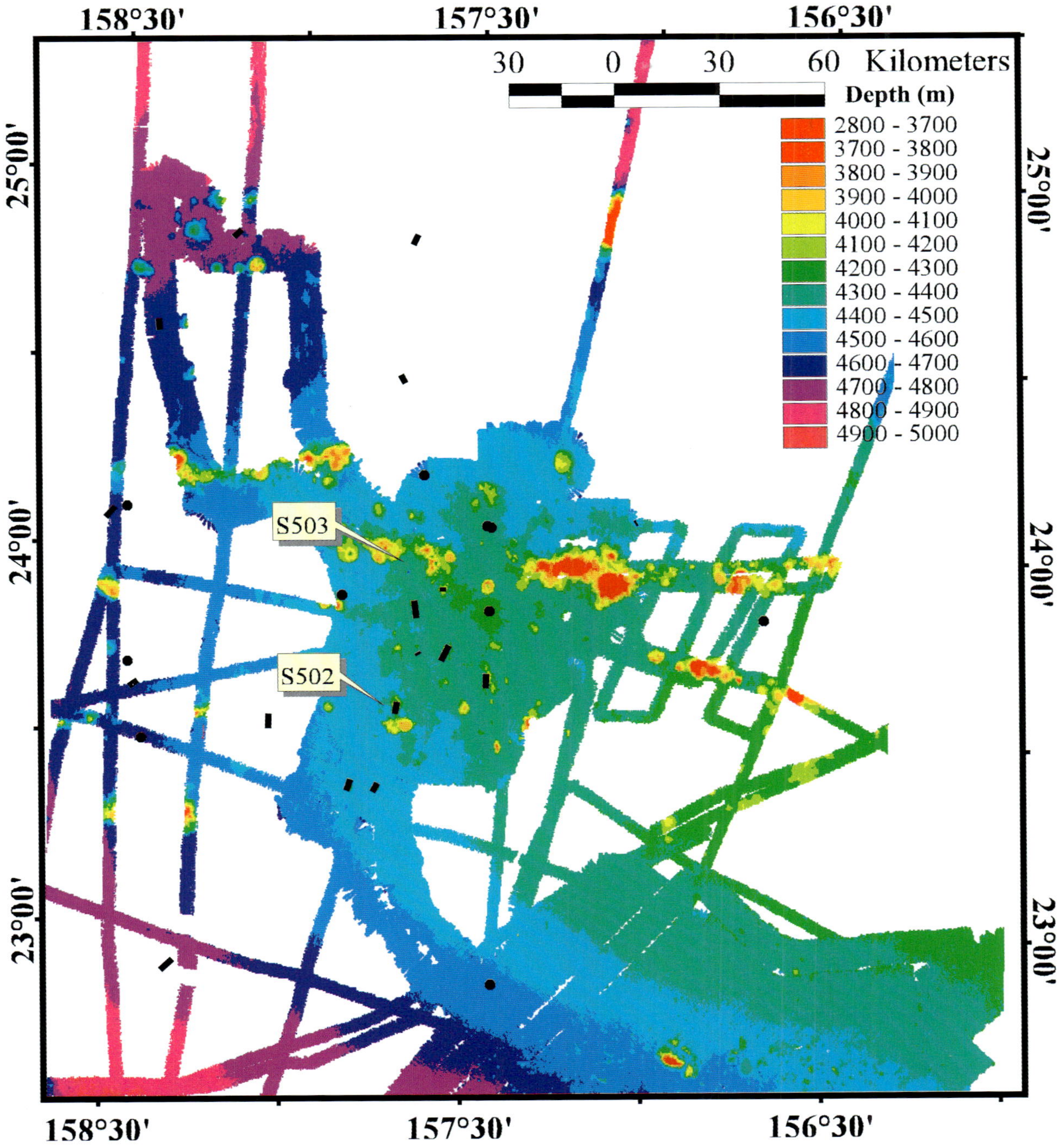

Plate 1. SeaBeam bathymetric coverage of the North Arch Volcanic Field showing that the North Arch Volcanic Field straddles the Hawaiian flexural arch. The area shown is the same as in Figure 1. Most of the data shown were collected from the *R/V Yokosuka* during the JAMSTEC program in 1999. The other narrow lines are transit lines through the region from the National Geophysical Data Center. Most of the largest volcanic cones and ridges (in red) are Cretaceous features.

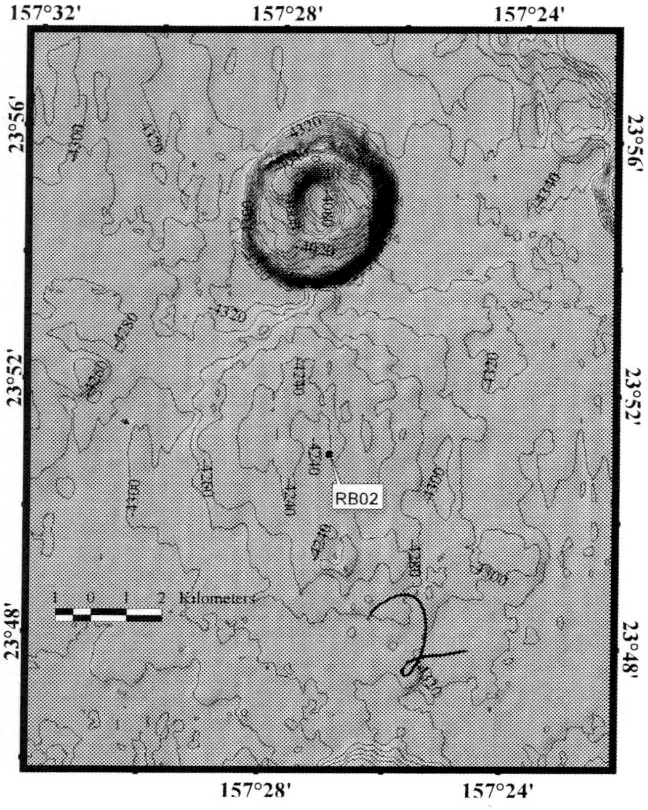

Figure 2. Shaded relief image of the SeaBeam bathymetry showing a low-relief shield volcano first surveyed with SeaBeam in 1988 [*Detrick et al.*, 1988] centered at 157°27'W, 23°51'N. The data shown are the newer data collected by the *R/V Yokosuka* in 1999. Contour interval is 20 m. The location of dredge RB02 [*Clague et al.*, 1990] is shown. The flat-topped volcanic cone with a crater in the upper part of the image is part of the Cretaceous seafloor beneath the North Arch flow field.

existing GLORIA data, which we utilized in our analysis. All the images shown are illuminated from the northwest.

The region also includes numerous older seamounts and volcanic ridges that can be distinguished from the much younger North Arch vents and flows by their low backscatter, which is interpreted as thicker sediment cover on these older features [*Bridges*, 1997]. Many of these older volcanoes are circular flat-topped cones, commonly with craters or calderas and are similar in form to seamounts formed near mid-ocean ridge spreading centers [e.g., *Clague et al.*, 2000c]. Many, but not all, of these volcanic cones are aligned in three roughly east-west trending lineaments that are approximately perpendicular to the orientation of the Cretaceous seafloor horst-graben fabric. The southernmost of these lineaments, at about 24°N, is located along the axis of the Hawaiian flexural Arch. Other low-relief areas within the high-backscatter flow field also have low backscatter and are interpreted to be old sediment covered seafloor surrounded by the young flows.

The regional bathymetry, coupled with the locations of the near-vent constructs, is the primary control on the distribution of the North Arch lava flows. The flow field is centered on the Hawaiian flexural arch about 250 km north of Oahu, so that greater depths occur north and south of the 24°N seamount lineament on the crest of the arch. There is also a strong east-west gradient in the depth of the flexural arch, particularly in the eastern part of the flow field. The depth to the flexural arch decreases from west to east from about 4625 m to 4240 m over a distance of about 150 km. This decrease in depth of the arch from west to east mirrors a complimentary increase in the depth of the Hawaiian Deep or Moat from west to east. This combination of shallower arch and deeper moat to the east results in much steeper gradients off the arch in the eastern, compared to the western, part of the flow field. This ~0.15° east-west slope, when coupled with the north-south slopes away from the axis of the arch, results in down-slope directions towards the north in the northwestern part of the flow field, towards the north-northwest in the northeastern part of the flow field (where only a few transit SeaBeam lines have been collected), but towards the southwest in the southern part of the flow field.

3.2 Near-Vent Constructs

The new bathymetric data delineate about 110 constructs ranging from low-relief shields to steep-sided cones. *Detrick et al.* [1988] described a single low shield volcano (Figure 2) based on the limited SeaBeam survey done in 1988. This shield was sampled (dredge RB02), and sheet flow fragments were recovered. *Clague et al.* [1990] described 7 small hills (50-200 m tall) that were identified based on their higher backscatter in the GLORIA imagery, even within the high backscatter region of the flows. These 7 hills were dredged, and pillow lava was recovered from 6 of them, with sheet flow fragments also present in 2 dredges. Hyaloclastite and volcanic breccia were also recovered from 2 of the hills that yielded pillow lava. The samples from the remaining hill were sheet flow fragments similar to those recovered from the flat lava flow terrain.

Based on the general recovery of vesicular pillow lava, volcanic breccia, and hyaloclastite from the hills, *Clague et al.* [1990] concluded that the hills were near-vent constructs in the form of pillow mounds, and that the eruptions of these volatile-rich alkalic lava had at times been Strombolian in nature, even at 4300 m depth. Little more could be determined without knowledge of the size, shape, and structure of the constructs. The new SeaBeam coverage of

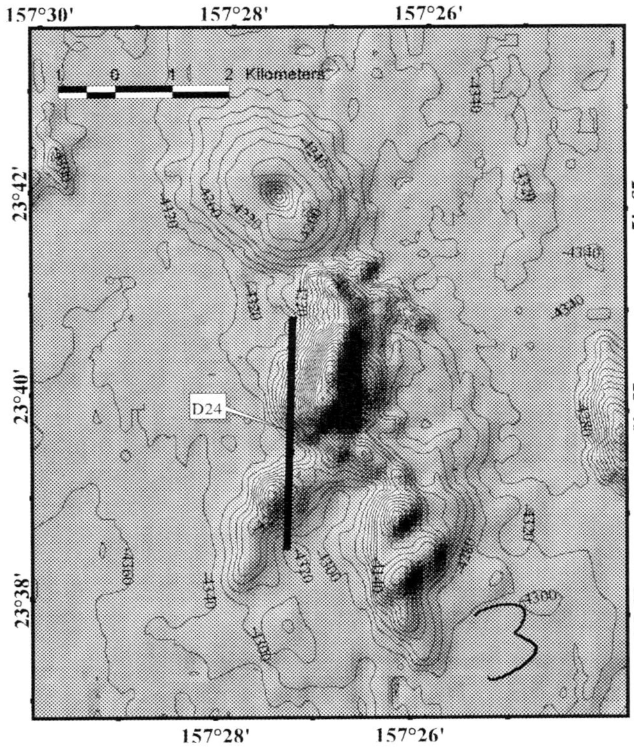

Figure 3. Shaded relief image of the SeaBeam bathymetry showing two different near-vent features nearby each other: a circular flat-topped cone with a summit crater to the north and a steep-sided elongate cone to the south. Contour interval is 20 m. The location of dredge D24 [*Clague et al.*, 1990] is shown.

part of the North Arch Volcanic Field has finally allowed us to evaluate what these and many other volcanic structures look like and how they formed. The detailed bathymetry has also revealed a few surprising features within the lava field and allowed us to quantitatively evaluate the emplacement of the lava flows.

3.2.1. Shield volcanoes. There are about 100 lava shields in the region covered by the SeaBeam bathymetry. The shields are less than 180 m tall and most are less than 50 m tall. Most have no obvious summit crater, although several of the largest shields are indented by craters as deep as about 100 m. One of the lowest-relief but largest of the shields is shown in Figure 2; it is located just south of a steep-sided flat-topped cone with a deep crater that has low backscatter and is inferred to be Cretaceous in age. The shield is about 10-12 km across, but only about 80 m tall and has average slopes of 0.9°. There are two 1-km across depressions in the summit, each only about 10-20 m deep. This is the first shield surveyed with Seabeam [*Detrick et al.*, 1988] and dredged in the North Arch (dredge 4 on cruise Roundabout Expedition Leg 2; called RB02 in *Clague et. al.*, [1990]).

3.2.2. Flat-topped circular volcanic cone. The only nearly-circular flat-topped cone found in the survey area is shown in Figure 3. It is about 4 km across and 140 m tall. The outer slopes average 8°. A 0.5-km diameter crater in its center is about 100 m deep. This cone is similar in shape and aspect ratio to the circular flat-topped cones on the slopes of the Hawaiian Islands. The top portions of several of the lava shields approach the shape of the flat-topped cone, but they are not as circular and are perched on top of irregular-shaped low-angle shields (Figure 4). We describe these composite vents in section 3.2.4.

3.2.3. Steep cones. We found only about 10 steep-sided volcanic cones. None are circular in plan view and most are elongate with the long axis aligned roughly parallel to the N-S horst-graben fabric of the underlying seafloor. Two types are shown with the first consisting of a steep-sided elongate hill. Figure 3 shows one such cone that is

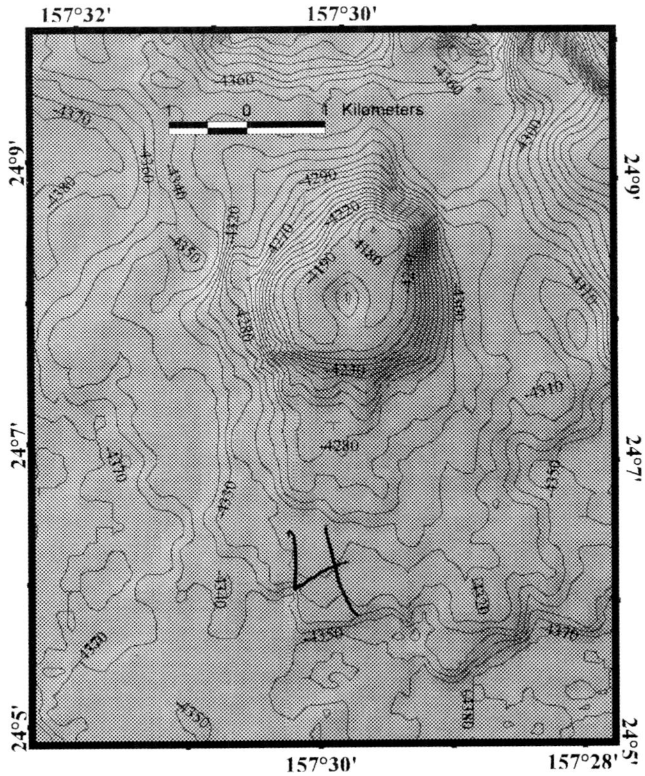

Figure 4. Shaded relief image of the SeaBeam bathymetry showing a composite volcano consisting of a roughly circular flat-topped cone with an off-center crater on an irregular-shaped low-relief shield. Contour interval is 10 m.

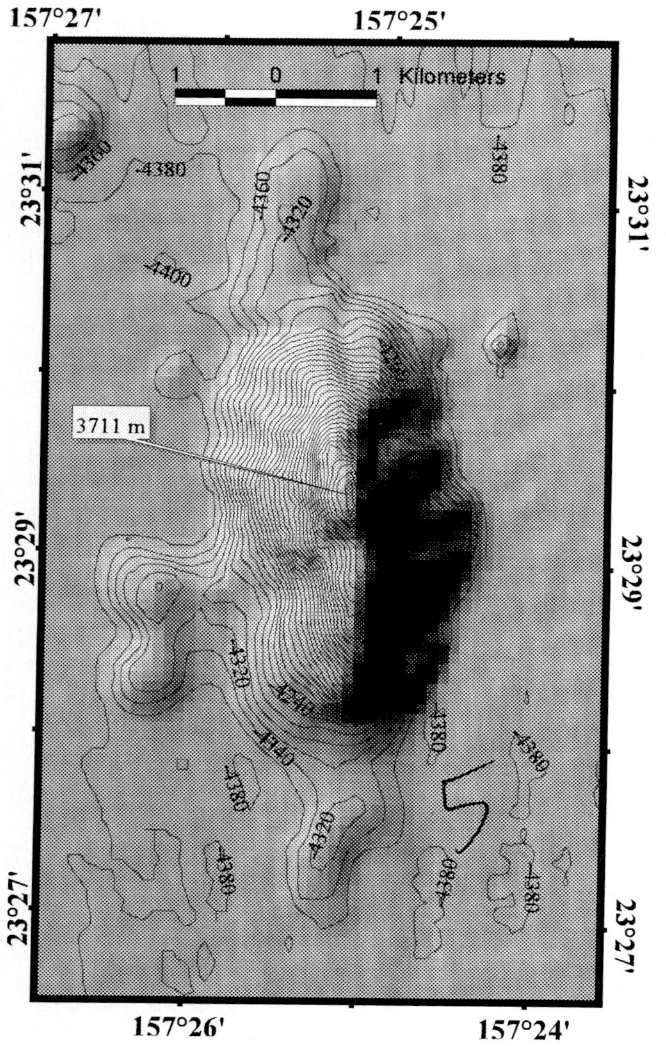

Figure 5. Shaded relief image of the SeaBeam bathymetry showing the largest steep-sided cone in the surveyed region. The contour interval is 20 m. The different gray-toned squares are the size of the bathymetric gridding and show the illumination from the northwest.

440 m tall. It has no summit crater. This steep cone was sampled by dredge D24 [*Clague et al.*, 1990] which recovered two distinct compositions of vesicular nephelinite. There are regions sloping gently away from the steep proximal slopes on its northwest and east flanks. They may be flows from more effusive eruption late in the history of the vent, or an earlier low-relief shield now almost completely buried by the overlying cone. A third possibility is that these aprons are debris flows from the cone, although no volcanic breccia or hyaloclastite was recovered from this particular cone.

A second steep-sided cone, also elongate N-S, is shown in Figure 5. This structure, at 670 m tall, is the tallest of the steep edifices in the survey area; it has no summit crater. There are areas of elevated relief extending away from the base of the cone to the north, south and west. These are probably pillow ridges above eruptive fissures, rather than fluid lava flows or debris flows, since they are 40-60 m thick. No sampling has been done at this cone. Other steep cones, not shown, have summit craters as deep as 200 m.

3.2.4. Composite volcanoes. There are several classes of composite volcanoes. The first type, which is transitional between the low-relief shields and the steep cones, is shown in Figure 4. This structure is about 4-5 km across, has gentle lower slopes (~1.5°), steeper upper slopes (~4.6°), and a nearly flat top that is about 1.5 km across and is indented by a shallow crater that is partly breached to the south. It is about 160 m tall and the crater is about 0.5 km across and 25 m deep. The shallowest point on the summit platform rises about 20 m above the northeastern part of the summit plateau. Another similar example is shown in Figure 6. The edifice is about 4 km across and also has gentle lower slopes (~2.1°) and steeper upper slopes (~6.2°). The nearly flat top is only about 0.6 km across and has two peaks near the edges of the platform. These may be the remnants of a rim that surrounded a small crater in the summit. This feature was sampled by dredge D21 [*Clague et al.*, 1990] which recovered 3 compositionally distinct types of alkalic basalt and nephelinite.

The second type of composite volcano consists of an earlier steep-sided cone or cluster of cones and subsequent construction of a low-relief shield. An example, the location of dredge D26 [*Clague et al.*, 1990] and dive S502 (*Shinkai 6500* dive 502), is shown in Figure 7. Steep-sided cones form a partial ring roughly 3 km across with a probable crater in the center that is roughly 1.5 km across. In the center of the ring of steep cones a low shield, only about 0.5 km across and 25 m tall, has filled the crater to the low points. In this case, a roughly ring-shaped collection of steep near-vent structures has had a subsequent low-relief shield volcano grow in its center. The steep cones consist of mainly pillow lava, some highly vesicular, with some hyaloclastite. The lava samples include a variety of compositions in the basanite to alkalic basalt range. Spherical glass beads found in some of the hyaloclastite suggest lava fountaining driven by exsolving magmatic volatiles [*Clague et al.*, 1990; *Dixon et al.*, 1997]. The ge-

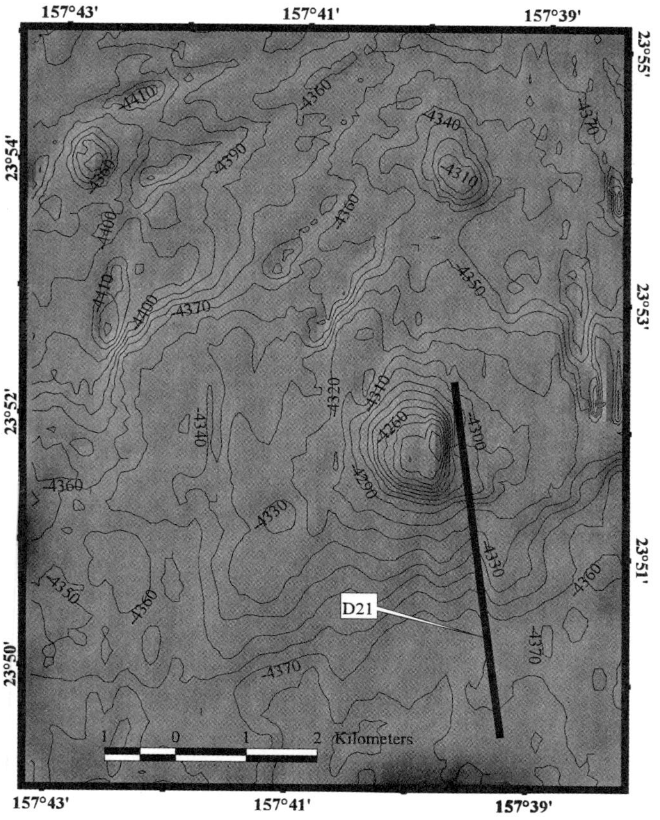

Figure 6. Shaded relief image of the SeaBeam bathymetry showing a composite volcano consisting of a small steep-sided nearly-circular cone on an irregular-shaped low-relief shield. Contour interval is 10 m. The location of dredge D21 [*Clague et al.*, 1990] is shown.

ology of this site, as observed from the *Shinkai 6500* submersible during dive S502, will be described in section 3.4.1.

3.2.5. Shield modified by pit crater. One shield in the surveyed area has been modified by formation of a pit crater on its flank (Figure 8). The shield is located west of the pit crater and is an irregular feature perhaps 1.5 km across. The shield is about 80 m tall. A pit crater, roughly 1.1 by 1.25 km across at the rim and about 300 m deep is centered about 0.8 km east of the summit of the shield. This pit crater has a volume of about 0.12 km^3, similar in size to the pit that formed on the southern rim of the summit of Loihi Seamount in 1996 [*Loihi Science Team*, 1997; *Davis and Clague*, 1998] and slightly smaller than Halema'uma'u Crater at Kilauea's summit. The North Arch pit crater was the location of dive S503 and the geology observed during the dive will be described in section 3.4.2. A retrospective examination of the GLORIA imagery shows the pit as a shadow; no other similar features were found in the GLORIA imagery of the entire flow field, so this feature appears to be unique.

3.2.6. Inferred fissure eruption. The southeastern part of the flow field (Figure 9) has nearly 100% SeaBeam coverage, but no vent structures were found. The regional slope is down towards the southwest, and several lava lobes, such as the long thin one, flow in this direction before being entrained into a graben in the fabric of the underlying sea floor that constrains it to flow roughly due south. However, the widest part of the flow extends towards the south-southeast, nearly parallel to the contours. We think that this spatial distribution results from eruption along a 75-km long fissure oriented south-southeast, and that the main flows from this fissure traveled downslope towards the southwest. This hypothesized fissure is not detected in the

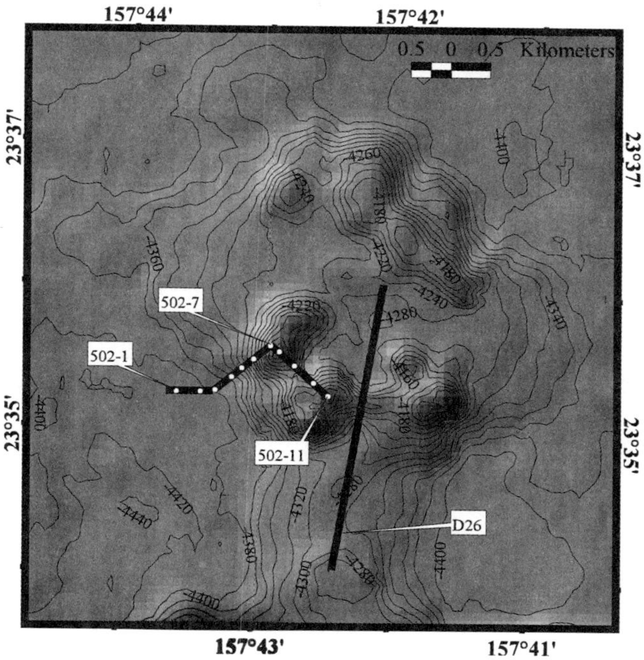

Figure 7. Shaded relief image of the SeaBeam bathymetry showing a composite volcano consisting of an incomplete ring of small steep-sided cones with a low shield in the center of the ring. Contour interval is 20 m. The different grey-toned squares are the size of the bathymetric gridding and show the illumination from the northwest. The locations of dredge D26 [*Clague et al.*, 1990] and *Shinkai 6500* dive S502 are shown. Some sample numbers are indicated along the dive track.

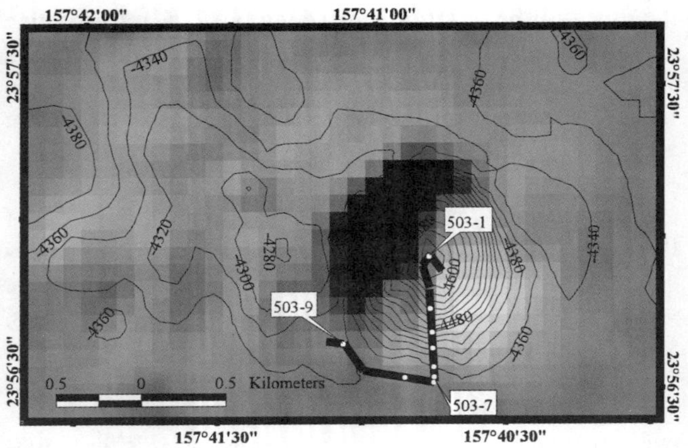

Figure 8. Shaded relief image of the SeaBeam bathymetry showing a low-relief shield modified by a large pit crater on its eastern flank. Contour interval is 20 m. The different grey-toned squares are the size of the bathymetric gridding and show the illumination from the northwest. The location of *Shinkai 6500* dive S503 is shown. Some sample numbers are indicated along the dive track.

SeaBeam data, suggesting that ramparts that may exist along the fissure have a relief <10 m.

3.3 Lava Flows

The clearest example of a long lava flow occurs in the northwestern corner of the flow field (Figure 10). The flow is constrained to flow almost due north in a graben in the fabric of the underlying seafloor. In the GLORIA imagery this flow can be traced for about 90 km, assuming the vent is located adjacent to a small Cretaceous seamount at about 158°20'W, 24°28'N. Just to the south of this seamount, the GLORIA data is degraded by a nadir, although there are hints that the flow may start near 158°20'W, 24°20'N. The new SeaBeam bathymetry does not show any vent structure near the Cretaceous seamount and the flow appears to originate at a series of small elongate hills at about 158°22'W, 24°22'N, just south of the nadir in the GLORIA mosaic. The flow is therefore 108 km long, generally about 2 km wide, and advanced down a gradient of 1.9 m/km (0.1°). The proximal 40 km of this flow has low marginal levees and the distal 50 km is a roughly 10-m tall lobe. The flow spread out and partially surrounded the Cretaceous seamount, filling in a shallow moat on all but the northeast margin of the seamount. At about 24°47'N, the flow spread out to the east and flooded a region between two lineaments of Cretaceous seamounts. The bathymetric data do not delineate an eastern boundary to this flow, although lava recovered in dredge D16 is compositionally distinct from that in D15 [*Clague et al.*, 1990] and is presumably from a different lava flow. It is also unclear if the eastern "breakout" was emplaced before, during, or after the northern 55 km of the flow.

The southern flows, discussed above because they define the 75-km long eruptive fissure, are also long flows (Figure 9). However, because we can only infer the location of the eruptive fissure, the exact lengths of these flows are less certain. The reconstruction of which flows travelled which paths is also less certain because the entire region does not have continuous bathymetric data. However, if we assume that the southwestern flow lobe erupted from our proposed fissure system, then this flow advanced about 110 km, first towards the southwest and then towards the south.

The flows from the vast majority of the 110 known vents are superposed on top of one another; determining which flow erupted from which vent is probably not possible without detailed sampling, or perhaps with near-bottom geophysical surveys that could accurately map the thickness (<3-4 m) of sediment covering the flows of different ages. In the northern part of the flow field, which was not surveyed in 1999, there are fewer flows and many can be traced several tens of km to specific vents in the GLORIA images.

3.4. Observations From the Shinkai 6500 *Submersible*

The two 1999 *Shinkai 6500* submersible dives in the North Arch Volcanic Field enabled observations of the bottom characteristics and collection of rocks from specific outcrops. Electron microprobe analyses of basaltic glass from these samples are presented in Table 1 and shown in Figure 11. The new samples are geochemically similar to those previously dredged and recovered in cores [*Clague et al.*, 1990; *Dixon et al.*, 1997; *Frey et al.*, 2000]. Further discussion about their chemistry is beyond the scope of this paper.

3.4.1. Dive S502. The first of these dives visited the composite volcano sampled by dredge D26 [*Clague et al.*, 1990]. The location of the dive track, shown on Figure 7, has been translated slightly to better fit the bathymetry. The dive began at 4403 m depth southwest of this feature on a sediment-covered bottom. Loose basaltic gravel was seen sporadically, but the only outcrop in the area was an enormous split tumulus that projected through the sediment by several m. The tumulus indicates that the flows beneath the sediment are tube-fed flows, with the tumulus above a tube. The dive proceeded to the western slope of the steeper cone and encountered the pillow flows that form the cone above about 4371 m depth. The highly-vesicular pillow lava (Figure 12A) constructing the cone is interbedded with thin layers of hyaloclastite (Figure 12B). The

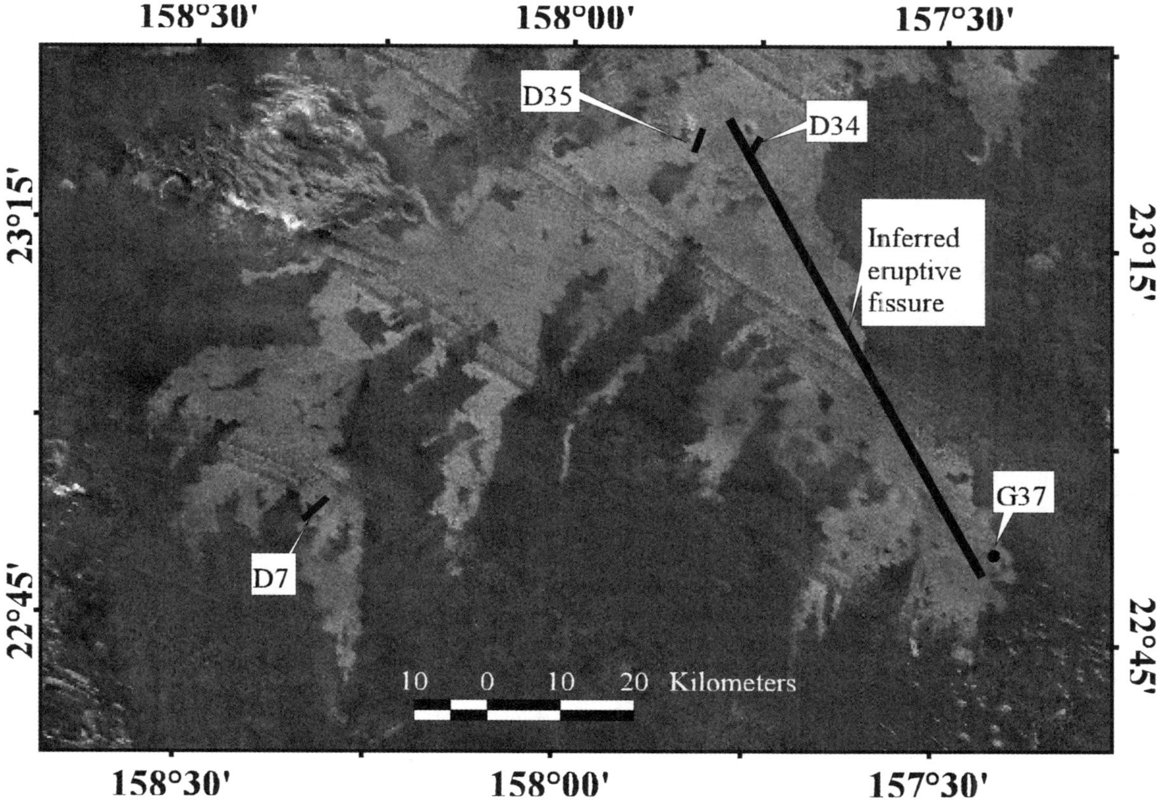

Figure 9. GLORIA image of the southern part of the North Arch Volcanic Field. The approximate location of the inferred eruptive fissure that fed lava to these flows is shown. Locations of several cores and dredges that recovered volcanic samples [*Clague et al.*, 1990] are shown. Regional contours for this area are shown in Plate 1.

most vesicular lava (sample S502-R7), with over 50% vesicles, occurs beneath a layer of hyaloclastite (samples S502-R6) at 4224 m depth. A sample from this hyaloclastite layer (sample S502-R7C) contains abundant thin, curved, dense glass bubble-wall fragments, now mainly altered (Figure 12C). These limu o Pele fragments are identical to those described from Loihi Seamount [*Clague et al.*, 2000a] that formed during small hydrovolcanic explosions when seawater, entrained in the erupting lava, expanded upon boiling [*Mattox and Mangan*, 1996]. Hyaloclastite (sample S502-R8) from 4038 m depth, near the summit of the cone, contains only angular dense to slightly vesicular glass fragments. Beyond the eastern base of the western cone the bottom is apparently composed of sheet flows, although none could be collected. The dive ended on a second steep cone composed dominantly of highly-vesicular pillow lava (samples S502-R9 and R10, collected at 4210 and 4114 m depth) with a thin veneer of hyaloclastite near the summit (sample S502-R11 at 4065 m depth). The hyaloclastite from near the summit contains angular moderately-vesicular glass fragments. The lavas, all apparently erupted from this single composite vent structure, have a wide range of compositions (43.6-47.0% SiO_2 and 5.6-10.2% MgO).

3.4.2 Dive S503. Dive S503 started in the pit crater shown in Figure 8. Once again, the dive track has been translated to better fit the bathymetry. The bottom of the pit at 4684 m depth is sediment covered with some angular basalt fragments and gravel in the sediment. The bottom where the submersible landed sloped about 10° to the south, but was still mainly sediment covered with random dense angular basalt blocks, some as large as 5 m across, scattered on the surface. The first outcrops of basalt were encountered only 122 m below the rim of the pit crater, so talus and sediment obscure the lower ~220 m of the crater wall. The accumulated post-Cretaceous flows in the area of the pit are therefore a minimum of 122 m thick. Most of the flows from 4460 m to 4348 m depth are slightly- to moderately-vesicular pillow lava, some with hollow interiors, but no tubes were observed. A thin deposit of hyaloclastite was recovered on pillow fragment S503-R5 at 4382

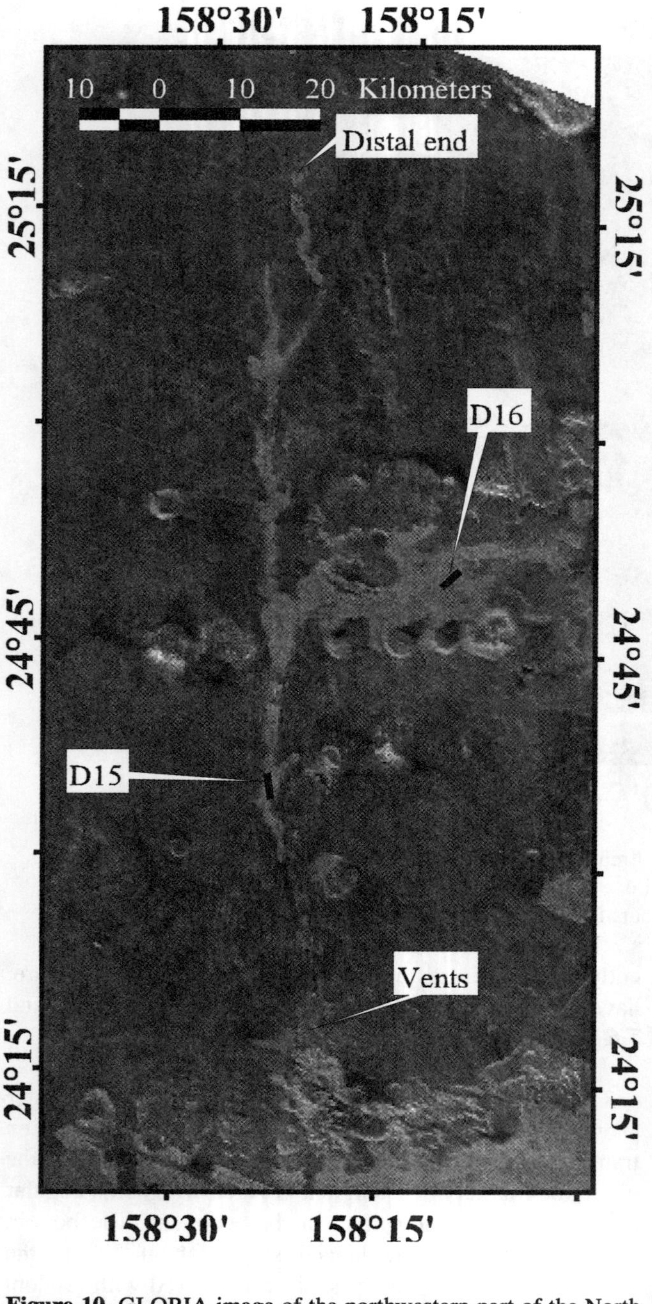

Figure 10. GLORIA image of the northwestern part of the North Arch Volcanic Field. The narrow flow erupted from a group of vents aligned N-S and flowed northward for 108 km, mainly within a graben in the underlying Cretaceous seafloor fabric. The average slope from the vents to the distal end of the flow is 0.1°. The flows spilled out of the graben to the east just north of 24°45'N and flooded a region between Cretaceous seamounts. Locations of two dredges from *Clague et al.* [1990] are shown. Regional contours for this area are shown in Plate 1.

m depth. The glass fragments are dense and angular. The uppermost flow is a massive 1.5- to 3-m thick sheet flow that has an elaborate drapery-folded surface (Figure12D) buried beneath only a few cm of sediment. Sample S503-R7 from this flow, however, had moderately thick palagonite rinds (~0.6 mm, although highly variable), suggesting that the flow is not as young as it appears from the thin sediment cover. The rim of the pit truncates flow channels and striations in the sheet flows and clearly shows that the pit formed, or has been extensively modified by slumping, after emplacement of at least the surficial lava flows. The abundant angular lava blocks and talus within the pit also attest to collapse of the pit walls, but breccia fragments were not seen around the rim indicating that collapse of the pit was not associated with hydrovolcanic explosions. Fractures concentric to the pit break up the surface sheet-flow within about 20 m of its rim. A few lava channels in the sheet flows have levees about one m high. Some of the sheet flows are lineated, as described by *Chadwick et al.* [1999] from the Juan de Fuca Ridge. Closer to the summit of the low-relief shield just west of the pit, pillow lava with delicate horn-like buds becomes more common. The glass rinds on lavas from the pit have a moderate compositional range (Table 1; 45.0-46.6% SiO_2 and 6.65-9.0% MgO). The entire exposed section may have been emplaced during a single eruption since no sediment layers were observed in the wall.

4. DISCUSSION

4.1. Volume of the Flows

The edges of the southern flows must be thin (probably <10 m thick) because we could not map the flow margins based on bathymetry collected by the SeaBeam system and because the flows enclose numerous islands of low-relief sediments, even near the proposed eruptive fissure. This complex flow covers an area of about 3,600 km^2 and has a volume of 36-72 km^3, assuming that it averages 10-20 m thick.

The elongate flow in the northwest part of the flow field (Figure 10) extends for 108 km, averages about 2 km wide, and fills a graben estimated from seismic profiles as 30 m deep, giving it a volume of about 6.5 km^3. This volume is probably a minimum because the SeaBeam data indicate that an unknown volume of lava also spilled out of the graben to the east at about 24°47'N. The total volume of this flow could be on the order of 10 km^3.

The entire flow field covers about 25,000 km^2 and has a minimum volume of 250 km^3, assuming only 10 m for the

Table 1. Electron Microprobe Glass Analyses

Sample	502-1B	502-2	502-3A	502-3B	502-4	502-5	502-5H	502-6A	502-6B-1
Depth	4403m	4401	4399	4399	4371	4312	4312	4224	4224m
# Avg.	5	4	4	4	4	6	2	14	1
SiO_2	46.9	47.0	46.9	46.8	45.5	44.1	43.6	44.1	46.3
TiO_2	2.01	2.03	1.96	1.95	2.19	2.39	2.31	2.40	1.69
Al_2O_3	16.60	16.60	15.86	15.86	14.44	14.40	14.18	14.42	14.78
FeO	11.93	11.92	11.64	11.71	11.88	12.14	11.82	12.11	11.84
MnO	0.19	0.18	0.16	0.17	0.19	0.19	0.17	0.18	0.19
MgO	5.57	5.67	6.26	6.26	8.30	7.64	7.40	7.57	7.70
CaO	11.72	11.73	12.30	12.39	12.17	12.84	13.02	12.80	11.95
Na_2O	3.43	3.41	3.26	3.29	3.45	3.80	3.77	3.73	3.09
K_2O	0.65	0.64	0.60	0.59	0.76	1.08	1.10	1.09	0.52
P_2O_5	0.33	0.35	0.32	0.33	0.38	0.56	0.55	0.54	0.27
S	0.175	0.171	0.166	0.167	0.161	0.122	0.140	0.126	0.116
Cl	0.093	0.081	0.059	0.061	0.086	0.101	0.108	0.096	0.086
Total	99.61	99.75	99.45	99.53	99.47	99.34	98.10	99.12	98.46

Sample	502-6B-2	502-6B-3	502-6B-4	502-7A	502-7B	502-7C	502-8	502-9	502-11
Depth	4224m	4224	4224	4091	4091	4091	4038	4210	4065m
# Avg.	1	1	5	7	7	4	13	3	14
SiO_2	46.3	46.3	46.1	44.2	44.2	44.4	44.2	44.3	44.0
TiO_2	1.64	1.60	1.53	2.36	2.35	2.42	2.37	2.48	2.39
Al_2O_3	14.33	14.16	13.86	14.46	14.44	14.59	14.58	14.83	14.36
FeO	11.90	11.85	11.78	12.18	12.13	12.01	12.14	12.14	12.07
MnO	0.18	0.23	0.16	0.17	0.18	0.20	0.17	0.17	0.19
MgO	9.23	9.61	10.24	7.41	7.42	7.23	7.44	6.82	7.61
CaO	11.03	10.81	11.07	12.90	12.80	13.08	12.76	13.25	12.84
Na_2O	3.09	3.01	3.01	3.81	3.82	3.72	3.76	3.86	3.74
K_2O	0.50	0.47	0.46	1.07	1.07	1.10	1.10	1.12	1.10
P_2O_5	0.25	0.28	0.28	0.57	0.56	0.54	0.52	0.58	0.57
S	0.136	0.135	0.135	0.120	0.132	0.135	0.137	0.145	0.123
Cl	0.078	0.065	0.044	0.104	0.101	0.100	0.098	0.107	0.098
Total	98.59	98.52	98.69	99.31	99.16	99.48	99.27	99.79	99.10

Sample	503-1B	503-4	503-5H-1	503-5H-2	503-5H-3	503-7	503-8	503-9A	503-9B
Depth	4684m	4421	4382	4382	4382	4342	4342	4330	4330m
# Avg.	4	6	4	6	3	7	6	4	3
SiO_2	45.5	45.5	45.0	46.6	45.2	45.02	45.0	45.1	45.6
TiO_2	2.71	2.14	2.31	1.83	2.67	1.90	1.82	1.89	2.79
Al_2O_3	14.53	14.31	13.92	15.06	14.49	14.29	14.29	14.31	14.77
FeO	12.94	11.89	11.95	11.42	12.56	11.62	11.58	11.58	12.89
MnO	0.17	0.18	0.18	0.17	0.19	0.18	0.18	0.15	0.18
MgO	7.13	8.20	8.97	7.77	7.66	8.71	8.86	8.92	6.65
CaO	11.05	11.99	11.81	12.09	11.18	12.18	12.12	12.35	11.17
Na_2O	3.81	3.47	3.23	3.23	3.61	3.48	3.52	3.48	3.81
K_2O	1.12	0.75	1.03	0.66	1.03	0.74	0.75	0.74	1.18
P_2O_5	0.47	0.39	0.42	0.35	0.44	0.45	0.47	0.51	0.50
S	0.122	0.162	0.165	0.164	0.155	0.148	0.149	0.147	0.122
Cl	0.059	0.086	0.074	0.064	0.065	0.074	0.085	0.083	0.075
Total	99.61	99.06	98.99	99.35	99.22	98.78	98.77	99.26	99.70

502 and 503 are *Shinkai 6500* dives S502 and S503, respectively. All samples are rock samples denoted in text with an R before sample number. H indicates hyaloclastite, with distinct glass compositions denoted by number after the H. #Avg. indicates the number of analyses averaged to obtain the analysis presented.

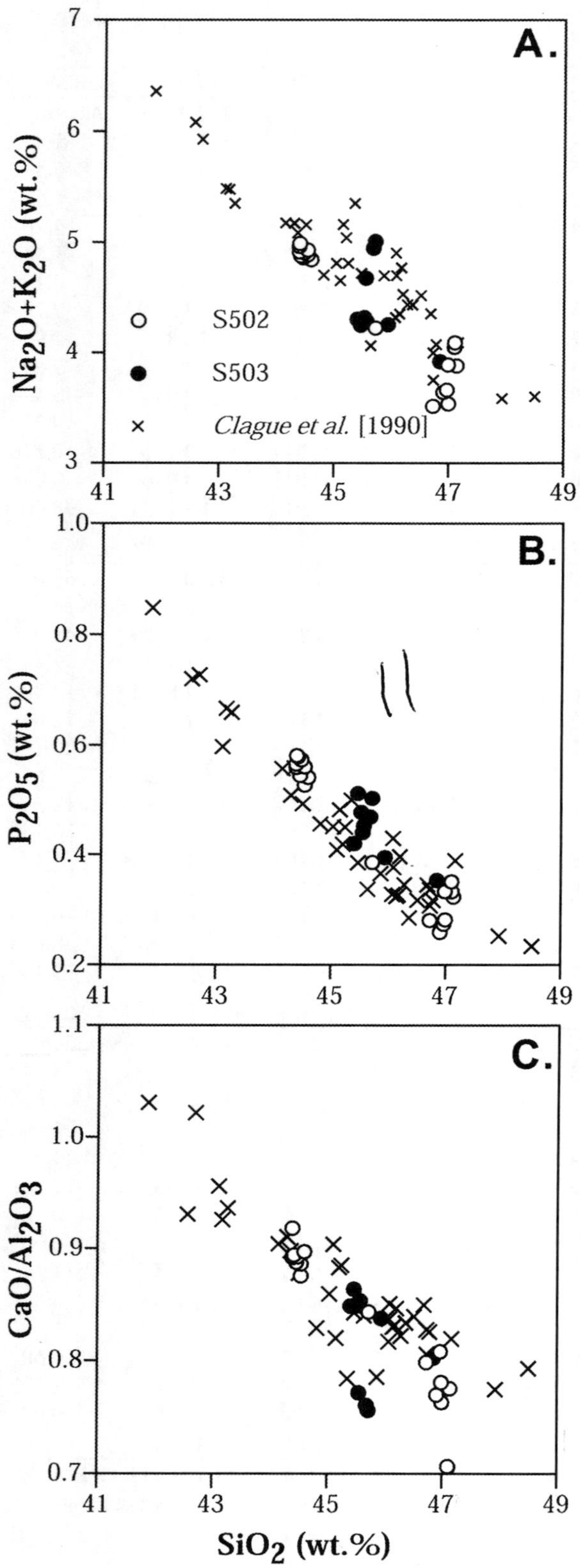

average flow thickness. A more realistic estimate of the average flow thickness and of the total volume of the flow field is based on the measured 122 m minimum thickness of flows in the wall of the pit crater and the lack of kipukas in the central part of the flow field. We estimate that the flows, on average, are about 40-50 m thick and the total volume is roughly 1,000-1,250 km^3.

This total volume has been erupted from a minimum of 110 vents, and more likely from about 200-300 vents, suggesting a volume per source vent that averages 3.3-6.2 km^3.

4.2. Age of the Flows

Clague et al. [1990] and *Dixon et al.* [1997] estimated that the flows comprising the North Arch Volcanic Field ranged from about 0.5 to 1.15 Ma. Using their estimation of the rate of palagonite formation, the flow at the top edge of the pit, with about 0.6 mm of palagonite, would be ~0.4 Ma, or slightly younger than the previous estimate of the youngest flows. The thin sediment cover on this flow would also imply that the sedimentation rate has not been constant and that the rate was higher when the older parts of the flow field erupted and has decreased since. Palagonite on a single glass chip from a core located east of the main North Arch Volcanic Field is consistent with an age of about 1.5 Ma [*Clague et al.*, 2001], suggesting that activity in the region began at least this long ago. This 1.5 Ma age of the flow is also consistent with the observed stratigaphic relations since the lava flows at this site underlie km-sized blocks in the distal Wailau landslide [*Moore et al.*, 1989] that originated on East Molokai Volcano where the tholeiitic shield lavas are about 1.5 Ma [see *Clague and Moore*, 2001]. *Uto et al.* [2000] report three preliminary Ar-Ar ages determined on samples dredged in 1988 from three steep near-vent structures [*Clague et al.*, 1990] that range from 1.0-1.4 Ma. These ages are at the old end of the range estimated previously and indicate that

Figure 11. Chemical plots of data from Table 1 showing the range of compositions of the volcanic glasses recovered by *Shinkai 6500* dives S502 and S503. A. Alkali-silica diagram showing that the new samples are alkalic basalt and basanite compositions and fall within the range of previously collected samples. B. P_2O_5 as a function of SiO_2 content, showing that the lower-SiO_2 melts have higher P_2O_5, consistent with their formation by lower melt fractions. C. CaO/Al_2O_3 as a function of SiO_2 content showing that the higher-SiO_2 lower percentage melts have lower CaO/Al_2O_3 consistent with the greater incompatibility of Al_2O_3 at low melting extents [*Frey et al.*, 2000].

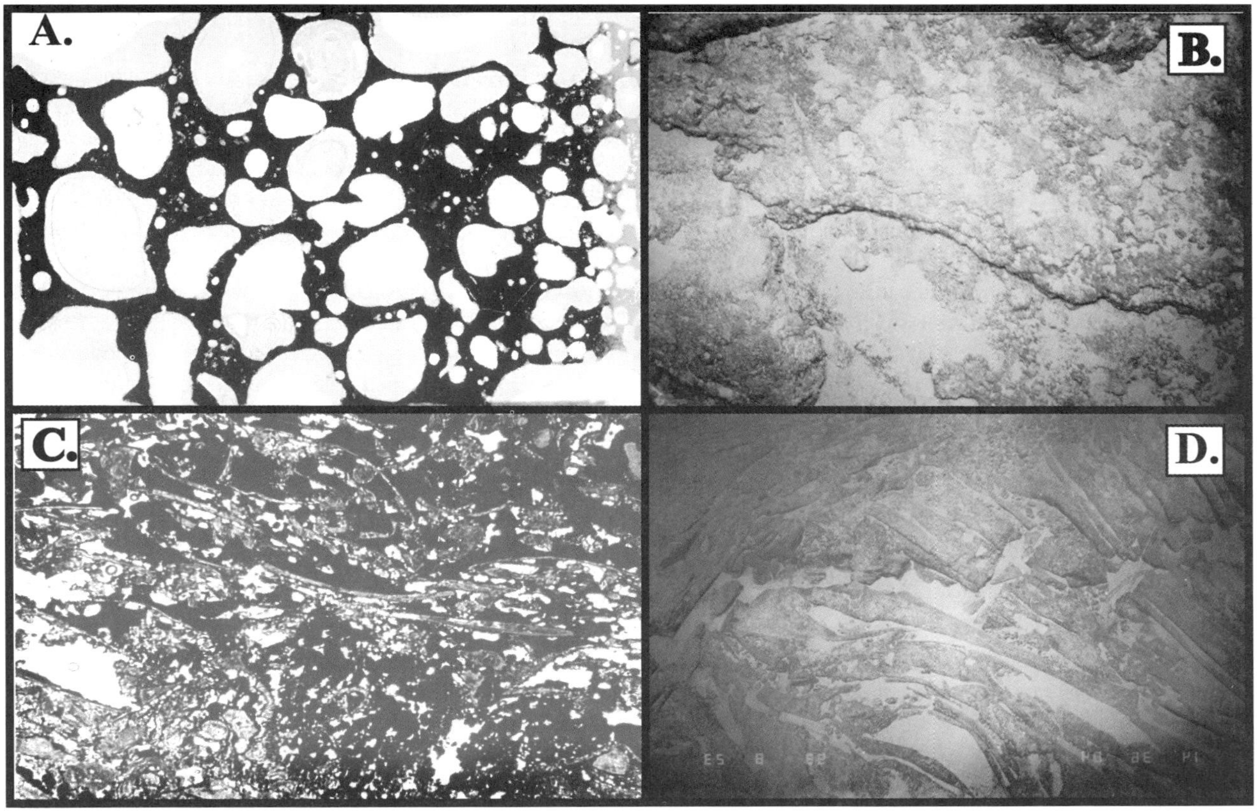

Figure 12. A. Photomicrograph of highly vesicular lava sample S502-R7C (location shown in Figure 7) collected from a steep-sided cone during *Shinkai 6500* dive S502. Image is 3.5 cm wide. B. Photograph of slabby hyaloclastite outcrop draping the slope; photograph taken near S502-R6 site (Figure 7) during dive S502. Field of view is about 2-3 m across. C. Photomicrograph of hyaloclastite sample S502-R6B (location shown in Figure 7) showing altered limu o Pele (bubble-wall) fragments formed during mild steam explosions. Image is about 1 cm wide. D. Photograph of drapery-folded sheet flows exposed on the rim of the pit crater near S503-R7 location in Figure 8; photograph taken during dive S503. Field of view is about 4-5 m across.

some of the steep near-vent structures are slightly older than the age range estimated for the sheet flows based on palagonite thickness and paleomagnetic study of the overlying sediment.

The known age range, when combined with the number of vents mapped in 1/3 of the flow field and estimated for the remainder, suggests an eruption frequency of one every 3.7-5.5 ka. The eruption rate, again averaged over the lifespan of the flow field, is roughly 0.9-1.1km^3/ka, about 1/50 of the recently estimated magma supply rate at Kilauea Volcano [*Cayol et al.*, 2000].

4.3. Eruption Styles and Rates

A wide range of eruption styles are recorded in the North Arch Volcanic Field, probably reflecting in part the high initial and dissolved contents of H_2O, CO_2, and S [*Dixon et al.*, 1997]. A small percentage of the cones formed during low-flux Strombolian eruptions that produced hyaloclastite and volcanic breccia. Even these features appear to be comprised mainly of vesicular pillow lava that erupted effusively, also probably at low eruption rates. The pillow lavas are in many cases more vesicular than the fragments in the hyaloclastite, indicating that fragmentation was not caused solely by high bubble content and its consequent increase in the surface area of the magma that could facilitate magma-water interaction. Thus fragmentation appears to be triggered by a combination of bubble content and eruption rate with the Stombolian hyaloclastite-forming eruptions being more vigorous than those forming vesicular pillow lavas. In any case, highly gas-charged magmas, apparently only in small volumes and erupted at low rates, usually were available at the beginning of an eruption, and produced irregularly-shaped cones of small volume that consist of pillows and hyaloclastite. *Clague et al.* [2000b] discussed the conditions under which steep sided, as opposed to flat-topped truncated, cones form on the flanks of Hawaiian volcanoes and noted the importance of high volatile contents and eruption of highly vesicular lavas. The data presented here are consistent with their interpretation.

Most of the volcanic constructs are low-relief shields comprised of sheet flows. These sheet flows may subsequently bury the early pillow/hyaloclastite deposits, as suggested by the outcrops in the wall of the pit crater. This sequence, with fluid, less-vesicular, effusive lava postdating early more-volatile-rich eruptions has also been observed in Hawaii [*Rowland and Walker*, 1990], particularly at Kilauea during the on-going Pu'u 'O'o eruption, which began with 47 high-volume fountain phases occurring roughly monthly and then evolved to quiet effusion of lava after the first three and a half years [see summaries in *Clague and Heliker*, 1993; *Thornber*, 2001]. The ca. 1800 Kaupulehu alkalic basalt eruption on Hualalai Volcano is interpreted to follow a similar pattern [*Kauahikaua et al.*, 2001]. As at Kilauea, the effusive phase in the North Arch appears to last far longer than the Strombolian phase and to produce most of the erupted volume. However, the eruption rate of the early fountain phase at Kilauea is far greater than the subsequent effusive phase, in contrast to the North Arch field where the eruption rates of the early Strombolian phase are small compared to those of the later effusive phase.

The North Arch shields formed during steady effusive eruptions of low-viscosity, low-bubble content, alkalic lava. The irregular form of the vast majority of these vents indicates that their eruption rates were greater than those that form flat-topped volcanic cones on the submarine rift zones of Hawaiian volcanoes so that levees could not form to impound lava into a pond above the vent [*Clague et al.*, 2000b]. These eruption rates also had to be sufficient to maintain the molten core of the flow from the vent to the distal end of the flow, some 100-plus km away. In addition, the presence of a pit crater also indicates that at least one of the eruptions was long-lived in order to form an underlying crustal magma reservoir. If the rates were 2-10 times as great as at Kilauea, then individual eruptions may have lasted for decades to perhaps as long as several centuries.

The pattern for the North Arch Volcanic Field is of eruptions of several km^3 to several tens of km^3 with durations of decades to centuries, and an average recurrence interval of about 5 ka.

4.4 Lava Flow Emplacement

The lava flows in the North Arch Volcanic Field are among the longest lava flows on Earth's surface and are the longest known submarine flows [*Cashman et al.*, 1998; *Gregg and Fornari*, 1998]. The basic constraints required to form long lava flows are that they must travel long distances before solidifying and that sufficient lava must be erupted to cover a large area. *Keszthelyi and Self* [1998] suggest that most long flows are limited by the volume of the flow rather than by cooling of the flow. Two competing models for emplacement of long flows are rapid emplacement and insulated emplacement, where the insulation occurs either by flow within tubes [e.g., *Peterson et al.*, 1994; *Kauahikaua et al.*, 1998] or as inflated sheets [*Hon et al.*, 1994]. Some special conditions, mainly concerning the rapid formation of an insulating glass rind, arise because of the submarine emplacement of these flows, but in general they behave similarly to subaerial lava transported in tubes [*Gregg and Formari*, 1998].

These long submarine lava flows advanced down slopes with gradients of 1.5-3.5 m/km (0.08-0.2°) and formed flows that are thin enough that their margins can rarely be detected in the Seabeam bathymetry. Most of these flows are not particularly narrow, however, with even the longest, narrowest, flow averaging about 2 km wide. A large tumulus observed near the beginning of *Shinkai 6500* dive 502 near the shield feeding lava to one such flow suggests that near the vents, the flow becomes narrowly defined in tubes, above which tumuli form. Farther downslope, the width of the flows suggests that advancement occurs as sheet or lobate flows that inflate beneath the crust in a manner analogous to that described by *Hon et al.* [1994] for subaerial Kilauea pahoehoe flows. The advancing edge of the North Arch flows presumably has a wide core of molten lava inflating the already formed insulating crust. Behind that flow front, the wide core of molten lava constricts to form tubes. The margins of the flows display numerous small breakouts that advance for a few km before stalling. The numerous branches of the flows are reminiscent of the pattern produced by the sequential breakouts from the 1983 Mauna Loa \`a\`a flows [*Lipman and Banks*, 1987], but are probably formed in a manner more analogous to the advance of pahoehoe flows at Kilauea [*Mattox et al.*, 1993].

Gregg and Fornari [1998] suggested that long submarine flows have low viscosities because the high ambient pressure prevents volatile exsolution and vesiculation. We agree that low vesicle contents characterize the long submarine North Arch flows, but note that these strongly alkalic lavas initially contain large inventories of volatiles which form vesicles. However, as submarine eruption of these lavas progresses, slow ascent rates through the mantle or residence in crustal storage reservoirs allows for loss of these bubbles and subsequent eruption of lava with low vesicle contents. The low viscosities of these alkalic lavas enhance the escape of bubbles, which rise and escape through the seafloor, and the settling of olivine phenocrysts and xenocrysts. In those rare cases where the initial eruptive products are preserved on the surface, the lavas are highly vesicular pillow lava and hyaloclastite that commonly contain xenocrysts of mantle olivine [*Frey et al.*, 2000]. Low bubble content lava appears to be a requirement to form long lava flows, but low bubble content can occur either through low initial volatile contents and lack of vesiculation (as in mid-ocean ridge basalt) or to loss of bubbles exsolved from high-volatile content magmas (like the North Arch alkalic lavas).

Brief, small volume eruptions (0.018 km^3 for the <1 month-long 1996 Gorda Ridge eruption) along the mid-ocean ridge system are known to trigger formation of "megaplumes" or event plumes [*Baker*, 1998] in the water column. The large volume (tens of km^3) of some of the North Arch flows, coupled with the rapid transfer of heat to seawater as the glass rind forms (in <1 sec; *Gregg and Fornari*, [1998]) suggests that equally large and long-lived event plumes might be generated during emplacement of submarine flows in the North Arch. Each km^3 of basalt would heat 1000 km^3 of seawater by 1°C [*Shaw and Moore*, 1988]. The largest single flow in the North Arch would therefore heat 36,000 to 72,000 km^3 of seawater by 1°C. Such a large heat input is 0.3-0.6 times that associated with El Niño sea surface temperature anomalies, although the heat is probably introduced over a longer time period, suggesting that emplacement of such voluminous submarine lava flows may influence ocean circulation and climate, as originally proposed by *Shaw and Moore* [1988].

4.5. Formation of the Submarine Pit Crater

The formation of pit craters is generally attributed to collapse of the surface into shallow magma storage reservoirs as they are drained [*Jaggar*, 1947, *Holcomb*, 1987], usually during an eruption or intrusion farther down the rift zone. On Kilauea, angular debris formed during phreatic eruptions is associated with pit crater formation [*Holcomb*, 1987]. Under the hydrostatic pressure at 4.3 km water depth where the single North Arch pit is located, no phreatic explosivity is expected, nor was any observed, since the expansivity of seawater as it is heated is relatively small [*Clague et al.*, 2000a]. Pit craters are now thought to form due to stoping over an underlying wide rift zone fracture [*Okubo and Martel*, 1998]. In the North Arch Volcanic Field, however, there are no rift zones and so this mechanism appears unlikely. We propose that this large pit crater formed by a reduction in the volume of magma stored underground. Since the magma in the reservoir consists of the sum of melt, crystals, and bubbles; the magma volume could be reduced simply by the escape of bubbles. Lava from the steep cones has high volatile contents [*Dixon et al.*, 1997] and many of the lava samples are highly vesicular (>50% bubbles, see Figure 12A). These bubbles consist mainly of exsolved CO_2 due to its low solubility in basaltic magma compared to that of other volatile species [*Dixon et al.*, 1997]. We propose that the reduction of magma volume in a reservoir, probably located 1-2 km below the surface, was caused by the escape of mainly CO_2 bubbles from the magma. While the eruption was ongoing, the supply of new magma maintained the pressure in the reservoir to support the lid, but once recharge ended, the stored magma quickly lost its complement of bubbles which rose through the magma and escaped from the top of the reservoir and into the water column. The roughly 0.12 km^3 volume of the pit would require loss of this volume of bubbles. If the stored magma contained 50% bubbles, then the volume of stored magma could have been as small as 0.25 km^3. On the other hand, smaller percentages of bubbles would require proportionally larger volumes of stored magma.

The presence of the pit crater also suggests that at least some of the magma delivered to these long-lived eruptions is stored at shallow crustal depths where it can cool and fractionate prior to eruption. Previously, the fractionated character of nearly all these lavas peripheral to the Hawaiian plume were attributed to cooling and fractionation only during their transit through the old, cold ocean lithosphere [*Clague and Dixon*, 2000]. The wide range of lava compositions erupted from these monogenetic vents suggests that little magma mixing occurs during either transport or storage in the crust. A similar wide range of compositions, although extending to more fractionated compositions, has been documented in a single historical (1949) eruption on La Palma Island in the Canary Islands [*Kluegel et al.*, 2000].

5. CONCLUSIONS

The North Arch Volcanic Field covers 24,000 km^2 and has a probable volume of 1,000-1,250 km^3. It consists of alkalic basalt, basanite, and nephelinite. More than 100 vent structures have been identified in the combined GLORIA side-scan and SeaBeam 2100 bathymetric data. Volcanic constructs are mainly low shields, some with small summit craters. Others are steep sided cones or rings of cones consisting of vesicular pillow lava and hyaloclastite. Some individual vent structures, though apparently monogenetic, erupted a range of alkalic lava compositions. One shield has been modified by a 300-m-deep, 1.1x1.25 km pit crater, probably formed by collapse due to loss of gas bubbles from magma stored in a crustal magma reservoir. The southernmost flows cover about 3,600 km^2 and have a volume of 35-75 km^3. They appear to have erupted from a single 75 km-long fissure near the eastern margin of the flow field. Fluid flows travel as far as 108 km down slopes as low as 1.9 m/km. The main control determining vent structure appears to be the volatile content of the magmas. Those with high volatile contents erupt vigorously and form steep sided cones of vesicular pillow lava and hyaloclastite whereas those with lower volatile contents erupt steadily and form sheet flows. The youngest sheetflows in the North Arch Volcanic Field, located adjacent to the pit crater, are somewhat younger than the previously estimated 0.5-1.6 Ma.

Acknowledgments. The authors thank JAMSTEC, and particularly J. Naka and E. Takahashi, the chief scientists on the *R/V Yokosuka*, Captain H. Tanaka, and Commander Y. Imai of the *Shinkai 6500* Team for their support of our program to map and sample the North Arch Volcanic Field. The entire shipboard science party assisted during and after the dives and their help was appreciated. DAC thanks Jim Moore for helpful comments on the manuscript and for discussions spread over many years about the eruption and emplacement of submarine lava and the formation of submarine vents in general and the North Arch lava flows in particular. K. Cashman, R. Fiske, and S. Takarada, provided constructive reviews that improved the manuscript significantly.

REFERENCES

Baker, E.T., Patterns of event and chronic hydrothermal venting following a magmatic intrusion: new perspectives from the 1996 Gorda Ridge eruption, *Deep-Sea Res. II, 45*, 2599-2618, 1998.

Ballard, R.D., R.T. Holcomb, and T.H. van Andel, The Galapagos Rift at 86°W. 3. Sheet flows, collapse pits, and lava lakes of the rift valley, *J. Geophys. Res., 84*, 5407-5422, 1979.

Batiza, R., P.J. Fox, P.R. Vogt, S.C. Cande, N.R. Grindlay, W.G. Melson, and T. O'Hearn, Morphology, abundance, and chemistry of near-ridge seamounts in the vicinity of the Mid-Atlantic Ridge ~26°S, *J. Geol., 97*, 209-220, 1989.

Batiza, R., and J.D.L. White, Submarine lavas and hyaloclastite, in Encyclopedia of Volcanoes, ed. by H. Sigurdsson, Academic Press, San Diego, 361-381, 2000.

Bridges, N.T., Characteristics of seamounts near Hawaii as viewed by GLORIA, *Mar. Geol., 138*, 273-301, 1997.

Cashman, K.V., H. Pinkerton, and J. Stephenson, Introduction to special section: Long lava flows, *J. Geophy. Res. 103*, 27,281-27,289, 1998.

Cayol, V., J.H. Deterich, A.T. Okamura, and A. Miklius, High magma storage rates before the 1983 eruption of Kilauea, Hawaii, *Science, 288*, 2343-2346, 2000.

Chadwick, W.W., Jr., T.K.P. Gregg, and R.W. Embley, Submarine lineated sheet flows: a unique lava morphology formed on subsiding lava ponds, *Bull. Volcanol., 61*, 194-216, 1999.

Clague, D.A., and G.B. Dalrymple, Age and petrology of alkalic postshield and rejuvenated stage lava from Kauai, Hawaii, *Contrib. Mineral. Petrol., 99*, 202-218, 1988.

Clague, D.A., A.S. Davis, J.L. Bischoff, J.E. Dixon, and R. Guyer, Lava bubble-wall fragments formed by submarine hydrovolcanic explosions on Loihi Seamount and Kilauea Volcano, *Bull. Volcanol., 61*, 437-449, 2000a.

Clague, D.A., A.S. Davis, J.E. Dixon, F.A. Frey, E. Hauri, M.D. Kurz, J. Mahoney, K.-H. Park, M.R. Rhodes, J. Sinton, and H. Staudigel, The distribution of alkalic lavas around Hawaii and the geometry of the Hawaiian plume, *J. Petrol.*, submitted, 2001.

Clague, D.A., and J.E. Dixon, Extrinsic controls on the evolution of Hawaiian ocean island volcanoes, *Geochem. Geophysics, Geosystems, 1*, 1999GC000023, 12 p., 2000.

Clague, D.A., and F.A. Frey, Petrology and trace-element geochemistry of the Honolulu Volcanics, Oahu: Implications for the oceanic mantle beneath Hawaii, *J. Petrol., 23*, 447-504, 1982.

Clague, D.A., and C. Heliker, The ten-year eruption of Kilauea Volcano. *Earthquakes and Volcanoes*, *23*, 244-254, 1993.

Clague, D.A., R.T. Holcomb, J.M. Sinton, R.S. Detrick, and M.E. Torresan, Pliocene and Pleistocene alkalic flood basalts on the seafloor north of the Hawaiian Islands, *Earth Planet. Sci. Lett., 98*, 175-191, 1990.

Clague, D.A., R.T. Holcomb, M.E. Torresan, and S. Ross, Shipboard report for Hawaii GLORIA ground-truth cruise F11-88-HW, 25 Oct.-7 Nov., 1988, *U.S. Geol. Surv. Open-file Report, 89-109*, 33 p., 1989.

Clague, D.A., and J.G. Moore, The headwall of the giant Wailau landslide, Molokai, Hawaii, *J. Volcanol. Geotherm. Res.*, submitted, 2001.

Clague, D.A., J.G. Moore, and J.R. Reynolds, Formation of submarine flat-topped volcanic cones in Hawai'i, *Bull. Volcanol., 62, 214-233*, 2000b.

Clague, D.A., J.R. Reynolds, and A.S. Davis, Near-ridge seamount chains in the northeastern Pacific Ocean, *J. Geophys. Res., 105*, 16,541-16,561, 2000c.

Clague, D.A., K. Uto, K. Satake, A.S. Davis, and J. Naka, Eruption style and flow emplacement in the submarine North Arch Volcanic Field, Hawaii, *EOS, Am. Geophys. Union, 81*, WP256, 2000d.

Dartnell, P., and J.V. Gardner, Sea-floor images and data from multibeam surveys in San Francisco Bay, Southern California, Hawaii, the Gulf of Mexico, and Lake Tahoe, California-Nevada, *U.S. Geol. Surv. Digital Data Series, DDS-55*, 1999.

Davis, A.S., and D.A. Clague, Changes in the hydrothermal system at Loihi Seamount after the formation of Pele's Pit in 1996, *Geology, 26*, 399-402, 1998.

Detrick, R.S., J. Sinton, and D.A. Clague, Volcanism on the Hawaiian flexural Arch: Results from SeaBeam, seismic reflection surveying, and dredging., *EOS, Trans. Am. Geophys. Union, 69*, 1444, 1988.

Dixon, J.E., D.A. Clague, P. Wallace, and R. Poreda, Volatiles in alkalic basalts from the North Arch Volcanic Field, Hawaii: Extensive degassing of deep submarine-erupted alkalic series lavas, *J. Petrol., 38*, 911-939, 1997.

Frey, F.A., D.A. Clague, J.J. Mahoney, and J.M. Sinton, Volcanism at the edge of the Hawaiian Plume: petrogenesis of submarine alkalic lavas from the North Arch Volcanic Field, *J. Petrol., 41*, 667-691, 2000.

Fink, J.H., and R.W. Griffiths, Radial spreading of viscous-gravity currents with solidifying crust. *J. Fluid Mech., 221*, 485-509, 1990.

Fornari, D.J., W.B.F. Ryan, and P.J. Fox, The evolution of craters and calderas on young seamounts: insights from Sea Mark I and Sea Beam sonar surveys of a small seamount group near the axis of the East Pacific Rise at ~10°N, *J. Geophys. Res., 89*, 11069-11083, 1984.

Gregg, T.K.P., and J.H. Fink, Quantification of submarine lava-flow morphology through analog experiments, *Geology, 23*, 73-76, 1995.

Gregg, T.K.P., and D.J. Fornari, Long submarine lava flows: Observations and results of numerical modeling, *J. Geophys. Res., 103*, 27,517-27,531, 1998.

Gregg, T.K.P., D.J. Fornari , M.R. Perfit, R.M. Haymon, and J.H. Fink, Rapid emplacement of a mid-ocean ridge lava flow on the East Pacific Rise at 9° 46'-51' N, *Earth Planet. Sci. Lett., 144*, E1-E7, 1996.

Gregg, T.K.P., D.J. Fornari , M.R. Perfit, W.I. Ridley, and M.D. Kurz, Using submarine lava pillars to record mid-ocean ridge eruption dynamics, *Earth Planet. Sci. Lett., 178*, 195-214, 2000.

Griffiths, R.W., and J.H. Fink, Solidification and morphology of submarine lavas: a dependence on extrusion rate, *J. Geophys. Res., 97*, 19729-19737, 1992.

Griffiths, R.W., and J.H. Fink, Effects of surface cooling on the spreading of lava domes and flows, *J. Fluid Mech., 252*, 667-702, 1993.

Holcomb, R.T., Eruptive history and long-term behavior of Kilauea Volcano, *U.S. Geol. Surv. Prof. Paper, 1350*, 261-350, 1987.

Holcomb, R.T., J.G. Moore, P.W. Lipman, and R.H. Belderson, Voluminous submarine lava flows from Hawaiian volcanoes, *Geology, 16*, 400-404, 1988a.

Holcomb. R.T., M. Holmes, R.P. Denlinger, R.C. Searle, and W.R. Normark, Submarine Hawaiian North Arch volcanic field, *EOS, Trans. Am. Geophys. Union, 69*, 1445, 1988b.

Hon, K., J. Kauahikaua, R. Denlinger, and K. Mackay, Emplacement and inflation of pahoehoe sheet flows: Observations and measurements of active lava flows on Kilauea Volcano, Hawaii, *Geol. Soc. Am. Bull., 106*, 351-370, 1994.

Jaggar, T.A., Origin and development of craters, *Geol. Soc. Am. Memoir, 21*, 1-508, 1947.

Kauahikaua, J., K.V. Cashman, T.N. Mattox, C.C. Heliker, K.A. Hon, M. Mangan, C.R. Thornber, Observations on basaltic lava streams in tubes from Kilauea Volcano, island of Hawaii, *J. Geophys. Res., 103*, 27,303-27,323, 1998.

Kauahikaua, J., K.V. Cashman, D.A. Clague, D. Champion, and J. Hagstrum, Geology of the most recent eruptions of Hualalai Volcano, Hawaii, *Bull. Volcanol.*, submitted, 2001.

Keszthelyi, L., and S. Self, Some physical requirements for the emplacement of long basaltic lava flows, *J. Geophys. Res., 103*, 27,447-27,464, 1998.

Kluegel, A., K.A. Hoernle, H.-U. Schmincke, and J.D.L.White, The chemically zoned 1949 eruption on La Palma (Canary Islands): Petrologic evolution and magma supply dynamics of a rift-zone eruption, *J. Geophys. Res., 105*, 5997-6016, 2000.

Lipman, P.L., and N.G. Banks, Aa flow dynamics, Mauna Loa, 1984, *U.S. Geol. Surv. Prof. Paper, 1350*, 1527-1567, 1987.

Lipman, P.L., D.A. Clague, R.T. Holcomb, and J.G. Moore, Newly identified young lava flows on the sea floor south of the Hawaiian ridge, *Geology, 7*, 611-614, 1989.

Loihi Science Team, Researchers rapidly respond to submarine activity at Loihi volcano, Hawaii, *EOS, Trans. Am. Geophys. Union, 78*, 231-234, 1997.

Lonsdale, P., Laccoliths[?] and small volcanoes on the flank of the East Pacific Rise, *Geology, 11*, 706-709, 1983.

Mattox, T.N., C.C. Heliker, J. Kauahikaua, and K. Hon, Development of the 1990 Kalapana flow field, Kilauea Volcano, *Bull Volcanol., 55*, 407-413, 1993.

Mattox, T.N., and M.T. Mangan, Littoral hydrovolcanic explosions: a case study of lava-seawater interaction at Kilauea Volcano, *J. Volcanol. Geotherm. Res., 75*, 1-17, 1996.

MBARI Mapping Team, MBARI Hawaii multibeam survey, *Monterey Bay Aquarium Research Institute Digital Data Series, No. 2*, 2000.

Menard , H.W., *Marine geology of the Pacific*, McGraw-Hill, New York, N.Y., 271 p., 1964.

Moore, J.G., Mechanism of formation of pillow lava, *Amer. Scientist, 63*, 269-277, 1975.

Moore, J.G., D.A. Clague, R.T. Holcomb, P.W. Lipman, W.R. Normark, and M. Torresan, Prodigious submarine landslides on the Hawaiian Ridge, *J. Geophys. Res., 94*, 16465-17484, 1989.

Moore, J.G., W.R. Normark, and R.T Holcomb, Giant Hawaiian underwater landslides, *Science, 264*, 46-47, 1994.

Moore, J.G., R.L. Phillips, R.W. Grigg, D.W. Peterson, and D.A. Swanson, Flow of lava into the sea, 1969-1971, Kilauea Volcano, Hawaii, *Geol. Soc. Am. Bull., 84*, 537-546, 1973.

Normark, W.R., and G.G. Shor Jr., Seismic reflection study of the shallow structure of the Hawaiian Arch, *J. Geophys. Res., 73*, 6991-6998, 1968.

Normark, W.R., R.T. Holcomb, R.C. Searle, M.L. Somers, and C.E. Gutmacher, Cruise report, Hawaiian GLORIA Legs 3 and 4, F3-88-HW and F4-88-HW, *U.S. Geol. Surv. Open-file report 89-213*, 56 p., 1989.

Okubo, C.H., and S.J. Martel, Pit crater formation on Kilauea Volcano, Hawaii, *J. Volcanol. Geotherm. Res., 86*, 1-18, 1998.

Perfit, M.R., and W.W. Chadwick Jr., Magmatism at mid-ocean ridges: constraints from volcanological and geochemical investigations. *Am. Geophy. Union Monograph, 106*, 59-116, 1998.

Peterson, D.W., R.T. Holcomb, R.I. Tilling, R.L. Christiansen, Development of lava tubes in the light of observations of Mauna Ulu, Kilauea Volcano, Hawaii, *Bull. Volcanol., 56*, 343-360, 1994.

Rappaport, Y., D.F. Naar, C.C. Barton, Z.J. Liu, and R.N. Hey, Morphology and distribution of seamounts surrounding Easter Island, *J. Geophys. Res., 102*, 713-728, 1997.

Rowland, S. K., and G.P.L. Walker, Pahoehoe and aa in Hawaii: Volumetric flow-rate controls the lava structure, *Bull. Volcanol., 52*, 615-628, 1990.

Shaw, H.R., and J.G. Moore, Magmatic heat and the El Niño cycle, *EOS, Trans. Am. Geophys. Union*, 69, 1553 and 1564-1565, 1988.

Smith, D.K., Shape analysis of Pacific Seamounts, *Earth Planet. Sci. Lett., 90*, 457-466, 1988.

Smith, D.K., and J.R. Cann, The role of seamount volcanism in crustal construction at the Mid-Atlantic Ridge (24°-30°N), *J. Geophys. Res., 97*, 1645-1658, 1992.

Smith, D.K., and T.H. Jordan, Seamount statistics in the Pacific Ocean, *J. Geophys. Res., 93*, 2899-2918, 1988.

Spiess, F.N., B.P. Luyendyk, R.L. Larson, W.R. Normark, and J.D. Mudie, Detailed geophysical studies of the northern Hawaiian Arch using a deeply towed instrument package, *Marine Geol., 7*, 501-527, 1969.

Thornber, C.T., Olivine-liquid relations of lava erupted at Kilauea Volcano from 1994-1998: Implications for shallow magmatic processes associated with the ongoing East-Rift-Zone eruption, *Canadian Mineral., 39*, 239-266, 2001.

Uto, K., D.A. Clague, O. Ishizuka, and J. Naka, $^{40}Ar/^{39}Ar$ geochronology of the North Arch Volcanic Field, Hawaii, *EOS Trans. Am. Geophys. Union, 81*, WP256, 2000.

Wallin, B.H., The northern Hawaiian Deep and Arch: Interpretation of geologic history from reflection profiling and echo-character mapping, MS Thesis, Univ. of Hawaii, Honolulu, 133p., 1982.

David A. Clague, Alicé S. Davis, Monterey Bay Aquarium Research Institute, 7700 Sandholdt Road, Moss Landing, CA 95039-9644

Kenji Satake, Kozo Uto, Geological Survey of Japan, 1-1-3 Higashi, Tsukuba, Ibaragi 305-8567 Japan

Emplacement and Inflation Structures of Submarine and Subaerial Pahoehoe Lavas From Hawaii

Susumu Umino[1], Sumie Obata[2], Peter Lipman[3], John R. Smith[4], Tsugio Shibata[5], Jiro Naka[6], and Frank Trusdell[7]

Features of subaerial pahoehoe tumuli from Kilauea and Mauna Loa Volcanoes in Hawaii and subaqueous flow lobes from Loihi Seamount off Hawaii and north of Oahu Island document the controlling factors of flow-lobe formation. Studied subaerial flow-lobe tumuli consist of uplifted pahoehoe crust, formed from coalesced flow lobes. The south rift zone of Loihi has abundant conical lava mounds and terraces with flat tops. Steep flanks (>4°) of these mounds are covered with elongate pillows, but the flat tops are overlain by lobate sheets and inflated pahoehoe flows. The pahoehoe lobes have an elongate dome-like shape 2 × 0.5 m to 20 × 20 m and up to 5 m in height. Hollow pahoehoe lobes, which are rare among subaerial pahoehoe flows on coastal plains in Hawaii Island, are common on Loihi. Similar hollow lobes and a subaqueous tumulus were found on a 2200-m deep terrace offshore of Oahu Island. In contrast to corrugated pillow lobes, these subaqueous flows have smooth surfaces covered with fine streaks of thin glass flakes. We suggest that subaqueous flow lobes grow like a slowly advancing subaerial pahoehoe flow. The plastic crust stretches, thins, and ruptures at the flow front, where new lava is exposed, and subsequently is quenched by water to form thin flaky glass. The Graetz number, giving a dimensionless temperature distribution within lava, varies flow by flow but remains almost constant for subaqueous flow lobes. This means that the lateral extension of a flow lobe is cooling limited, which results in a linear correlation between the rate of lava supply and the volume of lobes. Flow lobes stop advancing due to heat loss to the environment, as brittle crust thickens and hinders the lobes from deforming freely.

[1]Department of Biology and Geosciences, Shizuoka University, Shizuoka-shi, Shizuoka, Japan
[2]Mitsubishi Materials Natural Resources Development Corporation Ltd., Tokyo, Japan
[3]Volcano Hazards Team, U.S. Geological Survey, Menlo Park, California, USA
[4]Hawaii Undersea Research Lab, University of Hawaii,1000 Pope Rd, MSB 303, Honolulu, USA
[5]Department of Earth Sciences, Okayama University, Okayama-shi, Okayama, Japan
[6]Deep Sea Research Department, Japan Marine Science and Technology Center (JAMSTEC), Yokosuka-shi, Kanagawa, Japan
[7]Hawaiian Volcano Observatory, U.S. Geological Survey, Hawaii National Park, Hawaii, USA

Hawaiian Volcanoes: Deep Underwater Perspectives
Geophysical Monograph 128
Copyright 2002 by the American Geophysical Union

1. INTRODUCTION

Pahoehoe lava, flowing as interconnected small lobes at low extrusion rates, sometimes inflates to form structures such as tumuli. Inflation of simple and coalesced com-pound lobes, common in fluidal basaltic flows on gentle slopes, are an important mechanism to form vast sheet-like lava flows in Hawaii [*Hon et al.*, 1994; *Kauahikaua et al.*, 1998; *Mattox et al.*, 1993]. Columbia River Basalts and some sheet flows on mid-ocean ridges, which were once thought to be emplaced as thick copious flows at extremely high discharge rates ($10^6 \sim 10^3$ m^3/s), are now interpreted to form by coalescence and inflation of lava lobes extruded at smaller rates (10^1-10^3 m^3/s) [*Gregg and Chadwick*, 1996; *Gregg et al.*, 1996; *Self et al.*, 1996, 1997; *Thordarson and Self*, 1998]. When inflation is confined to a semicircular area overlying a magma pathway, it forms a domal uplift called a tumulus [*Walker*, 1991]. Less common but ubi-quitous on flat costal plains in Hawaii are flow-lobe tumuli, where the entire upper crust of an individual lobe or coa-lesced lobes has been fractured and uplifted [*Rossi and Gudmundsson*, 1996]. The rate of lava supply and the period of lava extrusion can be estimated by examining such inflation-related structures [*Rossi and Gudmundsson*, 1996; *Self et al.*, 1996, 1997; *Thordarson and Self*, 1998; *Umino et al.*, 2000]. Flow-lobe tumuli are particularly useful for this purpose [*Rossi and Gudmundsson*, 1996]. Subaqueous tumuli have been found from the Juan de Fuca Ridge [*Appelgate and Embley*, 1992], on the northern slope offshore of Oahu Island [*Umino et al.*, 2000], and in the A.D. 864 flow of Fuji Volcano, Japan [*Obata and Umino*, 1999]. They usually appear in fields of subaqueous lobate sheet flows and pahoehoe lavas, which were previously interpreted as flattened pillow lavas that formed at higher extrusion rates than typical pillows [*Ballard and Moore*, 1977; *Griffiths and Fink*, 1992a]. These lobate sheet flows are now thought to be subaqueous equivalents of pahoehoe sheet flows formed by coalescence and inflation of smaller pahoehoe lobes [*Gregg and Chadwick*, 1996]. Hollow lobes and collapse pits are characteristic of lobate sheet flows [*Ballard et al.*, 1979; *van Andel and Ballard*, 1979; *Chadwick et al.*, 1999]. Estimates of lava supply rates based on the inflation structures of subaqueous lavas suggest a systematic difference between subaerial and subaqueous lobes [*Umino et al.*, 2000]. Here, we describe subaerial flow-lobe tumuli from Kilauea and Mauna Loa Volcanoes in Hawaii and subaqueous pahoehoe flow lobes from Loihi Seamount and north of Oahu Island, and discuss the mode of emplacement and factors governing the flow-lobe volumes.

2. EMPLACEMENT AND INFLATION OF FLOW-LOBE TUMULI AND PAHOEHOE LOBES

Tumuli are composed of uplifted pahoehoe crust, formed from coalesced flow lobes, that have many outflows through inflation cracks. Coalesced pahoehoe lobes can be deduced from outlines shown by the hummocks and ropy wrinkles on the surface of tumuli (Figure 1A). Traces of merged crust of flow lobes can be seen on some walls of inflation cracks as curved, poorly vesiculated gray bands [*Hon et al., 1994*; Figure 1B]. Walls of inflation cracks commonly have bands, one to tens of centimeters wide, which differ in color, vesicularity, and roughness (Figure 1C). The first band beneath the glassy surface crust has low vesicularity, a smooth surface, and columnar joints, indicating that this part of the crust fractured brittley. A second lower band is usually more vesiculated, having a rusty, reddish-purple color due to oxidation. This is underlain by highly vesicular, spinous bands formed by pull-apart of plastic crust [*Hon et al.*, 1994]. Shelves of glassy billowy lava are attached to some crack walls, which were squeezed up at the opening tip of the cracks [*Hon et al.*, 1994; *Walker*, 1991]. Such textural variations result from differing temperature-stress responses by the inflating tumuli crust [*Hon et al.*, 1994; *Ryan and Sammis*, 1981], indicating cyclic repetition of brittle fracturing and viscoelastic pull-apart in the opening crack tips [*Obata and Umino*, 1999]. The uppermost crust with columnar joints, fractured brittley, was below the glass transition temperature at which compressibility and thermal expansivity change drastically [*Ryan and Sammis*, 1981]. *Hon et al.* [1994] estimated the temperature to be <800 °C based on the direct measure-ments of temperature variations within cracks on active inflating lobes at Kalapana.

The brittle crust in circumferential cracks of a tumulus is thinner than that exposed in the axial cracks traversing the entire lobe. This suggests that the flow lobe fractured first along the periphery as the upper crust rose. As outflow of lava through the circumferential cracks diminished, the upper crust was buckled up through the axial crack. We observed the same order of fracturing of an active tumulus in March 2000, which was emplaced on the 1999 flow field near Waha'ula Heiau (Figure 1A), although this is not the only way of crack opening [*K. Hon, personal comm.*, 2001]. This succession is the same as for inflation of much larger pahoehoe sheet flows in the Kalapana flow field [*Kauahi-kaua et al.*, 1998].

2.1 Subaerial Flow-lobe Tmuli: Kilauea 1990-91 and Mauna Loa 1843 Lavas of Hawaii

Episode 48 of the on-going eruption of Kilauea, Hawaii, began in July 1986 and effused lava from Kupaianaha vent on the east rift zone for 5.5 years (Figure 2A). In 1990-91 three successive pahoehoe flows attaining 2×10^7 m^3 reached the coastal plain and buried the town of Kalapana [*Mattox et al.*, 1993; *Kauahikaua et al.*, 1998]. As the flow

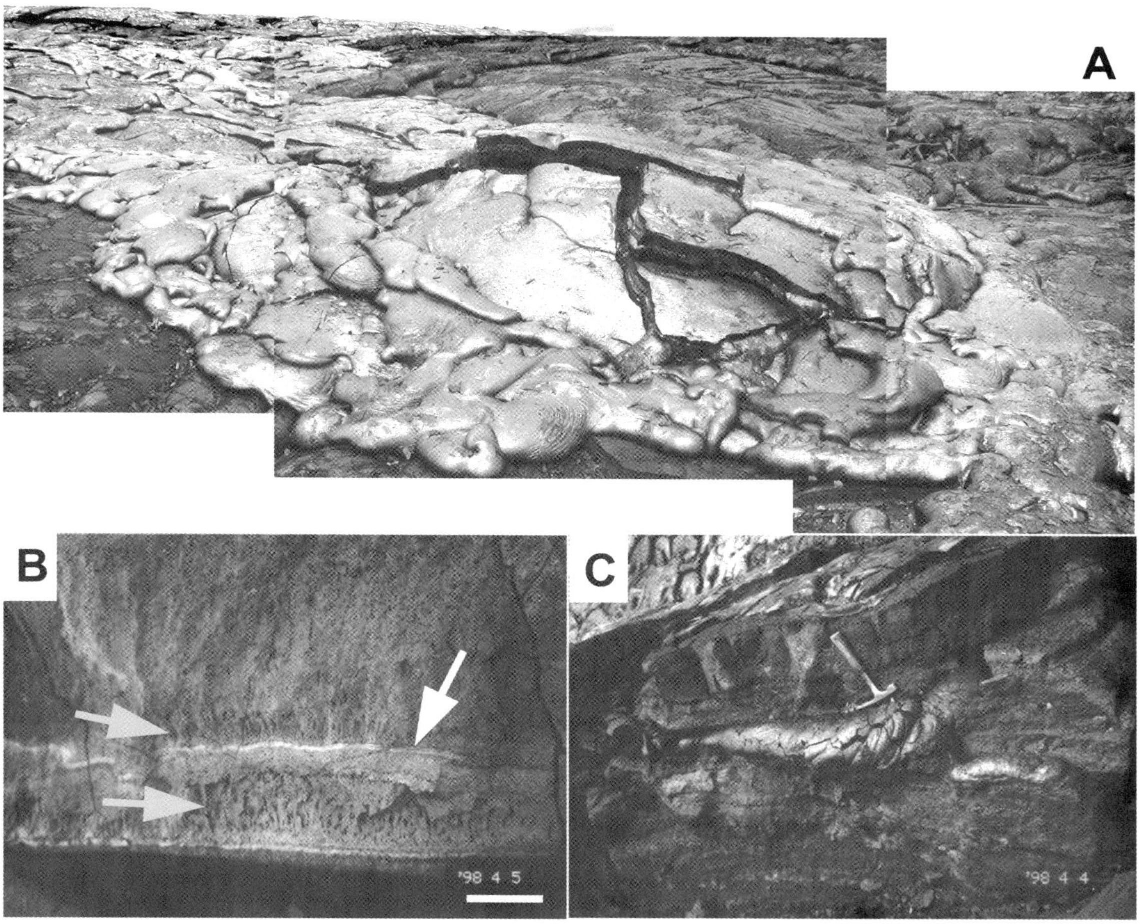

Figure 1. Features of inflated flow-lobe tumuli, southern flank of Kilauea. A: Growing tumulus near Waha'ula Heiau observed in March 2000. Width of the tumulus is 4-5 m. The crust of the tumulus, domed up in the center and tilted outwards, is faulted along the axial cracks and is being uplifted. Outflows of small pahoehoe toes surrounding the tumulus have stopped; they were fed through circumferential cracks opened prior to the axial cracks; B: Wall of inflation crack on a tumulus exposed on the coast cliff. Coalesced skin of two vertically stacked flow lobes is marked by a gray band (white arrow) parallel to the bottom of the lower lobe. Two rows of vertically oriented pipe vesicles (dark arrows) formed near the bottom of the flow lobes. Scale bar on bottom right is 20 cm long; C: Wall of an inflation crack on a tumulus at Kamoamoa, southern coast of Kilauea. The jointed uppermost band (hammer grip) shows the thickness of brittle crust when the crack opened. The hammer head (~20 cm across) rests on a lava shelf, which was squeezed out through the tip of the opening crack.

front widened, the advance rate of the front decreased, and smaller pahoehoe toes broke out along the front. These toes coalesced into a larger sheet through inflation. Surface outbreaks that leaked from cracks on the inflated sheets flowed a few meters to several tens of meters from the breakout points. Some outbreaks formed small flow-lobe tumuli on the master sheet flows late in the eruption. Four flow-lobe tumuli were chosen for morphometric study.

The 1843 flow of Mauna Loa erupted from fissures at 2900-3500 m elevation on the northern flank of the volcano, dividing at the saddle between Mauna Loa and Mauna Kea into two streams [*Barnard*, 1995; Figure 2A]. This lava consists of compound pahoehoe lobes bifurcating down-slope, that locally inflated to form large flow-lobe tumuli west of the saddle, some exceeding 100 m in length. A small flow-lobe tumulus, fed from a larger tumulus and terminating in a small basin, was chosen for the study.

The selected tumuli were mapped on a 1-m rectilinear grid (Table 1, Figure 3) using a measuring tape, wire, clinometer, and hand level. Heights of the tumuli were measured along cross sections spaced 1 m apart and are shown as the difference in height from an arbitrary X-Y origin (0,

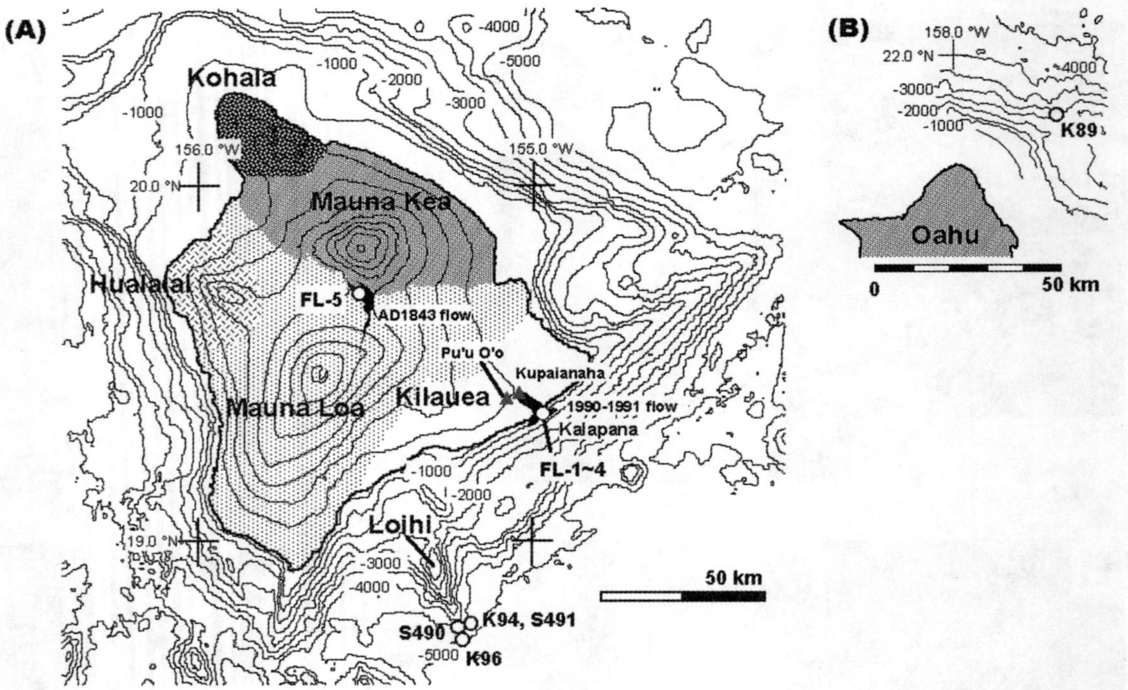

Figure 2. Locations of studied flow lobes (open circles). A: Kalapana 1990-91 flow field (FL-1~4) and Mauna Loa 1843 flow (FL-5) on Hawaii Island; *KAIKO* dive sites (K94, K96) and *SHINKAI* dive sites (S490, S491) on the south rift zone of Loihi Seamount; B: *KAIKO* dive site K89 off Oahu Island.

0) set at the lower left corner in the plan views (Figure 3). Because the Kalapana tumuli (FL-1-FL-4) were emplaced on flat and slightly inclined surfaces of the master sheets, thickness of the tumuli is readily known from the height data, which were used to calculate volumes (Table 1). Because the Mauna Loa tumulus (FL-5) filled a shallow basin between smaller pahoehoe lobes, the true base of the tumulus lies below the surrounding ground level. According, the base of this tumulus is assumed to be flat and below the ground level measured along its periphery. The calculated volume is the minimum, but the under-estimate should not exceed 30 %, judging from the curva-ture of the tops of adjacent flow lobes. Based on these maps, present volumes of the tumuli were calculated (Table 1). Pre-cracking lava volumes were calculated by closing all the axial inflation cracks so as to fit the jigsaw cracks on the brittle outer crust. Outflows of lava through the circum-ferential cracks were included in the pre-cracking volume estimates because most of them are considered to have leaked out before fracturing along axial cracks.

2.2 Subaqueous Lava Flow Lobes: Loihi Seamount and off Northern Oahu Island

Inflated pahoehoe flow lobes were observed on the south rift zone of Loihi Seamount and offshore of northern Oahu Island during submersible dives, by ROV *KAIKO* in 1998 and by *SHINKAI 6500* in 1999, operated by the Japan Marine Science and Technology Center (Figure 2). As a scale, we used parallel laser beams 10-cm apart projected on the object through a window of *SHINKAI 6500*, or scale bars on the *KAIKO* video display calibrated by the mani-pulator arms. The narrow views through video cameras and from the observer's window of *SHINKAI*, and limited mobi-lity of the submersibles allowed measurements of only length, width, height, and thickness of the upper crust of a selected lobe. When *KAIKO* or *SHINKAI* stoped for sam-pling and moved around a targeted lobe, views of the lobe from different angles enabled us to make measurements. In worst case, the moving submersible provided views of a targeted lobe that were considered adequate for our mea-surements. Because crustal thickness is critical in esti-mating the lava-supply rates, most thicknesses were also measured directly on collected samples.

Flow-lobe volumes were approximated by using a hemi-ellipsoidal geometric model (axial lengths l and width w, height h, and volume = $\pi lwh/6$) (Figure 4E), because most flow lobes have a flattened, elongate dome-like shape. The only exception is lobe 6K491-Pa-1, which is best approx-imated by a hemicylinder, for which volume is approxi-mated by $\pi lwh/4$ (Table 1). The hemiellipsoidal model is the minimum volume: many lobes are more bulbous than ideal hemiellipsoids but more flattened than hemicylinders. The hemiellipsoid approximation may underestimate lobe volumes, but by no more than 23%. Calculated volumes for selected lobes are 0.4-1050 m^3 (Table 1).

The dive offshore of Oahu Island (K89) was on the northern slope of a terrace 2200 mbsl (meters below sea level), which is most likely a landslide block from Koolau volcano [*Yokose et al.*, this volume] (Figure 2B). Of the studied slope from 2813 m to 2540 mbsl, the lower 0.5-km of the dive track has an average slope of 20°, and many outcrops of volcanic and tuff breccia are interbedded with smooth sheet and pillow flows. The upper 0.8 km is gentler (7°) and mostly covered with soft sediments, but the last 250 m exposes pahoehoe flow lobes among pillow lava. A tumulus found at the lower end of the pillow field at 2570 mbsl (Figure 2B; *Umino et al.* [2000]) is 10 m across and 2 m high (Table 1). Expansion cracks are 1 m wide at the upper end, filled with outflows of lava lobes. A hollow lobe nearby is 6 by 4.5 m across and 0.7 m high, with crust <15 cm thick.

The lower south rift zone of Loihi Seamount is fan-shaped with an apex at 2900 mbsl, broadening downslope to 4700-4900 mbsl (Figure 5). This broad rift zone has abundant lava terraces and lava mounds, with flat tops as wide as 1 km across. Steep flanks of the mounds are covered with elongate pillows, but the flat tops contain lobate sheets and pahoehoe lobes (Figure 4). We saw no tumuli similar to that north of Oahu. Some lobes have hollow interiors, thin roof crusts, and collapse pits (Figures 4B, 4F-H). Most flow lobes appear to be interconnected, and some merge upstream into a single wider sheet-like lobe (Figure. 4B). Collapse pits are common on such lobate sheet flows (dive K96). However, lobate sheet flows on the summit of a volcanic mound (dive K94) lacks collapse pits. Lava morphology gradually changes with increasing slope: from flat pahoehoe-like lobes, through narrow flat sheets of lava, to elongate pillows (> 4°). Preliminary estimates of supply rates of lava for pillow lobes (10^{-3}-10^{-1} m^3/min) are similar to the lower values of the lobate pahoehoe, suggesting that the underlying slope mainly controls the lava morphology [*Umino et al.*, 2000]. Elongate pillows preferentially form on steeper slopes, rather than pahoehoe flows; this relation is contrary to analogue experiments using polyethylene glycol wax by *Gregg and Fink* [2000].

Pahoehoe flow lobes are several decimeters to 20 m long and 0.3-2 m high (Table 1). Some have nearly equant cross sections, but most lobes are flattened like buns with a height/width ratio of 0.25-0.65 (averaging 0.53). Lobes pinch and swell, sometimes bifurcating into several lobes downstream (Figure 4A-D). Inflation cracks are irregular but ubiquitous, and displacement of the crust is small (< 10 cm) compared to the lobe height. This suggests that the flow lobes inflated when their crust was still plastic, or that they were emplaced as bulbous lobes at their current thickness. As on subaerial tumuli, the uppermost band on the inflation-crack walls has a smooth, polygonally jointed surface, which grades with depth into rougher bands.

Some pahoehoe lobes are hollow. They are similar to solid lobes in shapes, sizes, and presence of irregular cracks with small displacement. Lava drained out in the early stages of hollow-lobe formation, as shown by the thinness of remaining crust (5-20 cm) compared to lobe height (0.3-5 m) (Table 1; Figure 4F-H). Ceilings of the roof crust have glassy surfaces covered with lava curtains and small knobs (stalactites) up to 1 cm across, which indicates that the interiors were once filled with molten lava (Figure 4H). Lava of the hollow lobes is poorly vesiculated because of the large hydrostatic pressure, while the common shelly pahoehoe in proximal parts of subaerial flows are inflated due to high vesicularity and are emplaced as a foam [*Swanson*, 1973]. Hollow lobes are also present in subaerial tumuli on Hawaii Island [*Walker*, 1991], but are less common and have thicker roof crusts than underwater.

The solid and hollow lobes differ from tumuli in that they lack large open cracks, and from pillows in their lower height/width ratios and in lacking the streaks and fractures (i.e., corrugations and fault slivers of *Ballard and Moore* [1977]) formed during growth of pillow lobes. The height/width ratios of subaqueous pahoehoe lobes have an identical range to subaerial pahoehoe lobes (0.3-1.0) [*Wilmoth and Walker*, 1993].

2.3 Surface Textures of Flow Lobes

Pillow and pahoehoe flow lobes sampled during dives along the Loihi south rift zone are poorly to moderately vesiculated and have prominent surface structures (Figure 6). A sample of pahoehoe lobe crust (6S491-2; 4291 mbsl) has a well preserved original surface: the 4-mm thick glass rind has thin flakes and needle-like streaks subparallel to the direction of lobe growth. Streaks are up to 2 cm long and unevenly spaced every 1-5 mm (Figure 6B). The streaks consist of elongate flakes of glass standing 0.5-1 mm high and thinning toward the edges to <0.01 mm (Figures 6C, 6D). Some are oblique to others, but do not appear to interpenetrate. Some flakes are intensely folded, and others have curved or planar surfaces with sharp edges. Many olivine phenocrysts and sporadic spherical to elongated vesicles 0.2-0.5 mm in diameter are present within and between the flakes. Similar textures are common on surfaces of other sheet flows and pillow lobes. Crust of an inflated pahoehoe lobe (sample K94-2, 3970 mbsl) has a glassy selvage 10 mm thick, the upper 7 mm is flaky glass

Table 1. Subaerial and subaqueous flow lobes from Hawaii. Numbers in parenthases are those excluding outflows from pre-cracking volume estimates for subaerial tumuli.

					Pre-cracking			Observed			
		I.D.	Field	Morphology	Crust thickness* [m]	Height [m]	Volume [m³]	Maximum length [m]	Maximum width [m]	Maximum height [m]	Total volume [m³]
Subaerial Hawaii											
		Lobe-1	Kalapana 1990-91	Flow-lobe tumulus	0.07	0.20	5.81	6.50	5.50	1.28	17.35
		Lobe-2	Kalapana 1990-91	Flow-lobe tumulus	0.04	0.15	0.88 (0.67)	2.50	2.10	0.61	1.12
		Lobe-3	Kalapana 1990-91	Flow-lobe tumulus	0.06	0.21	31.41	21.00	8.00	1.20	82.34
		Lobe-4	Kalapana 1990-91	Flow-lobe tumulus	0.07	0.20	5.59 (4.18)	8.00	5.00	1.30	16.26
		Lobe-5	Mauna Loa 1843	Flow-lobe tumulus	0.03	0.50	11.49	6.40	4.30	1.47	21.45
Loihi Seamount and northern offshore of Oahu											
Dive#	time										
K89	13:56	Pa-1	North Oahu	Tumulus	0.20	-	-	10.00	10.00	2.00	104.72
K89	14:19	Pa-2	North Oahu	Hollow lobe	0.15	-	-	6.00	4.50	0.70	9.90
6K491	12:56	Pa-1	Loihi south rift	Hollow lobe	0.03	-	-	3.50	0.60	0.30	0.49
K94	12:42	Pa-1	Loihi south rift	Hollow lobe	0.15	-	-	20.00	20.00	5.00	1047.20
K94	12:53	Pa-2	Loihi south rift	Hollow lobe	0.04	-	-	4.40	0.80	0.40	0.74
K96	13:57	Pa-1	Loihi south rift	Inflated lobe	0.10	-	-	6.00	4.00	2.00	25.13
K96	13:58	Pa-2	Loihi south rift	Inflated lobe	0.10	-	-	3.40	2.60	1.00	4.63
K96	13:54	Pa-3-2	Loihi south rift	Inflated lobe	0.05	-	-	2.10	0.54	0.67	0.40
K96	13:58	Pa-4	Loihi south rift	Inflated lobe	0.08	-	-	3.14	2.78	0.98	4.48
K96	13:58	Pa-5	Loihi south rift	Inflated lobe	0.07	-	-	3.50	1.83	0.84	2.82

* Thickness of brittle crust for tumuli and inflated lobes, and that of the roof crust for hollow lobes.

** Time for surface crust to cool to glass transition temperature (Tg) for subaerial tumuli, and that to cool to solidus and Tg for hollow and inflated lobes, respectively.

*** Emplacement time is calculated as the observed lobe volume divided by the supply rate for subaerial tumuli, and taken to be equal to the solidification time of crust for submarine lobes

Table 1. (continued)

Time for crust to solidify** [min]	Supply rate [m^3/min]		Emplacement time*** [min]		1 / h^2 [m^{-2}]	F / V [min^{-1}]		Graetz number	
222.39	0.03		663.6		25.00	0.004		6.0	
86.89	0.01	(0.01)	110.8	(144.8)	44.44	0.012	(0.012)	8.6	(8.6)
181.66	0.17		476.3		22.68	0.006		8.1	
196.60	0.03	(0.02)	572.5	(764.9)	25.00	0.005	(0.005)	6.8	(6.8)
52.16	0.22		97.4		4.00	0.019		159.8	
						Average		37.9	
						(Excluding 159.8)		7.4	
845.01	0.12		845.0		0.25	0.001		157.8	
130.79	0.08		130.8		2.04	0.008		124.9	
3.71	0.13		3.7		11.11	0.270		809.5	
130.79	8.01		130.8		0.04	0.008		6371.6	
9.40	0.08		9.4		6.25	0.106		567.2	
212.10	0.12		212.1		0.25	0.005		628.6	
212.10	0.02		212.1		1.00	0.005		157.2	
53.45	0.01		53.4		2.23	0.019		280.0	
136.01	0.03		136.0		1.04	0.007		235.4	
104.28	0.03		104.3		1.42	0.010		225.5	
						Average Loihi		1159.4	
						(Excluding 6371.6)		414.8	

and the lower 3 mm is massive. Flakes are 5-10 mm long and up to 0.1 mm thick. The flakes are flat lying and are stacked subparallel to the surface like scales. The lower part of the flaky layer is sheared with apparent spacing of 0.2-0.3 mm between shear planes. Both open and trapped vesicles, 0.2-1 mm in diameter, are present in and between the flakes of glass.

Some fresh surfaces of subaerial pahoehoe flow lobes have similar structures (Figures 6E, 6F), but the surface glass forms highly stretched fibers up to 10 cm long, rather than short flakes. Their presence depends on lava vesicularity and shear or flow rates affecting the viscoelastic skin [*Hon et al.*, 1994]. *Wilmoth and Walker* [1993] distinguished two types of pahoehoe lobes: P-type is moderately vesiculated and characterized by the presence of pipe vesicles near the base; S-type is highly vesiculated, showing sponge-like textures [*Walker*, 1989]. P-type flow lobes have a silver-gray smooth glassy skin, without glass fibers or flakes. In contrast, more vesiculated S-type flow lobes have yellow-gold to brass luster like Pele's hair, inherited from thin fibers of vesicular glass covering the surface (Figure 6E). Vesicles in the surface skin are elongated (1:5 to 1:20) with intervening walls stretched to fibrous glass. Glass fibers are flat lying and form a relatively smooth surface. In contrast, highly deformed surfaces of ropy and shelly pahoehoe are rough and spinous (Figure 6F). The surface is covered with elongated flakes of glass as well as fibers, resulting from a large strain acted on the viscoelastic skin through twisting and folding of the surface crust.

3. LOBE EMPLACEMENT AND LAVA SUPPLY RATES

Growth time of a subaqueous flow lobe can be deduced from the thickness of roof crust for a hollow lobe and the polygonally jointed crust in inflation cracks. Because the ceilings of hollow lobes are draped with quenched lava dribbles, they were near the solidus temperature when the inner molten lava drained out. Therefore, growth time for hollow lobes can be approximated by the time for the roof crust to cool below the solidus. The lobe volume divided by the growth time gives the lava supply rate for hollow lobes. As displacements along subaqueous inflation cracks on non-hollow, solid lobes are small, the present lobe volume is almost equal to the volume before the cracks opened.

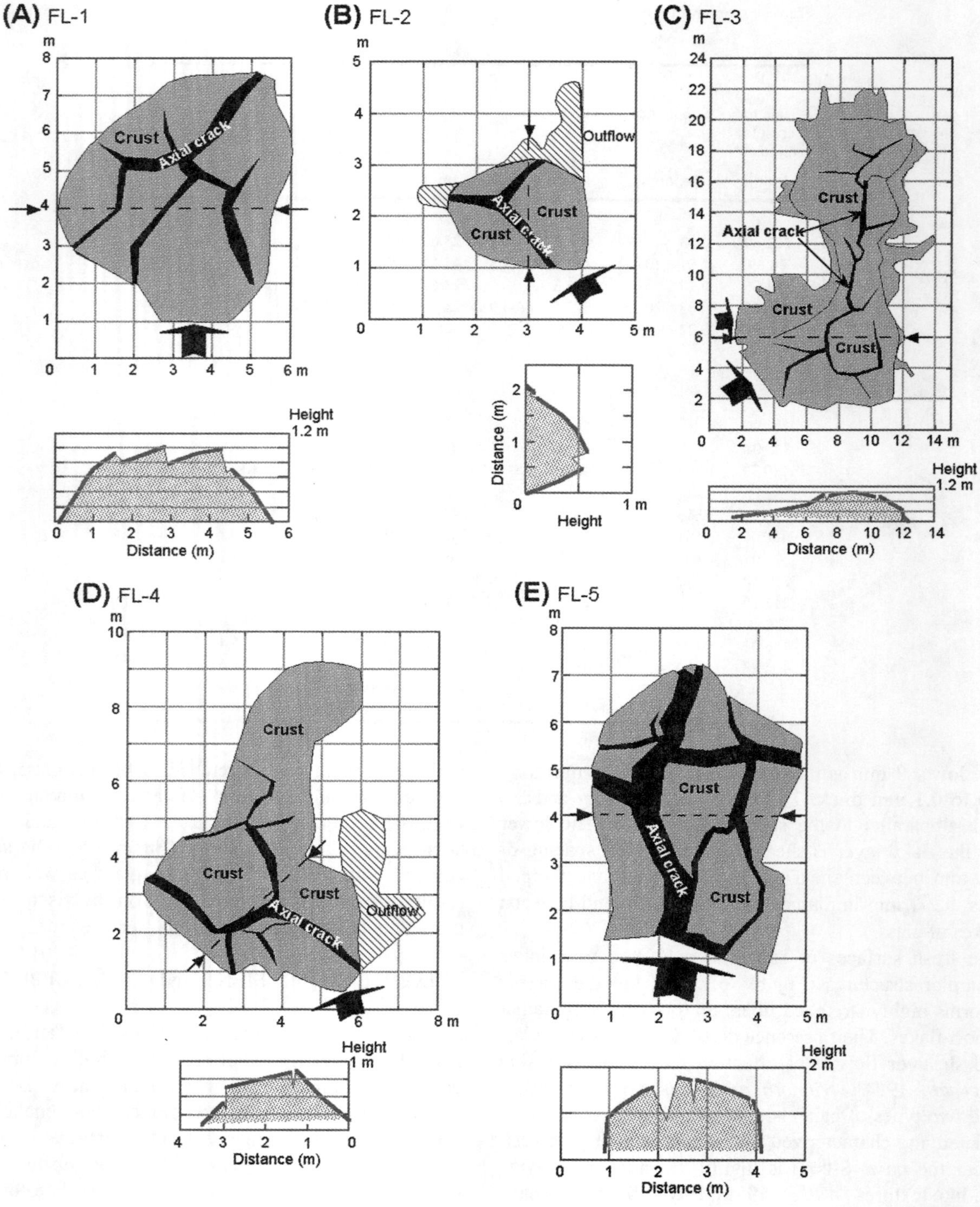

Figure 3. Maps and cross sections of subaerial tumuli from Kalapana flow field (FL-1~4) and Mauna Loa 1843 flow (FL-5). In plan views, gray and dark areas represent crust and inflation cracks. Outflows from circumferential cracks are shown by crosshatched pattern. Broad arrows denote conduits of lava supply. Locations of cross sections are indicated by thin arrows and dashed lines. In cross sections, thick lines represent crust of the lobes; vertical scale is approximately double horizontal in FL-1~4.

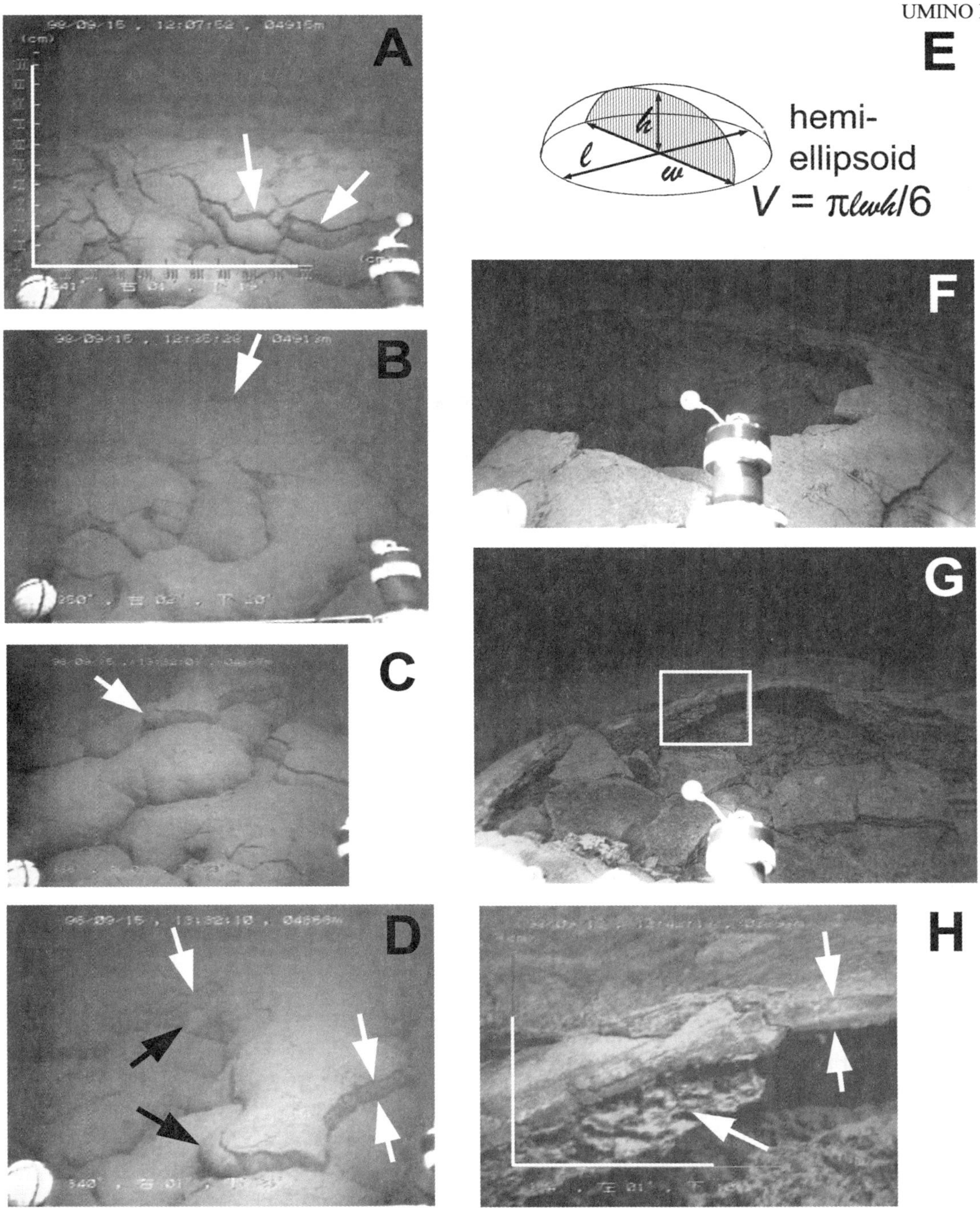

Figure 4. Inflated flow lobes at dive sites K96 (A-D) and K94 (F-H). A: Inflated flow lobes at 4915 mbsl. White arrows indicate inflation cracks. White bars are ~10 m long; B: Bulbous flow lobes merging upstream with a larger hollow lobe on top (white arrow indicates collapsed roof) at 4919 mbsl; C: Crust of the bulbous flow lobe in the center is buckled on the top, budding a small new flow lobe (white arrow) as an outflow through the inflation crack; D: Same lobe as in C, showing the cracks (white arrows) and new lobes (black arrows) coming out through the cracks; E: Flow-lobe volume is approximated by a hemiellipsoid with length l, width w and height h, giving a volume of $\pi lwh/6$; F: Collapsed hollow lobe (K94 Pa-1) at 3896 mbsl. Thin, dome-like roof has collapsed to the floor; G: Same lobe as in F. Floor of the flow lobe is covered with broken-up crust of the collapsed roof. Area of white rectangle is shown in H; H: The interior of the roof crust is coated with irregular curtains of lava, that dripped from the ceiling as lava drained. White arrows indicate the upper and lower surface of the roof crust. Scale bars are 1 m long.

These cracks opened brittley near the end of the lobe growth after the solid crust developed. The polygonally jointed band exposed at tops of the inflation cracks, underlain by rugged bands, gives the thickness of brittle crust when the lobe fractured. The time for this thickness of the crust to cool below the glass transition temperature is the growth time for the lobe. The supply rate of lava is given by the present lobe volume divided by the growth time.

Cooling of lava under was numerically simulated following the procedures of *Griffiths and Fink* [1992a, b], as described in *Umino et al.* [2000]:

The variation in temperature (T) due to heat conduction within lava is:

$$\delta T/\delta t = \kappa \delta^2 T/\delta x^2 \qquad (1)$$

where κ is thermal diffusivity of lava, t is time, and x is the distance from the lava surface. At the surface of lava, the heat flux Fc by conduction through lava is:

$$Fc = \lambda \delta Tc/\delta x \qquad (2)$$

where λ is thermal conductivity of lava. Fc is equal to the heat flux Fa by convection of water and radiation:

$$Fa = J(Tc - Ta)^{4/3} + \varepsilon\sigma(Tc - Ta)^4 \qquad (3)$$

where $J = \rho_a c_a \gamma\ (g\alpha_a \kappa_a^2/\nu_a)^{1/3}$, Tc and Ta are the surface-lava and ambient-water temperatures, and ε and σ are lava emissivity and Stefan-Boltzmman constant, respectively, ρ_a, c_a, α_a, κ_a, ν_a are density, specific heat, expansion coefficient, thermal diffusivity, and kinematic viscosity of seawater, respectively; g is acceleration of gravity, and γ is a semiempirical constant equal to 0.1 [*Griffiths and Fink*, 1992a, 1992b]. The physical parameters used in these equations are listed in Table 2. The lava samples from north Oahu and Loihi south rift are picritic tholeiite to transitional basalt, for which extrusion and solidus tem-peratures are assumed to be 1150 °C and 1050 °C, respec-tively [*Barth et al.*, 1994; *Montierth et al.*, 1995]. Ignoring latent heat of crystallization may underestimate the solidifi-cation time for thick crust with high crystallinity by as much as 35% [*Umino et al.*, 2000]. However, the hemi-ellipsoidal approximation for a flow lobe may also under-estimate the lobe volume by as much as 23%, which tend to offset effects of the latent heat of crystallization.

Banding on inflation-crack walls in subaerial sheet flows and tumuli results from differing temperature-stress res-ponses of stretching crust at the tip of an opening crack. Formation of such banding is mechanically identical to that seen on crack walls in growing lava domes ("crease struc-ture"; *Anderson and Fink* [1992]). The band width varies from band to band and also dome to dome. This indicates that crack growth cycled between brittle fracturing, visco-elastic stretching, and short repose periods, in response to stress accumulation at the crack tip. The depth-time curve for an opening crack is unique to the crack and the host lava dome [*Anderson and Fink*, 1992]. Similar stepwise opening has been demonstrated for inflation cracks on subaqueous tumuli from Fuji volcano, Japan [*Obata and Umino*, 1999]. Stepwise opening must also operate in Hawaii subaerial sheet flows and tumuli, as indicated by cyclically repeated smooth and spinous bands on crack walls [*Hon et al.*, 1994].

Although *Hon et al.* [1994] presented an empirical equation of growing crack depth with time based on direct measurements of an inflating sheet flow, such an empirical equation is only valid for the crack on which it was derived. Furthermore, a large thermal gradient ahead of the propa-gating tip of an opening crack lowers the solidifica-

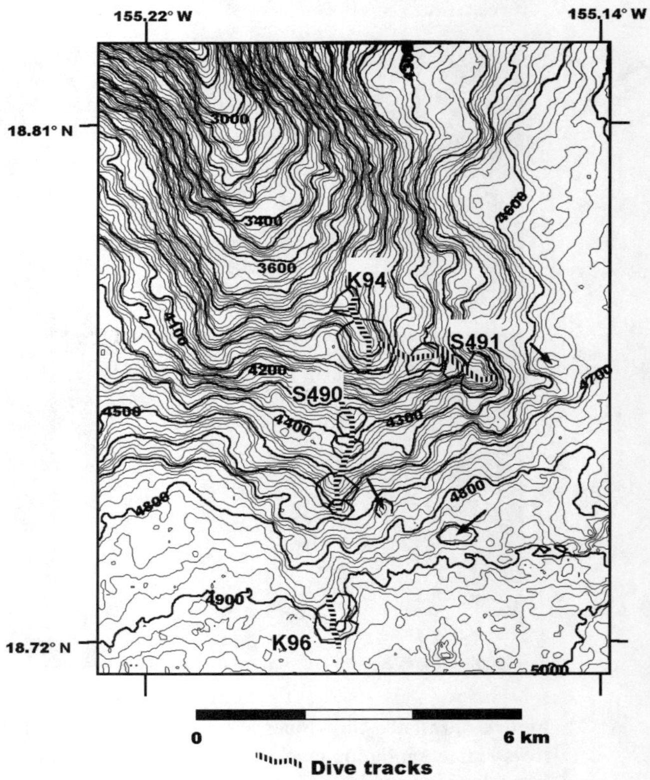

Figure 5. Lower south rift zone of Loihi Seamount. Gray lines encircle small terraces and conical landforms identified from bathymetric and slope maps. Arrows indicate depressions. Dive tracks are shown by hatchured lines. Bathymetric contours are 20 m, based on data collected by the SeaBeam 2112 system.

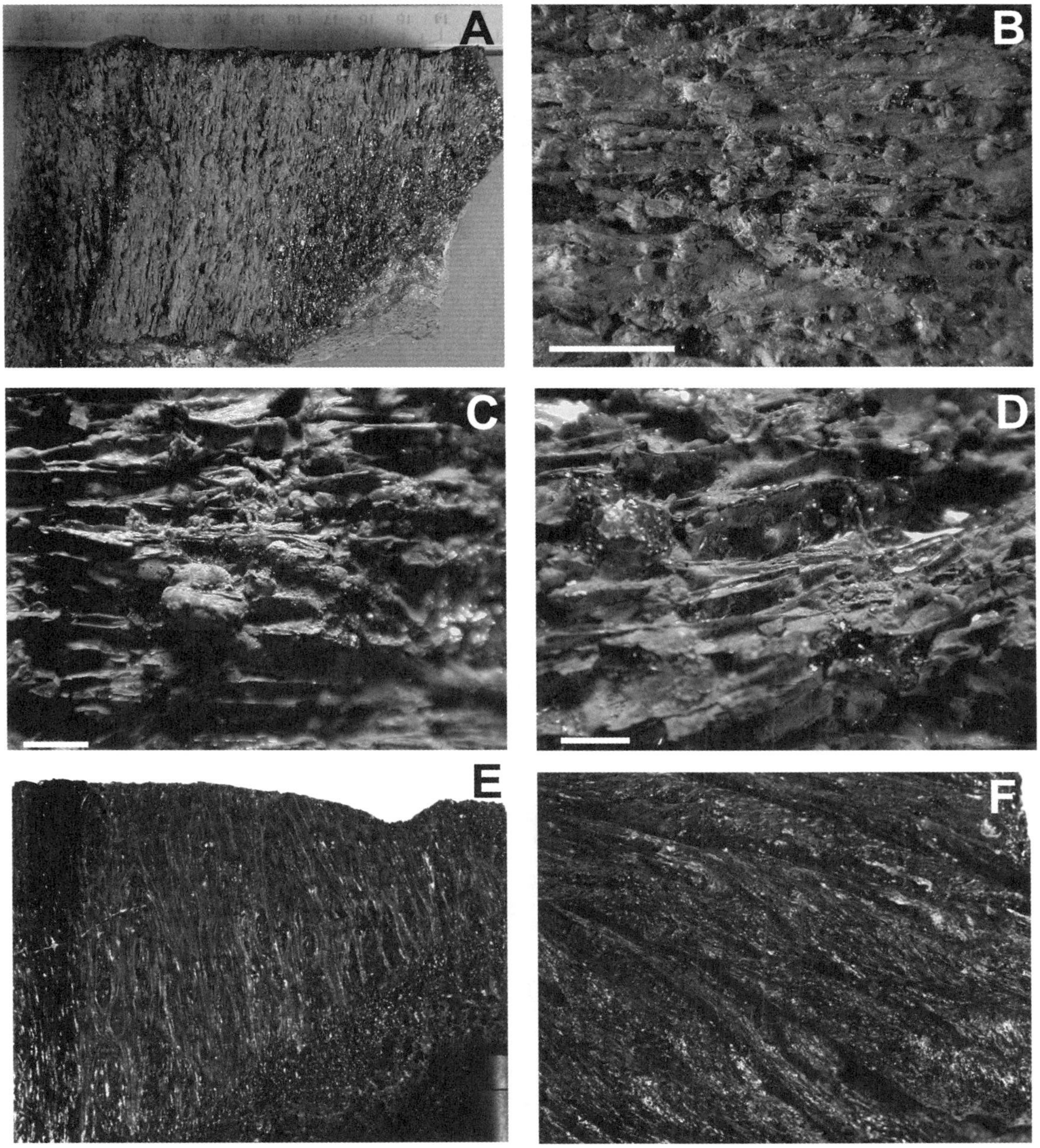

Figure 6. Surface textures on roof crust of a hollow lobe, south rift zone of Loihi Seamount (sample 6S491-2: A-D), and shelly pahoehoe at the foot of Pu'u O'o vent, Kilauea (E, F). A: Streaks on the surface of the flow lobe. Brownish color is from palagonitized glass. Width of the sample is 13 cm; B: Close-up of A, showing streaks of altered glass, knobby olivine crystals, and tiny bubbles; scale bar is 2 cm; C: Close-up of B showing sharp edges of the streaks consisting of thin plates of glass. Scale is 2 mm; D: Oblique view of C; scale is 2 mm; E: Smooth surface of shelly pahoehoe lava; width of sample is 6 cm; F: Folded surface of shelly pahoehoe lava; width of sample is 7 cm.

tion front below that away from the crack [*Obata and Umino*, 1999]. This effect is more pronounced for small lobes due to higher areal proportions of crack walls to pre-cracking surfaces than for flat sheet flows. Therefore, we modified the method of *Rossi and Gudmundsson* [1996] for esti-mating growth time of subaerial tumuli, using the thickness of brittle crust as an indication of the start of inflation of tumuli. We adopted the empirical equation on the

Table 2. Symbols and values used in the computation of cooling time of subaqueous lava after *Griffiths and Fink* (1992a, 1992b). The temperatures at the base of crust and glass transition are assumed to be 1323 K (1050 °C) and 1003 K (730 °C), respectively for tholeiitic basalt lobes observed on KAIKO dives at Loihi south rift (*Barth et al.*, 1994; *Montierth et al.*, 1995).

γ	constant	0.1	-
g	acceleration of gravity	9.8	$m^2 \cdot s^{-1}$
σ	Stefan-Boltzmman constant	5.67×10^{-8}	$kg \cdot s^{-3} \cdot K^{-4}$
ε	emissivity	0.9	-
Physical properties and symbols of basalt lava			
ρ_m	density	2600	$kg \cdot m^{-3}$
c_m	specific heat	1200	$m^2 \cdot s^{-2} \cdot K^{-1}$
κ_m	thermal diffusivity	5×10^{-7}	$m^2 \cdot s^{-1}$
λ_m	thermal conductivity	1.56	$m \cdot kg \cdot s^{-3} \cdot K^{-1}$
$T_{initial}$	initial temperature	1423	K
	solidus temperature	1323	K
Tg	glass transition temperature	1003	K
Tc	temperature at the surface	-	K
Physical properties of water at 1000-4000 m deep			
ρ_a	density	1000	$kg \cdot m^{-3}$
c_a	specific heat	4000	$m^2 \cdot s^{-2} \cdot K^{-1}$
κ_a	thermal diffusivity	1.0×10^{-7}	$m^2 \cdot s^{-1}$
ν_a	kinematic viscosity	1.0×10^{-6}	$m^2 \cdot s^{-1}$
α_a	expansion coefficient	1.0×10^{-2}	K^{-1}

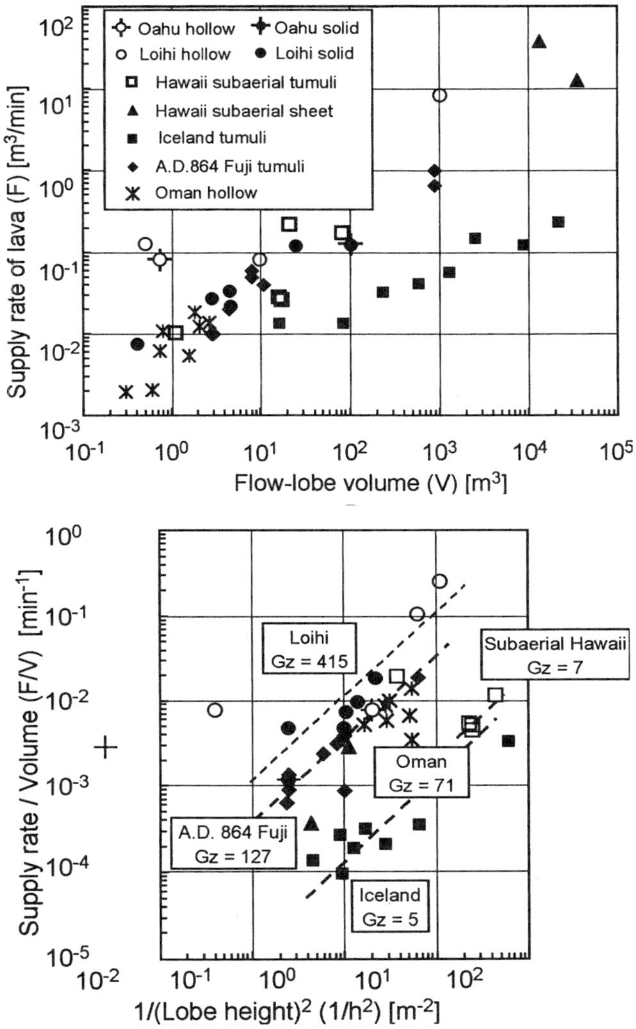

Figure 7. Relations among lobe height, lobe volume, and supply rate for pahoehoe lava lobes emplaced under subaerial and subaqueous conditions. A: Supply rate of lava versus volume of flow lobe. Data sources: subaerial tumuli from Iceland [*Rossi and Gudmundsson*, 1996] and Hawaii [this study], Hawaii pahoehoe sheet flows [*Hon et al.*, 1994], subaqueous tumuli of Fuji A.D. 864 flow [*Obata and Umino*, 1999; *Obata, unpublished data*], submarine flow lobes from Loihi Seamount and north off Oahu Island [this study], and submarine hollow lobes from the Oman Ophiolite [*Umino, unpublished data*]; B: Supply rate/volume versus 1/ (lobe height)^2 for flow lobes. Symbols are as in A.

800 °C isotherm depth-time relations by *Hon et al.* [1994] (equ-ation 7) with a correction for the short time range (<5 minutes) using the data of *Keszthelyi and Denlinger* [1996]. Brittle inflation along axial cracks is considered to be in the last stage during growth of a studied tumuli. The poly-gonally jointed uppermost band on the axial cracks (Figure 3) is underlain by a pull-apart band, which is continuous throughout the crack. This indicates that the initial cracking was immediately followed by opening of the entire crack as the lobe inflated. The jointed crust thickness (3-7 cm; Table 1) is uniform along the axial cracks on individual tumuli. This gives a minimum estimate for the emplacement of the tumuli before the axial cracks opened. The initial lobe volume, including the outflows through the circumferential cracks, was divided by the growth time.

As demonstrated by *Rossi and Gudmundsson* [1996] and *Umino et al.* [2000], supply rates correlate systematically with the flow-lobe volumes (Figure 7A). Hawaiian sub-aerial tumuli and pahoehoe sheet flows plot on the same trend as the subaqueous lobes. However, coalescing of flow lobes are much more common and complete on land than under water. In contrast, Icelandic subaerial tumuli [*Rossi and Gudmundsson*, 1996] have lava supply rates 1-2 orders of magnitude lower than those from Hawaii.

4. DISCUSSION

4.1 Growth Mechanism of Subaqueous Pahoehoe Lobes

Videos of growing pillows show that some pillow lobes ruptured near the lobe front continue to extrude new lava, but the site of rupturing is unstable and may be left behind as the lobe grows downslope [*Kjargaard*, 1996; *Moore*, 1975]. In the latter case, pillows tend to have subparallel surface corrugations, a few cm apart and up to 1 cm high, that are elongate in the growth direction of the lobes [*Ballard and Moore*, 1977]. As a pillow lobe grows, the solid crust of the lobe ruptures, and molten lava extrudes from beneath the crust, whose jagged edges rake the lava to make surface corrugations [*Ballard and Moore*, 1977; *Moore*, 1975].

In contrast, subaqueous pahoehoe lobes on gentle slopes and flat areas lack corrugations and have smooth surfaces except for the tiny glassy streaks described above. *Chadwick et al.* [1999] have proposed that "lineated sheet flows" on mid-ocean ridges form by grooving the surface of lava that is flowing from beneath solid crust, and the "lineation" diminishes as flow rate increases. This suggests that the presence of rigid crust and slow extrusion promotes formation of corrugated lava surfaces, and that pahoehoe lobes grow in a different manner from pillows and lineated sheets. A new flow lobe bud on some inflated lobes, where lava leaks from beneath the buckled crust of an older lobe to form a bulbous toe (Figures 4C, 4D). No corrugations are present on the new lobe. We suggest that such a subaqueous flow lobe grows just as a subaerial pahoehoe lobe slowly advances [*Keszthelyi and Denlinger*, 1996]. As a flow lobe expands forward, the thin viscoelastic crust is stretched and new lava comes out from the front of the lobe. The surface crust becomes thinner toward the front of the lobe and grades into molten lava extruding from beneath the crust. Solid jagged edges are absent along the ruptured front of the lobe, so no grooves form on the lava surface as it extrudes from the lobe. However, cooling by water reduces surface temperature of flowing lava below glass transition temperature in < 0.1 s [*Griffiths and Fink*, 1992a], and brittle crust forms. Stretching of the thin viscoelastic layer exposed on the surface of a growing lobe is immediately quenched by water, resulted in small glass flakes (< 1 cm in length) covering a subaqueous flow lobe.

In contrast, both direct observations and numerical model calculations suggest that subaerially extruded lava surface remains above the glass transition temperature at least for a few minutes [*Hon et al.*, 1994; *Keszthelyi and Denlinger*, 1996]. This is sufficient for the surface viscoelastic layer to stretch as much as 10 cm and subsequently quench to glass fibers, as seen on the surface of S-type pahoehoe.

4.2 What Determines the Volume of a Flow Lobe?

Topographic barriers may constrain flow-lobe volumes [*Rossi and Gudmundsson*, 1996], but such barriers can neither explain the correlation between the supply rates and flow-lobe volumes, nor the smaller flow-lobe volumes for subaqueous lavas compared to subaerial ones with identical supply rates (Figure 7A). Factors governing the lengths and volumes of lava flows are volume of supplied lava, rate of lava supply, underlying slope, and cooling of lava [*Malin*, 1980; *Pinkerton and Wilson*, 1994; *Walker*, 1973; *Wilson and Parfitt*, 1993]. The studied flow-lobe tumuli inflated because supply of lava continued after the flow lobes stopped advancing, which excludes the first possibility that the flow-lobe volume is determined by the amount of lava supplied. Under the same conditions of volumetric effusion rate and substrate slope, lavas stop flowing due to heat loss to the environment, as demonstrated by modeling and field observations of basaltic flows [*Gregg and Fink*, 2000; *Wilson and Parfitt*, 1993].

Cooling of lava can be parameterized by using a dimensionless Graetz number, as defined below [*Hulme and Fielder*, 1977]:

$$Gz = u \cdot h^2 / \kappa \cdot L \quad (4)$$

u, h and w are mean velocity, thickness and width of the flow, respectively. L is length of the flow when it ceased flowing, and κ is thermal diffusivity of lava (5×10^{-7} m^2/s). Gz represents the temperature distribution within a flow. Its meaning can be readily understood by replacing u with L/τ, where τ is time to flow the length L; Gz can be expressed as $h^2/k\tau$. Namely, Gz represents the state that conductive cooling of a semi-infinite half space by a sudden tem-perature decrease has proceeded to the depth of h at time τ. Gz has been applied to channelized aa and pahoehoe lava flows for the purpose of assessing the effect of cooling on flow length [e.g., *Hulme and Fielder*, 1977; *Pinkerton and Wilson*, 1994]. However, some flows with high discharge rates disrupt the surface crust, expose hot lava beneath the crust, and overturn lava between the flow interior and surface [*Kjargaard*, 1999]. For these flows, the assump-tions for applying Gz are not met. Alternatively, for slow emplacement of a pahoehoe lobe as discussed here, significant shearing and deformation take place only in a narrow zone of rupture where new lava comes out, while most lobe crust can be taken as a static slab above laminar flow of underlying lava. Therefore, the conditions for applying Gz are best fulfilled for steady pahoehoe lobe growth within a sufficiently short time (<1000 min) to produce a thin crust relative to horizontal extent of the lobe.

The volume (V) and supply rate of lava (F) are approximately:

$$V \sim L \cdot w \cdot h \quad (5)$$

$$F \sim u \cdot w \cdot h \quad (6)$$

From (4)-(6), we get

$$F / V \sim \kappa \cdot Gz / h^2 \quad (7)$$

For cooling-limited lava with the same physical properties and cooling conditions, all flows stop when Gz reaches a critical value. This critical Gz is constant for any individual flow in a compound lava field but varies among field. Therefore, we may expect a linear relationship between F/V and $1/h^2$, which gives a line with a slope of unity on a $\log(F/V)$-$\log(1/h^2)$ plot (Figure 7B). For pa-hoehoe lobes and tumuli, the width of a lobe varies downflow and cannot be defined as simply as for well-developed channelized flows. Instead, we redefined Gz by rewriting equation 7 as

$$Gz \sim (F / V)\, h^2 / \kappa \quad (8)$$

We use observed values of h for submarine lobes because the amount of inflation is negligible as indicated by the only small displacement along brittle fractures.

Despite the sizeable errors (up to a few tens percent) inherited from uncertainties of the lobe dimensions, most solid pahoehoe flow lobes from Loihi Seamount plot on a line with a slope of unity. Except for one lobe, the hollow lobes plot around the same line as the solid lobes, but are more dispersed. Subaqueous flow-lobe tumuli of the A.D. 864 flow of Fuji volcano [*Obata and Umino*, 1999; *Umino et al.*, 2000] also show a linear correlation between F/V and $1/h^2$. This shows that the lateral extension of these flow lobes was cooling limited, and flowage ceased when Gz of each lobe reached a critical value of ~415 for the Loihi lobes and ~127 for the A.D. 864 Fuji flow. Although the linear correlation is less clear, most subaqueous hollow lobes from the Oman Ophiolite [*Umino, unpublished data*] also plot around a line with a slope of unity, indicating a critical Gz of 71.

Hawaiian subaerial flow-lobe tumuli also plot near a line with a slope of unity (with one exception), but the correlation is less linear because of the limited range of data values. The average Gz for the Hawaiian tumuli on this line is 7, significantly smaller than for the subaqueous flows and published data on other subaerial flows (Gz ~300 [*Pinkerton and Wilson*, 1994]). This contrast partly arise from the different definition of Gz for the channelized Hawaiian flows, but may be largely due to overestimating of the thickness of the tumuli. We used the thickness of flow lobes before the axial cracks opened, but were unable to determine the initial thickness of the lobes when the flows came to a halt. Some tumuli have uninflated lobe margins 10-15 cm thick, which may approximate the initial thickness of flow lobes upon emplacement. Plots of mea-sured $F/V-1/h^2$ may significantly differ from values when the lobes initially stopped flowing and the calculated Gzs would be underestimated.

The plot for Icelandic subaerial tumuli [*Rossi and Gudmundsson*, 1996] also deviates significantly from a line with a slope of unity. As the thickness of the Icelandic tumuli is not given by *Rossi and Gudmundsson* [1996], it was estimated as the basin volume divided by the area. In addition, although all Icelandic tumuli are olivine tholeiite, the data include flows derived from several volcanoes. Any difference in chemical composition, extrusion temperature, crystallinity, or volatile content may change physical properties such as viscosity and yield strength of molten lava and crust, generating scatter for the Icelandic data.

Hollow lobes are common among the subaqueous Loihi pahoehoe flows, especially at the summit vent area of the K-94 conical mound. It is possible that these hollow lobes were formed by drainback of lava into the conduit. The common hollow lobes upstream of pillow flows reported from the Mid-Atlantic Ridge are interpreted as drainout of lava [*Ballard and Moore*, 1977]. However, much larger lobate sheets filling the majority of the K94 summit vent lack collapse pits, which are common in mid-ocean-ridge lobate sheet flows [*Ballard et al.*, 1979; *van Andel and Ballard*, 1979; *Chadowick et al.*, 1999]. Hollow lobes are also present in distal parts of a flow field (K-96). Therefore, drainback alone cannot explain all the hollow submarine lobes.

Inflation is likely to be a significant process in submarine lobes, because the critical inner pressure for lobes to inflate is lower for subaqueous lobes than for subaerial ones by as much as one to two orders in magnitude [*Umino et al.*, 2000]. If the inner pressure of lava were similar for both subaerial and subaqueous lobes, the latter would have much higher excess pressure. This would lead to rapid expansion of the lobes, which eventually permits the molten lava to break out, leaving hollow lobes behind. Umino et al. con-cluded that the lower critical pressures for subaqueous lobes result from shorter time lags between the em-place-ment and the initiation of inflation of the lobes. Effective cooling by water instantaneously forms brittle crust on the surface of subaqueous lava, which thickens twice as rapidly as that on subaerial lava for the first ten minutes. As brittle crust thickens, it hinders the lobes from deforming freely, and eventually flow lobes stop advancing and begin to inflate while the viscoelastic layer beneath the brittle crust is still thin. In contrast, the outer centimeter of subaerial lobes does not cool below the glass transition temperature for several minutes [*Hon et al.*, 1994]. Meanwhile, the lobes can flow freely and expand until their brittle crust thickens.

5. CONCLUSION

The Hawaii Kalapana 1990-91 flow field and Mauna Loa A.D. 1843 flow are associated with subaerial flow-lobe tumuli that consist of many flow lobes coalesced into a larger lobe. Submersible dives on the lower south rift zone of Loihi and offshore north of Oahu Island reveal that subaqueous flow lobes vary from pahoehoe-like lobes, through flat narrow flows, to elongated pillows as the underlying slope steepens. Among lobate sheet flows, inflated pahoehoe lobes are common on flat terraces and gentle slopes. In contrast with subaerial tumuli, they are exclusively single flow lobes. Hollow lobes are also commonly associated with inflated non-hollow lobes. Lack of corrugations on the surface of subaqueous pahoehoe lobes suggests that they formed by ballooning of the lava lobe and stretching of thin viscoelastic layer at the flow front.

The estimated lava supply rate correlates systematically with flow-lobe volume among both subaerial and sub-aqueous inflated flow lobes. As demonstrated by Graetz number of individual flows, cooling is the dominant factor governing the flow-lobe volumes for subaqueous lava. However, more quantitative data are needed for subaerial flows to assess the controlling factors of pahoehoe lobe emplacement and tumulus formation.

Acknowledgments. We thank the captains and crews of R/V *KAIREI* and *YOKOSUKA*, the *KAIKO* and *SHINKAI* Operation Teams, and the scientific parties of KR98 and YK99. Thanks also to K. Sugiyama for assistance and K. Johnson for providing comfortable accommodations during fieldwork. K. Hon, J. Kau-ahikaua, S. Self, and D. Swanson are appreciated for helpful sug-gestions and discussion on emplacement of flow lobes and model-ing their expansion. This paper has been benefited from critical reviews by M. Garcia, K. Hon and Th. Thordarson.

6. REFERENCES

Anderson, S.W. and Fink, J.H., 1992. Crease structures: Indicators of emplacement rates and surface stress regimes of lava flows. *Geol. Soc. Amer. Bull.*, *104*, 615 - 625.

Appelgate, B. and Embley, R.W., 1992. Submarine tumuli and inflated tube-fed lava flows on Axial Volcano, Juan de Fuca Ridge. *Bull. Volcanol.*, *54*, 447 - 458.

Ballard, R.D. and Moore, J.G., 1977. *Photographic Atlas of the Mid-Atlantic Ridge Rift Valley*. Springer-Verlag, New York, Inc., New York, U.S.A., 114p.

Ballard, R.D., Holcomb, R.T. and van Andel, T.H., 1979. The Galapagos Rift at 86°W: 3. Sheet flows, collapse pits, and lava lakes of the rift valley. *J. Geophys. Res.*, 84, 5407 - 5422.

Barnard, W.B., 1995. Mauna Loa Volcano: Historical eruptions, exploration, and observations (1779-1910). *In* Rhodes, J.M. and Lockwood, J.P. (eds.), *Mauna Loa Revealed.* AGU *Geophysical Monograph*, *92*, 1 - 19.

Barth, G.A., Kleinrock, M.C., and Helz, R.T., 1994. The magma body at Kilauea Iki lava lake: potential insights into mid-ocean ridge magma chambers. *J. Geophys. Res.*, 99, 7199 - 7217.

Chadwick, Jr., W.W., Gregg, T.K.P. and Embley, R.W., 1999. Submarine lineated sheet flows: a unique lava morphology formed on subsiding lava ponds. *Bull. Volcanol.*, *61*, 194 - 206.

Gregg, T.K.P. and Chadwick, Jr., W.W., 1996. Submarine lava-flow inflation: A model for the formation of lava pillars. *Geology*, *24*, 981 - 984.

Gregg, T.K.P., Fornari, D.J., Perfit, M.R., Haymon, R.M. and Fink, J.H., 1996. Rapid emplacement of a mid-ocean ridge lava flow on the East Pacific Rise at 9°46'-51'N. *Earth Planet. Sci. Lett.*, 144, E1 - E7.

Gregg, T.K.P. and Fink, J.H., 2000. A laboratory investigation into the effects of slope on lava flow morphology. *J. Volcanol. Geotherm. Res.*, *96*, 145 - 159.

Griffiths, R.W. and Fink, J.H., 1992a. Solidification and morphology of submarine lavas: a dependence on extrusion rate. *J. Geophys. Res.*, *97*, 19729 - 19737.

Griffiths, R.W. and Fink, J.H., 1992b. The morphology of lava flows in planetary environments: predictions from analog experiments. *J. Geophys. Res.*, *97*, 19739 - 19748.

Hon, K., Kauahikaua, J., Delinger, R. and Mackay, K., 1994. Emplacement and inflation of pahoehoe sheet flows: Observations and measurements of active lava flows on Kilauea Volcano, Hawaii. *Geol. Soc. Amer. Bull.*, *106*, 351 - 370.

Hulme, G. and Fielder, G., 1977. Effusion rates and rheology of lunar lavas. *Phil. Trans. R. Soc. London, Ser. A*, *285*, 227 - 234.

Kauahikaua, J., Cashman, K.V., Mattox, T.N., Heliker, C.C., Hon, K.A., Mangan, M.T. and Thornber, C.R., 1998. Observations on basaltic lava streams in tubes from Kilauea Volcano, island of Hawai'i. *J. Geophys. Res.*, *103*, 27303 - 27324.

Keszthelyi, L. and Denlinger, R., 1996. The initial cooling of pahoehoe flow lobes. *Bull. Volcanol.*, *58*, 5 - 18.

Kjargaard, J.I., 1996. Kilauea - Close up of an active volcano (30-min VHS video). Ka'Io Productions, Inc., Volcano, Hawaii, USA.

Kjargaard, J.I., 1999. Eruption update (63-min VHS video). Ka'Io Productions, Inc., Volcano, Hawaii, USA.

Malin, M.C., 1980. Lengths of Hawaiian lava flows. *Geology*, *8*, 306 - 308.

Mattox, T.N., Heliker, C., Kauahikaua, J. and Hon, K., 1993. Development of the 1990 Kalapana flow field, Kilauea Volcano, Hawaii. *Bull. Volcanol.*, *55*, 407 - 413.

Montierth, C., Johnston, A.D. and Cashman, K.V., 1995. An empirical glass-composition-based geothermometer for Mauna Loa lavas. *In* Rhodes, J.M. and Lockwood, J.P. (eds.), *Mauna Loa Revealed.* AGU *Geophysical Monograph*, 92, 207 - 217.

Moore, J.G., 1975. Mechanism of formation of pillow lava. *American Scientist*, *63*, 269 - 277.

Obata, S. and Umino, S., 1999. Morphology of A.D. 864 Aokigahara Lava Flow Exposed on the Coast of Motosuko Lake, Fuji Volcano (in Japanese). *Bull. Volcanol. Soc. Japan*, *44*, 201 - 216.

Pinkerton, H. and Wilson, L., 1994. Factors controlling the length of channel-fed lava flows. *Bull. Volcanol.*, *56*, 108 - 120.

Rossi, M.J. and Gudmundsson, A., 1996. The morphology and formation of flow-lobe tumuli on Icelandic shield volcanoes. *J. Volcanol. Geotherm. Res.*, *72*, 291 - 308.

Ryan, M.P. and Sammis, C.G., 1981. The glass transition in basalt. *J. Geophys. Res.*, *86*, 9519 - 9535.

Self, S., Thordarson, Th., Keszthelyi, L., Walker, G.P.L., Hon, K., Murphy, M.T., Long, P. and Finnemore, S., 1996. A new model for the emplacement of Columbia River basalts as large, inflated pahoehoe lava flow fields. *Geophys. Res. Lett.*, *23*, 2689 - 2692.

Self, S., Thordarson, Th. and Keszthelyi, L., 1997. Emplacement of continental flood basalt lava flows. *In* Mahoney, J.J. and Coffin, M. (eds.), *Large Igneous Provinces*, *Geophysical Monograph*, *100*, 381 - 410.

Swanson, D.A., 1973. Pahoehoe flows from the 1969-71 Mauna Ulu eruption, Kilauea Volcano, Hawaii. *Geol. Soc. Amer. Bull.*, *84*, 615 - 626.

Thordarson, Th. and Self, S., 1998. The Roza Member, Columbia River Basalt Group: A gigantic pahoehoe lava flow field formed by endogenous processes? *J. Geophys. Res.*, *103*, 27411 - 27446.

Umino, S., Lipman, P.W. and Obata, S., 2000. Subaqueous lava flow lobes, observed on ROV *KAIKO* dives off Hawaii. *Geology*, *28*, 502 - 506.

van Andel, T.H. and Ballard, R.D., 1979. The Galapogos Rift at 86°W: 2. Volcanism, structure, and evolution of the rift valley. *J. Geophys. Res.*, 84, 5390 - 5406.

Walker, G.P.L., 1973. Lengths of lava flows. *Phil. Trans. R. Soc. London, Ser. A*, *274*, 107 - 118.

Walker, G.P.L., 1989. Spongy pahoehoe in Hawaii: a study of vesicle-distribution patterns in basalt and their significance. *Bull. Volcanol.*, *51*, 199 - 209.

Walker, G.P.L., 1991. Structure, and origin by injection of lava under surface crust, of tumuli, "lava rises", "lava-rise pits", and "lava-inflation clefts" in Hawaii. *Bull. Volcanol.*, *53*, 546 - 558.

Wilmoth, R.A. and Walker, G.P.L., 1993. P-type and S-type pahoehoe: a study of vesicle distribution patterns in Hawaiian lava flows. *J. Volcanol. Geotherm. Res.*, *55*, 129 - 142.

Wilson, L. and Parfitt, E.A., 1993. The formation of perched lava ponds on basaltic volcanoes: the influence of flow geometry on cooling-limited lava flow lengths. *J. Volcanol. Geotherm. Res.*, *56*, 113 - 123.

Peter Lipman, Volcano Hazards Team, U.S. Geological Survey, MS 910, 345 Middlefield Road, Menlo Park, California 94025, USA

Jiro Naka, Deep Sea Research Department, Japan Marine Science and Technology Center (JAMSTEC), Natsushima-machi 2-15, Yokosuka, Kanagawa 37-0061, Japan

Sumie Obata, Mitsubishi Materials Natural Resources Development Corporation Ltd., Hamacyo 3-21, Nihonbashi, Cyuo-ku, Tokyo 103-0007, Japan

Tsugio Shibata, Department of Earth Sciences, Okayama University, Tsushima-naka 3-1-1, Okayama 700-12, Japan

John R. Smith, Jr., Department of Oceanography, University of Hawaii at Manoa, 1000 Pope Road, MSB 205, Honolulu HI 96822, USA

Frank Trusdell, Hawaiian Volcano Observatory, U.S. Geological Survey, P.O. Box 51, Hawaii National Park, HI 96718, USA

Susumu Umino, Department of Biology and Geosciences, Shizuoka University, Ohya 836, Shizuoka 422-8529, Japan

Introduction to Section 2
Climax-Stage Magmatism:
Growth History Of Kīlauea Volcano And Its Instability

Peter W. Lipman

U. S. Geological Survey, Menlo Park, California

On the south flank of Hawai'i Island, frequent eruptions, abundant earthquakes, and rapid ground deformation mark the current locus of volcanism along the Hawaiian Ridge. Kīlauea and Mauna Loa volcanoes are in a tholeiitic shield-building stage, erupting mainly on land. South of Kīlauea, Lō'ihi Seamount has erupted alkalic and transitional basalts that mark the growth of Hawai'i's youngest volcano. Kīlauea is the most active volcano on Earth, and its summit caldera and two rift zones characterize the typical shield stage of Hawaiian volcanoes. Kīlauea's south flank, between the rift zones, is subject to sustained and episodic seaward displacements associated with frequent earthquakes and expressed on land by the Hilina fault system.

Despite numerous studies, the early eruptive histories of Kīlauea and other Hawaiian volcanoes remain largely unknown, because older rocks are in deep water and largely concealed by younger subaerial lava. The deep offshore region (2500-5500 m) south of Kīlauea was explored and sampled for the first time by submersible and ROV (Remotely Operated Vehicle) dives during joint Japan-USA cruises in 1998-99. In addition, the offshore south flank has been the focus of recent multi-channel seismic-reflection studies, and the submarine extension of Kīlauea's east rift zone, the Puna Ridge, has been imaged by deep-tow sidescan-sonar surveys and a high-resolution photo system. The new data show that complex morphology, stratigraphy, structure, and eruptive products of submarine Kīlauea result from diverse volcanic processes, many of which were previously not well understood or in some cases even recognized.

Hawaiian Volcanoes: Deep Underwater Perspectives
Geophysical Monograph 128
This paper not subject to U.S. copyright
Published in 2002 by the American Geophysical Union

The submarine south flank of Kīlauea is characterized by a prominent mid-slope bench bounding a closed depression at about 3000 m depth. The upper slope steepens toward the shoreline, and a lower scarp descends from the lip of the mid-slope bench down to the 5000-m-deep Cretaceous sea floor. New seismic-reflection data provide important information on the subsurface stratigraphy and structure [*Hills and Morgan*, this chapter]. These authors show that the closed mid-slope depression contains as much as 1.5. km of sedimentary fill. This geometry must result from continuing active deformation, or else the basin would be completely filled by sediment from the Kīlauea shoreline. They also suggest that seaward sliding of the mobile south flank of Kīlauea results in overthrusting and progressive accretion of volcaniclastic strata to the toe of the flank. The authors recognize at least four laterally continuous thrust faults that are traced for up to ~60 km along strike. Making comparisons with accretionary prisms of subduction zones, they estimate that imbricate thrusting has produced as much as 30 km of cumulative shortening. Alternatively, others suggest that the south flank deformation may result from more localized compressive shortening at the toe of a large south flank slump structure, expressed on land as the Hilina fault system. Additional detailed stratigraphic, structural, and petrologic data from underwater observations and samples will be needed to resolve such uncertainties.

The crest of the Puna Ridge (submarine continuation of Kīlauea's east rift zone) is dominated by primary volcanic constructs; these include flow fields, pillow ridges, cratered cones, and terraces [*Smith et al.,* this chapter]. Three basic flow morphologies (pillow, lobate, and sheet flows) have developed, in response to varying eruption rates and topographic slopes along the Puna Ridge. Lava terraces along the ridge crest are interpreted by *Smith et al.* to grow upward and outward by inflation from within, by processes analogous to uplift of subaerial pahoehoe sheet flows. In contrast, others have interpreted the ridge-crest terraces as ponded lava lakes. Some morphologically similar terraces low on the flanks of the ridge seem likely due to gravitational spreading and slumping, as observed with the Shinkai 6500 submersible (dive S506).

Basaltic glasses collected along the crest of the Puna Ridge by dredging and during a dive are all tholeiitic. All these glasses have compositions similar to recent on-land eruptions of Kīlauea [*Johnson et al.*, this chapter], consistent with previous studies. Some new samples, from the south flank of the Puna Ridge are high in Al_2O_3 and CaO relative to the axial lavas, features interpreted as due to suppression of plagioclase crystallization in the presence of high water concentrations. New samples (dive S495) from a large lava flow at the deep south base of the Puna Ridge are alkalic basalt, as briefly reported from previous sampling. This appears to be a young flow of highly fluid low-viscosity lava, as indicated by only minor sediment accumulation on its low-relief upper surface. The source vent and relation of the alkalic flow to Kīlauea's east rift zone remains unknown.

In contrast to the dominantly constructional volcanic morphology along the crest of the Puna Ridge, the offshore south flank has more diverse morphologic, structural, and petrologic features. Visual observations during the 1998-99 JAMSTEC cruises, combined with sample characterization, indicate that the 3-km-deep mid-slope bench and lower scarp consist entirely of well-bedded volcaniclastic rocks: massive sandstone, debris-flow breccia, siltstone, and mudstone [*Lipman et al.*, this chapter]. South-flank breccia clasts are chiefly alkalic and transitional basalt, as are many glass grains in the interbedded sandstones [*Sisson et al.*, this chapter]. High-S compositions of many alkalic basalt clasts and some sandstone glasses indicate derivation from a submarine "Lō'ihi" stage of ancestral Kīlauea, prior to growth of its tholeiitic shield. In contrast, all observed outcrops above and east of the mid-slope bench, (dives K95, S504) are massive pillow lava and broken pillows of transitional basalt, which define the initial submarine flank of subalkaline Kīlauea. Fractures, shears, slickensided clasts, and open folds low in the lower scarp indicate widespread deformation, but upward-decreasing proportions of alkalic materials define a gross stratigraphy. These features indicate that ancestral Kīlauea was a sizeable submarine alkalic volcano, where recurrent slope failures generated a flanking apron of clastic debris that was subsequently overgrown by the modern tholeiitic Kīlauea shield.

Compositions of submarine-erupted glass from the lower scarp [*Sisson et al.,* this chapter], which encompass a range far greater in silica, alkalis, and other elements than that known for Loī'ihi Seamount, record magma-generation processes during early stages of volcanic growth, prior to tholeiitic shield-stage Kīlauea. The general compositional sequence on the south flank of Kilauea and on other Hawaiian volcanoes is compatible with genetic models of increasing degrees of partial melting in the mantle source, but the diverse compositions preserved in the volcaniclastic sediments of the south flank cannot have been derived entirely from a homogeneous source. The volcaniclastic apron on the south flank of Kilauea thus provides a broad petrologic sampling of an early stage in Hawaiian hotspot volcanism, much like that sought for Mauna Kea volcano by the in-progress Hawai'i Scientific Drilling Project. Recently acquired underwater and drill-hole samples have also documented greater compositional diversity than previously known for other Hawaiian volcanoes, such as Mauna Loa and Ko'olau. Compositional differences among tholeiitic basalts, previous considered characteristic of individual Hawaiian volcanoes (e.g., Kīlauea, Mauna Loa, Ko'olau), now appear more likely to represent a general evolutionary sequence sampled to different extents during growth of Hawaiian shields.

The offshore observations, in conjunction with previously reported subaerial geologic information, provide new perspectives on the growth of Kīlauea and other ocean-island volcanoes. Many previous studies have interpreted on-land features of Kīlauea's south flank and adjacent offshore region as products of large-scale block-slumping and volcano spreading since growth of the tholeiitic shield. The new data obtained by submersible operations suggest a more complex compositional evolution and prolonged history of slope failures. This interpretation involves widespread recurrent deposition of debris flows during early growth of an ancestral submarine alkalic volcanic edifice, succeeded relatively recently by modest-scale block-fault subsidence of the growing tholeiitic shield of Kīlauea.

In addition, the mid-slope bench would seem an exceptionally informative site for deep-ocean drilling, to investigate the stratigraphic, petrologic, and structural evolution of a major ocean-island hotspot volcano. Meanwhile, the success of JAMSTEC cruises in obtaining representative materials from the early alkalic growth stage of Kīlauea volcano suggest that similar sampling strategies may be applicable to other ocean-island volcanoes, especially where flanks are disrupted by slumping and landsliding associated with volcano spreading.

Structural Variability Along the Submarine South Flank of Kilauea Volcano, Hawai`i, From a Multichannel Seismic Reflection Survey

Denise J. Hills

Department of Geology and Geophysics, University of Hawai`i, Honolulu, Hawai`I

Julia K. Morgan

Department of Geology and Geophysics, Rice University, Houston, Texas

Gregory F. Moore and Stephen C. Leslie

Department of Geology and Geophysics, University of Hawai`i, Honolulu, Hawai`I

Recent analysis of marine multichannel seismic reflection data collected across the south flank of Kilauea volcano, Hawaii has confirmed that the distal flank records significant accretion and deformation of volcaniclastic strata, building a broad and laterally extensive midslope bench [*Morgan et al.*, 2000]. Further examination of adjacent reflection lines across the mobile flank of Kilauea demonstrates along-strike continuity in styles of deformation within the bench, as well as significant variability, which is commonly reflected in flank morphology. Correlation of reflections on 14 seismic lines allow us to identify at least 4 continuous thrust faults, several of which are traceable as far as ~60 km from the southwestern margin of the mobile flank to the subaerial to submarine transition of the East Rift Zone in the northeast. A landward thrust fault rises from the top of the Cretaceous oceanic crust, and displaces the primary volcanic edifice seaward onto folded and faulted bedded strata. The more seaward faults originate at a horizon slightly above the oceanic crust that appears to define a regional décollement at the base of the flank. To the southeast, imbrication of seaward thrust sheets builds a broad bench which ponds sediments behind it in a deep midslope basin. A lateral fault locally disrupts these structures, marking a transition in structural style to the northeast, where thrust sheets are stacked vertically defining an antiformal stack. The basin is absent in this region, and the primary volcanic edifice is overthrust directly upon accreted and recently deposited sediments building a narrow bench. A seaward change in décollement level from the top of oceanic crust beneath the volcanic edifice to a horizon within bedded strata beneath the bench, and the lateral change in thrust sheet geometries, suggest variations in sliding resistance beneath the mobile south flank of Kilauea, contributing to the complicated morphology evident in the offshore region.

Hawaiian Volcanoes: Deep Underwater Perspectives
Geophysical Monograph 128
Copyright 2002 by the American Geophysical Union

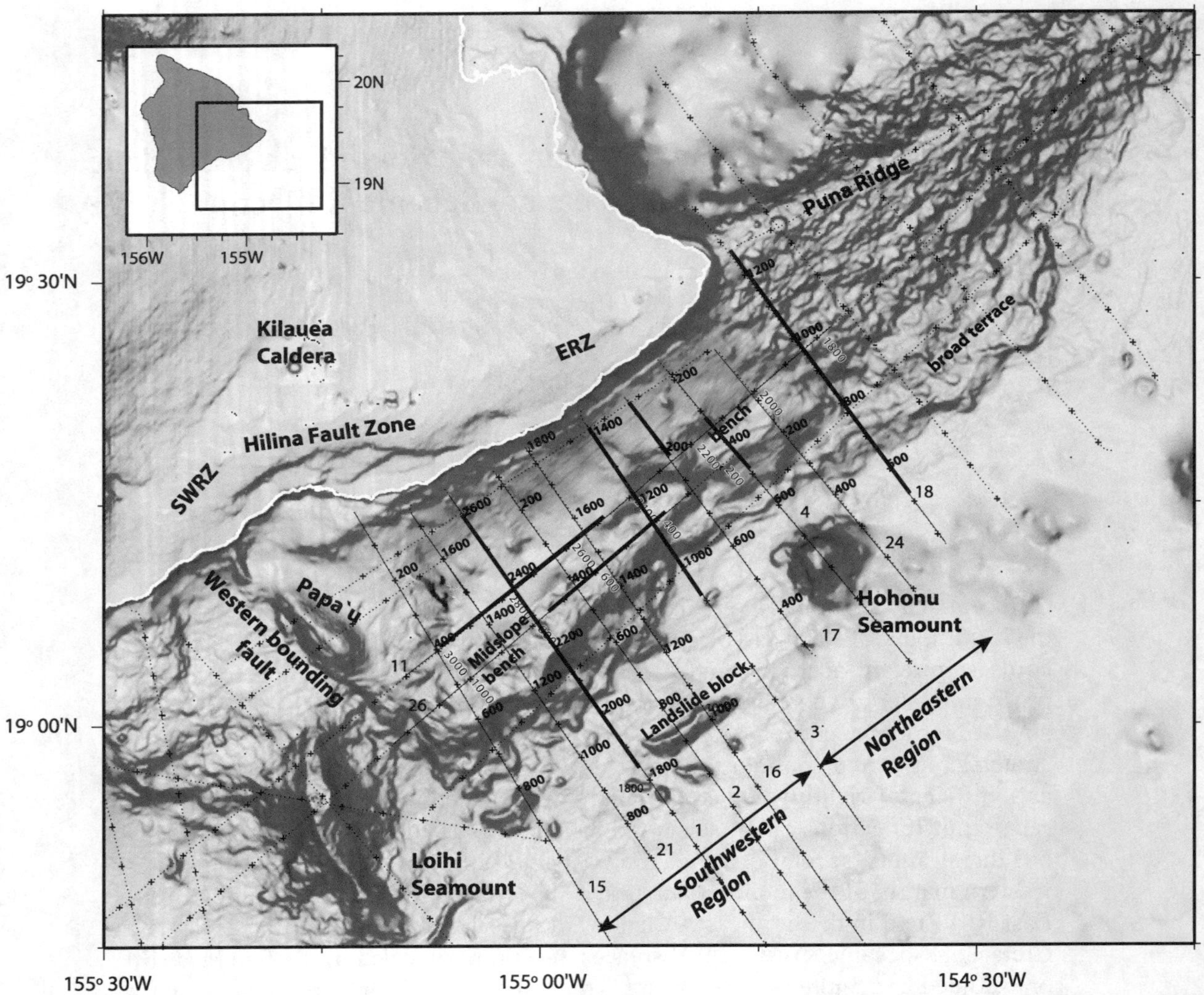

Figure 1. Slope map of bathymetry over the south flank of Kilauea, showing MCS lines collected during the survey. Heavy lines indicate MCS lines used in this study. Structural regions used in this paper are labeled. Detailed boundaries of morphologic features, such as the midslope bench and basin, and low terraces at the base of the flank, are shown in Figure 2. Data from *Smith et al.* [1994].

INTRODUCTION

The importance of deformation of the flanks of oceanic volcanoes during their evolution has been recognized around the world. Evidence for massive slope failures, in the form of chaotic deposits and large avalanche tracks, has been documented along submarine flanks in such places as the Marquesas [e.g., *Wolfe et al.*, 1994], the Canary Islands [e.g., *Carracedo*, 1994], La Réunion [e.g., *de Voogd et al.*, 1999], and the Hawaiian Islands [*Moore et al.*, 1989; *Smith et al.*, 1999; *Naka et al.*, 2000]. Intermittent displacements along near shore fault zones, and local tsunamis during recent earthquakes at Kilauea volcano, suggest that a large landslide, i.e., the Hilina slump, may break the south flank of the volcano [*Tilling et al.*, 1976; *Lipman et al.*, 1985]. This interpretation is consistent with existence of an extensive set of on-land normal faults, and a broad offshore bench along the distal edge of the flank (Figure 1) [e.g., *Lipman et al.*, 1985; *Moore et al.*, 1989; *Smith et al.*, 1999). This offshore bench has been interpreted to mark the uplifted toe of the deep-seated, largely submarine Hilina slump [e.g., *Lipman et al.*, 1985; *Moore et al.*, 1989; *Smith et al.*, 1999]. The active volcano also yields a record of ongoing rift-zone extension and ground displacements of up to 8 cm/yr, documented by long-term geodetic measurements [e.g., *Owen et al.*, 1995; 2000; *Delaney et al*, 1998]. Deep seismicity, such as the 1975 Kalapana M 7.1 earthquake, suggests that entire flank may slide seaward along the base of the edifice [e.g., *Crosson and Endo*, 1982; *Got et al.*, 1994]. Unstable slope movements therefore may be superimposed on this steady seaward motion at Kilauea, so the two processes contribute jointly to the formation of the offshore bench [*Denlinger and Okubo*, 1995], defining a complicated structural and dynamic picture of volcanic evolution.

Using recently acquired multichannel seismic reflection data over Kilauea's south flank, *Morgan et al.* [2000] have argued that the offshore bench reflects overthrusting and accretion of volcaniclastic material at the toe of the mobile south flank of Kilauea (Figure 2). They note that reconstructed flank displacements on the order of 15-24 km are far in excess of that expected due to intermittent movement of a surficial landslide, such as the Hilina slump; bench uplift instead is largely the result of cumulative displacements of the mobile flank as the volcano grew. This opens the possibility that more detailed analysis of the subsurface structure of Kilauea's distal flank can provide important information about the controls on and consequences of submarine volcano flank deformation and volcanic evolution, leading to a better understanding of oceanic volcanoes around the world.

Here we report on results of a marine multichannel seismic (MCS) reflection survey carried out over Kilauea's mobile south flank in 1998 on board the R/V *Maurice Ewing*. We integrate *Morgan et al.*'s [2000] work on a single line with the analysis of several additional lines across the distal flank to examine variability in subsurface structure as suggested by irregular flank morphology. We document evidence that volcaniclastic strata are involved in the formation of the bench structure across the flank, but significant along strike variations in intensity, style, and geometry of deformation exist. We then consider possible explanations for the distribution of flank deformation and implications for the evolution of Kilauea volcano.

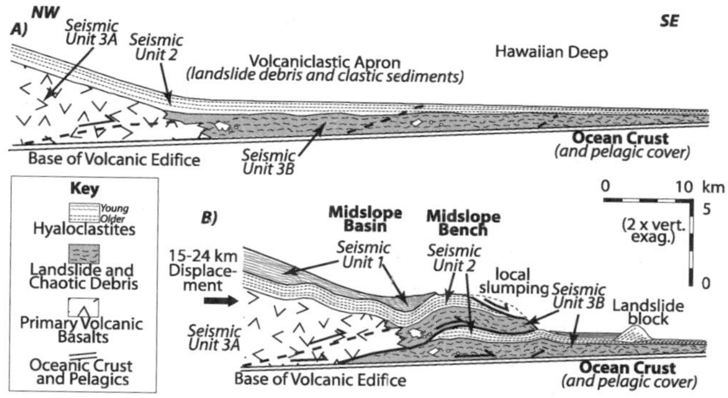

Figure 2. Model for origin of midslope bench along Kilauea's submarine south flank, showing overthrusing and accretion of the distal volcaniclastic apron near MCS Line 2, showing flank displacements of 15 to 24 km. Stratigraphic units and onlap relationships are indicated. Vertical exaggeration is 2 times. (Modified from *Morgan et al.* [2000]).

GEOLOGIC SETTING

Kilauea is a broad shield volcano built up on the southeastern flank of Mauna Loa (Figure 1). Eruptions are concentrated at the summit region and along two radial rift zones. The southwest rift zone (SWRZ) is poorly defined, lacking abundant craters or vents, although faults are numerous. The currently more active east rift zone (ERZ) is marked by vent cones and faults, can be followed to the northeastern point of Hawai`i Island, and continues more than 50 km underwater as the Puna Ridge (Figure 1). The Hilina fault zone, an arcuate system of seaward-dipping normal faults, breaks the surface just south of the summit area and seaward of the rift zones.

The onland limit of the ERZ defines the current subaerial to submarine lava transition. To the northeast, the submarine Puna Ridge lacks sediment cover and shows rough surface topography (Figure 1) interpreted to be submarine erupted lavas, including pillow basalts, lava benches and ponds [*Holcomb et al.,* 1988; *Lonsdale,* 1989; *Clague et al.*, 1994; *Moore and Chadwick*, 1995] and probably small debris flows. Southwest of this transition, subaerial flows enter the ocean, and fragmental basalt and hyaloclastite are generated at the shoreline by explosive interaction of lava and seawater [*Moore and Fiske*, 1969]. Debris cascades down slope, blanketing and smoothing the flank. As Kilauea's ERZ grows, the subaerial to submarine transition migrates along the Puna Ridge to the northeast [*Holcomb et al.*, 1988], leading to a progressive thinning of the sediment cover. Debris has collected along the base of the southeastern flank, forming a volcaniclastic apron that appears to thin to the northeast. This debris is thought to derive from slumps, small debris flows, and hyaloclastic sediment from the slope, intercalated with lava flows at the base of the Puna Ridge [*Clague et al.*, 1994; *Holcomb et al.*, 1988; *Leslie et al.,* in press].

Even with relatively little knowledge of the offshore morphology, early workers, such as *Stearns and Clark* [1930] and *Moore and Krivoy* [1964], postulated the existence of a major submarine landslide along Kilauea's south flank. With improved bathymetric [*Chadwick et al.*, 1993; *Smith et al.*, 1994] and side-scan sonar mapping [e.g., *Smith et al.*, 1994] of Kilauea's submarine south flank, the submarine morphology and shallow structure have become more clearly resolved. Detailed descriptions have been presented by *Moore and Chadwick* [1995] and *Smith et al.* [1999], which we briefly summarize here. A midslope bench is evident across much of the south flank, at depths of 2500-3000m (Figure 1). The bench is bounded to the southwest by the western limit of the mobile south flank, thought to define a fault that accommodates right-lateral strike slip motion [*Denlinger and Okubo*, 1995; *Smith et al.*, 1999]. The bench generally narrows and deepens from southwest to northeast, losing definition completely over the Puna Ridge (Figure 3). A 10-km wide sedimentary basin has developed in the southwestern part of the study area, landward of the midslope bench (Figure 1). The seaward portion of the bench in this area is characterized by very rugged surface morphology. An outer bench high (called the "transition zone" by *Smith et al.,* [1999]) has a shallow seaward dip and separates the midslope basin from the steep outer scarp, which has been incised by small landslides. This basin cannot be identified in the northeastern region (Figures 1 and 3). In contrast to the southwestern region, sediment cover smoothes the outer slope of the narrow northeastern bench. Several elongate ridges seaward of the bench (Figures 1, 2 and 3), are thought to be either compressional features at the toe of a slump [e.g., *Borgia and Treves*, 1992], or landslide blocks shed from the slopes of Kilauea or Mauna Loa [*Moore et al.*, 1989; *Moore and Chadwick*, 1995; *Smith et al.*, 1999; *Morgan et al.*, 2000].

Interpretations of the structure and dynamics of the south flank vary. Morphologically, the flank resembles a massive slow-moving landslide, named the Hilina Slump. By this interpretation, the active on-land portion of the Hilina fault zone marks the headwall of the slump, and the offshore bench reflects the slump's uplifted toe [*Lipman et al.*, 1985; *Moore et al.*, 1989, 1994; *Denlinger and Okubo*, 1995]. However, the importance of gravitational settling (or "volcanic spreading" [*Borgia et al.*, 1990]) in the formation of continuous rift zones as the volcano grows has been appreciated for some time [*Fiske and Jackson,* 1971]. This process may also result in uplift and deformation at the toe of the spreading flank [*Borgia et al.*, 1990; *Denlinger and Okubo,* 1995]. Steady seaward motion of the entire sub-aerial flank [e.g., *Owen et al.*, 1995; 2000; *Delaney et al*, 1998] and seismicity near the base of the volcanic edifice, for example the 1975 Kalapana earthquake, also sup-

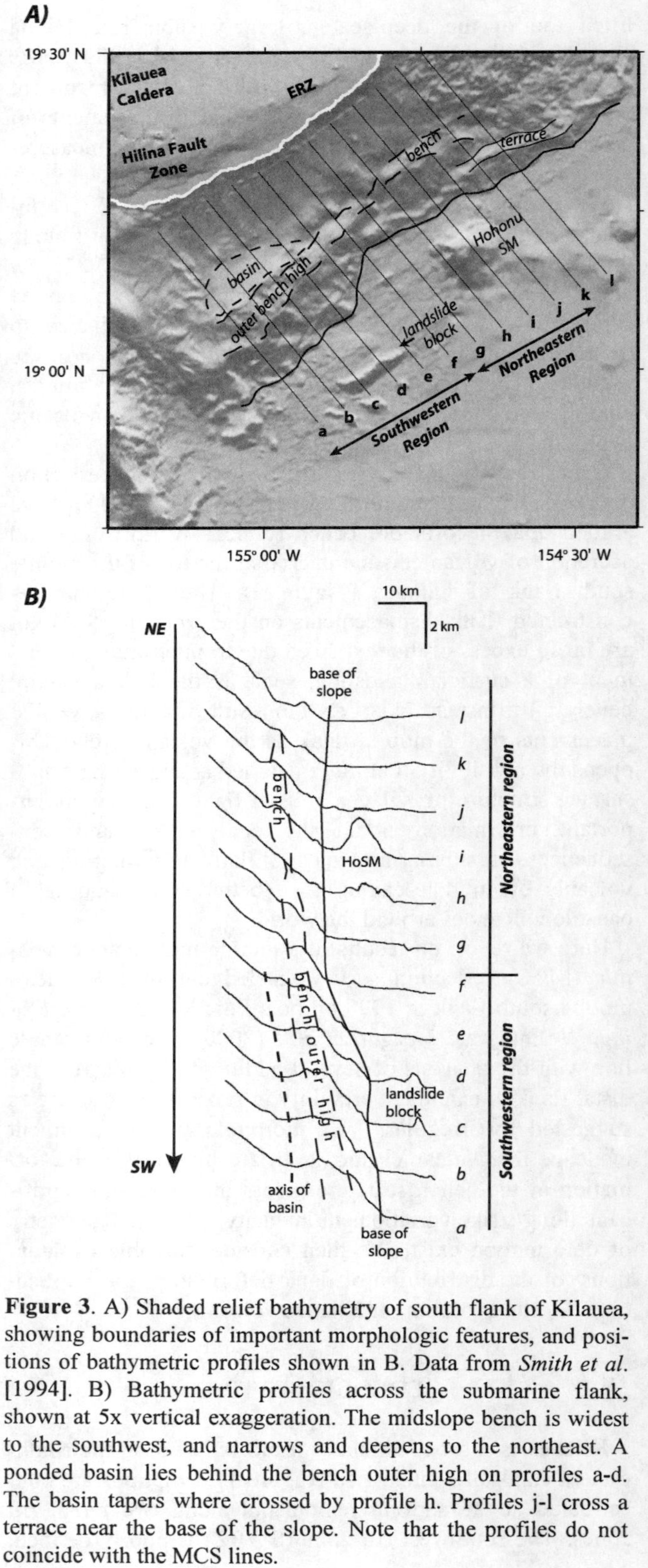

Figure 3. A) Shaded relief bathymetry of south flank of Kilauea, showing boundaries of important morphologic features, and positions of bathymetric profiles shown in B. Data from *Smith et al.* [1994]. B) Bathymetric profiles across the submarine flank, shown at 5x vertical exaggeration. The midslope bench is widest to the southwest, and narrows and deepens to the northeast. A ponded basin lies behind the bench outer high on profiles a-d. The basin tapers where crossed by profile h. Profiles j-l cross a terrace near the base of the slope. Note that the profiles do not coincide with the MCS lines.

Table 1. Processing sequence used on the Kilauea south flank MCS data.

Process	***Parameter***
Sort to common midpoint (CMP) bins	12.5 m
Edit noisy traces	
Velocity analysis	~ every 100 CMPs
Common offset F-K dip moveout (DMO) partial migration	
Minimum phase bandpass filter	4-8-72-80 Hz
Mute	Top (NMO stretch) Bottom (inside for multiple)
CMP ensemble stack	
Low pass filter for multiple removal	0-35 Hz; applied from the multiple to the bottom of the section
F-K migration	
F-K Filter to remove steeply-dipping migration noise	
Depth conversion	
Amplitude scaling (Automatic Gain Control)	200 or 500 ms

port the idea that the south flank of Kilauea volcano is deforming under more distributed tectonic stresses due to gravitational settling of the volcano edifice [*Delaney et al.*, 1998], or forceful intrusion of magmas at the rift zones [*Swanson et al.*, 1976]. Pelagic sediments sandwiched between the volcano and oceanic crust may provide a sliding surface along which volcanic spreading takes place [*Nakamura*, 1980; *Leslie et al.,* in press]. Focal mechanisms and relocations calculated for the 1975 Kalapana earthquake (M 7.2), and ongoing deep seismicity, imply a slip surface that corresponds in both depth and orientation to the oceanic plate beneath Kilauea's south flank [*Furumoto and Kovach,* 1979; *Zucca and Hill*, 1980; *Crosson and Endo*, 1982; *Got et al.*, 1994], further suggesting that this surface serves as a décollement along which the mobile flank may slide.

SEISMIC DATA ACQUISTION AND PROCESSING

Data Acquisition

We collected MCS data on the R/V *Maurice Ewing* with a 4-km, 160-channel streamer and a tuned 4336-in^3 (71 L) air gun array. Shots were fired every 50 m with a receiver spacing of 25-m to yield 40-fold coverage at 12.5-m CMP interval.

Data Processing

Time processing. A standard processing sequence (Table 1) was carried out on all seismic data using Landmark's ProMAX software. Data processing was complicated by the irregular morphology across the submarine south flank. The rough seafloor and steep slopes caused scattering of energy, and introduced out-of-plane reflections. Shallow depths of the upper flanks created a strong sea floor multiple that obscures some of deeper reflections on the landward-most ends of the lines. We attempted several multiple suppres-sion techniques, including deconvolution (pre- and post-stack), frequency-wavenumber (F-K) filter (pre- and post-stack) and pre-stack radon filtering. Radon filtering improved the velocity analyses and allowed more accurate velocity picks to be made [*Taylor et al.*, 1998], but did not appreciably attenuate the multiples. Although deconvolution and radon filtering somewhat improved the shallow portions of the data, these processes also severely attenuated the oceanic crust reflection, making it very difficult to distinguish in the small-scale plots that we are able to present in this paper. Pre-stack F-K filtering was unable to attenuate the multiple without also attenuating the ocean crust reflection because, in the time sections, they both dip in the same direction with only a few ms/trace difference in dip. A combination of band-pass filter to reduce high frequencies within the multiple signal, standard near-trace inside mute and post-stack F-K filtering had the most success at eliminating multiple energy. We used several versions of the stacked and migrated data to produce our final interpretations, but here present versions with post-stack F-K filtering without deconvolution (e.g., Figure 4).

All stacked sections were migrated with an F-K algorithm that uses *Stolt's* [1978] stretching technique to account for vertical and lateral velocity variations. Velocities were determined by iterative migrations to produce the best imaged sections.

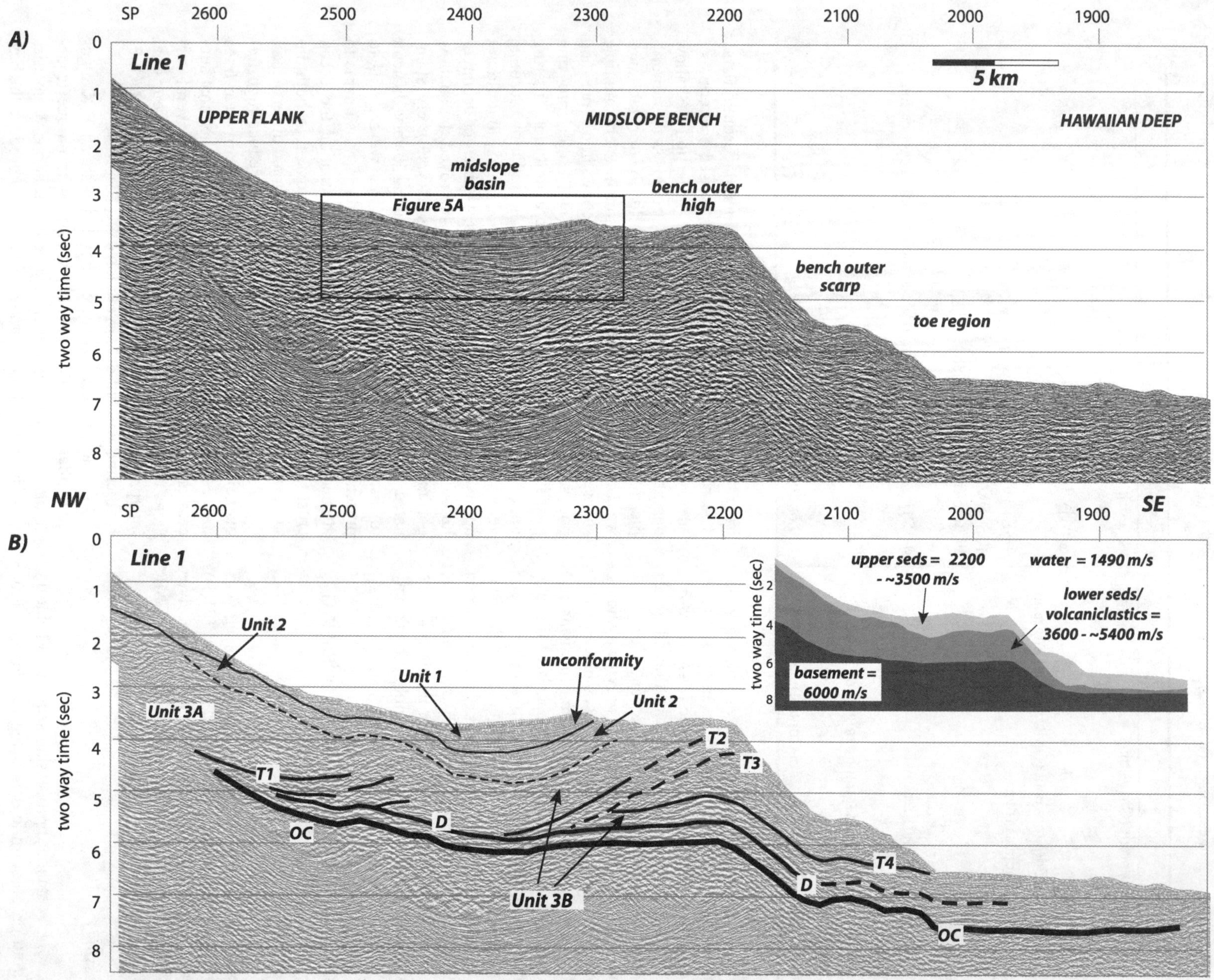

Figure 4. MCS Line 1; location indicated in Figure 1. See text for discussion of seismic stratigraphic units and interpreted reflectors. Uninterpreted (A) and interpreted (B) time sections, shown with 2x vertical exaggeration at the seafloor. Uninterpreted (C) and interpreted (D) depth converted sections, returning reflections to more realistic geometries, shown with 1.5x vertical exaggeration. Inset in (B) shows velocity model used for depth conversion.

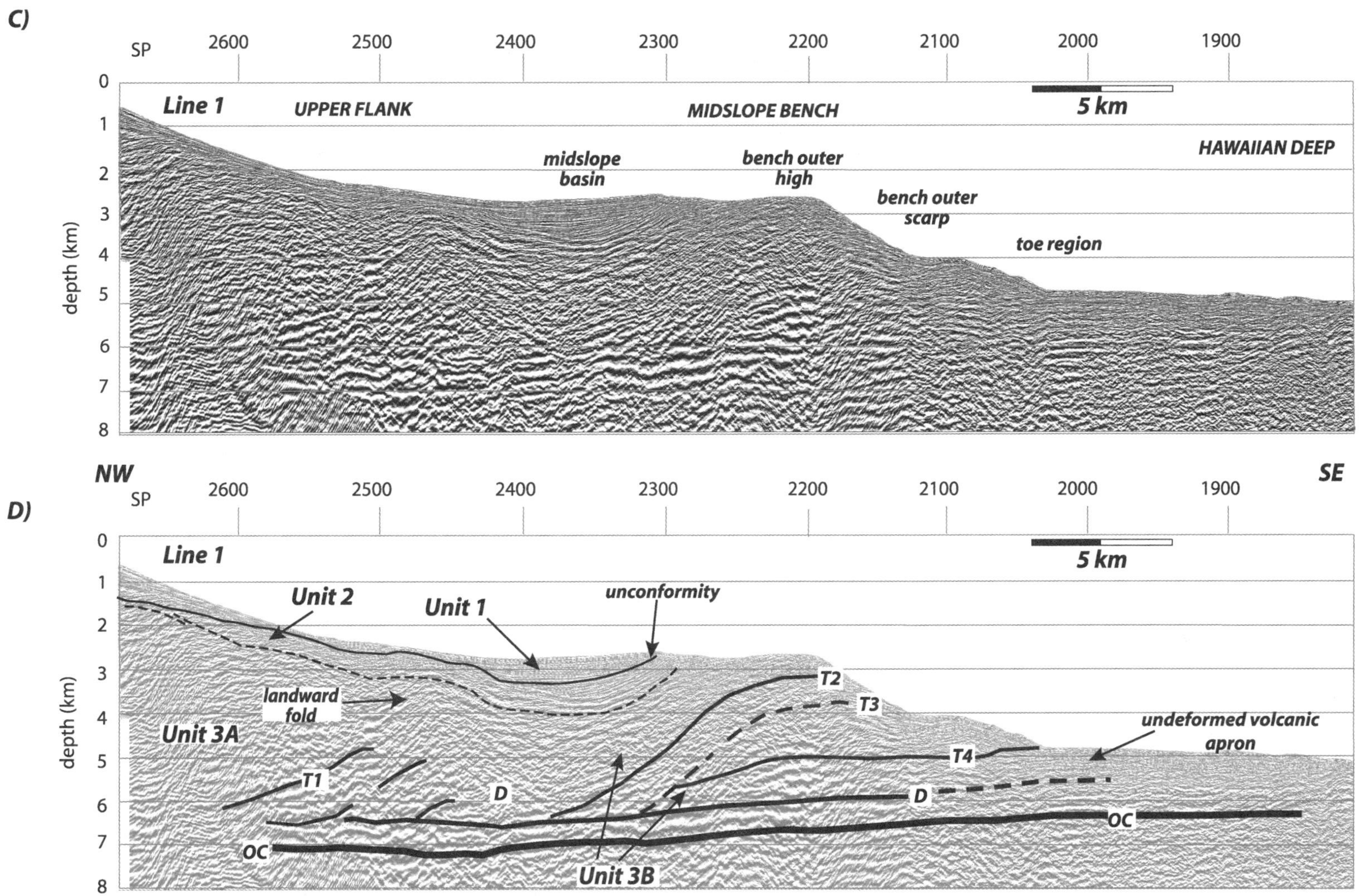

Figure 4. MCS Line 1; location indicated in Figure 1. See text for discussion of seismic stratigraphic units and interpreted reflectors. Uninterpreted (A) and interpreted (B) time sections, shown with 2x vertical exaggeration at the seafloor. Uninterpreted (C) and interpreted (D) depth converted sections, returning reflections to more realistic geometries, shown with 1.5x vertical exaggeration. Inset in (B) shows velocity model used for depth conversion.

Depth processing. Because of the steeply sloping seafloor and relatively high seismic velocities of the volcanic rocks and sediments, structural interpretations are very difficult to make on the time sections. In order to improve our interpretations, we converted the seismic time sections to depth. We were not able to collect refraction velocity data on our cruise, so we used published velocities for the depth conversions.

We identify a distinctive reflection that we interpret to be the top of oceanic crust (OC on Figure 4b). This reflection is present on all of our lines [*Morgan et al.*, 2000]. We are able to estimate the depth of this reflection with reasonable confidence seaward of the base of slope, where relatively thin sediments overlie the oceanic crust. Based on previous seismic refraction data and flexural studies [e.g., *Zucca and Hill*, 1980; *Hill and Zucca*, 1987; *Thurber and Gripp*, 1988; *Got et al.*, 1994], the oceanic crust dips landward under the south flank at an angle of 2-3°. Assuming this dip and a relatively planar surface for the top of oceanic crust reflection, we are able to construct a simple, three-layer velocity model that yields a reasonable representation of the oceanic crust and overlying units. Our velocity ranges are consistent with those determined from previous studies around Hawai`i [e.g., *Hill and Zucca*, 1987; *ten Brink and Brocher*, 1987; *Watts and ten Brink*, 1989; *Lindwall*, 1988]. Velocities within each layer increase with depth. Layers are defined, from shallowest to deepest, as (a) shallow sedi-ments, including sediments in the Hawaiian Deep (2200 - 3500 m/s); (b) deeper sediments and primary volcanics (3600 - 5400 m/s); and (c) oceanic basement (6000 m/s; this velocity was used to maintain a smooth gradient). The velocity model for Line 1 (Figure 4b-inset) generates the depth converted section for Line 1 shown in Figure 4c. Although the chances for error in these depth conversions are considerable, and our depth sections should therefore be considered to be approximations, we believe, however, that the depth sections are accurate enough to portray the structural styles of Kilauea's south flank.

RESULTS AND INTERPRETATION

Seismic Tectono-stratigraphic Units

We define four seismic units that allow us to delineate the regional structural geology of the south flank. These units are distinguished primarily by seismic character (i.e., frequency and continuity of reflections, external geometry, internal character, and boundaries). The units are illustrated on Line 1 (Figure 4) and our structural interpretation (Figure 2).

Unit 1 (upper sediments) is characterized by high frequency, laterally continuous reflections within the shallow section of the upper flank (e.g., Figure 4b, SP 2300 - 2675, seafloor to ~0.25 s - 0.75 s below the seafloor). These well-bedded sediments generally parallel the slope along the upper flank (Figure 4b, SP 2400 - 2675, seafloor to 0.25 s - 0.5 s below seafloor) and fill the midslope basin (Figures 5a and 4b, ~ SP 2300-2400, ~ 0.75 s thick at SP 2375). We interpret this unit as primarily hyaloclastic sediments generated along the shoreline of Kilauea by fragmentation of basalt flows as they enter the ocean that are then transported down the slope and into the basin by mass flow processes.

Unit 2 (lower sediments) is a regionally mappable unit that unconformably underlies Unit 1 (Figure 4b). Internal reflections are usually parallel and continuous, but locally irregular. Along the flanks of basement highs (e.g., Line 11, SP 2500-2600, Figure 5b), Unit 2 is on-lapped by Unit 1. In other areas (e.g., Line 11, SP 2700-2800, Figure 5b,), the boundary between Units 1 and 2 is concordant. In several locations, Unit 1 has either not been deposited or was subsequently eroded, and Unit 2 is exposed at the surface (e.g., Line 11, SP 2480-2525, Figure 5b; Line 1, SP 2225-2275, Figures 4a and 5a). The base of Unit 2 is locally difficult to identify due to low contrast with the underlying units. Based on the continuity of reflections, we interpret Unit 2 as older volcaniclastic sediments deposited on the slope and adjacent oceanic crust prior to major deformation of the south flank (Figure 2).

Unit 3, underlying the shallow sediments and continuing downward to reflector OC (oceanic crust), comprises two subunits that make up what we will refer to as the volcaniclastic edifice (Figure 4), following the usage of *de Voogd et al.* [1999]; in our usage, this term encompasses the primary volcanic edifice composed of submarine erupted lavas and breccias, Unit 3A, and layered volcaniclastic strata accumulated within the broad volcanic apron, Unit 3B. Deposition was contemporaneous and transitional, leading to interfingering relationships (Figure 2a). Unit 3A builds the upper flank, and is directly overlain by Unit 2; reflections are chaotic and discontinuous (Figure 4b, SP 2675, 1.9 s - 4.0 s, to SP 2400, 4.25 s - 5.75 s). Internal reflections are generally lower frequency than those in Unit 1 or 2, and reverberant (Figure 4b, SP 2550, ~ 4.75 and 4.9 s). The reflective character is consistent with what might be expected from volcanic breccias, pillow lavas, and other submarine extrusives that compose Puna Ridge, the modern analog for primary volcanic edifice [e.g., *Moore and Chadwick*, 1995].

Unit 3B lies seaward of, but at the same stratigraphic level as, Unit 3A. Unit 2 overlies Unit 3B in the midslope bench (Figure 4). Reflections within Unit 3B tend to be more continuous than those beneath the upper flank, and are mid- to low-frequency events (Figure 4b, SP 2400, 4.75

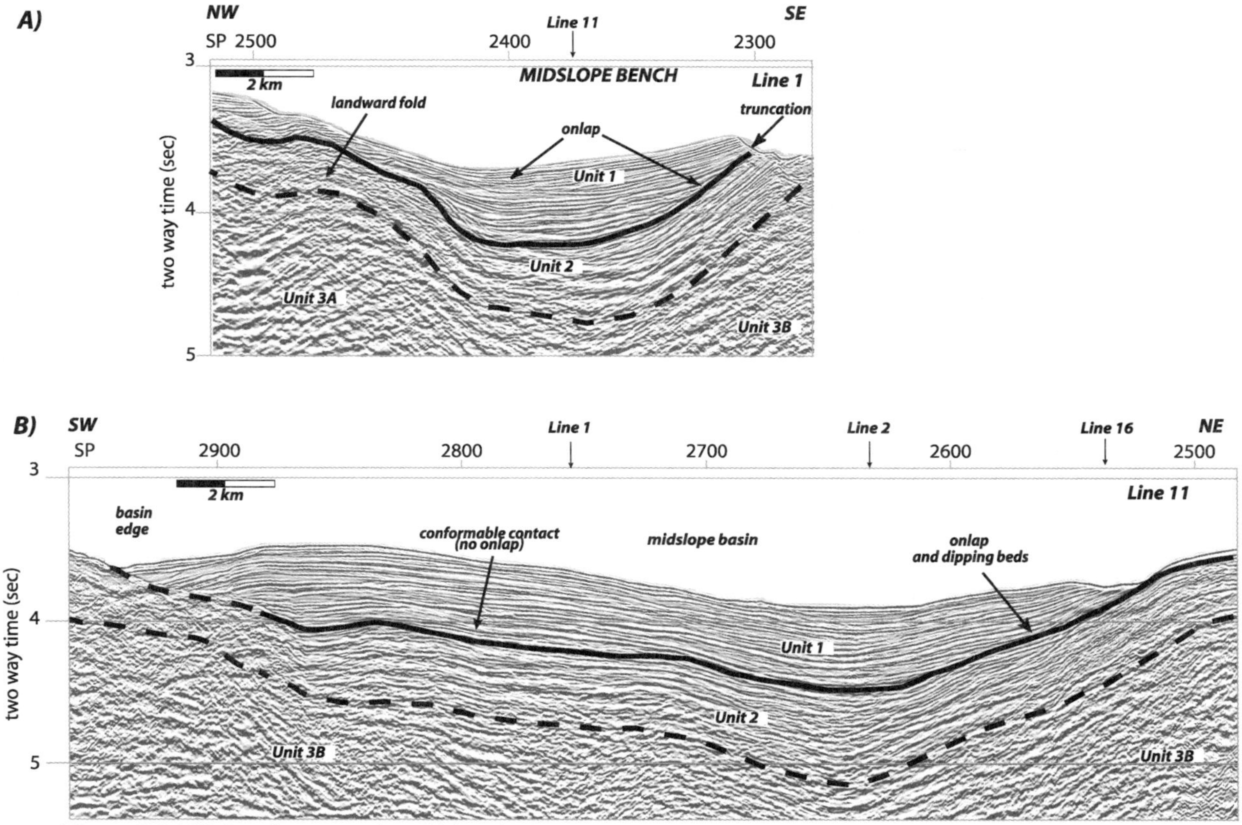

Figure 5. MCS time lines over the midslope basin. A) Enlargement of region on MCS Line 1 shown in Figure 4A. See text for discussion. B) Midslope basin as seen on MCS Line 11, follows the axis of the basin (location indicated on Figure 1).

s – 5.75 s, to SP 2000, 6.5 s - 7.25 s). Thus, we interpret Unit 3B to consist of layered volcanic strata, probably volcaniclastic breccias intercalated with hyaloclastic sands as documented by submersible dives along the outer bench scarp [*Naka et al.*, 2000]. The contact between Units 3A and 3B is indistinct on the seismic lines, suggesting complicated interfingering of proximal lava flows, pillow basalts and breccias, with avalanche deposits, slump debris, and turbidites making up an early volcaniclastic apron (Figure 2). Relatively horizontal, undeformed strata ~1 s thick lie at the base of the bench outer scarp at SP 2030 (Figure 4) and extend seaward into the Hawaiian Deep. We interpret these undifferentiated units to comprise the recent volcanic apron composed of hyaloclastites, volcaniclastic breccias, and debris flow deposits.

Structure of the South Flank

Our study of along-strike variations in south flank structure takes advantage of 14 seismic lines collected over this region (Figure 1). Morphologically, the flank can be subdivided into several regions. Our primary focus here is on the origin of the bench, so we distinguish a south-western region which contains the broad midslope basin and bench, and a northeastern region characterized by a narrower and deeper bench lacking a ponded basin, and which loses expression to the northeast (Figures 1 and 3b).

The basic structure across the southwestern bench and midslope basin is represented well on Line 1 (Figure 4), and is similar to that summarized by Morgan et al. [2000] for the adjacent Line 2. The midslope basin is a prominent feature also shown well on Line 11 that parallels the basin axis. High frequency, parallel reflections defining seismic Unit 1 onlap the steep landward (Figure 5a, Line 1, SP 2425) and southwestern (Figure 5b, Line 11, SP 2900) edges of the basin. The contact between Units 1 and 2 marks the base of the basin; Unit 2 is followed seaward to where it rises onto the bench (Figure 5a, Line 1, SP 2300-2350) and to the northeast onto the upper flank (Figure 5b, Line 11, SP 2500-2550). In these areas, Unit 1 beds onlap the dipping beds of Unit 2, and are tilted as well.

The bench rises about 4 km above the OC reflector, and is composed primarily of Unit 3B derived from the volcanic apron. Strong, landward-dipping reflections occur within the bench and bound packets of parallel and continuous reflections (Figure 4c and d, Line 1, SP 2150-2350, 3-6 km depth). These continuous reflections are recognized be-neath the basin and are traced laterally within the northeastern flank on Line 11 (Figure 5b). Both stratigraphic and structural interfaces can generate strong reflections such as we observe; however several standard criteria [e.g., Boyer and Elliot, 1982] allow us to identify these as thrust faults disrupting strata within the bench. These include (a) the divergence of reflections, e.g., from the OC or a shallower detachment horizon (e.g., Figure 4, T1), and (b) the termination and offset of reflections, indicating stratal truncation against a fault (e.g., Figure 4b and d, SP 2275, 5.5 s/5.6 km; reflections terminate against T3). In addition, we are able to correlate reflections along strike to adjacent lines, defining smoothly varying stratal packets. In this way, ambiguous reflections on one line can be correlated with distinct fault discontinuities on adjacent lines.

We recognize four continuous fault reflections across the study area, referred to as T1-T4 (Figure 4). Although not well expressed on Line 1, T1 rises from a deep reflection close to the OC horizon, landward of the midslope basin reflections (Figure 4d, SP 2475-2625, 4.5-6 km depth), slightly distorting reflections above (e.g., Figure 5a, SP 2400-2520, 4-5 s). The same structure and overlying fold were noted on Line 2 by *Morgan et al.* [2000]. Faults T2, T3, and T4 map out a broad fold within the bench (Figure 4d, SP 2100-2300, 3-6 km). In the landward direction, faults T2-T3 sole into a reflection near the OC horizon (Figure 4d). This consistent relationship between the thrust faults and subhorizontal horizon suggests a basal décollement beneath the basin and bench complex, in this case lying along a stratigraphic horizon up to 1 km above the OC. Labeled D (Figure 4d), the décollement is recognized on adjacent lines, with the thrust faults rising from it. These structures are traced along strike on Line 26, which traverses the axis of the bench outer high (Figure 6).

As shown on Line 26, the midslope bench steps down abruptly to the northeast (Figure 6, SP 550), marking the transition between the southwestern and northeastern regions. Similar reflections are recognized on either side of the step, but are clearly offset across a dipping region extending from the seafloor at ~SP 600, to 7 km at ~SP 450. On the northeastern side of this structure, reflections are warped, defining a monoclinal fold in the shallow sediments comprising Unit 2 (Figure 6, SP 400-550). Unit 1 strata are onlap the folded beds near SP 400. This disturbance of the bench reflectors defines a lateral fault, L, allowing for relative seaward displacement of the broad southwestern portion of the midslope bench. The seaward limit of the fault coincides with a recess in the toe of the flank, recognized just where Line 3 crosses the toe (Figure 1). The landward manifestation is the gradual rise from the basin onto a prominent upper flank, noted on Line 11 (Figure 5b, northeast of SP 2600) and south flank bathymetry (Figure 1). The change in morphology and presence of this lateral fault suggests a change in submarine structure of the flank across the transition.

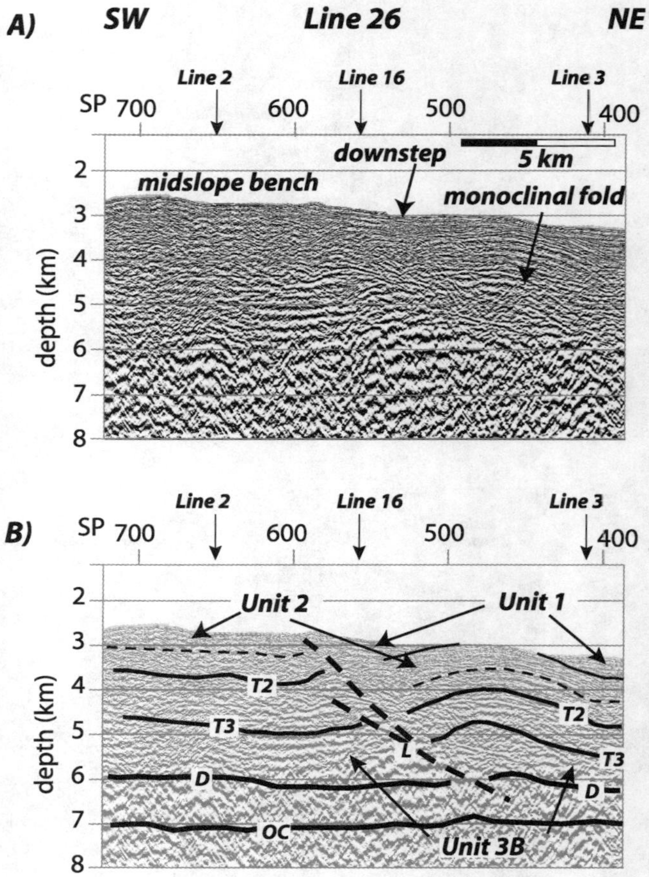

Figure 6. Portion of depth section of MCS Line 26, uninterpreted (A) and interpreted (B), plotted with 1.5 x exaggeration. An east dipping lateral transfer fault, L, offsets continuous thrust reflections, T2-T4. A fold is developed in Unit 2 to the northwest of L; Unit 1 onlaps the folded Unit 2. See text for discussion.

Line 3 (Figure 7) traverses the narrow bench and prominent upper flank just northeast of the transition (Figure 1). A pronounced deep reflection marks the OC beneath the upper flank (e.g., Figure 7, Line 3, SP 1300 and 1120, ~ 6.6-7.2 km depth). A strong set of reflections rises from the OC near SP 1400, approaching the seafloor near the base of the scarp landward of the midslope bench. Unit 2 is offset across this structure and folded within the scarp, supporting

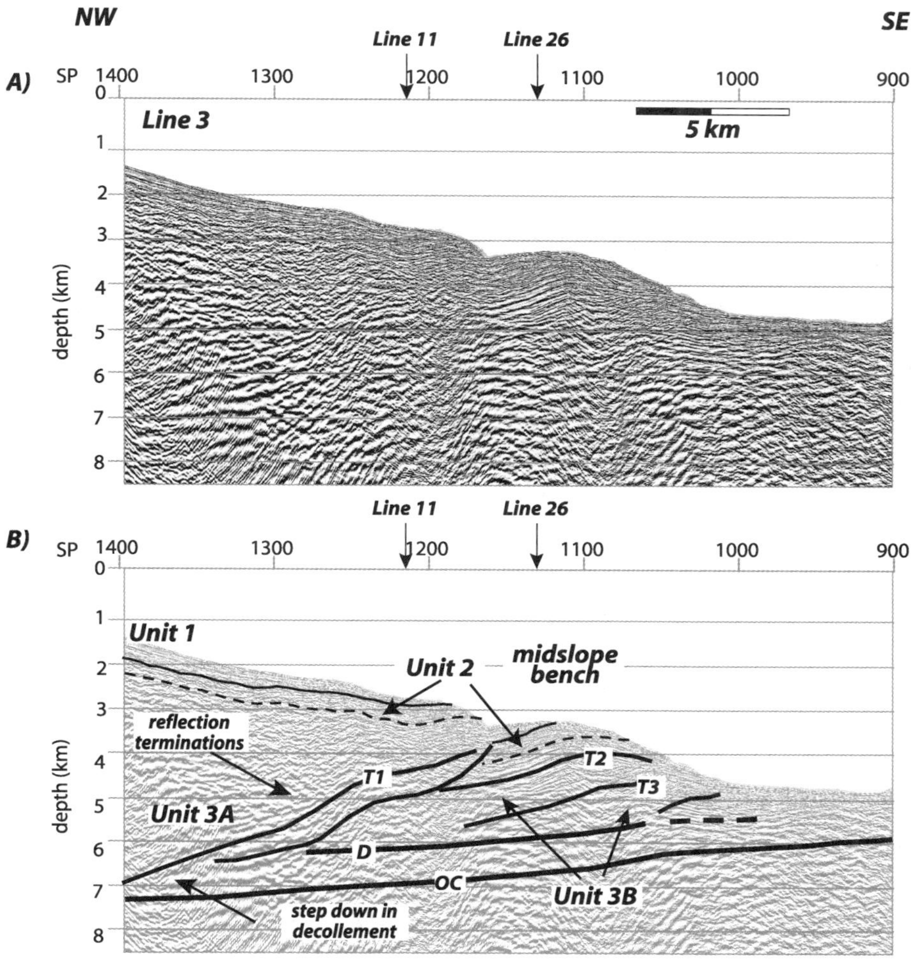

Figure 7. Depth sections of MCS Line 3, uninterpreted (A) and interpreted (B), plotted with 1.5 x vertical exaggeration. See text for discussion.

the interpretation that these reflections define a major thrust fault. The fault can be traced laterally, and appears to correlate with T1 on Line 1 (Figure 4d, ~SP 2550, 5.5 km), the most landward of the identified thrust faults on Line 1. The southwestern Line 1 revealed T1 behind the midslope basin and responsible for relatively little deformation of the overlying strata (Figure 4d, SP 2500-2600). We propose that displacement decreases along T1 to the southwest, and loses expression at the seafloor as the tip of the fault is buried by young slope sediments in the basin.

The more seaward thrust faults crossed by Line 26, T2-T3 (Figure 6), are also evident within the bench on Line 3 (Figure 7, SP 1050-1200, 4-6 km). Reflector geometry is more difficult to discern, but suggests that thrust sheets are stacked on top of each other, in contrast to imbricated geometry observed on Lines 1 and 2 (e.g., Figure 4d). Although not shown in detail, this interpretation is borne out on adjacent lines as well. Thrust T4 is not evident in this profile, possibly because of its incipient stage. As with Line 1, Line 3 also shows a prominent stratigraphic horizon, D, from which T3 rises. In contrast to Line 1 and similar lines in the southeastern region [e.g., *Morgan et al.*, 2000], D appears to terminate against the pronounced reflection T1. Reflectors seaward of T1 are continuous, implying bedded strata of Unit 3B; reflectors landward of T1 are characteristically discontinuous, with occasional through going horizons, consistent with primary volcanic material in Unit 3A. Based on reflector geometries, we propose that T1 displaces the primary volcanic edifice and overlying slope sediments above faulted and folded volcaniclastic strata in

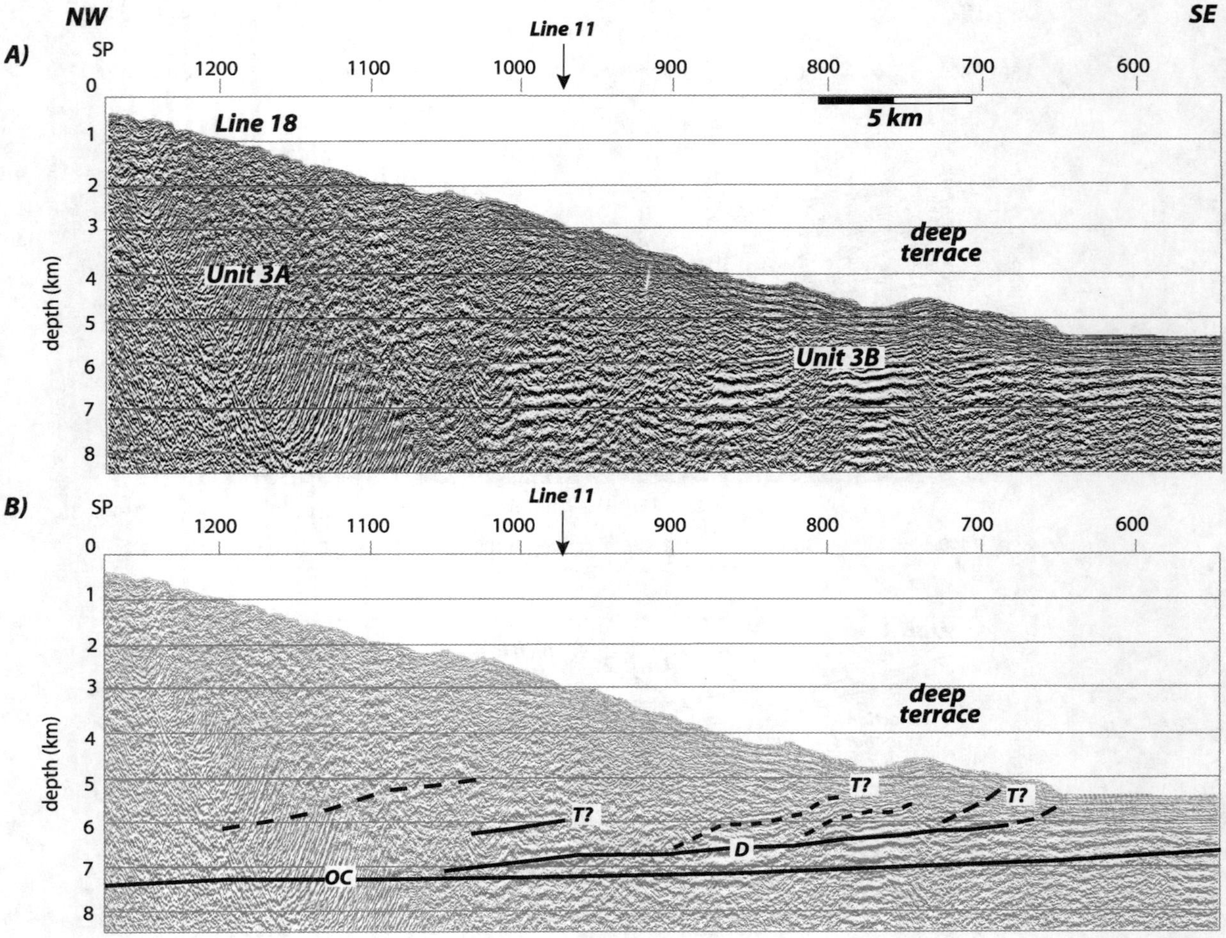

Figure 8. Depth sections of MCS Line 18, uninterpreted(A) and interpreted (B), plotted with 1.5 x vertical exaggeration. See text for discussion.

the bench. The décollement horizon, D, for the bench structures, does not extend landward of T1 (Figure 7, SP 1280, ~6 km). The geometries and offsets of reflections and structures observed in the bench on Line 3 are repeated on lines to the northeast, but becomes progressively less pronounced as displacement decreases.

Line 18 (Figure 8) lies just northeast of the limit of fragmental basalt deposition (Figure 1) and technically falls within the Puna Ridge region. This region lacks the pronounced midslope bench observed to the southwest. The flank is rougher than observed elsewhere, as expected for slope dominated by submarine erupted lavas. The seismic character of the entire flank is consistent with that of Unit 3A - short, irregular reflections assumed to denote primary pillow basalts and breccias. Despite this irregular surface, the OC reflection can be followed to the northwest beneath the flank, to the strong seafloor multiple (Figure 8, SP 1100, 7.2 km). Little disruption of the slope is evident. The toe of the flank is characterized by a broad terrace, which extends to the northeast along the base of Puna Ridge and southwest to Line 24. Subtle reflections beneath the terrace may be analogous to sub-bench reflections to the southwest, possibly indicative of incipient uplift of a small volcanic apron; a small basin has ponded bedded sediments behind the terrace (Figure 8, SP 760).

A veneer of well-bedded sediments of seismic Unit 1 drapes the upper flank regions across the study area, overlying Unit 2, which underlies the midslope basin and drapes the bench (e.g., Figure 4). Sediment thickness varies along and across the slope, and is thickest to the southwest, in particular in the vicinity of Line 16 near the current ocean entry (Figure 1). It thins rapidly to the northeast, and is nearly absent along Line 24 which lies just southwest of the edge of the fragmental basalt slope cover. In places, the slope sediments show landward rotation, bedding truncation, and/or chaotic reflections, indicating intermittent disruption and slope failure (e.g., Figure 9). Incipient slope collapse is suggested by the occurrence of several normal faults that break the uppermost strata along the well-sedimented northeastern midslope bench. We hypothesize that local failure of bedded strata that accumulated along previous bench configurations to the southeast resulted in deposition of the elongate landslide blocks in front of the bench outer slope (Figure 1).

The western boundary of the mobile south flank is not directly addressed by this study, but it helps to constrain the kinematics of the submarine flank. It can be identified as a southeast trending cliff with up to 1000 m of relief along its length (Figure 1), and is interpreted to result from right-lateral strike-slip displacement of Kilauea's flank relative to Mauna Loa [e.g., *Denlinger and Okubo*, 1995; *Moore and Chadwick*, 1995]. Line 15 crosses the midslope bench southwest of the main basin near the western margin and Papa'u seamount. These data are not discussed in detail here; however they have been considered in the preparation of the structural interpretations.

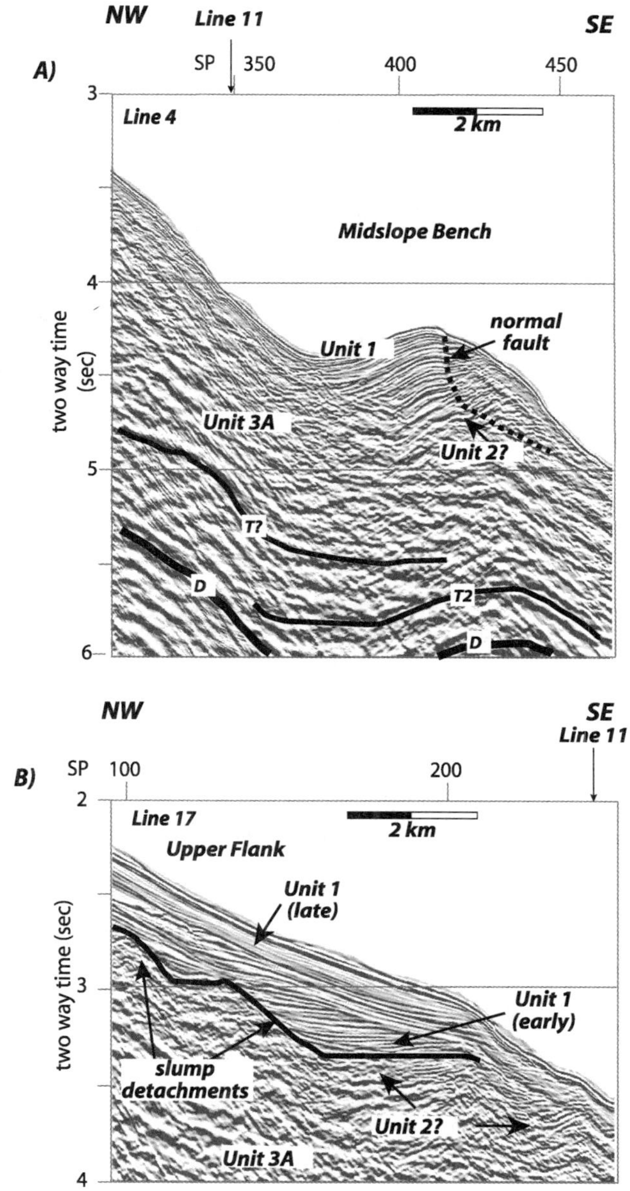

Figure 9. A) Narrow midslope bench along time section of MCS Line 4. A seaward dipping normal fault cuts the shallow, well-bedded, landward dipping sediments that cap the bench. Continuous reflections within the bench define the OC and landward dipping thrust faults that can be correlated on adjacent lines. B) Shallow slumps break the upper flank along time section of MCS Line 17. Sediments within the blocks have been rotated, erosionally truncated (arrows), and subsequently buried by younger, slope parallel hyaloclastic sediments.

DISCUSSION

Structural Summary

Despite variations in surface morphology, the seismic structure across the entire midslope bench region of the south flank proves to be very similar. We recognize pronounced, landward-dipping reflections that we interpret to be thrust faults that bound packets containing conformal bedding reflections. Many of these faults can be correlated along strike, extending up to 60 km within the midslope bench. Shallow strata on the benches are landward-tilted and on-lapped by younger fill. As noted by *Morgan et al.* [2000] reverse faulting and overthrusting primarily control the morphology of the distal flank, which has uplifted and folded strata in the midslope bench and ponded slope sediments behind it.

In places, uplifted and offset strata caused by the intersection of thrusts with the seafloor create slope breaks at the surface that can be traced across the outer scarp. Seismic stratigraphic and structural depth interpretations are shown for representative dip and strike lines in Figures 10 and 11, and summarized on the structural map of the mobile flank in Figure 12.

Dip lines crossing the southwestern and northeastern regions show contrasts in structural geometries that suggest a transition along strike. Key features of the southwest region (Figure 10a and b) include the intersection of thrust T1 with a décollement horizon above the OC, presence of the deep, wide basin and buried fold, and the imbricated arrangement of thrust sheets within the midslope bench. In

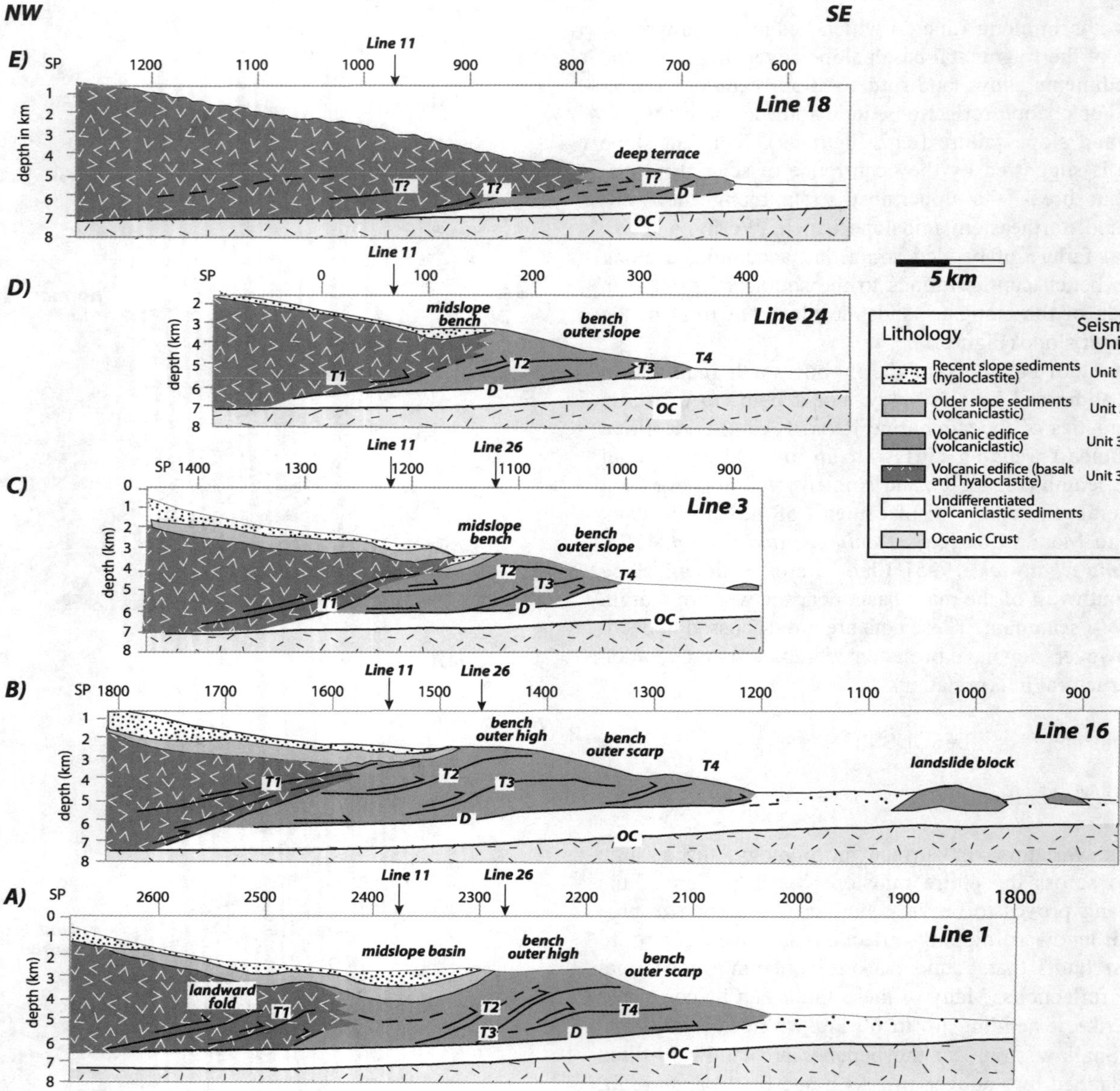

Figure 10. Selected interpreted depth sections of dip lines, shown with no vertical exaggeration. (A) MCS Line 1; (B) MCS Line 16; (C) MCS Line 3; (D) MCS Line 24; (E) MCS Line 18. Long thrust sheets are imbricated in the southeast, and a deep basin is developed behind the resulting bench. To the northeast, thin thrust sheets are vertically stacked, the upper flank is overthrust upon the bench, and the basin is absent. A low terrace near the base of Line 18 is the only disruption of the slope along the near Puna Ridge lines.

contrast, in the northeast region (Figure 10c-d), thrust T1 extends to the OC and truncates the shallower décollement horizon, D, the basin is lacking and replaced by a prominent uplift behind the narrow bench, and thrust faults define an antiformal stack within the bench. Within the Puna Ridge region (Figure 10e), the bench and basin are absent and a low terrace lies near the base of the flank. Along strike, Lines 11 and 26 (Figure 11) indicate dis-ruption and modest offset of the structurally lower faults, T2-3, along an east dipping lateral fault, L. The thrust sheet and units overlying fault T1 and its correlative horizon appear to extend unbroken above L, suggesting that the lateral fault is restricted to the bench strata. This meshes with the contrast in thrust geometry within the bench from southwest to

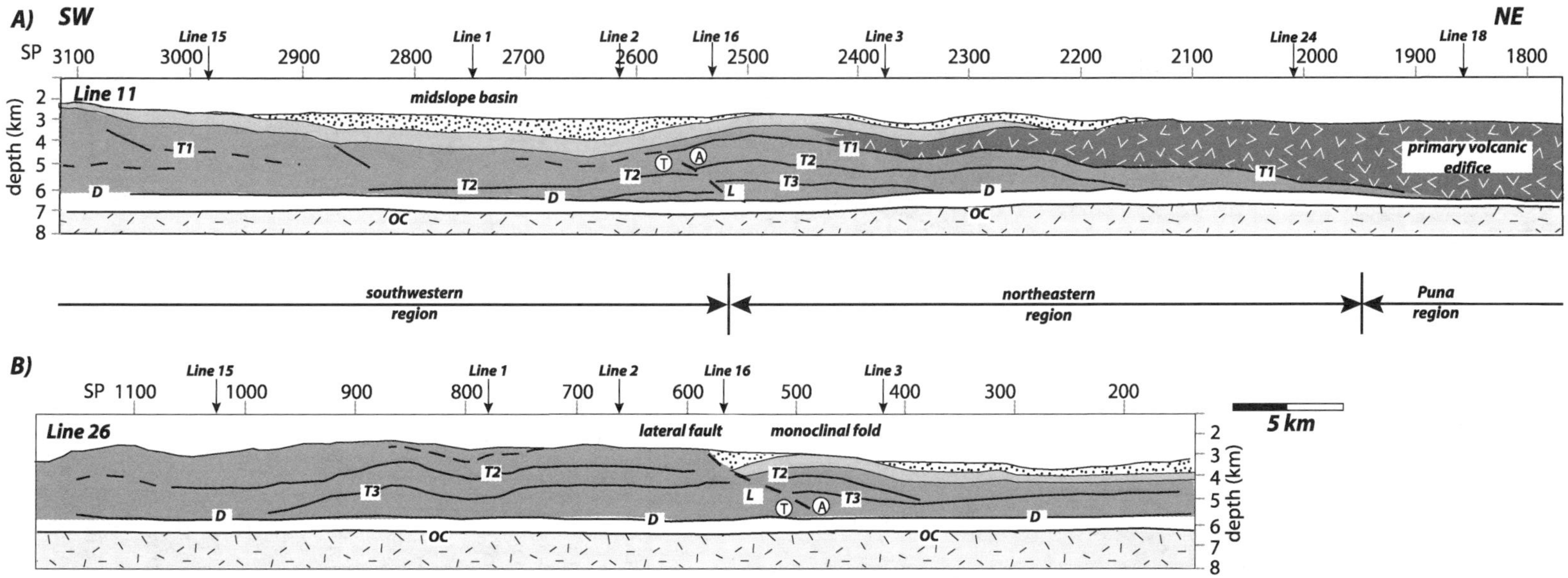

Figure 11. Interpreted depth sections of the strike lines across the study area, shown with no vertical exaggeration. (A) MCS Line 11; (B) MCS Line 26. Generally, thrust sheet thickness decreases to the northeast. To the northeast, Line 11 intersects a thickening wedge of primary volcanic material, Unit 3A, which has been overthrust upon a thinning wedge of volcaniclastic material, Unit 3B. Slope sediments also thin to the northeast. Symbols as on Figure 10.

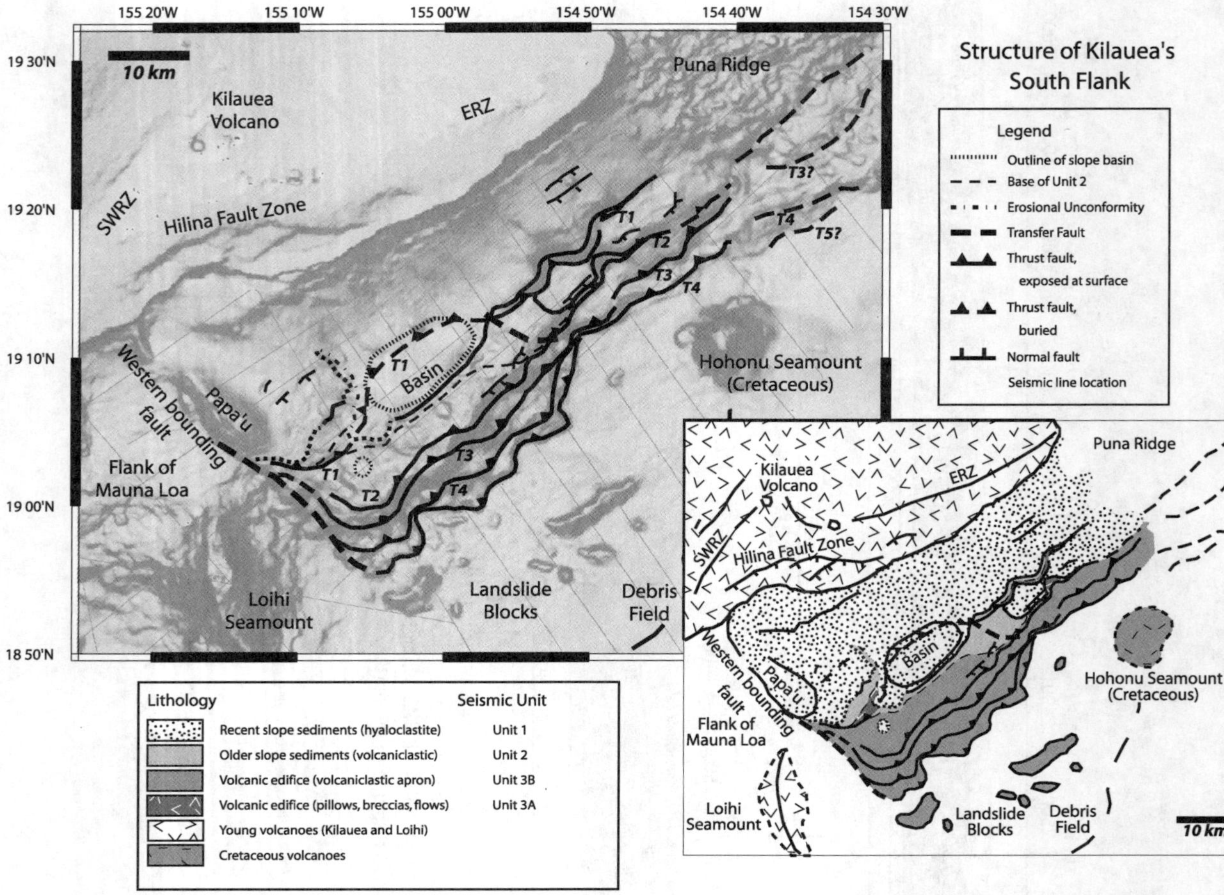

Figure 12. Structural map of the submarine south flank. Landward dipping thrust fault T1 is buried behind the basin to the southwest and carries the upper flank above the midslope bench to the northeast. Thrust faults, T2-T4, responsible for uplift of the bench, break the bench outer scarp and slope along its length. The upper flank is covered by hyaloclastic slope sediments, and the bench is dominated by volcaniclastic strata; primary volcanic edifice is exposed in few locales to the northeast [e.g., *Lipman et al.*, this volume], near the base of structurally higher thrust sheets. Large landslide blocks in front of the flank derive from local collapse of the bench outer high and scarp.

northeast. Horizon D extends seaward into strata flooring the Hawaiian Deep (Figure 10).

Thickness of thrust sheets also varies across strike. The structurally highest thrust sheet in the bench, overlying fault T2, is thinnest at 0.75 km where the basin is deepest (Line 11, near SP 2630, Figure 11a), and thickens to the south-west. The across-strike length for this upper thrust sheet is shown to be ~8-10 km on Line 1 (Figures 10a and 4d); this defines a minimum length, due to erosional truncation along the bench outer scarp. The correlative thrust sheet over-lying thrust T2 to the northeast (Line 24, Figure 10d) is no greater than 5 km in length, and has not been truncated along the outer scarp. The deeper thrust sheets appear to be thinner; the structurally lowest (most seaward) thrust sheets cannot be resolved on the seismic data, but are inferred to be ~0.1 km thick based on the height of sinuous terraces that form above them at the base of the flank.

The estimated lengths and numbers of thrust sheets within the midslope bench argue for very high cumulative displacements for the mobile south flank, of at least 30 km in the southeastern region (Figure 10a), where 3 thrust sheets are identified, and on the order of 5-10 km to the far northeast, e.g., Line 24 (Figure 10d). Assuming that each thrust sheet is composed of volcaniclastic debris, this would imply the presence of an extremely broad volcanic apron at the time of deformation, and the apron has subsequently become accreted to the distal portion of the sliding flank of Kilauea volcano.

Controls on Deformation Intensity and Surface Morphology

The high cumulative displacements of the flank recorded by the deformed flank and uplifted bench, here estimated at as much as 30 km, are inconsistent with shortening at the toe of a gravitationally driven (surficial) slump [*Morgan et al.*, 2000]. Instead, it seems that deformation along the distal flank is driven by more regional volcano-tectonic stresses. We hypothesize that such stresses result from volcanic spreading, the gravitational relaxation of the entire volcanic edifice, causing extension at the rift zones and shortening at the toe [*Denlinger and Okubo*, 1995; *Delaney et al.*, 1998]. The magnitude of this driving force will vary with the elevation and volume of the volcanic edifice, leading to high rates of displacement near the summit area relative to elsewhere on the volcano, as has been observed geodetically [e.g., *Owen et al.*, 2000]. Moreover, as the volcano elongates through lateral extension of the rift zones (for example, Puna Ridge), younger regions of the edifice will become involved in spreading process, but will reflect relatively lower cumulative displacements than do the more mature regions. These factors probably account in large part for the relative decrease in shortening and deformation from southwest to northeast along Kilauea's south flank.

The present deformational geometries of the south flank have also certainly been influenced by pre-existing morphology of the flank. A major recess may have been present landward of the current basin location. However, the broad influence of gravitational stresses driving seaward displacements of the flank will result in regional structures such as the laterally continuous thrust faults that we identify (Figure 12). Continuous displacements along such fault zones will therefore result in greater uplift and flank thickening at the edges of the recess relative to the center, in this case helping to preserve the midslope basin. Additionally, the thickness and extent of volcaniclastic debris in the surrounding volcanic apron will influence the length and thickness of thrust sheets accreted to the sliding flank. We recognize thrust sheets up to 2 km thick in the southwest region, and estimated displacements of at least 30 km suggest that an extremely wide apron was scraped up to form the broad, 4 km high midslope bench. Thinner thrust sheets to the northeast may explain the much lower profile in that region. However, the décollement level within the apron will also play a role; a very shallow décollement hori-zon will result in seaward sliding of the flank without further accretion, as may presently be the case to the north-east or along the Puna Ridge.

The strength of the sliding surface and deformability of flank strata will also influence final deformation geometries and slope morphology. The seaward step-up of the décollement horizon onto layered strata recognized seaward of thrust fault T1, suggests that the volcanic apron deposited in front of the flank provides a more favorable sliding surface than the top of the oceanic crust, presumably reflecting lower internal strength. Such changes in shear strength at the base of the volcano and its accreted material may also account for lateral variations in structural geometries. The lateral transfer zone reflects seaward offset of structures of the southwest region relative to the northeast. An increase in sliding resistance beneath the northeastern flank, possibly due to a thinning sedimentary sequence, rougher seafloor, presence of intercalated lava flows, or even absence of high pore fluid pressures, may have led to stacking of thinner thrust sheets in this region rather than seaward pushing and imbrication of thrust sheets in the southwestern region.

Relative Timing of Deformation

The landward dips of sediments at the seaward edge of the midslope basin provide a clear record of the sequence of deformation within the midslope bench. The top of Unit 2 marks a distinct unconformity, most apparent at the base of the midslope basin (Figure 5). The basin fill of Unit 1 onlaps dipping beds in Unit 2 along the seaward edge of the fold behind the basin; this depositional relationship implies that Unit 1 was laid down after tilting of the substrate, which is attributed to displacement along thrust faults T1 and T2. Continued displacement along these, or more recent thrust faults geometrically constrained to lie seaward of T2, led to further uplift and landward tilting of both Units 1 and 2.

The stratigraphic evidence for distinct episodes of deformation and overthrusting may help to explain the irregular upper surface of the bench outer high in the southwestern region. During an early stage of bench formation, the present upper surface would have defined the bench outer slope, presumably subject to slope failure just as has occurred along the current scarp. Subsequent displacements along more seaward faults would have caused landward rotation of the outer slope, generating the scalloped upper surface we see today (Figure 1). This model also suggests the origin of the elongate landslide blocks that currently lie seaward of the deformed flank. We speculate that they derived from the collapse of the previous bench outer slope, explaining some of the similarities in reflective character to that observed in the midslope bench.

CONCLUSIONS

We have analyzed the structure of Kilauea's mobile south flank using MCS data. Based on reflector divergence, stratal truncation, and along strike correlations, we identify 4 significant thrust faults that intersect the seafloor along the upper and outer slopes of the midslope bench, and several smaller ones near the base of the slope. The structurally higher thrust faults within the midslope bench extend up to ~60 km along strike, revealing a common origin for the midslope bench across the entire distal flank (Figure 12). More landward faults displace and fold the primary volcanic flank over the bench. Concordant reflections within the thrust sheets within the midslope bench indicate that they are composed of layered strata.

Analysis of the complete data set helps to refine the model proposed by *Morgan et al.* [2000]. The bench is constructed of volcaniclastic strata derived from successive flank failures, and deposited within a broad apron around the volcano. Seaward sliding of the mobile south flank results in overthrusting and progressive accretion of the volcaniclastic strata to the toe of the flank. From changes in thrust sheet thickness, we argue that the apron was thickest downslope of the summit region of Kilauea, which would have contributed to the apron for the longest time, and the apron tapered in the seaward and northeastward directions.

Thrust faults within the distal flank appear to sole into one of two detachment surfaces. The most landward thrust T1 extends to the OC reflector; the more seaward faults within the bench intersect a stratigraphically shallower horizon, D, within sediments beneath the midslope bench. This relationship indicates a step up in décollement level beneath the bench, most likely due to changes in the sliding resistance at the base of the flank. The primary flank composed of submarine erupted lavas is probably quite strong and resistant to internal deformation. The sedimented interface with the oceanic plate will define a relatively weak surface, localizing slip at that boundary. In contrast, volcaniclastic strata in front of the flank could be relatively weaker, allowing the décollement to step to a lower strength horizon within the unit. The ease with which the distal sedimentary unit deformed to build the uplifted benches is consistent with a seaward change in strength.

The lateral transfer zone marks a significant structural change in structural style, from imbricated thrust sheets to the southwest to antiformally stacked thrust sheets to the northeast. An increase in sliding resistance beneath the northeastern flank, possibly due to a thinning sedimentary sequence or rougher seafloor, may have led to stacking of thinner thrust sheets rather than seaward pushing and imbrication in the southwest region.

Acknowledgements. Numerous people assisted with the success of this study, especially the friendly and professional shipboard and technical crew and the scientific party of the R/V *Maurice Ewing,* who insured the collection of high quality data. Conversations with Brian Taylor and Michael Garcia greatly improved the clarity of the concepts presented. We also thank Taylor, Garcia, A. Taira and J-O. Park for critical reviews that helped to significantly improve this paper. This work was funded by National Science Foundation grant OCE-9711715. We also thank Landmark's University Grant Program for providing ProMAX processing software. SOEST contribution #5534.

REFERENCES

Borgia, A., J. Burr, W. Montero, L.D. Morales, G.E. Alvarado, Fault propagation folds induced by gravitational failure and slumping of the central Costa Rica volcanic range; implications for large terrestrial and Martian volcanic edifices *J. Geophys. Res., 95*, 14357-14382, 1990.

Borgia, A., and B. Treves, Volcanic plates overriding the oceanic crust: structure and dynamics of Hawaiian volcanoes, in *Ophiolites and their modern oceanic analogues*, edited by L.M. Parson, B.J. Murton, and P. Browning, pp. 277-299, The Geological Society, London, 1992.

Boyer, S.E., and D. Elliott, Thrust systems, *AAPG Bull., 66*, 1196-1230, 1982.

Carracedo, J. C., The Canary Islands; an example of structural control on the growth of large oceanic-island volcanoes, *J. Volc. Geotherm. Res.*, *60*, 225-241, 1994.

Chadwick, W.W., Jr., J.R. Smith, Jr., J.G. Moore, D.A. Clague, M.O. Garcia, C.G. Fox, Bathymetry of south flank of Kilauea volcano, HI, *U.S. Geol. Surv. Misc. Field Studies Map MF-2231*, 1993.

Clague, D. A., K. A. Hon, J. L. Anderson, W. W. Chadwick, Jr., C. G. Fox, Bathymetry of Puna Ridge, Kilauea Volcano, Hawaii, *Miscellaneous Field Studies Map - U. S. Geological Survey, MF-2237*, 1 sheet, 1994.

Crosson, R.S., and E.T. Endo, Focal mechanisms and locations of earthquakes in the vicinity of the 1975 Kalapana earthquake aftershock zone 1970-1979: Implications for tectonics of the south flank of Kilauea volcano, island of Hawai`i, *Tectonics*, *1*, 495-542, 1982.

Delaney, P.T., R.P. Denlinger, M. Lisowski, A. Miklius, P.G. Okubo, A.T. Okamura, and M.K. Sako, Volcanic spreading at Kilauea, 1976-1996, *J. Geophys. Res.*, *103*, 18003-18024, 1998.

Denlinger, R. and P. Okubo, Structure of the mobile south flank of Kilauea volcano, Hawai`i, *J. Geophys. Res.*, *100*, 24499-24507, 1995.

de Voogd, B., S.P. Palomé , A. Hirn, P. Charvis, J. Gallart, D. Rousset, J. Dañobeitia, and H. Perroud, Vertical movements and material transport during hotspot activity: Seismic reflection profiling offshore La R union, *J. Geophys. Res.*, *104*, 2855-2874, 1999.

Fiske, R.S., and E.D. Jackson, Orientation and growth of Hawaiian volcanic rifts; regional structure vs. the edifice effects, *EOS, Trans. AGU*, *54*, 372, 1971.

Furumoto, A.S., and R.L. Kovach, The Kalapana earthquake of November 29, 1975: an intraplate earthquake and its relation to geothermal processes, *Phys. Earth and Planet. Interiors*, *18*, 197-208, 1979.

Got, J.-L., J. Fréchet, and F.W. Klein, Deep fault plane geometry inferred from multiplet relative relocation beneath the south flank of Kilauea, *J. Geophys. Res.*, *99*, 15375-15386, 1994.

Hill, D.P., and J.J. Zucca, Geophysical constraints on the structure of Kilauea and Mauna Loa volcanoes and some implications for seismomagmatic processes, *U. S. Geol. Surv. Prof. Pap.*, *1350*, 1987.

Holcomb, R.T., J.G. Moore, P.W. Lipman, and R.H. Belderson, Voluminous submarine lava flows from Hawaiian volcanoes, *Geology*, *16*, 400-404, 1988.

Leslie, S.C., G.F. Moore, J.K. Morgan, and D.J. Hills, Seismic Stratigraphy of the Frontal Hawaiian Moat: Implications for Sedimentary Processes at the Leading Edge of an Oceanic Hotspot Trace, *Marine Geology*, in press.

Lindwall, D.A., A two-dimensional seismic investigation of crustal structure under the Hawaiian Islands near Oahu and Kauai, *J. Geophys. Res.*, *93*, 12,107-12,122, 1988.

Lipman, P.W., J.P. Lockwood, R.T. Okamura, D.A. Swanson, and K.M. Yamashita, Ground deformation associated with the 1975 magnitude-7.2 earthquake and resulting changes in activity of Kilauea volcano, Hawai`i, *U. S. Geol. Surv. Prof. Pap.*, *1275*, 1985.

Lipman, P.W., T.W. Sisson, T. Ui, J. Naka, J.K. Morgan, and J.R. Smith, Ancestral submarine growth of Kilauea Volcano and instability of its south flank, this volume.

Lonsdale, P., A geomorphological reconnaissance of the submarine part of the East Rift Zone of Kilauea Volcano, Hawai`i, *Bull. Vol.*, *51*, 123-144, 1989.

Moore, J.G., and H.L. Krivoy, The 1962 flank eruption of Kilauea Volcano and the structure of the East Rift Zone, *J. Geophys. Res.*, *69*, 2033-2045, 1964.

Moore, J.G., and W.W. Chadwick, Offshore geology of Mauna Loa and adjacent areas, Hawai`i. in *Mauna Loa Revealed: Structure, Composition, History, and Hazards*, edited by J.M. Rhodes and Lockwood, J.P., pp. 21-44, Am. Geophys. Union, Washington, D.C., 1995.

Moore, J.G., and R.S. Fiske, Volcanic substructure inferred from dredge samples and ocean-bottom photographs, Hawaii, *Geol. Soc. Amer. Bull.*, *80*, 1191-1201, 1969.

Moore, J.G., D.A. Clague, R.T. Holcomb, P.W. Lipman, W.R. Normark, and M.E. Torresan, Prodigious submarine landslides on the Hawaiian Ridge, *J. Geophys. Res.*, *94*, 17,465- 17,484, 1989.

Moore, J.G., W.R. Normark, and R.T. Holcomb, Giant Hawaiian Landslides, *Ann. Rev. of Earth and Plan. Sci.*, *22*, 119-144, 1994.

Morgan, J.K., G.F. Moore, D.J. Hills, and S.C. Leslie, Overthrusting and sediment accretion along Kilauea's mobile south flank, Hawai`i: Evidence for volcanic spreading from marine seismic reflection data, *Geology*, *28*, 667-670, 2000.

Naka, J., and Shipboard Scientific Group, Tectono-magmatic processes investigated at deep-water flanks of Hawaiian volcanoes, *EOS, Trans. AGU.*, *81*, 221, 226-227, 2000.

Nakamura, K., Why do long rift zones develop in Hawaiian volcanoes; a possible role of thick oceanic sediments, *Bull. Volc. Soc. of Japan*, *25*, 255-269, 1980.

Owen, S., P. Segall, J. Freymueller, A. Miklius, R. Denlinger, T. Arnadottir, M. Sako, and R. Burgmann, Rapid deformation of the south flank of Kilauea volcano, Hawai`i, *Science*, *267*, 1328-1331, 1995.

Owen, S.P., P. Segall, M. Lisowski, A. Miklius, R. Denlinger, and M. Sako, Rapid deformation of Kilauea Volcano: Global Positioning System measurements between 1990 and 1996, *J. Geophys. Res.*, *105*, 18,983-18,998, 2000.

Smith, J.R., A.N. Shor, A. Malahoff, and M.E. Torresan, Southeast flank of Island of Hawaii, SeaBeam multibeam bathymetry, HAWAII MR1 sidescan sonar imagery, and magnetic anomalies, Sheet a4. Hawaii Institute of Geophysics and Planetology, Honolulu, 1994.

Smith, J.R., A. Malahoff and A.N. Shor, Submarine geology of the Hilina slump and morpho-structural evolution of the Kilauea volcano, Hawai`i, *J. Volc. Geotherm. Res.* *94*, 59-88, 1999.

Stearns, H.T., and W.O. Clark, Geology and water resources of the Kau District, Hawaii (including parts of Kilauea and Mauna Loa volcanoes), *U.S. Geol. Surv., Water Supply Pap.*, *616*, 194 pp., 1930.

Stolt, R.H., Migration by Fourier transform, *Geophysics*, *43*, 23-48, 1978.

Swanson, D.A., W.A. Duffield, and R.S. Fiske, Displacement of the south flank of Kilauea volcano: The result of forceful intrusion of magma into the rift zones, *U. S. Geol. Surv. Prof. Pap.*, *963*, 1976.

Taylor, B., G.F. Moore, and S. Leslie, Attenuation of reverberation using radon transforms: Example of R/V *Ewing* data from offshore Hawai`i, *EOS Trans. AGU*, *79*, 1008, 1998.

ten Brink, U.S., and T.M. Brocher, Multichannel seismic evidence for a subcrustal intrusive complex under Oahu and a model for Hawaiian volcanism, *J. Geophys. Res.*, *92*, 13,687-13,707, 1987.

Thurber, C.H., and A.E. Gripp, Flexure and seismicity beneath the south flank of Kilauea volcano and tectonic implications, *J. Geophys. Res.*, *93*, 4271-4278, 1988.

Tilling, R.I., R.Y. Koyanagi, P.W. Lipman, J.P. Lockwood, J.G., Moore, and D.A. Swanson, Earthquake and related catastrophic events, Island of Hawaii, November 29, 1975; a preliminary report, *U.S.Geol. Surv. Circular*, *C 0740*, 32, 1976.

Watts, A.B., and U.S. ten Brink, Crustal structure, flexure, and subsidence history of the Hawaiian Islands, *J. Geophys. Res.*, *94*, 10,473-10,500, 1989.

Wolfe, C.J., M.K. McNutt, and R.S. Detrick, The Marquesas archipelagic apron: Seismic stratigraphy and implications for volcano growth, mass wasting, and crustal underplating, *J. Geophys. Res., 99*, 13591-13608, 1994.

Zucca, J.J., and D.P. Hill, Crustal structure of the south east flank of Kilauea Volcano, Hawai`i, from seismic refraction measurements, *Bull. Seis. Soc. Am., 70*, 1149-1159, 1980.

D.J. Hills, 1680 East West Road, Dept. of Geology and Geophysics, University of Hawaii, Honolulu, Hawaii, 96822, *djhills@soest.hawaii.edu*

S.C. Leslie and G.F. Moore, 1680 East West Road, Dept. of Geology and Geophysics, University of Hawaii, Honolulu, Hawaii, 96822

J.K. Morgan, Dept. of Earth Science, MS-126, 6100 Main St., Rice University, Houston, TX, 77005-1892

Volcanic Morphology of the Submarine Puna Ridge, Kilauea Volcano

Deborah K. Smith[1], Laura S. L. Kong[2], Kevin T. M. Johnson[3], Jennifer R. Reynolds[4]

Flow fields, pillow ridges, cones, and terraces are constructed on the Puna Ridge. Three basic flow morphologies are observed: pillow, lobate, and sheet flows. All three are found in each of the seven camera sites 660-3700 m deep. Pillow ridges are 500 m to a few kilometers long, and their heights reach to a few tens of meters. Cones are circular in plan, up to ~140 m high, and stand above surrounding topography on all sides. Terraces are symmetric in plan, have horizontal flat tops, and are up to several hundred meters high on their down-slope flanks. Terrace height-to-diameter ratios are commonly half that of cones (0.05 and 0.1, respectively). Other conditions being equal, feeder systems must have a larger excess magma pressure to build cones upwards, compared to terraces. Skylights mark the locations of lava tubes on terrace tops. Tumuli indicate that hydrostatic pressure was large enough to crack the overlying crust of the terrace tops. Small pillow cones (< 100 m high) are likely rootless vents, fed from lava that flowed through the cracks. Terraces are interpreted to grow upwards and outwards by inflation from within. A few simple trends in overall morphology are a function of water depth. Only two fissure vents are identified, both on the shallow ridge crest (<1100 m depth). Large central-vent eruptions dominate farther down the ridge crest. Large cratered cones are observed only at crestal depths between 600-2200 m, perhaps because of longer residence time and buildup of volatiles within a magma reservoir beneath this section of the ridge.

1. INTRODUCTION

Kilauea Volcano, Hawai`i (Figure 1) has two rift zones: the southwest rift zone (SWRZ) and the east rift zone (ERZ). Most studies have focused on the more active ERZ. Observations of recent eruptions [e.g., *Swanson*, 1973; *Swanson et al.*, 1976a; *Tilling et al.*, 1987; *Wolfe et al.*, 1987; *Kauahikaua et al.*, 1998], geologic mapping [e.g., *Holcomb*, 1987; *Moore and Trusdell*, 1991], and geophysical studies [e.g., *Klein et al.*, 1987; *Delaney et al.*, 1998] have provided a good understanding of the magmatic and volcanic processes along the 55 km of the ERZ from the summit to the shore. In contrast, relatively little is known about magmatic and volcanic processes along the 75-km-long submarine section called Puna Ridge.

Here we report the results of a high-resolution study of the morphology of the crest of the Puna Ridge using a near-bottom side-scan sonar and a photoimagery system. We describe the volcanic and tectonic landforms, and examine patterns in their formation to interpret dike emplacement and eruption mechanisms.

[1]Department of Geology and Geophysics, Woods Hole Oceanographic Institution, Woods Hole, Massachusetts

[2]Hawaii Institute of Geophysics and Planetology, University of Hawaii, Honolulu, Hawaii

[3]Department of Natural Sciences, Bishop Museum, Honolulu, Hawaii

[4]School of Fisheries and Ocean Sciences, University of Alaska, Fairbanks, Alaska

Hawaiian Volcanoes: Deep Underwater Perspectives
Geophysical Monograph 128
Copyright 2002 by the American Geophysical Union

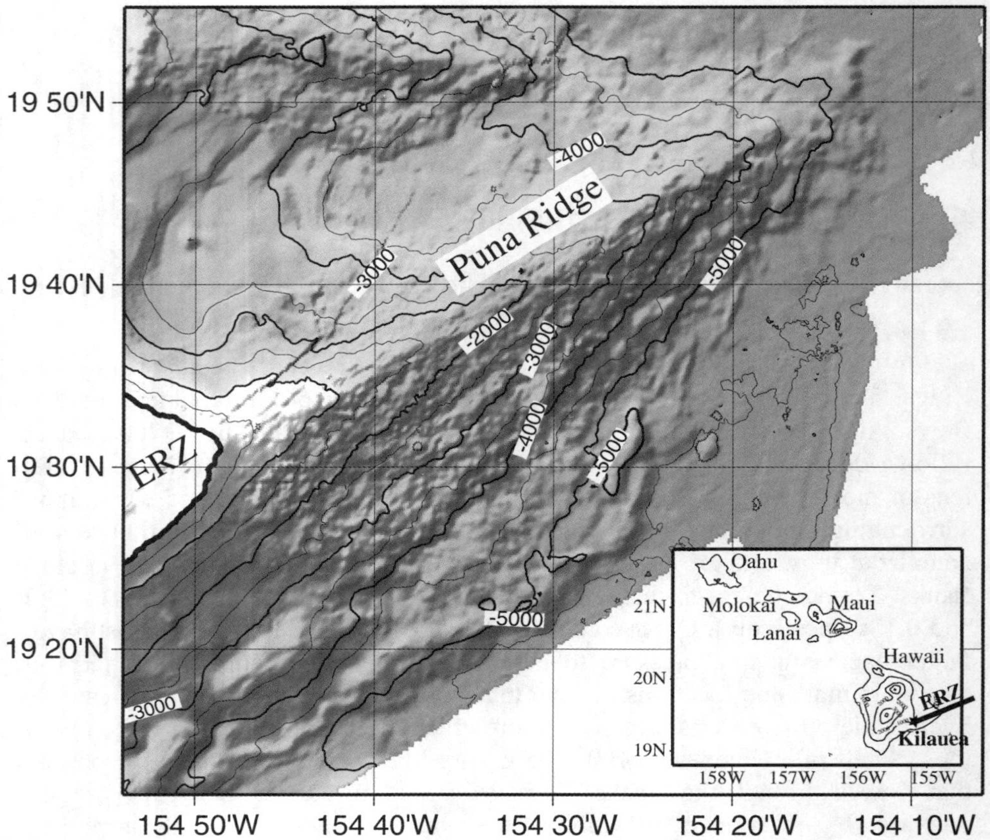

Figure 1. Shaded relief bathymetry of the Puna Ridge, 500-m contours [*Clague et al.*, 1994; R. T. Holcomb, unpublished data] Inset shows the summit of Kilauea Volcano (star), and the East Rift Zone (line).

2. BACKGROUND

Existing studies provide a broad framework within which our high-resolution data can be interpreted. Multibeam bathymetry data for the entire Puna Ridge [*Clague et al.*, 1994; *Monterey Bay Aquarium Research Institute (MBARI)*, 2000], along with limited high resolution side-scan sonar images [*Lonsdale*, 1989], photographic imagery [*Moore and Fiske*, 1969; *Clague et al.*, 1988; *Lonsdale*, 1989], and submersible dive observations [*Fornari et al.*, 1978; *Clague et al.*, 2000] verify that the Puna Ridge crest is the location of dike intrusions and fissure eruptions. GLORIA side-scan sonar data provide a regional framework of the Puna Ridge [*Holcomb et al.*, 1988]. Sea-surface magnetic data show a normally polarized anomaly centered over the ridge axis, consistent with the presence of a 11-km wide, 70-km long, nearly-vertical magnetic source, presumably describing a dike complex at depth [*Malahoff and McCoy*, 1967].

Relatively few dike intrusions have been recorded from the summit of Kilauea to the Puna Ridge in historic time. Submarine lavas from the crest and flanks of the Puna Ridge appear to be older than most subaerial lavas. *Holcomb* [1987] estimated that 70% of the subaerial portion of Kilauea is younger than ~500 years. Based on palagonite thicknesses, *Clague et al.* [1995] estimated that dredged lavas from the Puna Ridge are 700-24,000 years old, mostly 2000-7000 years old. The most recent submarine eruptions are thought to have occurred in 1790, 1884, and 1924 [*Stearns and Macdonald*, 1946]. The 1884 eruption was witnessed just offshore at 20 m water depth. In 1790 and 1924, magma withdrawal from the summit reservoir is inferred to have fed submarine eruptions on the Puna Ridge [*Stearns and Macdonald*, 1946]. *Klein et al.* [1987] noted a lack of seismicity along the ridge since 1960, suggesting recent inactivity.

The slope of the subaerial crest of the ERZ is fairly constant at ~23 m/km, but the Puna Ridge is much steeper [*Lonsdale*, 1989]. The Puna Ridge reaches a water depth of ~5400 m over the length of 75 km, yielding an average gradient of ~73 m/km. The gradient increases from ~51 m/km above ~2700 m depth, to ~95 m/km below that depth. From sea level to ~2100 m depth, the ridge axis trends 065° [*Lonsdale*, 1989], in line with the lower ERZ. The axis has a broad, 3-4 km wide crest, with a small (~1 km) right-stepping offset near a depth of 2000 m. Volcanic constructs, to 140 m high, are scattered along the ridge crest, especially at the shallower end. From about 2100-3400 m depth, the axis of the Puna Ridge bends to 045° [*Lonsdale*, 1989], and the crest narrows. At ~2700 m depth, the axis has a right-stepping offset, and the along-axis slope increases. Using a towed side-scan instrument, *Lonsdale* [1989] observed tightly clustered fissures and collapse pits concentrated in an inflated portion of the ridge above the offset, and suggested that intrusive rather than extrusive magmatism predominates. Extrusives were primarily pillow flows. Below 3400 m depth, the ridge crest is poorly defined. The axis is composed of 'steps' that *Lonsdale* [1989] interpreted as flow fronts steepened by mass wasting. GLORIA images [*Holcomb et al.*, 1988] and ground-truth sampling and photography [*Clague et al.*, 1988] indicate that sheet flows cover large areas north and southeast of the ridge base.

The flank slopes of the Puna Ridge are 160-275 m/km [*Fornari et al.*, 1978; *Smith and Cann*, 1999], steeper than on the subaerial ERZ (~50 m/km). The slopes on the south flank of the ridge do not change significantly with distance from the shoreline until the crest is deeper than 3500 m. The slopes are steepest on the north flank between ~1200-2000 m depth, and multibeam bathymetry data suggest a landslide scarp [e.g., *Swanson et al.*, 1976b; *Moore and Chadwick*, 1995].

3. DATA COLLECTION

In October 1998 the Woods Hole Oceanographic Institution Deep Submergence Lab's DSL-120 deep-towed vehicle obtained side-scan sonar data along the entire crest of the Puna Ridge and part of its south flank (Figure 2). The DSL-120 vehicle is towed at an average speed of 1.25 knots (1 knot = 0.5 m/s) at ~100 m above the seafloor. Pixel size of the images is typically less than 5 m. Side-scan images provide detailed morphological and structural information that can be used to map faults and fissures, volcanic landforms, and regions of mass wasting. A triaxial magnetometer on the DSL-120 vehicle obtained high-resolution magnetic data [*Smith et al.*, 2001]. To supplement the side-scan images, we collected ARGO II photographic images at seven sites (Figure 2). ARGO II is towed at altitudes of ~3-15 m above the seafloor at a speed of ~0.5 knots. The ARGO II data are useful for evaluating the detailed morphology of volcanic features [e.g., *Parfitt et al., 2001*] and determining the extent of sediment cover and landslide debris. Both video and electronic still-camera

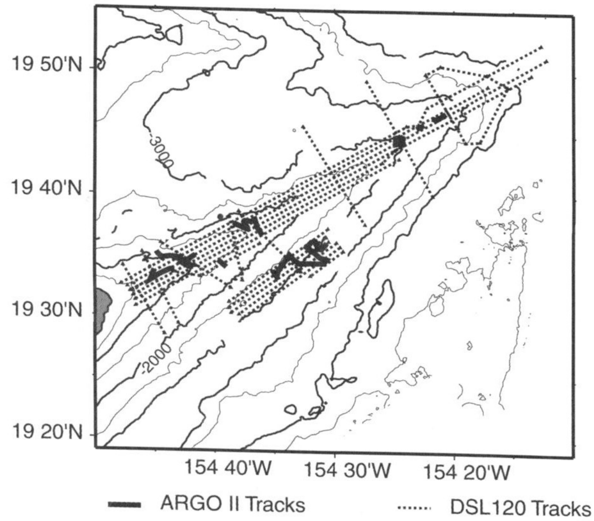

Figure 2 Surveys on the crest and flank of the Puna Ridge. Tracklines of the DSL-120 kHz side-scan sonar, shown as dotted lines. Tracklines of the ARGO II photoimagery survey, shown as thick solid lines.

images were collected. High-resolution bathymetric profiles of the seafloor were generated by combining the depth of the ARGO II vehicle, determined by a pressure gauge, and the altitude of the vehicle above the seafloor, obtained from a 100 kHz down-looking altimeter. The ARGO II imagery aided in the selection of sites for wax-core and dredge sampling for petrologic study [*Johnson et al.*, this volume]. All surveying and sampling were navigated within a network of bottom-moored, long-baseline transponders with horizontal relative errors of <5 m.

Existing SeaBeam and EM300 multibeam bathymetry [*Clague et al.*, 1994; *MBARI*, 2000] were used to constrain the shapes and volumes of volcanic features. The grid spacing is ~150 m for the SeaBeam data, ~30 m for the EM300 data.

4. VOLCANIC AND TECTONIC FEATURES ON THE CREST OF THE PUNA RIDGE

Detailed bathymetry (Figure 3) and geologic map interpretation (Plate 1) show that flow fields, pillow ridges, cones, and terraces are present on the crest of the Puna Ridge.

4.1. Flow Fields

Large areas with little or no relief or distinct morphological features are called flow fields. Bottom photographs show that these areas are covered by one to three lava morphologies [e.g., *Fox et al.*, 1988]: (1) Sheet flows are

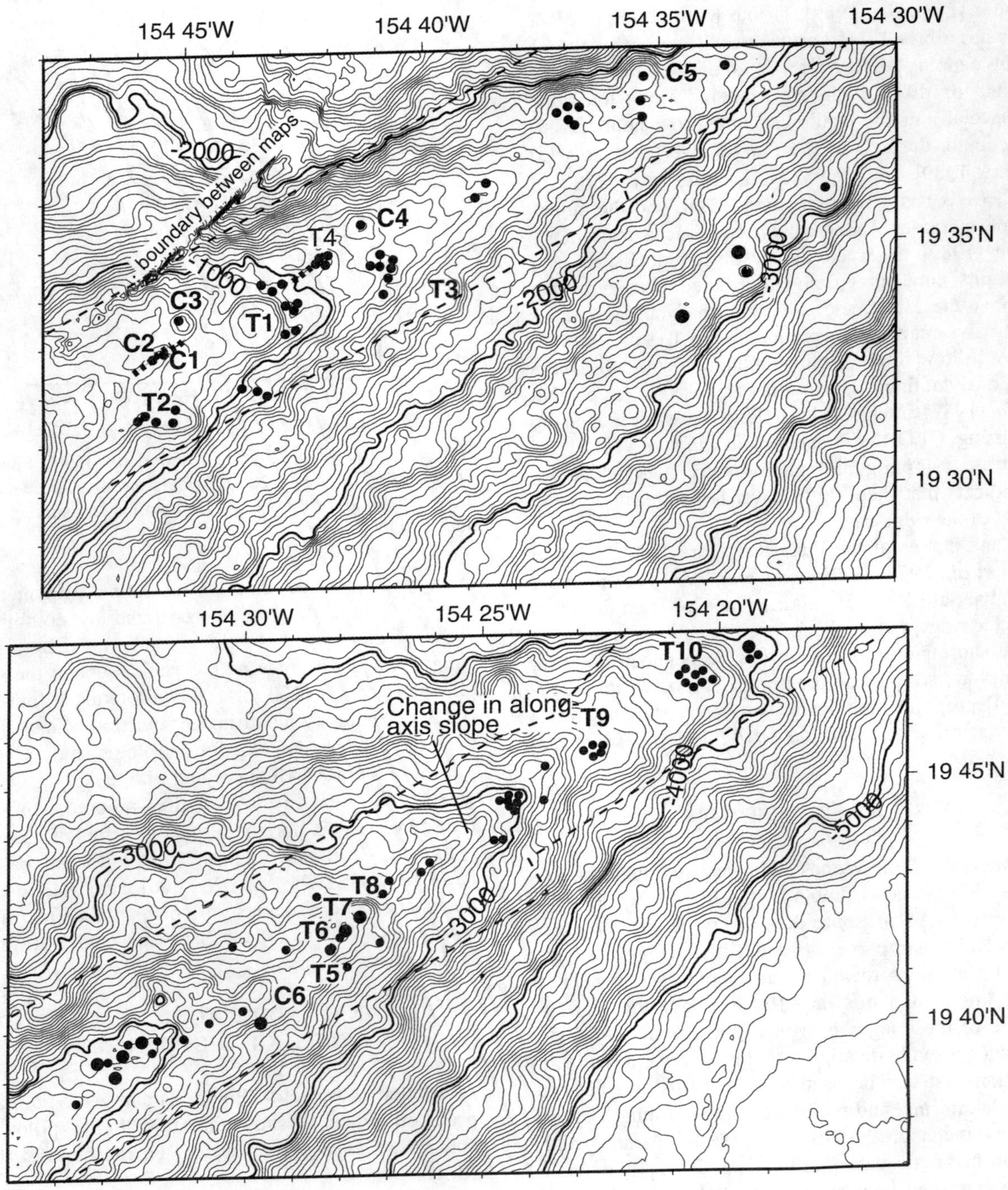

Figure 3. Bathymetric map of the Puna Ridge contoured at 50-m intervals [*Clague et al.*, 1994; R. T. Holcomb, unpublished data]. The map is in two parts: shallow and deep. The boundary between multibeam data to the southeast and lower-resolution single-beam data to the northwest is marked in the upper part. Dashed lines outline the extent of the DSL-120 survey along the ridge crest. C, cone; T, terrace. Black circles, locations of craters. Thick dashed lines, fissure vents (see Figures 5 and 10).

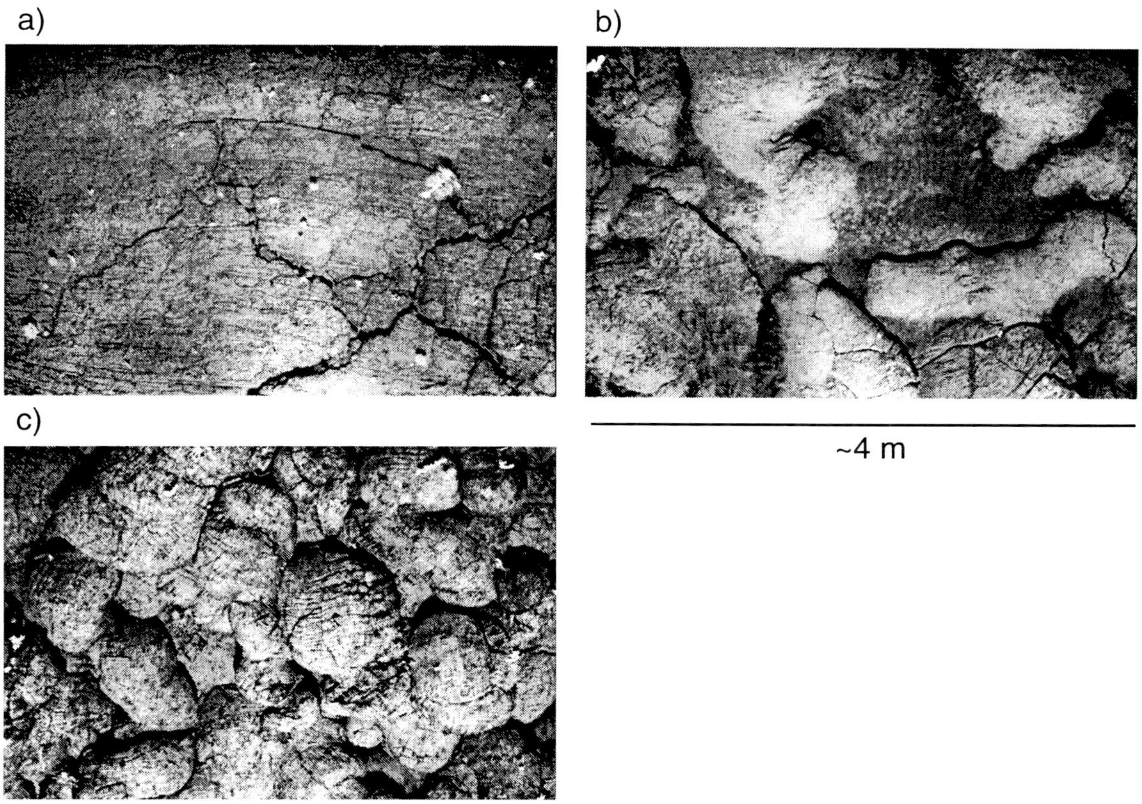

Figure 4. ARGO II photoimages of lava morphologies on the Puna Ridge, each ~4 m in width. (a) Sheet flow, smooth lava flow with a relatively featureless surface. (b) Lobate flow, low relief flow consisting of interconnected lobes. (c) Pillow flow, bulbous and elongate tubes with extrusion marks on their surfaces.

generally smooth lava flows with linear flow features (Figure 4a). They can be disrupted into jumbled, irregular, blocky pieces. (2) Lobate flows are interconnected low-relief lobes of lava (Figure 4b). (3) Pillow flows can be bulbous, elongate, or tubular (Figure 4c). Pillow flows are distinguished from lobate flows by extrusion marks on the surfaces. These lava morphologies are present at all water depths, unrelated to distance along the rift zone.

4.2. Pillow Ridges

Pillow ridges are composed of pillow flows along fissures [*Fornari et al.*, 1978]. Pillow ridges are morphologically similar to spatter ramparts on the subaerial ERZ, although the eruption dynamics differ [*Parfitt et al., 2001*]. Pillow ridges are 500 m to a few kilometers long, up to a few tens of meters wide, and a few to tens of meters high [*Fornari et al.*, 1978]. When sides are steep, they are called pillow walls.

Two distinctive pillow ridges at the shallow end of the Puna Ridge are associated with surface cracks paralleling the strike of the ridge (Figure 5). The two pillow ridges are 40-200 m wide and 9-20 m high [*Parfitt et al., 2001*]. They are composed primarily of bulbous pillows, in abrupt contact with underlying older flows. No other pillow ridges are observed deeper along the ridge, although *Fornari et al.* [1978] reported pillow walls during submersible dives at depths of 1500 m and 1850-1990 m.

4.3. Cones

Cones are circular in plan, and stand above the surrounding terrain on all sides. They are up to ~140 m high (Table 1), and cone height is about one-tenth of the basal diameter, typical of volcanoes in the Pacific basin [*Smith*, 1988]. Cone profiles range from pointed to flat-topped. The flatness, defined as the ratio of the top to the basal diameter, varies from 0 (pointed) to 0.7 (flat-topped).

Six large cones on the ridge crest have heights greater than 100 m (C1-C6, on Figure 3, Plate 1). All have flat tops and craters. For example, the C1 cone is ~100 m high (Table 1), with an ~50 m deep summit crater. Flanks are

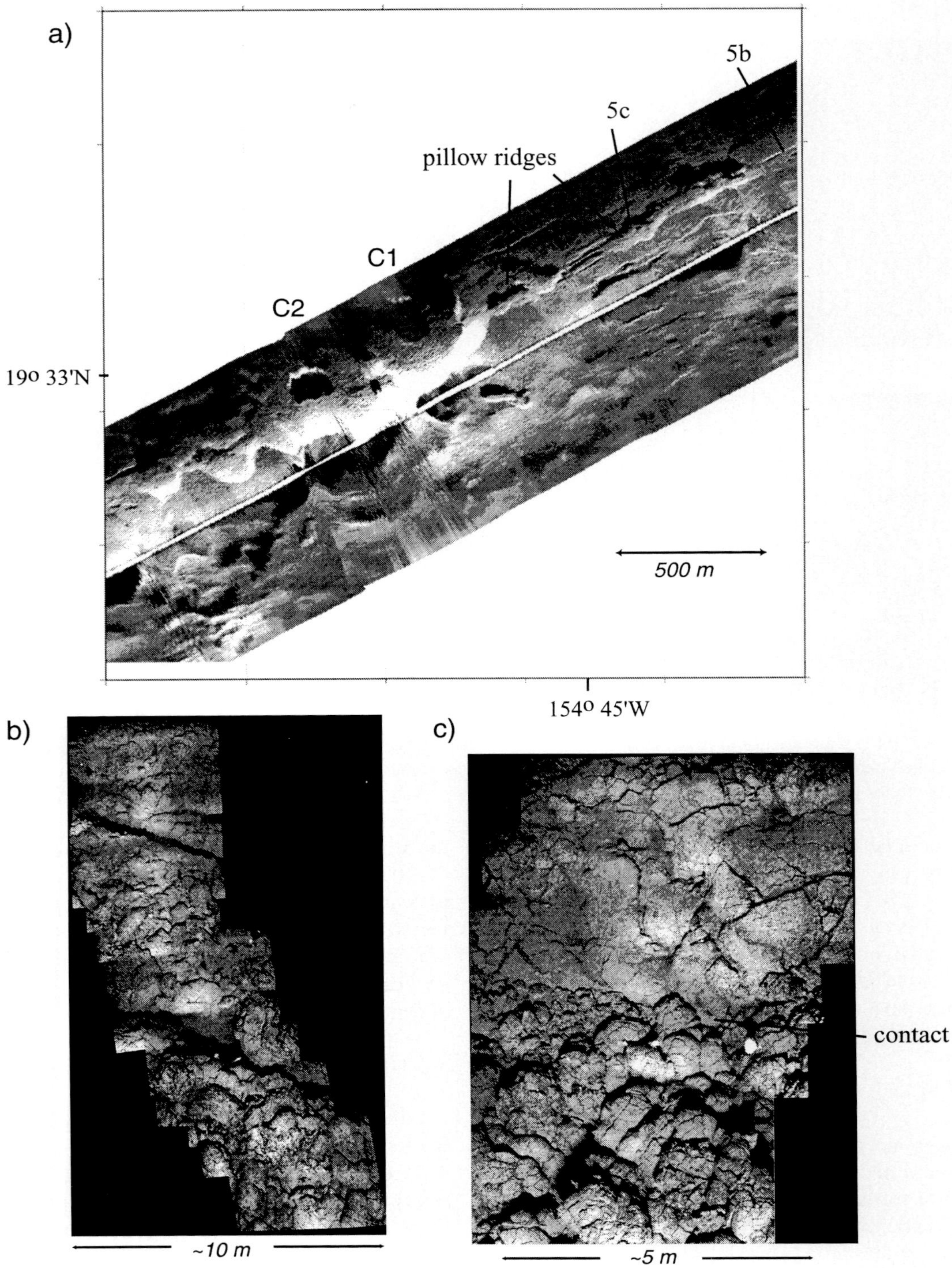

Figure 5. Products of a fissure eruption at the shallow end of the Puna Ridge. (a) Side-scan sonar image. Light is a reflection, dark a shadow. Fish track is in swath center. Craters marked as in Figure 3 and Plate 1. The deposit is ~3 km long and may consist of a number of flows associated with surface cracks paralleling the ridge. Locations of photomosaics in (b) and (c) are shown. (b) ARGO II photomosaic of ridge-parallel cracks. (c) Pillow ridges composed of bulbous pillows; contact between the pillows and underlying older flows is labeled.

Plate 2. Three-dimensional, perspective view of the shallow Puna Ridge, using EM300 swath bathymetry [*MBARI*, 2000]. Labeled craters and terraces correspond to those on Figure 3 and Plate 1.

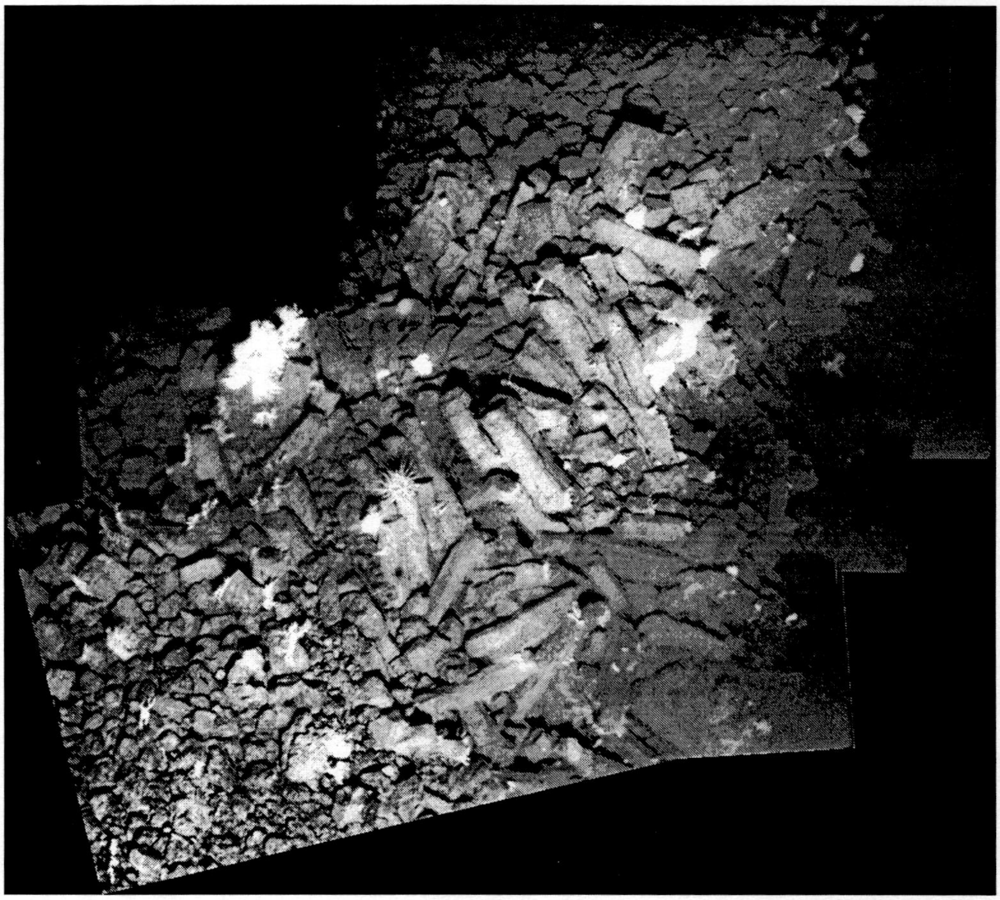

~15 m

Figure 6. ARGO II photomosaic of columnar joints that have tumbled out of a breach in the crater on cone C3 (Figure 3, Plates 1-2).

covered by rubble shed from its steep upper slopes. Mixed lobate, pillow, and sheet flows surround the crater on the bench. The C3 cone (Figure 3, Plates 1-2) is ~140 m high and has a crater similar in size to cone C1 (~50 m deep). On this cone rocks with well-formed columnar joints were photographed near a breach on the northeast side of the crater, and columns a few meters long have tumbled down the northeast flank (Figure 6). The columns indicate that flows at least a few meters thick cooled within the crater. The sides of the cone are covered with rubble, sediment, and perhaps hyaloclastite indicative of a phreatomagmatic eruption [*Lonsdale and Batiza*, 1980]. The largest volume cone identified on the Puna Ridge is C5 (Figure 3, Plate 1). Its volume is ~100 x 10^6 m^3 (Table 1) and it has a crater ~20 m deep. Relatively deep craters on all the large cones suggests that lava drained from the edifices after the eruption stopped.

A second class of cones is small and pointed in profile. Heights of these cones are <100 m and flatnesses are <0.5. Elongate and tubular pillows and rubble commonly cover their flanks (Figure 7). Because of their small size, we interpret them to be rootless vents, marking sites where excess pressure caused lava to break through the crust. Hummocky or lumpy topography, on a length scale of hundreds of meters and heights of a few to tens of meters, marks regions covered by pillow cones.

4.4. Terraces

Terraces are near-circular flat-topped features up to several kilometers wide and a few hundreds of meters high. Terraces are common along the crest and the flanks of the Puna Ridge. They have little or no relief on their up-slope side. Terraces have height-to-diameter ratios about half

Table 1. Sizes of Selected Cones and Terraces

Feature	Height (*H*)*, m	Basal diameter (*D*), m	Top diameter (*d*), m	Flatness (*d/D*)	Volume, 10^6 m^3	Height-to-diameter ratio (*H/D*)
C1	100	826	556	0.7	21	0.12
C2	80	800	558	0.7	19.2	0.1
C3	140	1000	500	0.5	64	0.14
C5	120	1300	800	0.7	101	0.09
T1	130	2400	1600	0.7	414	0.05
T2	100	1800	1400	0.8	200	0.05
T4	40	1800	1500	0.8	28	0.5
T5	35	550	350	0.6	6	0.06
T6	70	1150	650	0.6	45	0.06
T7	75	110	875	0.8	57	0.07

* average heights given where feature formed on a pre-existing slope.

that of the cones (0.05 compared to 0.1 for cones, Table 1). Terraces have flatness values >0.5; in places they form staircases, one on top of the next. Two styles of terraces are observed: (1) terraces whose tops are flat or slightly domed and have fine-scale inflationary features, and (2) terraces with one or more summit craters.

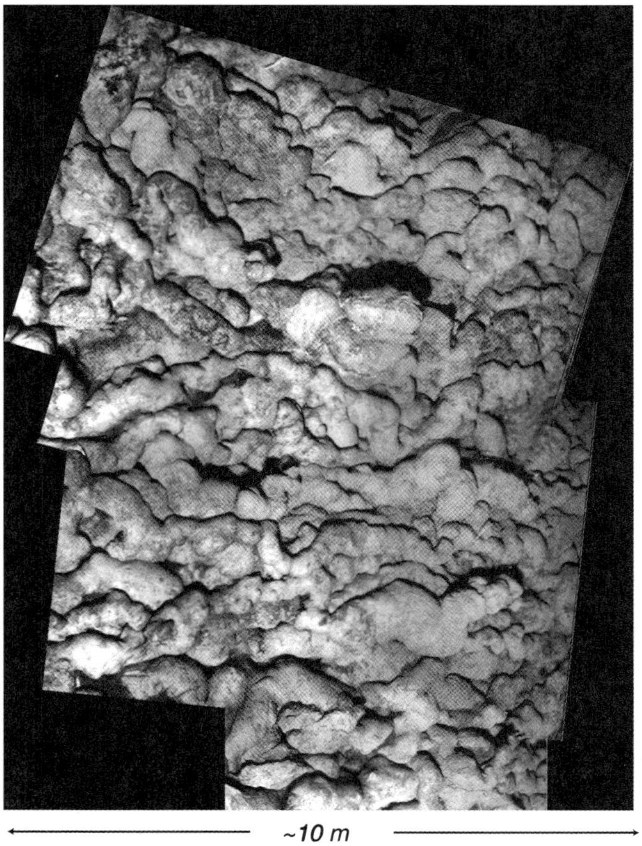

Figure 7. Flanks of a pillow cone, covered with tubular pillows. ARGO II photomosaic; downslope is to the right.

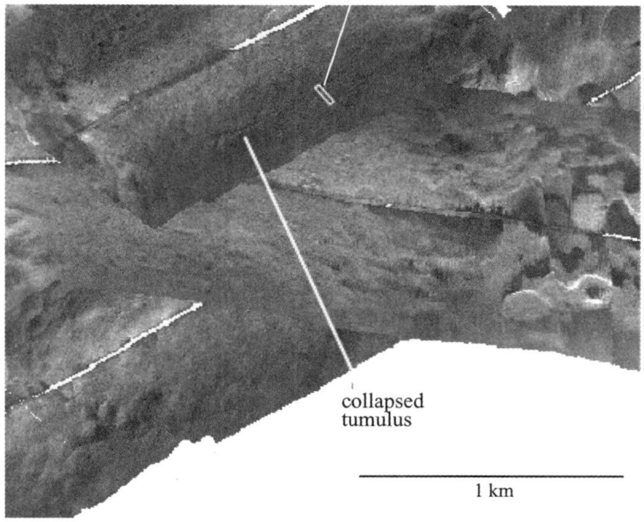

Figure 8. Top of Maka Nui (terrace T1; Figure 3, Plates 1-2), slightly domed with a large crack (white line) near its center. Interpreted as a collapsed tumulus (see text for details).

A distinctive terrace on the ridge crest at a depth of about 900 m (we refer to it as Maka Nui--Big Eye) is about 60-80 m high on its up-rift side and 150 m high on its down-rift side (Figure 3, Plates 1-2, Table 1). The center of Maka Nui is domed, rising ~20 m above its relatively flat top covered by pillow, lobate, and sheet flows. Several minor collapse features appear to be skylights, marking a network of tubes. A large crack (5-6 m wide, ~400 m long) has overhanging walls near the center of the dome (Figure 8). Because the crack occurs on a topographic high, we interpret it to be a tumulus that has collapsed into a tube as described on inflated subaerial flows [*Kauahikaua et al.*, 1998]. Tumuli (Figure 9) form when the pressure within a lava tube or flow is sufficient to bow up and crack the overlying lava crust [*Walker*, 1991].

A similar flat-topped terrace (T2, Figure 3, Plates 1-2), along the break in slope between the crest and the south flank at a depth of ~900 m, is similar to Maka Nui in that it has no summit crater and the top has inflation features. Two lines of small, cratered cones oriented 090°-100° lead from the crest of the Puna Ridge to the top of the terrace (Plates 1-2). They may mark primary eruptive fissures, but their orientations differ from those of fissures on the ridge crest (045°). In some places fissure orientations do deviate from the strike of a rift zone; for example, along rift zones of Mauna Loa [*Lockwood and Lipman*, 1987]. Because no similar features are known elsewhere on the Puna Ridge, however, we interpret the cones to be secondary vents built above lava tubes that fed the terrace. Tube-fed terraces have been mapped at the slow spreading Mid-Atlantic

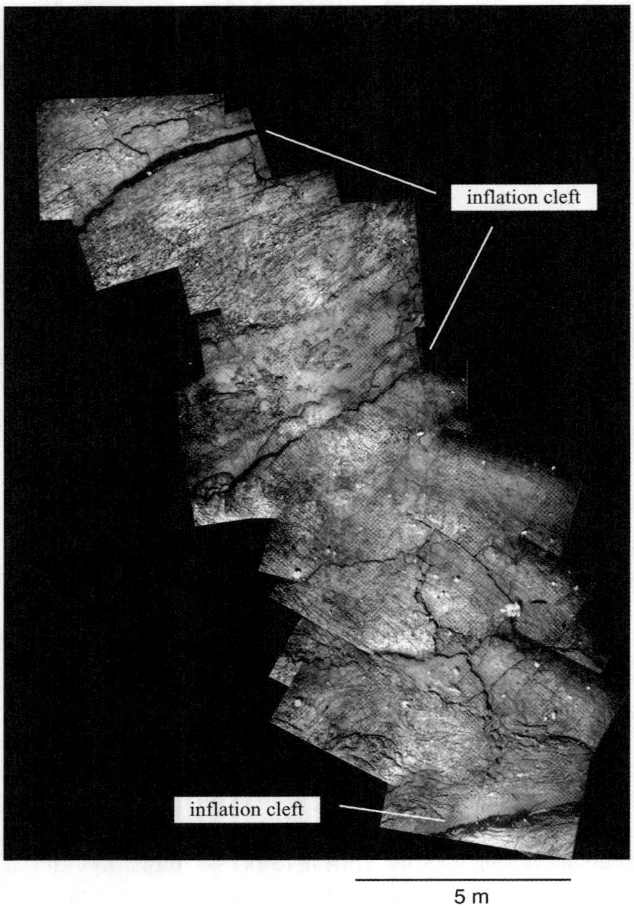

Figure 9. ARGO II photomosaic of inflation clefts on tumuli. Three are marked. Lava has squeezed up in the center cleft. Location labeled on Figure 8.

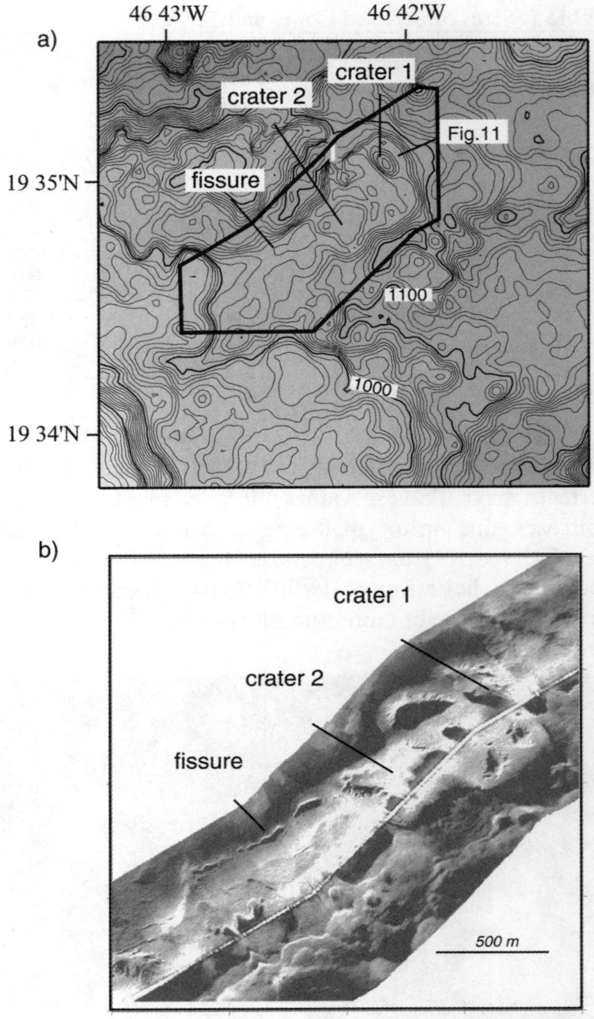

Figure 10. Terrace T4 (Figure 3, Plate 1), with 6 craters on its top. (a) EM300 multibeam bathymetry, contoured at 10-m intervals. (b) Side-scan sonar image. Photographs were obtained in the two labeled craters. A possible eruptive fissure runs along a topographic high, and flow fronts face away from the fissure.

Ridge [*Bryan et al.*, 1994]. There, a 60-m-high, cratered, flat-topped volcano is interpreted to have built over a lava tube that originated from a nearby volcanic ridge. Terrace T3 (Figure 3, Plate 1), located just below the break in slope between the crest and the south flank, is similar in shape and surface morphology to T2 and is assumed to be a tube-fed terrace as well.

Just as commonly, terraces have one or more craters on their summits. Some craters appear to be eruptive centers because lava tubes radiate from them. Bottoms of some craters are covered by rubble, although one has flows within it that cover the rubble in places. Terrace T4 (Figure 3, Plate 1), on the ridge crest at ~1100 m depth, has six craters (Figure 10), two of which were photographed. Crater 1 (Figure 10) is ellipsoidal with dimensions of ~150 m x ~200 m; it is ~30 m deep. The crater has steep overhanging walls and a rubble-filled bottom. A skylight, marking the path of a lava tube within surface flows, was photographed just to the northeast of crater 1 (Figure 11). Crater 2, on the southwest top of the terrace (Figure 10) is ~300 m wide, 15-20 m deep, and roughly circular in plan. Pillow lavas on the gentle slope into the crater appear to have flowed towards the crater floor which is covered with similar low-relief pillows. Another crater (~100 m wide and < 10 m deep) is nested within the larger one and has rough, jagged topography. It contains lava pillars, pedestals, and arches similar to features mapped in the axial zone of the East Pacific Rise, where lava has drained from the axial trough [*Fornari et al.*, 1998].

Figure 11. Skylights in a lava tube, surface flows on top of terrace T4. Location shown in Figure 10a. Black arrow marks a large skylight.

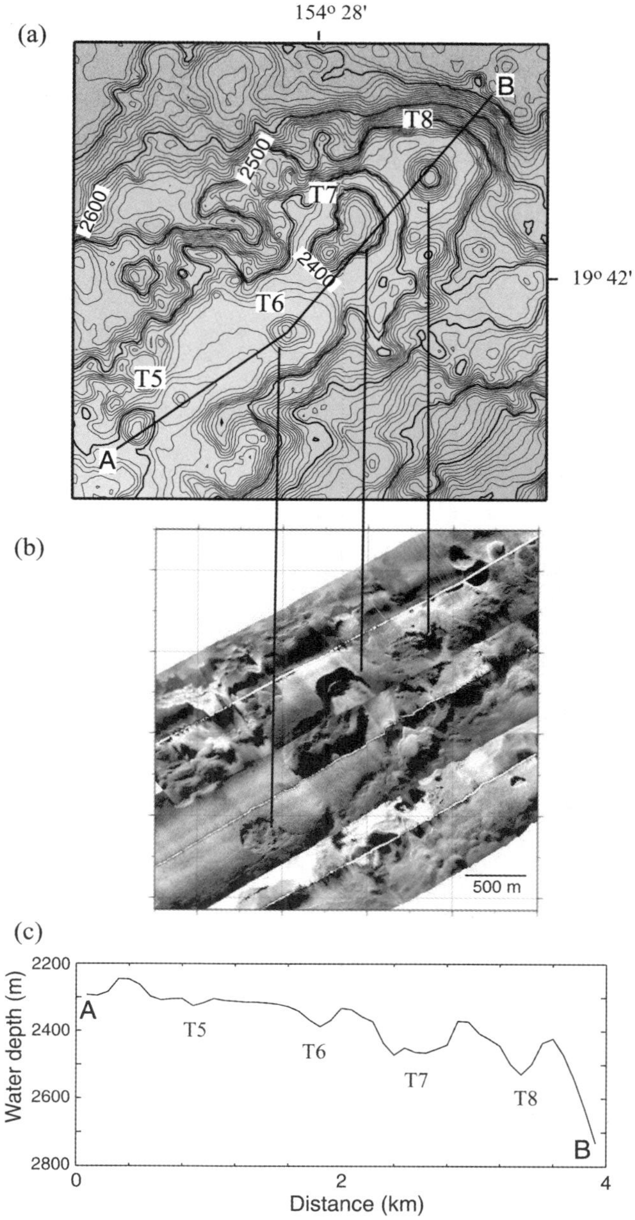

Figure 12. Staircase of terraces, at 2300 – 2500 m depths. (a) EM300 bathymetry data, 10-m contours. Terrace numbers as on Figure 3 and Plate 1. (b) Three mosaicked side-scan swaths. Vertical lines tie features in (a) and (b). Swaths are 1 km wide. (c) Bathymetric profile along the line on (a), showing terrace locations.

At ~2300 m depth, four terraces in a spectacular staircase (T5-T8; Figures 3, 12, Plate 1) increase in volume down-rift (Table 1), and each has a summit crater. The uppermost terrace (T5) has a crater 100-150 m wide and 10-20 m deep. T6 has a crater 300-400 m wide and 40-50 m deep. Two coalesced craters on T7 produce a dogleg plan shape, ~850 m long, 450-600 m wide, and ~100 m deep. The calculated volume of lava that drained away is ~15.5 x 10^6 m^3, ~27% of the terrace volume. The lowermost terrace (T8) has a crater that is ~500 m wide, ~100 m deep, and a volume of ~12 x 10^6 m^3, or ~12.5% of the volume of the terrace. Lava tubes appear to be associated only with the crater on T8. The radiating network of tubes suggests that lava flowed out onto the horizontal summit from the crater of T8.

The distal Puna Ridge below about 2700 m is a staircase of lava terraces, called deep steps by *Lonsdale* [1989], up

Figure 13. ARGO II photomosaic of drapery folds in lava on terrace T9 (Figure 3, Plate 1) at ~3300 m depth. This morphology is similar to that observed on the subaerial ERZ.

to 4 km or more wide measured perpendicular to the ridge. The terraces contain numerous craters and collapse features, except for the deepest large step at ~4100 m. The T9 step at 3360 m depth (Figure 3, Plate 1) has skylights on its top surface. Sheet flows on this step have fine examples of drapery folds (Figure 13). The top of the T10 step at 3700 m depth (Figure 3, Plate 1) is covered by pillow, lobate, and sheet flows. Abundant craters up to a few hundreds of meters wide are scattered across the top (Figure 14). One photographed crater has lava tubes leading from it, suggesting it was an eruptive center. The deep step at 4100 m (T11, Plate 1) appears hummocky or lumpy on the side-scan sonar images. Submersible-dive observations by one us [K. Johnson] show that this terrace is covered with numerous small pointed pillow cones.

4.5. Faults and fissures

One section of the ridge crest, at ~1300-1800 m depth, is highly faulted and fissured (Figure 3, Plate 1). The cracks and fissures are a few meters to 20 m wide, features consistent with a dike intruding a volcanic rift zone [e.g., *Parfitt and Head*, 1993]. At 1500-1600 m depth, small flows overlie the fissured and faulted terrain. *Smith et al.* [2001] estimate that these flows are 10-25 m thick, assuming that the topography was relatively flat at the time they were emplaced. Farther down the crest, the cracks and fissures become completely covered by lava flows starting at about 1800 m depth.

Multibeam bathymetry data show the head of a landslide (amphitheater) on the upper north flank at ~1200-2000 m depth along the ridge crest [e.g., *Moore and Chadwick*, 1995]. The scarp is identified by its high reflectivity in the DSL-120 images (Plate 1). Lavas coat the landslide scarp, suggesting that eruptions have occurred since the landslide. These flows may be starting the process of rebuilding the north flank [*Smith et al.*, 2001]. The relationship between the landslide scarp and the faulted and fissured section of the crest is unclear; collapse of the flank may have decreased the horizontal stress within the Puna Ridge, favoring dike intrusion and eruption at this location [*Smith et al.*, 2001].

5. DISCUSSION

The styles of volcanic features on the crest and flanks of Kilauea's ERZ change significantly as the subaerial rift zone dips below sea level. Along the crest of the subaerial ERZ cones are commonly eroded by subsequent eruptions. Lavas transported down the flanks of the subaerial rift zone form smooth, low-angle slopes. Lavas that cross the shoreline are believed to feed submarine debris flows [e.g., *Moore et al.*, 1973]. By contrast, the crest and flanks of the submarine Puna Ridge are covered by cones and terraces that have diameters of 1 km or more and sides several hundreds of meters high. Many of them have craters in their tops that are large enough to be resolved by multi-

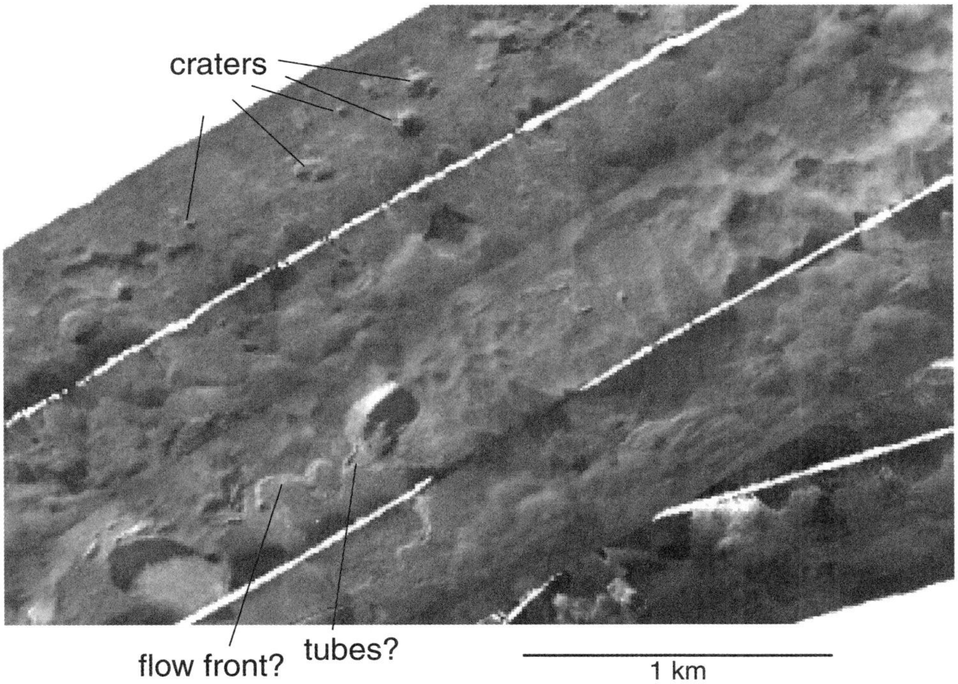

Figure 14. Mosaic of side-scan sonar swaths over terrace T10 (see Plate 1), at 3700 m depth. Each swath is ~1 km wide. The top is covered with a multitude of craters, only some marked. A large crater has apparent collapsed lava tubes leading from it, suggesting that lava flowed from the crater.

beam bathymetry. Cones and terraces are common on other submarine rift zones of Hawaiian volcanoes as well [*Clague et al.*, 2000].

Recent studies [*Smith and Cann*, 1999; *Clague et al.*, 2000] have focused on understanding how submarine terraces are constructed. Their large volumes (to 1 km^3), flat tops (that are horizontal rather than parallel to the local topographic slope), and near-circular plan shapes make them distinctive. The terraces are thought to form during single, long-lived eruptions with low to moderate effusion rates. *Smith and Cann* [1999] suggested that terraces grow upwards and outwards through inflation from within, and lava occasionally coats the top and flanks of the terrace. In contrast, Clague et al. [2000] proposed that terraces form as large, ponded lava lakes behind levees that stop the advance of the lava. Recent numerical studies show that high-viscosity lavas, gentle regional slopes, and low effusion rates favor the formation of terraces [W. Zhu, pers. comm.]. Also important in the formation of terraces is enhanced cooling in the submarine environment.

The common skylights, tumuli, and small pillow cones on the tops of terraces imply that a network of tubes develop within the flows. Pressure within the tubes and under the brittle surface of the terrace can be high enough to crack the surface and bow it up. In places, lava appears to have erupted through the cracks to form small pillow cones. Similar features (skylights, tumuli, rootless vents) are common on inflated pahoehoe flows observed along the subaerial ERZ [e.g., *Hon et al.*, 1994; *Kauahikaua et al.*, 1998]. Pressure build up at the cooling front of the advancing lobes of a subaerial pahoehoe sheet flow is translated throughout the flow, causing the still malleable top to raise as a single horizontal unit. Some flows on the subaerial ERZ have inflated by as much as 10-15 m. The overlying crust of the flow frequently splits, generating tumuli. In places lava spills out of the crack and drapes the tumulus, giving it the look of an eruptive vent rather than a pressure feature. When a flow stops inflating and advancing, the tubes drain, tumuli may collapse, and skylights form [*Kauahikaua et al.*, 1998]. The tumuli, skylights, and pillow cones on the top of the Puna Ridge terraces support the interpretation that the submarine terraces form mainly by inflation.

Notably, submarine terraces can be an order of magnitude larger than their subaerial counterparts. Terraces on the flanks and crest of the Puna Ridge have volumes to ~1 km^3 [*Smith and Cann*, 1999]. Volumes of many terraces (50-100 x 10^6 m^3) are similar to those of sustained eruptions from the subaerial ERZ (~160 x 10^6 m^3 at Mauna Ulu during 1972-1974 [e.g., *Tilling et al.*, 1987]). If eruption rates are similar on the subaerial ERZ and the Puna Ridge, the larger terraces are the product of an eruption lasting a few years, and most or all of the lava is incorporated into the single edifice.

Lava tube formation is common at the subaerial ERZ, and transporting lava through tubes allows flows to travel for long distances (tens of km) without losing significant heat [e.g., *Swanson*, 1973]. Do lava tubes transport large amounts of lava to sites of deposition far from the primary vents on the Puna Ridge? Collapsed tubes are common on surface flows, but these are different from the "master" tubes that develop at the primary vent and feed large secondary vent structures. Our high-resolution data do not provide the evidence necessary to distinguish between features that are tube fed, versus those that are primary vents fed directly from an underlying dike.

The idea that terraces and staircases of terraces are tube fed has been proposed for the terraces on the flanks of the Puna Ridge [*Smith and Cann*, 1999]. These authors argued that the width of dike intrusion along the Puna Ridge should not change significantly from the 1.5-3 km wide zone of eruptive fissures on the ERZ near the shoreline [*Holcomb*, 1987; *Moore and Trusdell*, 1991]. Gravitational spreading along Hawaiian rift zones [*Fiske and Jackson*, 1972] should not transport volcanic features constructed at the crest of the ERZ and Puna Ridge down the flanks. Instead, such features will subside along axial graben and be covered by subsequent flows. This suggests that the terraces and associated features on the rift flanks, beyond the zone of dike emplacement, are secondary vent structures.

Most volcanic features at the crest of the Puna Ridge are assumed to be primary vents, as they overlie the axis of dike intrusion. However, the staircase of terraces along the ridge crest at 2300-2500 m depth (T5-T8, Figure 12) could have alternate interpretations. One possibility is that the terraces were fed from vents that evolved from a long-lived fissure eruption. In this case, their heights might be expected to rise to the same water depth, based on hydraulic-head arguments [e.g., *Smith and Cann*, 1992]. Another possibility is that each terrace was built from its own eruption, fortuitously located to create the striking staircase. Finally, the terraces might have originated as secondary vent features, fed by lava tubes from a primary vent uprift. Craters could form when still-fluid lava drained into the tube system, perhaps as a blocked tube cleared. The mechanism by which terraces T5- T8 formed, and other volcanic features on the ridge crest, will likely remain unanswered until the geochemistry of their lavas is compared. Since lava is commonly transported in tubes along the crest of the subaerial ERZ [e.g., *Peterson et al.*, 1994] we assume that this happens on the crest of the Puna Ridge as well.

A few simple trends in the volcanic morphology of the ridge crest are a function of water depth. Only two fissure vents have been identified (Figures 5, 10); both are on the crest of the Puna Ridge. They occur on the shallow ridge above about 1100 m depth (red lines on Figure 3). One of them has been described in detail by *Parfitt* [2001]. Large central-vent eruptions appear to dominate the eruptive activity farther down the ridge crest, where the eruptions likely cover their own fissures.

Large cratered cones on the crest of the Puna Ridge are only observed in depths from ~600 to ~2200 m. As mentioned previously, the feeder systems for cones must have a larger excess magma pressure than for terraces, to overcome the hydraulic resistance necessary to build them higher. The hydraulic resistance includes both the frictional resistance to flow and the gravitational potential related to the elevation of the edifice above the source. If the topographic gradient and depth to the magma reservoir beneath the rift zone are fairly constant, as along the subaerial ERZ [*Rubin and Gillard*, 1998], then the excess magma pressure may be related to longer residence time and volatile build up within a reservoir beneath this section of the Puna Ridge.

The increase in along-axis slope from 51 m/km to 95 m/km at ~2700 m depth is accompanied by a changed style of volcanism [*Lonsdale*, 1989], to a series of large steps. Sizes of the steps tend to increase with increasing distance along the ridge. Photographs show that flat tops of most steps are strewn with craters (Figure 14), mostly too small (< ~50 m wide) to be defined in the bathymetry data. These differ from the terraces shallower than 2700 m, which commonly have single craters (e.g., terraces T5-T8) and/or larger ones (e.g., terrace T4) on their tops, which probably form by lava drainback. The many small craters on the deep steps may record stoping of crust where many pockets of fluid lava remain within the terrace [*Walker*, 1988].

Whether the distal steps are primary or secondary vents is not known. *Lonsdale* [1989] suggested that dikes rarely propagate past the change in slope at 2700 m. In that case, the steps would have been fed by lava tubes able to transport several cubic kilometers of magma for 15-20 kilometers from eruptive vents. The submarine tubes would have to be mechanically and thermally stable over these distances and times necessary for lava transport. *Smith and Cann* [1999] determined that in deep water (>2500 m) lava tubes should be mechanically stable under normal internal pressures. They suggested that the principal constraint on how far lava can flow in a tube is probably the critical length beyond which lava crystallizes. Using the formulation of *Lister and Dellar* [1996] for an axisymmetric channel embedded in an infinite solid, and assuming that lava tubes are thermally efficient [e.g., *Swanson*, 1973], the tubes can be ~0.4 - 4 km in length for volume fluxes of 1-10 m^3/s [*Smith and Cann*, 1999]. If this model is correct, submarine lava tubes would be too short to feed the most distal steps from vents near the slope break at 2700 m (distances >10 km), implying that some dikes propagate most of the length of the Puna Ridge.

Many questions remain unanswered about how the ERZ zone works along its entire length. One key will be to merge knowledge of the small-scale morphology with the information obtained from detailed sampling of the Puna Ridge. This work is currently on-going.

Acknowledgments. We thank WHOI's Deep Submergence Operating Group for their hard work on cruise TN084 to obtain these data. We also thank the captain and crew of the R/V Thompson. We had several fruitful discussions with Joe Cann. The manuscript benefited from the thorough reviews of W. Chadwick, M. Perfit, and P. Lipman. The work was supported by NSF grant OCE-9618226. WHOI contribution number 10455.

REFERENCES

Bryan, W.B., S.E. Humphris, B. Thompson, and J. Casey, Comparative volcanology of small axial eruptive centers in the MARK area, *J. Geophys. Res*, 99, 2973-2984, 1994.

Clague, D.A., R.T. Holcomb, M. Torresan, and S. Ross, Shipboard report for Hawaii GLORIA ground-truth cruise 11-88-HW, U.S. Geol. Surv. Open File Rep., 89-103, 1988.

Clague, D.A., K.A. Hon, J.L. Anderson, W.W. Chadwick, and C.G. Fox, Bathymetry of Puna Ridge, Kilauea Volcano, Hawaii, in U.S. Geol. Surv. Misc. Invest. Ser. Map, MF-2237, 1994.

Clague, D.A., J.G. Moore, J.E. Dixon, and W.E. Friesen, Petrology of submarine lavas from Kilauea's Puna Ridge, *J. of Petrol.*, 36, 299-349, 1995.

Clague, D.A., J.G. Moore, and J.R. Reynolds, Formation of submarine flat-topped volcanic cones in Hawai'i, *Bull. Volcanol.*, 62, 214-233, 2000.

Delaney, P.T., R.P. Denlinger, M. Lisowski, A. Miklius, P.G. Okubo, A.T. Okamura, and M.K. Sako, Volcanic spreading at Kilauea, 1976-1996, *J. Geophys. Res.*, 103, 18003-18023, 1998.

Fiske, R.S., and E.D. Jackson, Orientation and growth of Hawaiian volcanic rifts: The effect of regional structure and gravitational stresses, Proc. Roy. Soc. London, 329A, 299-326, 1972.

Fornari, D.J., R.M. Haymon, M.R. Perfit, T.K.P. Gregg, and M.H. Edwards, Axial summit trough of the East Pacific Rise 9°-10°N: Geological characteristics and evolution of the axial zone on fast spreading mid-ocean ridges, *J. Geophys. Res.*, 103, 9827-9855, 1998.

Fornari, D.J., A. Malahoff, and B.C. Heezen, Volcanic structure of the crest of the Puna Ridge, Hawaii: Geophysical implications of submarine volcanic terrain, *Geol. Soc. Am. Bull.*, 89, 605-616, 1978.

Fox, C.G., K.M. Murphy, and R.W. Embley, Automated Display and Statistical Analysis of Interpreted Deep-Sea Bottom Photographs, *Marine Geology*, 78, 199-216, 1988.

Holcomb, R.T., Eruptive history and long-term behavior of Kilauea volcano, U. S. Geol. Surv. Prof. Pap. 1350, 261-350, 1987.

Holcomb, R.T., J.G. Moore, P.W. Lipman, and R.H. Belderson, Voluminous submarine lava flows from Hawaiian volcanoes, *Geology*, 16, 400-404, 1988.

Hon, K., J. Kauahikaua, R. Denlinger, and K. Mackay, Emplacement and inflation of pahoehoe sheet flows; observations and measurements of active lava flows on Kilauea Volcano, Hawaii, *Geol. Soc. Am. Bull.*, 106, 351-170, 1994.

Kauahikaua, J., K.V. Cashman, T.N. Mattox, C. Heliker, K. A. Hon, M. T. Mangan, and C.R. Thornber, Observations on basaltic lava streams in tubes from Kilauea Volcanic, Island of Hawai'i, *J. Geophys. Res.*, 103, 27,303-27,323, 1998.

Klein, F.W., R.Y. Koyanagi, J.S. Nakata, and W.R. Tanigawa, The seismicity of Kilauea's magma system, U. S. Geol. Surv. Prof. Pap., 1350, 1019-1185, 1987.

Lister, J.R., and P.J. Dellar, Solidification of pressure-driven flow in a finite rigid channel with application to volcanic eruptions, *J. Fluid Mech.*, 323, 267-283, 1996.

Lockwood, J.P., and P.W. Lipman, Holocene eruptive history of Mauna Loa Volcano, U. S. Geol. Surv. Prof. Pap.. 1350, 509-535, 1987.

Lonsdale, P., A geomorphological reconnaissance of the submarine part of the East Rift Zone of Kilauea Volcano, Hawaii, *Bull. Volcanol.*, 51, 123-144, 1989.

Lonsdale, P., and R. Batiza, Hyaloclastite and lava flows on young seamounts examined with a submersible, *Geol. Soc. Am. Bull*, 91, 545-554, 1980.

Malahoff, A., and F. McCoy, The geologic structure of the Puna submarine ridge, Hawaii, *J. Geophys. Res.*, 72, 541-548, 1967.

Monterey Bay Aquarium Research Institute (MBARI), Monterey Bay Aquarium Research Institute Hawaii Multibeam Survey, in Digital Data Series 2, Moss Landing, Calif., 2000.

Moore, J.G., and W.W. Chadwick, Offshore geology of Mauna Loa and adjacent areas, Hawaii, in *Mauna Loa Revealed: Structure, Composition, History, and Hazards*, Geophys. Monogr. Ser., vol. 92, edited by J.M. Rhodes, and J.P. Lockwood, pp. 21-42, Washington, D. C., 1995.

Moore, J.G., and R.S. Fiske, Volcanic substructure inferred from dredge samples and ocean-bottom photographs, Hawaii, *Geol. Soc. Am. Bull.*, 1191-1201, 80, 1969.

Moore, J.G., R.L. Phillips, R.W. Grigg, D.W. Peterson, and D.A. Swanson, Flow of lava into the sea, 1969-1971, Kilauea Volcano, Hawaii, *Geol. Soc. Am. Bull.*, 84, 537-546, 1973.

Moore, R.B., and F.A. Trusdell, Geologic map of the lower east rift zone of Kilauea Volcano, Hawaii, U.S. Geol. Surv. Misc. Invest. Ser. Map, I-2225, 1991.

Parfitt, E., T. Gregg, and D. Smith, A comparison between subaerial and submarine fissure eruptions at Kilauea Volcano, Hawaii, *J. Vol. Geotherm. Res.*, accepted, 2001.

Parfitt, E.A., and J.W. Head, Buffered and unbuffered dike emplacement on Earth and Venus: Implications for magma reservoir size, depth, and rate of magma replenishment, *Earth Moon Planets*, 61, 249-281, 1993.

Peterson, D.W., R.T. Holcomb, R.I. Tilling, and R.L. Christiansen, Development of lava tubes in the light of observations of Mauna Ulu, Kilauea Volcano, *Hawaii, Bull. Volcanol.*, 56, 343-360, 1994.

Rubin, A.M., and D. Gillard, A reinterpretation of seismicity associated with the January 1983 dike intrusion at Kilauea Volcano, Hawaii, *J. Geophys. Res.*, 103, 10,003-10,015, 1998.

Smith, D.K., Shape analysis of Pacific seamounts, Earth Planet. Sci. Lett., 90, 457-466, 1988.

Smith, D.K., and J.R. Cann, The role of seamount volcanism in crustal construction at the Mid-Atlantic Ridge (24°-30°N), *J. Geophys. Res.*, 97, 1645-1658, 1992.

Smith, D.K., and J.R. Cann, Constructing the upper crust of the Mid-Atlantic Ridge: A reinterpretation based on the Puna Ridge, Kilauea Volcano, *J. Geophys. Res.*, 104, 25,379-25,399, 1999.

Smith, D.K., M.A. Tivey, P. Gregg, and L.S. Kong, Magnetic anomalies along the submarine Puna Ridge, Kilauea Volcano, Hawaii,, *J. Geophys. Res.*, in press, 2001.

Sterns, H.T., and G.A. Macdonald, Geology and ground-water resources of the Island of Hawaii, in *Hawaii Division of Hydrography*, pp. 363, 1946.

Swanson, D., Pahoehoe flows from the 1969-1971 Mauna Ulu eruption, Kilauea Volcano, Hawaii, *Geol. Soc. Am. Bull.*, 84, 615-626, 1973.

Swanson, D.A., D. B. Jackson, R.Y. Koyanagi, and T.L. Wright, The February 1969 East Rift eruption of Kilauea Volcano, Hawaii, U. S. Geol. Surv. Prof. Pap., 891, 1-30, 1976a.

Swanson, D.A., W.A. Duffield, and R.S. Fiske, Displacement of the south flank of Kilauea volcano: The result of forceful intrusion of magma into the rift zones, U. S. Geol. Surv. Prof. Pap., 963, 1-39, 1976b.

Tilling, R.I., R.L. Christiansen, W. Duffield, and A. Others, The 1972-1974 Mauna Ulu eruption, Kilauea volcano: An example of quasi-steady-state magma transfer, U. S. Geol. Surv. Prof. Pap., 1350, 405-469, 1987.

Walker, G.P.L., Three Hawaiian calderas: an origin through loading by shallow intrusions?, *J. Geophys. Res.*, 93, 14773-14784, 1988.

Walker, G.P.L., Structure, and origin by injection of lava under surface crust, of tumuli, "lava rises", "lava-rise pits", and "lava-inflation clefts" in *Hawaii, Bull. Volcanol.*, 53, 546-558, 1991.

Wolfe, E.W., M.O. Garcia, D.B. Jackson, R.Y. Koyanagi, C.A. Neal, and A.T. Okamura, he Pu'u 'O'o eruption of Kilauea volcano, episodes 1-20, January 3, 1983, to June 8, 1984, U. S. Geol. Surv. Prof. Pap., 1350, 471-508, 1987.

K. Johnson, Department of Natural Sciences, Bishop Museum, 1525 Bernice St., Honolulu, HI, 96817.

L. Kong, USDOT, Federal Highway Administration, Honolulu, HI 96850.

J. Reynolds, School of Fisheries and Ocean Sciences, University of Alaska, PO Box 757520, Fairbanks, AK 99775

D. Smith, Department of Geology and Geophysics, Woods Hole Oceanographic Institution, MS 22, Woods Hole, MA 02543.

---

[2]Now at USDOT, Federal Highway Administration, Honolulu, Hawaii

Petrological Systematics of Submarine Basalt Glasses From the Puna Ridge, Hawai'i: Implications for Rift Zone Plumbing and Magmatic Processes

Kevin T. M. Johnson[1], Jennifer R. Reynolds[2], Denys Vonderhaar[1], Deborah K. Smith[3], and Laura S. L. Kong[4*]

We report systematic chemical variations in volcanic glasses collected along the submarine East Rift Zone of Kilauea Volcano (Puna Ridge), Hawaii during a comprehensive mapping and sampling effort. Electron microprobe major and minor element data for 193 volcanic glass samples, and S, Cl, and F concentrations in 80 samples from 56 glass cores, 22 dredges, and two Shinkai 6500 submersible dives along the axis and south flank of the Puna Ridge are discussed. All Puna Ridge samples are tholeiitic basalts and most of the lavas contain olivine, augite ± plagioclase, but some contain only olivine and are modal picrites. The main suite of lavas define a fractionation trend similar to subaerial Kilauea lavas, with Al_2O_3 concentrations peaking at 6.3 wt% MgO. Several dredges collected on the south flank of the Puna Ridge and at the deep, distal end of the Puna Ridge axis are enriched in Al_2O_3 relative to other axial lavas at simi-lar Mg/(Mg+Fe). These liquids are also high in chlorine relative to most other Puna Ridge samples and we attribute their unique compositions to suppression of plagioclase crystallization either at higher pressure or in the presence of higher water concentrations. The high Cl concentrations are probably due to assimilation of altered oceanic crustal wall rocks into the fractionating magma chamber at the base of the rift zone. Six samples collected from a 70 km^2 flood basalt flow in the moat on the southern margin of the Puna Ridge are alkalic with 0.95 wt% K_2O. These lavas are aphyric, very fresh, and have distinctly higher K/Ti ratios than Puna Ridge lavas at similar Mg/(Mg+Fe) precluding a common mantle source for the two populations.

1. INTRODUCTION

Kilauea Volcano on Hawai'i Island is a typical Hawaiian shield volcano. It has a pronounced summit caldera and two rift zones, the 30-km-long Southwest Rift Zone (SWRZ) and the 125-km-long East Rift Zone (ERZ). The submarine portion of the ERZ, called the Puna Ridge, is 50% longer (~75 km) than the subaerial ERZ (~50 km; Figure 1). As in other Hawaiian volcanoes, the primary orientation and dimensions of the rift zones are controlled by gravitational stresses in the volcano edifice and by the buttressing effect of adjacent, preexisting shields [*Fiske and Jackson*, 1972].

[1]Department of Geology, Bishop Museum, Honolulu, Hawaii
[2]School of Fisheries and Ocean Sciences, University of Alaska, Fairbanks, Alaska
[3]Department of Geology and Geophysics, Woods Hole Oceanographic Institution, Woods Hole, Massachusetts
[4]Hawaii Institute of Geophysics and Planetology, University of Hawaii, Honolulu, Hawaii
* now at Federal Highway Administration, Honolulu, Hawaii

Hawaiian Volcanoes: Deep Underwater Perspectives
Geophysical Monograph 128
Copyright 2002 by the American Geophysical Union

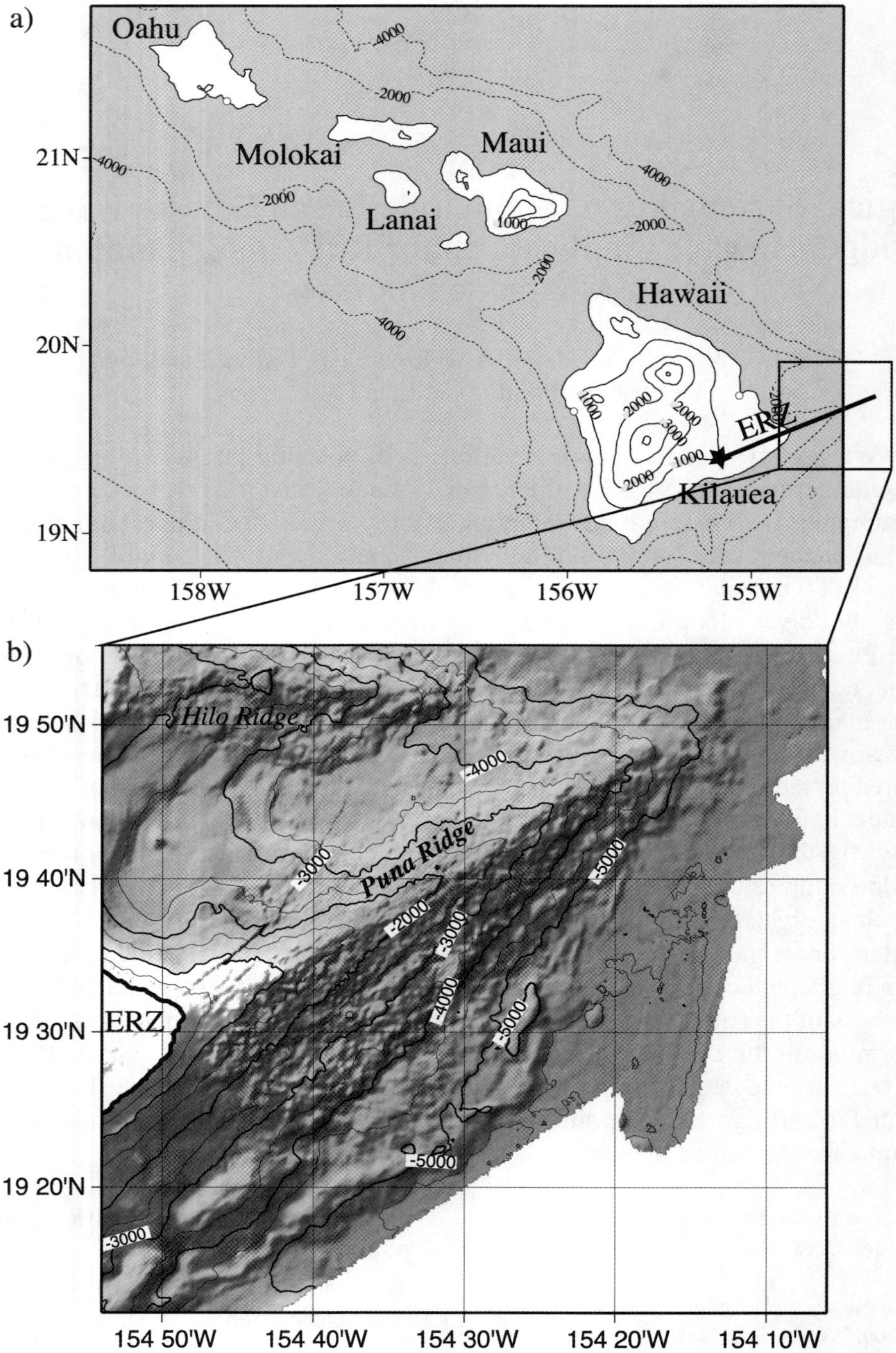

Figure 1. Location map showing the Puna Ridge extending off-shore of the Big Island of Hawai'i. The Puna Ridge is the submarine portion of the East Rift Zone (ERZ) of Kilauea Volcano. It is ~77 km long and drops from sea level to about 5400 m along its length.

The latter control probably accounts for the disparity in the lengths of Kilauea's two rift zones: the SWRZ is pinched between Mauna Loa's southeast flank and Loihi's growing edifice whereas the ERZ is buttressed only to the north, by Mauna Loa's northeast flank and by the submarine Hilo Ridge.

Rift zones are fundamental components of Hawaiian volcanoes. At the surface, the ERZ is 3-4 km wide and consists of tensional cracks, collapse pits, eruptive fissures, and a variety of constructional volcanic landforms such as cinder cones and spatter ramparts [*Macdonald et al.*, 1983] fed by dikes. Beneath the surface, rift zones are steeply dipping networks of dikes emanating from a shallow summit magma reservoir at 3-7 km depth [*Fornari*, 1987; *Klein et al.*, 1987; *Ryan et al.*, 1981]. However, although the most active portion of the subaerial ERZ is at about 3 km depth [*Klein et al.*, 1987], the base of the rift zone may extend to 8 km below sea level (8-9 km below the surface) near to where Kilauea rests on the Pacific plate [*Decker*, 1987]. Although extensive mixing between magmas equilibrated dominantly at low pressure has been documented in Puna Ridge lavas (e.g., *Clague et al.* [1995]), evidence for magmas equilibrated in the deeper portions of the rift system is sparse.

Several geological and geophysical data sets have been collected at the Puna Ridge prior to the studies reported in this paper. Swath bathymetry data covering the summit and flanks of the ridge [*Clague et al.*, 1994; *Clague et al.*, 2000b; *MBARI*, 2000], deep-tow side-scan sonar images [*Lonsdale*, 1989], photographic imagery [*Lonsdale*, 1989; *Moore and Fiske*, 1969], submersible dive observations [*Fornari et al.*, 1978b], and dredged rocks [*Clague et al.*, 1988; *Moore*, 1965; *Moore and Fiske*, 1969] all show that fissures and lava flows mark the crest of the Puna Ridge.

To obtain a better understanding of dike intrusion and lava emplacement in the submarine environment, we recently completed a detailed geological, geophysical, and geochemical investigation of the Puna Ridge. We used the deep-towed DSL-120 side-scan sonar system to map the entire length and width of the summit and part of the south flank (Figure 2). We also used the ARGO II photo-imagery system to obtain high-resolution electronic still camera images and video of the seafloor in selected regions within the side-scan sonar survey. Details of the DSL-120 and ARGO data collection techniques are reported elsewhere [*Smith et al.*, 2000]. Following the detailed deep-tow survey, we conducted a dense rock sampling program consisting of 22 dredges, 56 rock cores, and 1 submersible dive along the ridge crest and flank. The samples were collected along the entire 75 km length of the Puna Ridge at depths from 350 m to 5000 m in 1998 on cruise TN084 of the R/V *Thomas Thompson* and in 1999 from the Shinkai 6500 submersible on cruise YK99 of the R/V *Yokosuka* operated by JAMSTEC. In addition, we carried out one submersible dive on a broad, low-relief lava flow south of the Puna Ridge at a depth of 5565 m. In this paper we present the results of electron microprobe analyses of volcanic glass recovered in the dredges, cores and dives and discuss their implications with respect to existing models of magma transport within the rift zone.

2. MORPHOLOGY OF THE PUNA RIDGE

The along-axis slope of the subaerial ERZ is relatively constant at ~23 m/km [*Lonsdale*, 1989], whereas the upper part of the Puna Ridge is more than twice as steep (~51 m/km). Fialko and Rubin [*Fialko and Rubin*, 1998; *Fialko and Rubin*, 1999] suggested that the longitudinal slope of a rift zone may be an important factor in driving dike intrusion along the length of the rift and that the distribution of lava flows along the ridge acts to maintain an optimal slope. Their model, which accounts for the buoyancy effect of the water, predicts that the ratio for the initial change in slope immediately offshore should be ERZ/PR = 0.6. The along-axis slope of the Puna Ridge is slightly steeper than predicted by Fialko and Rubin (51 m/km observed, 38 m/km predicted) yielding an observed ratio of ~0.45. According to the ideas of Fialko and Rubin [*Fialko and Rubin*, 1998; *Fialko and Rubin*, 1999] eruptions should preferentially occur far from the shoreline to flatten the along-axis slope. Below 2700 m the along-axis slope of the Puna Ridge steepens to ~95 m/km. The cause of this increase in slope is not understood, although the steepening may indicate the limit of dike propagation [e.g., *Lonsdale*, 1989] or the effect of pressure on the critical point of seawater and the role of boiling in submarine eruptions [Moore and Chadwick, 1995].

The styles of volcanic features on the summit and flanks of the rift zone also change significantly just as the crest of the rift dips below sea level [*Clague et al.*, 2000b; *Moore and Chadwick*, 1995; *Smith and Cann*, 1999]. Lavas erupted from the subaerial ERZ form smooth, low-angle slopes, except where interrupted by faults. Gently dipping low-relief lava flows mantle the slopes. Large volcanic edifices are not commonly constructed. By contrast, on the Puna Ridge circular cones and semicircular flat-topped volcanic terraces that have diameters ranging up to 1 km or more and sides ranging up to several hundreds of meters high cover its summit and flanks. The volumes of individual volcanic terraces and cones on the Puna Ridge are large (0.1-1 km^3) and similar to the volumes of lava flows that

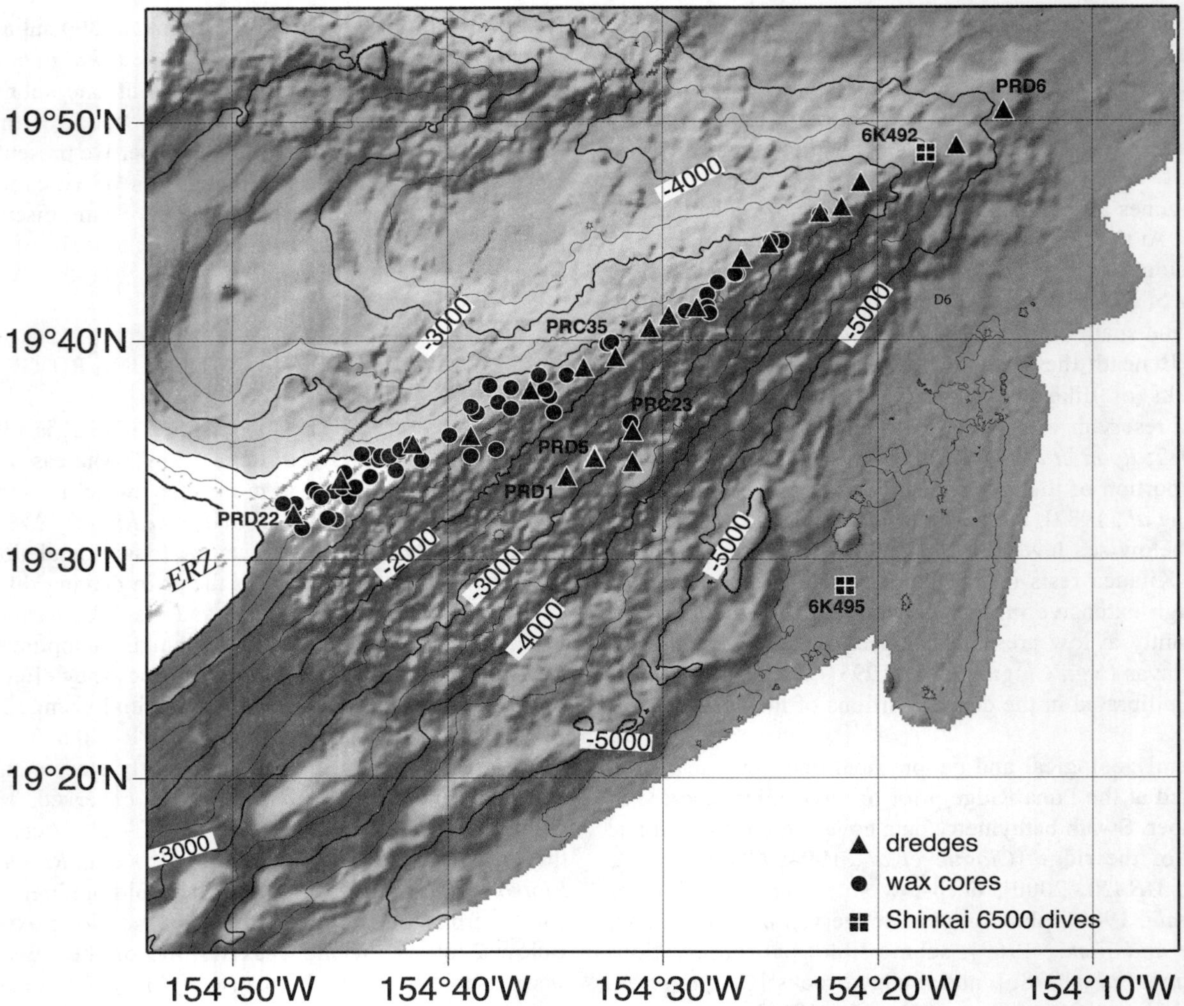

Figure 2. Map of the Puna Ridge showing sampling sites. Also shown are station numbers for stations discussed in the text.

are broadly distributed during eruptions at the subaerial ERZ [*Smith and Cann*, 1999].

The flank slopes of the Puna Ridge range between 160 and 275 m/km [*Fornari et al.*, 1978; *Smith and Cann*, 1999] and are steeper than on the subaerial slopes of the ERZ (~50 m/km). The slopes on the Puna Ridge south flank do not change significantly with distance from the shoreline until the crest is deeper than 3500 m. [*Smith and Cann*, 1999] argue that the south flank is resurfaced regularly and evenly as the ridge grows and that flows must be transported laterally 10-15 km from the ridge summit. Data to be presented below will have a direct bearing on this interpretation.

3. ROCK SAMPLING

3.1 Site Selection and Transponder Navigation of Sampling Devices

Prior to the TN084 cruise, the entire sampling effort for the survey area consisted of 19 stations on the Puna Ridge axis at an average spacing of more than 7 km; we collected samples from 78 sites at an average spacing of 1-2 km. The distribution of sites is shown in Figure 2 and tabulated in Tables 1-3.

Site selection was based on a combination of multi-narrow beam bathymetric survey maps from previous work on the Puna Ridge, and DSL-120 sidescan sonar data and ARGO II bottom photographs collected on this cruise. Nearly 100% coverage of much of the ridge axis was achieved on the TN084 cruise, which was very useful in selection of sample sites. Transponder navigation was used for the DSL-120 and ARGO surveys and we found that this enabled us to sample very small targets that would not be possible on a conventional dredge cruise. Precision site location was enhanced by attaching a relay transponder to the hydro-wire at 100m above the corer and to the trawl wire at 200-250 m above the dredge enabling us to precisely track the location of the sampling device while it was on the bottom. We quickly became accustomed to choosing, and hitting, targets from the DSL-120 and ARGO II lines that were as small as 100 meters in size for dredging and as small as 10-20 meters for coring in water depths of over 3400 meters. We noted that the typical scatter of fixes from a core in 2000-3000 meters of water was a circular cloud of points 10-30 meters in diameter, largely a function of the local current. The ability to sample with this precision enabled us to more closely approximate subaerial geological mapping and sampling and to infer relationships otherwise restricted to the land geologist or submersible vehicles.

3.2 Dredging

During the TN084 cruise, we conducted 22 dredge deployments and collected samples from 21 of these sites. Most were distributed along the axis; but five were located in the south flank survey area (PRD 1 to 5) (Table 1; Figure 2). The dredge assembly consisted of a rectangular steel dredge with a ~1 m-long steel chain bag and a heavy-gauge nylon mesh bag attached inside the chain bag. A 15 kg weight was placed inside a burlap bag and tied into the bottom of the mesh bag on each dredge. This burlap weight bag served the dual purposes of holding the net bag inside the chain bag and catching glass fragments during the dredge. On two occasions the burlap bag contained the only sample in the dredge.

3.3 Wax Rock Coring

We carried out 56 core deployments on TN084 (Figure 2). The core consisted of a 1.8 meter long, 4-inch diameter steel pipe with 600 pounds of steel weights secured on the top portion, which was rimmed by seven core tips. The base consisted of an 11-inch diameter steel plate with six core tips attached. The core tips were each 2 inches in diameter and were filled with wax. Basaltic glass from the surface of lava flows was sampled by lowering the corer to within 100 m of the bottom at low speed, stopping the hydro-winch, then lowering the device at high speed until impact. Most core locations were distributed along the axis of the ridge; one was located in the south flank survey area (PRC 5) and three elsewhere on the uppermost south flank (PRC 22, 38, 45). All of the rock cores recovered volcanic glass. Some also recovered rock chips, Mn crust, pelagic sediment, white sediment (ash?), and various biological material.

3.4 Submersible Sampling

During cruise YK99 of the R/V Yokosuka, the submersible Shinkai 6500 conducted two dives on or near the Puna Ridge. Dive 6K492 collected 6 samples along the axis on two large lava terraces at depths of 3800-4200 m (Figure 2). Dive 6K495 was carried out south of the Puna Ridge on a large, sidescan-sonar reflective lava flow (Figure 2) that was previously shown on a location map, but not discussed [*Clague et al.*, 1995]. The flow covers an area of some 70 km^2 with bathymetric relief of <5 meters over most of the 1.7 km dive track suggesting eruption of a highly fluid lava. Six samples were collected from this low-relief feature at depths of 5564-5568 m.

3.5 Sample Classification

3.5.1 Dredges. Dredge samples were sorted into groups based on observation of hand specimen characteristics including overall morphology, crystallinity, mineralogy, and extent of weathering or alteration. Ideally, each group would represent an individual lava flow, but a deliberate policy of forming the maximum number of likely groups resulted in more groups than were determined geochemically. Most of the axial sites recovered moderately fresh to very fresh glassy lavas. In addition, off-axis sites PRD 1 and 4 recovered very fresh glassy lavas or glass. Lava freshness did not appear to correlate with distance along axis or on- versus off-axis location; very fresh samples were collected from all areas of the ridge that we sampled.

3.5.2 Wax Cores. In contrast to dredge hauls, which were washed clean to a great extent during recovery, the wax cores contained a mixture down to sand and dust size of whatever lay on the sea floor. The wax core samples were examined under a binocular microscope and described in detail. They contained volcanic glass and rock; hyaloclastite with white matrix (ash?); altered and/or secondary material (e.g., authigenic manganese); sediment in the form of brown ooze; and biological material such as foraminifera and sponge spicules. The glass was further subdivided into six morphological categories, five of which represent energetic fragmentation of the magma:

Table 1. Representative electron microprobe analyses of glasses from dredge samples (wt%).

Sample	Lat (°N)	Lon (°W)	SiO_2	TiO_2	Al_2O_3	FeO	MnO	MgO	CaO	Na_2O	K_2O	P_2O_5	S	F	Cl	Total
PRD1-1	19.56172	154.57477	51.40	2.74	14.07	11.15	0.18	5.71	10.35	2.48	0.50	0.27				98.87
PRD1-2	19.56172	154.57477	50.57	2.61	14.28	10.57	0.16	5.94	10.71	2.53	0.48	0.24				98.08
PRD1-3	19.56172	154.57477	50.97	2.53	14.07	10.47	0.15	6.27	11.05	2.38	0.45	0.25	0.0457	0.0448	0.0714	98.75
PRD1-4	19.56172	154.57477	50.76	2.90	14.26	11.37	0.18	5.06	9.95	2.59	0.54	0.29	0.1022	0.0481	0.0627	98.12
PRD1-5	19.56172	154.57477	50.64	2.73	14.22	10.98	0.18	5.54	10.27	2.54	0.52	0.28	0.1122	0.0414	0.0619	98.10
PRD1-6	19.56172	154.57477	50.61	2.52	14.04	10.27	0.16	6.26	10.91	2.41	0.45	0.25				97.89
PRD3-1	19.59658	154.52358	50.81	3.03	13.79	11.22	0.16	6.01	10.42	2.45	0.55	0.28	0.0481	0.0472	0.0197	98.85
PRD4-3	19.57238	154.52368	50.86	3.12	13.72	11.35	0.17	5.54	10.01	2.62	0.61	0.34	0.0907	0.0532	0.0325	98.52
PRD4-6	19.57238	154.52368	49.79	3.79	12.85	13.82	0.19	4.67	9.22	2.71	0.67	0.40	0.1939	0.0614	0.0240	98.39
PRD5-1	19.57591	154.55271	51.04	2.87	14.06	11.36	0.16	5.44	10.11	2.55	0.55	0.30	0.0555	0.0547	0.0660	98.60
PRD5-2	19.57591	154.55271	51.02	2.84	14.21	11.11	0.18	5.34	10.05	2.58	0.53	0.31	0.0450	0.0443	0.0748	98.35
PRD6-1	19.84115	154.23403	50.87	3.03	14.12	10.70	0.16	5.81	10.37	2.47	0.56	0.32	0.0544	0.0534	0.0241	98.55
PRD6-2	19.84115	154.23403	50.91	2.88	13.97	11.29	0.17	5.38	10.05	2.61	0.56	0.30	0.0320	0.0316	0.0452	98.23
PRD7-1	19.81347	154.27157	51.56	3.24	13.60	11.33	0.15	5.56	9.71	2.59	0.61	0.34	0.0529	0.0520	0.0233	98.82
PRD8-1	19.76697	154.36057	50.74	2.80	13.75	11.15	0.15	6.00	10.48	2.47	0.51	0.28	0.1443	0.0458	0.0185	98.54
PRD9-1	19.76252	154.37706	50.65	2.99	13.61	11.58	0.17	5.62	10.16	2.49	0.55	0.28				98.12
PRD9-3	19.76252	154.37706	50.66	2.65	14.04	10.48	0.14	6.24	10.67	2.45	0.46	0.27	0.1264	0.0462	0.0187	98.26
PRD10-1	19.78518	154.34556	51.28	2.48	13.81	10.12	0.13	6.60	10.91	2.41	0.44	0.24	0.0981	0.0366	0.0187	98.57
PRD12-1	19.72808	154.43870	50.39	2.84	13.89	10.83	0.18	6.16	10.72	2.47	0.52	0.29	0.1206	0.0474	0.0175	98.47
PRD13-1	19.69079	154.47411	50.69	2.73	13.81	10.78	0.16	6.16	10.74	2.39	0.50	0.26	0.1371	0.0394	0.0155	98.41
PRD14-1	19.68356	154.49545	51.34	2.75	13.60	10.83	0.14	5.94	10.43	2.46	0.51	0.30	0.0927	0.0457	0.0202	98.47
PRD14-5	19.68356	154.49545	51.48	2.65	13.81	10.39	0.17	6.42	10.85	2.45	0.49	0.27				99.00
PRD15-1	19.67485	154.50972	51.08	2.77	13.88	10.88	0.17	6.15	10.66	2.41	0.51	0.29	0.1323	0.0495	0.0201	99.01
PRD17-1	19.64551	154.56151	50.42	3.15	13.46	11.97	0.17	5.69	10.32	2.47	0.59	0.31	0.1624	0.0451	0.0263	98.79
PRD17-2	19.64551	154.56151	50.73	3.46	13.18	12.61	0.17	5.28	9.72	2.61	0.67	0.35	0.1622	0.0572	0.0243	99.02
PRD18A	19.62687	154.60235	51.46	3.44	13.14	12.45	0.16	4.96	9.40	2.63	0.70	0.34				98.68
PRD19-2	19.59310	154.64926	51.07	2.69	13.33	10.71	0.15	7.24	10.87	2.33	0.50	0.26	0.0852	0.0428	0.0181	99.29
PRD20-1	19.58739	154.69458	50.88	3.07	13.54	11.89	0.18	5.71	10.22	2.51	0.58	0.30	0.1290	0.0554	0.0179	99.08
PRD21-1	19.55978	154.75045	51.43	3.09	13.44	11.64	0.16	5.77	10.34	2.52	0.55	0.33	0.0873	0.0441	0.0186	99.42
PRD22-1	19.53346	154.78614	50.90	2.32	12.90	10.87	0.15	9.16	10.28	2.20	0.36	0.23				99.38

Complete data set available on accompanying CD-ROM or from senior author.

1. Glass chips interpreted as fragments of lava flows. These were typically angular, and resembled glass from dredged rocks.

2. Pumice -- frothy glass with vesicularity greater than ~30%.

3. Spatter -- vesicular glass with fluidal shapes.

4. Spindles -- fluidal glass with elongated shape, grading into Pele's tears.

5. Pele's tears -- long, hair-thin glass fragments.

6. Limu o Pele -- curved bubble wall fragments, very thin and fragile.

Uncertain cases were assigned to the lava flow category. These categories conform to the definitions and descriptions of *Mattox and Mangan* [1993], *Batiza and White* [2000], *Clague et al.* [2000a] and *Clague et al.* [2001].

Particular care was taken to identify and sub-sample the explosive-type glasses, because monitoring of historical eruptions on Kilauea's subaerial East Rift and Southwest Rift has shown that such glasses can be compositionally distinct. The subaerial rift eruptions tend to follow a pattern that starts with energetic fissure eruption of gas-rich lava, often by fire-fountaining or spattering, and over time the eruption settles down to a less energetic outpouring of lava flows (e.g., *Rowland and Walker* [1990]). Sampling of the early erupted lavas has shown that the composition of the lava also often changes near the beginning of the eruption (e.g., *Garcia et al.* [1992]; *Helz and Wright*, [1992]; *Wright and Helz* [1996]), so we anticipated that fragmented glasses sampled from the submarine rift zone might also have a different compositional range compared with the submarine lava flows. Some may also be generated from those same or nearby lava flows, by non-explosive spalling of the glass surface or by entrainment of seawater bubbles into the lava (e.g., *Batiza and White* [2000]).

The depth of most Puna Ridge core samples with explosive-type glasses ranged from 450m to 2793m. Only two were deeper, at 2987m and 3422m. This depth range is comparable to that of hyaloclastite deposits on ocean ridges and near-ridge seamounts [*Batiza and White, 2000*].

4. ANALYTICAL METHODS

Microanalyses of glasses were done with a Cameca SX50 electron microprobe at the University of Hawai'i. Major and minor element geochemical data listed in Tables 1-3 are averages of 5 – 9 individual analyses for each sample. Microprobe operating conditions were 15 kV accelerating voltage, 10 nA beam current and spot size of ~10 µm. Counting times in seconds are: Na-20, Mg-30, Al-30, Si-30, P-85, K-20, Ca-20, Ti-30, Mn-30, Fe-50 with background counting times half of these values. Data were corrected by the PAP-ZAP method.

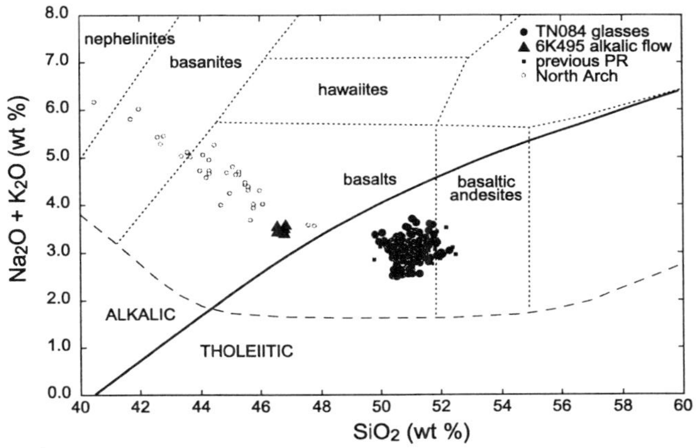

Figure 3. Alkali-silica diagram showing the cloud of data points from the Puna Ridge (large filled circles) and the alkalic flood basalt flow south of the Puna Ridge (6K495 - triangles). Previously published data from Puna Ridge (small squares [*Clague et al.*, 1995]), and data from the North Arch volcanic field (small open circles [*Clague et al.*, 1990]) are also shown for comparison.

For the volatiles, the beam current was 80.35 nA, beam size was 10 µm, accelerating voltage was 15.20 kV. Each analysis in Tables 1-3 represents an average of 20 measurements done on a single point for 20 seconds each, with the Faraday cup inserted for 10 seconds between each 20 second measurement. Fluorine analyses were done with the PCO crystal. Standards were phlogopite for F, troilite for S, and scapolite for Cl. Background counting times were 200 seconds.

5. RESULTS

5.1 Petrology of Puna Ridge Glasses

All of the analyzed volcanic glasses from TN084 are tholeiites that plot in the general field occupied by other Kilauea lavas on an alkali-silica diagram (Tables 2 and 3; Figure 3). Lavas are predominantly sparsely olivine-augite-plagioclase phyric tholeiites, but picritic mineralogy is also common.

Six lava samples collected by Shinkai 6500 submersible on dive 6K495 from a large, low-relief lava flow south of the Puna Ridge (Figure 2) are mildly alkalic basalts (Table 3 and Figure 3). The lavas are intermediate in composition and plot within the composition range of alkalic lavas from Loihi between Puna Ridge tholeiites and alkalic lavas from the North Arch volcanic field and rejuvenation-stage lavas (Honolulu Series and Koloa Volcanics) [*Clague and Frey*, 1982; *Clague and Dalrymple*, 1988; *Clague et al.*, 1990; *Frey et al.*, 2000] (Figure 4).

Table 2. Representative electron microprobe analyses of glasses from wax cores

Sample	Lat (°N)	Lon (°W)	SiO_2	TiO_2	Al_2O_3	FeO	MnO	MgO	CaO	Na_2O	K_2O	P_2O_5	S	F	Cl	Total
PRC1gl1	19.53345	154.78697	50.70	2.26	12.89	10.79	0.15	9.28	10.20	2.17	0.37	0.21	0.0858	0.0359	0.0134	99.16
PRC1ltpum	19.53345	154.78697	50.64	3.28	13.72	11.43	0.18	5.79	10.23	2.46	0.64	0.31	0.0170	0.0597	0.0176	98.77
PRC2	19.56222	154.72740	50.69	2.74	13.98	10.91	0.17	6.39	10.94	2.40	0.47	0.23				98.92
PRC3	19.56625	154.70785	51.45	2.61	13.79	10.71	0.14	6.66	10.91	2.42	0.47	0.25				99.41
PRC4	19.52897	154.75395	50.42	3.07	13.60	11.85	0.17	5.71	10.43	2.58	0.60	0.33				98.76
PRC5	19.75998	154.37680	50.80	2.68	14.27	10.50	0.16	6.35	10.81	2.42	0.44	0.26				98.66
PRC6	19.56572	154.74742	50.97	3.49	13.31	11.88	0.19	5.42	9.87	2.69	0.71	0.40				98.93
PRC7	19.55382	154.74893	51.09	2.58	13.91	10.86	0.16	6.15	10.97	2.43	0.49	0.26				98.89
PRC8	19.55277	154.74897	50.86	2.55	13.82	10.91	0.18	6.23	10.93	2.41	0.48	0.26				98.62
PRC9	19.55147	154.75615	51.02	2.92	13.85	11.28	0.17	6.05	10.53	2.46	0.52	0.26				99.07
PRC11	19.54167	154.78500	51.08	3.91	12.81	13.10	0.20	4.55	8.84	2.88	0.80	0.45	0.0679	0.0716	0.0297	98.78
PRC12	19.54988	154.76868	51.02	3.89	12.90	13.08	0.19	4.55	8.79	2.89	0.81	0.44	0.0887	0.0704	0.0301	98.76
PRC13	19.53050	154.75972	50.03	3.44	12.95	12.77	0.17	5.18	9.74	2.70	0.71	0.38				98.06
PRC14	19.54593	154.74312	49.94	3.09	13.18	11.54	0.17	5.49	10.00	2.60	0.62	0.33				96.96
PRC16	19.57923	154.72692	51.02	3.11	13.23	12.31	0.19	5.46	10.01	2.44	0.52	0.33				98.62
PRC17	19.57477	154.72077	50.94	3.05	13.39	11.97	0.17	5.55	10.22	2.52	0.58	0.33				98.73
PRC18	19.57733	154.71855	51.03	3.58	12.94	12.83	0.18	4.87	9.25	2.70	0.68	0.39	0.1188	0.0538	0.0222	98.65
PRC19	19.60867	154.64583	50.54	2.77	13.92	11.10	0.17	6.36	11.02	2.43	0.47	0.25				99.03
PRC21	19.61433	154.64933	51.15	2.80	14.00	10.94	0.19	6.00	10.63	2.54	0.49	0.26				98.98
PRC22	19.57710	154.64953	51.20	2.57	14.02	10.33	0.15	6.82	11.08	2.37	0.44	0.25				99.22
PRC23	19.60187	154.52530	51.29	2.96	13.62	11.15	0.17	5.70	10.22	2.62	0.59	0.31				98.65
PRC23spat	19.60187	154.52530	50.99	2.77	14.16	10.57	0.15	5.79	10.41	2.52	0.48	0.25	0.0569	0.0451	0.0310	98.23
PRC23pum2	19.60187	154.52530	50.47	2.33	12.96	11.05	0.17	8.54	10.67	2.16	0.39	0.22	0.0294	0.0282	0.0104	99.04
PRC24gl1	19.57910	154.73337	50.37	3.19	13.88	11.32	0.16	5.92	10.52	2.53	0.62	0.31	0.0907	0.0612	0.0252	98.99
PRC24gl2	19.57910	154.73337	50.91	3.21	13.26	12.42	0.17	5.37	9.92	2.43	0.52	0.29	0.1194	0.0473	0.0177	98.69
PRC26	19.73978	154.41586	51.32	2.77	14.40	10.07	0.14	6.14	10.56	2.60	0.56	0.27	0.1125	0.0451	0.0243	99.00
PRC28	19.71502	154.44333	51.71	3.51	13.49	11.21	0.17	5.15	9.31	2.83	0.75	0.43	0.1208	0.0685	0.0274	98.76
PRC35	19.66283	154.54273	50.74	2.88	13.58	11.58	0.17	5.81	10.35	2.53	0.54	0.30	0.1527	0.0479	0.0208	98.70
PRC35spat	19.66283	154.54273	50.50	2.49	13.21	10.77	0.15	7.92	10.66	2.26	0.42	0.24	0.1100	0.0444	0.0242	98.79
PRC35froth	19.66283	154.54273	50.50	2.57	13.45	10.49	0.16	6.89	10.57	2.30	0.45	0.24	0.0339	0.0397	0.0422	97.74
PRC39	19.62708	154.59198	51.27	3.45	13.10	12.12	0.20	4.91	9.23	2.71	0.71	0.40	0.1141	0.0466	0.0191	98.28
PRC39b	19.62708	154.59198	50.48	2.70	14.14	10.63	0.16	6.57	11.22	2.41	0.51	0.24				99.06
PRC41	19.62872	154.61735	51.39	2.65	14.02	10.37	0.12	6.64	11.12	2.43	0.47	0.24	0.0847	0.0476	0.0189	99.62
PRC44gl1	19.61322	154.61720	52.25	2.97	13.34	11.11	0.19	5.84	10.20	2.50	0.53	0.31				99.23
PRC45gl	19.58278	154.63017	50.71	2.97	13.68	11.63	0.17	5.88	10.42	2.40	0.49	0.24	0.1454	0.0495	0.0189	98.80
PRC46gl	19.59278	154.66505	51.24	2.91	13.90	10.90	0.16	6.13	10.48	2.50	0.51	0.27	0.0933	0.0498	0.0177	99.16
PRC47gl1	19.57500	154.68700	51.28	3.11	13.45	11.58	0.17	5.65	9.87	2.59	0.55	0.31	0.0755	0.0516	0.0231	98.72
PRC47tear1	19.57500	154.68700	50.94	2.50	13.48	10.57	0.17	7.00	11.03	2.23	0.43	0.25	0.1166	0.0456	0.0305	98.79
PRC48gl	19.58072	154.70425	50.56	3.04	13.54	11.84	0.16	5.66	10.12	2.47	0.56	0.28	0.1198	0.0594	0.0199	98.43
PRC49gl1	19.57791	154.71197	51.44	2.89	13.94	10.86	0.16	6.21	10.67	2.48	0.50	0.28	0.0864	0.0505	0.0155	99.58
PRC50	19.58250	154.70333	50.77	3.08	13.58	11.90	0.16	5.62	10.09	2.54	0.58	0.29				98.61
PRC51gl1	19.58727	154.69525	50.80	3.00	13.69	11.76	0.17	5.76	10.18	2.51	0.54	0.28				98.68
PRC52gl	19.54610	154.76572	51.07	2.99	13.74	11.64	0.16	5.83	10.25	2.50	0.54	0.28				98.99
PRC53gl	19.55455	154.74897	50.43	3.21	13.73	11.36	0.17	5.83	10.36	2.58	0.64	0.32				98.63
PRC54gl	19.55382	154.73840	51.34	2.83	14.00	10.91	0.17	6.14	10.61	2.48	0.51	0.27				99.27
PRC55gl1	19.54223	154.79496	50.82	4.27	12.61	13.58	0.15	4.34	8.61	2.75	0.79	0.41	0.0898	0.0706	0.0262	98.52
PRC56gl1	19.52297	154.78042	50.69	3.58	13.06	12.76	0.18	5.18	9.75	2.67	0.66	0.33	0.0861	0.0710	0.0239	99.04

gl – lava flow glass; limu – limu-type ejecta; dkpum, ltpum – darkor light pumice; spind – spindle ejecta; tear – tear-shape ejecta; spat – spatter; froth – frothy ejecta. Complete data set available on accompanying CD-ROM or from senior author.

5.1.1 Compositional ranges, MgO. Previous sampling on Puna Ridge fell into two general categories: lavas recovered in dredge hauls primarily along the crest of the ridge, and glass grains from turbidite sands recovered by box core at the base of the ridge [e.g., Garcia et al., 1989; Clague et al., 1995]. Although bottom photographs demonstrated the presence of a range of lava morphologies, including sheet flows [Lonsdale, 1989], the dredge hauls recovered only pillow lavas. The single sheet flow sample from Puna Ridge was collected during a *SeaCliff* submersible dive [Clague et al., 1995]. Geochemically, these samples had typical Kilauea-type compositions. However, there was a very pronounced division between compositions of the dredged lavas with a narrow, relatively evolved MgO range of 5.20%-6.96%, versus turbidite glasses that were dominated by compositions over 9% MgO (full range 0.98-15.0% MgO). It was clear that there was a sampling bias at work in both the dredge and box core/turbidite sample collections.

The results from our dredging and rock coring program also show distinct compositional populations for 1) dredged lavas, 2) lavas sampled by rock coring, and 3) fragmental glasses in the rock cores (Figure 5). The dredged lavas are similar to previous dredge collections from Puna Ridge, with MgO clustered between 5.0-7.3% and outliers at 4.67% and 9.2% MgO. Lava flows sampled by rock coring extend to significantly more evolved compositions. The overall range is 4.3-6.6%, with an outlier at 9.3% MgO, and about one third of them fall between 4.2-5.2 % MgO. The reason for this bias toward more evolved compositions may be related to our sampling strategy of using the pinpoint capabilities of rock coring to sample small cones. These

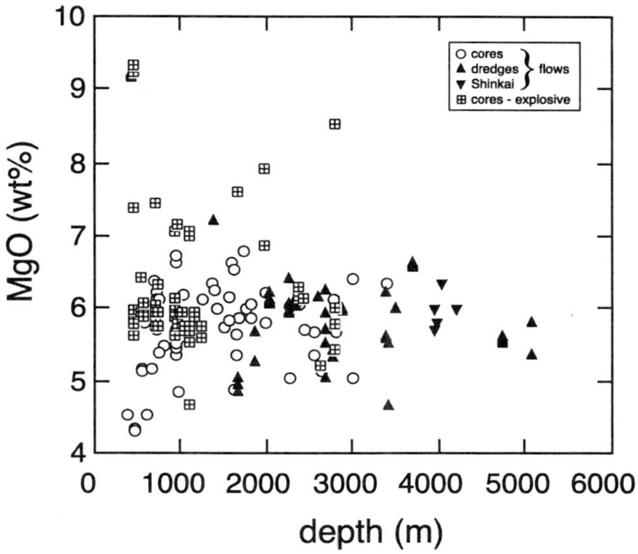

Figure 5. MgO versus depth for samples in this study grouped according to sampling method and eruptive style. Dredged samples generally plot within the MgO range of previous studies with MgO clustered between 5.0-7.3%, but with the addition of outliers at 4.67% and 9.2% MgO. The plot shows that lava flows sampled by rock coring extend to more evolved compositions than those sampled by dredging alone. The overall range is 4.3-6.6%, with an outlier at 9.3% MgO, and about one third of them fall between 4.2-5.2 % MgO (see text for discussion).

cones will have a greater tendency to be made of more evolved, viscous lavas.

Fragmental glasses recovered in the rock cores cover the widest range of compositions, a continuous trend from 4.7% to 9.3% MgO. Extremely primitive and extremely evolved compositions as in the turbidite sands are not represented in the new sample set, despite the more diverse flow types sampled by rock coring.

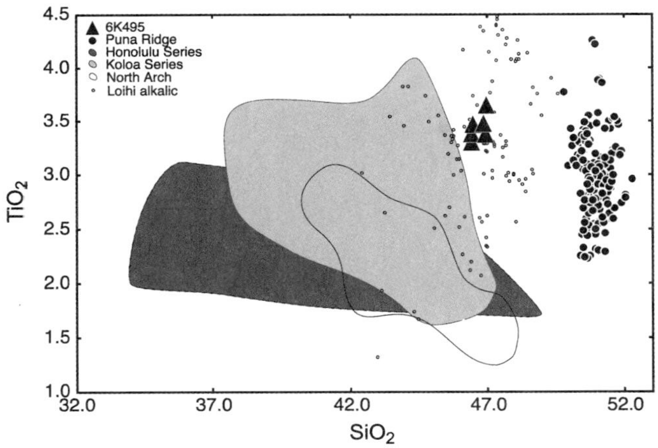

Figure 4. SiO_2 – TiO_2 plot showing the composition of the alkalic flood basalt relative to other alkalic lavas from Hawaii and lavas from the Puna Ridge.

5.1.2 Volatile Elements. Concentrations of volatile elements Cl, F, and S were determined on a subset of the samples employing long count-times on the electron microprobe. Fluorine and chlorine are plotted against MgO (Figure 6g-h) and both exhibit systematically incompatible behavior. The minor Cl enrichment in the frothy glass from wax core PRC35 may be due to alteration. However, the anomalous Cl concentrations above 0.04 wt% in several dredges probably indicate assimilation of a Cl-rich phase [*Michael and Cornell,* 1998; *Kent et al.,* 1999; *Dixon and Clague,* 2001]. These samples are from dredges PRD1 and PRD5 in the South Flank survey area, and PRD6 at the deep, distal end of Puna Ridge.

Sulfur in these glasses does not correlate with any single

Table 3. Electron microprobe analyses of glasses from Shinkai 6500 samples (wt%).

Sample	Lat (°N)	Lon (°W)	SiO_2	TiO_2	Al_2O_3	FeO	MnO	MgO	CaO	Na_2O	K_2O	P_2O_5	S	F	Cl	Total
6K492-1	19.80833	154.29567	51.41	2.89	14.07	10.34	0.16	5.99	10.44	2.55	0.54	0.29	0.0927	0.0458	0.0187	98.84
6K492-3	19.80833	154.29567	51.36	2.74	13.99	10.30	0.13	6.34	10.68	2.46	0.53	0.30				98.85
6K492-4	19.80833	154.29567	51.19	2.84	14.09	10.30	0.16	5.98	10.38	2.52	0.56	0.30				98.31
6K492-5	19.80833	154.29567	51.51	2.95	13.98	10.40	0.15	5.78	10.20	2.58	0.60	0.30				98.44
6K492-6	19.80833	154.29567	51.56	3.00	13.96	10.51	0.14	5.68	10.04	2.57	0.58	0.33	0.0931	0.0481	0.0209	98.54
6K495-1	19.47604	154.35500	46.54	3.46	13.60	12.01	0.16	6.87	11.20	2.56	0.96	0.39	0.1401	0.0800	0.0322	98.01
6K495-3	19.47604	154.35500	46.79	3.48	13.64	11.98	0.18	6.57	11.33	2.57	0.97	0.39				97.90
6K495-4	19.47604	154.35500	46.78	3.43	13.49	11.99	0.16	6.94	11.20	2.56	0.95	0.39				97.89
6K495-5	19.47604	154.35500	46.53	3.43	13.49	12.02	0.15	7.07	11.15	2.52	1.00	0.40				97.76
6K495-6	19.47604	154.35500	46.54	3.38	13.42	12.08	0.17	7.14	11.14	2.49	0.94	0.40				97.70
6K495-7	19.47604	154.35500	46.78	3.56	13.30	12.34	0.18	7.31	10.99	2.49	0.90	0.38				98.25

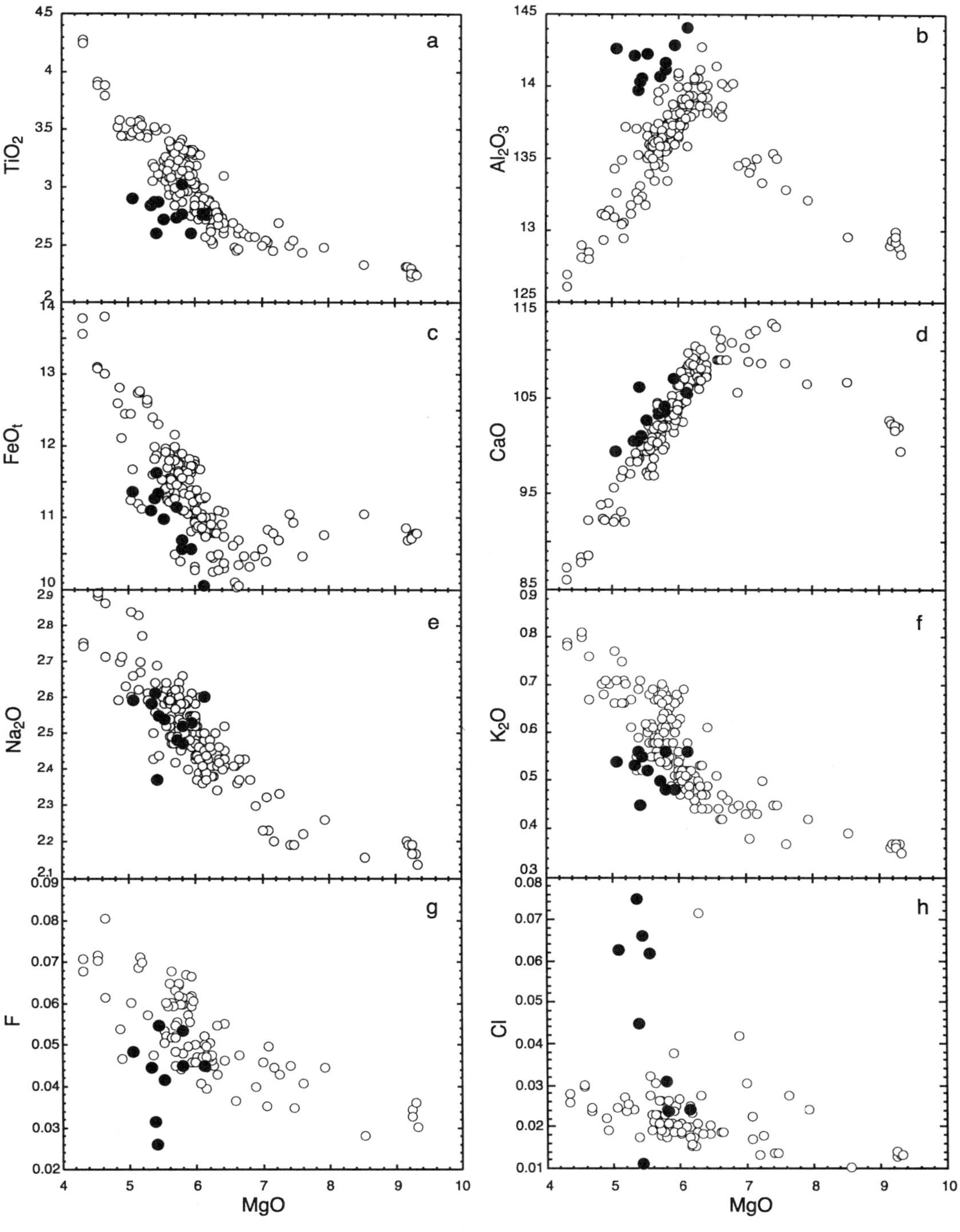

Figure 6. MgO variation diagrams of glasses in this study. In general, the trends are consistent with low pressure crystal fractionation of olivine, pyroxene, and plagioclase (open circles). However, several samples "overshoot" the liquid line of descent defined by the bulk of the Puna Ridge samples in Figure 6b. These samples are discussed in detail in the text and are plotted as solid circles in this figure.

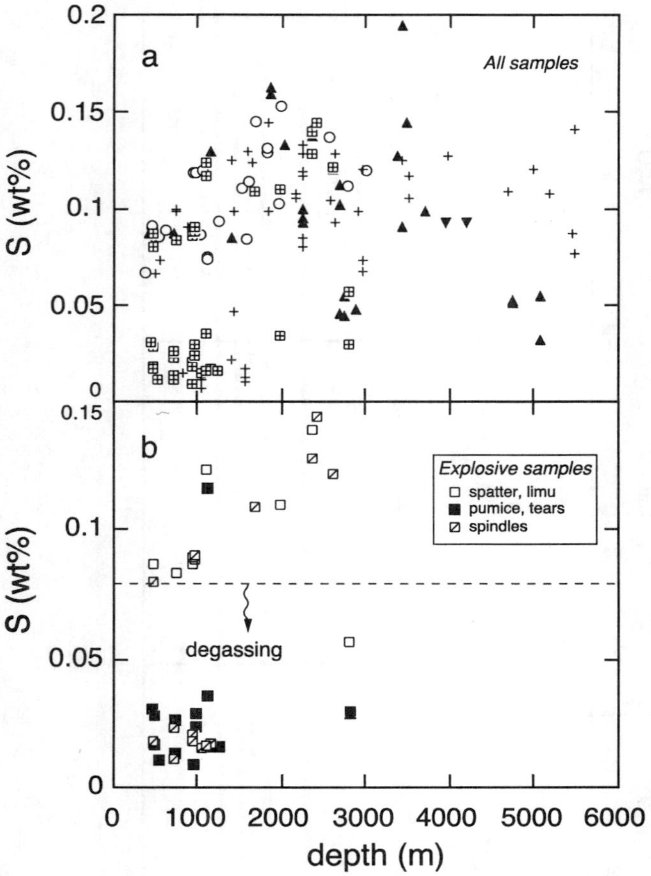

Figure 7. Concentration of sulfur (S) as a function of depth along the Puna Ridge. (a) Plot of all samples discussed in this study (symbols as in Figure 5) along with glasses from previous studies plotted as plusses [*Garcia et al.,* 1989; *Clague et al.,* 1995]. There is no correlation of S concentration with depth in the dredged samples, although a bimodal distribution is suggested. (b) Plot of only the explosive samples from wax rock cores grouped by their morphologies. The bimodality from (a) shows up clearly as a function of morphology with degassed samples represented by pumice and tears and undegassed samples represented by spatter and limu. Spindles are not diagnostic.

element. It is strongly affected by degassing at low confining pressures, but as in previous dredge collections from Puna Ridge, S also does not correlate with eruption depth in dredge samples in this study (Figure 7a). Basalts with <0.06 wt% S, a clear sign of low-pressure degassing, were dredged from water depths as great as 5000m (Table 1, Figure 7a). Dixon et al. [*1991*] attributed this apparent contradiction to magma mixing involving a component that was degassed in a shallow reservoir in the subaerial edifice, recirculated back into the deeper system, and erupted out the submarine rift zone.

Interpretation of the S contents of fragmental glasses in the rock cores is more problematic because these fragments are small enough to have been erupted elsewhere, including subaerial Kilauea, and transported to the sampling site. Indeed, when explosive glass fragments recovered in wax rock cores are subdivided by their morphology, S concentrations in the glasses correlate with the morphologic groups (Figure 7b). The 16 pumice and tears with S analyses have <0.04 wt% S (except PRC47tears1). Because of this consistency, these can be tentatively identified as subaerial glasses, possibly from fire fountaining, that were transported as far as 32 km offshore. The 10 spatter and limu samples with S data have >0.08 wt% S (except PRC23spat). These can be identified with confidence as submarine erupted lavas. Moreover, the undegassed subset of explosive samples does show a correlation of increasing S concentration with depth (Figure 7b). Tholeiitic submarine spatter and limu have been reported elsewhere [*Batiza and White, 2000*], whereas pumice and tears are known only from subaerial eruptions. The spindle category is not diagnostic; the 14 spindle samples with S data range from 0.0112 to 0.1440% S.

5.1.3 Fractionation trends. In general, glass compositions from dredges collected on TN084 and the submersible dive follow well-defined low-pressure crystal fractionation trends involving olivine±clinopyroxene/orthopyroxene± plagioclase that have been described previously for subaerial and submarine East Rift Zone lavas (Figure 6, [*Clague et al.*, 1995; *Garcia et al.*, 1989]). In detail, however, a significant deviation from the simple low-pressure fractionation model is observed in the Puna Ridge data set.

Figure 6b is a plot of MgO versus Al_2O_3 for the TN084 and YK99 glasses. The liquid line of descent reflects onset of plagioclase fractionation at around 6.4 wt% MgO with the concomitant decrease in Al_2O_3. A subset of the data does not turn over at 6.4% MgO, but continues to increase in Al_2O_3. These data points are dredges 1 and 5 and wax core 23 from the south flank off-axis zone (Figure 2). Dredge 6 from the distal end of the Puna Ridge axis also exhibits this Al_2O_3 "overshoot" to a lesser degree. Although not as apparent as the Al_2O_3 overshoot, these same samples plot off the trends of the other elements as well, plotting at the bend of several other plots (Figures 6a, 6c, and 6f-h).

6. DISCUSSION

The large new suite of samples discussed in this paper, which include a large number of wax rock cores, expands the compositional range of glasses reported from the Puna

Ridge. This expanded range provides a more continuous fractionation trend from 9 to 4.5 % MgO.

As discussed above, the morphology of wax rock core samples is diagnostic of their eruptive mode. The pumice, tears and all but six spindle samples that have S ≤ 0.04 wt% were erupted and degassed subaerially and are excluded from the following Puna Ridge fractionation models. The remaining samples are believed to be erupted in the submarine environment and are used in the modeling.

6.1 Fractionation models

The observation that several samples from the south flank and the distal end of the Puna Ridge fractionate beyond the inflection in MgO-Al_2O_3 is significant. The enrichment in Al beyond the trend defined by the majority of Puna Ridge lavas is probably a result of suppression of plagioclase fractionation in these magmas. Suppression of plagioclase fractionation occurs at higher pressures or at higher H_2O contents in the melt. To investigate the effects of each of these variables on melt compositions, we ran several fractionation models using the MELTS program [*Ghiorso and Sack*, 1995]. Figure 8 summarizes the results of these models. Figure 8a is a plot of the same data as shown in Figure 6b with the addition of other Puna Ridge samples [*Clague et al.*, 1995; *Garcia et al.*, 1989], Pu'u O'o [*Garcia et al.*, 1992], and the subaerial East Rift Zone plotted for comparison. The overshoot samples stand out as anomalous relative to most other samples from Kilauea.

Because we did not analyze water contents in our samples, we tested the effects of pressure and water on the liquid line of descent using one of our primitive lavas, PRD22-3, as the starting model composition at different pressures and with different initial water contents (Figures 8b-c). To evaluate the effect of pressure at a constant starting H_2O content of 0.1 wt%, the results of MELTS models at 1.0, 2.0, and 3.0 kb are plotted in Figure 8b. The plot clearly shows the effect of increasing pressure on liquid line of descent. The overshoot samples are consistent with higher pressures of crystallization, in the range of 2.0-3.0 kb, than the bulk of the lavas from Kilauea, which are consistent with crystallization at 1.0-2.0 kb. Puna Ridge glasses generally contain 0.3-0.5 wt% water [*Clague et al.*, 1995; *Garcia et al.*, 1989], and the modeled compositions have about 0.2 wt% H_2O in the 5.0-6.0 wt% MgO range, which is within the uncertainties of the analyses and MELTS calculations.

The results summarized in Figure 8b, specifically the elevation of Al_2O_3 concentrations at higher pressures are identical to the experimental results of *Presnall* [1999]. Presnall showed through high pressure experiments and

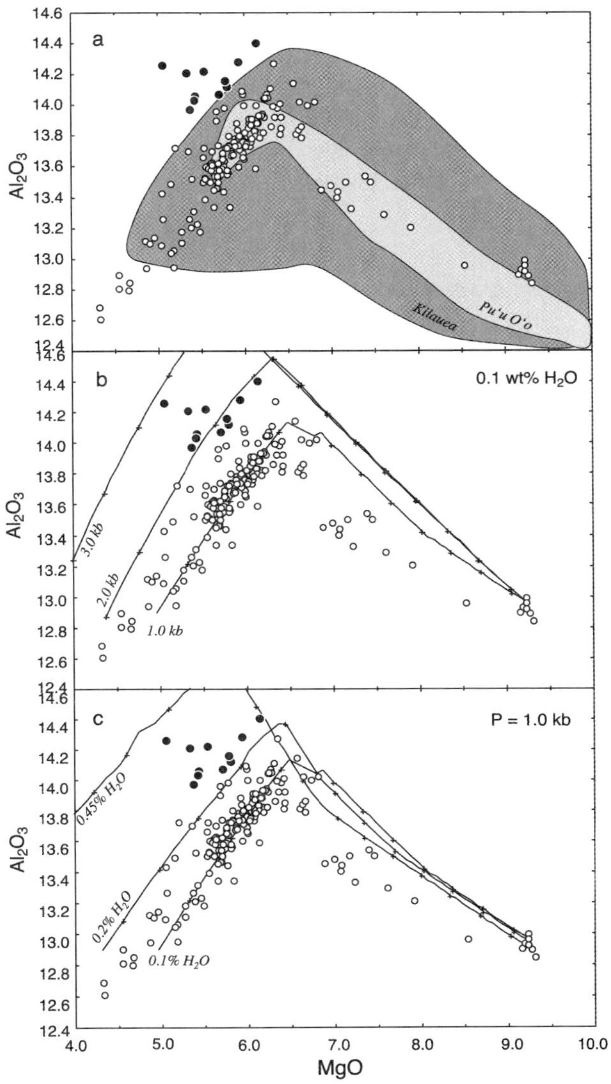

Figure 8. (a) MgO – Al_2O_3 plot of the Puna Ridge data showing the "overshoot" samples discussed in the text (solid circles) relative to all other samples from this study and other Kilauea whole rock and glass samples (open circles; Pu'u O'o field from [*Garcia et al.*, 1992], large field of Kilauea data from many sources). The samples that "overshoot" the trend are from the south flank and the tip of the Puna Ridge. (b) MELTS [*Ghiorso and Sack*, 1995] fractional crystallization model at a constant 0.1 wt% H_2O in a starting melt of primitive composition (PRD22-3) at 1.0, 2.0, and 3.0 kb. The bulk of the data set can be fit by fractional crystallization at ~1.0 kb and the anomalous samples can be fit by fractionation at 2.0-3.0 kb. (c) MELTS models at constant P=1.0 kb and various H_2O concentrations and same starting composition as in (b). The effects of increasing water in the melts are clearly evident and the data can be explained by fractional crystallization at ≤1.0 kb with varying amounts of H_2O in their parental magmas. See text for discussion.

careful geometric analysis of the resulting phase diagrams that the onset of plagioclase crystallization is inhibited by increasing pressure [*Presnall*, 1999]. The suppression of plagioclase crystallization with increasing pressure causes Al_2O_3 concentrations in the residual melts to continue increasing beyond the low pressure MgO-Al_2O_3 inflection (see Figure 8b).

6.2 *Effect of volatiles on fractional crystallization*

Water can also suppress plagioclase crystallization. Figure 8c shows the results of MELTS models at constant pressure (= 1.0 kb) with differing amounts of H_2O in the starting composition. The effect of increasing water in the magma is clearly seen in the model results. The results show that if fractional crystallization of the parental magmas takes place at 1.0 kb (~ 3 km depth), a range of 0.1 – 0.4 wt% H_2O is needed to explain the data.

The general trend of most Puna Ridge glass data (open circles in Figure 8) can be explained by crystal fractionation at ≤ 1 kb with 0.1 – 0.2 wt% H_2O in the parental magma while the overshoot samples (solid circles) require 0.2 – 0.4 wt% H_2O in the parental melts. Assuming a common source for Puna Ridge lavas, how might water be differentially introduced into the Kilauea plumbing system?

6.3 *Crustal assimilation as a potential source of volatile enrichment*

We noted above that most of the overshoot samples contain high concentrations of Cl and plot off the fractional crystallization trend for the bulk of Puna Ridge glasses (Figures 6 and 7). We also note that Cl positively correlates with H_2O in Puna Ridge glasses [*Clague et al.*, 1995]. Hence, in the absence of water analyses for the samples, we infer that the overshoot samples probably contain higher concentrations of H_2O than the "normal" samples. This leads to the conclusion that the suppression of plagioclase fractionation inferred from higher Al_2O_3 concentrations and deviation from the liquid line of descent in the flank and distal samples may be due to increased water in those melts. Thus, the magmas producing the flank and distal samples need not have fractionated at high pressures, but may have evolved at the same low pressures as the other magmas but contained more water (and chlorine). Why do they contain more water and chlorine?

One possibility is that water and chlorine were introduced into the melts as they flowed over the surface or in tubes. Although this may have happened, it can't be the cause of the increased Al_2O_3 concentrations because suppression of plagioclase crystallization must have occurred prior to eruption.

The higher chlorine contents of the flank and distal samples is most likely due to assimilation of crustal material that has been previously altered by seawater [*Michael and Schilling*, 1989; *Clague et al.*, 1995; *Michael and Cornell*, 1998]. The altered crust is also hydrated and contributes H_2O to the magma causing the suppression of plagioclase in the fractionating assemblage and the observed overshoot in Al_2O_3.

Assimilation of altered crust may occur by stoping of wall and roof rocks surrounding the older and distal submarine rift zone magma reservoir as the rift propagates through the crust. As the Puna Ridge grows and spreads out due to repeated intrusions, the axial zone pushes its older flanks and the seawater-altered crustal substructure farther out and away from the zone of present-day active magma intrusion. Dredge 6 was collected near the tip of the Puna Ridge where it is adjacent to Pacific oceanic crust. Thus, the end of the rift system probably both receives a smaller supply of magma than the upper sections of the rift zone and may be in contact with Cretaceous oceanic crust, resulting in continued exposure of the rift zone magma chamber to surrounding altered crustal wall rocks.

The flank samples were collected from prominent lava terraces surmounted by craters. These terraces are 10-11 km away from the present axial zone of intrusion on the ridge based on magnetic interpretations (*Smith et al.* [2000] and this volume). It is unclear whether the lava terraces on the flank were formed by flow in surface lava tubes as on subaerial Kilauea, or whether they were directly fed by intrusions. The magnetic information [*Smith et al.*, 2000] and analogy with subaerial lava emplacement mechanisms argues against a direct conduit from rift zone to flank eruptive site. However, the high chlorine concentrations in flank lavas argues for a magma source that has assimilated altered crustal material *while* fractionating. Furthermore, lavas collected directly upslope from the flank terraces, where putative sources for tube-fed flows would exist, do not show elevated chlorine contents or elevated Al_2O_3. This leads to the conclusion that some flank eruptions on the Puna Ridge are fed directly from portions of the deep rift zone that are in contact with old, altered crust and that their plumbing conduits are distinct from those of the main rift zone.

6.4 *Origin of alkalic lavas*

The 70 km^2 lava flow field south of the Puna Ridge is situated on the boundary between Puna Ridge and the flexural moat surrounding the island. Its young age, inferred from glass and rock freshness, argues against it representing an early stage of Puna Ridge volcanism. The composition of the glass is not as alkalic or as undersaturated as that of rejuvenation volcanism or North Arch lavas, but it is distinct from Kilauea and Puna Ridge glass compositions (Figures 3 and 4), suggesting a different source and mode of formation from these localities. The compositional similarity with alkalic lavas from Loihi suggests derivation from a similar source, by a similar degree of melting, or both (Figure 4).

It appears to be one of the youngest lava flows in the area. Its location at the junction between Hawaii Island and the oceanic plate suggests that it may be a product of a small degree of partial melting of the upper mantle resulting from decompression due to rupture of the oceanic lithosphere by the load from the island.

7. CONCLUSIONS

Lavas from the Puna Ridge generally define a liquid line of descent observed in other Kilauea lavas that can be explained by fractional crystallization of olivine±pyroxene± plagioclase at 1-2 kb. However, 10 samples from the south flank and the distal end of the Puna Ridge plot off the trend defined by the other lavas in MgO-Al_2O_3 space and continue to higher alumina contents at lower MgO. These lavas also contain elevated chlorine contents and are inferred to have higher water than other Puna Ridge samples. The "overshoot" in Al_2O_3 in these lavas can be best explained by suppression of the onset of plagioclase fractionation due to either increased pressure of crystallization or elevated H_2O in the parental magma. In the former scenario, crystallization of magma bodies at depths of 6-9 km causes elevated Al_2O_3 concentrations in the residual liquids that then erupt to form the flank and distal lava flows. This scenario implies eruption directly from deep portions of the rift system, bypassing the shallow rift zone plumbing that is the holding zone for magmas parental to most Kilauea lavas.

In the latter scenario, higher water contents of these melts might be caused by assimilation of altered oceanic crustal or older Puna Ridge wall rocks (containing elevated chlorine concentrations) into the magma chamber while the magma is fractionally crystallizing at depths of 3-6 km. A model whereby melts reside in portions of the rift zone surrounded by or in contact with altered wall rocks and erupt away from the main axis of the rift zone may explain the data. At present, it is not possible to choose between these two scenarios.

Six rock samples collected by Shinkai 6500 submersible at 5565 m depth on a 70 km^2 flood basalt flow south of the Puna Ridge are alkalic. They plot with alkalic basalts from Loihi between tholeiitic lavas from the Puna Ridge and strongly alkalic basalts from rejuvenation-stage volcanism and the North Arch volcanic field on K/Ti – Mg# and SiO_2-TiO_2 diagrams. Their source appears to be more enriched than that of the Puna Ridge, but less enriched than that of the North Arch lavas and rejuvenation stage volcanism on the islands and they are probably products of low degrees of melting induced by decompression due to lithospheric rupture from loading of Hawai'i Island on the Pacific oceanic plate.

Acknowledgments. We thank WHOI's Deep Submergence Operations Group for their hard work on cruise TN084 to obtain these data. We also thank the captain and crews of the R/V *Thompson*, the R/V *Yokosuka*, and the Shinkai 6500 for their hard work during the cruises. Discussions with Dave Clague and Jackie Dixon and constructive and thoughtful reviews by Dean Presnall and Junichi Kimura improved the manuscript. The work was supported by NSF grant OCE-9618226.

REFERENCES

Batiza, R., and J.D.L. White, Submarine lavas and hyaloclastite, in *Encyclopedia of Volcanoes*, edited by H. Sigurdsson, pp. 361-381, Academic Press, San Diego, CA, 2000.

Clague, D., and G.B. Dalrymple, Age and petrology of alkalic postshield and rejuvenated-stage lava from Kauai, Hawaii, *Contributions to Mineralogy and Petrology*, *v. 99*, p. 202-218, 1988.

Clague, D.A., A.S. Davis, J.L. Bischoff, J.E. Dixon, and R. Guyer, Lava bubble-wall fragments formed by submarine hydrovolcanic explosions on Loihi Seamount and Kilauea Volcano, *Bull. Volcanol.*, *61*, 437-449, 2000a.

Clague, D.A., and F.A. Frey, Petrology and trace element chemistry of the Honolulu volcanics, Oahu: implication for the oceanic mantle beneath Hawaii, *J. Petrol.*, *23*, 447-504, 1982.

Clague, D.A., R.T. Holcomb, J.M. Sinton, R.S. Detrick, and M.E. Torresan, Pliocene and Pleistocene alkalic flood basalts on the seafloor north of the Hawaiian Islands, *Earth Planet. Sci. Lett.*, *98*, 175-191, 1990.

Clague, D.A., R.T. Holcomb, M. Torresan, and S. Ross, Shipboard report for Hawaii GLORIA ground-truth cruise F11-88-HW, 25 Oct.-7 Nov., 1988, in *US Geol. Survey Open-file Rep.*, pp. 33, 1988.

Clague, D.A., K.A. Hon, J.L. Anderson, W.W. Chadwick, Jr., and C.G. Fox, Bathymetry of Puna Ridge, Kilauea Volcano, Hawaii, U.S. Geological Survey Map MF-2237, 1994.

Clague, D.A., J.G. Moore, J.E. Dixon, and W.B. Friesen, Petrology of submarine lavas from Kilauea's Puna Ridge, Hawaii, *Journal of Petrology*, *36* (2), 299-349, 1995.

Clague, D.A., J.G. Moore, and J.R. Reynolds, Formation of submarine flat-topped volcanic cones in Hawai'i, *Bull. Volcanol.*, *62*, 214-233, 2000b.

Clague, D.A., K. Uto, K. Satake, and A.S. Davis, Eruption style and flow emplacement in the submarine North Arch Volcanic Field, Hawaii, *J. Geophys. Res. (submitted)*, 2001.

Decker, R.W., Dynamics of Hawaiian volcanoes: an overview, in *Volcanism in Hawaii, v. 2*, edited by R.W. Decker, T.L. Wright, and P.H. Stauffer, pp. 997-1018, U.S. Geol. Surv. Prof. Paper, 1987.

Dixon, J.E. and D.A. Clague, Volatiles in basaltic glasses from Loihi seamount, Hawaii: evidence for a relatively dry plume component, *J. Petrol., 42*, 627-654, 2001.

Fialko, Y.A., and A.M. Rubin, Thermodynamics of lateral dike propagation: Implications for crustal accretion at slow spreading mid-ocean ridges, *J. Geophys. Res., 103*, 2501-2514, 1998.

Fialko, Y.A., and A.M. Rubin, What controls the along-strike slopes of volcanic rift zones?, *J. Geophys. Res., 104*, 20,007-20,020, 1999.

Fiske, R.S., and E.D. Jackson, Orientation and growth of Hawaiian rifts: the effects of regional structure and gravitational stresses, *Proc. Royal Soc. Lond., Series A, 329*, 299-326, 1972.

Fornari, D.J., The geomorphic and structural development of Hawaiian submarine rift zones, in *Volcanism in Hawaii, v. 1*, edited by R.W. Decker, T.L. Wright, and P.H. Stauffer, pp. 125-132, U.S. Geol. Surv. Prof. Paper, 1987.

Fornari, D.J., A. Malahoff, and B.C. Heezen, Volcanic structure of the crest of the Puna Ridge, Hawaii: Geophysical implications of submarine volcanic terrain, *Geol. Soc. of Am. Bull., 89*, 605-616, 1978.

Frey, F.A., D.A. Clague, J.J. Mahoney, and J.M. Sinton, Volcanism at the edge of the Hawaiian Plume: Petrogenesis of submarine alkalic lavas from the North Arch Volcanic Field, *J. Petrol., 41*, 667-691, 2000.

Garcia, M.O., D.W. Muenow, K.E. Aggrey, and J.R. O'Neil, Major element, volatile, and stable isotope geochemistry of Hawaiian submarine tholeiitic glasses, *J. Geophys. Res., 94*, 10525-10538, 1989.

Garcia, M.O., J.M. Rhodes, E.W. Wolfe, G.E. Ulrich, and R.A. Ho, Petrology of lavas from episodes 2-47 of the Pu'u O'o eruption of Kilauea Volcano, Hawaii: evaluation of magmatic processes, *Bull. Volcanol., 55*, 1-16, 1992.

Ghiorso, M.S., and R.O. Sack, Chemical mass transfer in magmatic processes. IV. A revised and internally consistent thermodynamic model for the interpolation and extrapolation of liquid-solid equilibria in magmatic systems at elevated temperatures and pressures, *Contrib. Mineral. Petrol., 119*, 197-212, 1995.

Helz, R.T., and T.L. Wright, Differentiation and magma mixing on Kilauea's east rift zone: a further look at the eruptions of 1955 and 1960 Part I. The late 1955 lavas, *Bull. Volcanol., 54*, 361-384, 1992.

Kent, A.J.R., M.D. Norman, I.D. Hutcheon, and E.M. Stolper, Assimilation of seawater-derived components in an oceanic volcano: evidence from matrix glasses and glass inclusions from Loihi seamount, Hawaii, *Chemical Geology, 156*, 299-319, 1999.

Klein, F.W., R.Y. Koyanagi, J.S. Nakata, and W.R. Tanigawa, The seismicity of Kilauea's magma system, in *Volcanism in Hawaii, v. 2*, edited by R.W. Decker, T.L. Wright, and P.H. Stauffer, pp. 1019-1185, U.S. Geol. Surv. Prof. Paper, 1987.

Lonsdale, P., A geomorphological reconnaissance of the submarine part of the East Rift Zone of Kilauea Volcano, Hawaii, *Bull. Volcanol., 51*, 123-144, 1989.

Macdonald, G.A., A.T. Abbott, and F.L. Peterson, *Volcanoes in the Sea-The Geology of Hawai'i*, University of Hawai'i Press, Honolulu, HI, 1983.

Mattox, T.N., and M.T. Mangan, Littoral hydrovolcanic explosions: a case study of lava-seawater interaction at Kilauea Volcano, *J. Volcanol. Geotherm. Res., 75*, 1-17, 1993.

MBARI, MBARI Hawaii Multibeam Survey CD, Monterey Bay Aquarium Research Institute, 2000.

Michael, P.J., and W.C. Cornell, Influence of spreading rate and magma supply on crystallization and assimilation beneath mid-ocean ridges: Evidence from chlorine and major element chemistry of mid-ocean ridge basalts, *J. Geophys. Res., 103*, 18325-18356, 1998.

Michael, P.J., and J.-G. Schilling, Chlorine in mid-ocean ridge magmas: evidence for assimilation of seawater-influenced components, *Geochim. Cosmochim. Acta, 53*, 3131-3143, 1989.

Moore, J.G., Petrology of deep-sea basalt near Hawaii, *Am. J. Sci., 263*, 40-52, 1965.

Moore, J.G., and W.W.J. Chadwick, Offshore geology of Mauna Loa and adjacent areas, Hawaii, in *Mauna Loa revealed: structure, composition, history, and hazards*, edited by J.M. Rhodes, and J.P. Lockwood, pp. p. 21-44, American Geophysical Union, Washington, DC, 1995.

Moore, J.G., and R.S. Fiske, Volcanic substructure inferred from dredge samples and ocean bottom photographs, Hawaii, *Geol. Soc. of Am. Bull., 80*, 1191-1203, 1969.

Presnall, D.C., Effect of pressure on the fractional crystallization of basaltic magma, in *Mantle Petrology: Field Observations and High Pressure Experimentation: A Tribute to Francis R. (Joe) Boyd*, edited by Y. Fei, C.M. Bertka, and B.O. Mysen, pp. 209-224, The Geochemical Society, Houston, TX, 1999.

Rowland, S.K., and G.P.L. Walker, Pahoehoe and aa in Hawaii: Volumetric flow-rate controls the lava structure, *Bull. Volcanol., 52*, 615-628, 1990.

Ryan, M.P., R.Y. Koyanagi, and R.S. Fiske, Modeling the three-dimensional structure of macroscopic magma transport systems: application to Kilauea Volcano, Hawaii, *J. Geophys. Res., 86*, 7111-7129, 1981.

Smith, D.K., and J.R. Cann, Constructing the upper crust of the Mid-Atlantic Ridge: A reinterpretation based on the Puna Ridge, Kilauea Volcano, *J. Geophys. Res., 104*, 23,579-25,399, 1999.

Smith, D.K., M.A. Tivey, P. Gregg, and L.S.L. Kong, Magnetic anomalies at the Puna Ridge, a submarine extension of Kilauea Volcano: Implications for lava deposition, *J. Geophys. Res., in press*, 2000.

Wright, T.L., and R.S. Fiske, Origin of the differentiated and hybrid lavas of Kilauea Volcano, Hawaii, *Journal of Petrology, v. 12* (no. 1), p. 1-65, 1971.

Wright, T.L., and R.T. Helz, Differentiation and magma mixing on Kilauea's east rift one: a further look at the eruptions of 1955 and 1960. Part II. The 1960 lavas, *Bulletin of Volcanology*, v. *57*, p. 602-630, 1996.

Yang, H.-J., F.A. Frey, M.O. Garcia, and D.A. Clague, Submarine lavas from Mauna Kea Volcano, Hawaii: implications for Hawaiian shield stage processes, *J. Geophys. Res.*, *99*, 15577-15594, 1994.

Kevin T. M. Johnson, Department of Geology, Bishop Museum, 1525 Bernice St., Honolulu, HI 96817-2704, USA. E-mail: kevinj@soest.hawaii.edu

Laura S. L. Kong, U. S. Department of Transportation, Federal Highway Administration-Hawaii Division, 300 Ala Moana Blvd. Ste. 3-306, Box 50206, Honolulu, HI 96850, USA.

Jennifer R. Reynolds, University of Alaska Fairbanks, P.O. Box 757220, 213 O'Neill Building, Fairbanks, AK 99775, USA.

Deborah K. Smith, Department of Geology and Geophysics, Woods Hole Oceanographic Institution, Woods Hole, MA 02543, USA.

Denys Vonderhaar, Department of Geology, Bishop Museum, 1525 Bernice St., Honolulu, HI 96817-2704, USA.

Ancestral Submarine Growth of Kīlauea Volcano and Instability of its South Flank

Peter W. Lipman[1], Thomas.W. Sisson[1], Tadahide Ui[2], Jiro Naka[3], and John R. Smith[4]

Joint Japan-USA cruises in 1998-99 explored and sampled the previously unstudied deep offshore region south of Kīlauea. Bathymetric features, dive observations, and recovered samples indicate that the 3-km-deep mid-slope bench, bounded seaward by a 2-km-high lower scarp, is underlain by massive turbidite sandstone and interbedded debris-flow breccia. Debris-flow clasts are submarine-erupted (high-S) alkalic basalt, distinctive fine- to coarse-grained alkalic gabbro and nephelinite (some containing phlogopite), and subordinate transitional basalt. No recovered clasts are similar to recent Kīlauea tholeiite. Primary volcanic deposits (pillow basalt, hyaloclastite breccia) are absent. The sandstone and breccia matrix, a large fraction of the volcaniclastic apron, are mainly subaerially erupted (low-S) tholeiitic glass sand generated by shoreline processes on pre-Kīlauea volcanoes. Fractures, shears, slickensided clasts, and open folds indicate widespread deformation low in the lower scarp; upward-decreasing proportions of alkalic materials define a gross stratigraphy. Alkalic high-S compositions of many basalt clasts and some sandstone glass indicate derivation from the submarine "Lō'ihi" stage of ancestral Kīlauea, prior to growth of its tholeiitic shield. Slopes (3300-2800 m depth) above the mid-slope bench contain submarine-erupted (S >750 ppm) pillow lava of transitional basalt, defining the initial flank of subalkaline Kīlauea. The geometry and diverse constituents of the bench and lower scarp require initial landsliding during alkalic volcanism at 200-300 ka, prior to shield growth at Kīlauea and inception of Hilina faults. The active Hilina slump structures on Kīlauea's south flank are in an early growth stage, thus possibly posing greater potential for future large-scale landsliding and tsunamis.

[1]Volcano Hazards Program, U.S. Geological Survey, Menlo Park, California

[2]Graduate School of Science, Hokkaido University, Sapporo, Japan

[3]Japan Marine Science and Technology Center, Yokosuka, Japan

[4]School of Ocean and Earth Science and Technology, University of Hawai'i, Honolulu, Hawai'i

Hawaiian Volcanoes: Deep Underwater Perspectives
Geophysical Monograph 128
This paper not subject to U.S. copyright
Published in 2002 by the American Geophysical Union

1. INTRODUCTION

Despite numerous above-sea-level studies, the early eruptive histories of Hawaiian volcanoes are largely unknown, because younger subaerial lavas conceal older rocks. Most subaerial lavas that enter the ocean advance only short distances beyond shoreline [*Moore and Chadwick*, 1995]; therefore, older submarine-erupted lavas —farther and deeper offshore—may escape burial by shoreline-crossing flows and remain exposed on steep submarine flanks. Large submarine landslides and slumps might also expose products of early volcano growth. Potential sites for such exposures are landslide headwall

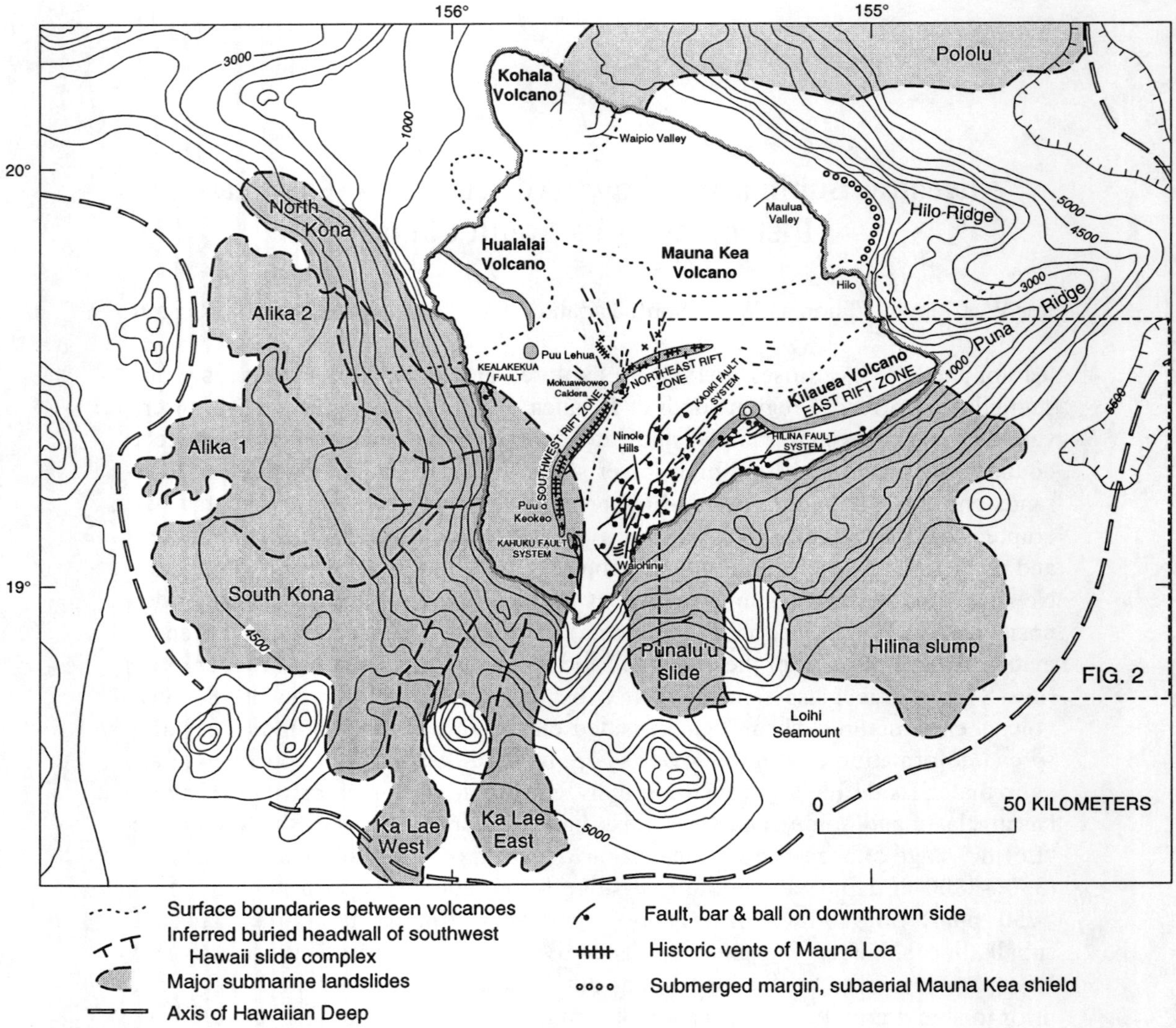

Figure 1. Map showing major structural features of Kīlauea volcano in relation to other volcanoes of Hawai'i Island. Contour interval, 500 m. Modified from *Lipman et al.* [1990]

scarps, steep lower flanks compressionally thickened by gravitational subsidence and volcano spreading, and distal run-outs of far-traveled (to 200 km) landslides [*Lipman et al.*, 1985, fig. 21; *Borgia and Treves,* 1992; *Moore et al.*, 1989].

The south flank of Hawai'i Island is the main locus of recent volcanism along the Hawaiian Ridge, including shoreline-crossing lavas from Kīlauea and Mauna Loa, and submarine eruptions at Lō'ihi Seamount. Morphologically complex deep-water bathymetry adjacent to topographically simpler slopes of on-land Kīlauea and Mauna Loa [*Chadwick et al.,* 1993; *Moore and Chadwick,* 1995] suggests that the submarine slopes might record early events in the growth of these volcanoes. Before the present study, however, no samples or camera images have been available for the south flank area at depths greater than about 2000 m. Prior submersible-based studies of the region between Kīlauea's east rift zone and Mauna Loa's southwest rift zone have focused on upper parts of Lō'ihi and Pāpa'u Seamounts [*Fornari et al.,* 1979; *Malahoff,* 1987; *Garcia et al.*, 1995a].

The deep offshore region south of Kīlauea was explored and sampled for the first time by submersible and ROV (Remotely Operated Vehicle) dives during joint Japan-USA cruises in 1998-99 [*Naka et al.*, 2000; *Lipman et al.*, 2000], supported by the Japan Marine Science and Tech-

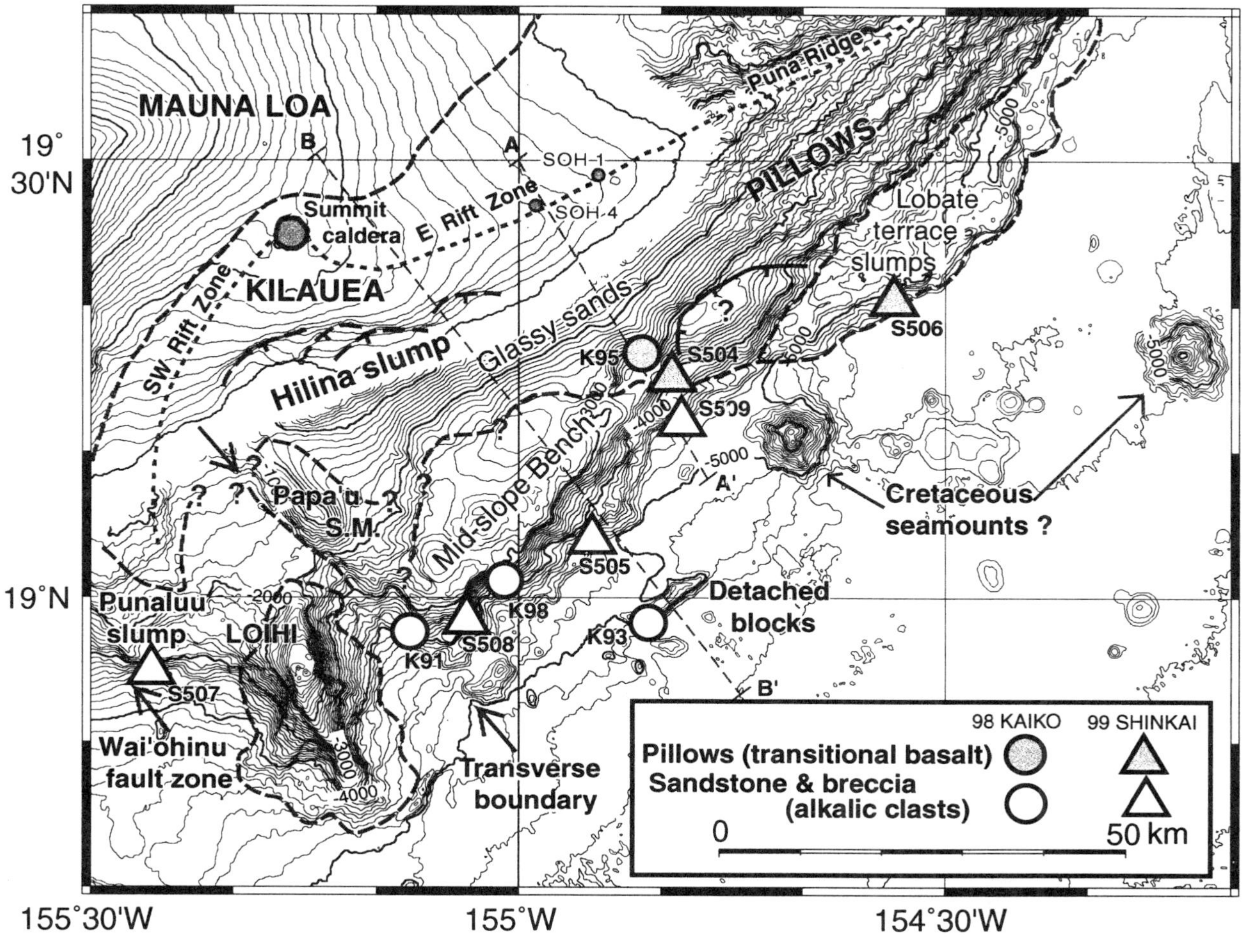

Figure 2. South-flank bathymetry (from JAMSTEC SeaBeam surveys in 1998-99 [*Smith and Satake*, this volume]), dive sites, and general distribution of breccia-clast and pillow compositions. Contour interval, 100 m. Shaded pattern, subaerial Hawai'i Island; geologic features generalized from *Wolfe and Morris* [1996]. Dashed lines, boundaries between inferred structural domains on south flank of Kīlauea. Hachured lines, fault scarps. Lines A-A' and B-B' are locations of cross sections in Plate 1. SOH 1, SOH 4: drill holes along east rift zone [Quane et al., 2000]. Dive sites on Lō'ihi not shown.

nology Center (JAMSTEC). Goals of this work were to better characterize compositions, style, and volume of early submarine Hawaiian volcanism, and to better understand potentially catastrophic submarine mass-wasting processes in several parts of the island chain. The southern slope of Kīlauea was studied because of the clearly documented historical deformation in this region, including past and potential future slumps and landslides along the Hilina fault system and related structures (Figure 1). The new data show that ancestral Kīlauea was a sizeable submarine alkalic volcano, where recurrent slope failures generated a flanking apron of clastic debris that was subsequently overgrown by the modern tholeiitic Kīlauea shield.

2. GEOLOGIC SETTING

Kīlauea is the world's most active volcano, and about 90% of its subaerial surface is less than 1100 years old [*Holcomb*, 1987; *Tilling and Dvorak*, 1993]. Two rift zones radiate from the approximately 1200 m elevation summit caldera (Figure 2). The east rift zone runs east-northeast for 50 km to the shoreline and then continues for another 60 km beneath the sea as the Puna Ridge, ending at about 5500 m depth at the Cretaceous seafloor. The morphology of the sediment-poor, pillow-lava-dominated Puna Ridge is in striking contrast to the smooth upper submarine slope offshore from the island [*Chadwick et al.*, 1993], which is blanketed by coastal-derived glassy sand [*Lee et al.*, 1994]. The less conspicuous southwest rift zone runs 30 km to the southwest from the summit caldera to the shoreline and then another ~20 km as an indistinct submarine ridge.

Kīlauea's offshore south flank lies between these rift zones and is characterized by a prominent mid-slope bench bounding a closed depression at about 3000 m depth [*Moore and Peck*, 1965; *Chadwick et al.*, 1993; *Smith et al.*, 1999]. The upper slope steepens toward the shoreline, and a lower scarp ('outer scarp' of *Smith et al.* [1999]) descends from the lip of the mid-slope bench down to the 5000-m-deep Cretaceous sea floor (Figure 2). Scattered along the base of the lower scarp are isolated blocks and elongate ridges [*Chadwick et al.*, 1993; *Smith et al.*, 1999], some as much as 10 km long, that are interpreted as slide blocks from the lower scarp, based on bathymetry and compositions of recovered samples. A prominent northwest-trending transverse scarp, aligned with Pāpa'u Seamount ("transverse boundary," in Figure 2), bounds the bench and lower scarp to the southwest [*Moore and Chadwick*, 1995; *Denlinger and Okubo*, 1995]. Southwest of this boundary are the active Lō'ihi Seamount, containing alkalic and transitional basalts that record growth of Hawai'i's youngest volcano [*Moore et al.*, 1982; *Malahoff*, 1987; *Garcia et al.*, 1995a], and the smaller Punalu'u slump derived from Mauna Loa [*Lipman et al.*, 1990].

While little direct observational data have been available previously, underwater slopes of large subaerial Hawaiian volcanoes have been inferred to contain an early pillow-lava core, built while the edifice was entirely submarine [*Moore and Fiske*, 1969; *Moore and Chadwick*, 1995, Figure 6]. These authors inferred that the pillow-lava core is extensively overlain and mantled by angle-of-repose hyaloclastite breccias resulting from fragmentation during shoreline entry of lava erupted above sea level. Such deposits on the south flank of Hawaii appear to have strikingly different bathymetric expressions (Figure 2; also *Chadwick et al.* [1993]). The irregular lobate contours of the Puna Ridge mark submarine-erupted pillow lava [*Clague et al.*, 1995], emplaced beyond the influence of shoreline-derived fragmental debris. In contrast, the smooth offshore flank downslope from subaerial parts of the south flank is blanketed by quenched fragments of subaerially erupted basalt [*Lee et al.*, 1994; *Moore and Chadwick*, 1995]. The hyaloclastite blanket on the upper submarine slope of Kīlauea has been estimated to be as much as 1 km or more thick from seismic-reflection profiling [*Morgan et al.*, 2000; *Hills et al.*, this volume], but JAMSTEC dives show that submarine-erupted pillow lavas crop out in places along this slope. Recent submersible observations along the west flank of Mauna Loa suggest that pillow lavas derived from subaerial eruptions may also be abundant constituents on upper submarine slopes of other Hawaiian volcanoes [*Garcia and Davis*, 2001].

The south flank of Kīlauea also contains the active Hilina fault system, a series of arcuate down-to-the-south normal faults (Figure 2) where multi-meter seaward displacements and large tsunamis were generated concurrently with magnitude 7.9 and 7.2 earthquakes in 1868 and 1975 [*Wood*, 1914; *Tilling et al.*, 1976; *Lipman et al.*, 1985; *Wyss*, 1988]. Between such large infrequent events, seaward motions of 5-10 cm/yr along the south-flank coastline have been geodetically documented for most of the 20th century [*Swanson et al.*, 1976; *Delaney et al.*, 1993, 1998; *Owen et al.*, 2000]. On land, individual fault scarps of the Hilina system are up to 200-m high, widely draped by younger lavas, and have cumulative overall relief of as much as 800 m. Offsets along these scarps largely or entirely postdate the deposition of ash beds, interleaved with Kīlauea basalt flows [*Easton*, 1987], that are younger than about 39 ka [*Beeson et al.*, 1996].

The Hilina fault system was early recognized from geologic evidence to define a region of seaward motion of Kīlauea's south flank, resembling a giant landslide [*Stearns and Clark*, 1930, p. 97; *Moore and Krivoy*, 1964]. Combined geologic, geodetic, and geophysical studies in the aftermath of the 1975 Kalapana earthquake (M=7.2) demonstrated that coastal seaward motions of as much as 8 m horizontally and subsidence of as much as 3.5 m [*Lipman et al.*, 1985] were accommodated by slip along a gently dipping fault at depths of 7-10 km, at or near the base of the Hawai'i Island volcanic edifice [*Ando*, 1979; *Crosson and Endo*, 1982; *Thurber and Gripp*, 1988; *Got et al.*, 1994; *Denlinger and Okubo*, 1995]. Forceful dike injection along rift zones [*Swanson et al.*, 1976; *Dvorak*, 1994], gravitational spreading of the overall volcanic edifice [*Borgia and Treves*, 1992], density-driven flowage of olivine-rich cumulates at magmatic temperatures [*Clague and Denlinger*, 1994], and combined effects of such processes have been inferred to cause the slumping and seaward motions of Kīlauea's south flank. Displacement along a basal decollement is likely facilitated by thick sediment at the interface between the volcanic pile and prevolcanic oceanic crust [*Nakamura*, 1980; *Crosson and Endo*, 1982]. Landward first motions of the tsunami generated by the 1975 earthquake require uplift of a large area offshore [*Hatori*, 1976; *Ando*, 1979; *Ma et al.*, 1999], interpreted as involving the mid-slope bench. The Hilina faults have been interpreted as having a shallow rotational slump geometry [*Moore and Fiske*, 1969, fig. 3; *Swanson et al.*, 1976; *Dvorak et al.*, 1986; *Riley et al.*, 1999] or, alternatively, having deeper connections with the basal decollement surface [*Lipman et al.*, 1985; *Denlinger and Okubo,* 1995; *Moore et al.*, 1995; *Okubo et al.*, 1997].

Based on the low-resolution bathymetry available at the time of the 1975 earthquake (Figure 1), submarine benches and scarps on the south flank were interpreted as down-dropped segments of the Kīlauea shield, accompanied by local compression and uplift at the distal submarine toe of the sliding edifice [*Lipman et al.,* 1985; *Moore et al.,* 1989]. Critical evidence for the distal compression and uplift of the seafloor was the landward first arrival of the 1975 tsunami. With access to higher-resolution surveys [*Chadwick et al.,* 1993; *Smith and Satake,* this volume], the mid-slope bench and a closed basin behind it (Figure 1) have increasingly been interpreted as due dominantly to major compressional uplift and thrusting along the lower south flank of the actively spreading Kīlauea volcano [*Borgia and Treves,* 1992; *Denlinger and Okubo*, 1995; *Smith et al.*, 1999]. Recently collected multi-channel seismic-reflection data have been interpreted to indicate that the bench and lower scarp mark sites of major imbricate thrusting that has accommodated 15-24 km of displacement [*Morgan et al.,* 2000; *Hills et al.,* this volume].

3. STRATIGRAPHY AND STRUCTURE OF THE SUBMARINE SOUTH FLANK

Video-camera images, visual observations, and samples obtained by the ROV *Kaiko* (four unmanned dives in 1998) and *Shinkai 6500* submersible (six dives in 1999) provide the first direct geologic evidence on the stratigraphy and structure of the deep south flank of Hawai'i Island. The geology of this region was previously interpreted solely from bathymetric and seismic data (except for a single camera station [*Moore and Fiske,* 1969] showing mud-covered bottom at 2930 m depth). The new observations and samples (Figure 2) are from the lower Hilina slump scarp (dives K91, K98, S505, S508, S509), lower parts of the upper Kīlauea slope (K95, S504), a deep outlying northeast-trending block (K93), the southern base of the proximal Puna Ridge (S506), and a deep traverse on the Punalu'u slump (S507), west of Lō'ihi Seamount. Many features of the dives and samples are summarized in Table 5 (CD ROM); additional documentation is available from on-board cruise reports [*Shipboard Scientists*, 1998, 1999].

In addition to direct observations by scientists in the *Shinkai 6500*, Super VHS tapes were recorded continuously by two video cameras for all ROV and submersible dives, and color transparencies were obtained by 35-mm camera for many important features. Rock samples (85 total) were collected with the manipulator arms--most directly from outcrops, some from uniform talus at cliff bases. Weakly indurated bottom sediment and breccia matrix were sampled by push core [*Naka et al.*, this volume]. At two sites, additional useful samples were obtained (S507-R7, S509-R7) when the submersible gently grazed steep breccia outcrops, and small breccia clasts fell into the metal-mesh sample tray.

The ocean-bottom sampling process by ROV (monitored by video images on the surface ship), and even by submersible with a scientist observer on board, was challenging in a region as structurally and lithologically complex as the south flank. Even for these relatively youthful rocks, which lack thick palagonite or manganese coatings, lithologies at some sites were initially misidentified from video images. Some angular fragments of well-indurated sandstone resemble prismatic pillow fragments, and none of the gabbro-nephelinite sites (five localities) was recognized prior to surface arrival and examination of samples on the support ship. Nevertheless, the numerous rock samples obtained during the two cruises have yielded substantial redundancy and repetitions among sample types (Table 5, Appendix), suggesting that the major lithologies of the lower south flank have now been characterized.

Dive traverses were targeted at scarps, ridge crests, and other relatively steep segments of the lower south flank in order to optimize rock exposures. The overall slopes of the submarine south flank average about 10 degrees, and the lower scarp about 15-20 degrees [*Moore and Mark,* 1992; *Chadwick et al.,* 1993]. Despite the relatively gentle overall slopes, near-vertical cliffs alternated with sediment covered slopes during most dives. All flatter portions of the lower scarp are heavily mantled by young sediment composed of basaltic glass sand derived from coastal environments, intermixed with abyssal clays and radiolarian tests [*Naka et al.*, this volume]. In places, indurated slope-mantling sediments were difficult to distinguish unambiguously from older strata of the lower scarp that they unconformably overlie.

Extensive visual observations, combined with sample characterization, indicate that the 3-km-deep mid-slope bench and lower scarp consist entirely of well-bedded volcaniclastic rocks. Massive sandstone is interbedded with breccia, siltstone, and mudstone (Figure 3A). Unexpectedly, no primary volcanic deposits (pillow basalt, sheet flow, hyaloclastite pillow breccia) were observed anywhere along the lower scarp.

The sandstone beds (Figure 3B) commonly are massive and well indurated, form cliffs, and are interbedded with coarse breccia. The sandstone is brown to black in hand sample, typically well sorted, and consists mainly of fresh basalt glass, variably rimmed by palagonite (commonly

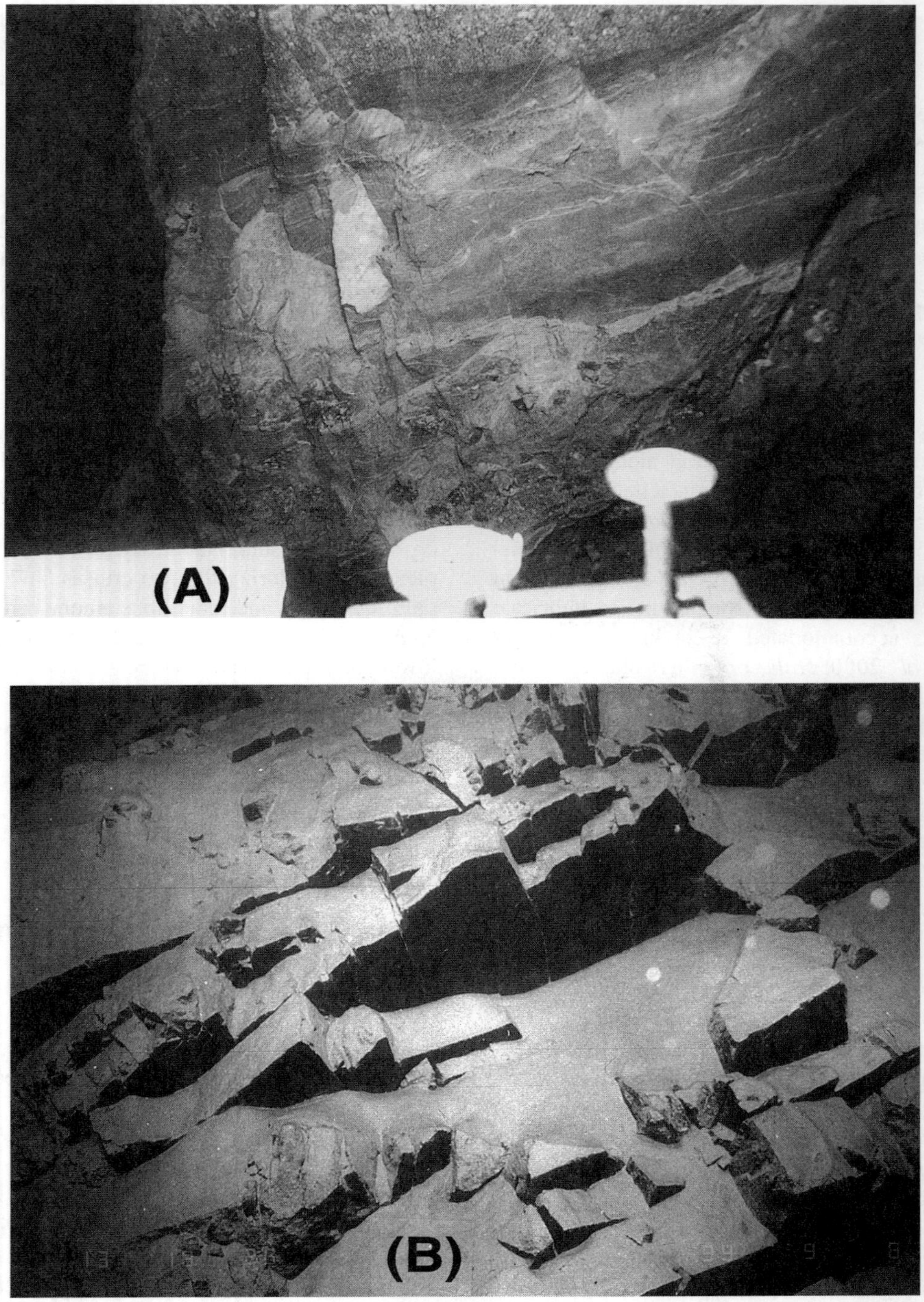

Figure 3. Characteristic rock types observed during south-flank dives.
(A) Bedded basaltic-glass sandstone and breccia below mid-slope bench. Dive K91, at 4300 m depth, width of view ~2 m.
(B) Massive well-indurated turbidite sandstone, consisting of well sorted grains of basaltic glass, cut by rectilinear joints. Dive S508, 3735 m depth (sample site S508-R5), width of view ~2 m.
(C) Massive fine-grained breccia, consisting mainly of small clasts of devitrified alkalic basalt, cut by a steeply dipping narrow shear zone. Dive S509, at 4515 m (near sample site S509-R2B; see Figure 4C), width of view ~1 m.
(D) Prismatically jointed pillow lava in cross section above mid-slope bench. Dive K95, at 3100 m depth, width ~1 m.

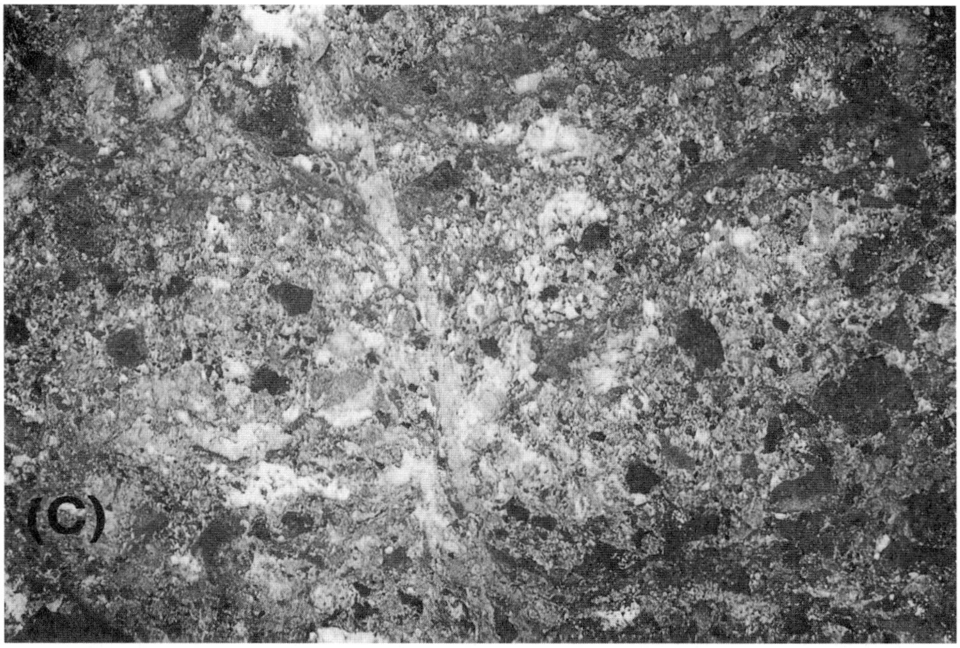

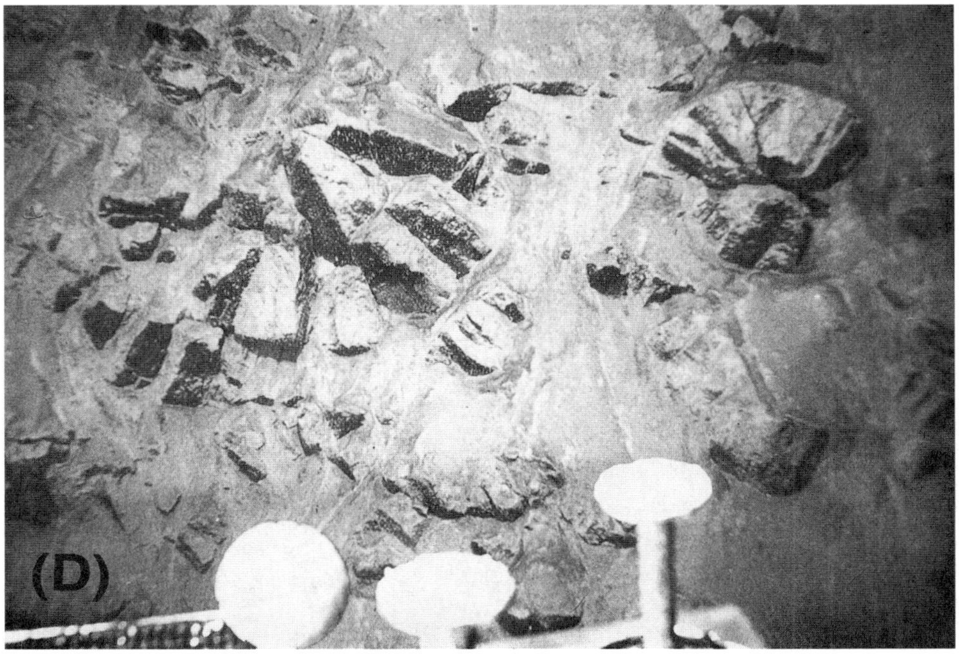

0.05-0.5 mm rinds). Sandstone clasts from near the base of the lower scarp are generally more palagonitized than samples from higher on the scarp, but some samples collected sequentially in vertical sections have variably thick rinds on glass grains. Such variations indicate that permeability and other fluid-flow variations must significantly influence palagonitization rates in this area. Variable combinations of completely palagonitized small grains cement the sandstones and, in some samples, zeolites fill pore space. Bedded sandstones (Figure 3A), where successfully sampled, are commonly graded, fining upward. The well-sorted sandstones (Figure 4A), some forming massive beds 10 m

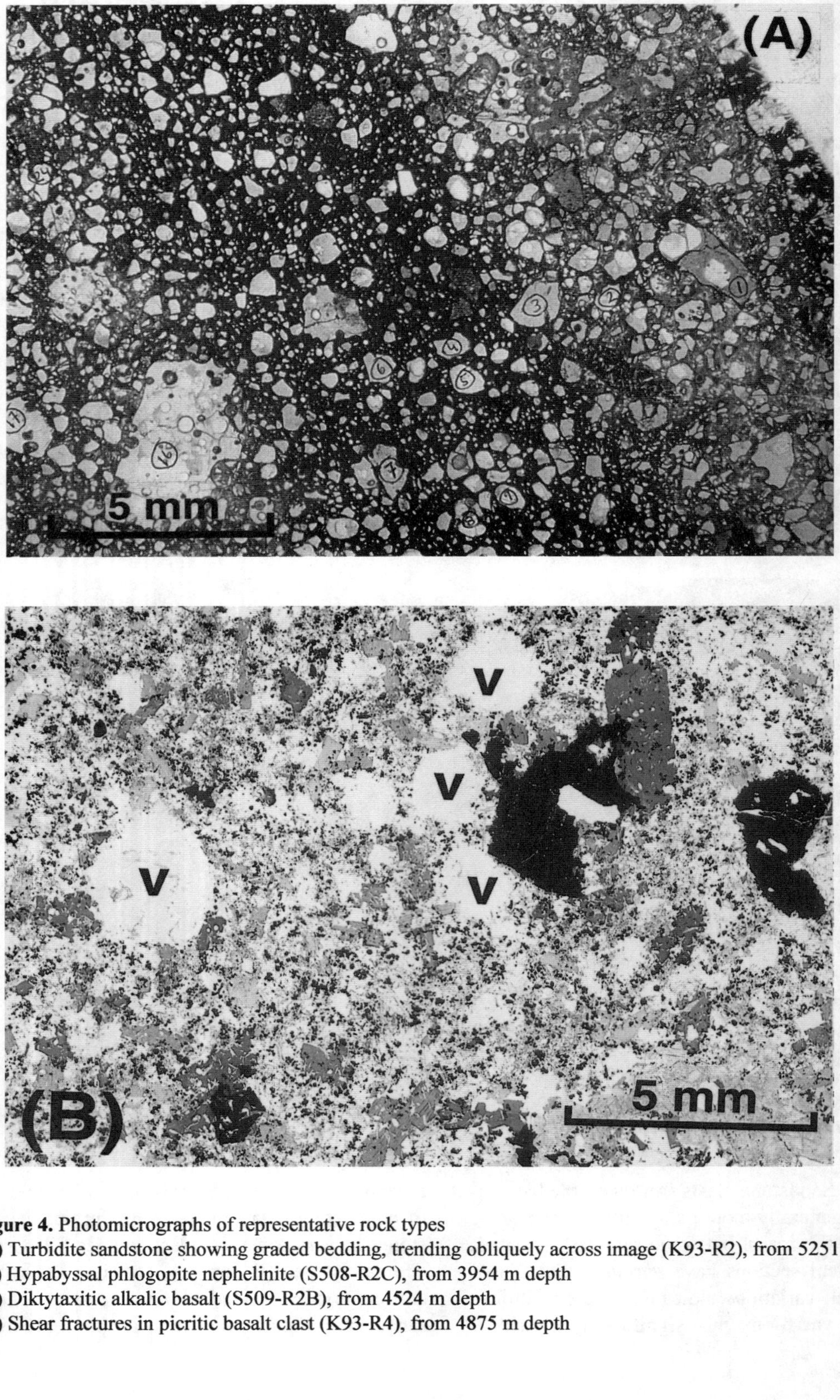

Figure 4. Photomicrographs of representative rock types
(A) Turbidite sandstone showing graded bedding, trending obliquely across image (K93-R2), from 5251 m depth
(B) Hypabyssal phlogopite nephelinite (S508-R2C), from 3954 m depth
(C) Diktytaxitic alkalic basalt (S509-R2B), from 4524 m depth
(D) Shear fractures in picritic basalt clast (K93-R4), from 4875 m depth

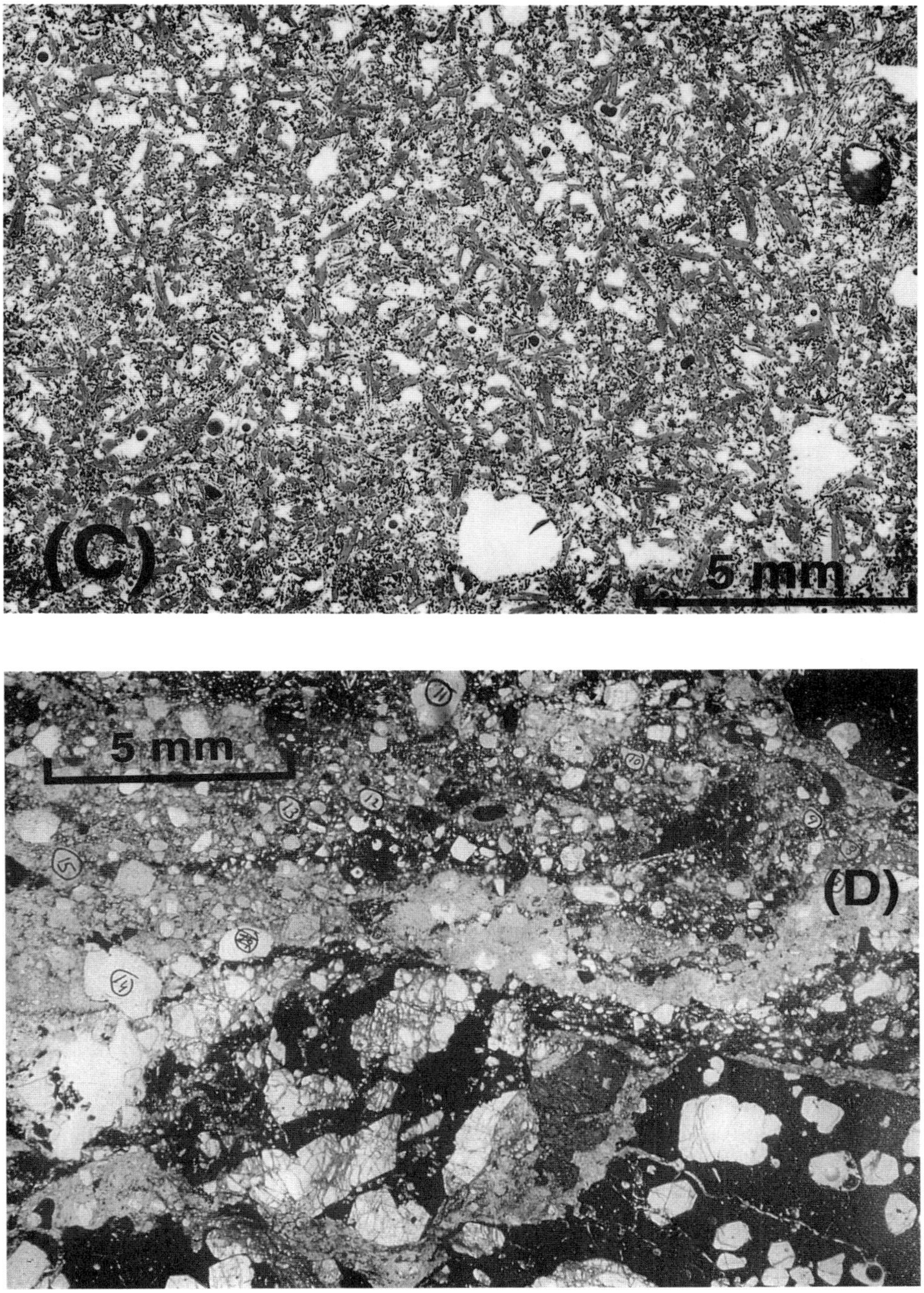

or more thick, are interpreted as indurated turbidite deposits, probably largely derived from subaerial lavas entering the sea (discussed further in section 8.1).

Breccias are bedded clast-supported deposits composed of generally angular clasts 5-25 cm in diameter (Figure 3C); larger clasts are rare or absent. Some breccia exposures are massive, forming near-vertical outcrops tens of meters thick without obvious internal bedding, that are suggestive of deposition on proximal flanks of the source volcano. Where possible, two or more clasts of differing

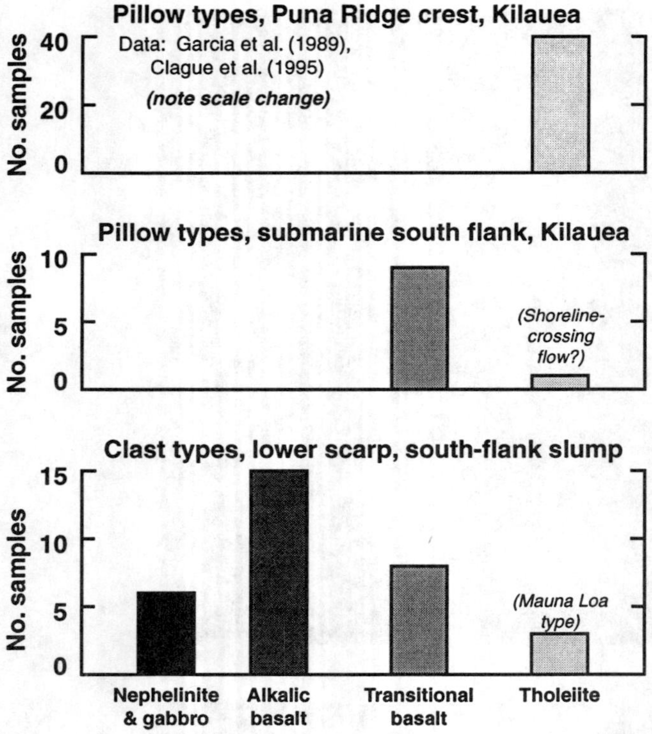

Figure 5. Basalt types in pillow and breccia-clast samples, offshore south flank of Hawai'i Island. Alkalic clasts are dominant along the lower scarp, while tholeiitic compositions resembling present-day Kīlauea are rare except along the crest of the Puna Ridge (submarine east rift zone).

appearance were collected from individual breccia sites; such multiple samples typically show diverse textures and/or compositions, indicating that many of the breccias are heterogeneous and were not derived from fragmentation of single lava flows. The dominant clast type, especially low on the lower scarp, is dense sparsely porphyritic alkalic basalt; two dives (S505, S508) also yielded clasts of fine-grained hypabyssal hawaiite or phlogopite nephelinite (Figure 4B). No coarse breccias were discovered that consist largely or entirely of glass fragments (classic coastal-generated "hyaloclastite breccia" [e.g., *Moore and Fiske*, 1969; *Yokose et al.*, this volume; *Clague et al.*, this volume]. In the lower scarp, clasts vary from angular, especially for dense sparsely porphyritic basalts, to subrounded where crystals are more abundant (picrite, nephelinite). Many basalt clasts contain a glass rind on one surface and are readily identifiable as pillow fragments. High sulfur contents (>750 ppm) in glass rinds document that all but one these (S509-R6B) are fragments of submarine-erupted lava [*Sisson et al.,* this volume]. Matrix of the breccias typically is well-sorted coarse heterolithologic sand, dominated by glass grains of diverse compositions but including texturally variable grains of devitrified basalt.

Variable dips, open folds, fractures, shears, and slickensided clasts are conspicuous in some outcrops [*Ui et al.,* 2000], especially deep along the lower scarp (dives S505, S509). Where sufficiently exposed for measurement, dips of bedding vary from subhorizontal to about 25 degrees, and changes in attitude were locally detected within short vertical distances during dive traverses, documenting weak open folding or juxtaposition of structural domains. In places, divergent attitudes of discontinuously exposed older substrate were difficult to distinguish from slope-parallel bedding in the relatively young sediments that mantle surfaces, leading to some uncertainty in the orientations and intensity of folding and other deformation affecting rocks of the lower scarp. Most of the massive sandstones are cut by well-developed rectilinear fractures (Figure 3B) that probably represent primary joints related to diagenesis and lithification. Other fractures are obviously tectonic in origin, especially closely spaced subparallel or anastamosing cracks associated with shears showing small offsets and occupied by finely comminuted material. Most such tectonic fractures observed during the dives dip steeply; no preferred orientation was recognized initially [*Ui et al.,* 2000], but more detailed geometric analysis of the dive video tapes might reveal more diagnostic information about deformation processes. Several clasts from deep sites near the base of the lower scarp (dives S505, S509) and an outlying elongate ridge (dive K93) have slickensided or faceted surfaces, indicative of shearing. The sheared surfaces of some glass-rich sandstones are black, resembling massive fresh glass. The intensity of deformation appears to decrease upward, in accord with the widely proposed inference that seaward motion and deformation along a basal decollement are major components in the spreading of Kīlauea [*Nakamura*, 1980; *Lipman et al.,* 1985; *Borgia and Treves*, 1992; *Denlinger and Okubo*, 1995; *Morgan et al*., 2000].

Proportions of alkalic materials, both in numbers of sampled breccia clasts and abundance of glass types in the breccia matrix, seemingly define a gross stratigraphy (Figures 5-6), without obvious reversals or repetitions. Alkalic materials, as clasts, clast matrix, and well-sorted sandstone, were especially abundant in the lower samples collected during dives S505 and S509. The absence of obvious reversals in the decrease of alkalic constituents up section suggests a gradual change in the volcanic source, with transitional and tholeiitic compositions increasing with time. Such changes are unlikely to have been completely systematic, because compositionally diverse lavas likely were erupted concurrently, as is known for present-day

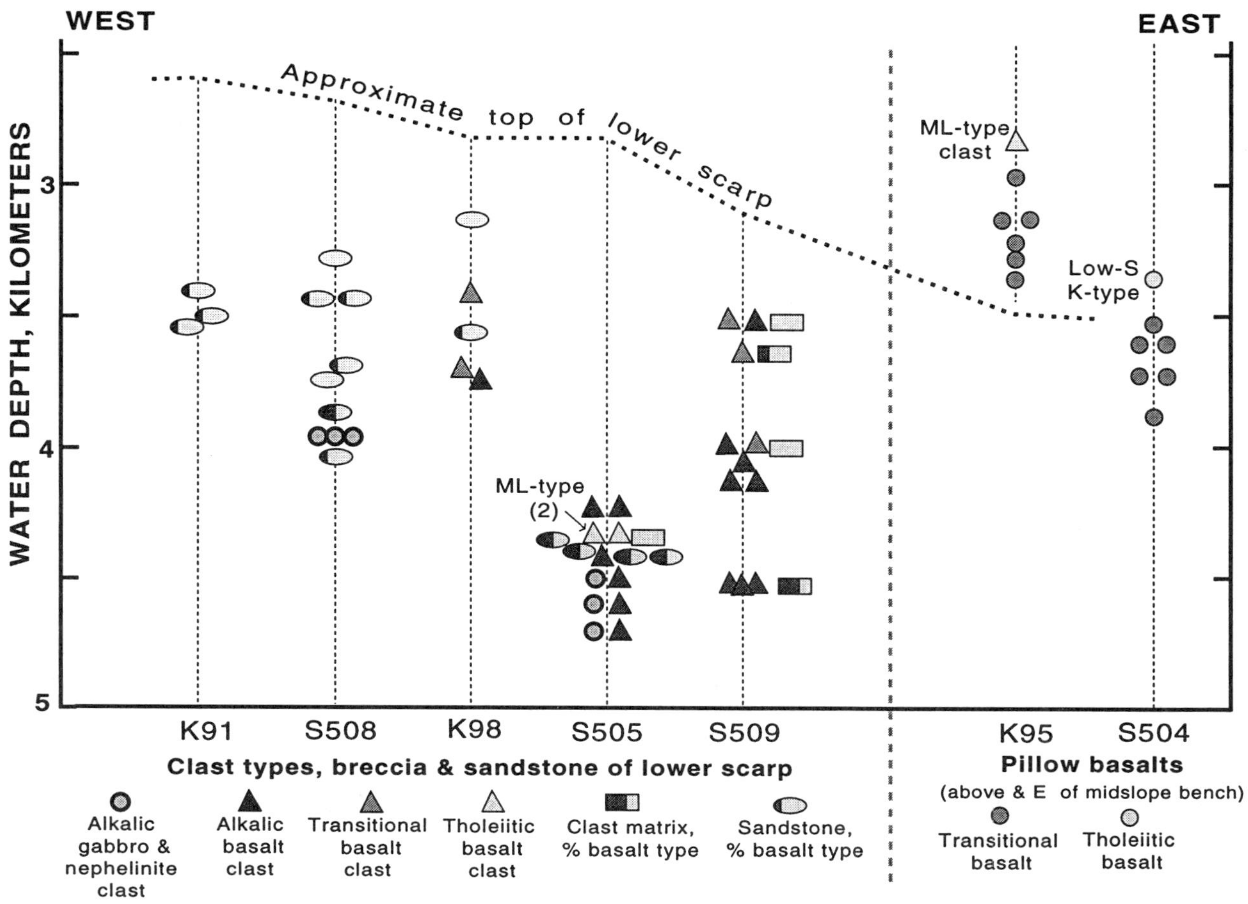

Figure 6. Composition vs. depth relations, clasts and sandstone samples from the lower south flank of Hawai'i Island. Submarine-erupted (high S) alkalic clasts are dominant in the deepest volcaniclastic samples from the lower scarp below mid-slope bench. Clasts of transitional basalt become more common upward, without obvious compositional reversals that might indicate major imbricate fault repetitions. All tholeiitic clasts are Mauna Loa (ML) types, distinct from Kīlauea-like compositions. In the breccia matrix and sandstones, grains of alkalic glass are again most common in deep samples; tholeiite grains, which are common throughout, are overwhelmingly derived from subaerial lavas (low S) erupted from pre-Kīlauea volcanoes. Above and east of mid-slope bench, most samples are submarine-erupted pillows of transitional basalt. A single tholeiitic pillow basalt of degassed Kīlauea (K) type, from the eastern margin of the mid-slope bench (top of dive S504), is interpreted as a relatively young subaerially erupted flow.

Lō'ihi [*Moore et al.,* 1982; *Garcia et al.*, 1995a], and successive debris flows probably came from different parts of the growing volcanic edifice. The observed progression provides no evidence for structural repetitions, such as might occur if intense thrusting had deformed the strata exposed in the lower scarp as inferred from seismic-reflection data [*Morgan et al.*, 2000], but the number of samples may be insufficient to provide a robust test for such hypotheses.

Similar bedded volcaniclastic strata, showing structural features indicative of mild deformation, characterize the steep southeast slope of a large elongate ridge outboard from the lower scarp (dive K93). Similarities in clast compositions, degree of palagonitization, and sedimentary character are consistent with its emplacement as a large slide block derived from high on the nearby lower scarp. Two clasts collected from massive breccia near the base of the elongate ridge are fragments of palagonitized turbidite sandstone (K93-R1, -R2), suggesting brecciation near the base of the ridge. Such brecciation is consistent with interpretation of this and adjacent large outlying ridges as slide blocks [*Moore et al.*, 1989; *Smith et al.*, 1999].

In contrast, during two dives (K95, S504) above and east of the mid-slope bench, all observed outcrops were

Table 1. Representative Compositions Of Submarine-Erupted Rocks, South Flank Of Hawai'I Island

	Basanite	Alkalic basalt	Phono-tephrite	Phlogopite nephelinite	Transitional basalt	Tholeiite (Puna Ridge)
Sample type	Sand	Clast	Sand	Clast	Pillow	Pillow
Sample no.	508-R3-24	S509-R7-3	S505-R6B-15	S508-R2A	K95--R5	S506-R5a
SiO_2	41.12	45.73	50.25	37.35	48.59	51.90
TiO_2	2.72	4.39	2.31	6.42	3.17	2.36
Al_2O_3	14.66	14.45	17.66	10.98	14.59	14.25
FeOt	12.99	13.41	12.05	17.92	11.83	11.05
MnO	0.20	0.17	0.21	0.23	0.16	0.18
MgO	6.07	5.97	2.47	9.40	6.39	6.43
CaO	14.02	10.67	5.12	12.25	11.86	10.89
Na_2O	5.84	3.42	5.63	2.52	2.52	2.34
K_2O	1.88	1.14	2.70	2.03	0.60	0.38
P_2O_5	0.50	0.64	1.60	0.92	0.29	0.22
S, ppm	2080	1810	1180	--	1190	430
Cl, ppm	2350	450	610	--	400	100

(Glass analyses, except for nephelinite)

massive pillow lava and broken pillows (Figure 3D) (excepting a talus block of gabbroic cumulate: K95-R1), without any exposed sandy volcaniclastic interbeds. Both dive tracks were along steep spurs, protruding through the sedimented upper slope, that probably define lateral margins of small slumps or landslides. These sites were chosen to optimize exposures and minimize potential for drape by younger shoreline-crossing lavas. Along both dive tracks, equant and cylindrical pillows are abundant and well exposed, showing both outer depositional surfaces and broken pillows in cross section. No bedded reworked pillow breccia or finely comminuted hyaloclastite was identified, although sizable covered areas between outcrops, even on the steepest slopes, could be sites for such material, which might be less indurated and less well exposed than areas of massive pillows. Most exposed pillows are dense, containing only 5-10% vesicles and lacking hollow cores, and have well-defined palagonite rinds on their outer glassy surfaces. Along with high sulfur contents (S>750 ppm), as discussed in section 4, these textural features indicate old submarine eruption of undegassed lava at high confining pressures of the water column. An exception is the highest sampled flow on dive S504, which mantles a gentle break in slope that marks the eastern margin of the mid-slope bench. This sample (S504-R5) is more vesicular, less palagonitized, and low in S (100 ppm)—features suggestive of emplacement as a young subaerially erupted Kīlauea flow that crossed the shoreline. Shoreline-crossing lavas have been emplaced frequently at Kīlauea *[Moore et al.,* 1973; *Tribble,* 1991], and compositional contrasts *[Sisson et al.,* this volume], summarized next, support such an interpretation for this site.

4. COMPOSITIONS OF BRECCIA CLASTS, BRECCIA MATRIX, AND SANDSTONES

Compositions of the basalt fragments and glass sands, in addition to constraining melting sources and magmatic evolution for a Hawaiian volcano [*Sisson et al.*, this volume], provide critical evidence bearing on depositional processes during sedimentation and early growth of the source volcano. As part of the present study, pillow fragments (19 samples) and large breccia clasts (26) were analyzed for major and minor elements by x-ray fluorescence and instrumental neutron activation. Glassy rims of pillows, large clasts, small glass clasts, and glass sand grains from breccia matrix and sandstone blocks (27) were analyzed by electron microprobe (Table 1) [*Sisson, et al.* this volume]. More than 700 separate glass samples were analyzed from the Hilina slump region, each at least in triplicate.

South-flank breccia clasts are chiefly alkalic basalt or basanite (glass SiO_2, 41-47%), transitional basalt (glass SiO_2, 47-50%), hawaiite, or at one site mafic phlogopite-nephelinite (S508-R2A, -R2B, -R2C). Only two of the individually sampled large clasts are tholeiite (S505-R9A, -R9B); these, and a single similar talus block from the pillow section above the mid-slope bench (K95-R7), plot along an olivine-control trend more compatible with Mauna Loa compositions than with Kīlauea (arrow, Figure

7A). Olivine phenocrysts and Cr-rich spinel microphenocrysts are common in the transitional basalts; most alkalic basalts are aphyric or nearly so. Microphenocrysts and microlites of olivine, plagioclase, and clinopyroxene are common, especially for evolved compositions (<6.5% MgO). Margins (0.2-0.5 mm) of many glass pillow rinds and sand grains are palagonitized, but cores are unaltered (e.g., coherent plots for mobile vs. immobile elements such as glass K/P ratios [*Sisson et al.*, this volume]), reflecting prolonged storage in cold ocean-bottom waters (<1° C). As emphasized by *Clague et al.* [1995], such glass analyses, even for small sand grains, provide valid data for magmatic liquids, more reliable than subaerially preserved materials for eruptions more than a few thousand years old.

High sulfur (S>750 ppm) in the clast glasses (rinds of pillow fragments) with alkalic and transitional compositions indicates that they were erupted in deep water without having degassed at low pressure [*Sisson et al.*, this volume]. The submarine-erupted (high-S) glass sands (Figure 7C) include a highly varied assemblage of alkali basalt, hawaiite, basanite, transitional basalt, sparse typical tholeiite, and rare highly alkalic phonotephrite and nephelinite (to 10.8% total alkalis). These greatly enlarge the spectrum of alkalic compositions defined by the breccia clasts (Figure 7A) and document eruption of magma types not previously recognized for Hawaiian volcanoes [*Sisson et al.*, this volume]. Sparse grains of variably degassed glass of alkalic and transitional basalt indicate complex additional sources (Figure 7B). The glass compositions suggest that the ancestral alkalic edifice of Kīlauea approached or even reached sea level, allowing for syneruptive degassing. Alternatively, perhaps some of the degassed glass came from shallow-water or subaerial parts of Mauna Kea.

In contrast, the dominant glass grains in debris-flow matrix and sandstone are subaerially erupted (S<400 ppm) tholeiitic basalt, similar to recent Kīlauea and Mauna Loa lavas (Figure 7B). The high proportion of glass grains, relative to devitrified basalt grains, and low sulfur contents point to seawater quenching of subaerial lavas as they

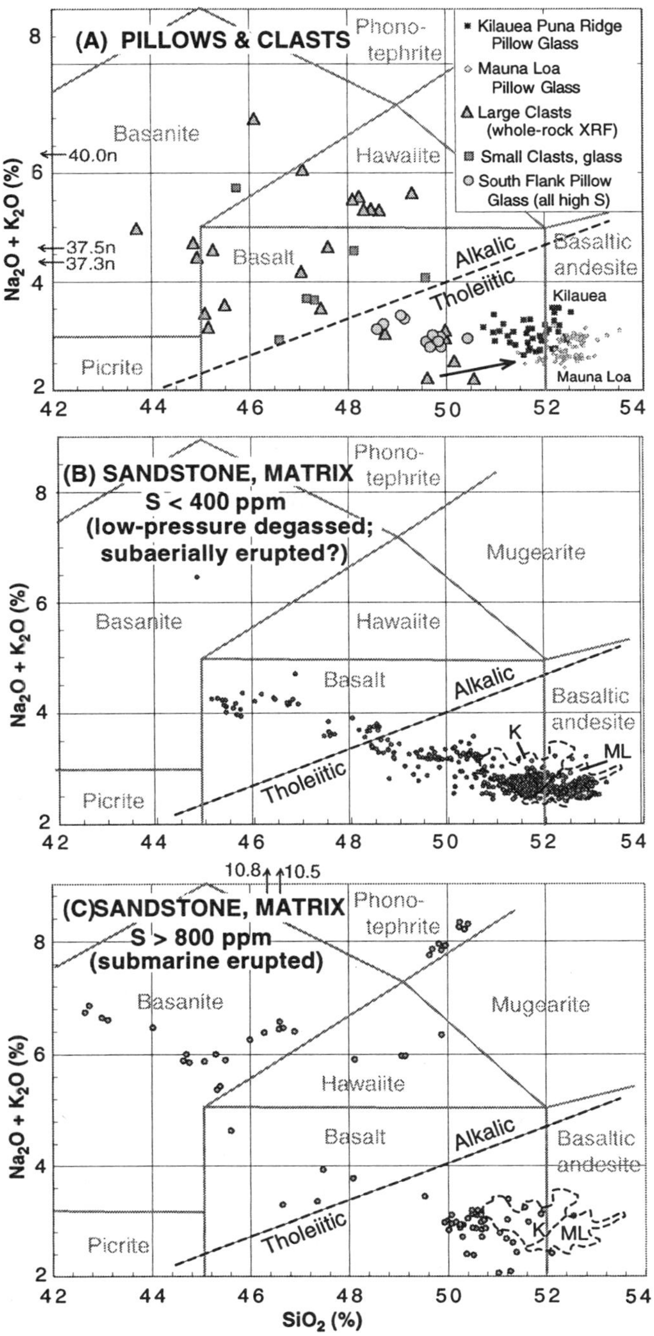

Figure 7. Alkali-silica diagrams, showing glass compositions for pillows, breccia clasts, sandstones, and breccia matrix from lower south flank (normalized volatile-free, all Fe as FeO). *Macdonald-Katsura* [1964] alkalic-tholeiitic boundary (diagonal dashed line) and IUGS rock classification are shown for referencing alkalic compositions, although not fully appropriate for glass compositions.

(A) Pillow glasses from above mid-slope bench and breccia clasts (some low-MgO whole-rock analyses) from debris-flow deposits of lower scarp. Plotted for comparison are glass-rind data for pillows from Kīlauea (*Garcia et al.*, 1989; *Clague et al.*, 1995) and Mauna Loa (various sources). Three clasts lie along olivine-control trend for Mauna Loa tholeiite (large arrow). Small arrows/n, phlogopite-bearing nephelinite (off chart).

(B) Sandstone and breccia-matrix glass. Black dashed line, compositional fields of Kīlauea and Mauna Loa pillow glass, from (A).

(C) Sandstone and breccia-matrix glass.

reached the shoreline, similar to the voluminous production of glass sand during the current long-lived east-rift eruption of Kīlauea [*Mattox and Mangan*, 1997]. Although unlikely as the primary source for the abundant low-S glass sands, some grains could represent products of variable degassing as magma traveled at shallow depths along subaerial parts of a rift zone prior to submarine eruption. Such a process is documented for some young pillow lavas along the Puna Ridge [*Dixon et al.*, 1991; also Shinkai dive S506].

A recurring problem with interpreting the glass compositions, especially for the abundant turbidite sandstones with tholeiitic compositions, is distinguishing among potential sources lying along the "Kea" trend of volcanoes [*Dana*, 1849], which includes Kīlauea, Mauna Kea, Kohala, and Haleakalā (Maui). As discussed more fully in section 8.1, the abundant glass sands of degassed Kea-type tholeiite in the lower-scarp breccia matrix and sandstone probably came from pre-Kīlauea volcanoes such as Mauna Kea and/or Kohala. In contrast to ready distinction from 'Loa'-trend compositions (e.g., Mauna Loa vs. Kīlauea [*Wright*, 1971; *Rhodes*, 1996]), no major-or trace-element criteria have yet been identified that distinguish reliably among tholeiites from the Kea-trend volcanoes. Reliable compositions are sparse for the older volcanoes because of subtle post-emplacement alteration in subaerial environments. Needed for the older volcanoes are more glass analyses from submarine-emplaced samples, such as those determined during the current south-flank studies [*Sisson et al.*, this volume] and from drill holes [*Quane et al.*, 2000; *DePaolo et al.*, 2001].

In contrast to the picritic glass (to 14% MgO) in some turbidite sands north of Kīlauea's Puna Ridge [*Clague et al.*, 1991, 1995], all analyzed tholeiitic glass sands on the south flank are low MgO (mostly in the range 5-7%). These include indurated lower-scarp sandstones, breccia matrix, and turbidite layers in the distal piston cores. The highest MgO glasses yet found in the submarine Hilina rocks are widespread but compositionally uniform sandstone grains of transitional basalt with 9.5–10.7% MgO, 46.2–46.6% SiO_2, and intermediate S contents (400-650 ppm). Sparse high-MgO glass grains were found only in surficial sands from two dives (S495, S506) along the south base of the Puna Ridge [*Naka et al.*, this volume, appendix]. While a variety of petrologic evidence documents the role of picritic primary magmas during tholeiitic activity at Hawaiian volcanoes [e.g., *Garcia et al.*, 1995b; *Rhodes*, 1995], our additional sampling of turbidite glass sands south of Hawai'i Island indicates that picritic liquids rarely escape modification by crustal-level magma-chamber processes.

Pillow lava and pillow breccia, which form slopes (2800-3300 m depth) above the mid-slope bench toward Kīlauea's shoreline and along the south base of the Puna Ridge, appear to define the submarine flank of initial Kīlauea subalkaline volcanism. Most pillows (Figure 7A) from above and immediately east of the mid-slope bench (dives K95 and S504) are submarine-erupted, non-degassed (1050-1350 ppm S; 5 of 6 glass rinds) olivine basalt and picrite with low-MgO glass. The glass compositions are distinctly lower in SiO_2 (47.9-49.8% SiO_2) than glass from subaerial Kīlauea or the crest of the Puna Ridge (50.5-52.5% SiO_2). Such compositions occur sparsely on subaerial Kīlauea, where they have been termed "transitional basalt" [*Wolfe and Morris*, 1996]. These compositions (Figure 7A) plot just within the tholeiitic field of *Macdonald and Katsura* [1964], comparable to the 'tholeiitic' pillow lavas on Lōi'ihi Seamount [*Moore et al.*, 1982; *Garcia et al.*, 1995a]. The transitional basalt pillows above the mid-slope bench do not overlap in composition with the oldest tholeiite flows exposed on land along scarps of the Hilina fault system [*Chen et al.*, 1996] or those penetrated in deep drill holes along the east rift zone [*Quane et al.*, 2000]. In contrast, a single pillow sample from high in the section (S504-R5), having a present-day Kīlauea glass composition (51.7 % SiO_2: Figure 7A), is low in sulfur (100 ppm) and likely represents a relatively young shoreline-crossing flow. No alkalic basalt, basanite, nephelinite, or hawaiite were found with the transitional pillows on the south flank.

Contrasts with the sandstone and breccia below the mid-slope bench preclude prior suggestions [*Lipman et al.*, 1985; *Moore et al.*, 1989], based on low-resolution bathymetry, that the bench might be part of the Kīlauea shield that slumped along Hilina-style faults. Nevertheless, compositional similarities suggest that all the south-flank submarine-erupted rocks are part of the Kīlauea magmatic system [*Sisson et al.*, 2000, this volume]. Breccia clasts of transitional basalt from the lower scarp have major-element compositions similar to the pillow basalts that overlie the bench, and both plot with Kīlauea tholeiites on many trace-element diagrams. Alkalic basalt clasts are enriched in incompatible elements relative to other samples but share ratios of highly incompatible elements with transitional basalts (e.g., La/Ta: 12.4±0.4, both below and above the bench).

An additional dive (S506), at the south base of the Puna Ridge, downslope from the near-shore ridge crest, transected variably slumped and brecciated pillow lavas. Of seven analyzed samples, only three have present-day Kīlauea compositions (high SiO_2, low TiO_2 and K_2O, etc.); the others are transitional basalt similar to those sampled at shallower levels above and east of the mid-slope bench (dives K95, S504). Of five samples with glass rinds (all with palagonite rims), the three transitional basalts have high sulfur contents (>1100 ppm) indicative of deep submarine eruption without degassing. Both Kīlauea tholeiite glasses have much lower sulfur contents (400, 450 ppm) that likely resulted from low-pressure degassing during transport at shallow levels along the east rift zone. Similar intermediate to low sulfur contents characterize many pillow lavas along the crest of the Puna Ridge, all of which have major-oxide compositions similar to historical Kīlauea lavas [*Garcia et al.*, 1989; *Dixon et al.*, 1991; *Clague et al.*, 1995; *Johnson et al.*, this volume]. All the newly collected glass samples have substantial palagonite rinds (0.2-0.5 mm thick) indicative of relatively old eruption ages. Thus, these deep samples along the base of the Puna Ridge (dive S506) provide a compositional (and age?) bridge from the transitional basalt sampled at shallower levels above and east of the mid-slope bench (dives K95, S504) to younger Kīlauea lavas on the Puna Ridge crest and the subaerial volcanic edifice.

5. SUBMARINE GROWTH OF ANCESTRAL KĪLAUEA

The offshore observations and compositional data, in conjunction with previously reported subaerial geologic information, provide new perspectives on the early growth of Kīlauea. Scaled geologic cross sections through the south flank (Plate 1) are particularly useful to summarize geometric relations between successive petrologic and depositional sequences. The distribution of submarine rock types along the lower scarp, especially the clasts of highly diverse alkalic rocks at low structural and stratigraphic levels overlain by non-degassed pillows of transitional basalt (Figure 6), constrains the first possible deposition of Kīlauea tholeiitic lavas with present-day compositions to above about 2600 m depth along this segment of the south flank. The JAMSTEC-dive samples, together with documentation of modern tholeiitic Kīlauea compositions to >1700 m depth (1050 m below paleo-sea level) in the SOH1 drill hole (Figure 2) along Kīlauea's east rift zone [*Quane et al.*, 2000] and inference of a modest seaward slope of the volcanic edifice (Plate 1a), appear to provide a broadly continuous sequential record of Kīlauea's growth, deformation, and magmatic evolution. The evolution from transitional basalt to present-day tholeiitic compositions is also bracketed at greater depth along the deep southern base of the Puna Ridge (dive S506). There, the presence of both compositional types contrasts with the sole presence of tholeiites along the crest of the rift zone, both subaerial and submarine [*Garcia et al.*, 1989; *Clague et al.*, 1995].

Compositions of submarine-erupted glass from the lower scarp, which encompass a range far greater than that known for Lō'ihi Seamount, record the alkalic early stages of volcanic growth, prior to tholeiitic shield-stage Kīlauea (Table 2). The source volcano must have been northwest of the present-day mid-slope bench, toward modern Kīlauea (Plate 1), but no primary lavas have been found, only volcaniclastic debris derived from these early eruptions of ancestral Kīlauea. Early submarine alkalic volcanism evolved, through an intermediate stage characterized by pillow lava of transitional basalt, into present-day subaerial tholeiitic shield-stage Kīlauea. Although tholeiites now dominate Kīlauea's magmatic system, at least as far back as the oldest lavas exposed on land [*Chen et al.*, 1996] and in deep drill holes along the east rift zone [*Quane et al.*, 2000], compositionally diverse magmas continue to erupt in small volume. Examples include a relatively recent alkalic basalt at the south base of the distal Puna Ridge [*Clague et al.*, 1995; Shinkai dive S495], volumetrically minor transitional basalt in the A.D. 600-1000 Kulanaokuaiki tephra [*Dzurisin et al.*, 1995; *Fiske et al.*, 1999] and an associated local lava flow [*Wolfe and Morris*, 1996, table 1, sample DS-32], and local subaerial lavas of Mauna Loa-like tholeiite [*Rhodes et al.*, 1989].

The general compositional sequence on the south flank of Kīlauea and on other Hawaiian volcanoes (Table 2) is compatible with genetic models of increasing degrees of partial melting of a relatively homogeneous mantle source [*Garcia et al.*, 1989; *DePaolo and Stolper,* 1996; *Sisson et al.,* this volume]. The volcaniclastic apron on the south flank of Kīlauea provides a broad petrologic sampling of an early stage in Hawaiian hotspot volcanism, much like that sought for Mauna Kea volcano by the in-progress Hawai'i Scientific Drilling Project [*DePaolo et al.*, 2001]. Deep in the Hawai'i Scientific Drill Hole (below 2500 m), submarine-erupted pillow lavas of similar transitional composition are sparsely interlayered with the dominant tholeiitic lavas [*Sherman et al.*, 2000; *Rhodes,* 2000]. This suggests a stratigraphic and magmatic sequence similar to that for Kīlauea on the south flank, although no submarine-erupted alkalic rocks have been penetrated thus far in the drill hole. Recently acquired underwater and drill-hole samples have also documented greater compositional diversity than previously known for other Hawaiian volcanoes (Table 2), such as Mauna Loa [*Garcia et al.*, 1995b] and Ko'olau [*Haskins and Garcia*, 2000; *Tanaka et al.*, this volume]. Compositional differences among tholeiitic basalts, previous considered characteristic of individual Hawaiian volcanoes (e.g., Kīlauea, Mauna Loa, Ko'olau [*Wright*, 1971; *Frey and Rhodes*, 1993], now appear more

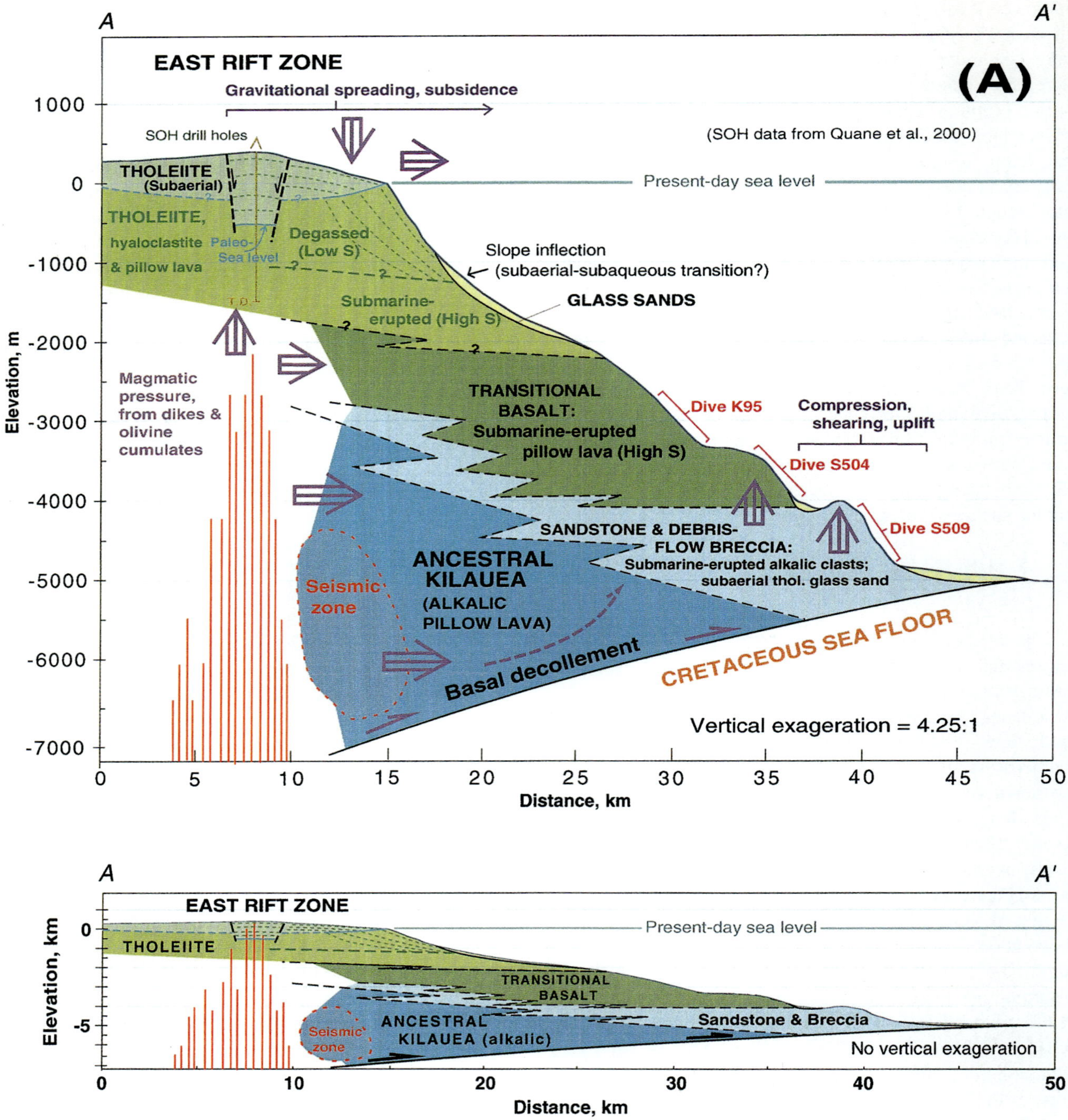

Plate 1. Northwest-trending cross sections (A-A' and B-B'), showing inferred stratigraphy and structure of the south flank as constrained by dive results, bathymetry, and on-land geology. Locations of sections shown on Figure 2. Dashed lines within shield-stage tholeiite unit indicate generalized geometry of depositional contacts. Large open arrows indicate inferred deformation directions and processes. Deformation is thought to result from combined effects of lateral spreading of the volcanic edifice, dike intrusion along rift zones, and gravitational slumping along oversteepened flanks. Data for SOH drill holes (Scientific Observations Holes) from *Guillou et al.* [1997] and *Quane et al.* [2000].

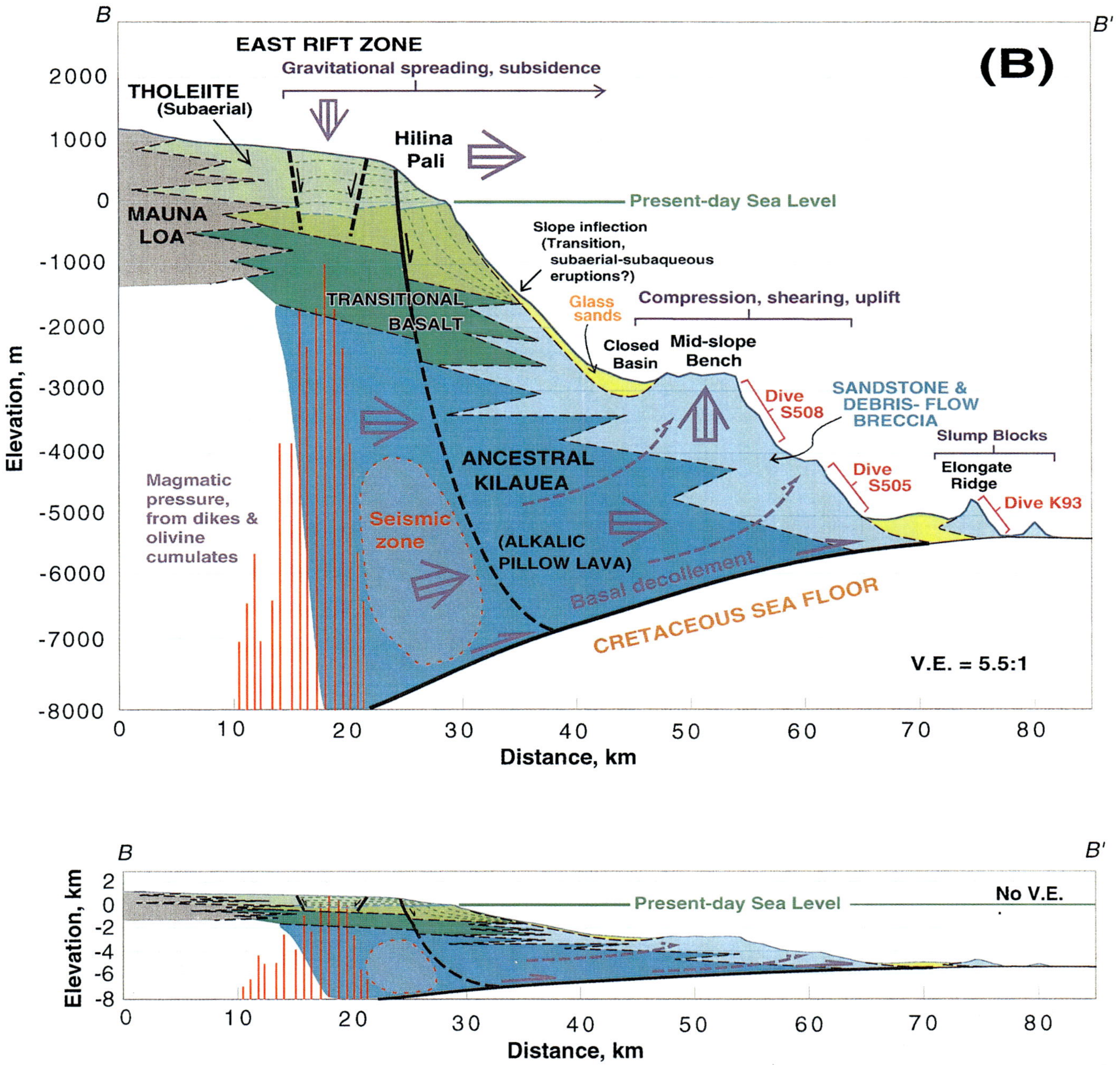
B
B'
(B)
EAST RIFT ZONE
Gravitational spreading, subsidence
THOLEIITE
(Subaerial)
Hilina
Pali
Present-day Sea Level
MAUNA
LOA
Slope inflection
(Transition,
subaerial-subaqueous
eruptions?)
TRANSITIONAL
BASALT
Glass
sands
Compression, shearing, uplift
Closed
Basin
Mid-slope
Bench
Dive
S508
SANDSTONE &
DEBRIS- FLOW
BRECCIA
Slump Blocks
Elongate
Ridge
Dive
S505
Dive K93
ANCESTRAL
KILAUEA
Magmatic
pressure,
from dikes &
olivine
cumulates
Seismic
zone
(ALKALIC
PILLOW LAVA)
Basal decollement
CRETACEOUS SEA FLOOR
V.E. = 5.5:1
Elevation, m
Distance, km
No V.E.
Elevation, km
Distance, km

Table 2. Hawai'i Basalt Types: Associated Volcanoes

	Early alkalic	Transitional basalt	Shield tholeiite, Low Si	Shield tholeiite, Silicic	Shield tholeiite, High Si	Post-shield alkalic	Post-erosional alkalic
Typical volcano	(Lō'ihi)	(Lō'ihi)	(Kīlauea)	(Mauna Loa)	(Ko'olau)	(Mauna Kea)	(Ko'olau)
Lō'ihi	⊗	⊗					
Kīlauea	#	#	⊗	x			
Mauna Loa	#?		x	⊗			
Huālalai				⊗		⊗	
Mauna Kea		x?	⊗			⊗	
Kohala				⊗		⊗	
Ko'olau (Oahu)			#	#	⊗		⊗

⊗, Dominant exposed phase, and x, minor phase, of historically active or geologically mapped volcanoes (see text for data sources)
#, Newly recognized rock type/volcano association

likely to represent the general evolutionary sequence during growth of Hawaiian shield volcanoes.

In addition to the compositional and depositional sequences, the geometry of the mid-slope bench and exposed pillow lavas of transitional basalt near its east end provide confirming evidence for volcano spreading along a basal decollement and compression of the deep distal flank. The height of the lower scarp and the thickness of truncated gently-dipping volcaniclastic sandstone and breccia exposed there are interpreted most reasonably as due to oversteepening by compression, whether primarily simple shortening [*Lipman et al.,* 1985; *Clague and Denlinger,* 1994] or major thrust faulting [*Morgan et al.*, 2000; *Hills et al.,* this volume]. Increased shearing and fracturing seen during dives along deep parts of the lower scarp probably reflect proximity to the basal decollement, the existence and geometry of which were initially inferred from seismic data [*Ando*, 1979; *Zucca and Hill*, 1980].

Outcropping ribs of pillow lavas just above the mid-slope bench (dive K95) show that the hyaloclastite apron derived from ocean entry of subaerially erupted lavas is thinner and less continuous than previously inferred in generalized models for the south flank [*Moore and Fiske,* 1969; *Moore and Chadwick,* 1995, Figure 6B]. From seismic-reflection profiles [*Morgan et al.*, 2000; *Hills et al.,* this volume], thick volcaniclastic sediment has also been interpreted for the area where pillow lavas crop out, as seen in dives K95 and S504. Volcaniclastic sediment may have ponded deeply in landslide-scar reentrants between ribs of pillow lava, and the thickness of such sediment accumulation must be highly variable. The volume of volcaniclastic sediment on the upper slope and mid-slope basin, which is associated with emplacement of Kīlauea's shield-stage tholeiite on the south flank (seismic unit 1, from *Hills et al.*, this volume), is roughly calculated to be less by a factor of about 5 than that of the sandstone and debris-flow breccia (seismic units 2-3) derived from the alkalic and transitional basalt of ancestral Kīlauea.

Details of the south flank stratigraphy and structure, as depicted in Plate 1, remain uncertain at depth. For example, does the contact between transitional basalt and tholeiite dip seaward, parallel to the present-day water-rock surface, as might be expected if the south flank were widely draped by hyaloclastite as depicted in many previous models for underwater flanks of Hawaiian volcanoes [e.g., *Moore and Chadwick*, 1995]? Alternatively, could this contact be back-rotated and landward-dipping, due to combined effects of rotational listric faulting along the Hilina system [*Swanson et al.*, 1976; *Riley et al.*, 1999] and subsidence of the island under isostatic loading by the volcanoes [*Moore*, 1970, 1987]? Or does uplift associated with magmatic inflation and endogenous growth (vertical arrows in Plate 1) due to dike injection along the rift zones [*Swanson et al.*, 1976; *Dvorak et al.*, 1986; *Dieterich*, 1988] counterbalance the component of landward tilt?

Other questions include the location and size of the concealed ancestral alkalic volcanic edifice. The source for the voluminous submarine-erupted basalt debris in sediments of the lower scarp is interpreted as related to Kīlauea and entirely submarine in Plate 1. Alternatively, could these sediments be far-traveled debris from ancestral Mauna Kea, rather than related to Kīlauea? This seems unlikely, because of the greater transport distance required for coarse clasts, and inappropriately young ages indicated by isotopic dating and stratigraphic correlations with piston cores (discussed in section 6). Deep parts of the subaerial tholeiitic shield of Mauna Kea have yielded ages as old as 400 ka from the pilot hole of Hawai'i Scientific Drilling Project [*Sharp et al.*, 1996]. Also, Mauna Kea appears to lack large shield-stage rift zones definable by pronounced potential-field anomalies [*Hildenbrand et al.*, 1993; *Kauahikawa et al.*, 2000]. The absence of conspicuous rift zones suggests that Mauna Kea may contain only a modest-size tholeiitic shield that was built high on the flanks of older volcanoes, perhaps without a sizable pre-shield edifice.

While such features of the three-dimensional constructional geometry of the south flank remain uncertain, the general configuration and distributions depicted schematically in Plate 1 (large concealed alkalic edifice, mantled by its debris; overlain successively by subalkaline transitional basalt, then capped by tholeiite) are thought to be broadly valid. This geometry would require the thickness and volume of Kīlauea's tholeiitic shield to be smaller than previously envisaged, as also indicated by the estimated sediment volumes for successive growth stage on the south flank. The overall volume of Kīlauea and its alkalic core remain difficult to estimate with rigor because of (1) the lack of identified in-place alkalic lavas, (2) uncertainties concerning the amount of structural thickening of the alkalic volcaniclastic sequence exposed on the lower scarp, and (3) the unknown regional onlap versus interfingering relations with adjacently active Mauna Loa (Plate 1B) [*Rhodes et al.*, 1989; *Lipman*, 1995; *Quane et al.*., 2000]. If the summit of Kīlauea initial grew near paleo-sealevel, on the flank of a large existing Mauna Loa edifice, the primary alkalic edifice could have been much smaller than if Kīlauea were initiated in deep water, near the base of old landslide-slump scarps along the Kao'iki- Wai'ōhinu fault zone (Figure 1). Additional volatile studies, especially CO_2 determinations on glassy margins of pillow clasts [e.g., *Dixon et al.*, 1991], might provide better controls on eruptions depths for early alkalic Kīlauea.

The three clasts with Mauna Loa-type compositions (S505-R9A, -R9B, K95-R7) show that the south flank of this volcano was sufficiently steep to shed debris during the early submarine growth of Kīlauea, but no consistent stratigraphic or structural relations emerged from the limited underwater sampling. In particular, no evidence was found that supports prior speculation that deep exposures along the lower scarp might largely be the south flank of Mauna Loa or landslide debris from this flank, projecting seaward beneath Kīlauea [*Lipman*, 1995, p. 66]. In contrast, 12 analyzed samples of breccias and lavas from steep lower slopes of the Punalu'u slump (dive S507), which morphologically resembles the lower scarp of the Hilina area, all have Mauna Loa-type compositions (Table 5; *T Sisson*, unpubl. data).

6. AGES OF EARLY EVENTS: ISOTOPIC DATING AND PISTON-CORE STRATIGRAPHY

The inception ages and life spans of individual volcanoes such as Kīlauea represent some of the least understood aspects of Hawaiian volcanism [*Moore and Clague*, 1992; *Lipman*, 1995; *DePaolo and Stolper*, 1996], but radiometric dating of the new submarine samples provides important controls on the timing of early volcanism on the south flank (Table 3; *M. Lanphere*, written commun., 2001). Three alkalic basalt clasts of submarine-erupted pillow fragments, collected from debris-flow breccias along the lower scarp, have yielded reliable ^{40}Ar-^{39}Ar plateau and isochron ages of 220 to 275 ka. A fourth basalt clast, the shallowest sample collected during dive S505, yielded a younger age of 130 ka. Phlogopites from two clasts of hypabyssal nephelinite (S508-R2A, R2C) have well defined ^{40}Ar-^{39}Ar plateau ages of 234 and 239 ka. These results suggest that early alkalic volcanism at ancestral Kīlauea was underway during the 220-275 ka interval. Some alkalic eruptions may have continued for an additional 100,000 years, probably overlapping with the inception of shield-stage tholeiitic eruptions.

Additional age information on events during the growth of Kīlauea and older Hawaiian volcanoes is suggested by possible correlations with features in piston cores P6 (120 km southeast of the Kīlauea shoreline) and P5 (250 km southeast) [*Naka et al.*, this volume; *Kanamatsu et al.*, this volume]. Both cores contain multiple turbidite sand layers of basaltic glass (8-10 in P6, 4 in P5) and disseminated glass grains in the intervening pelagic clay. Glass sands in both cores are dominated by degassed tholeiite compositions [*Naka et al.*, this volume; *T. Sisson*, unpubl. data]. The proximal P6 core is entirely normally magnetized, indicating that its turbidite layers are all younger than 780 ka [*Kanamatsu et al.*, this volume]. Sparse grains of alkalic basalt and basanite in core P6 are submarine-erupted (high S) and compositionally similar to the south-flank alkalic clasts. The alkalic grains are strongly concentrated at depths between 1.75 to 3.3 m, which can be correlated with the upper part of core P5 above the 780 ka Brunhes-Matuyama paleomagnetic boundary [*Naka et al.*, this volume]. If sedimentation rates were generally constant, such depths would correspond to a range from about 180 to 330 ka for the period of early alkalic volcanism (Table 3). Alkalic grains are absent from turbidites in the distal P5 core, indicating that the turbidites sampled in core P6 were relatively low energy. A pulse of submarine-erupted transitional-basalt sand of Kea-type composition at a depth of 0.6 m in core P5 [*Naka et al.*, this volume], corresponding to an age of about 200 ka (Table 3), may record the inception of subalkaline eruptions at Kīlauea.

Such relatively young time spans for the bulk of early deep submarine alkalic volcanism from ancestral Kīlauea (~330-180 ka), as well as the initial appearance of subalkaline compositions at >200 ka, are compatible with previous volume-based estimates for the inception of Kīlauea volcanism [*Moore and Clague*, 1992; *Lipman*, 1995]; they are inconsistent with some recently published isotopic ages for other Kīlauea rocks. Unspiked K-Ar dates as old as 436 ka have been reported for Kīlauea tholeiites having compositions indistinguishable from modern lavas, at depths near paleo sea level in drill holes SOH4 and SOH1 along the east rift zone [*Guillou et al.,* 1997; *Quane et al.,* 2000]. Anomalously old ages for samples like these, which are characterized by low yields of radiogenic argon, might be due to excess mantle-derived argon, a problem long recognized in dating submarine-erupted tholeiite lavas [*Dalrymple and Moore,* 1968]. Our new age determintions for inception of Kīlauea are also inconsistent with the models presented by *DePaolo and Stolper* [1996, p. 11,652], who infer an inception age of about 600 ka for growth of Kīlauea, and that by *Holcomb et al.* [2000, figure 5], who infer younger inception of Kīlauea volcanism at about 100-150 ka. The *Holcomb* estimate for all of Kīlauea is, alternatively, roughly consistent with implications, from the new isotopic and piston-core ages, that the bulk of tholeiitic shield-stage volcanism at Kīlauea is more recent than about 130 ka. If the tholeiitic carapace of modern Kīlauea averages about 2 km thick for the overall edifice (thicker near the summit, thinner along the distal Puna Ridge), as suggested by the south-flank cross sections and supporting data at mid distances from the summit (Plate 1), then the erupted volume of tholeiitic Kīlauea is ~8,000 km^3. This amount of tholeiite could have accumulated in ~80,000 years at the inferred long-term magma-supply rate of ~0.1 km^3/yr [*Dzurisin et al,*, 1984; *Dvorak and Dzurisin,* 1993]. In comparison, total edifice volume may be ~12,000-20,000 km^3 [*Lipman,* 1995; *DePaolo and Stolper,* 1996]; or even larger depending on the size of deep intrusions and cumulates [*Fiske and Wright,* 1997].

Abundant degassed Loa-type tholeiite sands sampled at varying depths in both piston cores [*Naka et al.*, this volume] were likely derived from subaerial Mauna Loa, prior to interference with sediment-transport directions by inception of the ancestral Kīlauea edifice. Somewhat speculatively, an exceptionally thick (>1.3 m) turbidite sand layer at the bottom of core P6, which is dominated by low-S Loa-type glass sand, may represent a distal facies of the Punalu'u slide at piston-core depths corresponding to an age of about 410 ka (Table 3). Sparse low-S tephra of ve-

Plate 2. Schematic cross sections through south flank and summit area, showing inferred growth of Kīlauea. Geometric relations are simplified, because volcaniclastic material was likely generated from a broad swath along the south flank, not just upslope along the line of section.

(A) Early submarine alkalic stage constructs a compositionally diverse edifice, including distinctive basanite and rare phonotephrite, that are more compositionally extreme than known Lö'ihi lavas. Lavas of this early alkalic stage may have locally reached sea level, indicated by sparse low-S glass in lower-scarp sandstones (Figure 7B). Submarine slides deposit breccia clasts of alkalic basalt and sparse gabbro/nephelinite (from intrusive core). Turbidites deposit massive graded beds of glass sand, derived from shoreline-crossing lavas of Mauna Loa (ML) and/or Mauna Kea (MK). Sparse alkalic tephra erupt from Mauna Kea or possibly Kohala volcanoes.

(B) Submarine-erupted transitional basalt marks beginning of subalkaline stage, forming slope above present-day mid-slope bench, and continuing to deposit clasts on flanking debris apron. Abundant sand grains of degassed transitional basalt (Figure 7B) suggest the beginning of a subaerial shield, but absence of tholeiitic clasts indicates that this composition remained volumetrically minor. Hilina fault system may have begun to develop at this stage.

(C) Present-day shield stage: tholeiite dominates, but sparse transitional basalt continues (e.g., A.D. ~600-100 Kulanaokuaiki tephra). Hilina faults begin to accommodate active slumping and volcano spreading along basal decollement, possibly at accelerating rates. Compression along toe of slump uplifts mid-slope bench, thickening the volcaniclastic sections along the lower scarp and triggering secondary landslides to generate isolated slump ridges along basal south flank (Figure 2).

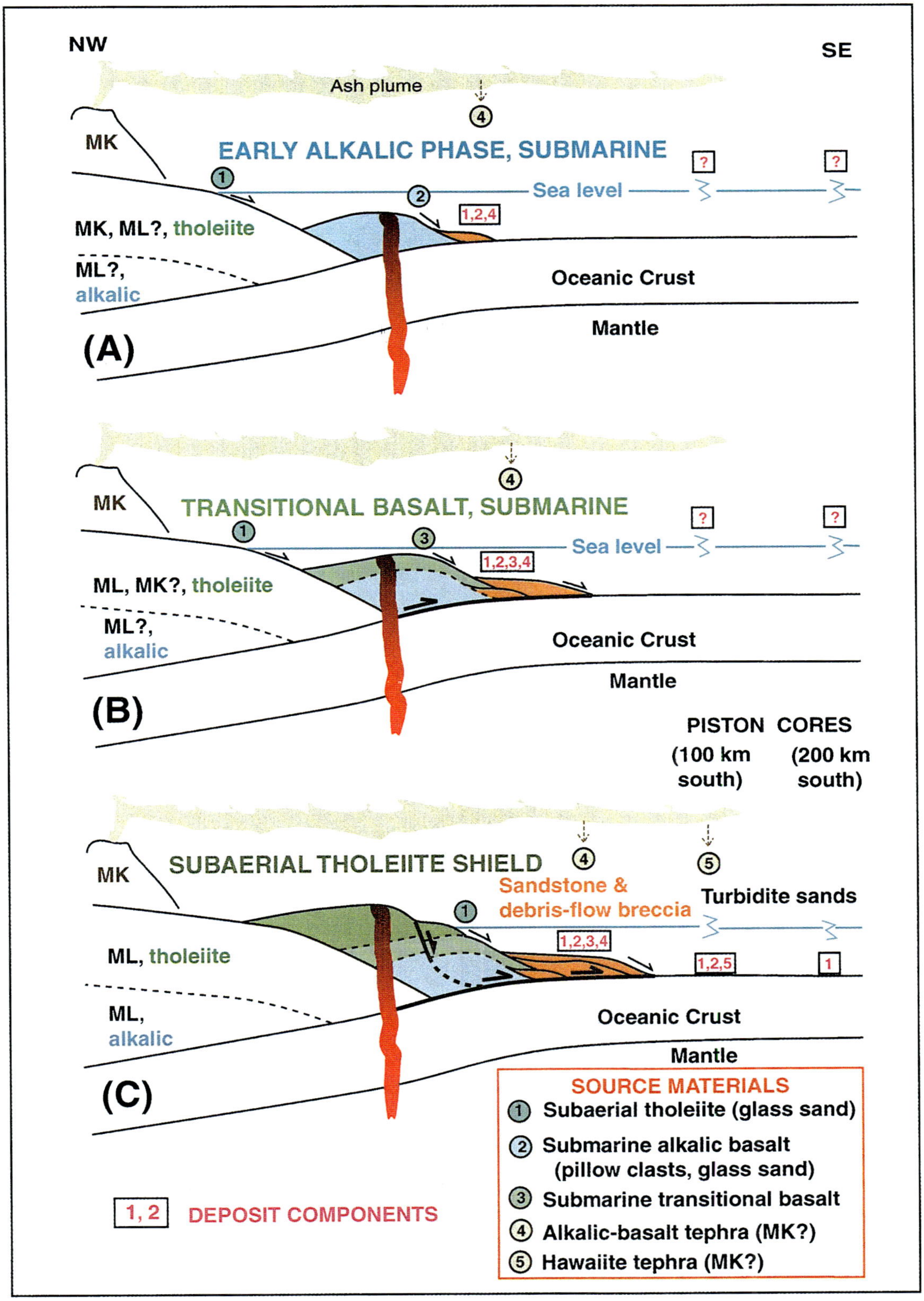

NW
SE
Ash plume
MK
EARLY ALKALIC PHASE, SUBMARINE
Sea level
1,2,4
?
MK, ML?, tholeiite
ML?, alkalic
Oceanic Crust
Mantle
(A)
MK
TRANSITIONAL BASALT, SUBMARINE
Sea level
1,2,3,4
ML, MK?, tholeiite
ML?, alkalic
Oceanic Crust
Mantle
(B)
PISTON CORES
(100 km south)
(200 km south)
MK
SUBAERIAL THOLEIITE SHIELD
Sandstone & debris-flow breccia
Turbidite sands
1,2,3,4
1,2,5
ML, tholeiite
ML, alkalic
Oceanic Crust
Mantle
(C)
SOURCE MATERIALS
1 Subaerial tholeiite (glass sand)
2 Submarine alkalic basalt (pillow clasts, glass sand)
3 Submarine transitional basalt
4 Alkalic-basalt tephra (MK?)
5 Hawaiite tephra (MK?)
1, 2 DEPOSIT COMPONENTS

sicular hawaiite and benmoreite in some turbidite sandstones from the lower scarp compositionally resemble the Laupahoehoe Volcanics of Mauna Kea, which are younger than about 75 ka [*Wolfe and Morris*, 1996]. This association probably does not have age significance, because rare grains of benmoreite tephra have been found in all the turbidite layers in cores P5 and P6 [*T.W. Sisson*, unpubl. data]. This distribution indicates that evolved alkalic tephra have been erupted and dispersed widely from various Hawaiian volcanoes over the last 2 Ma (Table 3).

Because the more distal P5 core is magnetically reversed below a depth of 2.3 m, the ages of its lower three turbidite layers, which are dominated by Kea-type tholeiite grains, must be greater than 0.78 Ma; the lowermost layer may be as old as 2 Ma (Table 3). Tholeiitic materials of these ages seem impossible to have originated from Kīlauea. Even shield-stage Mauna Kea is likely to have lacked a large subaerial edifice before 780 ka [*DePaolo and Stolper,* 1996]. Only subaerial Kohala, Haleakalā (Maui) at paleo-shoreline distances of 350-450 km, or the volcanoes on O'ahu at distances of >600 km, could be sources for such energetic long-traveled turbidites, which seem most plausibly related to large landslide events (Table 3). If ever feasible, distinguishing among these Kea-trend tholeiite sources, especially in small samples such as glass sands, could greatly improve understanding of the growth histories of Hawaiian volcanoes.

7. OTHER POSSIBLE SITES OF EARLY ALKALIC VOLCANISM

Kīlauea now joins Lō'ihi as the only Hawaiian volcanoes with a documented early alkalic stage, and the diversity of alkalic compositions found thus far is much greater for Kīlauea. Our success in obtaining analyzable and datable basaltic samples from the sedimentary deposits along the south flank of Hawai'i Island suggests that morphologically similar submarine slopes of other volcanoes might expose analogous materials, especially as the compositional diversity of other Hawaiian volcanoes becomes more clearly documented (Table 2). Perhaps a greater range of alkalic compositions might be found on lower flanks of the steep east and west slopes of Lō'ihi, which have been dissected by landsliding. To date, only the upper parts of these flanks (<2000 m depth) have been sampled [*Garcia et al.*, 1995a].

Where gravitational slumping and landsliding have been most extensive, the early submarine record of volcano growth is most likely to be preserved at distal localities in deep water. For the Nu'uanu and Wailau slides, which have been other major foci for the recent Japan-USA cooperative studies [*Naka et al.*, 2000; *Takahashi et al.*, 2000], all dive sites to date have been at proximal and mid-slide distances. No alkalic and only rare submarine-erupted (high S) rock fragments were recovered [*Clague et al.*, this volume; *Shinozaki et al.*, this volume]. Perhaps more compositionally diverse samples might be exposed on small distal blocks, some as much as 200 km from present-day shoreline. Other attractive sites to search for early submarine-erupted alkalic phases of volcano growth could be along distal basal parts of the South Kona slump on the west side of Hawai'i Island [*Lipman et al.,* 1988; *Moore et al,* 1995a] for evidence of the early history of Mauna Loa. Another potentially informative area is the newly recognized Laupāhoehoe slump on the northeast side of Hawai'i Island [*Smith et al.,* this volume], where a mid-slope bench and lower scarp (pre-Mauna Kea) appear morphologically similar to the offshore lower south flank and might expose early rocks from Kohala Volcano.

8. PAST SLOPE FAILURES AND FUTURE STABILITY OF THE SOUTH FLANK

Previous studies of Kīlauea's south flank have interpreted features of the Hilina fault system and adjacent offshore bathymetry as products of large-scale block-slumping and volcano spreading since growth of the Kīlauea shield [e.g., *Moore and Krivoy*, 1964; *Swanson et al., 1976*; *Lipman et al.*, 1985; *Borgia and Treves,* 1992; *Moore and Chadwick*, 1995; *Delaney et al.*, 1998]. In contrast, our new data obtained by submersible operations suggest a more prolonged and complex history of slope failures. This interpretation involves widespread recurrent deposition of debris flows during early growth of an ancestral submarine alkalic volcanic edifice (Plate 2), succeeded relatively recently by modest-scale block-fault subsidence of the growing tholeiitic shield of Kīlauea.

8.1. Significance of Compositionally Diverse Volcaniclastic Rocks

The compositionally diverse high-S volcaniclastic rocks in breccia beds exposed deep along the south flank, and absence of Kīlauea-like tholeiite clasts in these deposits, are strong evidence for recurrent landsliding from the slopes of an active submarine alkalic volcano. Fairly uniform clast sizes, absence of large clasts (megablocks), and well developed meter to multi-meter bedding thickness of these breccias interstratified with well-bedded sandstones, indicate repeated deposition from submarine debris flows that were individually of modest size, rather than by chaotic landsliding or large-scale block slumping. Such processes probably also characterize recent growth of Lō'ihi

Table 3. Growth of Hawaiian volcanoes and landslide/turbidite events, recorded in south-flank volcaniclastic sediments and distal piston-cores P5 and P6

Deposit/feature	Core depth (in meters)	Age (method)	Glass chemistry (4, 5)	Inferred volcanic significance
S495, Lava flow	--	Recent	Alkalic basalt	Deep submarine flow, S base Puna Ridge
P6: Tholeiite grains	0.0-1.75	0-175 ka	Low-S K>>L type	Coastal sands, subaerial Kilauea shield
P6: Sparse alkalic grains	0.3	30 ka (2)	High CaO	Alkalic volcanism, Lō'ihi Seamount
Ash bed, HSDP	--	39 ka (7)	--	Correlative with Pahala Ash; may predate inception of subaerial Hilina fault system
S505-10B: alkalic clast	--	130 ka (3)	High S alk b	Late alkalic pillow fragment, Kīlauea
P6-1 Turbidite	1.75-1.85	180 ka (2)	Low-S K>L type Some hi-S alk b	Coastal slide (Kil/MK > Mauna Loa?) Late submarine-erupted alkalic Kīlauea glass
P6-2 Turbidite	1.85-1.95	190 ka (2)	Low-S K>L type Some hi-S alk b	Coastal slide (Kil/MK > Mauna Loa?) Submarine-erupted alkalic Kīlauea glass sand
P5: Tholeiite grains	0.6	200 ka (1)	High S	Early submarine transitional basalt, Kīlauea
K95, S504: pillows	--	~130-220 ka	Hi-S trans. basalt	Upper submarine slope of Kīlauea
P6-3 Turbidite	2.15-2.30	220 ka (2)	Low-S K>L type Some hi-S alk b	Coastal slide (Mauna Kea? > Mauna Loa?) Submarine-erupted alkalic Kilauea glass sand
S505-2B: alkalic clast	--	220 ka (3)	High S alk b	Submarine-erupted pillow, ancestral Kīlauea
P6-4 Turbidite	2.30-2.40	230 ka (2)	Low-S K>L type Some hi-S alk b	Coastal slide (Mauna Kea? > Mauna Loa?) Submarine-erupted alkalic Kīlauea glass sand
HSDP drillcore	--	235 ka (6)	--	End of tholeiitic shield-building, Mauna Kea
S508: mica nephelinite	--	235 ka (3)	--	Shallow intrusion, ancestral alkalic Kīlauea
S509-4-B: alkalic clast	--	~275 ka (3)	High S alk b	Submarine-erupted pillow, ancestral Kīlauea
S509-2A: alkalic clast	--	270 ka (3)	High S alk b	Submarine-erupted pillow, ancestral Kīlauea
P6-5 Turbidite	3.0-3.05	300 ka (2)	Low-S K>L type Some hi-S alk b	Subaerial slide (Mauna Kea? > Mauna Loa?) Submarine-erupted alkalic Kīlauea glass sand
P6-6 Turbidite	3.2-3.3	330 ka (2)	Low-S K>L type Some hi-S alk b	Subaerial slide (Mauna Kea? > Mauna Loa?) First submarine-erupted alkalic Kilauea glass
P6-7 Turbidite	3.55-3.65	360 ka (2)	Low-S K>L type No hi-S alk b	Subaerial slide (Mauna Kea? > Mauna Loa?) Predates early alkalic Kīlauea?
P6-8 Turbidite	4.05-	410 ka (2)	Low-S L-type thol.	Punalu'u slide (large subaerial Mauna Loa)?
P5: High mag. SUS	1.2	410 ka (1)	Low-S L-type thol.	Punalu'u slide (large subaerial Mauna Loa)?
P5-1 Turbidite	1.8-1.9	625 ka(1)	K-type thol.	Subaerial slide (Kohala?)
P5: Polarity reversal	2.3	780 ka (2)	--	Brunhes-Matuyama boundary
P5-2 Turbidite	4.35-4.55	1.5 Ma (1)	K-type thol.	Wailau slide? (Molokai'i)
P5-3 Turbidite	5.10-5.25	1.7 Ma (1)	K-type thol.	Large subaerial slide (O'ahu or Molokai'i)?
P5-4 Turbidite	6.05-6.30	2.0 Ma (1)	K-type thol.	Nu'uanu slide (Ko'olau)?

Notes:

1. Ages interpolated from pelagic sedimentation rate (2.95 mm/ka) in P5 core [*Naka et al.*, this volume]
2. Ages interpolated from correlation of P6 core with upper part of P5 core, based on paleo-magnetic susceptibility, microfossil, and glass compositions [*Kanamatsu et al.*, this volume; *Naka et al.*, this volume]
2. Documented magnetic chron boundary
3. $^{40}Ar/^{39}Ar$ plateau age on biotite [*M. Lanphere*, written commun., 2001]
4. Glass-sand chemistry [*Naka et al.*, this volume; *T. Sisson*, unpubl. data]
5. K-type, subaerially erupted (low S) "Kea-trend" tholeiite (SiO_2<52%); L-type, subaerially erupted (low S) "Loa-trend" tholeiite (SiO_2>52%)
6. $^{40}Ar/^{39}Ar$ ages [*Sharp et al.*, 1996] and rock chemistry [*Rhodes*, 1996]
7. Age on ash bed in Hawai'i Scientific Pilot Drill Hole, by ^{14}C [*Beeson et al.*, 1996]

Seamount, where the steep east and west slopes of the volcano have been interpreted as scarps remnant from submarine landsliding [*Malahoff*, 1987; *Garcia et al.*, 1995a; *Moore and Chadwick*, 1995]. At ancestral Kīlauea, the early debris-flow deposits on the south flank and turbidite sands in the proximal piston core P6 suggest that in its alkalic stage this volcano was a similarly steep sided submarine edifice, capable of generating frequent landslides. These slide events appear to have been modest in scale, because none is recorded in the more distal P5 core [*Naka et al.*, this volume].

The diverse compositions of turbidite sands interbedded with the debris flow-breccias and forming matrix of the breccia clasts imply multiple sources and diverse slope-failure mechanisms. The most common compositional types in the massive sandstone samples and breccia matrix from the lower scarp of the south flank are subaerially erupted (low S) glass sands with tholeiitic compositions similar to present-day Kea- and Loa-trend magmas. Comparably thick and voluminous turbidite sandstones are absent on the proximal submarine flanks of Mauna Kea, where sampled by drilling [*DePaolo et al.*, 2001], or in the large landslide blocks of the Nu'uanu landslide that sample the north flank of Ko'olau [*Yokose*, this volume; *Clague et al.*, this volume]. The turbidite sandstones on the south flank of Hawai'i thus appear to represent more distal deposits from large coastal-failure events. These sands are largely derived from coastal environments, where subaerially erupted lavas quench to glass as they enter the ocean, processes well documented for the long-lived current eruption of Kīlauea [*Mattox and Mangan*, 1997]. Intermittent gravitational collapse of growing coastal lava deltas during such eruptions [*Tribble*, 1991] likely generates coarse debris that would be deposited in proximal environments and abundant glass sand that could be transported longer distances in turbidity flows.

Although the inception age and early growth histories of individual Hawaiian volcanoes remain poorly known, tholeiitic coastal sand from shield-stage Mauna Loa should have been readily available at 200-300 ka, during early growth of Kīlauea. In contrast, the source volcanoes for the common Kea-trend (Kīlauea-like) tholeiitic glass sand in the turbidite and breccia beds below the mid-slope bench are more difficult to identify. Despite the abundant subaerially erupted tholeiite as glass sand, the absence of Kea-type tholeiite clasts in the south-flank sediments is strong evidence against their derivation from any proximal (Kīlauea) source. Kīlauea probably could not have erupted voluminous subaerial tholeiitic lavas from its summit area before, or even while, submarine-erupted pillow lavas of transitional basalt were accumulating several thousand meters downslope from vents along the middle east rift zone (30-40 km from the present day summit: Figure 2).

More likely, the Kīlauea-type glass sand in volcaniclastic deposits on the lower scarp is derived from another Kea-trend volcano—Mauna Kea, or even Kohala if its shield stage was as large as recent evidence suggests [*Holcomb et al.*, 2000; *Kauahikaua et al.*, 2000]. Deposition of some tholeiitic glass sand from even more distant Kea-trend volcanoes, perhaps as far west as O'ahu, is documented by compositionally similar turbidite sands older than 0.78 Ma, probably back to 2 Ma or older (Table 3), in piston core P5 [*Naka et al.*, this volume].

The intermixed subaerially erupted sand of degassed tholeiite from other volcanoes represents a large volumetric fraction of the volcaniclastic apron on ancestral Kīlauea, and submarine volcanoes adjacent to large islands are likely to contain clastic material from multiple sources. Again, Lō'ihi Seamount provides a useful analog; degassed glass sand derived from shoreline-crossing Kīlauea and Mauna Loa lavas have been identified both on the flanks and at the very summit of this submarine volcano [*Clague et al.*, 2000; *Naka et al.*, this volume]. Mixing of coastal-derived glass sand and larger fragments of degassed hyaloclastite with submarine erupted pillow lavas in deeper parts (below 2500 m) of the Hawaii Scientific Drill Hole [*Sherman et al.*, 2000] may be interpretable analogously as recording sources from multiple volcanoes, respectively the subaerial shield of Kohala and submarine early Mauna Kea.

8.2. Northwest-Trending Transverse Structures

Additional evidence for early landslides as important processes in generating the complex features of the offshore south flank comes from the growth history of the conspicuous northwest-trending scarp aligned with Pāpa'u Seamount (Figure 2). The ā scarp bounds the Hilina submarine slump terrain all the way to the base of the south flank. Its transverse geometry suggests origin as a tear fault with about 15 km of right-lateral separation and 1500 m of bathymetric relief; relative components of horizontal versus vertical motion are unconstrained. The transverse submarine scarp has no subaerial topographic or geologic expression in the tholeiitic shield-stage lavas and vents of present-day Kīlauea [*Wolfe and Morris,* 1996] or geophysically imaged displacement of well-defined gravity and magnetic trends along the southwest rift zone [*Kauahikaua et al.,* 2000; *Hildenbrand et al.*, 1993]. The subaerial obscurity of the transverse structure seems unlikely to have resulted from deep burial of a young or active structure by lavas erupted from the relatively weak southwest rift zone.

Elsewhere on subaerial Kīlauea and Mauna Loa, even small structurally controlled topographic highs have persisted for long periods, as lavas divert around them (e.g., small horsts within the Hilina system).

In comparison, a parallel submarine scarp, which bounds the Punalu'u slump from Mauna Loa (Figure 2), continues on land as the Wai'ōhinu fault that moved 2-3 m laterally during the 1868 earthquake [*Wood*, 1914; *Lipman*, 1980]. The Wai'ōhinu fault bounds the Ka'ōiki-Honu'apo fault system on Mauna Loa (Figure 1), interpreted as a Mauna Loa south flank analogue of the Hilina fault system on Kīlauea [*Stearns and Clark*, 1930; *Lipman et al.*, 1990]. In contrast to the lack of topographic, structural, or geophysical expression of the subaerial projection of the Pāpa'u scarp, the Wai'ōhinu fault coincides with a conspicuous southward topographic bulge along the subaerial south flank of Mauna Loa, a prominent aeromagnetic feature (Anomaly C, in *Hildenbrand et al.* [1993, Pl. 2]), with offsets in a large positive gravity anomaly [*Kinoshita et al.*, 1963; *Kauahikaua et al.*, 2000], and with seismic-tomography velocity gradients [*Okubo et al.*, 1997]. The gravity and seismic-tomography deflections may mark a deep zone of dense olivine-rich cumulates, associated with prolonged seaward spreading of Mauna Loa and growth of the Ka'ōiki-Honu'apo fault system [*Lipman et al.*, 1990]. The large anomalies suggest processes on a larger scale but analogous to the dense dikes and deep olivine cumulates inferred to play an important role in destabilizing the upper south flank of present-day Kīlauea [*Clague and Denlinger,* 1994]. These contrasts in subaerial morphology and structure, along with dominance of alkalic clasts in the lower scarp, suggest that the south flank and its transverse (Pāpa'u) boundary were actively deforming prior to growth of Kīlauea's tholeiitic shield and inception of the current Hilina faults. Slumping of the Kīlauea shield along the Hilina fault system appears to be relatively young, postdating some of the deformation within the volcaniclastic wedge underlying the Hilina bench. Encroachment of the shield onto the clastic wedge derived from ancestral Kīlauea and associated old landslide scars on the submarine edifice may have localized the currently active Hilina slumping.

In contrast, seaward spreading of the unbuttressed south flank of Mauna Loa would have been especially active prior to, and during early growth of Kīlauea. If initially at rates comparable to current south-flank deformation at Kīlauea (5-10 cm/yr), seaward motion on Mauna Loa's south flank may have been substantial (10-15 km?) since the birth of Kīlauea at ~300 ka, even if declining as a growing Kīlauea stabilized the larger edifice. Seaward spreading of Mauna Loa probably contributes to the location of the present-day depositional contact between these volcanoes, in close proximity to the summit of Kīlauea (Figure 2), and also allows for an unexpectedly large volume of submarine-erupted ancestral Kīlauea lavas. If the south flank of Mauna Loa had already grown to near its present size and slope when ancestral Kīlauea began, any initial Kīlauea magma source at the site of the present caldera would necessarily have vented in shallow water, or even subaerially. Lavas from these two volcanoes probably interfinger complexly at depth, due to combined depositional and deformation processes, and the overall orientation is likely to be much steeper than a simple depositional onlap.

8.3. Prior Shield-Stage Slope Failures?

In contrast to the abundant evidence for submarine landsliding during the early alkalic growth of Kīlauea, the absence of clasts or large slide blocks resembling present-day Kīlauea tholeiite within, atop, or beyond the mid-slope bench and lower scarp suggests that no major slope failures have occurred along the submarine south flank of this volcano since commencement of its shield stage. Consistent with such an inference is the absence of any identifiable turbidite sand bed in proximal piston core P6 that could have been derived from a subaerial tholeiitic shield stage of Kīlauea [i.e., <175 ka, table 3; *Naka et al.*, this volume].

The only clear evidence of subsidence associated with submarine extensions of present-day Hilina faults is a shallow bench, about 500 m below sea level north of Pāpa'u Seamount [*Chadwick et al.*, 1993]. This bench is bounded by seismically active normal faults that have been traced across the shore line [*Denlinger and Okubo*, 1995]. Total exposed displacements along the Hilina fault system, amounting to about 800 m, could have been generated by historical rates of faulting (2-3 m/100 yr earthquake) and sustained seaward spreading (5-10 cm/yr since beginning of the 20th century) in 30-40,000 years. Such a brief time interval for Hilina faulting is also roughly compatible with the age of the oldest ash beds offset along these faults [~39 ka; *Beeson et al,* 1996].

Accordingly, rapid extension along the subaerial south flank and displacement at historical rates along a basal decollement [*Swanson et al.*, 1976;*Crosson and Endo*, 1982; *Lipman et al.*, 1985; *Thurber and Gripp,* 1988] may be relatively recent events in the growth of Kīlauea. The recent extension at Kīlauea may be rejuvenating and sustaining deformation of similar style that was more active earlier along the south flank of Mauna Loa during the peak of its shield-buiding stage [*Lipman et al.,* 1990; *Lipman*, 1995].

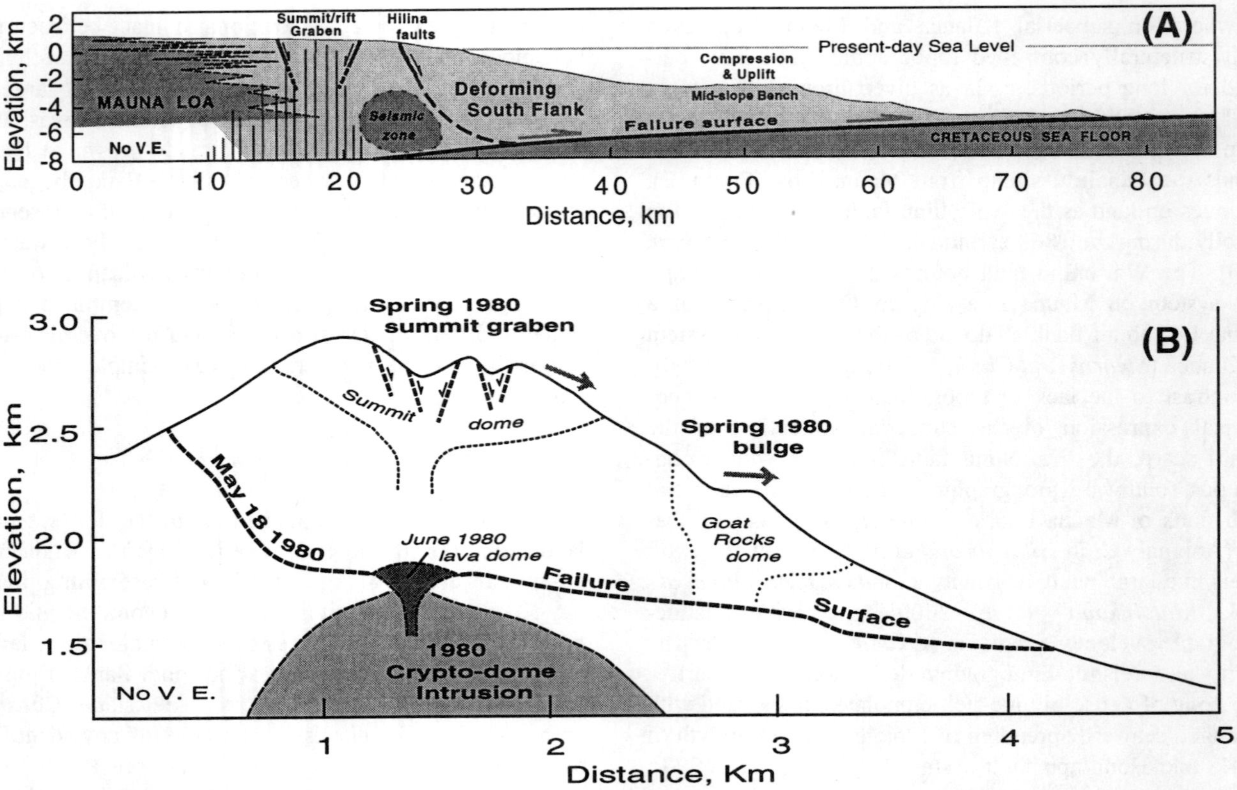

Figure 8. Possible future geometry of major south-flank landslide at Kīlauea **(A),** in comparison to the geometry of Mount St. Helens in 1980 before its flank failure on May 18, 1980 **(B).**

8.4. Compressional Deformation: Imbricate Thrusts Versus Localized Uplift

Shearing, gentle folding, complex fracturing, and other deformation observed along the lower scarp confirm prior interpretations of compressional deformation on the lower south flank. The deformation involves uplift of the bench and steepening of the lower scarp, and accommodating seaward spreading higher on the south flank [*Lipman et al.*, 1985; *Borgia and Treves*, 1992; *Denlinger and Okubo*, 1995; *Morgan et al.*, 2000]. Large-scale imbricate thrust faulting (15-24 km shortening) has also been inferred for the lower Hilina slump from seismic-profiling data [*Morgan et al.*, 2000; *Hills et al.*, this volume], but no confirming geologic evidence, such as repetition of the compositional sequence from alkalic to transitional basalt, was obtained from dive observations and samples. The meter- to decimeter-scale interbedding of massive debris-flow breccias with turbidite sandstone, observed widely in the lower scarp, seems more plausibly interpreted as an assemblage in the proximal volcaniclastic apron, rather than requiring that such breccias have been initially deposited in distal parts of the apron, as much as 30 km from their present locations [*Hills et al.*, this volume] The gently dipping beds of massive well-indurated turbidite sandstone, interlayered with less lithified mudstone and debris-flow breccia, could provide well defined seismic reflectors that would be difficult to distinguish from bedding-parallel thrust faults. Even in well exposed subaerial desert outcrop exposures, distinction among coherent stratigraphic sequences versus bedding-plane faults can be challenging and controversial [e.g., discussions in *Dickinson*, 1984; *Lipman and Hagstrum,* 1992].

As an alternative to proposals for long-traveled imbricate thrust faults [*Morgan et al.*, 2000; *Hills et al.*, this volume], could compressional shortening and uplift by open folding and small-scale reverse faulting have formed the mid-slope bench during growth of the Hilina fault system? At geodetically constrained recent deformation rates of about 10 cm/yr at the coastline [*Delaney et al.*, 1998; *Owen et al., 2000*], Kīlauea would have spread about 4 km since inferred inception of the Hilina faults at about 40 ka. The pre-deformation morphology of the south flank probably was a thick volcaniclastic apron, similar to that recon-

Table 4. Comparisons: Deformation and Slope Stability

	Mt. St. Helens 1980	Kīlauea South Flank
Timing (relatively recent phase of both volcanoes)		
Inception	March, 1980	< ~20-30 ka (Hilina faults)
Prior history	No previous major failure	No major failure, tholeiitic shield?
Deformation geometry & style		
Headwall/extension	Slump, summit graben	Rift graben, Hilina faults
Toe/compression	N slope bulge	Mid-slope bench, lower scarp
Motion	Subhorizontal	Subhorizontal
Displacement rate	~ 1.5 m/day	~10 cm/yr at S.L.
Seismicity	M=4-5 quakes; 1-3/day	M=5-6 quakes, ~2 /10 yr M-7-8 quake, ~1 /100 yr
Failure event		
Date	May 18, 1980	Future?
Trigger	M=5.1 quake	(M=7-8 quake?)
Event	Debris avalanche	(Debris avalanche?)
Post/pre failure areas	Much larger	(Much larger?)
Immediate precursor	None	???
Causes	Cryptodome intrusion Gravity spreading	Dike intrusions Gravity spreading

structed for the north flank of O'ahu prior to the Nu'uanu slide [*Yokose*, this volume; *Moore and Clague*, this volume]. About 1-1.5 km of uplift, above the axis of the mid-slope basin, would be needed. In a simple geometric model, a prism of volcaniclastic rocks on the lower flank averaging 10 km wide x 3 km thick (cross-sectional area = 30 km^2) compresses to 6-7 km wide x 4+ km high. The primary volcanic edifice acts as a relatively rigid wedge, pushing weak volcaniclastic rocks up over its toe. Such an alternative, involving local compression and uplift, would greatly reduce the distance required for the debris-flow breccias to have traveled from the ancestral alkalic edifice. There would be no requirement for long-traveled debris from a relatively small ancestral Kīlauea, in comparison with absence of comparably distal debris generated during the larger and more voluminous tholeiitic shield stage. This alternative would also be consistent with the apparent simple upward compositional change from dominantly alkalic to transitional and tholeiitic compositions, without obvious imbricated repetition (Figure 6).

Despite such uncertainties, our new data confirm the importance of compression, uplift, and oversteepening of the mid-slope bench in the evolution of submarine flanks of Hawaiian volcanoes. Such dynamic processes may eventually lead to large-scale catastrophic landslides and debris avalanches (e.g., Nu'uanu, Wailau, Alikä). Reconstruction of the paleo-geometry of these slide areas suggests the presence of a pre-slide mid-slope bench, or "bulge," especially for Nu'uanu [*Moore and Clague*, this volume; *Yokose*, this volume]. These older large-scale slope failures probably were also associated with flank deformation and volcano spreading, as exemplified by present-day processes along the Hilina fault system and the adjacent offshore south flank.

8.5. Evolution of the Hilina Fault System and Future Slope Stability

The inception and early history of the Hilina fault system and associated block slumping on the south flank remain poorly known, but the interpretations developed here suggest that the active Hilina slump structures on the south flank may be only in an early growth stage. These faults may have initiated relatively recently at Kīlauea, as the subaerial tholeiitic shield has grown larger and increasingly unstable. In this scenario, the modern Hilina faults, block slumping, and associated volcano spreading have been partly localized by older geologic units and structures formed during ancestral growth of an early alkalic edifice. If the Hilina system represents only a relatively recent evolutionary stage of Kīlauea's growth, and no major gravitational failure of the subaerial tholeiitic shield has occurred previously, the potential for large-scale landsliding (e.g., Alikā or Nu'uanu slides) may be increasing.

In such a perspective, the present geometry and kinematics of the Hilina fault system and offshore slump on the south flank are reminiscent of the situation on the north flank of Mount St. Helens in 1980 [*Lipman and Mullineaux,* 1981; *Moore and Albee*, 1981], before the eruption-triggering slope failure on the morning of May 18 (Figure 8). Features in common, despite the highly different volcanic settings, include: (1) the presence of actively opening graben structures high on the volcanic edifice, at the headwall of a large deforming flank area, (2) compression and bulging in a distal toe area, (3) roughly steady-state displacement rates, (4) subhorizontal motions in which downward pull of gravity is roughly balanced by upward pressure from magma, and (5) association with large earthquakes (Table 4).

Perhaps along the south flank (as at Mount St. Helens in 1980), eventual catastrophic failure could be triggered by an earthquake no larger than previous events, without additional precursor warnings. Although Kīlauea's south flank seems poised to fail catastrophically, such failure may not take place for tens of thousands of years, until the edifice reaches some critical size. The Hilina fault system has been active for about 40 ka [*Beeson et al,* 1996], and piston-core stratigraphy [*Naka et al.,* this volume] confirms that no large failures have thus far occurred from the Kīlauea shield. Nevertheless, recurrent earthquakes on the Hilina fault system and their associated tsunamis demonstrate a persistent threat to life and property that grows as Hawai'i's population increases. Relatively modest displacements of the Hilina fault system, too small to produce any recognized deep marine deposits, are a continuing local hazard, potentially punctuated by catastrophic slope failures at intervals of tens to hundreds of thousands of years.

Acknowledgments. We thank Captain O. Yukawa of the RV *Kairei* and Captain H. Tanaka of the RV *Yokosuka*, T. Fukui and Y. Imai (commanders of the ROV *Kaiko* and *Shinkai 6500*), the operation teams and marine crews for their skilled support, and the scientific team for shipboard assistance and subsequent discussion. D. Siems provided rapid-turnaround XRF rock analyses, and M. Lanphere provided critical geochronologic data, confirming Kīlauea as the source of the alkalic basalt clasts. The present manuscript has benefited from review comments by J. Morgan, D. Swanson, R. Tilling, M. Walter, and H. Yokose.

REFERENCES

Ando, M, The Hawaii earthquake of November 29, 1975,: Low dip angle faulting due to forceful injection of magma, *J. Geophys. Res., 84,* 7616-7626, 1979.

Beeson, M.H., D.A. Clague, and J.P. Lockwood, Origin and depositional environment of clastic deposits in the Hilo drill hole, Hawaii, *J. Geophys Res., 101,* 11,617-11,630, 1996.

Borgia, A. and B. Treves, Volcanic plates override the ocean crust: Structure and dynamics of Hawaiian volcanoes, *Geol. Soc. Spec. Publ. London, 60,* 277-299, 1992.

Chadwick, W. W. Jr., J.R. Smith Jr., J.G. Moore, D.A. Clague, M.O. Garcia, and C.G. Fox, Bathymetry of south flank of Kilauea volcano, Hawaii, *U.S. Geol. Survey Misc. Field Studies Map MF*-2231, 1993.

Chen, C.-Y., F.A. Frey, J.M. Rhodes, and R.M. Easton, Temporal geochemical evolution of Kilauea Volcano: Comparison of Hilina and Puna Basalt, *Am. Geophys. Union Mon. 95,* 161-181, 1996.

Clague, D.A., A.S. Davis, J.L. Bischoff, J.E. Dixon, and R. Geyer, Lava bubble-wall fragments formed by submarine hydrovolcanic explosions on Lō'ihi Seamount and Kīlauea Volcano. *Bull. Volc.*, 61, 437-449, 2000.

Clague, D.A., and R.P. Denlinger, The role of cumulus dunite in destabilizing the flanks of Hawaiian volcanoes, *Bull. Volc., 56,* 425-434, 1994.

Clague, D.A., J.G. Moore, and A.S. Davis, Volcanic breccia and hyaloclastite in blocks from the Nuuanu and Wailau landslides, Hawaii, *this volume.*

Clague, D.A., J.G. Moore, J.E. Dixon, and W.B. Friesen, Petrology of submarine lavas from Kilauea's Puna Ridge, Hawaii, *J. Petrol., 36,* 299-349, 1995.

Clague, D.A., W. Weber, and J. E. Dixon, Picritic glasses from Hawaii, *Nature, 353,* 553-556, 1991.

Crosson, R.S. and E.T. Endo, Focal mechanisms and locations of earthquakes in the vicinity of the 1975 Kalapana earthquake aftershock zone 1970-1979: Implications for tectonics of the south flank of Kilauea volcano, island of Hawaii, *Tectonics, 1,* 495-542, 1982.

Dalrymple, G.B., and J.G. Moore, Argon-40: Excess in submarine pillow basalts from Kilauea Volcano, Hawaii, *Science, 161*, 1132-1135, 1968.

Dana, J.D., 1849, Geology, *United States Exploring Expedition, 10,* George P. Putnam, New York, 436 p.

Delaney, P.T., A. Miklius, T. Arnadottir, A.T. Okamura, and M.K. Sako, Motions of Kilauea volcano during sustained eruption from the Pu'u O'o and Kupai'anaha vents, 1983-1991, *J. Geophys. Res., 98,* 17,801-17,820, 1993.

Delaney, P.T., R.P. Denlinger, M. Lisowski, A. Miklius, P.G. Okubo, A.T. Okamura, and M.K. Sako, Volcano spreading at Kilauea, 1976-1996, *J. Geophys. Res., 103,* 18,003-024, 1998.

Denlinger R.P., and P. Okubo, Structure of the mobile south flank of Kilauea Volcano, Hawaii, *J. Geophys. Res., 100*, 24,499-24,507, 1995.

DePaolo, D. J., and E.M. Stolper, Models of Hawaiian volcano growth and plume structure: Implications of results from the Hawaii Scientific Drilling Project, *J. Geophys. Res., 101*, 11,643-11,654, 1996.

DePaolo, D.J., E.M. Stolper, and D.M. Thomas, Deep drilling into a Hawaiian volcano, *Eos Trans. AGU, 82,* 154-155, 2001.

Dickinson, W.R., Reinterpretation of Lime Peak thrust as a low-angle normal fault; implications for the tectonics of southeastern Arizona, *Geology, 12,* 610-613, 1984.

Dvorak, J.J., and D. Dzurisin, Variations in magma supply at Kilauea volcano, Hawaii, *J. Geophys. Res., 98,* 22,255-22,268, 1993.

Dzurisin, D., R.Y. Koyanagi, and T.T. English, Magma supply and storage at Kilauea volcano, Hawaii, *J. Volc. Geother. Res., 21,* 177-206.

Dieterich, J.H., Growth and persistence of Hawaiian volcanic rift zones, *J. Geophys. R, 93,* 4258-4270, 1988.

Dixon, J.E., D.A. Clague, and E.M. Stolper, Degassing history of water, sulfur, and carbon in submarine lavas from Kilauea Volcano, Hawaii, *J. Geol., 99,* 371-394, 1991.

Dvorak, J.J., An earthquake cycle along the south flank of Kilauea volcano, Hawaii, *J. Geophys. Res., 99,* 9533-41, 1994.

Dvorak, J.J. A.T. Okamura, T.T. English, R.Y. Koyanagi, J.S. Nakata, M.K. Sako, W.T. Tanigawa, and K.M Yamashita, Mechanical response of the south flank of Kilauea Volcano, Hawaii, to intrusive events along the rift systems, *Tectonophysics, 124,* 193-209, 1986.

Dzurisin, D., J.P. Lockwood, T.J. Casadevall, and M. Rubin, The Uwekahuna Ash: Product of large prehistoric phreatomagmatic eruptions at Kilauea volcano, Hawaii, *J. Volc. Geotherm. Res., 66,* 163-184, 1995.

Easton, M., Stratigraphy of Kilauea Volcano, *U.S. Geol. Surv. Prof. Pap.* 1350, 243-260, 1987.

Fiske, R. S., T. R. Rose, D.A. Swanson, and J. C. McGeehin, The Kulanaokuaiki tephra: product of newly recognized pyroclastic eruptions at Kilauea volcano, Hawai'i [abs.], *Eos Trans. AGU, 80,* F1196-1197, 1999.

Fiske, R.S., and T.L. Wright, Volume of Kilauea: Estimates and implications [abs.], *Eos Trans. AGU, 77,* , F644-F645, 1997.

Fornari, D.J., J. G Moore, and L. Calk, A large submarine sand-rubble flow on Kilauea Volcano. *J. Volc. Geotherm. Res., 5,* 239-256, 1979.

Garcia, M.O., and M.G. Davis, Submarine growth and internal structure of ocean island volcanoes based on submarine observations of Mauna Loa volcano, Hawaii, *Geology, 29,* 163-166, 2001.

Garcia, M. O., D. J. P. Foss, H. B. West, and J. J. Mahoney, Geochemical and isotopic evolution of Lō'ihi Volcano, Hawaii, *J. Petrology, 36,* 1647-1674, 1995a.

Garcia, M.O, T.P. Hulsebosch, and J.M. Rhodes, Olivine-rich submarine basalts from the southwest rift zone of Mauna Loa Volcano: Implications for magmatic processes and geochemical evolution, *in* Mauna Loa Revealed: Structure, composition, History, and Hazards, *Am. Geophys. Union Mon. 92,* 219-239, 1995b.

Garcia, M.O., D.W. Muenow, K.E. Aggray, and J.R. O'Neil, Major element, volatile, and stable isotope geochemistry of Hawaiian submarine tholeiitic glasses, *J. Geophys. R., 94,* 10,525-10,538, 1989.

Got, J.-L., J. Frechet, and F.W. Klein, Deep fault plane geometry inferred from multiple relative relocation beneath the south flank of Kilauea, *J. Geophys. R., 99,* 15,375-15,386, 1994.

Guillou, H., L. Turpin, F. Garnier, S. Charbit, D.M. Thomas, Unspiked K-Ar dating of Pleistocene tholeiitic basalts from the deep core SOH-4, Kilauea, Hawaii, *Chem. Geol., 140,* 81-88, 1997.

Haskins, E.H., and Garcia, M.O, Scientific drilling of a Hawaiian volcano: Does Ko'olau have a skin? [abs.], *Eos Trans. AGU, 81,* F1349, 2000.

Hatori, T., Wave source of the Hawaii tsunami in 1975 and tsunami behavior in Japan, *J. Seis. Soc. Jpn., 29,* 355-364, 1976.

Hildenbrand, T.G., J.G. Rosenbaum, and J.P. Kuahikaua, Aeromagnetic study of the Island of Hawaii, *J. Geophys. Res., 98,* 4099-4119, 1993.

Hill, D.P., and J.J. Zucca, Geophysical constraints on the structure of Kilauea and Mauna Loa volcanoes, and some implications for seismomagmatic processes, *U.S. Geol. Surv. Prof. Pap. 1350,* 903-917, 1987.

Hills, D., J. Morgan, and G. Moore, Structural variability along the submarine south flank of Kilauea volcano, Hawaii, from a multichannel seismic reflection survey, *this volume.*

Holcomb, R., Eruptive history and long-term behavior of Kilauea Volcano. *U.S. Geol. Survey Prof. Paper 1350,* 261-350, 1987.

Holcomb, R., B.K. Nelson, P. W. Reiners, and N.-L. Sawyer, Overlapping volcanoes: The origin of the Hilo Ridge, Hawaii, *Geology, 28,* 547-550, 2000.

Johnson, K.T, Reynolds, J.R., and Vonderhaar, D., Petrological systematics of submarine lavas from the Puna Ridge, Hawaii: Implications for rift zone plumbing and magmatic processes, this volume.

Kanamatsu, T., E. Herrero-Bervera, and G. M. McMurtry, Magnetostratigraphy of deep-sea sediments from areas adjacent to the Hawaiian Islands: Implication for ages of turbidites derived from submarine landslides, *this volume*

Kauahikaua, J., T. Hildenbrand, and W. Webring, Deep magmatic structures of Hawaiian volcanoes, imaged by three-dimensional gravity models, *Geology, 28,* 883-886, 2000.

Kinoshita, W.T., H.L. Krivoy, D.R. Mabey, and R.R. MacDonald, Gravity survey of the island of Hawaii, *U.S. Geol. Surv. Prof. Paper 475-C,* C114-C116, 1963.

Lee, H.J., M.E. Torresan, and W. McArthur, Stability of submerged slopes on the flanks of the Hawaiian Islands, a simplified approach, *U.S. Geological Survey Open-File Report 94-638,* 54 p, 1994.

Lipman, P.W., The southwest rift zone of Mauna Loa--Implications for structural evolution of Hawaiian volcanoes, Am. *J. Sci, 280-A,* 752-776, 1980.

Lipman, P.W., Declining growth of Mauna Loa during the last 100,000 years; Rates of lava accumulation vs. gravitational subsidence, *in* Mauna Loa Revealed: Structure, composition, History, and Hazards, *Am. Geophys. Union Mon. 92,* 45-80, 1995.

Lipman, P.W., and Hagstrum, J. T., Jurassic ash-flow sheets, calderas, and related intrusions of the Cordilleran volcanic arc in southeastern Arizona: Implications for regional tectonics and ore deposits, *Geol. Soc. Am. Bull., 104,* 32-39, 1992.

Lipman, P.W., J.P. Lockwood, R.T. Okamura, D.A. Swanson, and K.M. Yamashita, Ground deformation associated with the 1975 magnitude-7.2 earthquake and resulting changes in activity of Kilauea volcano, Hawaii, *U.S. Geol. Surv. Prof. Pap. 1276*, 45 pp, 1985.

Lipman, P.W., and D.R. Mullineaux, eds., The 1980 eruptions of Mount St. Helens, Washington, *U.S. Geol. Surv. Prof. Paper 1250,* 844 p., 1981.

Lipman, P.W., W.R. Normark, J.G. Moore, J.B. Wilson, and C.E. Gutmacher, The giant submarine Alika debris slide, Mauna Loa, Hawaii, *J. Geophys.. Res., 93,* 4279-4299, 1988.

Lipman, P.W., J.M. Rhodes, and G.B. Dalrymple, The Ninole Basalt - Implications for the structural evolution of Mauna Loa volcano, Hawaii, *Bull. Volc., 53,* 1-19, 1990.

Lipman, P. W., Sisson, T. Ui, T. and Naka, J, Ancestral growth of Kilauea volcano: submarine alkalic through tholeiitic shield-*stage development and landsliding on the offshore south flank of Hawaii Island, Geology, 28*, 1079-1082, 2000.

Ludwig, K.R., B.J. Szabo, J.G. Moore, and K.R. Simmons, Crustal subsidence rate off Hawaii determined from $^{234}U/^{238}U$ ages of drowned coral reefs, *Geology, 19,* 171-174, 1991.

Ma, K.-F, H. Kanamori, and K. Satake, Mechanism of the 1975 Kalapana, Hawaii, earthquake inferred from tsunami data, *J. Geophys. Res., 104,* 13,153-13,168, 1999.

Macdonald, G.A., and T. Katsura, Chemical composition of Hawaiian lavas, J. *Petrol., 5,* 82-113, 1964.

Malahoff, A., Geology of the summit of Lō'ihi Submarine volcano. *U.S. Geol. Survey Prof. Paper 1350,* 133-144, 1987.

Mattox, T.N., and M.T. Mangan, Littoral hydrovolcanic explosions: a case study of lava-seawater interaction at Kilauea Volcano, *J. Volc. Geotherm. Res, 75*, 1-17, 1997.

Moore, J.G., Relationship between subsidence and volcanic load, Hawaii, *Bull. Volc., 34,* 562-576, 1970.

Moore, J.G., Subsidence of the Hawaiian Ridge, *U.S. Geol. Surv. Prof. Pap. 1350*, 85-100, 1987.

Moore, J.G., and W.C. Albee, Topographic and structural changes, March-July 1980—photogrammetric data, *in* The 1980 eruptions of Mount St. Helens Washington, *U.S. Geol. Surv. Prof. Paper 1250*, 123-134, 1981.

Moore, J.G., W.B. Bryan, M.H. Beeson, and W.R. Normark, Giant blocks in the South Kona landslide, Hawaii, *Geology, 23, 125-128*, 1995.

Moore, J. G., and W. W. Chadwick, Jr., Offshore geology of Mauna Loa and adjacent areas, Hawaii, *in* Mauna Loa Revealed: Structure, composition, History, and Hazards, *Am. Geophys. Union Mon. 92,* 21-44, 1995.

Moore, J. G., and D.A. Clague, Volcano growth and evolution of the island of Hawaii, *Geol. Soc. Am. Bull., 104*, 1471-1484, 1992.

Moore, J. G., and D.A. Clague, Mapping of the Nuuanu landslide in Hawaii, this volume.

Moore, J. G., D. A. Clague, and W.R. Normark, Diverse basalt types from Lō'ihi seamount, Hawaii, *Geology, 10*, 88-92, 1982.

Moore, J.G., D.A. Clague, R.T. Holcomb, P.W. Lipman, W.R. Normark, and M.E. Torresan, Prodigious submarine landslides on the Hawaiian Ridge, *J. Geophys. Res., 94,* 17,465-17,484, 1989.

Moore, J.G., and R.S. Fiske, Volcanic substructure inferred from dredge samples and ocean bottom photographs. *Geol. Soc. Am. Bull, 80,* 1191-1201, 1969.

Moore, J.G., and H.L. Krivoy, The 1962 flank eruption of Kilauea volcano and structure of the East Rift Zone, *J. Geophys. Res., 69*, 2033-2045, 1964.

Moore, J.G., and R.K. Mark, Morphology of the Island of Hawaii, *GSA Today, 2,* 12, 1992.

Moore, J.G., and D.L. Peck, Bathymetric, topographic, and structural map of the south-central flank of Kilauea Volcano, Hawaii, *U.S. Geol. Survey Misc. Geol.. Map*, I-456, 1965.

Moore, J.G., R.L. Phillips, R.W. Grigg, D.W. Peterson, and D.A. Swanson, Flow of lava into the sea, 1969-1971, Kilauea Volcano, Hawaii, *Geol. Soc. Am. Bull., 84*, 537-546, 1973.

Morgan, J. K., G. F. Moore, D. J. Hills, and S. Leslie, Overthrusting and sediment accretion along Kīlauea's mobile south flank, Hawaii, from marine seismic reflection data. *Geology, 28,* 667-670, 2000.

Naka, J., and 34 others, Tectono-magmatic processes investigated at deep-water flanks of Hawaiian volcanoes. *Eos Trans. AGU, 81,* 221, 222-226, 2000.

Naka, J., T. Kanamatsu, P. Lipman, T. Sisson, N. Tsuboyama, J. Morgan, J. Smith,' and T. Ui, Deep-sea volcaniclastic sedimentation around the southern flank of Hawaii Island, *this volume*.

Nakamura, K., Why do long rift zones develop in Hawaiian volcanoes—A possible role of thick ocean sediments [in Japanese], *Bull. Volc. Soc. Japan, 25*, 255-267, 1980.

Okubo, P., H.M. Benz, and B.A. Chouet, Imaging the crustal magma sources beneath Mauna Loa and Kilauea volcanoes, Hawaii, *Geology, 25*, 867-870, 1997.

Owen, S., P. Segall, M. Lisowski, A. Miklius, R. Denlinger, and M. Sako, The rapid deformation of Kilauea Volcano: GPS measurements between 1990 and 1996, *J. Geophys. Res., 105,* 18,983-18998, 2000.

Quane, S.L., Garcia, M. O., Guillou, H., and Hulsebosch, T. P., Magmatic history of the east rift zone of Kilauea volcano, Hawaii, based on drill core from SOH 1: *J. Volc. Geotherm. Res., 102,* 319-338, 2000.

Riley, C.M., J.F. Diehl, J.L. Kirschvink, and R.L. Ripperdan, Paleomagnetic constraints on fault motion in the Hilina fault system, south flank of Kilauea Volcano, Hawaii, *J. Volc. Geotherm. Res., 94*, 233-250, 1999.

Rhodes, J.M, The 1852 and 1868 Mauna Loa picrite eruptions: Clues to parental magma compositions and the magmatic plumbing system, *in* Mauna Loa Revealed: Structure, composition, History, and Hazards, *Am Geophys. Union Mon. 92,* 241-262, 1995.

Rhodes, J.M., Geochemical stratigraphy of lava flows sampled by the Hawaii Scientific Drilling Project, *J. Geophys Res., 101,* 11,729-11,746, 1996.

Rhodes, J.M., Geochemistry of Mauna Loa and Mauna Kea lavas sampled by the Hawaii Scientific Drilling Project [abs.], *Eos Trans. AGU,* 81, F1342, 2000.

Rhodes, J.M., M. K. Wenz, C. Neal, J. Sparks, J., and J.P. Lockwood, Geochemical evidence for invasion of Kilauea's plumbing system by Mauna Loa magma, *Nature, 337,* 257-260, 1989.

Sharp, W.D., B.D. Turrin,and P.R. Renne, The $^{40}Ar/^{39}Ar$ and K/Ar dating of lavas from the Hilo 1-km core hole, Hawaii Scientific Drilling Project, *J. Geophys Res., 101,* 11,607-11,616, 1996.

Sherman, S. B., Garcia, M.O, Seaman, C., Baker, M., and Stolper E., Geochemistry of glasses from the submarine rocks of the Hawaiian Scientific Drilling Project: Dueling volcanoes? [abs.], *Eos Trans. AGU,* 81, F1342, 2000.

Shinozaki, K., Z-Y Ren, and E Takahashi, Geochemical and petrologic characteristics of Nu'uanu and Wailau landslide blocks, *this volume.*

Shipboard Scientists, Kairei cruise (KR98-09): Onboard report, *Japan Marine Sci. Tech. Center,* Yokosuka, 64 p., 1998.

Shipboard Scientists, Shinkai 6500 / Yokosuka cruises (YK99-07, -08) around Hawaiian Islands: Onboard report, *Japan Marine Sci. Tech. Center,* Yokosuka, unpaged, 1999.

Sisson, T.W., P.W. Lipman, and J. Naka, Submarine alkalic through tholeiitic shield-stage development of Kilauea volcano, *this volume.*

Smith, J. R., A. Malahoff, and A. N. Shor, Submarine geology of the Hilina slump and morpho-structural evolution of Kilauea volcano, Hawaii, *J. Volc. Geotherm. Res., 94,* 59-88, 1999.

Smith, J. R. and K. Satake, Deep water multibeam surveys along the southeastern Hawaiian Ridge, 1998-1999, explanation of CD-ROM, this volume.

Smith J.R., K. Satake, J.K. Morgan, P.W. Lipman, The Pololu, Laupahoehoe, and Hana Ridge landslides revealed by recent multibeam sonar surveys in Hawaii, this volume.

Stearns, H.T, and W.O. Clark, Geology and water resources of the Kau District, Hawaii, *U.S. Geol. Surv. Water-Supply Pap. 616,* 1-94, 1930.

Swanson, D.A., W.A. Duffield, and R.S. Fiske, Displacement of the south flank of Kilauea volcano: The result of forceful intrusion of magma into the rift zones, *U.S. Geol. Surv. Prof. Pap 963,* 30 pp., 1976.

Tanaka, R., Nakamura, E., and Takahashi, E., Geochemical characteristics of Koolau volcano and the Nuuanu and Wailau landslide deposits, Hawaii: Implication for the evolution of Hawaiian plume. this volume

Takahashi, E. J.G. Moore, J.O. Garcia, K. Shinozaki, H. Yokose, T. Kanamatsu, J. Naka, J.R. Smith, K. Satake, S. Takarada, and T. Ui, Towards reconstruction of the Nuuanu and Wailau giant landslides in the northeast of Oahu Island, Hawaii [abs.], *Eos Trans AGU, 81,* WP250-51, 2000.

Tilling, R.T, R.Y. Koyanagi, P.W. Lipman, J.P. Lockwood, J.G. Moore, and D.A. Swanson, Earthquake and related catastrophic events, island of Hawai'i, November 29, 1975: A preliminary report, *U.S. Geol. Survey Circ. 740,* 33 p., 1976.

Thurber, C.H., and A.E. Gripp, Flexure and seismicity beneath the south flank of Kilauea Volcano and tectonic implications, *J. Geophys. R, 93,* 4271-4278, 1988.

Tilling, R.I, and Dvorak, J.J., Anatomy of a basaltic volcano, *Nature, 363,* 125-133, 1993.

Tribble, G.W., Underwater observations of active lava flows from Kilauea Volcano, Hawaii, *Geology, 19,* 633-636, 1991.

Ui, T., S. Takarada, J.G. Moore, J.K. Morgan, P.W. Lipman, K. Satake, J.R. Smith, M. Yoshida, Fracture analysis of Hilina slump, Nuuanu and Wailau debris avalanches, Hawaii: Transport mechanism and composition with subaerial analogs. 1. Interpretation of video images [abs.], *Eos Trans AGU, 81,* WP237, 2000.

Wolfe, E.W., and J. Morris, Geologic map of the Island of Hawaii. *U.S. Geol. Survey Misc. Invest. Map I-* 2524, 1996.

Wood, H.O, On the earthquakes of 1868 in Hawaii. *Bull. Seis. Soc. Am., 4,* 169-203, 1914.

Wright, T.L., Chemistry of Kilauea and Mauna Loa in space and time, *U.S. Geol. Surv. Prof. Pap. 735,* 40 pp, 1971.

Wyss, M., A proposed source model for the great Kau, Hawaii, earthquake of 1868, *Bull, Seis. Soc. Am., 78,* 1450-62, 1988.

Yokose, H., Landslides on the windward flanks of Oahu and Molokai, Hawaii: SHINKAI 6500 submersible investigation, *this volume.*

Zucca, J.J., and D.P. Hill, Crustal structure of the southeast flank of Kilauea volcano, Hawaii, from seismic fraction measurements, *Seis. Soc. Am. Bull., 70,* 1149-1159, 1980.

Peter W. Lipman, MS 910, U.S. Geological Survey, 345 Middlefield Road, Menlo Park CA 94025, USA (plipman@usgs.gov)

Thomas W. Sisson, MS 910, U.S. Geological Survey, 345 Middlefield Road, Menlo Park CA 94025, USA

Tadahide Ui, Graduate School of Science, Hokkaido University, Sapporo 060-0810, Japan

Jiro Naka, Deep Sea Research Department, Japan Marine Science and Technology Center (JAMSTEC), 2-15 Natsushima-cho, Yokosuka, Kanagawa 237-0061, Japan

John R. Smith, School of Ocean and Earth Science and Technology, University of Hawai'i, Honolulu, Hawai'i 96822, USA

Submarine Alkalic Through Tholeiitic Shield-Stage Development of Kīlauea Volcano, Hawai'i

Thomas W. Sisson and Peter W. Lipman

Volcano Hazards Program, U.S. Geological Survey, Menlo Park, California, USA

Jiro Naka

Japan Marine Science and Technology Center, Yokosuka, Japan

The submarine Hilina region exposes a succession of magma compositions spanning the juvenile "Lō'ihi" through tholeiitic shield stages of Kīlauea volcano. Early products, preserved as glass grains and clasts in volcaniclastic rocks of the 3000 m deep Hilina bench, include nephelinite, basanite, phonotephrite, hawaiite, alkali basalt, transitional basalt, and rare alkali-poor Mauna Loa-like tholeiite. Transitional basalt pillow lavas overlie the volcaniclastic section and record an early phase of subsequent subalkaline magmatism. Rare degassed tholeiitic pillow lava and talus above the volcaniclastic section are products of subaerial shield volcanism. Major and trace element variations of clasts and pillow lavas point to a factor of 2 – 2.5 increase in degree of melting from juvenile alkalic to modern tholeiitic Kīlauea. Progressive changes in element ratios that distinguish Hawaiian shield volcanoes, without commensurate changes in elements fractionated by partial melting, also signal increased contributions from Mauna Loa-type source regions as Kīlauea matured from its juvenile alkalic to its tholeiitic shield stage. Ancestral Kīlauea basanites and nephelinites were not primitive magmas but might have evolved from plume-derived alkali picritic parents by lithospheric-level crystallization differentiation, or solidification and remelting, involving pyroxene and garnet, similar to the subcrustal differentiation origin of hawaiites [*Frey et al.*, 1990]. Low magmatic productivity early in Kīlauea's history sustained a poorly integrated trans-lithospheric conduit system in which magmas stalled and differentiated, producing evolved hawaiites, nephelinites, and basanites. This contrasts with shield-stage Kīlauea where high magmatic productivity flushes the conduit system and delivers primitive magmas to shallow levels.

Hawaiian Volcanoes: Deep Underwater Perspectives
Geophysical Monograph 128
Copyright 2002 by the American Geophysical Union

INTRODUCTION AND GEOLOGIC SETTING

Hawaiian volcanoes pass through a series of growth stages as the lithospheric plate on which they ride advances over and then moves beyond a mantle melting anomaly. Early volcanism is submarine and alkalic, as represented by Lō'ihi Seamount, the youngest Hawaiian volcano situated at the leading edge of the volcanic chain [*Moore et al.*, 1982]. Magmas become tholeiitic and production rate increases substantially as the volcano moves over the central part of the melting anomaly. Kīlauea and Mauna Loa, Earth's most active and largest single volcanoes respectively, are examples of this tholeiitic shield stage. The volcanoes die with reduced eruption rate and reversion to alkalic magmas. Mauna Kea and Kohala are well studied examples this late post-shield alkalic volcanism [*Lanphere and Frey*, 1987; *Frey et al.*, 1990]. Small volumes of extremely alkalic magmas can then erupt after a negligible or prolonged hiatus in volcanism [*Clague and Frey*, 1982; *Clague and Dalrymple*, 1988]. These extremely alkalic post-erosional magmas resemble alkalic magmas erupted along the submarine arch surrounding the Hawaiian Trough [*Frey et al.*, 2000] and may not be related directly to the extinct shield volcanoes on which they erupt.

The symmetric succession in magma types and eruption rates, from modest submarine alkalic, to voluminous shield-stage tholeiitic, to moderate post-shield alkalic likely reflects a symmetry in melting processes in the rising mantle plume. Margins of the plume may melt to lower degrees than the plume interior, thereby generating small volumes of alkali-rich liquids from the plume edges and greater volumes of relatively alkali-poor liquids from the plume core [*Moore et al.*, 1982; *Frey and Clague*, 1983; *Chen and Frey*, 1983]. Depth of melting, as well as margin-to-core zonation in composition of the plume, may also vary [*Hauri et al.*, 1994b]. Understanding of the development of Hawaiian volcanoes is limited, however, because Lō'ihi Seamount has been the only well studied and widely accepted example of the early pre-shield alkalic stage. Other Hawaiian volcanoes may have begun with different magma types than those erupted at Lō'ihi, derived from different sources and by different melting processes. Quantitative understanding of Hawaiian magma generation would benefit from additional examples of the pre-shield stage, as well as from examples of the complete transition from pre-shield alkalic to shield-building tholeiitic stages.

The joint Japan – US JAMSTEC dive program to the offshore south flank of Kīlauea Volcano (Figure 1) was designed to study presumed landslide and slump processes that generated the 3000 m deep Hilina mid-slope bench. It was also hoped that the lower scarp, below the mid-slope bench, might expose lavas from early in Kīlauea's growth. Instead, bedded volcaniclastic rocks form all examined parts of the 3000-m-deep Hilina bench and its lower scarp (to 5000 m depth), as well as at least one slide block south of the frontal scarp [*Lipman et al.*, this volume]. Clasts in debris-flow breccias are dominantly submarine-erupted (high-S in glass) alkalic and transitional basalts. Glass grains in bedded turbidite sandstones are chiefly subaerial erupted basalts, but include a substantial fraction of submarine-erupted (high-S) glasses of which alkalic types are abundant. Widespread submarine-erupted alkalic materials point to the presence of a submarine alkalic volcano that lay to the northwest beneath present day Kīlauea. Combined stratigraphic, petrologic, and age data [*Lipman et al.*, this volume] suggest that this submarine volcano was the pre-shield alkalic "Lō'ihi-stage" of modern Kīlauea.

The Hilina bench volcaniclastic section is overlain sharply to the northwest (Figure 1) by sand-mantled nondeformed pillow lava and pillow breccia that form a continuous slope up to Kīlauea's shoreline [*Lipman et al.*, this volume]. To the northeast, pillow lavas form a continuous slope from the crest of Kīlauea's submarine Puna Ridge down to the Cretaceous seafloor at 5000 m depth. The pillow lavas upslope toward Kīlauea's shoreline from the bench are products of volcanism younger than that which produced the alkalic materials in the bench volcaniclastic rocks. Continuity in slope, eruptive style, and rock types indicate that pillow lavas at the foot of the Puna Ridge, northeast of the volcaniclastic bench, are also part of this younger episode of volcanism that has encroached upon and buried the volcaniclastic section. Most samples from upslope and northeast the bench are submarine-erupted transitional basalts, but rare degassed tholeiites are like those from modern Kīlauea and Mauna Loa. No alkalic samples have been recovered upslope or northeast of the bench, indicating that alkalic magmatism had greatly abated by the time those pillow lavas erupted.

The overall stratigraphic succession in the Hilina bench – Puna Ridge – Kīlauea region suggests that (1) early episodes of submarine alkalic and transitional magmatism, represented by clasts and grains in the bedded volcaniclastic rocks, were followed by (2) submarine transitional basaltic volcanism, preserved as pillow lavas upslope and northeast of the mid-slope bench, and then by (3) modern tholeiitic volcanism, as represented by subaerial Kīlauea tholeiites and submarine tholeiites from the crest of the Puna Ridge [*Garcia et al.*, 1989; *Clague et al.*, 1995]. Thus, the submarine Hilina region provides a newly discovered example of the juvenile alkalic phase of Hawaiian volcanism and presents a generally complete sampling of the progression to mature tholeiitic magmatism. This paper presents initial results of geochemical and petrologic studies aimed at understanding the causes for Kīlauea's magmatic evolution.

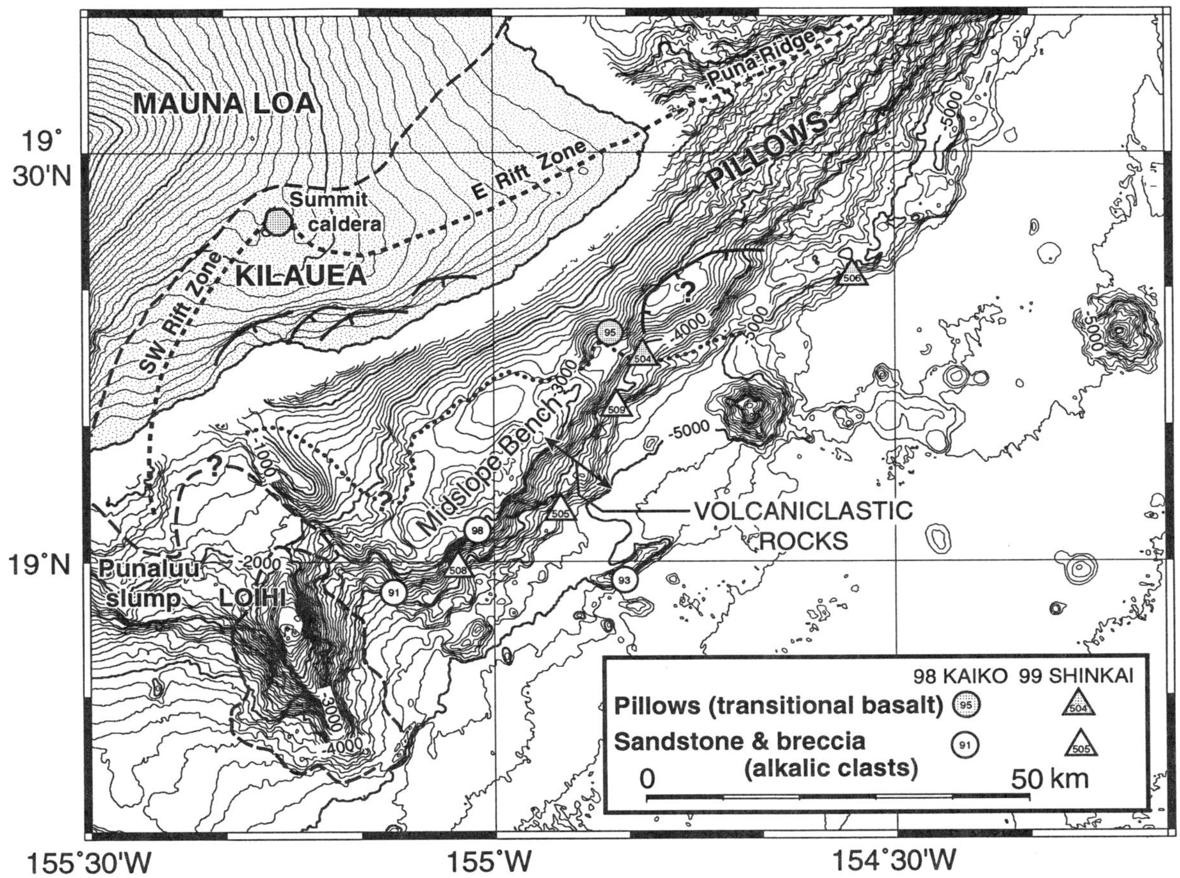

Figure 1. Bathymetric map of the southeast flank of Hawai'i Island, showing locations of major volcanic features, JAMSTEC dive sites, and dive numbers. Heavy dashed lines show approximate boundaries of major volcanoes. Short dashed lines show locations of Kīlauea's rift zones. Dotted line shows approximate contact between volcaniclastic rocks of the Hilina mid-slope bench and younger transitional basalt pillow lavas upslope to the north and east. Mapped (on land) and inferred (submarine) normal faults are shown as heavy solid lines with ticks on down-thrown side. On-land structure generalized from *Wolfe and Morris* [1996]. Contour interval: 100 m, heavy index contours: 1000 m. Dive sites on Lō'ihi Seamount are omitted for clarity.

ANALYTICAL TECHNIQUES

Submarine glasses preserve especially faithful records of the petrologic and geologic development of Hawaiian volcanoes [*Garcia et al.*, 1985; *Clague et al.*, 1995] because cold water storage limits alteration to the development of thin palagonite rinds that are easily avoided during electron microprobe analyses. Major element, S, and Cl concentrations of glass grains in volcanic glass turbidite sandstones, as well as glass from breccia clast margins, breccia matrix, pillow rinds, and glass grains in push and piston cores were measured by wavelength-dispersive techniques with a 5-spectrometer JEOL 8900R electron microprobe at the USGS in Menlo Park, California. Instrument conditions were 15 kV accelerating potential, 25 ηA beam current, and beam diameter of 20 μm. Na was measured first for 15 s to minimize loss of X-ray counts, Mn for 20 s; remaining elements were measured for 40 s. F was analyzed late in this study on an alkali basalt, a basanite, and two evolved nephelinite glasses at 30 ηA using an LDE crystal, with an empirical correction for interference by the Fe Kα peak. Concentrations were determined with the CIT-ZAF reduction scheme [*Armstrong,* 1995] utilizing the standards of *Clague et al.* [1995] to assure comparability to their extensive suite of tholeiitic glass analyses from Kīlauea's Puna Ridge and the alkalic North

Arch volcanic field [*Clague et al.*, 1990]. Instrument drift was monitored by 6 – 10 replicate analyses of basaltic glass standard VG-2 (MORB) and in some cases also A-99 (Kīlauea tholeiite) between each rock sample [*Jarosewich et al.*, 1979]. VG-2 is the primary standard for Na, Si, Al, Fe, Mg, and Ca, and, where necessary, drift outside of expected analytical precision was compensated by reprocessing analyses utilizing standard intensities determined at the beginning and end of the rock sample. Reported glass compositions are averages of 3 (sandstone, push core, and piston core grains), 5 (breccia matrix grains), or 10 – 15 (pillow and clast rinds) replicate analyses on non-overlapping locations, with background intensities determined on each grain.

Generally 25 separate glass grains were analyzed for each sandstone sample except where glass grains appeared to be unusually diverse (50 grains analyzed) or homogeneous (15 grains analyzed). Glass grains were selected on the basis of variations in color, shape, and backscattered electron brightness to characterize the diversity of melt compositions. Alkalic glasses differ from tholeiites in color, transparency, and backscatter brightness; they were analyzed in greater proportion than their true abundance.

Whole rock compositions were determined for clasts, pillows, and pillow fragments at the USGS in Denver, Colorado. Fresh appearing parts were sawn from hand samples, saw marks were removed with a diamond grinding wheel, and samples were crushed and powdered in alumina "chipmunk" jaw crushers and shatterboxes. Major element oxide concentrations were determined by wavelength dispersive X-ray fluorescence by David Siems, using equipment and techniques described in *Taggart et al.*, [1987]. Trace element concentrations were measured by energy-dispersive X-ray fluorescence (Rb, Zr, Sr, Y, Nb, Cr, V, Cu, Ni, Zn) [*Siems*, 2000] and by neutron activation analysis (REE, Cs, U, Th, Ta, Ba, Hf, Sc, Co) [*Baedecker and McKown*, 1987].

RESULTS

Hilina Bench Glass Compositions

Volcaniclastic rocks of the Hilina bench contain the earliest known potential products from ancestral Kīlauea [*Lipman et al.*, this volume]. Glasses from Hilina bench sandstones, debris-flow clasts, and debris-flow matrix form a continuous array of compositions spanning the tholeiitic – alkalic basalt boundary, and extending into the hawaiite, basanite, phonotephrite, and nephelinite compositional fields (Figure 2; representative analyses in Table 1; complete analyses on CD-ROM, this volume).

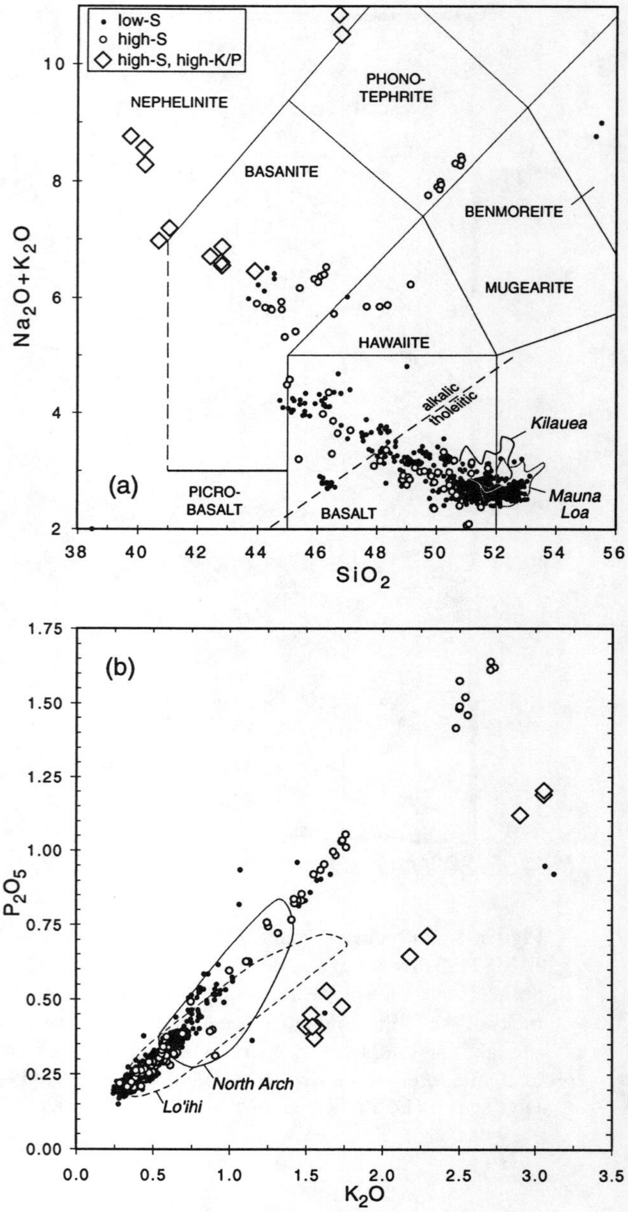

Figure 1. Bathymetric map of the southeast flank of Hawai`i Island, showing locations of major volcanic features, JAMSTEC dive sites, and dive numbers. Heavy dashed lines show approximate boundaries of major volcanoes. Short dashed lines show locations of Kīlauea's rift zones. Dotted line shows approximate contact between volcaniclastic rocks of the Hilina mid-slope bench and younger transitional basalt pillow lavas upslope to the north and east. Mapped (on land) and inferred (submarine) normal faults are shown as heavy solid lines with ticks on downthrown side. On-land structure generalized from *Wolfe and Morris* [1996]. Contour interval: 100 m, heavy index contours: 1000 m. Dive sites on Lō'ihi Seamount are omitted for clarity.

Table 1. Representative electron microprobe analyses of Hilina region glasses.

sample	type[a]	SiO_2	TiO_2	Al_2O_3	FeO	MnO	MgO	CaO	Na_2O	K_2O	P_2O_5	S[b]	Cl[b]	total %
Hilina bench sandstone high-S grains														
S508-R3-10	n	39.8	2.92	14.0	12.6	0.21	6.00	14.5	6.47	2.30	0.71	2600	2850	100.1
S508-R3-16	n	40.2	2.94	14.0	13.1	0.21	5.95	13.9	6.11	2.18	0.64	2350	1600	99.7
S508-R3-23	n	40.7	2.46	13.3	13.3	0.21	9.53	12.7	5.23	1.74	0.47	1850	2300	100.0
S508-R3-6	n	41.1	2.49	14.9	12.9	0.20	6.19	14.2	5.56	1.63	0.53	1650	1200	100.0
S508-R3-3	b	42.4	2.32	15.1	12.3	0.19	6.75	13.2	5.16	1.55	0.41	1250	1000	99.7
S508-R3-18	b	42.7	2.41	15.5	12.4	0.19	5.90	13.3	5.04	1.57	0.40	1400	1550	99.7
S508-R3-14	b	42.8	2.29	15.5	12.0	0.18	6.01	13.6	5.05	1.50	0.41	1200	1000	99.5
S508-R3-24	b	43.9	2.19	16.1	11.8	0.20	6.35	12.4	4.90	1.56	0.37	1000	800	99.9
S505-R7-37	b	44.4	5.35	15.4	13.4	0.18	4.95	8.88	4.39	1.43	0.82	1250	550	99.5
S505-R7-39	b	44.8	4.91	15.7	12.9	0.14	4.57	9.14	4.46	1.47	0.83	1300	600	99.1
S505-R7-35	b	46.3	4.09	15.9	12.8	0.21	3.96	8.57	4.77	1.76	1.05	1350	850	99.5
S505-R6B-14	b	44.8	4.77	15.7	12.6	0.17	4.63	9.13	4.36	1.44	0.83	1100	500	98.6
S505-R6B-10	b	46.1	4.14	15.8	12.8	0.18	3.98	8.60	4.64	1.74	1.03	1350	850	99.2
S509-R1B-18	b	44.9	4.83	16.2	12.4	0.16	4.68	9.83	4.07	1.24	0.76	1050	500	99.2
S509-R1B-17	b	45.3	4.79	16.2	12.5	0.16	4.72	9.88	4.16	1.25	0.74	1000	450	99.9
K98-R7-11	b	44.0	4.52	15.6	13.0	0.17	5.07	9.39	4.50	1.40	0.77	1050	500	98.5
S505-R7-43	pt	50.0	2.64	18.0	12.5	0.19	2.73	5.19	5.43	2.48	1.42	1100	550	100.8
S505-R7-45	pt	50.1	2.58	18.0	12.3	0.16	2.70	5.22	5.36	2.50	1.48	1100	500	100.6
S505-R7-38	h	47.6	3.65	16.2	11.6	0.17	4.08	8.99	4.38	1.47	0.85	1050	500	99.2
S505-R6B-27	h	48.3	3.42	16.1	11.2	0.18	3.81	8.50	4.28	1.59	0.94	1050	600	98.6
S505-R6B-29	ab	47.1	4.66	13.2	13.5	0.19	5.14	10.1	2.95	0.75	0.49	2000	500	98.2
S505-R7-44	ab	45.4	4.16	14.2	12.4	0.16	6.09	11.4	2.61	0.59	0.37	1400	950	97.5
S505-R6B-35	tr	49.3	3.00	13.6	12.5	0.19	5.74	10.7	2.47	0.59	0.31	1200	300	98.6
K93-R3-7	tr	49.4	2.88	14.1	11.6	0.17	6.22	11.2	2.43	0.55	0.30	1200	300	99.1
K93-R2-19	t	49.9	2.44	13.0	10.9	0.16	8.32	11.7	2.00	0.37	0.21	1200	200	99.2
K93-R4-2	t	51.0	2.49	12.8	10.9	0.19	7.87	12.4	1.62	0.44	0.23	1000	400	100.1
S508-R3-20	t	51.2	2.85	14.2	10.9	0.19	5.99	10.5	2.42	0.53	0.29	1250	250	99.2
Hilina bench sandstone low-S grains														
K98-R7-4	b	44.3	4.68	15.3	12.9	0.19	4.26	9.73	4.86	1.66	0.92	450	450	98.9
K98-R7-1	b	44.6	5.18	14.9	12.9	0.17	4.29	10.0	4.74	1.59	0.90	650	500	99.3
K91-R2-13	ab	46.4	5.48	12.3	16.2	0.22	4.67	9.50	3.28	0.84	0.58	300	500	99.4
S505-R7-21	ab	47.7	3.74	13.0	13.4	0.18	5.69	11.1	2.68	0.69	0.37	200	200	98.7
K91-R2-12	tr	46.2	2.77	14.1	12.2	0.17	9.46	10.7	2.50	0.38	0.25	600	300	98.8
K91-R2-14	tr	46.3	2.63	13.6	12.4	0.16	10.3	10.6	2.40	0.30	0.23	550	300	98.9
S505-R7-2	tr	48.1	3.74	13.4	13.2	0.18	5.69	11.2	2.64	0.68	0.40	150	150	99.3
K93-R3-11	tr	48.9	2.83	13.6	13.0	0.19	6.04	11.3	2.57	0.42	0.26	150	200	99.1
K93-R3-14	t	51.9	2.18	14.0	10.0	0.14	6.95	11.6	2.41	0.30	0.18	100	150	99.7
K93-R4-19	t	51.0	2.11	14.0	10.6	0.17	7.12	11.4	2.19	0.31	0.18	50	100	99.1
K93-R4-22	t	51.3	2.30	14.1	10.4	0.16	7.04	11.2	2.18	0.29	0.19	100	200	99.2
K93-R4-21	t	52.1	2.04	14.1	10.5	0.15	6.88	10.9	2.18	0.28	0.17	<50	50	99.3
K98-R7-5	t	52.1	2.24	13.2	12.0	0.19	5.95	10.3	2.30	0.38	0.21	100	100	98.8
K98-R7-2	t	52.4	2.71	13.3	12.2	0.18	5.79	10.3	2.43	0.43	0.29	200	100	100.1
K98-R7-3	t	52.5	2.18	14.0	10.3	0.13	6.65	11.1	2.27	0.31	0.20	100	100	99.7
S505-R9b-23	t	52.8	2.08	14.2	10.5	0.20	6.77	10.9	2.19	0.31	0.19	<50	<50	100.2
S505-R7-31	t	52.2	2.55	13.2	12.0	0.20	5.85	10.3	2.27	0.39	0.24	<50	100	99.3
S505-R7-41	t	52.6	2.38	13.6	11.2	0.18	6.22	10.5	2.30	0.39	0.24	50	50	99.7
S509-R7-5	t	53.0	2.35	13.8	11.4	0.17	6.16	10.5	2.44	0.38	0.24	<50	50	100.5
Hilina bench sandstone anomalous low-S grains														
S505-R7-15	bm	55.5	1.31	19.9	8.94	0.18	1.48	3.37	5.88	3.12	0.92	450	1450	100.8
S505-R6B-36	bm	55.3	1.33	19.7	9.06	0.16	1.53	3.36	5.71	3.06	0.95	550	1500	100.4
S508-R3-5	n	38.5	2.64	9.43	13.6	0.22	10.1	15.0	0.93	1.06	0.82	500	170	92.3
S509-R1B-7	ab	46.1	4.41	14.20	13.5	0.17	7.2	11.0	2.42	0.48	0.31	650	350	99.9
S508-R3-11	ab	46.7	3.20	14.0	12.1	0.19	6.44	11.9	3.06	1.62	0.45	<50	150	99.7

Table 1. (continued).

sample	type[a]	SiO_2	TiO_2	Al_2O_3	FeO	MnO	MgO	CaO	Na_2O	K_2O	P_2O_5	S[b]	Cl[b]	total %
Hilina bench debris flow clasts														
S505-R2B	ab	46.7	3.82	14.6	12.4	0.17	5.40	11.5	2.77	0.87	0.39	1500	850	98.9
S505-R10B	ab	46.2	3.82	14.6	12.2	0.17	5.37	11.2	3.09	0.89	0.40	1800	800	98.2
S509-R4B	h	46.5	4.38	16.5	12.9	0.17	4.29	8.71	4.40	1.31	0.72	1350	550	100.1
S509-R4A	tr	49.0	2.93	14.4	12.5	0.19	5.56	11.2	2.42	0.47	0.25	1550	250	99.1
S509-R6B	tr	48.1	4.05	13.6	13.1	0.18	5.57	11.3	2.84	0.82	0.44	100	150	100.0
Hilina bench debris flow matrix high-S grains														
K98-R1	n	40.2	5.09	13.7	12.0	0.21	5.22	11.1	5.67	2.90	1.12	3000	1350	97.6
S509-C2-5	n	46.8	2.48	16.6	11.5	0.31	3.51	7.50	7.46	3.05	1.18	2300	2100	100.8
S509-C2-6	n	46.7	2.47	16.6	11.5	0.29	3.54	7.61	7.80	3.05	1.19	2200	2100	101.2
S509-C2-1	ab	45.1	4.34	14.4	13.2	0.17	5.90	10.6	3.47	1.11	0.63	1650	450	99.1
S509-C2-2	ab	46.5	2.58	14.0	12.0	0.17	6.23	12.3	2.95	0.91	0.31	1850	600	98.3
S509-R7-3	ab	45.0	4.32	14.2	13.2	0.17	5.87	10.5	3.36	1.12	0.63	1800	450	98.6
basalts upslope and northeast of Hilina bench														
K95-R2	tr	48.1	3.41	14.3	12.4	0.16	5.78	11.1	2.61	0.56	0.33	1300	450	98.9
K95-R4	tr	49.1	3.00	14.3	12.1	0.18	6.01	11.0	2.37	0.48	0.29	1150	250	99.0
K95-R5	tr	47.9	3.13	14.3	11.7	0.16	6.30	11.7	2.49	0.59	0.29	1200	400	98.7
K95-R6	tr	49.0	2.77	14.3	11.3	0.16	6.22	11.5	2.28	0.48	0.25	1050	300	98.4
S504-R2B	tr	49.8	2.96	14.2	11.6	0.18	5.91	10.9	2.38	0.53	0.28	1350	150	98.9
S504-R5	t	51.7	3.15	13.5	12.1	0.17	5.57	10.3	2.51	0.61	0.33	100	150	99.9
S506-R8A	tr	48.8	2.64	14.7	11.3	0.17	5.98	11.4	2.34	0.50	0.25	1150	600	98.3
S506-R8C	tr	48.9	2.64	14.6	11.1	0.15	6.07	11.5	2.33	0.51	0.25	1150	600	98.2
S506-R2A	tr	48.9	2.71	14.8	11.4	0.18	5.77	11.1	2.41	0.51	0.25	1300	600	98.2
S506-R4A	t	51.9	2.43	14.2	11.0	0.17	6.32	10.8	2.33	0.38	0.24	400	150	99.8
S506-R5A	t	52.0	2.37	14.3	11.1	0.18	6.44	10.9	2.35	0.38	0.22	450	100	100.3

a: n - nephelinite, b - basanite, h - hawaiite, ab - alkali basalt, pt - phonotephrite, bm - benmoreite, tr - transitional basalt, t - tholeiite

b: S and Cl reported in ppm

Glass sulfur (S) content reflects a magma's shallow ascent history. Although there are no clear divisions, most subaerially-erupted Hawaiian basaltic magmas degas to < 300 ppm S in the melt [*Dixon et al.*, 1991]; mid-ocean ridge and Hawaiian magmas that erupt in deep water without previously experiencing low pressures have ~ 1000 ± 300 ppm S in the melt [*Jambon* 1994]; whereas glasses with intermediate S values may have erupted at shallow water depths, may have degassed incompletely as tephra in subaerial eruptions, or may have mixed with low-pressure degassed magma before erupting in deep water [*Dixon et al.*, 1991].

We divide the Hilina glasses on the basis of sulfur content into low and high-S (> 750 ppm S) groups to distinguish low-pressure-degassed from nondegassed (deep submarine-erupted) magmas. Our intent is to identify those compositions that *must* have originated from a deep-water juvenile volcano (ancestral Kīlauea) and to screen out grains from any mature shield volcano that may have degassed incompletely while transiting rift zones before erupting in deep water [*Dixon et al.*, 1991]. Many grains in the low-S group could also originate from ancestral Kīlauea if that volcano reached shallow water depths.

Alkalic compositions are present in both low- and high-S groups, but the two groups differ in that high-SiO_2, alkali-poor tholeiite glasses are abundant in the low-S population, whereas the strongly alkalic hawaiite, basanite, phonotephrite, and nephelinite glasses are almost entirely high S and likely were erupted at substantial water depths. Rare basanite grains in the low-S group (Figure 2) are partly degassed with 400 – 650 ppm S (Table 1, and appendix on CD-ROM, this volume) and may also derive from ancestral Kīlauea. The high-S population can be further subdivided on the basis of K_2O and P_2O_5 contents (Figure 2). Most high-S and low-S glasses have nearly constant K_2O/P_2O_5 of ~1.67 despite ranging by a factor of 8 in K_2O and P_2O_5 concentrations. Constant K_2O/P_2O_5 is widely observed for Hawaiian glasses and fresh lavas and has been interpreted as resulting from magma sources with uniform K_2O/P_2O_5 and melting in which neither P-rich nor K-rich minerals were residual [*Wright*, 1971; *Clague et al.*, 1995; *Frey et al.*, 2000]. A separate population of high-S glasses forms an array displaced to higher K_2O values (> 1.5 wt.%). These high-S, high-K/P glasses are all nephelinite and low-SiO_2 basanite grains; they were found only in sandstone sample S508-R3 and as matrix of debris flow breccias. High-K/P glasses are also distinguished by high Cl (> 500 ppm, Table 1) and by elevated P_2O_5, Na_2O, and CaO values for their MgO contents (Figure 3). Three low-S grains plot near the high-S high K/P trend. Two are evolved, vesicular benmoreites with low P_2O_5 and 400 – 500 ppm S that may have originated as partly degassed post-shield tephra from an older Hawaiian volcano. The third is an alkali basalt glass with no detectable S. Three high-K/P glasses and one high-S, normal K/P glass were analyzed for F to search for evidence of phlogopite or amphibole in the source region. Basanite glass S509-C2-1 has 550 ppm F, evolved nephelinite glasses S509-C2-5 and S509-C2-6 have 1500 and 1900 ppm F, and normal K/P alkali basalt glass S509-C2-2 has 200 ppm F.

Glass from tholeiites erupted along Kīlauea's Puna Ridge [*Garcia et al.*, 1989], as well as MgO-rich picritic and tholeiitic glass grains in sands collected near the ridge define a single trend with K_2O/P_2O_5 of ~1.62 [*Clague et al.*, 1995], indistinguishable from the main Hilina glass population. Some melt inclusions from Puna Ridge phenocrysts and xenoliths have anomalous K_2O/P_2O_5 [*Clague et al.*, 1995], but these deviate from the main Puna Ridge trend to low, not high, K_2O/P_2O_5 (to ~1). Submarine glasses from Lō'ihi Seamount, which include alkali basalts, as well as submarine glasses from the North Arch volcanic field, which include nephelinites and basanites, also do not match the Hilina high-K/P group (Figure 2).

Except for the high-K/P group, high- and low-S glasses define trends on MgO variation diagrams (Figure 3) that are familiar from other Hawaiian volcanoes [*Basaltic Volcanism Study Project*, 1981] and are mostly explained by fractional crystallization. Glass CaO values peak at 6 – 7 wt.% MgO, which marks the onset of abundant clinopyroxene crystallization. For most of the low-S glasses, Al_2O_3 values also peak at 6 – 7 wt.% MgO, marking the onset of abundant plagioclase crystallization. The high-S suite contains greater numbers of alkalic compositions and shows no well-defined Al_2O_3 maximum. In the low-S suite FeO* values rise continuously from 10 – 11 wt.% at 7 wt.% MgO to near 16 wt.% at 4.5 – 5 wt.% MgO. The generally more alkalic high-S suite fails to exceed 13.5 wt.% FeO*. The range in FeO* values at a given MgO content in both low and high-S groups is similar to that determined by *Clague et al.* [1995] for high-MgO submarine tholeiite glasses from Kīlauea's Puna Ridge. TiO_2 values peak at 5 – 5.5 wt.% at MgO of 4.5 – 5 wt.%, marking the onset of abundant Fe-Ti oxide crystallization.

Outliers on the MgO variation diagrams are listed separately in Table 1. The lowest MgO glasses are the benmoreite tephra grains noted previously. Their low P_2O_5 and TiO_2 contents signify extensive apatite and Fe-Ti oxide crystallization. Another outlier is an Fe-Ti-rich alkali basalt glass with 650 ppm S; its S content approaches our arbitrary division of 750 ppm, and this grain could belong to ancestral Kīlauea. A second alkali basalt outlier with high K_2O for its MgO value is the degassed, high-K/P grain noted previously. The remaining outlier is a strongly undersaturated glass distinguished by low SiO_2 and Na_2O, moderate S (500 ppm), and high CaO, MgO, and CaO/Al_2O_3. The glass differs in most respects from the submarine-erupted nephelinites and basanites, but it was found in sandstone sample S508-R3 that contains common high-S nephelinites. The low analytical total for the unusual grain may indicate that it is an altered nephelinite.

Some Hawaiian shield volcanoes can be distinguished by the compositions of their basaltic rocks and glasses. Magmas from Mauna Loa are generally richer in SiO_2 and poorer in alkalis and TiO_2 than those of Kīlauea, Mauna Kea, or Lō'ihi, [*Wright*, 1971; *Garcia et al.*, 1989, 1993, 1995a, 1995b; *Clague et al.*, 1995], although rare Mauna Loa-like magmas erupt from Kīlauea [*Rhodes et al.*, 1989] and Kīlauea-like magmas from Mauna Loa [*Garcia et al.*, 1995a]. Mauna Loa-like alkali-poor and relatively SiO_2-rich tholeiite glasses are abundant in the low-S suite, and probably originated from subaerial Mauna Loa lavas that entered the sea. Only six high-S grains, out of nearly 800 analyzed from the Hilina volcaniclastic rocks, have low alkali and TiO_2 contents like magmas erupted at Mauna Loa (Figure 2).

Most high-S glass grains have lower SiO_2 than is common even for Kīlauea liquids (Figure 2a). For convenience of discussion, glasses that are close to the tholeiitic – alkalic boundary and that have SiO_2 < 51 wt.% and Na_2O+K_2O > 2.5 wt.% will be referred to as transitional. This is in keeping with *Wolfe's* [1997] study of the geology and petrology of Mauna Kea, where "transitional" is used for basalts that plot near the alkalic – tholeiitic boundary, or between this and the fields defined by the dominant shield-stage tholeiites. The majority of high-S grains from the Hilina volcaniclastic section that fall on the tholeiitic side of the tholeiitic – alkalic boundary are this relatively SiO_2-poor transitional type. Most of these high-S transitional glasses have alkali contents equal to shield-stage Kīlauea glasses, yet are poorer in SiO_2 (Figure 2a), so they are not Kīlauea-type tholeiite liquids that have simply undergone less olivine fractionation. Glass in rocks named by other workers as tholeiite from Lō'ihi Seamount [*Moore et al.*, 1982; *Garcia et al.*, 1993, 1995b; *Clague et al.*, 2000] share these relatively low-SiO_2, high-alkali features and would be classified as transitional following our guidelines.

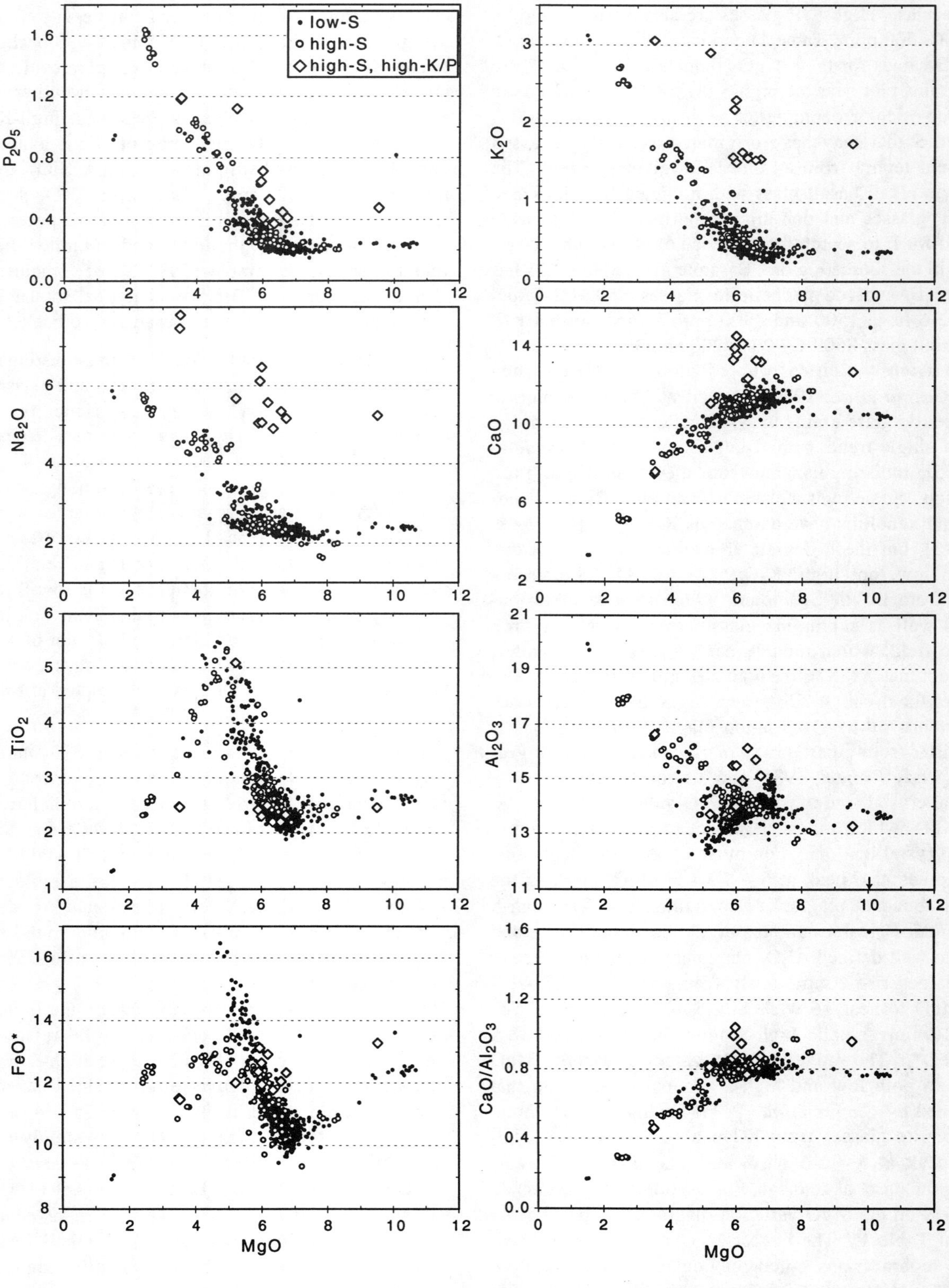

Figure 3. MgO variation diagrams for glasses from Hilina bench volcaniclastic rocks.

Hilina Bench Clasts – Descriptions and Major Element Compositions

As with glasses, Hilina bench debris-flow clasts span the alkalic – tholeiitic boundary and extend far into the alkalic field, including alkali basalts, hawaiites, basanites, and phlogopite nephelinites (Figure 4). Many clasts lack glass margins; of those with glass, all but one are high-S (Table 1) and were erupted under deep water conditions.

Alkali basalts are nearly aphyric to conspicuously olivine-phyric rocks with phenocrysts of olivine with Cr-spinel inclusions and sparse microphenocrysts of clinopyroxene. One alkali basalt (S509-R2B) that plots close to the basanite – alkali basalt boundary (Figure 4) closely resembles basanite samples with sparse olivine and abundant deep brown acicular clinopyroxene. Basanites and hawaiites are nearly aphyric rocks with microphenocrysts of olivine, plagioclase, Fe-Ti oxides, and brown clinopyroxene. The most evolved (low-MgO) hawaiite (S509-R1A, Table 2), is a densely crystalline flow interior cut by granular shear bands containing clay minerals; this rock may not closely preserve its magmatic composition. Three samples of fine-grained vuggy hawaiitic gabbro consist of acicular brown clinopyroxene, Fe-Ti oxides, and plagioclase, with both angular and spherical open vugs. Hawaiitic gabbros lack cumulate textures and probably solidified in shallow intrusions, perhaps as dikes or sills. Transitional basalts range from nearly aphyric, with microphenocrysts of clinopyroxene and plagioclase and lesser olivine, to conspicuously olivine-porphyritic. Two Mauna Loa-like clasts (S509-R9A, S509-R9B), distinguished by low alkali and TiO_2 contents for their SiO_2 and MgO values, are olivine porphyritic tholeiites.

Three distinctive phlogopite-bearing clasts have extreme low-SiO_2 contents (37 – 40 wt.%, Table 2) and merit special description. These clasts, S508-R2A, S508-R2B, and S508-R2C were collected from the same debris-flow breccia exposure. Though texturally variable, each consists of randomly oriented medium-grained acicular brown titaniferous diopside-salite, evenly distributed Fe-Ti oxides, and poikilitic titaniferous phlogopite (<0.4 wt.% F) set in an interstitial matrix of nepheline and its hydration products analcite and gonnardite (a zeolite found elsewhere as a reaction product after nepheline [*Mason*, 1957]). Accessory minerals include, in decreasing abundance, apatite, titanite, Fe-sulfide, and perovskite. Coarse grain sizes, compared to lava samples, and the poikilitic habit of the phlogopite grains indicate that these are intrusive rocks. Abundant open diktytaxitic vugs in all three samples, and round vesicles in one, suggest that the magmas solidified at shallow depths below the seafloor. The rocks lack cumulate textures and will be referred to as phlogopite nephelinites with the understanding that they are shallow intrusive rocks, not lavas.

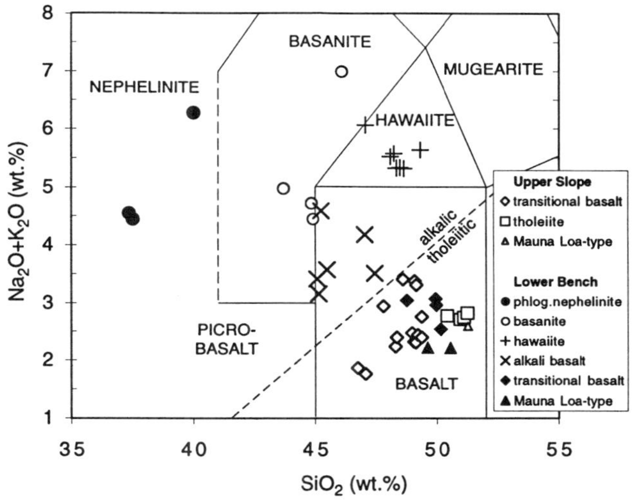

Figure 4. Alkali – silica diagram for clasts from the Hilina bench and pillow lavas and fragments from above and east of the bench. Classification fields from *LeMaitre* [1991]; alkalic – tholeiitic boundary from *Macdonald and Katsura* [1964].

Glasses and Whole Rocks Upslope and Northeast of the Hilina Bench – Descriptions and Major Element Compositions

Most samples collected upslope and northeast of the Hilina mid-slope bench are pieces of intact pillow lavas or are pillow fragments. Most samples were submarine-erupted (high-S) and have adhering glass with SiO_2 < 51 wt.% and Na_2O+K_2O > 2.5 wt.% (Table 1). In keeping with glass nomenclature, such samples are grouped as transitional basalts. No recovered samples are alkalic (Table 2). Phenocryst abundances range widely, from nearly aphyric rocks with sparse phenocrysts of plagioclase and olivine, to olivine-rich picrites. Complexly zoned equant clinopyroxene phenocrysts, to 2.5 mm diameter, accompany olivine in all but the nearly aphyric rocks; plagioclase phenocrysts are present but scarce. Many olivine phenocrysts are strained with subgrain boundaries, and mono- and poly-mineralic glomeroporphyritic clots are widespread, together suggesting that some phenocrysts were entrained and did not grow from their host magmas, as has been inferred for other Hawaiian basalts [*Helz*, 1987; *Clague et al.*, 1995; *Garcia*, 1996].

Table 2. Major oxide analyses of Hilina region whole rocks, normalized to 100% with all Fe as FeO

sample	type[a]	SiO_2	TiO_2	Al_2O_3	FeO*	MnO	MgO	CaO	Na_2O	K_2O	P_2O_5	total[b]
upslope and northeast of Hilina bench												
K95-R6	tr	49.0	2.44	12.3	11.1	0.17	10.7	11.6	1.97	0.51	0.24	99.3
K95-R5	tr	47.8	2.97	12.9	11.8	0.17	9.87	11.3	2.34	0.60	0.31	99.3
K95-R4	tr	49.4	2.78	14.1	11.7	0.17	7.47	11.4	2.20	0.56	0.29	98.9
K95-R2	tr	48.6	3.47	14.0	12.3	0.17	6.31	11.4	2.65	0.76	0.36	99.4
S504-R4	tr	49.1	3.17	14.1	12.2	0.17	6.19	11.3	2.57	0.80	0.39	98.5
S504-R3B	tr	49.1	2.17	11.6	11.0	0.17	11.9	11.5	1.87	0.45	0.26	99.4
S504-R3A	tr	49.4	2.20	12.2	10.8	0.16	10.9	11.7	1.88	0.51	0.26	99.4
S504-R2B	tr	49.1	2.24	12.4	10.8	0.16	11.3	11.4	1.87	0.44	0.27	99.1
S504-R2A	tr	49.2	2.25	12.6	10.6	0.17	10.8	11.7	1.96	0.47	0.28	98.1
S504-R1	tr	49.1	3.13	14.2	12.2	0.17	6.24	11.2	2.56	0.74	0.41	99.6
S506-R8A	tr	46.8	1.69	8.92	11.5	0.16	19.8	9.13	1.52	0.34	0.23	98.4
S506-R8C	tr	48.3	2.11	11.0	11.1	0.17	13.8	11.2	1.85	0.38	0.25	97.9
S506-R6A	tr	47.1	1.73	7.89	11.7	0.18	21.5	7.96	1.44	0.33	0.22	96.6
S506-R2A	tr	48.4	2.24	11.6	11.3	0.17	12.8	10.9	1.97	0.43	0.27	98.9
S504-R5	t low-S	50.4	2.57	12.7	11.6	0.18	9.14	10.3	2.29	0.48	0.32	100.5
S506-R4A	t low-S	51.0	2.32	13.6	11.1	0.17	8.28	10.6	2.34	0.37	0.29	100.1
S506-R5B	t	51.1	2.35	13.6	11.1	0.17	7.89	10.7	2.37	0.37	0.29	100.2
S506-R5A	t low-S	51.3	2.36	13.7	11.1	0.17	7.53	10.8	2.43	0.39	0.29	100.1
K95-R7	mlt	51.3	2.09	13.3	11.1	0.17	9.03	10.3	2.23	0.36	0.23	99.9
Hilina bench debris flow clasts												
S508-R2A	pn	37.3	6.42	11.0	17.9	0.23	9.40	12.2	2.52	2.03	0.92	95.6
S508-R2B	pn	40.0	5.65	13.0	14.3	0.21	7.18	12.4	4.33	1.94	1.04	95.5
S508-R2C	pn	37.5	6.35	10.5	18.1	0.23	9.95	12.1	2.69	1.75	0.84	96.7
K98-R1	b	46.1	4.24	15.9	12.3	0.18	4.36	8.97	5.25	1.75	0.99	97.9
S509-R2A	b	44.9	4.51	13.6	13.8	0.16	7.54	10.5	3.39	1.05	0.59	97.7
S509-R4B	b	43.7	5.43	15.1	15.5	0.18	6.40	7.97	3.69	1.29	0.77	95.7
R509-R7(1)	b	44.8	4.24	13.5	13.5	0.17	8.41	10.0	2.78	1.94	0.65	96.4
S509-R1A	h	47.1	4.17	16.5	14.5	0.22	1.91	7.32	3.96	2.10	2.24	91.3
S505-R1B	h	49.3	3.53	16.4	11.8	0.17	4.47	7.82	4.56	1.07	0.84	96.4
S505-R4	h	48.2	3.61	16.4	12.9	0.20	4.91	7.45	4.51	1.07	0.72	95.9
S505-R5A	h	48.1	3.57	16.4	13.0	0.20	4.93	7.55	4.37	1.15	0.74	96.0
S505-R1A	hg	48.6	3.68	16.3	13.1	0.20	5.11	6.98	4.38	0.94	0.68	96.0
S505-R2A	hg	48.5	3.70	16.3	13.2	0.20	5.08	7.04	4.38	0.95	0.68	98.6
S505-R3	hg	48.3	3.76	16.3	13.3	0.21	4.98	7.14	4.29	1.04	0.67	98.9
S505-R2B	ab	45.1	3.15	11.4	13.7	0.19	13.5	9.23	2.33	1.08	0.36	96.2
S505-R10A	ab	45.1	3.12	11.2	13.4	0.17	14.0	9.40	2.45	0.70	0.37	99.4
S505-R10B	ab	45.5	3.35	12.1	13.3	0.20	11.8	9.84	2.29	1.28	0.39	96.4
S509-R1B1	ab	47.0	4.37	15.1	12.7	0.14	4.66	11.3	3.34	0.84	0.54	98.7
S509-R2B	ab	45.3	4.43	13.8	13.1	0.15	7.73	10.3	3.50	1.08	0.61	98.3
S509-R6B	ab	47.4	3.48	13.8	12.3	0.16	7.79	11.0	2.79	0.71	0.49	99.8
K98-R6	tr	49.9	2.66	14.2	11.4	0.17	6.96	11.3	2.56	0.51	0.34	99.3
S509-R4A	tr	48.8	2.83	14.0	12.3	0.18	6.86	11.7	2.52	0.52	0.38	99.0
S509-R5	tr	50.0	2.83	14.3	11.4	0.17	6.65	11.3	2.45	0.51	0.37	99.1
S509-R6C	tr	50.2	2.41	12.5	11.4	0.19	8.50	12.0	2.07	0.47	0.28	98.5
S505-R9B	mlt	50.6	1.91	12.2	11.2	0.17	12.0	9.52	1.90	0.31	0.23	100.6
S505-R9A	mlt	49.6	1.76	11.2	11.3	0.17	14.6	8.89	1.95	0.27	0.25	99.0

a: tr - transitional basalt, t - tholeiite, mlt - Mauna Loa-type tholeiite, pn - phlogopite nephelinite,
b - basanite, h - hawaiite hg - hawaiitic gabbro, ab - alkali basalt

b: total gives original value

Besides the submarine-erupted transitional basalts, a single block of Mauna Loa-like alkali- and TiO_2-poor tholeiitic basalt (K95-R7) was collected upslope from the mid-slope bench. The sample is moderately olivine porphyritic (< 10 % phenocrysts) and lacks a glass rind, precluding assessment of subaerial versus submarine eruption. In addition, one pillow fragment from above the mid-slope bench (S504-R5) and three pillow samples from the south base of the Puna Ridge well east of the bench (S506-R4A, -R5A, -R5B) are classified as tholeiite, as opposed to transitional basalt, due to glass SiO_2 ~52 wt.%. The tholeiites have abundant phenocrysts of olivine, plagioclase, and clinopyroxene. Their glasses are low-S, and the samples may be shield-stage Kīlauea magmas that degassed in a low-pressure environment during subaerial eruption or during transit downrift, before eventual deposition (or eruption) in deep water.

Although rock samples generally appear fresh with palagonite-poor nonhydrated glass rinds, clay minerals partly infill vesicles in some samples and are present locally in groundmass. Major element analytical totals are also commonly low (Table 2), reflecting addition of clay minerals or hydration of evolved groundmass glass. For many samples, whole-rock K_2O values are similar to or slightly exceed those of rind glasses (Tables 1 and 2). Abundant olivine and clinopyroxene phenocrysts should lead to lower whole-rock K_2O than for glass, and the disparity in K_2O values may also result from secondary clays.

Whole-Rock Trace Element Results

Rare earth element (REE) abundance patterns correspond with rock type (Figure 5). Chondrite-normalized light REE abundance patterns $(La/Sm)_{cn}$ steepen in the series: Mauna Loa-type (1.1), tholeiite (1.2 – 1.4), transitional basalt (1.3 – 1.6), alkali basalt (1.6 – 1.9). Normalized middle–to-heavy REE patterns $(Sm/Yb)_{cn}$ also steepen in the same sequence: Mauna Loa-type (2.4 – 2.6), tholeiite (2.6 – 3.0), transitional basalt (3.0 – 3.6), alkali basalt (3.7 – 4.2). Transitional basalts upslope and northeast of the mid-slope bench cannot be distinguished from those of the bench based on REE (or other major or trace element) abundances or patterns. Hawaiites, basanites, and phlogopite nephelinites have substantially higher light-REE contents than other rocks (Figure 5). Six of seven hawaiite and hawaiitic gabbro samples have similar REE abundances, with normalized light REE slopes ($(La/Sm)_{cn}$: 1.5 – 1.6) at the upper end of transitional basalt values and with normalized middle-to-heavy REE slopes ($(Sm/Yb)_{cn}$: 3.1 – 3.3) in the midst of the transitional basalt group. The remaining hawaiite, altered sample S509-R1A, has low MgO and high REE and Y abundances, a steeply sloping light REE pattern ($(La/Sm)_{cn}$: 2.1), and an enormous negative Ce anomaly, recalling altered subaerial basalts from Oahu that have gained REE (except Ce) and Y [*Frey et al.*, 1994]. Henceforth, discussions of hawaiites refer collectively to hawaiites and hawaiitic gabbros and exclude the obviously altered sample. Basanites and the basanite-like alkali basalt (S509-R2B) are distinct with high REE concentrations, light REE relative abundances similar to alkali basalts ($(La/Sm)_{cn}$: 1.7 – 1.9), but with much more steeply sloping middle-to-heavy REE patterns ($(Sm/Yb)_{cn}$: 4.2 – 6.1). Phlogopite nephelinites have the most strongly fractionated REE patterns ($(La/Sm)_{cn}$: 2.2 – 2.6, $(Sm/Yb)_{cn}$: 9.5 – 9.9).

Concentrations of incompatible trace elements Zr and Sr increase with P_2O_5 and form simple positively sloping arrays (Figure 6). Phlogopite nephelinites have the highest incompatible element concentrations, followed by basanites and hawaiites, then by alkali basalts, by transitional basalts and tholeiites, and with picritic transitional basalts and Mauna Loa-type basalts having the lowest concentrations. For Ba and Nb the variations with P_2O_5 are similar but more complex, with phlogopite nephelinites and three alkali basalts (S505-R2B, S505-R10A, S505-R10B) deviating to higher Ba and Nb, and Mauna Loa-type basalts deviating slightly to lower Ba and Nb for their P_2O_5 values. Consistent differences in trace element ratios have been used to fingerprint tholeiites from certain Hawaiian shield volcanoes [*Frey and Rhodes*, 1993]. Low-S tholeiites upslope and northeast of the mid-slope bench have Kīlauea-like Sr/Nb (21 – 25), confirming that they are shield-stage Kīlauea lavas, whereas samples classified as Mauna Loa-type based on major element composition have higher Mauna Loa-like Sr/Nb (29 – 33). Hilina transitional basalt compositions overlap Kīlauea tholeiites with Sr/Nb (19 – 25). Abundances and variations of incompatible trace elements are discussed further in following sections.

Unlike other incompatible trace elements, Rb and Cs do not form simple arrays with P_2O_5 and instead form scattered positively sloping fields (not illustrated). Rb and Cs concentrations have probably been modified by hydration and by addition of clay, as was inferred for K_2O, and are not utilized for petrogenetic interpretations. Three transitional basalt pillow lavas (K95-R5, K95-R2, S504-R3B) collected upslope and immediately northeast of the mid-slope bench have anomalously elevated U relative to Th; these samples have among the highest Cs and K/Rb values (not illustrated), and U may have been added to the samples in vesicle-filling secondary minerals.

The elements Ni, Cr, and Co are highly compatible in olivine and Cr-spinel and form simple arrays of monotoni-

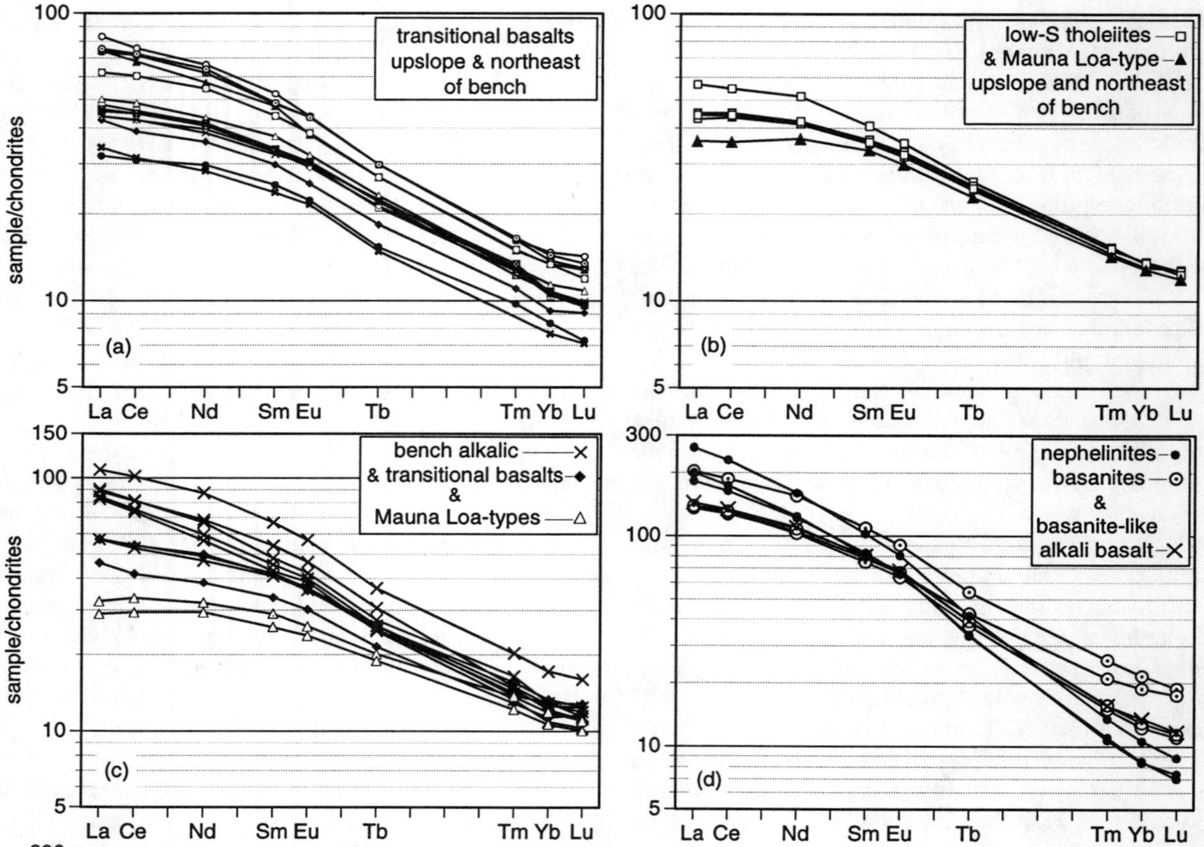

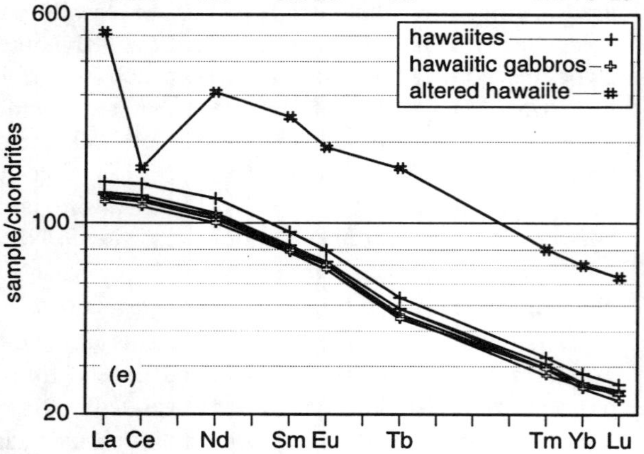

Figure 5. Chondrite-normalized REE for upper and east slope pillow lavas and fragments (a, b), and Hilina bench clasts (c, d, e). Chondritic normalizing values from *Boynton* [1984].

cally decreasing concentrations with decreasing whole-rock MgO, as would be anticipated from the common olivine phenocrysts with Cr-spinel inclusions in the alkali basalt – transitional basalt – tholeiite suite. The Ni array reaches nearly 850 ppm at 21.5 wt.% MgO and declines to < 20 ppm at 5 wt.% MgO. The Co array reaches nearly 90 ppm at 21.5 wt.% MgO and decreases to 40 – 55 ppm at ~ 6.5 wt.% MgO; hawaiites then continue to 29 ppm at 4.25 wt.% MgO. The Cr – MgO array is more scattered, but reaches ~ 1500 ppm Cr at 21.5 wt.% MgO and < 100 ppm at ~ 6 wt.% MgO. Hawaiites all have < 10 ppm Cr, basanites have 175 – < 10 ppm Cr, and phlogopite nephelinites have 75 – < 10 ppm Cr.

Vanadium is compatible in Fe-Ti oxides but not in olivine; its concentration in the basaltic rocks generally increases with decreasing MgO and olivine content, from 250 ppm V at 21.5 wt.% MgO to 500 ppm at 4.7 wt.% MgO. Hawaiites form a separate group at low MgO and low V (214 – 344 ppm). In contrast to the basaltic samples, V and MgO are positively correlated in the basanites and phlogopite nephelinites. Phlogopite nephelinites reach 700 ppm V at 10 wt.% MgO; V concentration drops to a low of 250 ppm at 4.7 wt.% MgO in basanite. Scandium is compatible in high-Ca pyroxene and in garnet, but it is incompatible in olivine. In the basaltic samples Sc concen-

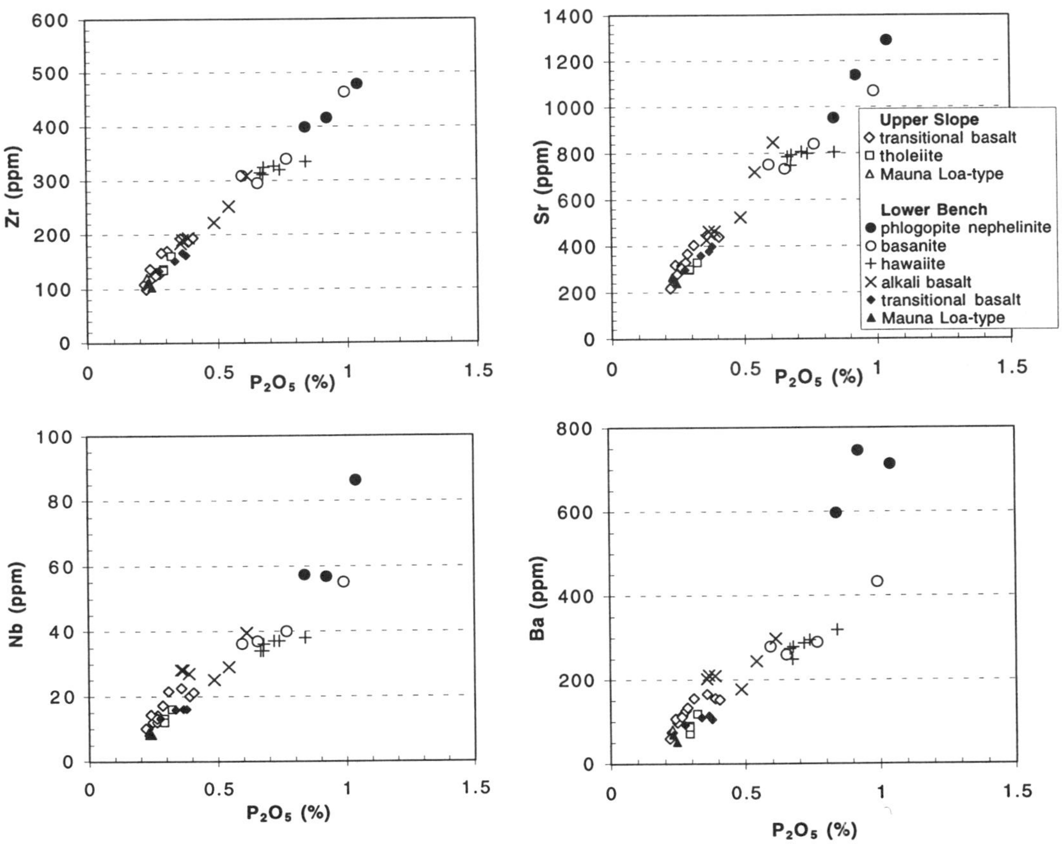

Figure 6. P_2O_5 - incompatible trace element variation diagrams for Hilina region samples. Altered hawaiite (S509-R1A, not illustrated) plots far off trend at high P_2O_5 for Nb, Zr, Sr, and Ba.

trations increase slightly as MgO and olivine content decline; the alkali basalts have slightly lower concentrations (24 – 30 ppm) than the transitional, tholeiitic, and Mauna Loa-type basalts (25 – 34 ppm). Scandium concentrations are lower in the basanites and phlogopite nephelinites (15 – 25 ppm) but are not correlated with MgO or rock-type. Hawaiites have a low and narrow range in Sc content (17 – 19 ppm).

INTERPRETATION AND DISCUSSION

Source Characteristics for Hilina – Kīlauea Magmas

New ^{40}Ar-^{39}Ar age measurements linking the alkalic rocks to Kīlauea [*Lipman et al.*, this volume], as well as the stratigraphic succession of dominant magma types, from alkalic to transitional to tholeiitic, suggest the progressive development of a single magmatic system. Alkalic and transitional basalt, hawaiite, and basanite clasts from the Hilina bench and transitional basalt pillow lavas from upslope and northeast of the mid-slope bench define common trends on plots of weathering- and alteration-resistant highly incompatible elements such as La versus Ta, and La versus Th (Figure 7). These trends project to near the origin, consistent with variable degrees of melting of a source with uniform La, Ta, and Th concentrations. Shield-stage low-S tholeiites from upslope and northeast of the mid-slope bench and Kīlauea reference sample BHVO-1 also fall on these trends, suggesting that magmas of the Hilina – Kīlauea region derived from sources with uniform La/Ta (12 – 12.5) and La/Th (12.5 – 13). Mauna Loa-like clasts plot at higher La/Ta (~14) and La/Th (20 – 22). *Garcia et al.* [1995b] also used uniform ratios of alteration- and weathering-resistant highly incompatible elements to argue that alkalic, transitional, and tholeiitic magmas of Lō'ihi Seamount were derived from melting of a nearly uniform source region (Lō'ihi: La/Ta ~15, La/Th ~14), although isotopic diversity in Lō'ihi magmas requires the Lō'ihi

Table 3. Whole rock trace element analyses (ppm) of submarine Hilina region rocks

		XRF										INAA																
sample	type	V	Cr	Ni	Cu	Zn	Rb	Sr	Y	Zr	Nb	Cs	Ba	Th	U	La	Ce	Nd	Sm	Eu	Tb	Tm	Yb	Lu	Hf	Ta	Sc	Co
basalts upslope and northeast of Hilina bench																												
K95-R6	tr	331	772	265	117	98	8	317	24	137	14	0.08	106	0.61	0.3	11.8	29.4	19.6	5.50	1.80	0.84	0.32	1.85	0.26	3.50	0.93	32.8	56.6
K95-R5	tr	382	589	283	121	112	10	404	28	170	22	0.15	154	1.07	1.6	17.4	41.0	26.0	7.00	2.12	0.98	0.36	2.18	0.31	4.20	1.43	30.9	58.5
K95-R4	tr	364	260	129	110	112	11	364	30	166	17	0.14	132	1.08	0.3	14.6	36.6	24.8	6.47	2.15	0.98	0.36	2.17	0.29	4.29	1.16	30.9	50.6
K95-R2	tr	431	92	82	98	120	12	445	31	193	22	0.14	164	1.42	2.7	19.5	45.7	30.0	7.73	2.46	1.08	0.39	2.39	0.35	4.93	1.56	30.7	49.2
S504-R4	tr	357	72	95	150	111	15	448	30	187	20	0.15	154	1.32	0.6	17.3	43.1	28.1	7.12	2.44	1.08	0.39	2.24	0.31	4.97	1.44	31.7	52.4
S504-R3B	tr	289	687	297	114	97	7	299	22	124	12	0.12	111	0.70	1.2	10.3	25.7	17.9	4.86	1.63	0.77	0.32	1.70	0.23	3.26	0.82	34.1	63.7
S504-R3A	tr	288	700	284	123	90	9	304	23	124	13	0.12	111	0.64	0.3	10.3	25.8	17.4	4.75	1.67	0.82	0.31	1.75	0.24	3.18	0.81	33.7	59.6
S504-R2B	tr	291	750	292	102	88	7	315	22	128	13	0.10	112	0.74	0.3	10.9	27.1	18.4	4.93	1.71	0.81	0.33	1.73	0.24	3.39	0.87	32.6	58.8
S504-R2A	tr	293	628	271	117	91	8	329	21	127	13	0.12	125	0.76	0.9	10.7	27.1	18.8	4.98	1.70	0.80	0.30	1.75	0.23	3.34	0.86	32.9	57.1
S504-R1	tr	395	71	86	143	117	12	437	30	193	21	0.12	152	1.34	0.5	17.7	43.6	28.9	7.18	2.43	1.08	0.40	2.34	0.33	4.92	1.45	30.4	50.8
S506-R8A	tr	247	1470	815	122	99	6	225	17	98	10	0.08	74	0.51	0.2	8.0	19.0	12.8	3.50	1.21	0.54		1.25	0.17	2.31	0.61	26.2	80.5
S506-R8C	tr	295	1340	477	136	98	7	278	21	119	12	0.08	98	0.70	0.2	10.0	23.5	16.2	4.38	1.43	0.67	0.27	1.50	0.22	2.89	0.74	31.0	63.8
S506-R6A	tr	259	1250	841	87	100	7	217	19	108	10	0.05	60	0.52	0.2	7.5	18.6	13.4	3.72	1.25	0.56	0.24	1.36	0.18	2.51	0.55	24.9	87.3
S506-R2A	tr	315	874	429	119	96	9	298	23	129	14	0.12	115	0.75	0.4	11.3	27.6	18.8	4.91	1.66	0.76	0.31	1.69	0.24	3.33	0.89	31.3	67.2
S504-R5	t	351	442	207	117	109	9	328	27	160	16	0.11	118	0.93	0.4	13.3	33.0	23.3	5.99	1.99	0.95	0.37	2.19	0.30	4.20	1.02	30.9	54.0
S506-R4A	t	380	381	203	136	106	7	297	28	136	13	0.11	71	0.61	0.4	10.1	26.3	18.7	5.19	1.78	0.89	0.35	2.12	0.31	3.47	0.77	31.7	48.9
S506-R5B	t	376	372	176	130	103	6	295	26	135	13	0.16	89	0.68	0.2	10.6	27.2	19.1	5.35	1.85	0.91	0.36	2.22	0.31	3.62	0.83	32.8	48.9
S506-R5A	t	371	349	151	138	104	6	299	26	134	12	0.07	87	0.62	0.2	10.4	26.7	19.1	5.26	1.81	0.90	0.37	2.20	0.30	3.54	0.80	32.4	46.9
K95-R7	mlt	335	393	166	93	100	5	264	25	121	9	0.10	82	0.41	0.1	8.5	21.6	16.6	4.92	1.67	0.84	0.35	2.09	0.29	3.13	0.60	30.2	50.7
Hilina bench debris flow clasts																												
S508-R2A	pn	640	<3	104	92	156	58	1140	24	416	57	0.54	746	3.19	0.9	46.9	105	56.3	11.9	3.74	1.21	0.27	1.38	0.17	9.75	4.35	17.1	64.0
S508-R2C	pn	702	75	144	106	167	47	953	25	399	57	0.47	597	2.59	0.8	42.9	98.9	55.2	12.2	3.73	1.23	0.26	1.36	0.18	9.96	4.32	22.3	69.1
S508-R2B	pn	478	<3	62	84	153	53	1290	29	479	86	0.48	713	4.43	1.3	61.7	138	72.5	15.1	4.51	1.51	0.33	1.72	0.21	11.3	6.62	14.7	48.6
K98-R1	b	250	<3	6	19	143	30	1070	53	463	55	0.31	432	3.81	1.7	47.8	113	70.4	16.1	5.07	1.96	0.62	3.50	0.46	10.5	3.79	16.0	31.5
S509-R2A	b	409	127	166	62	137	17	752	32	309	36	0.09	280	2.48	1.2	32.7	79.3	48.9	11.9	3.76	1.42	0.38	2.10	0.28	7.92	2.60	25.0	61.3
S509-R4B	b	431	8	55	30	138	16	840	42	340	40	0.22	289	2.41	0.8	31.9	76.8	47.6	11.5	3.71	1.55	0.51	3.05	0.42	8.16	2.61	17.3	55.0
S509-R7(1)	b	417	175	181	48	148	21	734	31	295	37	0.19	261	2.45	0.9	32.3	78.4	46.5	11.0	3.55	1.35	0.36	1.99	0.27	7.32	2.50	23.2	56.3
S509-R1A	h	344	8	16	14	186	34	387	171	313	43	0.81	212	3.54	1.1	121	96.8	139	36.5	10.7	5.80	1.94	11.4	1.54	6.73	2.79	22.9	15.7
S505-R1B	h	220	<3	7	14	143	11	803	55	334	38	0.08	318	2.44	0.8	33.4	84.5	55.7	13.6	4.46	1.94	0.78	4.57	0.63	8.50	2.73	19.3	28.6
S505-R4	h	224	<3	9	16	136	14	806	55	326	37	0.06	287	2.22	0.8	29.9	74.1	47.8	11.8	4.05	1.76	0.71	4.20	0.57	7.88	2.49	18.3	32.5
S505-R5A	h	230	8	8	14	140	15	797	52	320	37	0.21	294	2.29	1.5	30.5	76.3	49.2	12.2	4.04	1.76	0.74	4.25	0.59	8.09	2.58	18.4	33.0
S505-R1A	hg	214	8	14	19	113	11	746	51	310	34	0.12	248	2.14	0.7	28.1	69.3	45.2	11.5	3.79	1.62	0.71	4.05	0.55	7.41	2.39	17.4	33.1
S505-R2A	hg	220	8	4	17	115	14	792	54	324	36	0.10	278	2.18	0.8	29.0	72.2	46.6	11.7	3.93	1.65	0.67	4.13	0.58	7.66	2.44	18.3	34.6
S505-R3	hg	222	<3	8	13	121	15	784	52	314	34	0.08	273	2.23	0.8	29.2	72.5	48.3	11.9	3.94	1.68	0.71	4.19	0.58	7.73	2.50	18.7	35.3
S505-R2B	ab	378	784	451	103	129	16	424	29	183	28	0.15	201	1.33	0.4	19.4	44.4	25.6	6.29	2.17	0.90	0.32	1.77	0.25	4.35	1.62	24.3	75.9
S505-R10A	ab	411	706	437	95	134	11	462	28	190	28	0.14	211	1.19	0.4	19.9	45.5	27.0	6.66	2.24	0.94	0.36	1.98	0.27	4.61	1.65	25.7	79.9
S505-R10B	ab	402	678	363	84	129	15	462	28	193	27	0.19	209	1.44	0.5	21.2	49.3	30.2	7.07	2.35	0.99	0.40	2.10	0.28	4.85	1.83	25.8	69.4
S509-R1B1	ab	499	75	45	112	135	11	718	36	252	29	0.15	244	1.88	0.9	25.3	61.4	39.6	9.82	3.19	1.33	0.49	2.80	0.39	6.35	1.97	30.4	47.2
S509-R2B	ab	389	149	130	73	150	17	845	34	308	39	0.18	297	2.55	0.8	33.9	81.1	49.9	11.8	3.82	1.43	0.38	2.19	0.29	7.95	2.68	24.6	58.5
S509-R6B	ab	392	251	158	92	127	13	522	30	222	25	0.09	177	1.55	0.7	20.8	49.1	31.0	7.94	2.60	1.11	0.36	2.07	0.29	5.44	1.59	28.3	51.3
K98-R6	tr	380	140	81	126	110	8	359	27	151	16	0.15	109	0.88	0.3	13.5	32.6	22.6	6.10	2.00	0.94	0.38	2.17	0.30	3.83	1.05	30.7	44.6
S509-R4A	tr	392	52	77	117	116	9	397	30	161	16	0.10	105	0.93	0.3	13.4	31.9	21.3	6.01	2.02	0.91	0.34	2.06	0.30	3.98	1.00	31.6	48.2
S509-R5	tr	353	87	74	103	109	9	375	27	166	16	0.07	113	1.05	0.4	13.4	32.8	22.1	6.16	2.06	0.94	0.34	2.10	0.31	4.10	1.03	31.1	42.8
S509-R6C	tr	336	302	136	109	98	7	296	25	133	13	0.09	92	0.71	0.3	10.8	25.1	17.4	4.95	1.69	0.78	0.31	1.84	0.28	3.25	0.77	34.4	49.3
S505-R9B	mlt	339	671	413	130	103	4	261	22	110	8	0.10	68	0.50	0.2	7.6	20.2	14.5	4.27	1.45	0.73	0.33	1.94	0.27	2.89	0.51	29.0	62.3
S505-R9A	mlt	300	791	617	96	103	4	237	21	102	8	0.10	50	0.32	0.2	6.8	17.7	13.3	3.78	1.33	0.69	0.30	1.73	0.24	2.57	0.45	27.2	69.2

type: tr - transitional basalt, t - tholeiite, mlt - Mauna Loa-type tholeiite, pn - phlogopite nephelinite, b - basanite, h - hawaiite, hg - hawaiitic gabbro, ab - alkali basalt

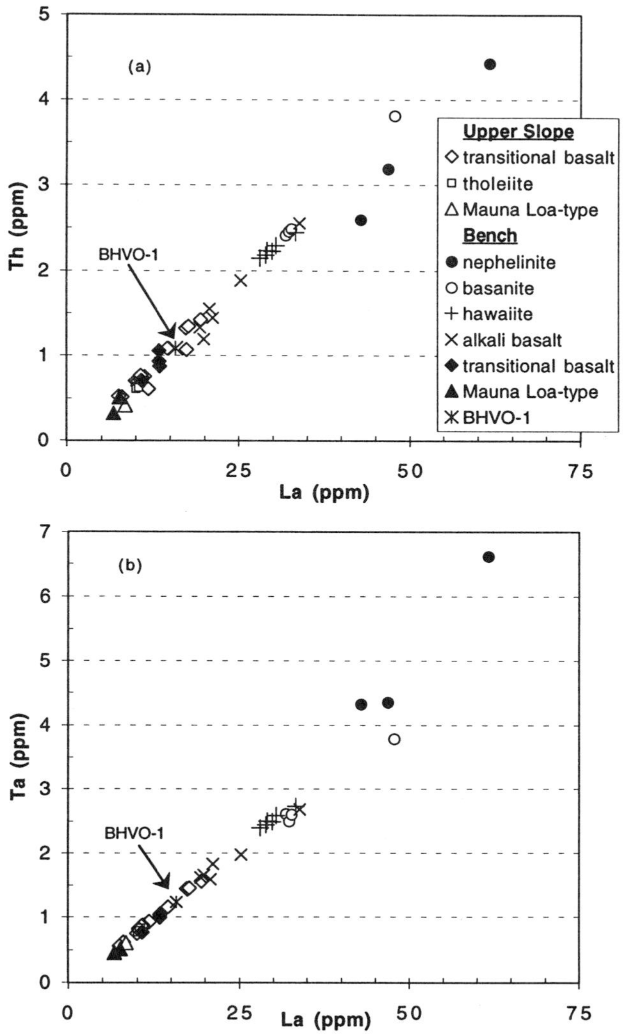

Figure 7. Representative variation diagrams of highly incompatible elements for Hilina bench clasts, and upper and east slope pillow lavas and fragments: (a) La vs. Th, (b) La vs. Ta. Altered hawaiite S509-R1A (not plotted) lies far off these trends at high La for its Th and Ta.

source to be heterogeneous in respects that do not correlate simply with magma composition [*Staudigel et al.*, 1984; *Clague and Frey*, 1984; *Garcia et al.*, 1993; *Garcia et al.*, 1995b].

Hilina phlogopite nephelinites plot off of the Hilina – Kīlauea trends at slightly lower La/Ta (9.3 – 10.8) and slightly higher La/Th (13.9 – 16.6), suggesting either a different source composition or petrogenesis than for the basaltic samples. Nevertheless, in addition to their location on the lower Hilina slope, origin of the phlogopite nephelinites as part of the Kīlauea magmatic system is supported by relatively young isotopic ages for these clasts (235 – 240 ka), too recent to have been derived from pre-shield volcanism at Mauna Kea or older Hawaiian volcanoes [*Lipman et al.*, this volume].

Uniform K_2O/P_2O_5 in all but the most strongly alkalic glasses (Figure 2) from the Hilina region is also consistent with a common Kīlauea source for the alkali basalts, hawaiites, transitional basalts, subaerial tholeiites, and the higher-SiO_2 basanites. Three possibilities could explain the high-K_2O/P_2O_5 trend for the nephelinite and lowest-SiO_2 basanite glasses (Figure 2) that projects to the K_2O axis: (1) The magmas could derive from a high-K_2O/P_2O_5 source, followed by crystallization differentiation involving a potassic mineral, thereby limiting enrichment of derivative liquids in K_2O relative to their enrichment in P_2O_5. (2) A potassic mineral could have been residual to melting, yielding high-K_2O liquids even as P_2O_5 was diluted by progressive melting. (3) The magmas could have originated from the same source as the "normal" K_2O/P_2O_5 Kīlauea magmas but at degrees of melting sufficiently low to fractionate K from P and produce high K_2O/P_2O_5 primary liquids, followed by crystallization differentiation involving a potassic mineral, as in possibility (1).

Fractional crystallization of apatite is unlikely as the cause of the high-K_2O/P_2O_5 suite: (1) Except for the two most evolved nephelinites, glasses that define the high K_2O/P_2O_5 array have high, not low, concentrations of P_2O_5 for their MgO values (Figure 3). (2) Only these two most evolved of the high K_2O/P_2O_5 glasses contain apatite, and for the rest, apatite saturation would require P_2O_5 concentrations higher than the glasses possess [*Harrison and Watson, 1984*]. (3) The high K_2O/P_2O_5 array persists to glass with 9.5 wt.% MgO, which would require crystallization of apatite at unusually high temperature.

In situ trace element and isotopic analyses of ancestral Kīlauea glass grains would help to decide among alternative interpretations for the high K_2O/P_2O_5 group. Relatively high Cl and F contents of the high-K_2O/P_2O_5 nephelinite and basanite glasses support the interpretation of melting of phlogopite- or amphibole-bearing sources. Phlogopite has not been found in any of the ancestral Kīlauea glass grains but is conspicuous in the phlogopite nephelinite clasts, and a kaersutite fragment (with 0.6 wt.% K_2O) was found in the sandstone sample (S508-R3) that contains nephelinite and low-SiO_2 basanite glass grains. These occurrences are evidence that phlogopite and amphibole participated in crystallization differentiation of some ancestral Kīlauea alkalic magmas. Our preferred interpretation is that the source for the high K_2O/P_2O_5 glasses contained phlogopite or amphibole, or both, and that these minerals may also have participated in subsequent crystallization differentiation.

Partial Melting Conditions

Variations in whole-rock REE contents are consistent with residual garnet in the sources of the Hilina region magmas. Following *Frey et al.* [2000], we plot Tb/Yb versus Th (Figure 8). The garnet/melt partition coefficient for Yb is greater than for Tb [*Shimizu and Kushiro*, 1975], whereas Yb and Tb partition nearly equally into clinopyroxene [*Hart and Dunn*, 1993; *Johnson*, 1998; *Green et al.*, 2000]. As a result, residual garnet leads to sharply declining Tb/Yb as degree of melting increases. Th can be used as a monitor of degree of melting, so that variable melting of a single source in the garnet peridotite stability field produces a positive sloping array on a plot of Tb/Yb against Th. Subsequent fractional crystallization of olivine – clinopyroxene – plagioclase can modify this simple relation by displacing liquids to higher Th and by displacing magmas with entrained or accumulated crystals to lower Th, both without substantial changes in Tb/Yb. Most Hilina region rocks define a positive slope on the Tb/Yb versus Th diagram (Figure 8), with apparent degree of partial melting increasing in the sequence nephelinites < basanites < alkali basalts < transitional basalts < tholeiites.

Hawaiites fall off this positive array with high Th for their Tb/Yb, consistent with their being evolved magmas produced by fractional crystallization, as has been inferred for post-shield hawaiites [*Frey et al.*, 1990]. Relative abundances of light REE in alkali basalts are higher ($(La/Sm)_{cn}$: 1.6 – 1.9) than for hawaiites ($(La/Sm)_{cn}$: 1.5 – 1.6), so fractional crystallization of alkali basalts would produce derivative liquids with steeper REE patterns than the hawaiites. Hawaiite REE patterns are more similar to those of the transitional basalts ($(La/Sm)_{cn}$: 1.3 – 1.6)), so transitional basalts are more probable parents for Hilina hawaiites. The relatively low SiO_2 and Sc contents of the hawaiites suggest fractional crystallization of a clinopyroxene-rich assemblage, and the low V of the hawaiites signals Fe-Ti oxide crystallization. Hilina hawaiites may result from stagnation of early transitional basalts at elevated pressure where the clinopyroxene stability field is enlarged, as has also been concluded for post-shield hawaiites [*Frey et al.*, 1990]. Seven transitional basalts from above and east of the Hilina bench cluster slightly off the main Tb/Yb array at low Th. These samples (Figure 8) also have high Ca/(Mg+Fe) and high Sc for their Mg/(Mg+Fe). Accumulation or entrainment of clinopyroxene can account for these features and has petrographic evidence in the prominent clinopyroxene plus olivine phenocrysts in many of the above-bench transitional basalts.

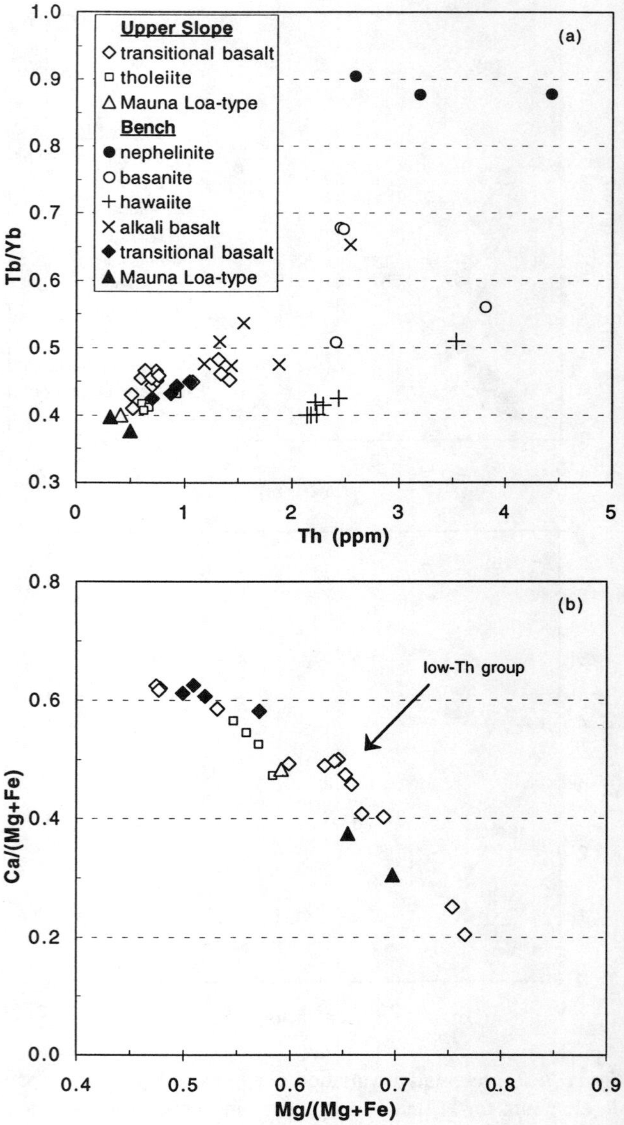

Figure 8. Process discrimination diagrams for Hilina bench clasts, and upper and east slope pillow lavas and fragments: (a) Tb/Yb vs. Th (ppm), (b) molar Mg/(Mg+Fe) vs. molar Ca/(Mg+Fe) for Hilina transitional, tholeiite, and Mauna Loa-type basalts. Points labeled low-Th group in (b) are transitional basalts that plot off the Tb/Yb vs. Th array in (a).

Melting in the upper part of the garnet peridotite stability field is also suggested by the major element composition of glasses in the Hilina bench volcaniclastic rocks. Except for the high-K_2O/P_2O_5 group, high-S glasses with MgO > 6.5

wt.% have CaO/Al_2O_3 in the range 0.8 – 1 (Figure 3). Partial melt CaO/Al_2O_3 contours have gentle positive slopes at pressures > 40 kbar in the garnet stability field and have steep negative slopes in the spinel stability field (Figure 9). Lherzolite partial melts yield CaO/Al_2O_3 values in the range 0.8 – 1 that are nearly insensitive to degree of melting, up to the loss of garnet, in the pressure interval 30 – 35 kbar in the shallow part of the garnet stability field. We infer that parent magmas for the Hilina basaltic suite were dominated by melts that were produced from, or last equilibrated with, garnet peridotite in this pressure interval, equivalent to 95 – 110 km depth. High-S transitional basalt glasses from above and east of the Hilina bench have slightly lower CaO/Al_2O_3 of 0.76 – 0.82, but these are minimum values for their parent magmas because the pillow glasses are low-MgO (< 6.5 wt.%) and may have been modified by clinopyroxene crystallization. For comparison, MgO-rich glasses from Lō'ihi Seamount generally have higher CaO/Al_2O_3 (0.85 – 1.1) [*Moore et al.*, 1982; *Garcia et al.*, 1993; *Garcia et al.*, 1995b], which could indicate derivation from slightly greater depths (40 kbar or 125 km) than for the Hilina suite (Figure 9). High-MgO tholeiitic and picritic glasses from Kīlauea's Puna Ridge and vicinity have CaO/Al_2O_3 of 0.74 - 0.84 [*Clague et al.*, 1995]. The presence of CaO/Al_2O_3 values < 0.8 in high-MgO glasses supports the interpretation of *Clague et al.* [1995] and *Wagner et al.* [1993] that some parent magmas for Kīlauea tholeiites continue to equilibrate with peridotite at depths shallower than the garnet stability field.

Hofmann et al. [1984] evaluated relative degrees of melting and general source characteristics for Kīlauea tholeiites by adjusting whole-rock compositions to constant MgO (12 wt.%) and thereby minimizing effects of olivine loss or accumulation. We apply this approach to the submarine Hilina rocks (Figure 10) to further evaluate if increasing partial melting can account for the progression from alkali basalts through modern tholeiites. The seven transitional basalts with accumulated clinopyroxene were treated in two ways: (1) with the conventional olivine addition-subtraction calculation, and (2) by first removing 4 – 13 wt.% clinopyroxene (using Kīlauea augite composition MP218AUG from *Wright* [1971]) to bring the compositions onto the main Ca/(Mg+Fe) vs. Mg/(Mg+Fe) trend of Figure 8, and then applying the olivine adjustment. Inconsequential differences result from the two approaches (Figure 10a).

Elements are ordered in Figure 10 so as to give smooth tholeiite-normalized enrichment patterns for the average transitional basalt from above and east of the mid-slope bench. General agreement between this element order and that of systematically increasing compatibility for oceanic

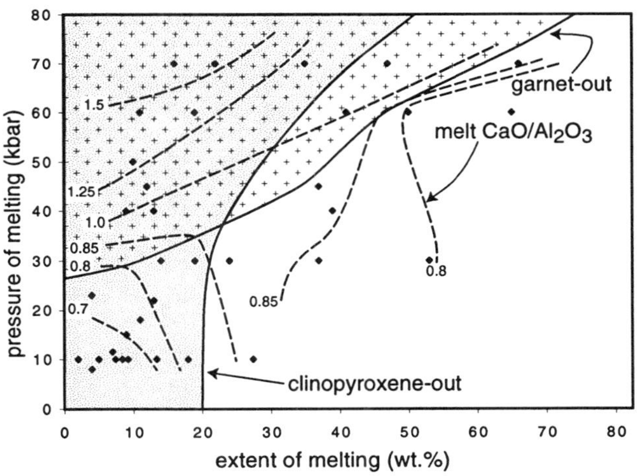

Figure 9. Pressure and extent-of-melting diagram, contoured for CaO/Al_2O_3 of peridotite partial melts. Contours are constrained by data (diamonds) from *Walter* [1998], *Kinzler* [1997], *Baker and Stolper* [1994], *Baker et al.* [1995], and *Hirschmann et al.* [1998]. Shaded field shows clinopyroxene stable; stippled field shows garnet stable.

basalts [*Hofmann*, 1988] (exceptions are relative positions of Nb vs. Ta and La, and Sm vs. Zr) is consistent with increasing partial melting as a dominant factor for the change from alkalic to exclusively tholeiitic magmas. Alkali basalt clasts have steeper but otherwise similar tholeiite-normalized enrichment patterns (Figure 10b), except for a positive Nb anomaly. This is due to the three alkali basalt samples that are enriched in Nb and Ba relative to P_2O_5 (Figure 10b omits sample S509-R2B that compositionally and texturally belongs with the basanites). The average enrichment pattern is more irregular for transitional basalt clasts from Hilina bench debris flow breccias (Figure 10c), probably due to the small number of these clasts.

Relative degrees of partial melting can be inferred from the fractionation-adjusted incompatible element enrichments. Of the elements considered, Ba is most enriched in alkali and transitional basalts relative to tholeiites, but we utilize Th and Ta because they are also highly incompatible, they were, arguably, uniformly distributed in the source region (Figure 7), and they are generally unmodified by alteration (greater range in normalized Th values is due, in part, to a 2 – 4 times lower measurement precision for Th than for Ta). If Th and Ta were perfectly incompatible, average enrichments of 1.5 and 1.4 for the transitional basalt pillows, and of 2.4 and 2.3 for alkali basalt clasts would correspond to average degrees of melting of ~70 % and ~40 % of that which yielded the Kīlauea tholei-

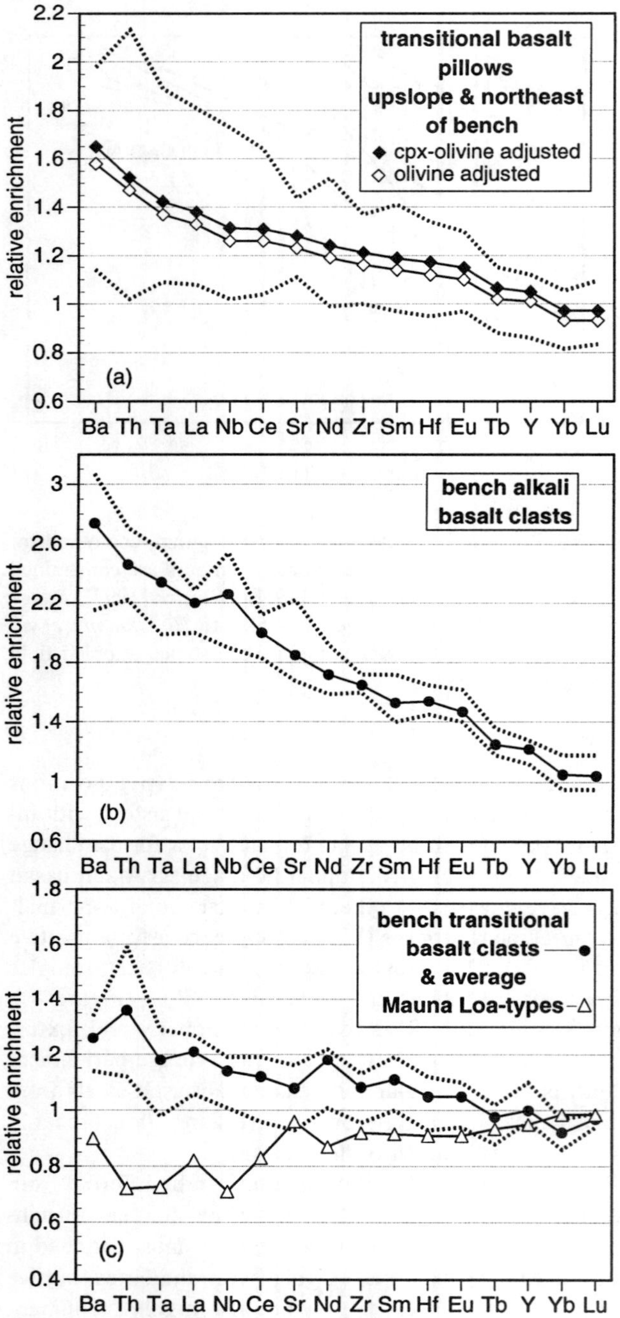

Figure 10. Trace element enrichments relative to average of modern-Kīlauea-like tholeiites (S506-R4A, S506-R5A, S506-R5B), all fractionation-adjusted (to 12% MgO). Dotted lines show ranges; lines with symbols show average compositions. (a) Transitional basalt pillow lavas upslope and east of the mid-slope bench. Open symbols show olivine-adjusted composition, filled symbols show pyroxene- and olivine-adjusted composition. (b) Alkali basalt clasts from debris-flow breccias, exclusive of basanite-like sample S509-R2B. (c) Transitional basalt clasts from debris-flow breccias (filled circles); average Mauna Loa–type tholeiite (triangles) for reference.

ites. True relative degrees of melting will be slightly less than these estimates, but by no more than 10 % so long as the bulk distribution coefficient values for Th and Ta are < 10 % of the melt fraction value, a condition satisfied if tholeiites form by more than about 5 % partial melting of peridotite. These estimates show that the degree of melting has increased by nearly a factor of 2.5 from the period in which alkali basalts were common. Comparing Th and Ta relative enrichments of the least enriched alkali basalt with those of the most enriched transitional basalt shows that alkali basalts no longer formed once degrees of melting generally exceeded 50 % of that which produces tholeiites.

The ancestral Kīlauea enrichment patterns add further support the interpretation that garnet was residual to melting, as was previously determined by *Hofmann, et al.* [1984] for shield-stage Kīlauea tholeiites. Evidence for residual garnet are the near-constant fractionation-adjusted Yb and Lu, signaling that bulk distribution coefficient values were near 1 for these elements, and the fractionation between middle and heavy REE, that is more pronounced in the alkalic, lower-degree partial melts, consistent with the relative partitioning behavior of REE in garnet [*Shimizu and Kushiro,* 1975].

Comparisons With Lō'ihi Seamount and a Progressive Change in Source Composition

Lō'ihi Seamount is the sole active juvenile Hawaiian volcano, and it is instructive to compare its magmas to those of ancestral Kīlauea. Limited information on the crystallization differentiation behavior of strongly alkalic magmas precludes adjusting basanite and nephelinite rock and glass analyses to primitive compositions. Instead, we consider trace element ratios that are not strongly modified by fractional crystallization of olivine, clinopyroxene, or spinel. Most Hilina region basaltic samples define restricted shallow sloping arrays on plots of Ce/Y versus Ba/Zr and Zr/Y versus Zr/Nb (Figure 11), and the position of samples in these arrays follows rock composition. For example, Zr/Nb generally increases in the series alkali basalt – hawaiite and transitional basalt – tholeiite, with Mauna Loa-type samples plotting at distinctly higher Zr/Nb, all with small decrease in Zr/Y; Ba/Zr varies in the reverse order. The basanites, basanite-like alkali basalt, and phlogopite nephelinites define a distinct steeply sloping array in each diagram, diverging from the main body of basaltic analyses to much higher Ce/Y and Zr/Y. Phlogopite nephelinites and basanites also define arrays that diverge from the basalt and hawaiite samples on plots of other trace element ratios, such as La/Sm vs. Sm/Yb, reaching high Sm/Yb values (not illustrated). Most Lō'ihi rocks plot along and extend the shallow-sloping "basaltic"

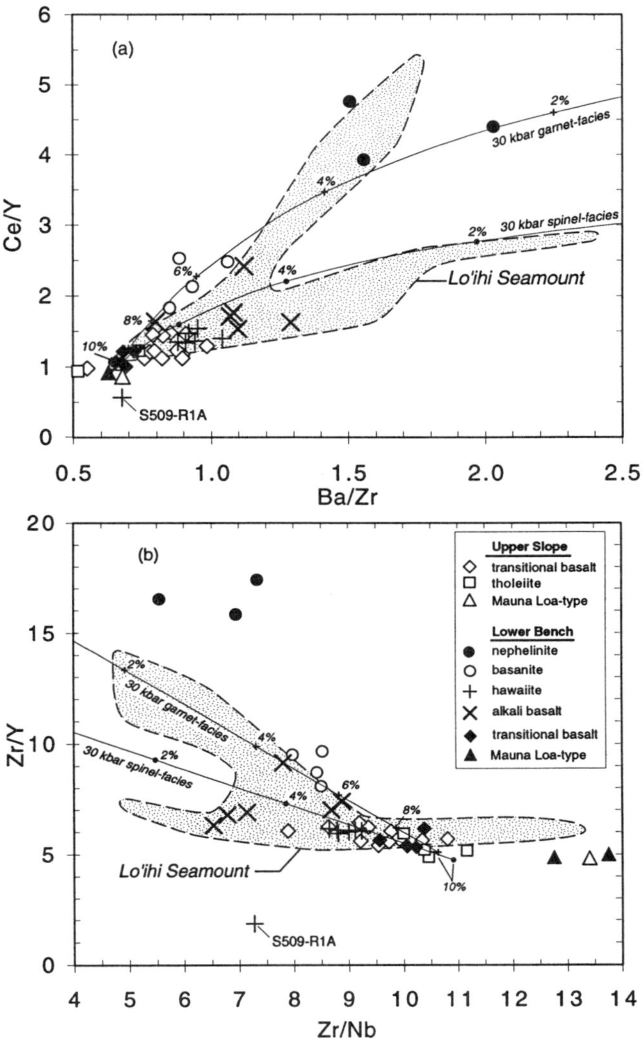

Figure 11. Incompatible element ratios that are insensitive to olivine – clinopyroxene – plagioclase crystallization. (a) Ce/Y vs. Ba/Zr. (b) Zr/Y vs. Zr/Nb. Curves show trajectories for batch partial melts of spinel (•) and garnet (+) peridotite, calculated with the non-modal partial melting equation of *Shaw* [1970] and parameters of Table 4. Tick marks on curves are 2% melting increments; curves are designed so that trace element abundances of 10% garnet-facies partial melt match the average Kīlauea tholeiite (S506-R4A, S506-R5A, S506-R5B), without dilution to primary compositions. Field for Lō'ihi Seamount from *Frey and Clague* [1983], *Garcia et al.* [1993], and *Garcia et al.* [1995b] (for which study Zr is calculated as 40 times reported Hf).

legs of the ancestral Kīlauea fields; the most alkalic Lō'ihi samples (basanites) studied by *Garcia et al.* [1995b] define the low-Zr/Nb, high-Ba/Zr endmember, and Lō'ihi tholeiites define the opposite ends of the arrays. In contrast, the most alkalic Lō'ihi samples studied by *Frey and Clague* [1983] (two basanites and an alkali basalt) diverge to high Ce/Y and Zr/Y and define subordinate steeply-sloping arrays similar to, but not as extreme as, the ancestral Kīlauea basanite – nephelinite array.

Garcia et al. [1995b] interpreted the Lō'ihi basaltic magmas that plot along the shallow trace element arrays in Figure 11 as partial melts of peridotite sources that were uniform with respect to most trace elements; degrees of melting were inferred to increase from alkali basalts to their tholeiites (comparable to our transitional basalts). Some important trace element ratios vary significantly and systematically between estimated parent magmas, however; their alkali basalt parent has low Hf/Nb and high Ba/Hf (comparable to low Zr/Nb and high Ba/Zr), and the reverse in their tholeiite parent. We have calculated trace element variations during progressive partial melting (Table 4), using recent experimentally determined reactions for melting deep in the spinel peridotite field [*Walter et al.*, 1995] and shallow in the garnet peridotite field [*Walter*, 1999]. The calculations are designed so that average trace element ratios of the three most depleted Hilina low-S tholeiites are produced as a 10% batch melt at 30 kbar in the garnet stability field to evaluate simple melting processes that might create shield-stage Kīlauea tholeiites. The choice of degree of melting is somewhat arbitrary, but greater degrees of melting would place tholeiites well beyond the point of garnet exhaustion using *Walter's* [1999] melting reaction and fertile peridotite mineralogy, which would conflict with evidence for residual garnet for shield-stage Kīlauea tholeiites [*Hofmann et al.*, 1984]. Ten percent melting is also consistent with the best-fit estimate of *Garcia et al.* [1995b] for production of Lō'ihi tholeiites. Calculated melt compositions grossly mimic the coupled variation in trace element ratio and magma type, with alkali basalts having formed at lower extents of melting, transitional basalts at intermediate degrees, and tholeiites at the highest extents of melting. More importantly, the calculated melting curves do not match the Hilina and Lō'ihi basaltic suites particularly well, and this mismatch is most severe for melting in the garnet facies. Lō'ihi and Hilina basaltic rocks vary much more in their Zr/Nb and Ba/Zr, without changing sufficiently in Ce/Y and Zr/Y, than would be expected for melting of a uniform garnet peridotite source. Reasonable alternative selections of mineral-melt partition coefficients do not change this result. Melting deeper in the garnet stability field, where garnet constitutes a greater fraction of the mineral assemblage, leads to larger changes in Ce/Y and Zr/Y with degree of melting and increases the disparity between calculated and observed trace element variations. Models of melting deep

Table 4. Peridotite melting parameters

	Olivine	Opx	Cpx	Gt	Sp	Co_{ppm}
40 kbar mode	0.533	0	0.36	0.11	0	
& melt reaction	0.11	0	0.56	0.33	0	
30 kbar garnet mode	0.531	0.177	0.27	0.019	0	
& melt reaction	0.08	-0.19	0.81	0.3	0	
30 kbar spinel mode	0.519	0.19	0.28	0	0.014	
& melt reaction	0.01	-0.56	1.43	0	0.12	
D Ba	1.3E-05	2E-04	0.01	4E-04	0	8.86
D Zr	0.0018	0.032	0.1	0.27	0	17
D Y	0.0027	0.095	0.41	4	0	5.53
D Nb	0.0012	0.001	0	0.002	0	1.29
D Ce	3.6E-05	0.007	0.08	0.02	0	3.07

Partition coefficients (D): garnet: Ce, Y - after *Shimizu and Kushiro* [1975], Zr - *Johnson* [1998], Nb - average of *Green et al.* [2000], Ba - *Hauri et al.* [1994a]; clinopyroxene: average of *Johnson* [1998], *Green et al.* [2000], and *Hart and Dunn* [1993]; orthopyroxene: *Green et al.* [2000]; olivine: *Skulski et al.* [1994], *Beattie* [1994].

in the spinel peridotite stability field more closely match the ranges of basaltic sample trace element ratios, but conflict with evidence presented previously and by other workers for residual garnet in the source of Lō'ihi and Kīlauea magmas.

Hawaiian magmatism is dominated by three sources, as identified by radiogenic isotopes [*Tatsumoto*, 1978; *Stille et al.*, 1986]. These are: (1) a depleted mid-ocean ridge-like source sampled most purely by post-erosional and North Arch magmas at low degrees of partial melting [*Clague and Frey*, 1982; *Stille et al.*, 1986; *Frey et al.*, 2000], (2) a source with radiogenic Sr and non-radiogenic Pb sampled most purely by alkali-poor high-Zr/Nb basalts and basaltic andesites of Ko'olau Volcano, Oahu [*Frey et al.*, 1994; *Roden et al.*, 1994], and (3) a "Kea" source end-member with high ^{206}Pb and intermediate $^{87}Sr/^{86}Sr$ sampled most purely in some historic tholeiites from Kīlauea [*Stille et al.*, 1986]. Figure 12 shows that alkalic basalts from both Lō'ihi and from ancestral Kīlauea approach the compositions of post-erosional and North Arch magmas, whereas tholeiitic magmas are more similar to Mauna Loa basalts, which approach Ko'olau magmas in their trace element and isotopic characteristics [*Frey and Rhodes*, 1993]. Our working interpretation is that temporally increasing degrees of melting at Kīlauea and at Lō'ihi have been accompanied by shifts from chiefly Kea and post-erosional-type sources to increasing contributions from Mauna Loa – Ko'olau-type source materials as magmas were tapped from further into the plume. Such a progression in source would account for the increasingly Mauna Loa-like Zr/Nb and Ba/Zr (and other) values as magmas became more tholeiitic, without commensurate changes in garnet-sensitive trace element ratios. Support for shifting source

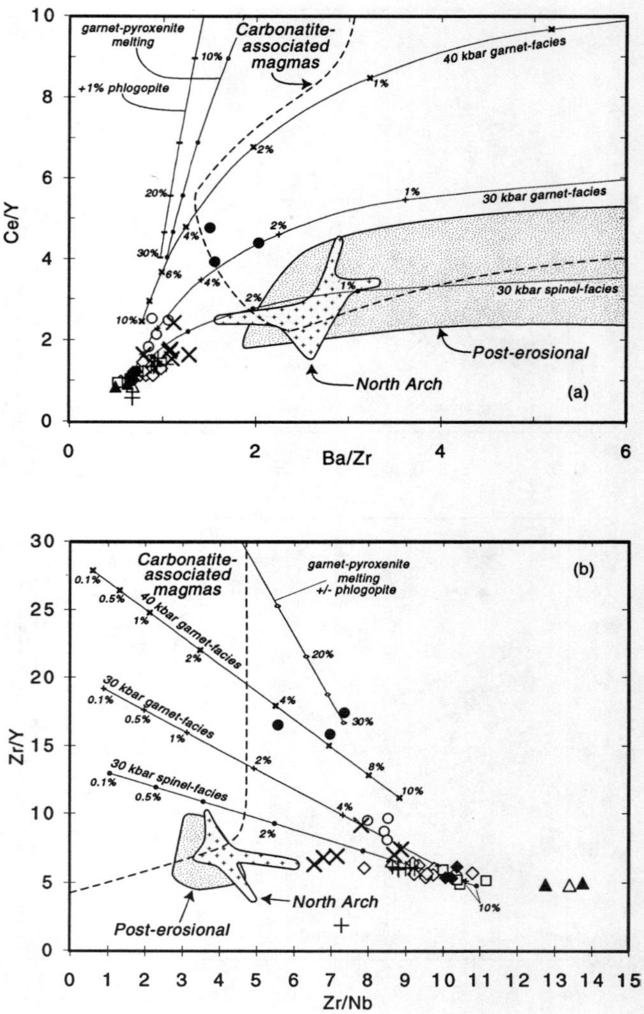

Figure 12. Diagrams of incompatible element ratios, showing fields for Hawaiian post-erosional and North Arch alkalic magmas [*Clague and Frey*, 1982; *Clague and Dalrymple*, 1988; *Frey et al.*, 2000], plus field of carbonatite-associated silicate magmas from other localities (heavy dashed line). Hilina region samples are symbols as in Figure 11. (a) Ce/Y vs. Ba/Zr. (b) Zr/Y vs. Zr/Nb. Melting curves for garnet peridotite and spinel peridotite at 30 kbar as in Figure 11. Partial melting calculation at 40 kbar based on melting reaction and source mode from *Walter* [1999], partition coefficients and trace elements abundances from Table 4. Garnet-pyroxenite melting curves are for batch partial melting of 70% clinopyroxene – 30% garnet, or 69.5% clinopyroxene – 30% garnet – 0.5% phlogopite. This latter source is based on the average ancestral Kīlauea transitional basalt, the partition coefficients of Table 4, and phlogopite from *Green et al.* [2000]. Field for carbonatite-associated magmas from *Barreiro et al.* [1987], *Schleicher et al.* [1991], *Keller and Spettel* [1995], *Johnson et al.* [1997], and *Dawson et al.* [1985].

proportions is given by preliminary *in situ* Pb isotope measurements of high-S ancestral Kīlauea glass grains [*N. Shimizu*, written communication]. Results for nephelinite (2) and basanite (1) glasses extend the Kīlauea isotopic field to even higher ^{206}Pb, demonstrating that some ancestral Kīlauea magmas sampled relatively pure "Kea" source materials with less contributions from Ko'olau – Mauna Loa-type sources than for modern Kīlauea tholeiites. One high-S alkali basalt grain has Pb isotope values similar to but more extreme than post-erosional magmas, confirming the involvement of post-erosional-type sources during the inception of Kīlauea. *Tanaka et al.* [this volume] and *Shinozaki et al.* [this volume] present evidence of a temporal shift from Kea-type to Ko'olau-type source compositions for shield-stage tholeiites from Ko'olau, Oahu, also consistent with a zoned plume. A decisive test for a zoned plume model will be to search for Ko'olau-like isotopic features in the abundant high-S glass grains from ancestral Kīlauea. These should be rare if the plume is well zoned with Ko'olau-type source materials in the interior, but the presence of scarce high-S low-alkali tholeiite glass grains from ancestral Kīlauea (Figure 2) suggests that they will not be completely absent.

Origin of Ancestral Kīlauea Nephelinites and Basanites

Nephelinites were formerly known only from late post-erosional stages of Hawaiian volcanism and from deep-water lava flows erupted on the crest of the Hawaiian Arch, leading to interpretations that they originate from melting processes localized to the plume margins. Their discovery as submarine-erupted glasses and shallow intrusive rocks from early Kīlauea enhances the apparent symmetry between juvenile and late-stage alkalic magmatism. An obvious feature of the comparison in Figure 12 is that the phlogopite nephelinites and basanites from ancestral Kīlauea are unlike post-erosional and North Arch strongly alkalic lavas. The basanite – nephelinite array attains high Ce/Y and Zr/Y at much lower Ba/Zr and higher Zr/Nb than the post-erosional or North Arch rocks. Post-erosional and North Arch alkalic magmas have high concentrations of compatible major and trace elements (to 480 ppm Ni, 600 ppm Cr, 29 ppm Sc), leading to the interpretation that they are low-degree peridotite partial melts little modified by fractional crystallization [*Clague and Frey*, 1982; *Frey et al.*, 2000]. In contrast, ancestral Kīlauea phlogopite nephelinites and basanites have moderate compatible element contents (to 180 ppm Ni, 175 ppm Cr, 25 ppm Sc), signaling either fractional crystallization or non-peridotitic source materials. An further noteworthy feature is that the nephelinite – basanite array merges with the main field of ancestral Kīlauea basaltic samples, in much the same was as the most alkalic Lō'ihi rocks studied by *Frey and Clague* [1983] merge with the main Lō'ihi field (Figure 11). This relationship, as well as preliminary Pb isotope results mentioned previously, are evidence that strongly alkalic magmas from ancestral Kīlauea, with fractionated REE and low Y, were petrogenetically linked to the more common basalts and were not simply melts of foreign source materials. Low Y and heavy-REE contents suggest a greater involvement of garnet in the petrogenesis of the phlogopite nephelinites and basanites than for the alkalic, transitional, and tholeiitic basalts that successively dominated Kīlauean magmatism. Greater involvement of garnet could have been through partial melting of source regions with higher garnet contents, because these sources were deeper and further in the garnet peridotite stability field, or because the sources were compositionally distinct. Greater involvement of garnet could also have been through high-pressure crystallization-differentiation from broadly basaltic parents, analogous to the high-pressure fractional crystallization of clinopyroxene and Fe-Ti oxides that creates hawaiites. Even more complex scenarios are possible, involving solidification of magmas from the plume interior as garnet pyroxenites within the lithosphere, followed by low-degree remelting as the lithosphere was heated by the impinging plume and its high-temperature magmas.

An origin as lower degree melts from deeper in the plume is explored using the 40 kbar peridotite melting experiments of *Walter* [1999] and the same partition coefficients and trace element relative abundances as for the 30 kbar calculations (Figure 12, Table 4). More abundant garnet at 40 kbar leads to melts with higher Zr/Y and Ce/Y (and more fractionated REE patterns) than at 30 kbar, but the 40 kbar melting curves do not closely match the steep nephelinite – basanite array. The results are permissive of an interpretation that the phlogopite nephelinites and basanites represent a spectrum of melts formed at greater depths than the basalts and over a wide pressure range. Incipient high-pressure partial melting of the Kīlauea plume source does not explain the high K_2O/P_2O_5 of the nephelinite and low-SiO_2 basanite glass grains. As discussed previously, high K_2O/P_2O_5 even in high-MgO, low-P_2O_5 glass suggests melting of a phlogopite or amphibole-bearing source, but the low thermal stability of phlogopite and amphibole [*Wendlandt and Eggler*, 1980; *Holloway*, 1973; *Konzett and Ulmer*, 1999] would preclude this from being the high-temperature source that yields Kīlauea picritic parent magmas by greater degrees of melting at shallower levels [*Clague at al.*, 1995]. Fluorine stabilizes amphibole and phlogopite to high temperatures [*Foley et al.*, 1986a], and although the F contents of the nephelinitic and basanitic glasses greatly exceed those of Hawaiian tholeiite glasses (550 – 1900 ppm vs. 20 – 150 ppm) [*Garcia et al.*, 1989], they are less than the high F contents of most magmas derived by melting of F-phlogopite bearing sources (2000 – 8000 ppm) [*Foley et al.*, 1986b]. Based on our present understanding, it seems unlikely that the nephelinites and low-SiO_2 basanites originated by incipient partial melting of the main plume source that now feeds Kīlauea.

High-pressure crystallization differentiation of garnet from basaltic or picritic plume magmas, or low-degree remelting of plume magmas that solidified as garnetiferous rocks in the deep lithosphere, may be viable interpretations for the ancestral Kīlauea nephelinite – basanite suite. To produce liquids with SiO_2 as low as in nephelinites from a basaltic or picritic parent magma would require separation of dominantly pyroxene in addition to garnet, and the pyroxene would have to be moderate or low in CaO to produce the high CaO and CaO/Al_2O_3 of nephelinitic liquids. Parent magmas with SiO_2 contents well below that of pyroxene, such as primitive alkali basalts or alkalic picrites, might also be essential to produce very low-SiO_2 derivative liquids. *Frey* [1980] presents trace element evidence that some garnet pyroxenite xenoliths erupted in post-erosional magmas on Oahu formed as cumulates from alkalic magmas transiting the lithosphere. These xenoliths have subcalcic clinopyroxene, some contain primary orthopyroxene, and most contain traces of phlogopite. The garnet pyroxenite xenoliths show that the appropriate cumulate assemblage forms at high pressure and likely derived from alkalic parent magmas. High-pressure fractional crystallization of a garnet – pyroxene assemblage might produce basanites with "normal" K_2O/P_2O_5 directly from plume-derived parent magmas. A more complex petrogenetic process may be necessary to produce the high-K_2O/P_2O_5 nephelinite – basanite suite. Low-degree remelting of plume magmas, which previously solidified in the lithosphere as phlogopite- or amphibole-bearing garnet pyroxenites, could have given rise to high-K_2O/P_2O_5 nephelinites and basanites. Trace element variations estimated for batch melting of a garnet clinopyroxenite (± phlogopite) source with composition of the average Hilina region transitional basalt are in qualitative agreement (Figure 12, Table 4) with a remelting interpretation (ad hoc adjustments in mineral proportions, melting reactions, and modal mineralogy can further improve the agreement).

This deep crystallization differentiation model is similar to the well-established interpretation of hawaiite petrogenesis [*Frey et al.*, 1990], except that it includes garnet, may require lower-SiO_2 alkalic parent magmas, and includes the possibility that early intrusions can be remelted. Remelting of clinopyroxenite or wehrlite has recently been shown to account for undersaturated ankaramitic liquids in subduction zone arcs [*Schiano et al.*, 2000]. Other Hawaiian nephelinites have been interpreted as direct low degree partial melts of peridotitic sources, in some instances modified by olivine and Cr-spinel fractional crystallization [*Clague and Frey*, 1892; *Frey, et al.*, 2000]. We accept this interpretation, due to the generally high contents of compatible major and trace elements in those rocks, but some may also have been modified by high-pressure garnet-pyroxenite crystallization differentiation. For example, two North Arch nephelinites with the most steeply fractionated REE patterns and highest Ce/Y values are anomalously depleted in Sc for their whole-rock MgO contents (samples 23-2 and 23-6 of *Frey et al.* [2000]), as would result from crystal fractionation of garnet or garnet plus pyroxene. Trace element analyses of ancestral Kīlauea alkalic glass grains will allow better development and testing of our high-pressure crystallization differentiation and remelting hypothesis. Crystallization differentiation in the lithospheric upper mantle signals a cold, poorly-integrated magma transport system that differs from the hot, well-flushed conduits that must exist to feed picritic liquids to mature, shield-stage Hawaiian volcanoes [*Clague et al.*, 1995].

Few magmas share the low-SiO_2, high Ce/Y, Zr/Y, and other characteristics of the Hilina nephelinites and basanites. Extreme compositions that lie on the extrapolation of trace element arrays for the Hilina phlogopite nephelinite to basanite suite are common in the undersaturated silicate lavas associated with carbonatite magmatism. Such lavas include nephelinites, melilitites, and phonolites (Figure 12). No carbonatites have yet been discovered in the Hawaiian Islands, but there is evidence for carbonatite liquids in the shallow oceanic mantle elsewhere [*Amundsen*, 1987; *Ionov et al.*, 1993; *Hauri et al.*, 1993; *Schiano et al.*, 1994]. Intrusive carbonatites exposed in the volcanoes of the Cape Verde Islands [*Gerlach et al.*, 1988] demonstrate that carbonatites can be part of ocean island magmatism. High CO_2 contents of glasses from the submarine North Arch volcanic field [*Dixon et al.*, 1997] establish that some Hawaiian plume-marginal magmas and their sources are carbon-rich. While it is possible that the Hilina nephelinite – basanite suite was associated with unrecognized carbonatite magmatism, a more certain point is that high CO_2 contents stabilize low-Ca pyroxene in alkalic liquids [*Eggler*, 1978], and could promote fractional crystallization of a pyroxene-rich assemblage, thus driving residual liquids to lower SiO_2 contents. High-CO_2 contents might be essential to produce nephelinites from alkali picritic or basaltic parent magmas.

Implications for the Inception and Development of Kīlauea Volcano and for the Structure of the Hawaiian Plume

The geology and geochemistry of the submarine Hilina region support general concepts about the inception of Hawaiian shield volcanoes, developed from Lō'ihi Seamount [*Moore et al.*, 1982]. At their beginning, the magmas form mainly by melting near the top of the garnet peridotite stability field at 95 – 110 km (Hilina/Kīlauea) to perhaps as much as 125 km depth (Lō'ihi) (Figure 13). Changes in magma type from alkalic and transitional, to transitional, to subaerial tholeiitic result from the combined effects of increasing degrees of melting and changing source composition, but these factors cannot be distinguished unambiguously with major and trace element data alone. Complicating this simple progression, rare submarine-erupted low-alkali tholeiite glasses show that intersecting magmatic plumbing systems [*Rhodes et al.*, 1989] or small domains of Mauna Loa source materials situated near the plume margin delivered traces of Mauna Loa–like magmas to ancestral Kīlauea.

Hilina and Lō'ihi magma compositions also provide evidence for why melting stops in the plume. Decompression causes plume melting, and melting can cease, creating a top on the melting zone, if that portion of the plume stops ascending and spreads beneath the lithosphere. Alternatively, melt productivity can greatly decline, giving the appearance of having stopped, if an important melt-contributing mineral phase is exhausted in the source, even if the source continues to ascend (assuming that melts segregate from the source). Garnet or clinopyroxene are the minerals likely to be exhausted first during melting of ascending lherzolite (Figure 9). Major and trace element features of ancestral Kīlauea and Lō'ihi magmas point to melt production in the shallow garnet stability field, with no clear evidence for shallower melting in the spinel stability field. This, and evidence for residual garnet left in the source, suggest that plume spreading, not exhaustion of an important melt-contributing mineral, creates the top of the melting zone. Hawai'i is built atop Cretaceous oceanic crust, and the thickness of Cretaceous oceanic lithosphere is on the order of 90 – 105 km [*Clague and Dalrymple,* 1987; *Leitch and Davies,* 2001]. The base of the lithosphere is close to the spinel peridotite – garnet peridotite transition (Figure 13), which may account for the sparse evidence for spinel peridotite melts in Hawai'i. Once plume material impinges on the base of the lithosphere, it spreads, perhaps as a wake behind the plume, and melting ends in those parts of the plume that are no longer rising. The onset of seismicity beneath Hawaiian shield volcanoes at 60 km depth signals that vigorous magma conduit systems extend from that depth to the volcanoes' summits [*Eaton and Murata*, 1960; *Koyanagi et al.*, 1987].

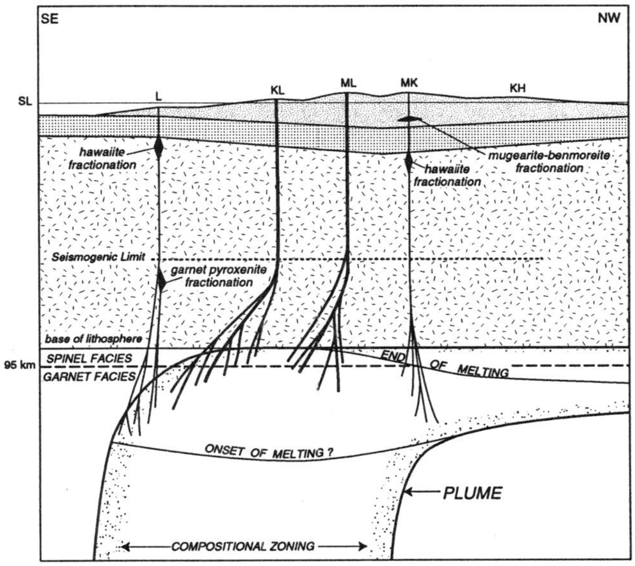

Figure 13. Conceptual present-day SE – NW cross section though Hawai'i Island and the Hawaiian mantle plume (no vertical exaggeration), with projected summit locations of major volcanoes: L – Lō'ihi Seamount, KL – Kīlauea, ML – Mauna Loa, MK – Mauna Kea, KH – Kohala (Hualalai lies far off line of section). Depth to plume top based on melt CaO/Al_2O_3 – pressure relations (Figure 9). Depths for onset of melting are not well known, but are shown as depressed in plume interior due to higher temperatures. End-of-melting curve indicates where plume transport becomes horizontal. Depths in the lithosphere of hawaiite generation (by clinopyroxene crystallization) and basanite to nephelinite generation (by garnet-pyroxenite crystallization) are unknown.

CONCLUSIONS

The 1998 – 1999 JAMSTEC cruises to the offshore Hilina region, on the south flank of Hawai'i Island, recovered submarine-erupted rocks and glasses that reveal the growth history and evolution of Kīlauea Volcano. Early volcanism was mainly alkalic and transitional basalt, similar to Lō'ihi Seamount, but included nephelinite, phlogopite nephelinite, phonotephrite, and basanite with geochemical characteristics unlike rocks recovered from Lō'ihi or elsewhere in Hawai'i. Alkalic volcanism greatly abated and was succeeded by transitional basaltic magmas that constitute much of the submarine south slope of Kīlauea Volcano. Rocks like subaerial shield-stage Kīlauea cap the submarine Puna Ridge and are restricted to relatively shallow depths on Kīlauea's submarine south flank. As such, they are a relatively young feature of Kīlauea magmatism [*Lipman et al.*, this volume].

The monotonic succession of dominant basaltic compositions, from alkalic to transitional to tholeiitic, is explained by a combination of increasing degree of partial melting and shift of source composition. Melting took place in the upper part of the garnet peridotite stability field, at roughly 95 – 115 km depth, and degree of melting has increased by up to a factor of 2.5 between early alkalic and modern tholeiitic shield-stages. Absence of features suggestive of melting in the spinel stability field suggest that the top of the melting zone in the plume margin lies at > 90 km depth, and the top of the melting zone probably results from plume spreading. Melting or melt-solid reaction in the plume interior may extend a short distance into the spinel stability field [*Clague et al.*, 1995; *Wagner et al.*, 1993], suggesting a dome-shaped top for the melting zone or melt-solid interaction region.

The peculiar submarine-erupted nephelinite – basanite glasses and phlogopite nephelinite – basanite clasts from early submarine Kīlauea may result from: (1) high pressure fractional crystallization of pyroxene and garnet from CO_2-rich alkali picritic parent magmas, and (2) low degree remelting of alkalic magmas and their cumulates that solidified in the lithosphere as phlogopite-garnet pyroxenites at shortly earlier times. These and common hawaiites from ancestral Kīlauea, that also result from high pressure crystallization, are products of a cold, poorly integrated, juvenile magma conduit system, unlike the vigorous trans-lithospheric magma supply systems that feed mature shield volcanoes.

Acknowledgments. We thank Captain Yukawa of the R.V. *Kairei* and Captain Tanaka of the R.V. *Yokosuka*, and the JAMSTEC operations teams and marine crews for their efficient and thorough support. N. Shimizu generously shared new Pb isotope analyses that substantially clarified our understanding of possible magma sources. D. Siems and J. Budahn analyzed whole rock compositions quickly and precisely. R. Oscarson provided assistance with the electron probe. We thank D. Clague, D. Sherrod, M. Walter, and H. Yokose for their constructive and insightful reviews, and E. Takahashi for his editorial guidance.

REFERENCES

Amundsen, H.E.F, Evidence for liquid immiscibility in the upper mantle, *Nature, 327,* 692-695, 1987.

Armstrong, J.T., CITZAF: a package of correction programs for the quantitative electron microbeam X-ray analysis of thick polished materials, thin films, and particles, *Microbeam Anal., 4*, 177-200, 1995.

Baedecker, P.A., and D.M. McKown, Instrumental neutron activation analysis of geochemical samples, in: P.A. Baedecker (ed.), Methods for geochemical analysis, *U.S. Geol. Surv. Prof. Pap., 1770*, H1-H14, 1987.

Baker, M.B., and E.M. Stolper, determining the composition of high-pressure mantle melts using diamond aggregates, *Geochim. Cosmochim. Acta, 58*, 2811-2827, 1994.

Baker, M.B., M.M. Hirschmann, M.S. Ghiorso, and E.M. Stolper, Compositions of near-solidus peridotite melts from experiments and thermodynamic calculations, *Nature, 375*, 308-311, 1995.

Barreiro, B.A., and A.F. Cooper, A Sr, Nd, and Pb isotope study of alkaline lamprophyres and related rocks from Westland and Otago, South Island, New Zealand, *Geol. Soc. Am. Spec. Pap., 215*, 115-126, 1987.

Basaltic Volcanism Study Project, *Basaltic volcanism on the terrestrial planets*, Pergamon Press, Inc., New York, 1286 pp, 1981.

Beattie, P., Systematics and energetics of trace-element partitioning between olivine and silicate melts: implications for the nature of mineral/melt partitioning, *Chem. Geol., 117*, 57-71, 1994.

Boynton, W.V., Cosmochemistry of the rare earth elements: meteorite studies, in: P. Henderson, ed., *Rare Earth Element Geochemistry*, Elsevier, New York, 1984.

Chen, C.-Y., and F.A. Frey, Origin of Hawaiian tholeiites and alkalic basalt, *Nature, 302*, 785-789, 1983.

Clague, D.A., and F.A. Frey, Petrology and trace element geochemistry of the Honolulu Volcanics, Oahu: implications for the oceanic mantle below Hawaii, *J. Petrol., 23*, 447-504, 1982.

Clague, D.A., and G.B. Dalrymple, The geology of the Hawaii – Emperor volcanic chain, *U.S. Geol. Surv. Prof. Pap., 1350*, 5-55, 1987.

Clague, D.A., and G.B. Dalrymple, Age and petrology of alkalic postshield and rejuvenated-stage lava from Kaui, Hawaii, *Contrib. Mineral. Petrol., 99*, 202-218, 1988.

Clague, D.A., J.G. Moore, J.E. Dixon, and W.B. Friesen, Petrology of submarine lavas from Kilauea's Puna Ridge, Hawaii, *J. Petrol., 36*, 299-349, 1995.

Clague, D.A., R.T. Holcomb, J.M. Sinton, R.S. Detrick, and M.E. Torresan, Pliocene and Pleistocene alkalic flood basalts on the seafloor north of the Hawaiian islands, *Earth Planet. Sci. Lett., 98*, 175-191, 1990.

Clague, D.A., A.S. Davis, J.L. Bischoff, J.E. Dixon, and R. Geyer, Lava bubble-wall fragments formed by submarine hydrovolcanic explosions on Lō'ihi Seamount and Kilauea Volcano, *Bull. Volc., 61*, 437-449, 2000.

Dawson, J.B., J.V. Smith, and A.P. Jones, A comparative study of bulk rock and mineral chemistry of olivine melilitites and associated rocks from east and south Africa, *Neues Jahr. Miner. Abh., 152*, 143-175, 1985.

Dixon, J.E., D.A. Clague, and E.M. Stolper, Degassing history of water, sulfur, and carbon in submarine lavas from Kilauea Volcano, Hawaii, *J. Geol., 99*, 371-394, 1991.

Dixon, J.E., D.A. Clague, P. Wallace, and R. Poreda, Volatiles in alkalic basalts from the North Arch volcanic field, Hawaii: extensive degassing of deep submarine-erupted alkalic lavas, *J. Petrol., 36*, 911-939, 1997.

Donaldson, C.H., J.B. Dawson, R. Kanaris-Sotiriou, R.A. Batchelor, and J.N. Walsh, The silicate lavas of Olddoinyo Lengai, Tanzania, *Neues Jahr. Miner. Abh., 156*, 247-279, 1987.

Eaton, J.P., and K.J. Murata, How volcanoes grow, *Science, 132*, 925-938, 1960.

Eggler, D.H., The effect of CO_2 upon partial melting of peridotite in the system Na_2O-CaO-Al_2O_3-MgO-SiO_2-CO_2 to 35 kb, with an analysis of melting in a peridotite-H_2O-CO_2 system, *Am. J. Sci., 278*, 305-343, 1978.

Frey, F.A., The origin of pyroxenites and garnet pyroxenites from Salt Lake Crater, Oahu, Hawaii: trace element evidence, *Amer. J. Sci., 280-A*, 427-449, 1980.

Frey, F.A., and D.A. Clague, Geochemistry of diverse basalt types from Loihi Seamount, Hawaii: petrogenetic implications, *Earth Planet Sci. Lett., 66*, 337-355, 1983.

Frey, F.A., and J.M. Rhodes, Intershield geochemical differences among Hawaiian volcanoes: implications for source compositions, melting process and magma ascent paths, *Phil. Trans. R. Soc. London. A, 342*, 121-136, 1993.

Frey, F.A., M.O. Garcia, and M.F. Roden, Geochemical characteristics of Ko'olau Volcano: Implications of intershield geochemical differences among Hawaiian volcanoes, *Geochim. Cosmochim. Acta, 58*, 1441-1462, 1994.

Frey, F.A., D. Clague, J.J. Mahoney, and J.M. Sinton, Volcanism at the edge of the Hawaiian plume: Petrogenesis of submarine alkalic lavas from the North Arch volcanic field, *J. Petrol., 41*, 667-691, 2000.

Frey, F.A., W.S. Wise, M.O. Garcia, H. West, S.-T. Kwon, and A. Kennedy, Evolution of Mauna Kea Volcano, Hawaii: Petrologic and geochemical constraints on postshield volcanism, *J. Geophys. Res., 95*, 1271-1300, 1990.

Foley, S.F., W.R. Taylor, and D.H. Green, The effect of fluorine on phase relationships in the system $KAlSiO_4 - Mg_2SiO_4 - SiO_2$ at 28 kbar and the solution mechanism of fluorine in silicate melts, *Contrib. Mineral. Petrol., 93*, 46-55, 1986.

Foley, S.F., W.R. Taylor, and D.H. Green, The role of fluorine and oxygen fugacity in the genesis of the ultrapotassic rocks, *Contrib. Mineral. Petrol., 94*, 183-192, 1986.

Garcia, M.O., Petrography and olivine and glass chemistry of lavas from the Hawaii Scientific Drilling Project, *J. Geophys. Res., 101*, 11701-11713, 1996.

Garcia, M.O., T.P. Hulsebosch, and J.M. Rhodes, Olivine-rich submarine basalts from the southwest rift zone of Mauna Loa volcano: Implications for magmatic processes and geochemical evolution, *Am. Geophys. Union Mon. 92*, 219-239, 1995a.

Garcia, M.O., D. Muenow, K. Aggrey, and J. O'Neil, Major element, volatile, and stable isotope geochemistry of Hawaiian submarine tholeiitic basalt, *J. Geophys. Res., 94*, 10525-10538, 1989.

Garcia, M.O., D.J.P Foss, H.B. West, and J.J. Mahoney, Geochemical and isotopic evolution of Loihi volcano, Hawaii, *J. Petrol., 36*, 1647-1674, 1995b.

Garcia, M.O., B.A. Jorgenson, J.J. Mahoney, E. Ito, and A.J. Irving, An evaluation of temporal geochemical evolution of Loihi summit lavas: results from Alvin submersible dives, *J. Geophys. Res., 98*, 537-550, 1993.

Gerlach, D.C., R.A. Cliff, G.R. Davies, M. Norry, and N Hodgson, Magma sources of the Cape Verdes archipelago: Isotopic and trace element constraints, *Geochim. Cosmochim. Acta, 52*, 2979-2992, 1988.

Green, T.H., Experimental studies of trace element partitioning applicable to igneous petrogenesis: Sedona 16 years later, *Chem. Geol., 117*, 1-36, 1994.

Green, T.H., J.D. Blundy, J. Adam, and G.M. Yaxley, SIMS determination of trace element partition coefficients between garnet, clinopyroxene and hydrous basaltic liquids at 2-7.5 Gpa and 1080-1200 °C, *Lithos, 53*, 165-187, 2000.

Harrison, T.M., and E.B. Watson, The behavior of apatite during crustal anatexis: Equilibrium and kinetic considerations, *Geochim. Cosmochim. Acta, 48*, 1467-1477, 1984.

Hart, S.R., and T. Dunn, Experimental cpx/melt partitioning of 24 elements, *Contrib. Mineral. Petrol., 113*, 1-8, 1993.

Hauri, E.H, N. Shimizu, J.J. Dieu, and S.R. Hart, Evidence for hotspot-related carbonatite metasomatism in the oceanic upper mantle, *Nature, 365*, 221-227, 1993.

Hauri, E.H., T.P. Wagner, and T.L. Grove, Experimental and natural partitioning of Th, U, Pb, and other trace elements between garnet, clinopyroxene, and basaltic melt, *Chem. Geol., 117*, 149-166, 1994.

Hauri, E.H., J.A. Whitehead, and S.R. Hart, Fluid dynamic and geochemical aspects of entrainment in mantle plumes, *J. Geophys. Res., 99*, 24275-24300, 1994.

Helz, R.T., Diverse olivine types in lava of the 1959 eruption of Kilauea volcano and their bearing on eruption dynamics, *U.S. Geol. Surv. Prof. Pap., 1350*, 691-722, 1987.

Hirschmann, M.M., M.B. Baker, and E.M. Stolper, The effect of alkalis on the silica content of mantle-derived melts, *Geochim. Cosmochim. Acta, 62*, 883-902, 1998.

Hofmann, A.W., Chemical differentiation of the Earth: the relationship between mantle, continental crust, and oceanic crust, *Earth Planet. Sci. Lett., 90*, 297-314, 1988.

Hofmann, A.W., M.D. Feigenson, and I. Raczek, Case studies on the origin of basalt: III. Petrogenesis of the Mauna Ulu eruption, Kilauea, 1969-1971, *Contrib. Mineral. Petrol., 88*, 24-35, 1984.

Holloway, J.R., The system pargasite-H_2O-CO_2: A model for melting of a hydrous mineral with a mixed-volatile fluid – I. Experimental results to 8 kbar, *Geochim. Cosmochim. Acta, 37*, 651-666, 1973.

Ionov, D.A., C. Dupuy, S.Y. O'Reilly, M.G. Kopylova, and Y.S. Genshaft, Carbonated peridotite xenoliths from Spitzbergen: implications for trace element signature of mantle carbonate metasomatism, *Earth Planet. Sci. Lett., 119*, 283-297, 1993.

Jackson, E.D., E.A. Silver, and G.B. Dalrymple, Hawaii-Emperor chain and its relation to Cenozoic circum-pacific tectonics, *Geol. Soc. Amer. Bull.*, *83*, 601-617, 1972.

Jambon, A., Earth degassing and large-scale geochemical cycling of volatile elements, in: M.R. Carroll and J.R. Holloway (eds.) Volatiles in Magmas, *Rev. Mineral.*, *30*, 479-517, 1994.

Jarosewich, E., J.A. Nelen, and J.A. Norberg, Electron microprobe reference samples for standard analysis, *Smithsonian Contrib. Earth Sci.*, *22*, 68-72, 1979.

Johnson, K.T.M, Experimental determination of partition coefficients for rare earth and high-field-strength elements between clinopyroxene, garnet, and basaltic melt at high pressures, *Contrib. Mineral. Petrol.*, *133*, 60-68, 1998.

Johnson, L.H., A.P. Jones, A.A. Church, and W.R. Taylor, Ultramafic xenoliths and megacrysts from a melilitite tuff cone, Deeti, northern Tanzania, *J. South African Earth Sci.*, *25*, 29-42, 1997.

Keller, J., and B. Spettel, The trace element composition and petrogenesis of natrocarbonatites, in: K. Bell and J. Keller (eds.), Carbonatite Volcanism, *IAVCEI Proceedings in Volcanology*, *4*, 70-86, 1995.

Kinzler, R.J., Melting of mantle peridotite at pressures approaching the spinel to garnet transition: application to mid-ocean ridge basalt petrogenesis, *J. Geophys. Res.*, *102*, 853-874, 1997.

Konzett, J., and P. Ulmer, The stability of hydrous potassic phases in lherzolitic mantle – an experimental study to 9.5 Gpa in simplified and natural bulk compositions, *J. Petrol.*, *40*, 629-652, 1999.

Koyanagi, R.Y., B. Chouet, and K. Aki, Origin of volcanic tremor in Hawaii - Part I, *U.S. Geol. Surv. Prof. Pap.*, *1350*, 1221-1257, 1987.

Lanphere, M.A., and F.A. Frey, Geochemical evolution of Kohala volcano, Hawaii, *Contrib. Mineral. Petrol.*, *95*, 100-113, 1987.

Leitch, A.M., and G.F. Davies, Mantle plumes and flood basalts: enhanced melting from plume ascent and an eclogite component, *J. Geophys. Res.*, *106*, 2047-2060, 2001.

LeMaitre, R.W., A classification of igneous rocks and glossary of terms, *Blackwell Scientific Publications*, Oxford, 193 pp., 1991.

Lipman, P.W., T.W. Sisson, T. Ui, J. Naka, and J. Smith, Ancestral submarine growth of Kilauea Volcano and instability of its south flank, *this volume*, 2001.

Macdonald, G.A., and T. Katsura, Chemical composition of Hawaiian lavas, *J. Petrol.*, *5*, 82-133, 1968.

Mason, B., Gonnardite (ranite) from Langesundsfjord, *Norsk Geol. Tidssk*, *37*, p. 435, 1957.

Moore, J.G., and D.A. Clague, Coastal lava flows from Mauna Loa and Hualalai volcanoes, Kona, Hawaii, *Bull. Volc.*, *49*, 752-764, 1987.

Moore, J.G, D.A. Clague, and W.R. Normark, Diverse basalt types from Loihi seamount, Hawaii, *Geology*, *10*, 88-92, 1982.

Moore, J.G., D.J. Fornari, and D.A. Clague, The 1877 submarine eruption of Mauna Loa, Hawaii, and the variation of palagonitization rate with temperature, *U.S. Geol. Surv. Bull.*, *1663*, 11 pp., 1985.

Nelson, D.R., A.R. Chivas, B.W. Chappell, and M.T. McCulloch, Geochemical and isotopic systematics of carbonatites and implications for the evolution of ocean island sources, *Geochim. Cosmochim. Acta*, *52*, 1-17, 1988.

Roden, M.F., T. Trull, S.R. Hart, and F.A. Frey, New He, Nd, Pb, and Sr isotopic constraints on the constitution of the Hawaiian plume: Results from Ko'olau Volcano, Oahu, Hawaii, USA, *Geochim. Cosmochim. Acta*, *58*, 1431-1440, 1994.

Rhodes, J.M., K.P. Wenz, C.A. Neal, J.W. Sparks, and J.P. Lockwood, Geochemical evidence for invasion of Kilauea's magmatic plumbing system by Mauna Loa magma, *Nature*, *337*, 257-260, 1989.

Schiano, P., R. Clocchiatti, N. Shimizu, D. Weis, and N. Mattielli, Cogenetic silica-rich and carbonate-rich melts trapped in mantle minerals in Kerguelen ultramafic xenoliths: implications for metasomatism in oceanic upper mantle, *Earth Planet. Sci. Lett.*, *123*, 167-178, 1994.

Schiano, P., J.M. Eiler, I.D. Hutcheson, and E.M. Stolper, Primitive CaO-rich silica undersaturated melts in island arcs: Evidence for involvement of clinopyroxene-rich lithologies in the petrogenesis of arc magmas, *Geochem. Geophys. Geosyst.*, 1, paper number 1999GC000032, 2000.

Schleicher, H., A. Baumann, and J. Keller, Pb isotopic systematics of alkaline volcanic rocks and carbonatites from the Kaiserstuhl, upper Rhine rift valley, *Chem. Geol.*, *93*, 231-243, 1991.

Shaw, D.M, Trace element fractionation during melting, *Geochim. Cosmochim. Acta*, *34*, 237-243, 1970.

Shimizu, N., and I. Kushiro, The partitioning of rare earth elements between garnet and liquid at high pressure: preliminary experiments, *Geophys. Res. Lett.*, *2*, 413-416, 1975.

Shinozaki, K., Z.-Y. Ren, and E. Takahashi, Geochemical and petrological characteristics of Nuuanu and Wailau landslide blocks, *this volume*, 2001.

Siems, D.F., The determination of 30 elements in geologic materials by energy-dispersive X-ray fluorescence spectrometry, *U.S. Geol. Surv. Open-File Report*, *00-475*, 13 pp.

Skulski, T., W.G. Minarik, and E.B. Watson, High-pressure experimental trace element partitioning between clinopyroxene and basaltic melts, *Chem. Geol.*, *117*, 127-147, 1994.

Staudigel, H., A. Zindler, S.R. Hart, T. Leslie, C. –Y. Chen, and D. Clague, The isotope systematics of a juvenile intraplate volcano: Pb, Nd, and Sr isotope ratios of basalts from Loihi Seamount, Hawaii, *Earth Planet. Sci. Lett.*, *69*, 13-29, 1984.

Stille, P., D.M. Unruh, and M. Tatsumoto, Pb, Sr, Nd, and Hf isotopic constraints on the origin of Hawaiian basalts and evidence for a unique mantle source, *Geochim. Cosmochim. Acta*, *50*, 2303-2319, 1986.

Taggart, J.E., J.R. Lindsay, B.A. Scott, D.V. Vivit, A.J. Bartel, and K.C. Stewart, Analysis of geologic materials by wavelength-dispersive X-ray fluorescence spectrometry, in: P.A. Baedecker (ed.), Methods for geochemical analysis, *U.S. Geol. Surv. Prof. Pap.*, *1770*, E1-E19, 1987.

Tanaka, R., E. Nakamura, and E. Takahashi, Geochemical evolution of Ko'olau volcano, Hawaii, *this volume*, 2001.

Tatsumoto, M., Isotopic composition of lead in oceanic basalts and its implication to mantle evolution, *Earth Planet Sci. Lett.*, *38*, 63-87, 1978.

Wagner, T.P, D.A. Clague, T.L. Grove, E. Hauri, and E. Takazawa, Melt-lithosphere interaction in the Hawaiian plume: reconciliation of trace element abundances and phase equilibria results, *International Workshop on Intraplate Volcanism: The Polynesian Plume Province*, Papeete, Tahite, p. 65, 1993.

Walter, M.J., Melting of garnet peridotite and the origin of komatiite and depleted lithosphere, *J. Petrol.*, *39*, 29-60, 1999.

Walter, M.J., T.W. Sisson, and D.C. Presnall, A mass proportionation method for calculating melting reactions and application to melting of model upper mantle lherzolite, *Earth Planet. Sci. Lett.*, *135*, 77-90, 1995.

Wolfe, E.W., The geology and petrology of Mauna Kea Volcano, Hawaii – A study of postshield volcanism, *U.S. Geol. Surv. Prof. Pap.*, *1557*, 129 pp.

Wolfe, E.W., and J. Morris, Geologic map of the Island of Hawaii, *US Geol. Surv. Misc. Invest. Map I-2524*, 1996.

Wendlandt, R.F., and D.H. Eggler, The origins of potassic magmas: 2. Stability of phlogopite in natural spinel lherzolite and in the system $KAlSiO_4$-MgO-SiO_2-H_2O-CO_2 at high pressures and high temperatures, *Amer. J. Sci.*, *280*, 421-458, 1980.

Wright, T.L., Chemistry of Kilauea and Mauna Loa lava in space and time, *U.S. Geol. Surv. Prof. Pap. 735*, 40 pp, 1971.

T.W. Sisson, MS 910, U.S. Geological Survey, 345 Middlefield Rd., Menlo Park CA 94025, USA

P.W. Lipman, MS 910, U.S. Geological Survey, 345 Middlefield Rd., Menlo Park CA 94025, USA

J. Naka, Deep Sea Research and Development, Japan Marine Science and Technology Center (JAMSTEC), 2-15 Natsushima-cho, Yokosuka, Kanagawa 237-0061, Japan

Introduction to Section 3
Giant Landslides Northeast of O‘ahu: When, Why and How?

Michael Garcia

University of Hawai‘i

The presence of giant landslides on the flanks of numerous ocean island volcanoes including Hawai'i, Reunion, and the Canary islands is well documented. The catastrophic failure associated with some of these giant landslides appears to be a common feature of ocean island volcanoes. The collapse of ocean island is now recognized as an important global erosional agent. The rapid growth (~1 to 1.5 million years) and enormous size of Hawaiian volcanoes (individually, up to 8.5 km of relief and 80,000 km^3 volume) causes them to be particularly unstable. The collapse of Hawaiian volcanoes has generated some of the largest landslides on Earth and colossal tsunami waves. Dozens of major landslides, some with volumes >1000 km^3 and large blocks >1 km in height, have been recognized along the Hawaiian Ridge. On average, a major landslide has been identified every 32 km along the ridge. Based on an average ~10 cm/year northwest motion of Hawaiian volcanoes, a giant landslide occurs about every 320 thousand years, and smaller slides must have occurred much more frequently. The wealth of information on Hawaiian landslides has led to Hawai'i becoming the type example for ocean island landslides.

A major obstacle in gaining a better understanding of when, why and how giant landslides form in Hawai'i has been the limited detailed bathymetric and side-scan sonar imagery coverage for Hawai'i. This dramatically changed following the surveying northeast of the island of O‘ahu during JAMSTEC cruises to Hawai'i in the summers of 1998 and 1999. The new bathymetric data allow for comprehensive reconstructions of the Nu‘uanu and Wailau landslides by *Moore and Clague* [this section], *Yokose* [this section] and *Satake et al.* [this section]. *Moore and Clague* [this section] also review the history of bathymetric surveys and geologic interpretations for these slides. *Moore and Clague* [this section] conclude that the volume inequities between the size of the "holes" left by these slides and the volume of the blocks indicate that there was long-term bulging on the slopes of Koolau volcano before the landslide. *Yokose* [this section] enhances our understanding of the products and processes related to both these landslides by combining the new bathymetric data with detailed submersible observations of landslide debris outcrops. The high rainfall on the windward side of O‘ahu and Moloka`i is interpreted by *Yokose* [this section] to have increased the importance of sedimentary processes in creating instabilities on the submarine flanks of Ko`olau and Moloka`i volcanoes. The exceptional size of the Nu`uanu and Wailua landslides are thought by *Yokose* [this section] to be related to the high rainfall on the windward sides of these islands.

How big were the tsunamis associated with the Nu`uanu and Wailau landslides? *Satake et al.* [this section] answer this question by determining the volume change associated which each landslide. Calculations for tsunami generation and propagation for these slides were computed using a kinematic landslide model. *Satake et al.* [this section] calculated that mammoth waves were created by the Nu`uanu and Wailua slides. For example, waves from the Nu`uanu slide would have been >100 m high in Hawaii and would have struck the islands within minutes. This tsunami was directed towards California. Waves up to 70 m high would have reached the shores of southern California in ~5 hours [*Satake et al.*, this section].

Another major obstacle to understanding giant Hawaiian landslides concerns the lack of sampling of landslide debris. This was due in part to the debris being in water depths of 3-5 km below sea level, which is outside the range of most submersibles. The JAMSTEC deep submergence vehicles *KAIKO* and *SHINKAI 6500,* however, are ideal for exploring landslide debris because they can explore the full depth range of the debris. The papers by *Yokose* [this section], *Clague et al.* [this section], *Shinozaki et*

Hawaiian Volcanoes: Deep Underwater Perspectives
Geophysical Monograph 128
Copyright 2002 by the American Geophysical Union

al. [this section] and *Tanaka et al.* [this section] examine the lithologic variation, volcaniclastic origin, petrology, and geochemistry of rocks recovered from the landslide debris.

In addition, four piston cores and many rock dredgehauls were made to sample the debris from the landslides. The piston cores contain fresh volcanic glasses in turbidite sands and silts. The geochemistry of these glasses [*Sherman et al.*, this section], in combination with paleomagnetic and paleontologic studies of the cores [*Kanamatsu et al.*, section 2 of this volume], help to determine the history of several large and numerous smaller landslides that have occurred northeast of O'ahu and Moloka`i islands. Many of the rock samples recovered are breccias and hyaloclastites. *Clague et al.* [this section] examined the origin of these monomict and polymict rocks and their implications for the tectonic history of these islands. They conclude that volcano spreading has occurred along low angle thrust faults within the tectonized core of the volcano. The petrology and geochemistry of the lavas collected from the rock dredges was combined with the results for the submersible-collected samples in papers by *Shinozaki et al.* [this section] and *Tanaka et al.* [this section]. Their results show that the volcanic rocks from the Nu'uanu landslide debris are petrographically and geochemically distinct from the geochemically unusual subaerial lavas that characterized Ko'olau Volcano. *Shinozaki et al.* [this section] conclude that the distinct geochemistry of subaerial Ko`olau lavas represents just the final stage of this volcano. *Tanaka et al.* [this section] provide trace element and Sr-Nd-Pb isotopic data for subaerial and submarine Ko`olau, Nu`uanu and Wailua landslide deposit lavas. These data reveal that Ko`olau volcano evolved from Kea-type to Ko`olau-type geochemistry, which spans nearly the entire compositional range for Hawaiian lavas. The results document the secular changes in the Hawaiian plume.

Mapping the Nuuanu and Wailau Landslides in Hawaii

James G. Moore

U. S. Geological Survey, Menlo Park, California

David A. Clague

Monterey Bay Aquarium Research Institute, Moss Landing, California

The development of concepts about the giant Nuuanu and Wailau landslides have been dependant on bathymetric surveys of the landslide region northeast of Oahu and Molokai. Improved technology of sonar surveying and of navigation through the years has increased the precision of these maps providing more details about the landslides. The landslides were mapped by a U.S. Navy-Scripps Institution single-beam sonar survey in 1953-54, but the resulting map was not of sufficient detail to permit recognition of the landslides. A more detailed U. S. Navy survey in 1958 utilized an improved radio navigation system and provided the bathymetric detail necessary for recognition of the landslides. The GLORIA side-scan sonar surveys during 1986 to 1991 produced an acoustic reflectance image of the seafloor of the entire Hawaiian area and these images clearly revealed the nature and size of the Nuuanu and Wailau landslides and identified an additional 66 landslides longer than 20 km in the Hawaiian region. In early 1998 Simrad multibeam high-resolution surveys utilizing GPS navigation revealed new detail in the proximal parts of the landslides. The 1998 and 1999 JAMSTEC multibeam surveys produced detailed bathymetric maps of the entire landslide area for the first time. In the Nuuanu landslide the exposed volume of the 54 blocks larger than 1 km^3 is 1400 km^3. Refitting of the blocks back to their apparent pre-slide positions shows that they do not fit easily into the reentrants of the host volcanoes. A pre-landslide seaward-extending bulge in the host volcano flanks could accommodate this excess volume.

1. INTRODUCTION

The giant submarine landslides northeast of Oahu were recognized only after the techniques of bathymetric mapping were able to provide accurate and detailed charts. Subsequently the increasing detail of new generations of bathymetric and sidescan maps has led to an improved understanding of these giant features. The development of ideas on these landslides parallels the history of technological advances made in sea floor mapping, which will be outlined in this paper.

Other types of oceanographic investigations, including geophysical measurements of gravity, magnetics, and seismic reflection studies, and ocean floor investigations, including dredging, bottom photography, coring, drilling, and manned submersible studies, have continued to enhance our understanding of the landslides, but seafloor mapping

Hawaiian Volcanoes: Deep Underwater Perspectives
Geophysical Monograph 128
Copyright 2002 by the American Geophysical Union

remains the most important investigative tool. The early recognition of such landslides in the Hawaiian region has stimulated studies leading to the discovery of similar large landslides on marine volcanoes the world over.

2. SEAFLOOR SURVEYS OF THE NUUANU AND WAILAU LANDSLIDES

2.1. Early Bathymetric Mapping in Hawaii

The advent of echo sounding in the 1920s greatly improved knowledge of the bathymetry of the abyss. Previously, bathymetric maps were necessarily generalized because the laborious process of lowering a weight to the ocean floor made wire soundings tedious, time consuming, and therefore widely spaced. As a result of the development of the sonic depth finder, the Hydrographic Office of the United States Navy began a systematic program of bathymetric mapping in regions of interest. The Hydrographer of the Navy, Captain G. S. Bryan [1940] noted that "The main highways of the seven seas have been fairly well covered, but there still exist many unsounded areas lying outside of the regular steamer lanes. With a view to filling in these vacant spaces on charts, the Hydrographic Office has systematically arranged to route naval vessels through them whenever it was possible to do so. The naval tanker *Ramapo*, for instance, has for years been periodically crossing the Pacific Ocean. Each time it has followed a different route in order to obtain soundings in previously unexplored areas. During this time it has crossed that ocean 50 times and taken over 35,000 soundings." A bathymetric map of the North Pacific Ocean [*U. S. Navy,* 1939] based on these echo soundings (Figure 1) remained the authoritative reference for this large region for years.

Betz and Hess [1942] recontoured the sounding data of the 1939 Navy chart and discovered important regional bathymetric features. The volcanic Hawaiian Ridge was found to be centered along the axis of a large-scale region of elevated topography some 1000 km (600 miles) wide, trending west-northwest, which they named the *Hawaiian Swell.* In addition, zones of apparent transcurrent faulting, identified by topographic irregularities, were mapped that paralleled the Hawaiian Ridge. Some of these were believed to act as fissure vents that enabled volcanic material to erupt on the sea floor where it constructed the Ridge and Swell.

Such early regional bathymetric maps combined with inshore and harbor charts were all that was available during the early phases of geologic mapping of the Hawaiian Islands begun in 1924 by Harold Stearns. During this period of active fieldwork on land, only the most rudimentary knowledge of submarine geology existed. The resources and methods had not yet been developed to undertake marine studies of the major part of the volcanoes that lay beneath the sea. The unattributed bathymetric map (Figure 2) utilized by *Stearns* [1946] in his discussion of the Hawaii Ridge was apparently derived from an early U. S. Coast and Geodetic Survey chart. This map (Figure 2) shows little detail in the landslide region northeast of Oahu and north of Molokai except for a northeast-trending ridge that may have been based on one or a few soundings falling on one of the landslide blocks. Stearns' map does, however, show one major improvement over the previous Navy map (Figure 1). It depicts the major rift zone ridge (Haleakala or Hana Ridge) extending east from the east cape of Maui, a feature missing from the Navy map.

2.2. Mid-Pacific Expeditions During the 1950s

About a decade after publication of the Navy bathymetric map of the Pacific, further improvement of mapping adjacent to Hawaii was facilitated by a series of fathograms obtained during the joint Scripps Institution of Oceanography-U.S. Navy Electronics Laboratory Mid-Pacific Expedition in 1950 using the Navy vessel *USS EPCE(R) 857* and the Scripps vessel *Horizon.* This refined map (Figure 3) outlined the depressed area on the Hawaiian Swell that flanks the central Hawaiian Ridge and was termed the *Hawaiian Deep* [*Dietz and Menard*, 1953] (later called the *Hawaiian Trough* by *Stearns* [1966]). The high region between the Deep and the outer margin of the Swell was called the Hawaiian Arch. *Dietz and Menard* suggested that the Swell may "have been produced by arching of the crust above a zone of divergence between two thermal convection cells in the viscous subcrust." The Ridge was presumably built by lava outpouring from fracture-fissures produced at the axis of the uparched crust. *Dietz and Menard* [1953] first proposed that the great weight of the newly erupted volcanic material comprising the Ridge caused the central zone of the Swell to subside isostatically producing the Deep flanking the Ridge.

Shortly after the 1950 Mid-Pacific expedition, six joint U.S. Navy Electronics Laboratory-Scripps Institution expeditions took place to the mid-Pacific area, and a number of crossings of the Hawaiian Deep and Arch were made. The resulting maps refined the length and character of the Hawaiian Deep and Arch. Part of these single-beam echo-sounding surveys conducted in 1953 and 1954 along a 8-km (5-mile) grid detailed for the first time the sea-floor topography northeast of Oahu and revealed a number of peaks in the Hawaiian Deep (Figure 4). The largest peak (later named Tuscaloosa Seamount), located near the axis of the Hawaiian Deep, displayed a basal diameter of 20-30 km and a height of 2000 m.

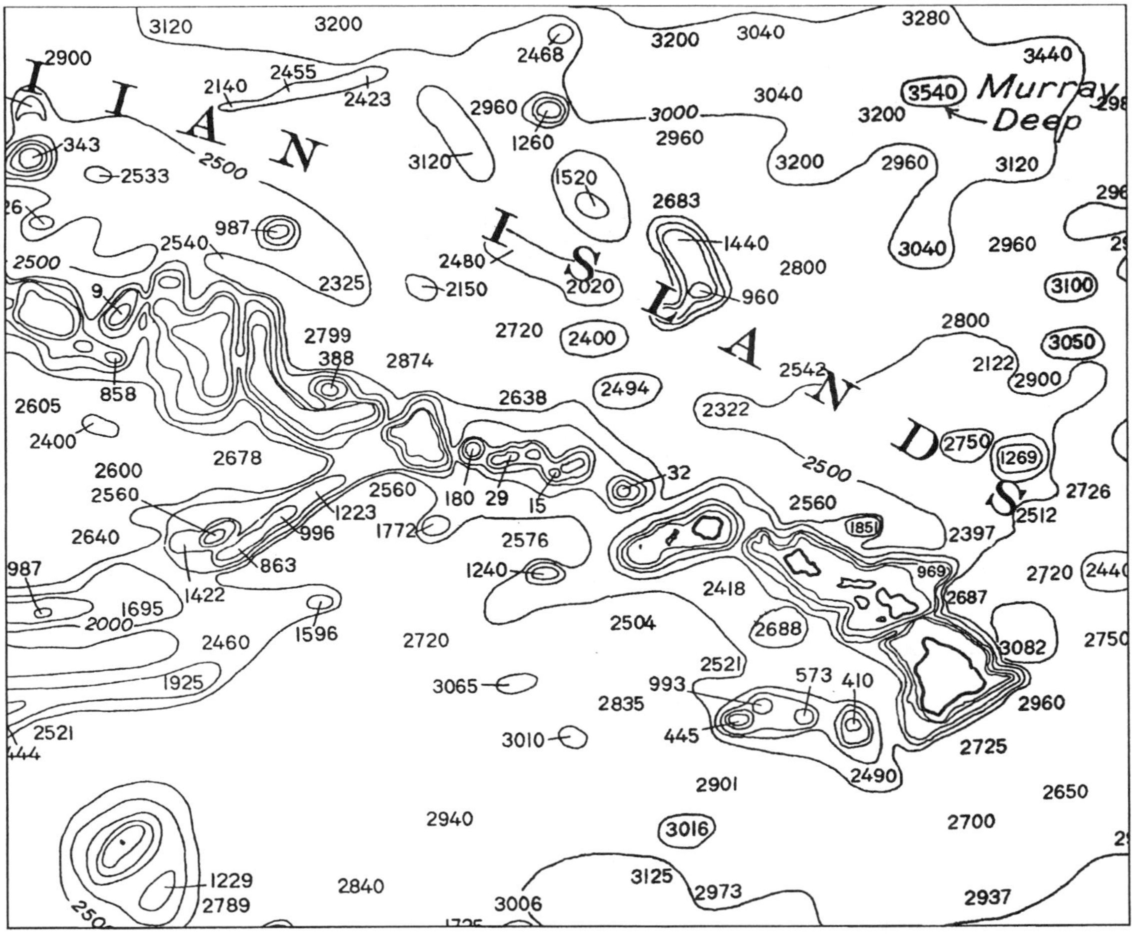

Figure 1. Part of U. S. Navy Hydrographic Office chart of southeast Hawaiian Ridge [*U. S. Navy*, 1939]. Contour interval 500 fathoms (914 m).

Hamilton [1957] concluded that these peaks were volcanic cones that had grown in their present position on the sea floor. He stated; "The volcanic character of these seamounts is indicated by their slopes, symmetrical profiles, oval shapes in plan, their association with volcanic features of the Hawaiian Ridge, and by the volcanic material cored from their slopes or the adjacent sea floor." The association he mentioned was the alignment of the offshore seamounts with the late volcanic eruptions in the southeast part of Oahu (the Honolulu Volcanics). G. A. Macdonald concurred that the offshore seamounts were along this secondary volcanic trend [*Hamilton*, 1957, p. 1019].

The contouring of bathymetric data points that are limited and widely spaced necessarily produces generalized curves that are rounded in shape. It was these rounded contours that suggested to Hamiliton that the bathymetric highs northeast of Oahu were volcanoes. When comparing this map with later more detailed maps of this same area (Figure 5), the overall shape and position of the Nuuanu and Wailau landslides can be discerned, but the representation is so poor that landsliding was not proposed at that time as an explanation for these seamounts.

The Hamilton map (Figure 4) also revealed a series of prominent submarine canyons in the insular shelf northeast of Oahu. The canyons are as much as 200 m deep and lead down to depths of 1000 to 2000 m. *Hamilton* [1957] concluded that these canyons were partly eroded by submarine processes because the depths of the lower parts of the canyons were below any previous estimates of submergence.

2.3. The Pioneer-Rehoboth Survey, 1957

Paralleling the advances in measuring the ocean depth by sound pulses were improvements in navigation that were basic to producing accurate maps. The exact position of a measured depth was just as important as the precision of

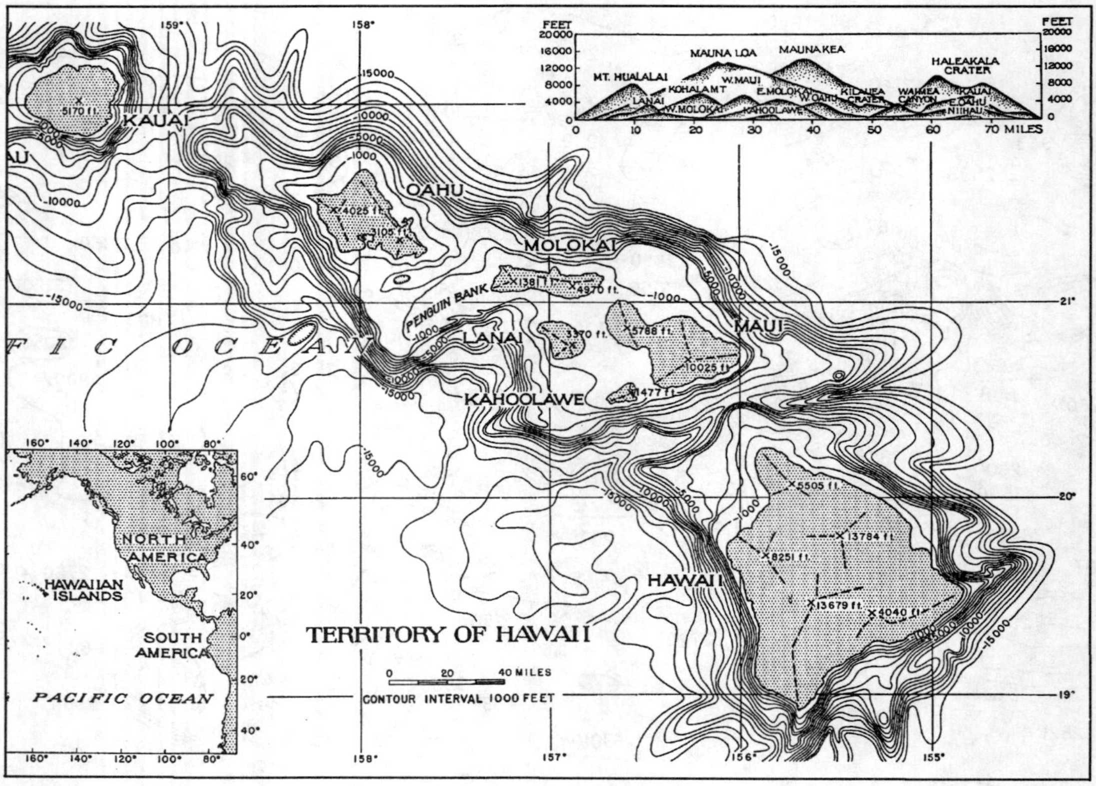

Figure 2. Bathymetry of the Hawaiian Ridge from Kauai to Hawaii, contour interval 1000 feet (304 m). Inset shows subaerial profiles of the volcanic peaks of the main islands [*Stearns,* 1946].

that depth. Prior to the Second World War navigation when out of sight of land was based on celestial observations and dead reckoning. Errors in position were commonly greater than several nautical miles.

A major step forward in navigation was the development of Loran (LOng range RAdio Navigation), a position-fixing system that consists of a series of stations on land that transmit radio signals picked up by receivers on ships. Determination of ship position is calculated by the time difference between reception of two or more modulated signals from widely spaced transmitters. By 1945 a chain of Loran-A transmitters was in operation and widely used for navigation. The range was 400-800 nautical miles during the daytime (groundwave) and as much as 1400 nautical miles at night (skywave) when radio waves were reflected from the upper ionized portion of the atmosphere. Accuracy was within about one nautical mile in the daytime and as poor as 6 nautical miles at night. An improved loran system (Loran-C) with greater accuracy and range became operational in 1957.

In 1957, the same year that the Hamilton map was published, the U. S. Navy and the U. S. Coast and Geodetic Survey embarked on a new mapping program of the Hawaiian Region. The Coast and Geodetic Survey ship *Pioneer* and the Navy ship *Rehoboth AGS-50* conducted this single-beam echo-sounding program—the Area Victor Exploration Survey— during April to November 1957. Navigation was controlled by a system of land-based radio beacons especially installed for the purpose. The resulting bathymetric map was released by the U. S. Navy early in the 1960s (Figure 5) and published in 1966 [*U. S. Navy,* 1966]. It was available for a research cruise utilizing the *Pioneer* off Hawaii in 1962 [*Moore and Reed,* 1963]. The map (scale, 1:1,000,000; contour interval, 100 fathoms = 600 feet or 183 m) displayed the bathymetry in far greater detail than was previously available, and was a marked improvement over the *Hamilton* [1957] map published a few years earlier (Figure 4).

The *Pioneer-Rehoboth* map depicted a prominent slope break at a depth of 3000-3600 ft (~1000 m) about 15 km offshore from the islands. This discontinuity apparently represents the position of the shoreline when shield building ceased on Koolau volcano (the eastern Oahu volcanic edifice). The map also shows better detail of the Hawaiian

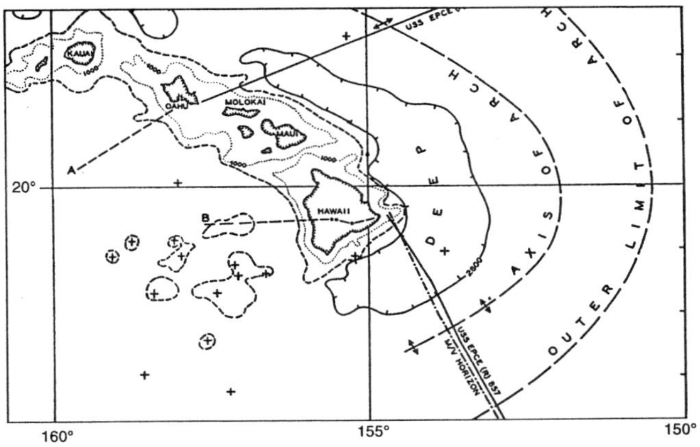

Figure 3. Regional chart of the southeast end of the Hawaiian Ridge showing the axis and outer limit of the Hawaiian Arch (which define the Hawaiian Swell). The 1000 fathom (dotted) and 2000 fathom (dashed) contours are shown, and the axis of the Hawaiian Deep is defined by the 2600 and 2800 fathom contours. Crosses indicate the summits of seamounts. [*Dietz and Menard*, 1953]

Deep on the northeast base of the Ridge in the vicinity of Oahu and Molokai (Figure 5).

The most arresting feature on the *Pioneer-Rehoboth* map, however, is the hilly terrain noted by Hamilton off northeast Oahu. The new survey showed more than twice as many hills and defined two blocky, weltlike masses that crossed the axis of the Hawaiian Deep. These masses appear to originate in embayments on the flank of the Hawaiian Ridge, one on the northeast flank of east Oahu, and one on the north flank of Molokai (Figure 5). The individual bathymetric highs are better resolved on the new map with several showing more angular and irregular shapes as compared to the Hamilton map (Figure 4). The tongue-like masses of hilly terrain were attributed to giant landslides [*Moore*, 1964] on the basis of this map. The landslides apparently transported blocky debris across the Deep and up onto the Arch. The larger landslide could be traced 160 km offshore and was up to 50 km wide. The largest block in the landslides, as identified by *Moore* [1964] corresponded to the volcano mapped by *Hamilton* [1957]. In the new map it is depicted as a nearly flat-topped seamount about 30 km long and two km high. This feature was later named Tuscaloosa Seamount (Figure 7).

Data from the *Pioneer-Rehoboth* bathymetric chart was quickly incorporated into oblique-viewing sea-floor maps. A physiographic map of the northeastern Pacific (Figure 6, scale 1:6,000,000, *Menard* [1964], and a relief map of the ocean floor surrounding the major Hawaiian Islands (Figure 7, scale 1:6,000,000, *Stearns* [1966] both utilized the new bathymetry. The latter chart contains new names of seamounts selected by Harold Stearns to honor scientists who worked on Hawaiian geology. Among the first planimetric bathymetric maps utilizing the data was published in 1970 as part of a series of maps covering the North Pacific [*Chase et al., 1970*].

2.4. The Spencer F. Baird Survey, 1962

In 1947, while working on an assignment for the U.S. Navy, Francis Shepard ran sounding lines along the northeast side of Oahu and the north shore of East Molokai and discovered several large submarine canyons on the island shelves. Returning in 1962 with the Scripps vessel *Spencer F. Baird*, bathymetric studies were made on canyons off the islands of Kauai and Molokai. Navigation of the echo-sounding surveys was by visual sights to land in the day-time and radar during darkness and cloudy weather.

The resulting map which extends 25 km offshore north of Molokai (Figure 8) indicates that submarine canyons occur offshore from the high and precipitous cliffs along the east part of the windward north coast, a region of high rainfall [*Shepard and Dill*, 1966]. Surveys east and west of this region revealed only rare small submarine canyons. Generally the canyons occur offshore from, and on line with, subaerial canyons on East Molokai Volcano. A prominent peninsula off the north coast, Kalaweo (also called Kalaupapa), was built by a volcano younger than the large sea cliff and has been dated at 455 ± 115 ka [*Clague et al.*, 1982].

Shepard and Dill noted that the courses of the upper (southern) parts of the canyons adjacent to Kalaupapa suggest that this volcano grew after initial canyon cutting, and partly filled the upper parts of the canyons. Their map shows that the canyon heads are consistently a few kilometers offshore, and extend about 13 km to the edge of the shelf at about 1200 m depth (Figure 8). Some of the canyons extend down the slope beyond the shelf edge to depths of at least 2000 m. The general size and shape of these canyons, and the fact that they appear to be extensions of on-land canyons, suggests that they represent the drowned portions of a canyon system that was largely cut subaerially and later subsided beneath the sea. However, the part of the canyons that extend beyond the lip of the 1200-m shelf down to water in excess of 2000 m deep must have been carved subaqueously [*Clague and Moore*, 2001].

2.5. The McArthur Surveys, 1968

A more extensive inshore survey (extending only 9 km offshore) was conducted on the north insular shelf of Molokai by the Coast and Geodetic Survey ship *McArthur* in

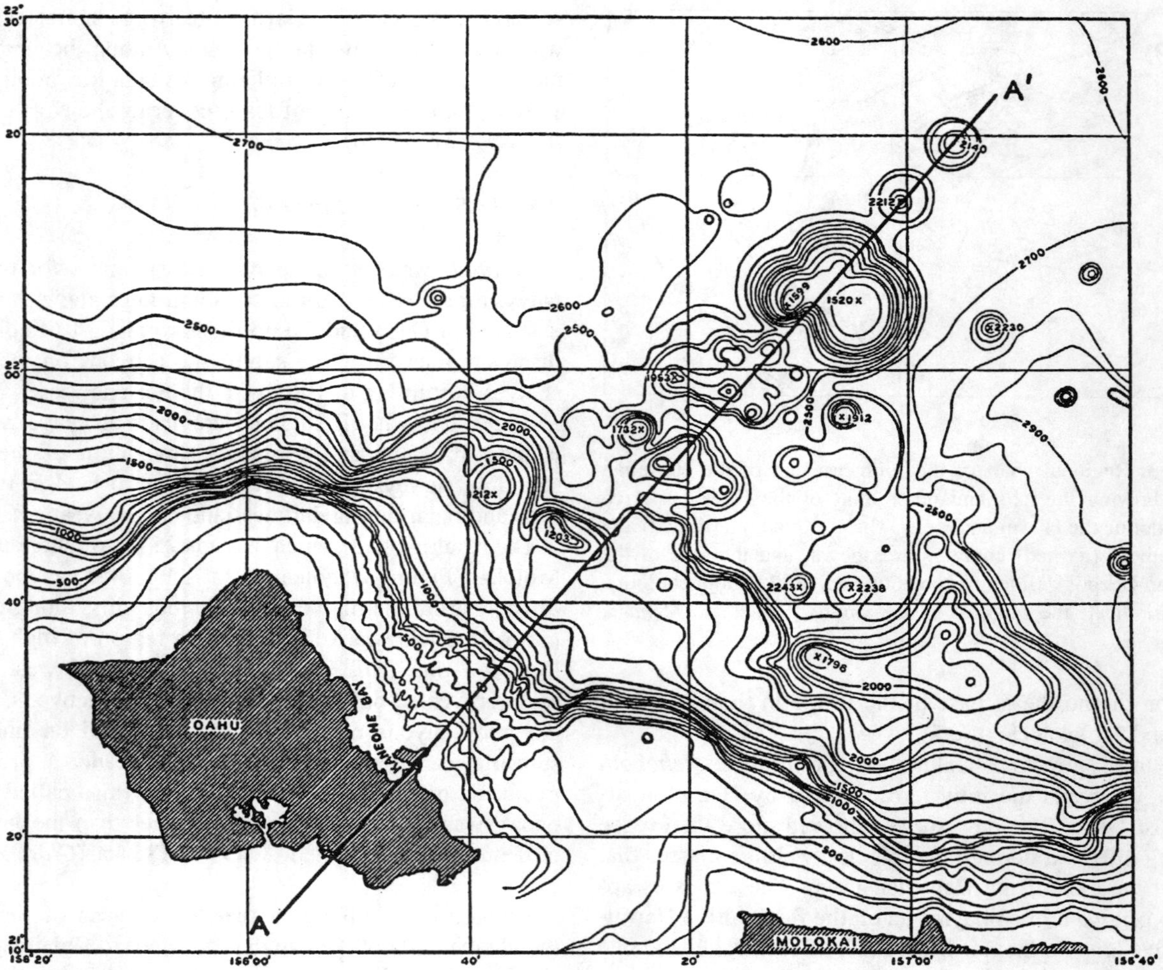

Figure 4. Bathymetric chart of an area northeast of Oahu; contour interval 100 fathoms (600 ft = 183 m). From *Hamilton* [1957]

1968. A single beam echosounder was used and navigation was by visual and radar fixes on land. The major parts of a dozen large canyons that cut into the shelf were then mapped in more detail. This mapping also supported the notion that, despite the 3 km gap where canyons are missing adjacent to the coast, the on land canyons can generally be linked with offshore canyons [*Mathewson*, 1970]. It was further concluded that this study "does not lend support to the contention" that the north flank of Molokai was modified by the action of a giant landslide.

Also in 1968 the *McArthur* remapped with a single beam survey about the same area off north Molokai as that mapped by the *Spencer F. Baird* in 1962. The resulting map was not published until 1989 [*Moore et al, 1989*]. It displays, in particular, more detail of the giant scarp descending from the 1200 m terrace, but does not reach far enough north to show any of the Wailau landslide blocks.

2.6. The Mahi Survey, 1969

Tuscaloosa Seamount was further surveyed during a 1969 cruise of the University of Hawaii research vessel *Mahi* utilizing a single beam echosounder, refined tables for sound velocity in seawater, and navigation based on 3-minute Hawaii Army National Guard radar fixes [*Langford and Brill*, 1972]. A detailed bathymetric map of the seamount contoured at 100 m interval clearly showed the flat top at a depth of 2700 m (Figure 9). This led to the notion that the seamount was a guyot rather than a giant landslide block and that the flat top was carved by marine coastal erosion at a time when the block stood above the sea [*Langford*, 1969; *Langford and Brill*, 1972]. This process, plus the alignment of the seamounts with the northeast-trending volcanic vents of east Oahu, and the huge size of the proposed landslide, led others to support the volcanic

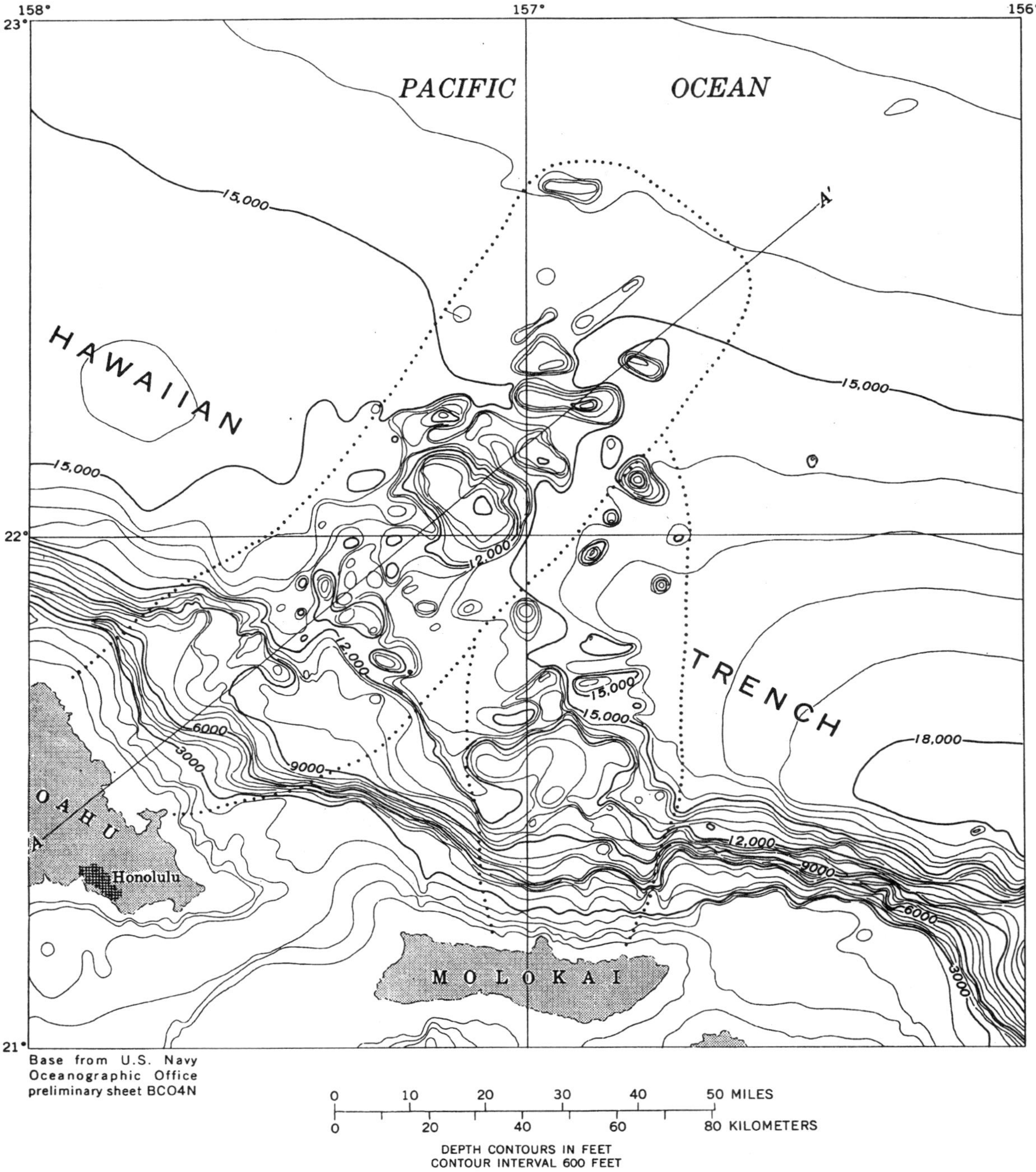

Figure 5. Part of *Pioneer-Rehoboth* 1957 survey map of submarine topography northeast of Oahu [*U. S. Navy*, 1966]. Contour interval 600 feet; inferred boundaries of giant landslides shown by dotted lines [*Moore*, 1964].

model for the anomalous seamounts northeast of Oahu [*Macdonald and Abbot*, 1970; *Andrews and Bainbridge*, 1972].

Also during 1969, the research vessel *Mahi* obtained reflection profiles using airgun and sparker sources across the canyons off the northeast coast of Oahu at the head of the Nuuanu landslide [*Andrews and Bainbridge*, 1972]. These profiles showed that the upper parts of some of the canyons were filled with up to 600 m of sediment and suggested that "the canyons were originally subaerial features

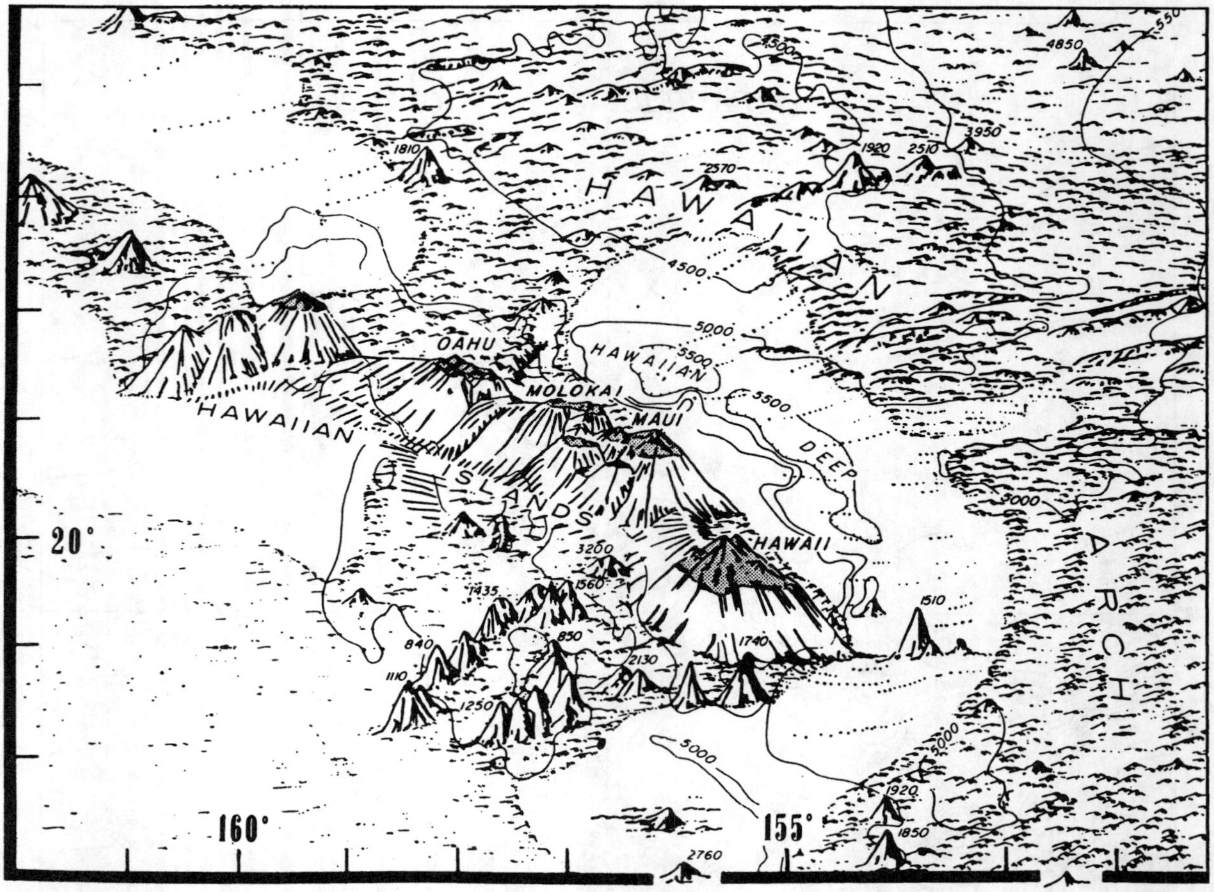

Figure 6. Physiographic map of the Hawaiian Region [*Menard*, 1964]. The map incorporates data from the *Pioneer-Rehoboth* survey; deep-water contours have 500 m interval.

which have drowned as the island ridge subsided." Further seismic reflection surveys in the early 1970s from the research vessel *Teritu* defined several sediment-filled submarine canyons off Oahu that clearly link the present subaerial canyons with the offshore submarine canyons across the 3-km-wide canyon-free nearshore zone. According to *Coulbourn et al.* [1974] "Results of sub-bottom profiling provide strong support to those who advocate a subaerial origin for Oahu's submarine canyons."

In the next several years, a series of somewhat detailed single beam charts were compiled that covered various parts of the Hawaiian region. These included a moderately detailed chart of the general region of the Hawaiian Islands [*Wilde et al.*, 1980, scale, 1:1,000,000; contour interval, 100 m], and a bathymetric atlas [*Campbell*, 1987, scale 1:200,000, contour interval 20 m] utilizing a large number of recent track lines adjacent to the Hawaiian Islands but only marginally including the region north of Oahu and Molokai.

2.7. The Farnella Surveys, 1986-91

A major advance in mapping of the ocean floor came with the invention of the side-scan sonar system utilizing a linear array of sonic sources in line with the ship that send sonic pulses out normal to the ship's track to intersect the ocean floor at different distances from the ship track. These signals are reflected back to receivers mounted on the ship's hull or on a towed vehicle. The point on the ocean floor from which they reflect is determined by the length of time taken to return to the receivers *assuming a flat ocean floor*. From the intensity of received reflections, the sonic reflectance of the ocean floor is recorded and from the timing of received reflections, the location on the ocean floor is fixed. As the ship moves along its track, and the process of sending and receiving signals continues, these data are assembled and plotted into a swath image or chart depicting the sonic reflectance or depth of the ocean floor. The advantage of this system is that it maps a swath several

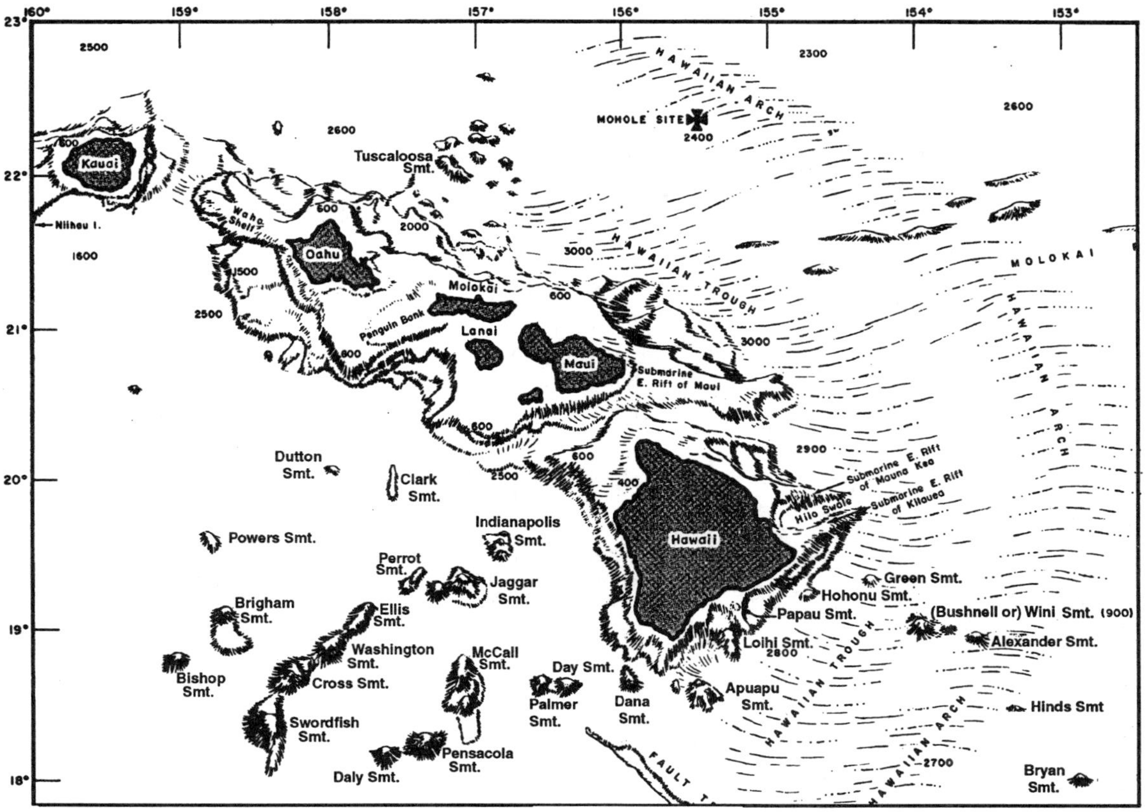

Figure 7. Relief map of the ocean floor surrounding the major Hawaiian Islands, depths in fathoms. Most of the seamounts have been named to honor deceased scientists who have contributed to Hawaiian geology [*Stearns*, 1966].

times wider than the ocean depth, rather than a single-track line directly beneath the ship. Hence, mapping can proceed at a rapid pace.

President Reagan, in 1983, proclaimed United States sovereignty over seabed resources in an Exclusive Economic Zone (EEZ) that extends 370 km (200 nautical mile) offshore from U.S. territory. The U.S. Geological Survey was assigned the task of mapping this vast area of some 13 million square km. A swath mapping system was employed using a side-looking sonar called GLORIA (for Geologic LOng-Range Inclined ASDIC, where ASDIC is an acronym for sonar), because a swath up to 30 km wide could be mapped from a single track, thus minimizing the time needed for the survey. This system, which produced an acoustic backscatter image (Figure 10), but not a contour map, was developed by the Institute of Oceanographic Sciences in the United Kingdom [*Somers et al.*, 1978].

Most of this survey utilized a new satellite-based navigation system capable of locating the ship more precisely than the Loran system. The Satellite Transit system relied on an array of 6 satellites, but required that only one satellite be above the horizon at a time; otherwise the satellite signals would interfere with one another. The position was determined by the Doppler shift of the satellite signal coupled with the orbital parameters of the satellite. The ship position could be fixed to within about 500 m. During the periods when no satellites were visible, dead reckoning and Loran-C signals were used to fix the ship position.

Later in the GLORIA survey the more robust GPS navigation system became operational. It employed a constellation of Earth-orbiting satellites that provided ranging information from each satellite so that a position required information from at least three satellites above the horizon. This system provided fixes more frequently with an accuracy of a few tens of meters.

The entire Hawaiian EEZ was mapped from 1986 to 1991 in 400 days at sea with the research vessel *Farnella* towing the GLORIA vehicle [*Groome et al.*, 1997]. The GLORIA images of the Hawaiian region revealed the astonishing fact that major landslides cover more than half of the slopes of the Hawaiian Ridge [*Moore, et al.*, 1989; *Moore, et al.*, 1994). Sixty-eight landslides more than 20 km long occur on the 2200 km of the Ridge imaged, and therefore represent one mapped landslide every 32 km,

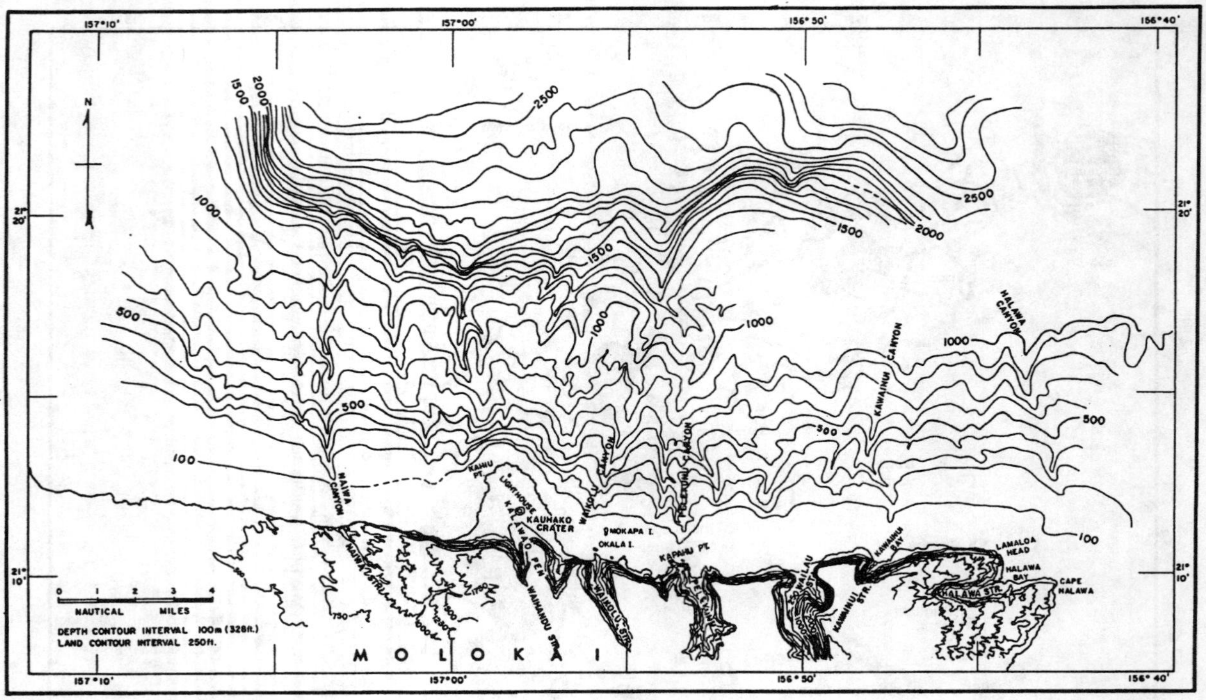

Figure 8. Bathymetric map off the north coast of Molokai showing submarine canyons and their apparent relation to land canyons [*Shepard and Dill*, 1966]. Kauhako Crater and the peninsula (Kalaupapa) on which it sits is a young post-landslide volcano 455 ± 115 years old [*Clague et al.*, 1982]. Contour interval 100 m.

which is equivalent to one about every 350,000 years. Actually many other landslides are no doubt covered by younger ones as well as by the propagating volcanoes of the Hawaiian Ridge so that this recurrence interval should be considered a maximum. Continued work on the estimation of past landslide frequency is important not only because of the hazard of catastrophic movement, but also because of the role of sliding as an integral process in the life history of the volcanoes.

Individual landslide blocks ranging in size from 30 km in length down to blocks occupying a single pixel 50 X 100 m in size were recorded on these images by their high reflectivity (Figure 10). The larger blocks are commonly angular and flat-topped, and their geometry invites attempts to fit them together in their pre-landslide configuration. Near the distal part of the Nuuanu landslide, pre-slide lava flows can be detected through the landslide, implying no more than a few tens of meters of debris covering the flows.

The GLORIA image of the Nuuanu landslide shows hundreds of landslide blocks more than 0.5 km in length. Their backscatter images comprise a characteristic polka-dot pattern, one of the hallmarks of the landslides in their

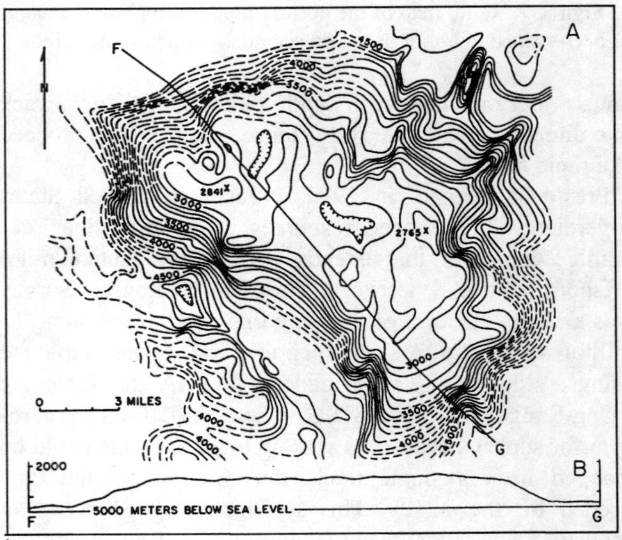

Figure 9. Bathymetric map of Tuscaloosa Seamount (22.1° N, 157.1°W), contour interval 100 m. From *Langford and Brill* [1972], and *Macdonald and Abbott* [1970].

Figure 10. Final processed GLORIA side-looking sonar image of Nuuanu and Wailau landslides, with lighter shade indicating higher acoustic reflectance. The apparent elongation of small blocks out near the distal ends of the landslides is an acoustic artifact; the elongation is parallel to the track line and caused by beam divergence and yawing of the towed GLORIA vehicle. The large 30-km block in the center is Tuscaloosa seamount. Previous unprocessed published images were of lower resolution [*Moore et al.*, 1989]

distal regions (Figure 10). This pattern is caused by the acoustic reflectance from the large fields of hummocky terrain formed by smaller landslide blocks. Such features, not shown by earlier maps, indicate that the landslides are bigger and longer than formerly supposed; they demonstrate that the Nuuanu landslide extends out 220 km northeast of Oahu.

2.8. Thomas G. Thompson and Laney Chouest Surveys, 1992

The next major advance in bathymetric mapping came with the multibeam sonar system, another swath system like GLORIA, but capable of accurately plotting ocean depth as well as variations in sonic backscatter. This system utilizes two hull-mounted arrays of sonic sources, one parallel with the axis of the ship, and the other at right angles. The angle of arrival of the signal reflected from the bottom is determined by the relative time of arrival across the cross array of receivers, and the slant distance to the bottom by the time difference between sending a pulse and receiving its reflection from the bottom. Hence this system is capable of mapping not only the intensity of backscatter (like GLORIA), but also the bottom topography over a swath which may be as wide on each side as twice the ocean depth.

Apparently the first effective multibeam system (named SASS for Sonar Array Survey System) was developed by the United States Navy for military purposes in the late 1960s. SASS swath maps were first released for general research purposes to facilitate the French American Mid-Atlantic Undersea Study (FAMOUS project) in 1973. More than a decade later, the first SASS multibeam bathymetric map in Hawaiian waters charted the summit area of Loihi Volcano [*Malahoff*, 1987].

Other multibeam mapping systems were quickly developed with similar technology to the SASS system. Two limited seafloor surveys, using such systems, were conducted on the upper reaches of the Wailau landslide in the summer of 1992 utilizing both the research vessels *Thompson G. Thompson* and *Laney Chouest*. Mark Holmes directed the *Thompson* survey utilizing a Krupp Atlas Electronik Hydrosweep multibeam system. R. T. Holcomb directed the *Laney Chouest* survey utilizing a SeaBeam I multibeam system. The combined surveys did not produce continuous overlapping swaths, but produced maps that showed far more detail of the giant canyons on the insular shelf than had heretofore been generated. The maps delimited for the first time the most proximal of the large landslide blocks at the head of the Wailau landslide. The surveys were never formally published but are available in the NOAA database.

2.9. The Ewing Survey, 1998

A multibeam bathymetric survey was made of part of the Nuuanu landslide in 1998 from the research vessel *Ewing*. In addition to a partial bathymetric survey of the landslide, two seismic lines, collected at the same time, cross the Nuuanu landslide and reveal much of its internal structure. The proximal slide blocks are separated by sediment ponds approximately 1.5-2 km thick that represent debris largely emplaced with the slide blocks. Post-slide sediments are apparently less than 100-250 m thick [*Moore et al.*, 1998].

2.10 The Ocean Alert Surveys, 1998

Early in 1998, the Monterey Bay Aquarium Research Institute and the U. S. Geological Survey collected detailed high resolution bathymetric and sidescan data from the vessel *Ocean Alert*. A 30 khz – Simrad EM300 multibeam system was utilized. The system produces both bathymetric contour maps, and acoustic backscatter images, at depths less than 3500 m.

A further improvement of the satellite-based GPS navigation system employed a non-moving base station on land. This differential GPS system—that continuously determined the location of the ship relative to the land station—provided ship position within a few meters. Areas were mapped in the inshore, shallower parts of both the Nuuanu [*Dartnell and Gardner*, 1999] and Wailau [*MBARI Mapping Team*, 2000] landslides (Figures 11 and 12)

These surveys show details of the steep scarps at the head of the landslides— scarps that notch into a prominent slope break at 1000-1500 m depth that apparently represents the shoreline at the end of the shield building phase of host volcano growth. This observation suggests that at the time of failure, the head of the landslides were close to, or perhaps a few km inshore from, the general coastal zone at that time [*Clague and Moore*, 2001]. Due to later subsidence, the heads of the landslides are now 10-15 km offshore.

The maps provide notable details of the canyons within and above the headwalls of the giant landslides. The shape and size of the upper part of the submarine canyons are so similar to those on land that they appear to have resulted from subaerial erosion prior to the subsidence of the island volcanoes. In contrast some features of the canyons depicted in these maps indicate that the canyons have been modified by flow of density currents down the drowned

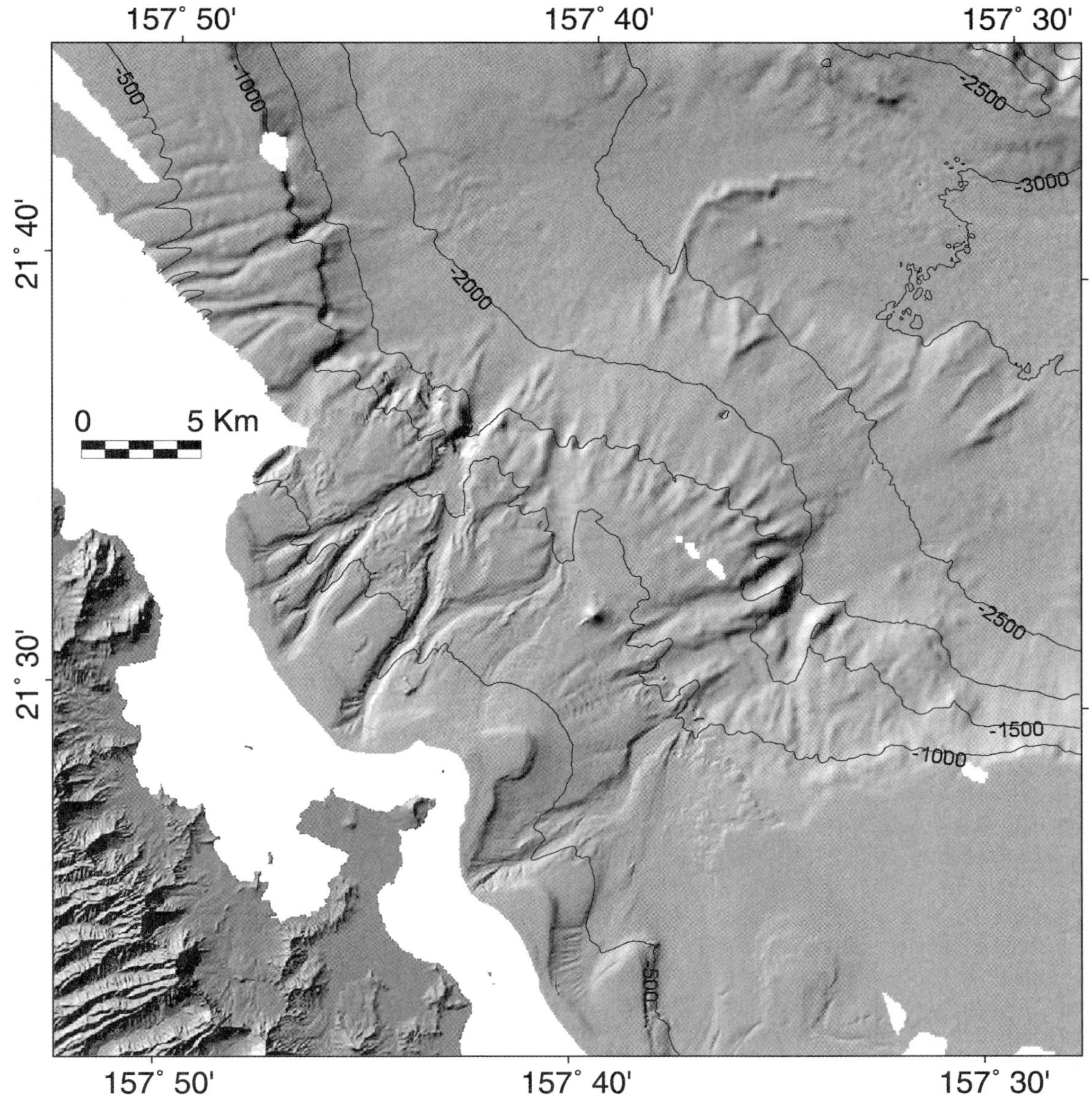

Figure 11. Shaded relief map showing submarine canyon system northeast of Oahu. Note dunes arranged normal to canyon 7 km northeast of prominent (Kaneohe) peninsula (21° 30' N lat, 157° 40' W long). Higher-resolution inboard Simrad image [*Dartnell and Gardner*, 1999] is joined to lower resolution SeaBeam image farther offshore [*Smith and Satake*, this volume].

canyons. Giant ripple marks or dunes were imaged on the shelf adjacent (and normal) to the canyons (Figure 11). These dunes were apparently deposited when density currents flowing down the canyons overflowed and dropped part of their load.

In addition, the canyons extend down the steep slope hundreds of meters below the primary slope break at 1000-1500 m (Figures 11 and 12). In one of these extended canyons, where the main slope break is at 1300 m, a large plunge pool 90 m deep occurs at 1600 m depth at the base of a near-vertical 300 m high scarp. Such a depression, reminiscent of the pools at the base of terrestrial waterfalls, occurs at such a depth that it was probably not carved subaerially. The submarine canyons, and their deep off-

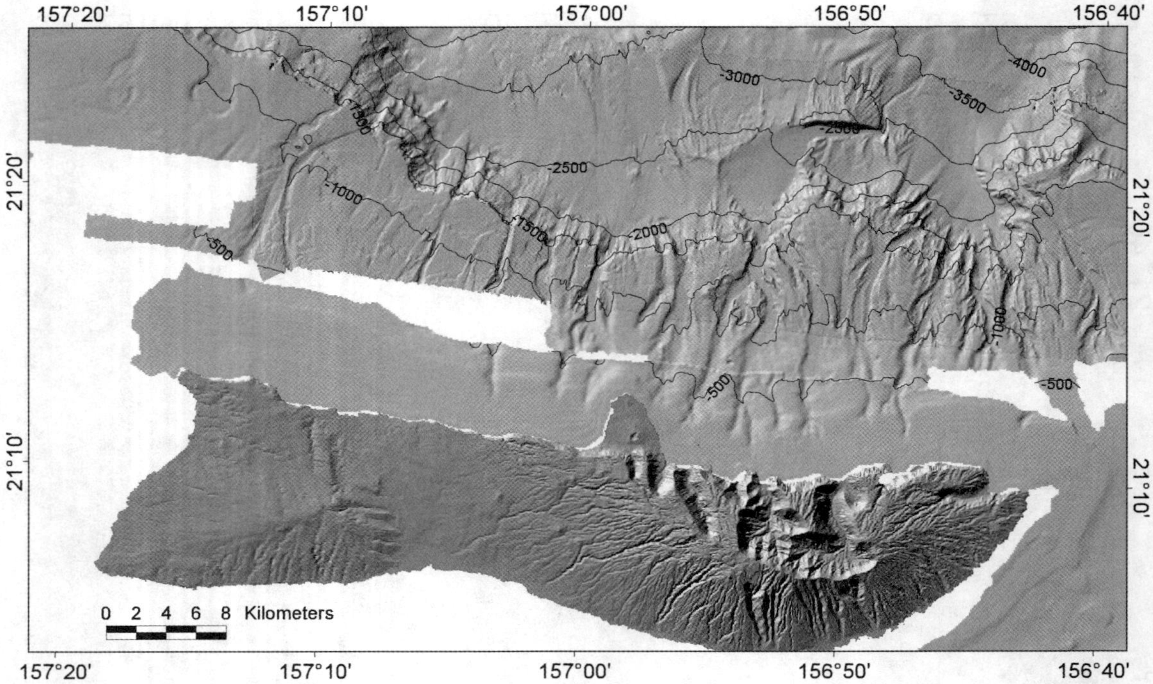

Figure 12. Shaded relief map of Molokai and its north offshore region. Simrad image [*MBARI Mapping Team*, 2000] at center of map is spliced to inshore single-beam survey [*Mathewson*, 1970] and offshore SeaBeam bathymetry [*Smith and Satake*, this volume]. Submarine canyons begin 3 km offshore, traverse shelf to its edge at ~1200m depth, carve down steep slope and can be traced in places below 2500 m depth. Most proximal block of Wailau landslide is visible in northeast part of map.

shore extensions, occur only offshore from the deepest subaerial canyons that head in the highest parts of the windward, high rainfall parts of the islands, suggesting that subaerial cutting is the primary mode of upper canyon formation. Once these canyons are established they funnel density currents down slope that promote submarine canyon erosion. The extensive fresh water discharges near the shoreline suggests that groundwater flow may be important in triggering the giant landslides, as evaluated by *Iverson* [1995].

2.11. The Kairei and Yokosuka Surveys, 1998 and 1999

In the summers of 1998 and 1999 a cooperative Japan-United States oceanographic program focussed on some key areas adjacent to the Hawaiian islands utilizing the research vessels *Kairei* and *Yokosuka* from the Japan Marine Science and Technology Center (JAMSTEC). In addition to mapping, dredging and coring, sea bottom investigations were made with the tethered-remotely-operated vehicle *Kaiko* and the manned submersible *Shinkai 6500*. Both ships utilized multibeam bathymetric mapping systems (SeaBeam 2100) and GPS navigation to produce detailed bathymetric maps of the ocean floor around the Hawaiian Islands [*Naka et al.*, 2000]. Surveys of the Nuuanu-Wailau landslide complex produced the most detailed regional map yet made (Figure 13), a marked advance over previous contour maps of the region (Figure 5). At the greater depths of the central and lower parts of the landslide, the digitized map has been gridded at 100 m and can be readily interpreted at a scale of 1:140,000 with a contour interval of 10 m.

The map shows the individual blocks in the landslide down to a few hundred meters in length, and from it the dimensions and volume of individual blocks and the areal distribution of blocks can be measured (Figure 14). The irregular shapes of the landslide blocks are well defined, and different models of re-assembly of the blocks back to their pre-movement condition can be made and compared.

3. DISCUSSION

3.1 Nature of the Landslide Blocks

The resolution of the JAMSTEC bathymetry is such that when plotted at 1:140,000 landslide blocks can be defined that are as small as 500 m in length where delimited by two contours (that is where exceeding 10 m in height). Irregu-

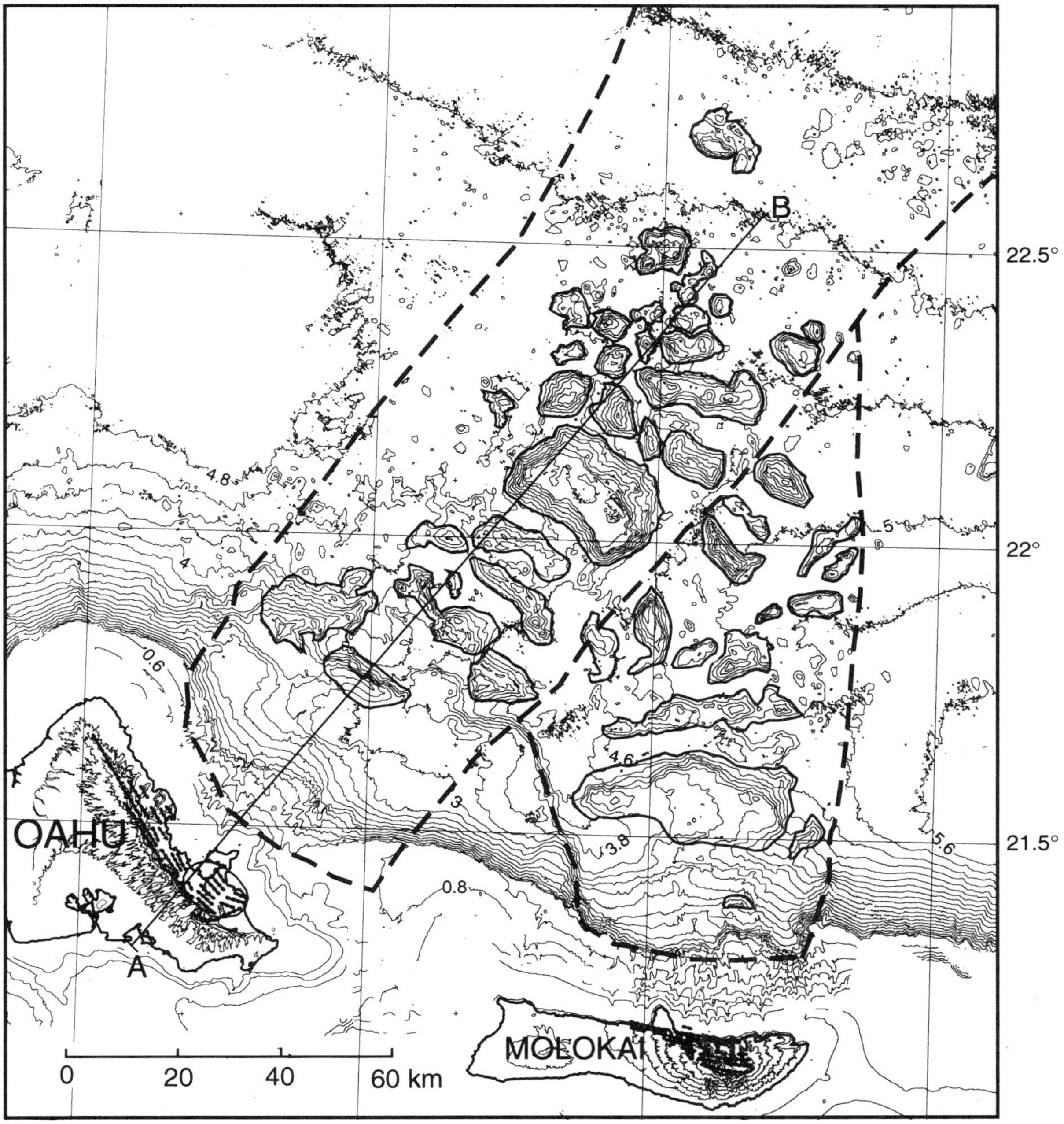

Figure 13. JAMSTEC map of the Nuuanu and Wailau landslides, contour interval 200 m, depth labels in km [*Naka*, 2000; *Smith and Satake*, this volume]. Dashed lines are tentative boundaries around the Nuuanu landslide on the west and the Wailau landslide on the east. Larger landslide blocks are outlined by solid lines. The dike complexes of the Koolau volcano on Oahu and of the East Molokai volcano on Molokai are shown by beaded lines, and the inferred position of the Koolau caldera by a shaded ellipse [*Walker*, 1987]. A-B is profile line for figure 16.

larities in the contours caused by instrumental effects make meaningful measurements on smaller blocks difficult.

Mapped blocks range in size from less than 500 m up to about 30 km for the largest block, Tuscaloosa seamount. Many smaller blocks, no doubt, are partly buried in sediment and are not defined by the mapping system. Bottom-towed bathymetric surveys in the younger Alika landslide off west Hawaii define landslide blocks down to 5 m in size

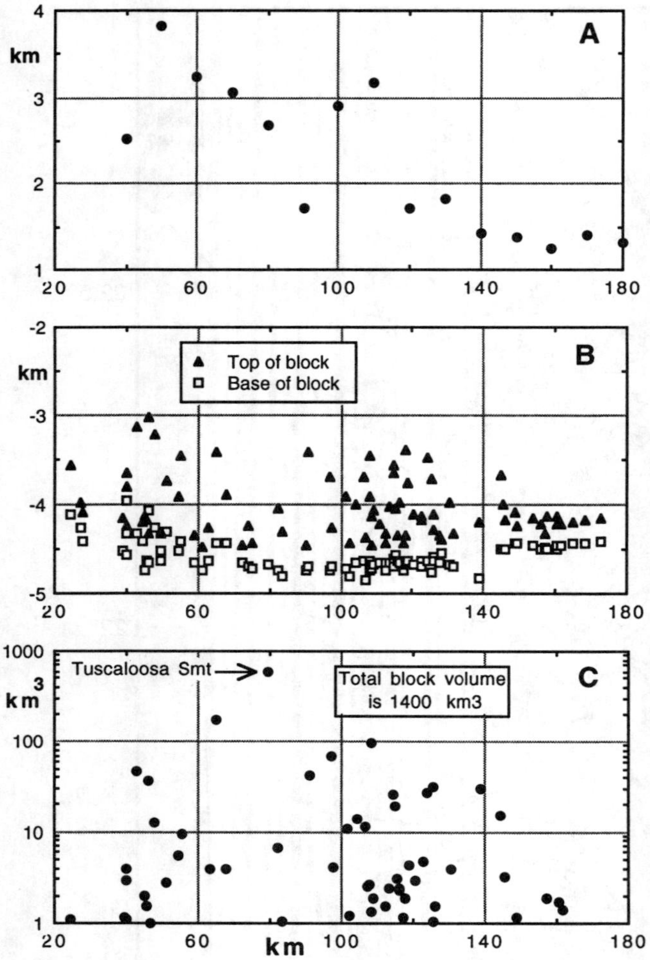

Figure 14. Nature of blocks in the Nuuanu landslide along a northeasterly trending line along the axis of the landslide (see figure 13).). A, Average length of blocks longer than 1 km in 10-km-wide zones. B, Depth of top and of average base of blocks larger than 2 km. C, Volume of blocks greater than 1 km^3.

and near-bottom photographs show blocks less than 1 m in size [*Moore et al.*, 1992]. Therefore it is likely that the landslides off Oahu contain a continuum of block sizes down to a small size. The average size of blocks systematically decreases outward from the head of the Nuuanu landslide. When all blocks larger than one km in length are measured, the average length of the blocks (grouped in 10-km-wide zones) shows a general decrease from 2-4 km near the headwall to less than 1.5 km near the distal end of the landslide at 180 km from the landslide head (Figure 14).

The height of the larger landslide blocks above their exposed bases is commonly about 1 km, and Tuscaloosa seamount rises 2 km above its base to a water depth of 2750 m (Figure 14). Beyond about 140 km from the headwall, the larger blocks generally are less than a few hundred meters in height. The exposed base of the blocks reaches a maximum depth of 5250 m at about 85 km from the headwall, and rises further northeast to 4400 m at 170 km (Figure 14). The region of maximum depth coincides with the axis of the Hawaiian Deep, which attains about 4900 m depth northwest of the landsiide and about 5700 m southeast. The average axis of the Deep beneath the landslide is inferred be about 5300 m if landsliding had not occurred and normal sedimentation ensued.

The exposed volume of the larger blocks in the Nuuanu landslide have been measured making use of the digital data derived from the JAMSTEC map survey. Fifty-four blocks have volumes greater than 1 km^3, and the largest (Tusaloosa Seamount) has a volume of nearly 600 km^3. The total volume (down to their exposed bases) of all the large blocks (which dominate the volume of all blocks in the landslide) is 1400 km^3 (Figure 14) and, hence, about 40 % of the total exposed volume of blocks in the landslide resides in Tuscaloosa Seamount. *Satake et al.* [this volume] have estimated that the volume of the debris deposited by the Nuuanu landslide is (3.0 ± 0.5) x 10^3 km^3. Hence the exposed part of the blocks larger than 1 km collectively comprise about one half of the landslide deposit.

The makeup of the landslide blocks can be defined by the petrology of available samples as discussed in other chapters in this volume, but several factors are summarized here. Most of the samples from the Nuuanu landslide are basaltic hyaloclastites with degassed clasts. Hence, the bulk of the blocks represent lava flows that originally erupted on land, crossed the shoreline and poured into the sea where the lava was quenched and shattered. No subaerially-cooled lava has been found in the blocks, and only a minor amount of the material represents lava that was submarine erupted.

The two landslide blocks sampled from the Wailau landslide contain abundant hyaloclastites also, but they do contain some lava that is non-degassed and hence apparently erupted under the elevated pressures of submarine conditions.

3.2. Geometry of Nuuanu and Wailau Landslides

Many features of the giant landslides are imperfectly known and, therefore, models that attempt to reconstruct the pre-landslide conditions are poorly constrained. The new JAMSTEC bathymetry, however, suggests that the irregularly shaped landslide blocks fit together, and that the blocks can be reassembled in jigsaw puzzle fashion. In order to generate a model of pre-landslide conditions, assumptions must be made on the following four factors:

3.2.1. Position of landslide head. The precise shape and position of the Nuuanu and Wailau headwalls immediately after landsliding is poorly known. A long history of post-landslide volcanism, sedimentation, and marine erosion modified the region extending out to the main slope break at about 1200 m. In addition to the extensive post-landslide alkalic volcanism of Koolau and East Molokai volcanoes, some tholeiitic shield-building volcanism probably also partly rebuilt the failed flanks after landslide movement and before alkalic volcanism. Post-landslide erosion of much of the subaerial Koolau volcano has provided a rich source of sediment that has buried much of the upper part of the landslide and its reentrant. The offshore shelf extending from about 1200 m depth up to the present shoreline has been occupied by shifting shorelines as the volcano subsided and as sea level rose and fell during its natural (ice-age induced) fluctuations over the past 2 million years. Such marine erosion caused the most extensive modifications on the windward (northeast) sides of the islands. Moreover when the Nuuanu landsliding took place the island of Molokai and its volcanoes were at an earlier growth stage and were smaller and lower [*Clague and Dalrymple*, 1987; *Moore*, 1987]. Part of the east side of the reentrant of the Nuuanu landslide has no doubt been modified by post-landslide accumulation of volcanic materials and sediment from the younger Molokai volcanic centers.

The apparent non-disruption of the regional Koolau dike complex [*Walker*, 1987] near and east of the axis of the steep Oahu cliff (Nuuanu Pali) indicates that the cliff or the axis of the rift zone (as defined by the dike complex) is not the head of the slide and that the head probably is located offshore to the northeast. *Clague and Moore* [2001] make a similar argument for the offshore headwall of the Wailau landslide.

The head of the Nuuanu slide probably is located at the irregular concave reentrant in the 15-km wide shelf that surrounds Oahu. The shelf edge is at about 1200 m depth and the reentrant is 30-40 km wide (Figures 14, 16). The shelf edge is believed to be a sea-level feature, and to have formed about in its present position when shield building of the east Oahu volcano (Koolau) ceased and lava no longer crossed the shoreline. The break-in-slope was formed (as is occurring today on Kilauea and Mauna Loa Volcanoes) by the tendency of lava flows that cross the shoreline to maintain a lesser gradient above sea level than below. Subsequently, post-landslide subsidence of the entire Hawaiian Ridge has lowered the volcano and its slope break to its present depth.

A series of submarine canyons incise the slope above the slope change and can be traced over its edge down to depths of 2000, and broader channels with flanking levees can be traced to 3000 m (Figure 11). A sedimented zone a few km wide offshore is free of the canyons, but they are present on land and appear to be on line with those on the submarine slope. Seismic sub-bottom profiling clearly shows buried continuations of the canyons from the shoreline to their appearance offshore [*Coulbourn et al.*, 1974]. The size and shape of the canyons on the gentle slope is similar to that of canyons on subaerial Hawaiian volcanoes, such as Kohala on the island of Hawaii. The canyons only occur offshore from the windward side of the highest part of the islands, the site where rainfall is highest. These factors suggest that much of the canyon cutting above 1200 m occurred when the gentle slope was above sea level before landsliding and before subsidence.

After landsliding and overall subsidence when the slope was submerged, cutting of the canyons continued both above and below the change in slope. Alternating sedimentation and marine erosion during the Pleistocene ice ages 125-m oscillation of sea level further modified the slope and canyons. The coastal parts of the canyons out to 3 km from the shoreline were filled in by sediment. The deeper parts on the shelf were occupied by brim-full turbidity currents that deposited dunes on their flanks, and currents below 1200 m depth actively eroded and deposited down to 3000 m depth (Figures 12, 13).

3.2.2. Position and nature of the pre-landslide slope. Recent deformation studies of the active Kilauea volcano indicates that the south flank of the volcano is creeping seaward (southward), and that the rift zone forms the north boundary of this creeping zone [*Swanson et al,* 1976; *Owens et al*., 1995]. Gravity-induced spreading of the volcano occurs at the rift zone axis, and the rift zone dikes (feeding fissure eruptive vents) are emplaced in this tensional zone. Seaward from the Kilauea east rift zone is a major submarine bench with low ridges near its outer margin that bound several closed depressions. This bench or "bulge" is tectonically active and is apparently related to major thrust faults in the lower compressional part of the Hilina slump [*Moore and Normark*, 1994].

No doubt the dikes of the Koolau and Wailau volcanoes formed in similar fashion [*Walker*, 1987] so that they mark the head of the zone that formerly creeped in the early, non-catastrophic phase of volcano spreading. Emplacement of rift zone dikes may not only play a role in affecting mechanical and fluid pressures along the failure zones at the base of the slide, but they may trigger landslide failure [*Elsworth and Day*, 1999]. The creeping phase of move-

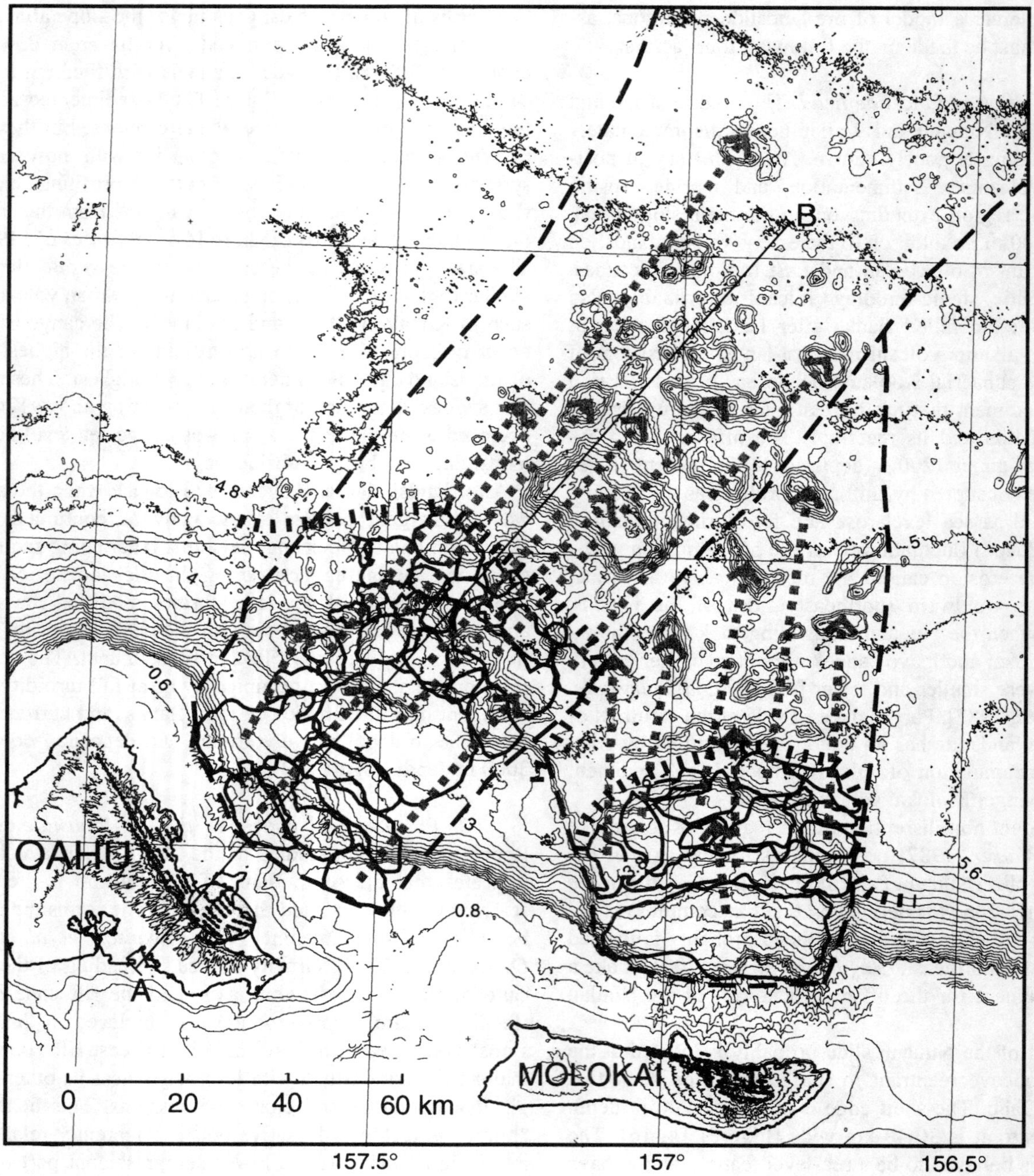

Figure 15. A hypothetical reassembly of blocks before movement of the Nuuanu and Wailau landslides showing the movement paths (dashed gray lines) of individual blocks whose final position are outlined in figure 13. Hachured lines are assumed pre-landslide position of bulged base of volcanic edifice (about 4600 m depth contour). Slide outlines, dikes, and caldera as in Figure 13.

ment is believed to have caused bulges on the northeast submarine flank of the Koolau volcano and the north flank of the East Molokai volcano, and it was these bulges that failed catastrophically to produce the Nuuanu and Wailau avalanches. Hence, the nature of the present Kilauea submarine bench [*Moore and Normark*, 1994] provides a model for the pre-landsliding deformation, and is used to estimate the shape and position of the pre-landslide bulging of the volcano flank (Figures 15 and 16). One difference between the Kilauea model and that proposed for Nuuanu

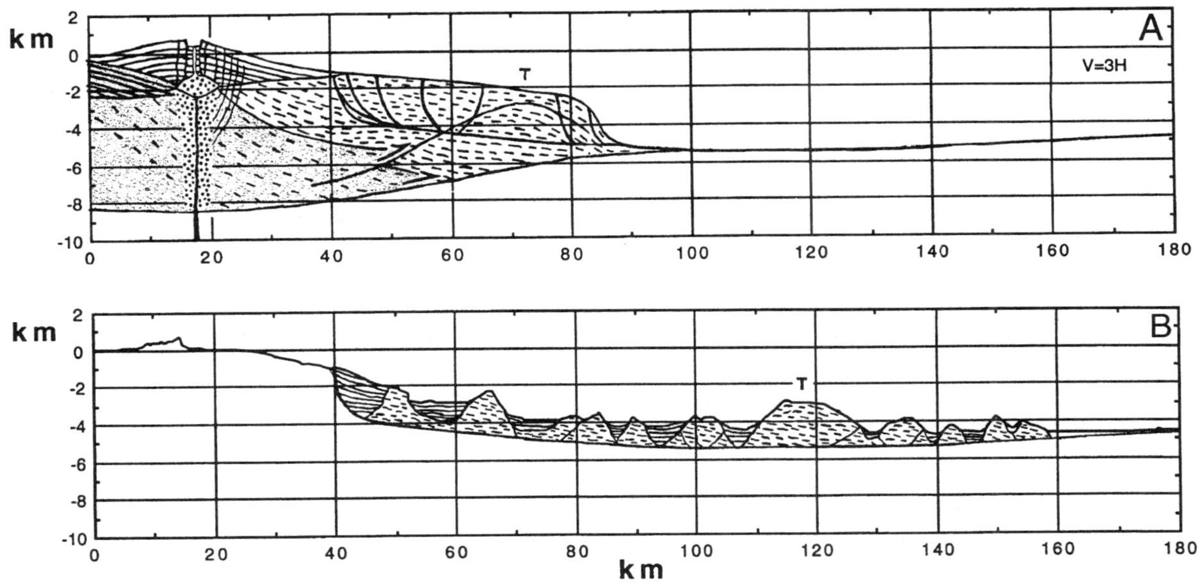

Figure 16. Profiles showing possible configuration of Nuuanu landslide before and after movement. (A) Profile before landsliding showing basal slip surface and fracture locations prior to failure of prominent bulge caused by spreading of volcano; such spreading would be similar to that presently occurring on the south flank of Kilauea volcano. Subaerial lava, lined; hyaloclastite, dashed; Waianai Volcano, shaded; (B) Final profile after landslide-induced deformation of blocks and subsequent sedimentation. Profiles are drawn with present, subsided state of volcano. The left (SW) side of profile A would have been about 1.2 km higher before and during landsliding. T indicates Tuscaloosa seamount; vertical exaggeration threefold.

is that Kilauea stands less than 1 km above sea level upslope from the Hilina slump whereas the Koolau and Wailau volcanoes presumably stood 2-3 km above sea level prior to landsliding (and later subsidence and erosion). Hence the bulging of the Koolau submarine flank would have been accordingly larger.

3.2.3. Thickness of the landslide. The Nuuanu landslide tongue is presumably thickest where it crosses the Hawaiian Deep. As mentioned above, the average depth of the Deep under the landslide is about 5300 m. Seismic reflection studies indicate that "The slide blocks are separated by sediment ponds approximately 1.5-2.0 sec (1.5-2 km) thick that probably represent debris emplaced with the slide blocks. Recent (post-slide) sediments appear to make up less than 100-250 m of the top of this unit..." [*G. Moore et al.*, 1998]. For the purposes of this reconstruction, we consider that the landslide extends to a maximum depth of 5300 m, and that the slide averages about 1.5 km thick along its axis out to about 120 km from the headwall (Figure 16). From 120-220 km out the overall average thickness decreases down to less than a few tens of meters.

3.2.4. Relationship of Nuuanu and Wailau landslides. Details of the relationship of the Nuuanu landslide blocks with those of the Wailau landslide are poorly known, and studies to date do not provide unambiguous geochemical methods of distinguishing rocks from the two source volcanoes [*Clague et al.*, this volume]. Manganese thickness of the blocks sampled are thinner on those of the Wailau landslide indicating that it is younger [*Naka et al.*, 2000; *Clague et al.*, this volume]. We have assumed that all the blocks directly offshore and in line with the apparent course of the Nuuanu landslide belong to that landslide as suggested in figure 14, and have moved over unobstructed terrain. Blocks that are assigned to the Wailau landslide are those that clearly came from, and are directly downslope from the reentrant on East Molokai Volcano. The Wailau landslide is considered to end where it abuts against the blocks assigned to the Nuuanu landslide [*Moore,* 1964]. Blocks from the Wailau landslide probably overran some of the area covered by the Nuuanu landslide and blocks from both landslides are most likely intermixed in this central portion.

3.3. Development of Nuuanu Landslide

Individual blocks of the Nuuanu landslide were defined on the map by drawing bounds around their exposed bases, and the blocks thus defined were moved back into the upslope reentrant as described above. The blocks were stacked one against another in both map plan (Figure 15),

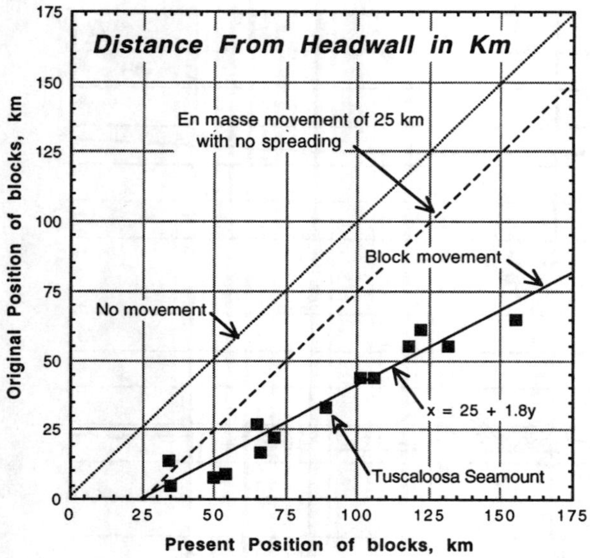

Figure 17. Relation of original inferred pre-landslide positions of blocks to their present post-landslide positions, all relative to the landslide headwall. Those blocks closest to the headwall have moved about 25 km, whereas blocks originally more distant from the headwall have moved further—indicating an overall spreading of the landslide debris that produced the spaces between the blocks.

and in profile section (Figure 16). The resulting reconstruction requires that the Tuscaloosa block moved about 55 km while those originally closer to the volcano moved less and those further out moved more. The many smaller blocks filled in regions between the larger blocks and apparently some formed by fragmentation of the larger blocks. Many smaller blocks moved farther out to the distal end of the landslide and formed the outer hummocky apron of landslide material so apparent in the GLORIA image (Figure 10).

Since the blocks that were originally closer to the headwall moved a shorter distance than those that were farther from the headwall (Figure 15), the landsliding has caused the original material to spread during movement. This spreading has generated spaces between the blocks.

When the inferred original position of each block is compared with its present position (Figure 17), we can get a notion of the relative distance each block moved relative to its neighbors and the amount of spreading of the landslide debris. Blocks near the headwall moved out about 25 km whereas those originally at 25 km from the headwall moved 45 km and those originally at 75 km moved 85 km (Figure 17). The blocks moved further than if they had all moved 25 km en masse. Each block moved not only 25 km, but an additional distance of about 1.8 times their original distance from the headwall. This 1.8 factor is a measure of the spreading of the landslide. This relation does not consider the small blocks and hummocks whose original position cannot be approximated. Some moved to the extreme toe of the landslide more than 200 km from the headwall (Figure 10).

The model highlights the fact that the large blocks reassembled to their prelandslide condition comprise a bulk some 70 km long (to the northeast) for the Nuuanu and 40 km (to the north) for the Wailau (Figure 15). These masses do not fit readily into the apparent landslide amphitheaters on Koolau and East Molokai volcanoes, which each extend shoreward only about 20-km from a reconstructed smooth slope. One scheme to overcome this problem, as suggested above, is that a pre-landslide seaward-extending bulge in the volcanic flank could accommodate the apparent excess volume of the block material.

The headwall of the landslides may have been originally closer to the islands, so as to increase the volume of the reentrant. A cover of post-landslide lava and sediment might have obscured the original position of the landslide heads. However, the distance between the assumed Koolau landslide head and the on-land high scarp (Nuuanu Pali) is about 15 km and the comparable distance for the Wailau landslide is about 13 km, so that this adjustment would help, but not overcome, the excess block volume problem.

If some of the blocks assigned to the Nuuanu landslide actually came from the Wailau landslide, then the volume inequities could be made more similar between the two landslides and their respective amphitheaters—but they would not be eliminated. Hence the sum of the evidence at this time suggests that some form of long-term, creep-induced bulges of Kilauea style presaged the catastrophic failure of both the Nuuanu and Wailau landslides.

Acknowledgments. The JAMSTEC multibeam bathymetry was generated by SeaBeam systems mounted on the research vessels *Kairei*, Captain O. Yukawa, and *Yokosuka*, Captain H. Tanaka, during 1998 and 1999. K. Satake and J. Smith managed the shipboard and on-shore processing of the SeaBeam data. J. Naka was chief scientist during both cruises. P. W. Lipman and M. O. Garcia headed the U.S. science group. E. Takahashi headed the Japanese group, and coordinated scientific efforts of both the American and Japanese teams. Scientific discussions with P. W. Lipman were particularly helpful. Thoughtful reviews of W. R. Normark, K. Satake, G. F. Moore, and S. J. Day improved the perspective and readability of the manuscript. We are pleased to record our appreciation of the help and support of these persons, of the participating scientists and crews, and of the funding agency—Japan Marine Science and Technology Center.

REFERENCES

Andrews, J. E., and C. Bainbridge, Submarine canyons off eastern Oahu, *Pac. Sci.*, *26*, 108-113, 1972.

Betz, F. Jr., and H. H. Hess, The floor of the North Pacific Ocean, *Geographical Rev.*, 32, 99-116, 1942.

Bryan, G. S., A bathymetric chart of the North Pacific Ocean, Ass. D'oceanographie Physique, Un. Geodesique et Geophys. Internat., Pub. Scientifique No. 8, 78-80, 1940.

Campbell, J. F., *Bathymetric atlas of the southeast Hawaiian Island,* Hawaii Inst. of Geoph. Tech. Rep. 87-1, 21 p., 1987.

Chase, T.E., H. W. Menard, and J. Mammerickx, Bathymetry of the North Pacific, Chart no. 8 of 10: Scripps Institution of Oceanography and Institute of Marine Resources, 1970.

Clague, D. A., G. B. Dalrymple, The Hawaiian-Emperor volcanic chain, Part I Geologic evolution, *U. S. Geol. Surv. Prof. Pap.*, *1350*, 5-54, 1987.

Clague, D. A., C. Dao-gong, R. Murnane, M. H. Beeson, M. A. Lanphere, G. B. Dalrymple, W. Friesen, and R. T. Holcomb, Age and petrology of the Kalaupapa Basalt, Molokai, Hawaii, *Pac. Sci.*, *36*, 411-420, 1982.

Clague, D. A., and J. G. Moore, The headwall of the giant Wailau landslide, Hawaii, *J. Volcanol. Geotherm. Res.*, in press 2001.

Clague, D. A., J. G. Moore, A. S. Davis, Formation of volcanic breccia and hyaloclastite in landslide blocks from the Wailau and Nuuanu giant landslides, Hawaii, this volume.

Coulbourn, W. T., J. F. Campbell, R. Moberly, Hawaiian submarine terraces canyons, and Quaternary history evaluated by seismic-reflection profiling, *Marine Geology*, 17, 215-234, 1974.

Dartnell, P., and J. V. Gardner, Sea-floor images and data from multibeam surveys in San Francisco bay, Southern California, Hawaii, the Gulf of Mexico, and Lake Tahoe, California-Nevada. *U.S. Geol. Surv., Digital Data Series* DDS-55, 1999.

Dietz, R. S. and H. W. Menard, Hawaiian Swell, Deep, and Arch, and subsidence of the Hawaiian Islands: *J. Geology*, 61, 99-113, 1953.

Elsworth, D. and S. J. Day, Flank collapse triggered by intrusion: the Canarian and Cape Verde archipelagoes, *J. Volcanol. Geotherm. Res.*, 94, 323-340, 1999.

Groome, M.G., C.E. Gutmacher, and A.J. Stevenson, Atlas of GLORIA sidescan-sonar imagery of the Exclusive Economic Zone of the United States: EEZ-View, *U.S. Geol. Surv. Open-File Rept.* 97-540, 1997.

Hamilton, E. L., Marine geology of the southern Hawaiian Ridge, *Geol. Soc. Am. Bull.*, *68*, 1011-1026, 1957.

Iverson, R. M. Can magma-injection and groundwater forces cause massive landslides on Hawaiian volcanoes? *J. Volcanol. Geotherm. Res.*, *66*, 295-308,1995.

Langford, S. A., The surface morphology of the Tuscaloosa Seamount, *Unpublished M.S. thesis, University of Hawaii,* 17 p., 1969.

Langford, S. A., and R. C. Brill, Giant submarine landslides on the Hawaiian Ridge: a rebuttal: *Pac. Sci.*, *26*, 254-258, 1972.

Macdonald, G. A., and A. T. Abbott, Volcanoes in the sea: *University of Hawaii Press*, 441p., 1970.

Malahoff, A., Geology of the summit of Loihi submarine volcano, *U. S. Geol. Surv. Prof. Pap.*, *1350*, 133-144, 1987.

Mathewson, C. C., Submarine canyons and the shelf along the north coast of Molokai Island, Hawaiian Ridge, *Pac. Sci.*, *24*, 235-244, 1970.

MBARI Mapping Team, MBARI Hawaii multibeam survey, Monterey Bay Aquarium Research Institute Digital Data Series No. 2, 2000.

Menard, H. W., *Marine geology of the Pacific,* McGraw-Hill, New York, 271 p. Contains in pocket map: Physiography of the Northeastern Pacific, scale 1:6,000,000, 1964.

Moore, G. F., J. K. Morgan, S. Leslie, B. Taylor, and C. Berndt, Morphology and structure of the Nuuanu debris avalanche, North of Oahu, Hawaii (abstract), *EOS, 79*, 1008, 1998.

Moore, J. G., Petrology of deep-sea basalt near Hawaii, *Am. J. Sci.*, *263*, 40-52, 1964.

Moore, J. G., and R. K. Reed, Pillow structures of submarine basalts east of Hawaii, *U. S. Geol. Surv. Prof. Pap.*, 475-B, B153-B157, 1963.

Moore, J. G., Subsidence of the Hawaiian Ridge, *U. S. Geol. Surv. Prof. Pap.*, *1350*, 87-100, 1987.

Moore, J. G., W. R. Normark, and C. E. Gutmacher, Major landslides on the submarine flanks of Mauna Loa volcano, Hawaii, *Landslide News, 6*, 13-16, 1992.

Moore, J. G., D. A. Clague, R. T. Holcomb, P. W. Lipman, W. R. Normark, and M. E. Torresan, Prodigious submarine landslides on the Hawaiian Ridge: *J. Geophys. Res.*, *94*, 17,465-17,484, 1989.

Moore, J. G., W. R. Normark, Giant Hawaiian Landslides: *Annu. Rev. Earth Planet. Sci, 22*, 119-144, 1994.

Moore, J. G., W. R. Normark, R. T. Holcomb, Giant Hawaiian underwater landslides, *Science, 264*, 46-47, 1994.

Naka, J. and 34 other authors, Tectono-magmatic processes investigated at deep-water flanks of Hawaiian volcanoes, *EOS, 81*, 221 and 226-227, 2000.

Owen, S., P. Segall, J. Freymueller, A. Miklius, R. Denlinger, T. Arnadottir, M. Sako, and R. Burgmann, Rapid deformation of the south flank of Kilauea volcano, Hawaii, *Science*, 267, 1328-32, 1995

Satake, K., Smith, J.R. and Shinozaki, K., Three-Dimensional reconstruction and tsunami model of the Nuuanu and Wailau giant landslides, Hawaii, this volume.

Shepard, F. P., and Dill, R. F., *Submarine canyons and other sea valleys*, Chicago, Rand McNally and Company, 381p., 1966.

Smith, J. R. and K. Satake, Bathymetry around Hawaiian islands, this volume.

Somers, M. L., R. M. Carson, J. A. Revie, R. H. Ege, B. J. Barrow, and A. G. Andrews, GLORIA II—an improved long range sidescan sonar: Proceedings of the IEE/IERE Subconference on Ocean Instruments and Communication, p. 16-24, Oceanology International, London, 1978.

Stearns, H. T., Geology of the Hawaiian Islands: *Hawaii Division of Hydrography*, Bulletin 8, 106 p., 1946.

Stearns, H. T., *Geology of the state of Hawaii*, Pacific Books, Palo Alto, California, 266 p., 1966.

Swanson, D. A., W. A. Duffield, and R. S. Fiske, Displacement of the south flank of Kilauea volcano: The result of forceful intrusion of magma into the rift zones, *U. S. Geol. Surv. Prof. Pap., 936*, 30 pp., 1976.

United States Navy, Hydrographic Office, Bathymetric Chart of the North Pacific Ocean, no. 5486, scale 1:13,600,000 at 40° North Latitude, 500 fathom contour interval, 1939.

United States Navy, Oceanographic Office, Contoured Position Plotting Sheet 1604LC, scale 1: ~1,000,000; contour interval 100 fathoms, 1966.

Walker, G. P. L., The dike complex of Koolau volcano, Oahu: internal structure of a Hawaiian rift zone, *U. S. Geol. Surv. Prof. Pap.*, 1350, 961-993, 1987.

Wilde, P., T. E. Chase, W. R. Normark, J. A. Thomas, J. D. Young, Oceanographic data off the southern Hawaiian Islands: LBL Pub. 359, Laurence Berkeley Laboratory, University of California, Berkeley, 1980.

Yokose, H., Landslides on the windward flanks of Oahu and Molokai, Hawaii: SHINKAI 6500 submersible investigations, this volume.

Landslides on the Windward Flanks of Oahu and Molokai, Hawaii: SHINKAI 6500 Submersible Investigations

Hisayoshi Yokose

Department of Earth Sciences, Faculty of Science, Kumamoto University, Kumamoto, Japan

The submarine geology of the Nuuanu (perhaps the largest landslide in the world) and Wailau landslides was investigated during 1999, using the SHINKAI 6500 manned submersible and its mother ship, R.V. YOKOSUKA. Observations were made at eight dive sites, and about 100 in-situ rock samples were collected from the landslide deposits at depths of 2500 to 4500 m. Based on video images and rock samples, the outcrops are subdivided into three basic types, each with several facies: (1) products of volcanic eruptions, (2) coarse volcaniclastic deposits, and (3) unconsolidated and reworked epiclastic sediments. The coarse volcaniclastic deposits are thought to have originated from debris flows in alluvial-fan systems. Unconsolidated muddy gravel is abundant in submarine channels and piedmont slopes; this facies is considered to be the product of reworked sediment in submarine channel systems and small-scale slope failures. These facies comprise most of the megablocks and the apron of Oahu. The steep cliffs of many megablocks can be reasonably fitted together like a jigsaw puzzle. The megablocks can be graphically re-joined to form a huge slab that can be extrapolated back to the submarine eastern flanks of Oahu, at ocean depths of 1000 to 3000 m. Growth of the pre-failure volcaniclastic apron has been enhanced by debris supplied by subaerial erosion from the windward, rainy sides of Oahu and Molokai. Thus, the exceptional size of the landslides described here may have been influenced by heavy trade-wind rainfall.

1. INTRODUCTION

Submarine landslides have been identified on the flanks of many volcanic islands [*Lipman et al.*, 1988; *Moore et al.*, 1989; *Holcomb and Searle,* 1991; *Stillman,* 1999; *Labazuy,* 1996; *Carracedo,* 1999] and in front of submarine deltas on continental shelves [e.g., *Embley and Jacobi*, 1978; *Scheidegger,* 1982; *Prior and Coleman,* 1982]. Recognition of submarine landslides is based mainly on bathymetry. A headwall commonly marks the source area; more detailed understanding comes from direct observations and in-situ samples. Submarine landslides can be destructive because they break submarine cables, collapse oil well platforms, and produce tsunamis [e.g., *Prior and Coleman,* 1982; *Saxov,* 1982; *Smith and Shepherd*, 1996; *McGuire,* 1996].

Hawaiian Volcanoes: Deep Underwater Perspectives
Geophysical Monograph 128
Copyright 2002 by the American Geophysical Union

Among the many submarine landslides around the Hawaii islands, the Nuuanu slide northeast of Oahu is perhaps the world's largest, with a volume exceeding 5000 km^3 [*Moore et al.,* 1994]. Geologic understanding of the Nuuanu and Wailau landslides has been poor, because most of the deposits are located in the heretofore-inaccessible deep ocean (>2000 m). The purpose of this paper is to report new submersible observations of deep submarine outcrops of the Nuuanu and Wailau landslides,

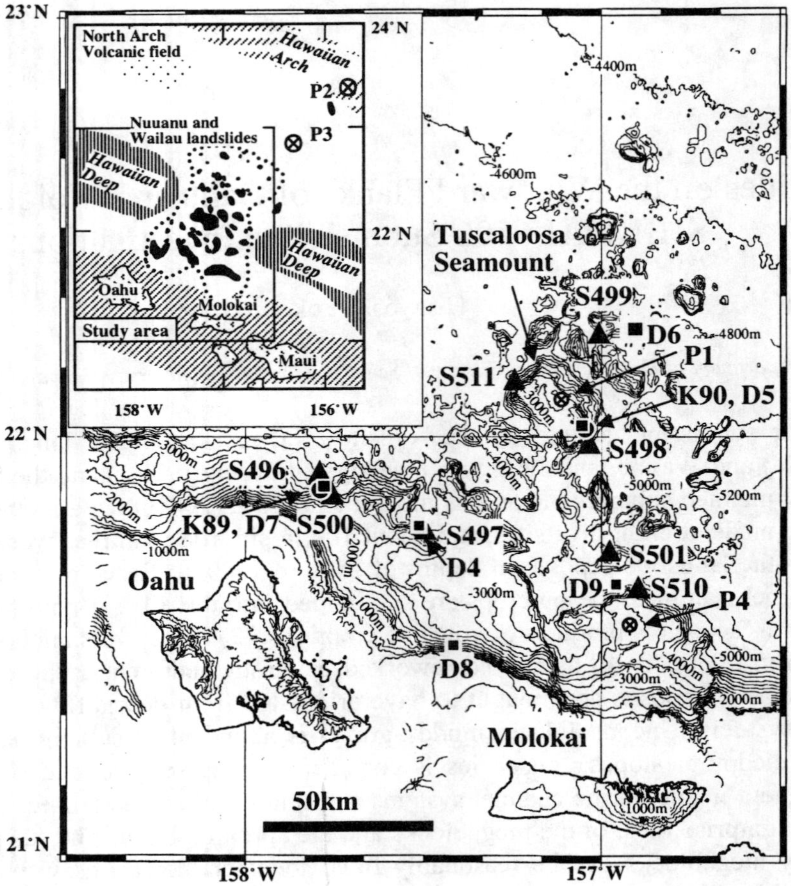

Figure 1. Location map of the dredges (D) and piston cores (P) by KAIREI in 1998, and dives by remotely operated vehicle KAIKO in 1998 (K) and SHINKAI 6500 in 1999 (S) north and northeast of Oahu and Molokai. Inset shows principal structural features of this area. Bathymetric contour interval is 200m.

and to discuss mechanisms for the evolution of these giant features.

2. DIVE LOCATIONS

Eight dives were made on the proximal and distal blocks of the Nuuanu and Wailau landslides using the SHINKAI 6500 submersible of the Japan Marine Science and Technology Center (JAMSTEC) in August and September 1999 (Fig. 1). Continuously recording video images and about 100 in-situ rock samples were collected during these dives. Dives S500 and S496 were on the relatively undisturbed northern flank of Koolau volcano. Dive S497 visited a proximal megablock of the Nuuanu landslide. Dives S511 and S498 were on the northern and southern scarps of the Tuscaloosa Seamount, the largest block in the Nuuanu landslide. Dive S499 examined a distal megablock. Dives S501 and S510 investigated megablocks north of Molokai, which likely belong to the Wailau landslide.

3. OUTCROP CLASSIFICATION

Although video images and rock samples are normally obtained during submersible dives, both the quality and quantity of the information obtained is limited when compared with that obtained in land-based studies. Such limitations exist because: (1) the field of view is narrow, (2) color of outcrops is degraded due to selective color absorption by water and the near-ubiquitous presence of manganese oxide coatings, (3) small-relief details on rock surfaces are masked by manganese coatings, (4) sample collection is difficult. (5) the strike and dip of strata cannot be measured accurately in video images, and (6) dive time is short in deep water, chiefly because of battery-power constraints. These problems are less serious where the geological setting is simple, but Nuuanu and Wailau landslides have complex geologic features and extend over a wide area. Therefore, great care had to be taken to characterize the various outcrop types.

Large geologic structures can be difficult to interpret underwater. In such cases, computer-joined images can provide panoramas. In-situ sampling is a powerful tool for submarine investigations since it helps to overcome problems (2) and (3) above. In-situ samples both verify different outcrop types, and also provide a basis for interpreting outcrops that could not be sampled. Based on such data, the deposits in the Nuuanu and Wailau landslides can be divided into three main rock units: (1) products of volcanic eruptions (lobate, massive, and autobrecciated lavas); (2) volcaniclastic rocks (breccia, sandy facies); and (3) unconsolidated and reworked epiclastic sediment (muddy gravel facies, siltstone, and mud).

3.1. Products of Volcanic Eruptions

3.1.1. Lobate lava flows. Pillow lava, which indicates submarine emplacement, was observed at two sites: dive S500 at a depth of 2650 m (Fig. 2A) and dive S510 at about 3700 m (Fig. 2B). These outcrops occur on the edges of terraces near Oahu and Molokai, respectively. Most pillows are 30 to 60 cm in diameter, with well preserved outer surfaces and internal structures. The surfaces are smooth; none are striated or lineated. These are interpreted as lobate lava flows, based on surface morphology [*Batiza and White,* 2000]. Such lobate lavas are considered to form at higher eruption rates than pillow lavas having striated glassy surfaces [*Griffiths and Fink*, 1992]. High eruption rates are inferred to have occurred near dive S500, where hollow pillows were observed [*Umino et al.,* 2000]. The lavas are aphyric basalt with low vesicle content (Fig. 2C and 2D), features commonly associated with lavas erupted at considerable depth. Based on the sulfur content in glass from pillow rims, the eruption depths at dives S500 and S510 are estimated to be about 1000 m [*Shinozaki et al., this volume*].

3.1.2. Massive lava and dike. Massive lavas were observed during dive S497 at a depth of 3100 m and dive S498 at 4400 m, but relations with other rock facies are unknown. A dike, observed on dive S501 at a depth of 4200 m, intrudes autobrecciated lava (Fig. 3A); exposed thickness of the lava and dike is about 1 m. The dike may represent basalt injected from the flow interior, alternatively, it may be a vent feeder. In video images, massive lavas superficially resemble the sandy volcaniclastic facies, but they commonly have columnar joints. The outcrop areas of this unit are too small to show on the bathymetric map (see route maps in accompanying CD-ROM, this volume).

3.1.3. Autobrecciated lava. Autobrecciated lava was observed during four dives: S499, S500, S501, and S510 (see route maps in accompanying CD-ROM, this volume). It can be distinguished from epiclastic facies by the following criteria: (1) randomly oriented monolithologic angular clasts (Fig. 4B); (2) matrix of fine particles of lava or volcanic glass, altered to palagonite; and (3) consolidated outcrops (Table 1).

Not all autobrecciated lavas are associated with lava flows but some are. Brecciated lava observed on dive S510 is overlain by a lobate lava flow. Although talus covers some intervals, the lava flow observed on dive S500 (at 2700 m) may underlie autobrecciated lava. Clasts in the autobrecciated lavas, both on dives S500 and S510, consist of the same black aphyric basalts as the overlying lobate lava flows. Similar associations of lava flows and autobrecciated lava have been reported elsewhere [*Jones and Nelson,* 1970; *Lonsdale and Batiza,* 1980; *Yamagishi,* 1991]. Thus, these Hawaiian brecciated lavas may represent fragmented parts of the associated lava flows. Again, these units are too small to depict on bathymetric maps.

3.2. Volcaniclastic Rocks

3.2.1. Volcaniclastic breccia. Volcaniclastic breccias were observed on all dives except S499. Representative occurrences are shown in Fig. 5 and characteristics summarized in Table 1. The breccias consist mainly of matrix and clast-supported lapilli to blocks. The clasts are subangular to subrounded and are polymictic (e.g., picritic, olivine, and aphyric basalts). The matrix generally makes up less than 30% of the deposits. At dive site S497, which examined a proximal megablock, the volcaniclastic breccia is well bedded (Figs. 5A, 5B), but this facies is massive in the other dives (Fig. 5C). This facies is widely interbedded with sandy volcaniclastic facies (dives S497, S498, and S511). At dive sites S498 and S511 on Tuscaloosa Seamount, layers of volcaniclastic breccia are < 1 m thick and massive.

In the Nuuanu landslide, the maximum thickness of beds and size of clasts tend to decrease with distance from Oahu. Subrounded clasts are more abundant in distal megablocks. At dive site S497, near Oahu, this facies has a weak fabric and inversely graded bedding (Figs. 5A and 5B). Accord-

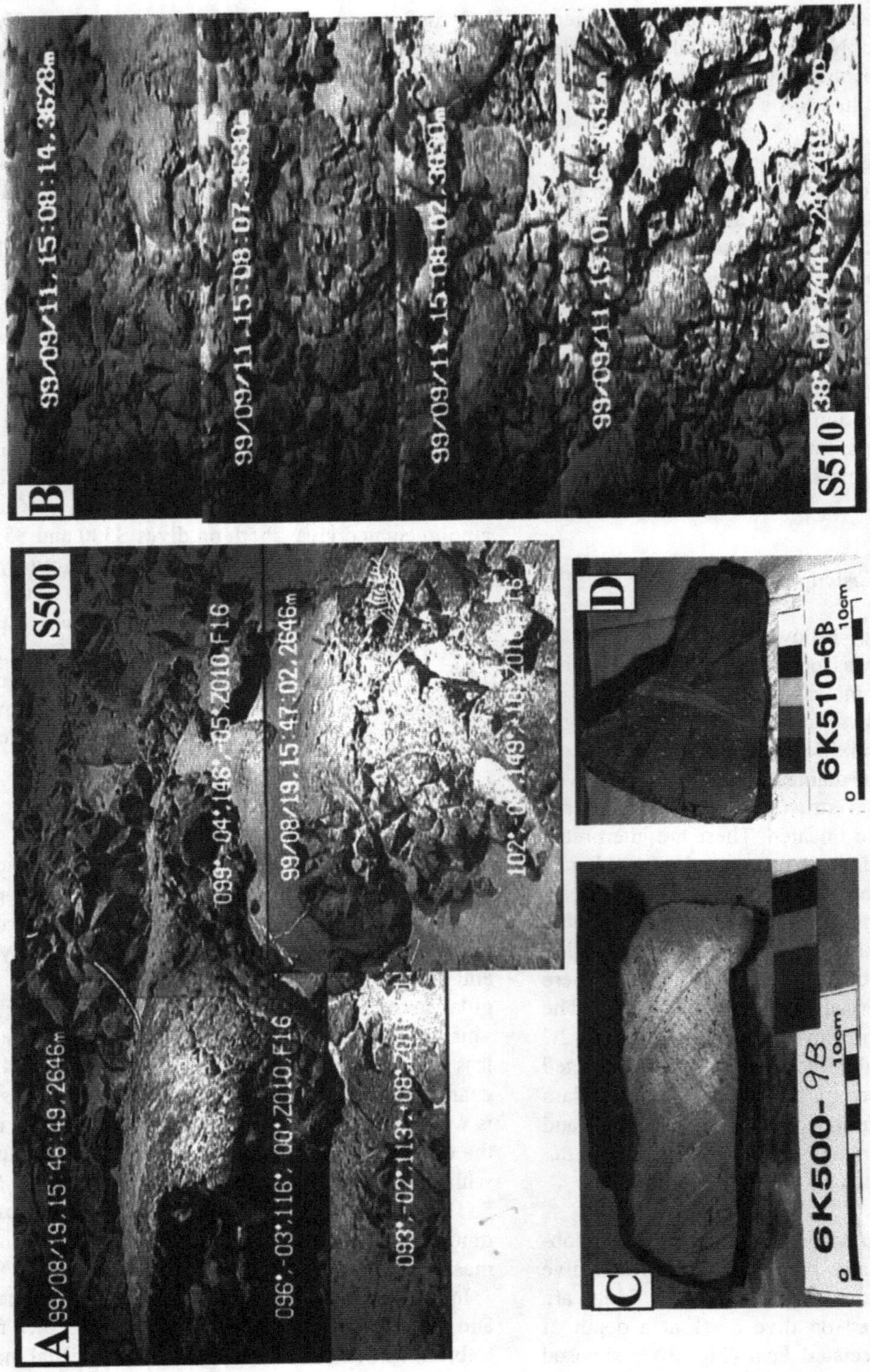

Figure 2. Pillow lavas. Representative pillow lavas capping landslide blocks, as seen in dive S500 (A) and S510 (B). The field of view is about 3 m wide. Morphologically, the lavas are lobate flows; they consist of dense aphyric basalt with few vesicles (C and D).

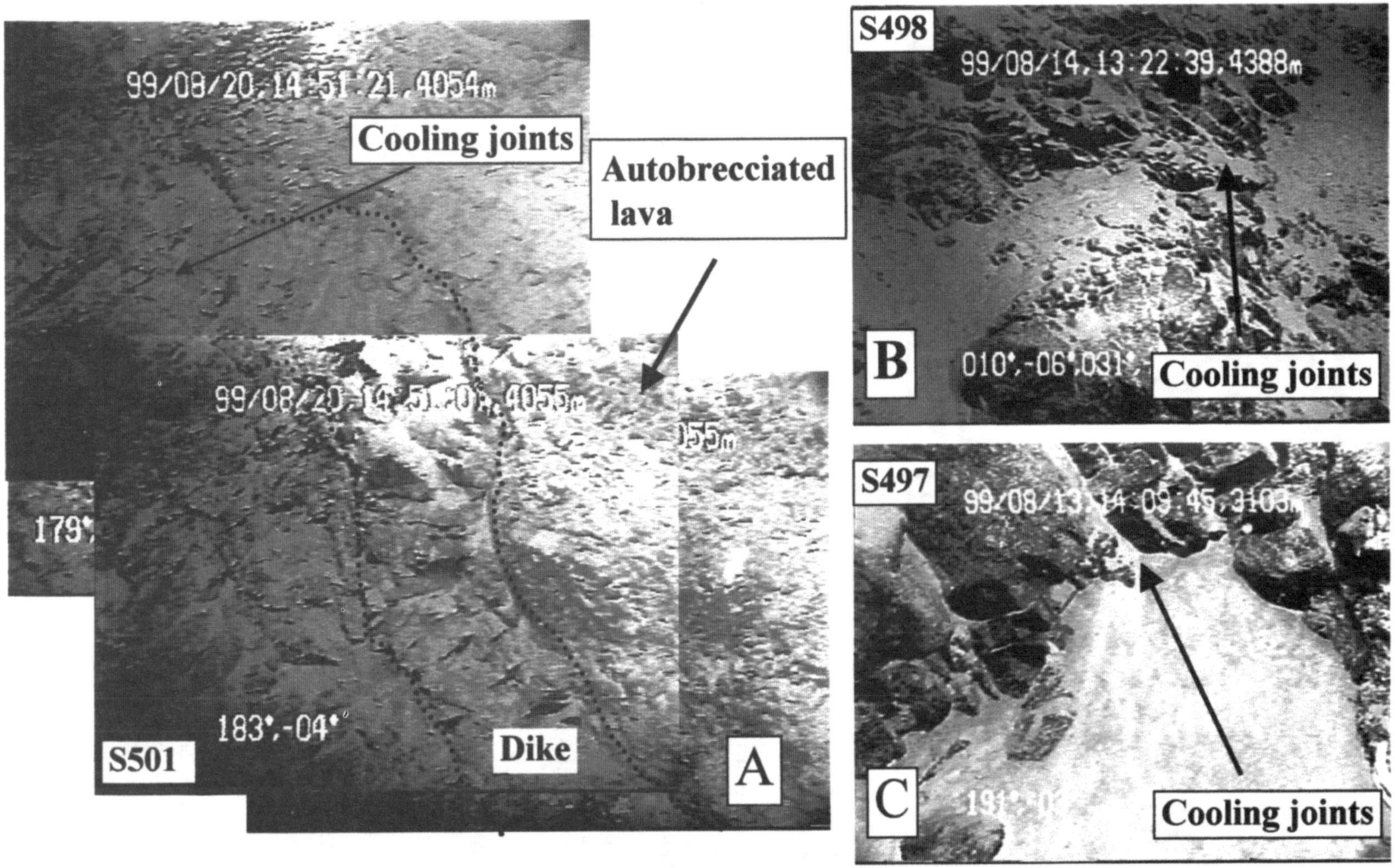

Figure 3. (A) Dike with crude columnar jointing intrudes autobrecciated lava; field of view is about 3 m wide. (B) and (C) Massive lava flows, both less than 2 m thick; the field of view is about 3 m and 1.5 m, respectively.

Figure 4. Autobrecciated lava (A) and rock sample (B) of the same lithology. The field of view in (A) is about 3 m wide. The rock sample is highly fractured and is not abraded.

ing to the classification of debris flow deposits of *Walker* [1975], the volcaniclastic breccia outcrops at dive S497 correspond to an "inverse to normally graded model" and a "graded-stratified model", while at the other localities they more resemble a "disorganized-bed model". The sedimentary characteristics of volcaniclastic breccias are similar to those of debris-flow deposits on alluvial fans [*e.g., Blair and McPherson,* 1994; 1999], or a submarine delta [e.g., *Sohn,* 2000].

3.2.2. Sandy volcaniclastic facies. Representative occurrences of the sandy volcaniclastic facies are shown in Fig. 6, and characteristics are summarized in Table 1. Sandy volcaniclastic rocks crop out at all dive sites between depths of 2500 to 4300 m. This facies displays distinct bedding that makes it easily distinguishable from other facies. In video images, this facies resembles massive sandstone. In general, grading and fine bedding cannot be confirmed in video images because manganese coatings obscure surfaces.

Common characteristics of these rocks are: (1) matrix-supported texture, (2) well sorted clasts, (3) subangular to subrounded clast shapes, (4) highly vesiculated and reddish clasts. These characteristics are consistent with deposition in the submarine environment.

The rocks are polymictic, and vesicles in the clasts are commonly filled with zeolites. The matrix mainly consists of palagonite. The thickness of beds ranges from 6 to 70 cm, and does not vary outcrop scale (Fig. 6). Bed thickness decreases with distance from the source volcano (Fig. 6B). These rocks tend to lie beneath the volcanic breccia, or if more distal, laterally adjacent to it. This facies resembles the bedded material in Papa'u Seamount [*Fornari et al.,* 1979] and Hilina bench [*Lipman et al., this volume*]. The clasts are inferred to have formed as hyaloclastite, near the shoreline when lava flowed into the sea. In that environment, the fragments of hyaloclastite and lava rubble would be abraded by wave action. Finally, these clasts were likely transported as grain flows and deposited on volcanic aprons.

Because video images provide only two-dimensional data, the true dips of the observed strata are uncertain. Estimates of strike and dip can be made, however, using the angles of the video camera (see route maps in accompanying CD-ROM, this volume). Some dips are greater than 30 degrees due to landsliding. On the basis of the available data, a major part of the landslide blocks is made of volcaniclastic breccia and sandy volcaniclastic rocks.

3.3. Unconsolidated Reworked Epiclastic Sediments

3.3.1. Muddy gravel facies. To maximize chances of finding outcrops, dives were planned mainly to traverse the steep cliff faces of landslide blocks. Muddy gravel deposits occur as slope-fan sediments at the base of cliffs and within gullies on megablock slopes. Muddy gravel facies

Table 1. Features of volcaniclastic rocks of the Nuuanu and Wailau landslides.

Characteristics	Autobrecciated lava	Volcaniclastic breccia	Sandy volcaniclastic facies	Muddy gravel facies
Fragments composition	monomictic	polymictic	polymictic	polymictic
Rock types	fine grained basalts	aphyric to picritic basalts	aphyric to picritic basalts	aphyric to picritic basalts
Grain size	pebble to boulder	pebble to boulder	granule to pebble	pebble to boulder
Roundness	fragmented	subangular to subrounded	subrounded to rounded	subangular to subrounded
Fabric	clast supported	variably clast to matrix supported	clast supported	variably clast to matrix supported
Grading	ungraded	inversely graded to ungraded	not observed	inversely graded
Imbrication	random orientation	slightly organized (p)	random orientation	slightly organized (p)
	massive	planar-bedded	planar-bedded	panar-bedded
Bed thickness	50 cm <	30 - 100 cm	5 - 60 cm	20 - 50 cm
Matrix	fine tuff	not observed	palagonite (tuff)	silt to sand
Lithified or unlithified	lithified	lithified	lithified	unlithified
Sorting	polymodal	polymodal	unimodal	bimodal
Nomenclature (Folk et al., 1970)	breccia	conglomerate	sandy conglomerate	muddy gravel
Nomenclature (Fisher , 1966)	pyroclastic breccia	volcanic breccia	lappillistone	pyroclastic breccia, tuff breccia
Interpretation	autobrecciated lava	debris flow deposits	debris flow deposits	debris flow deposits

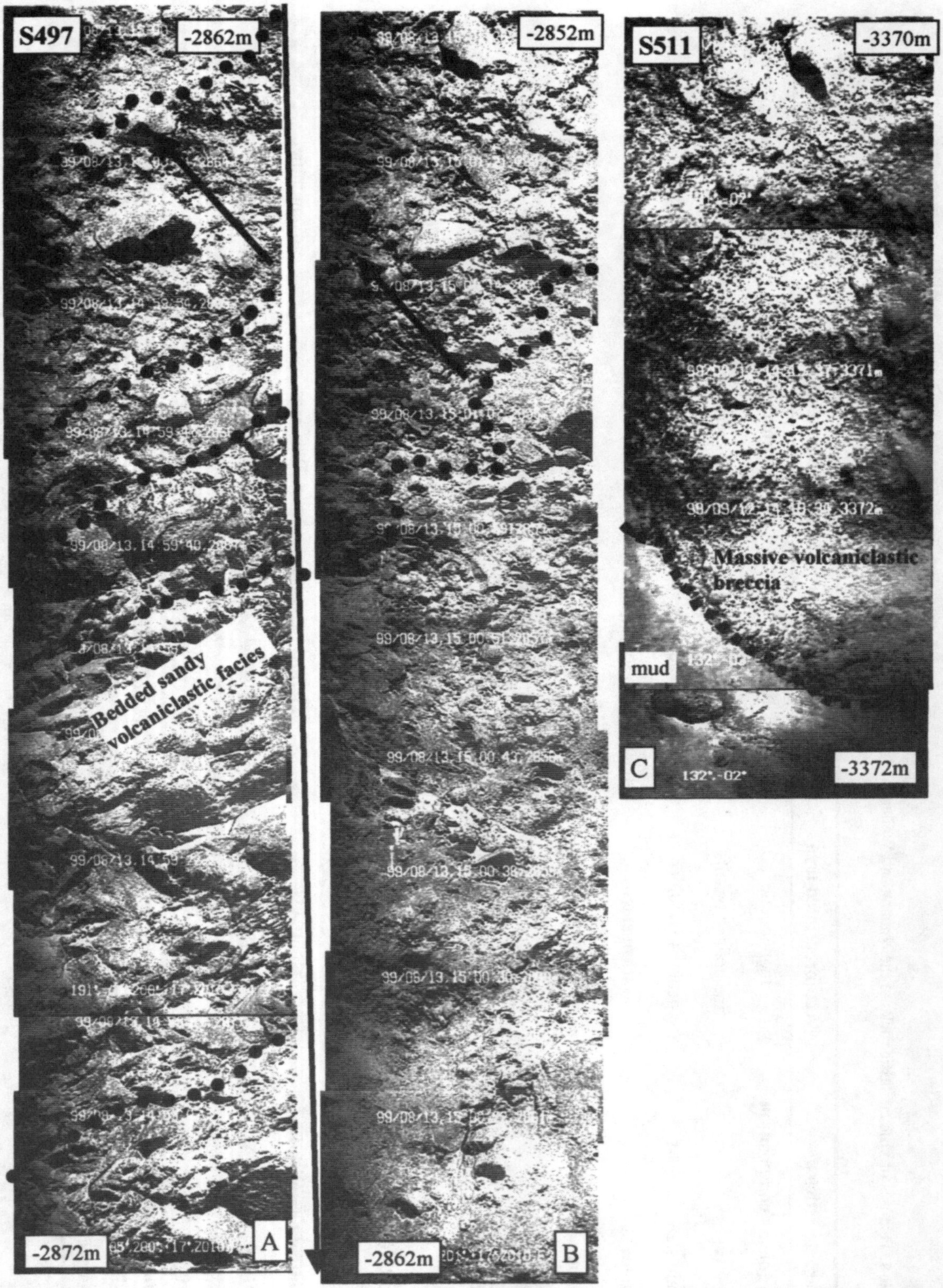

Figure 5. Representative volcaniclastic breccia. Crudely bedded breccia at dive S497 (A and B) and massive breccia at dive S511 (C). The fields of view are about 2 m wide. Thick dotted lines show bedding contacts. The arrows show inverse grading

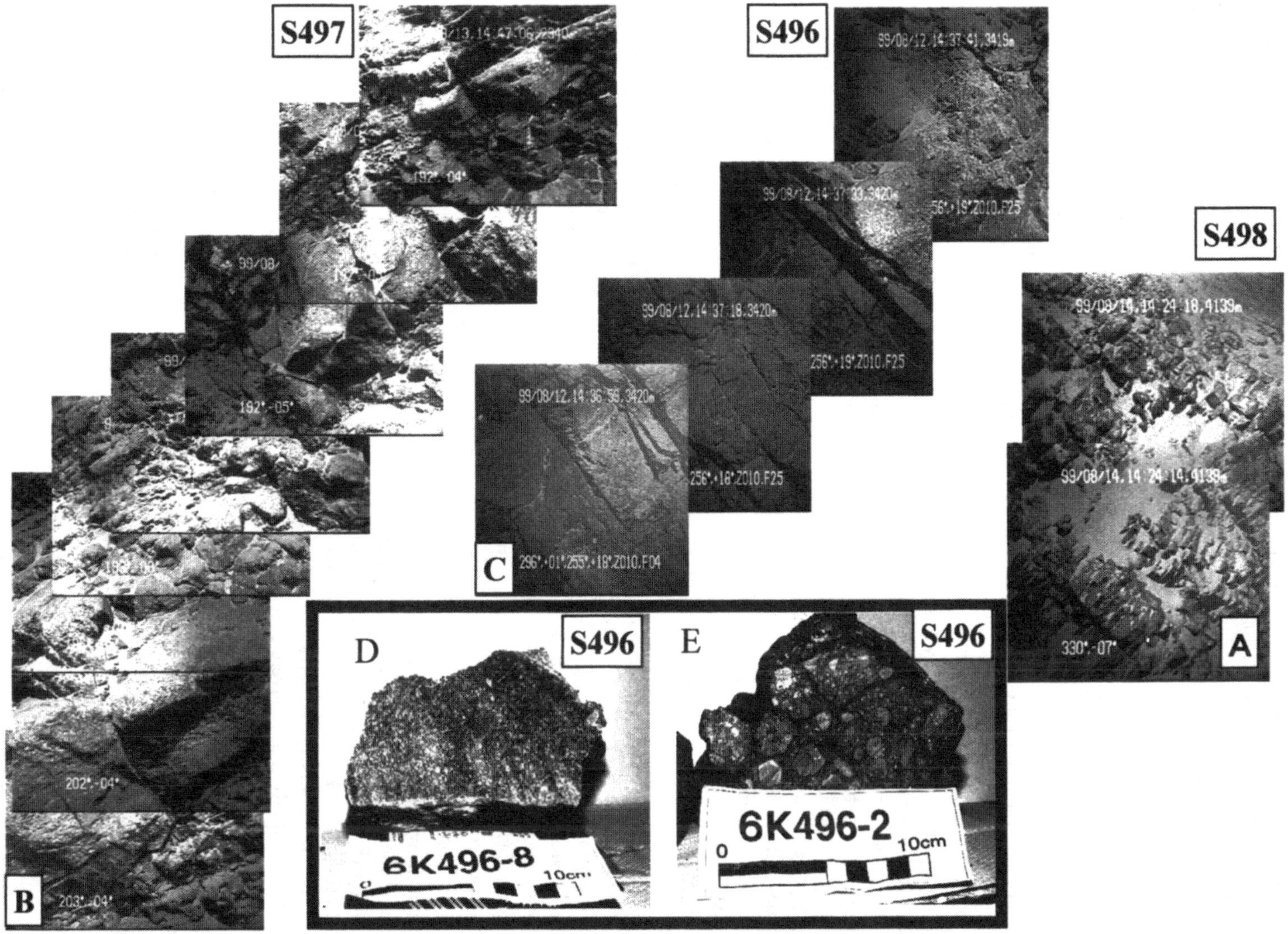

Figure 6. Outcrops (A-C) and samples (D-E) of sandy volcaniclastic rocks. Individual beds are 0.1 to 1m thick. The view in (A) is about 2 m wide. This facies resembles coarse grained sandstone. Bedding observed at several localities dips about 30 degrees. The samples are well sorted, contain abraded clasts.

are planar-bedded cobble and pebble gravel that is clast to matrix supported, rhythmically interstratified with sand and mud (Fig. 7). The matrix of this facies ranges from sand to mud, similar to that currently mantling the surrounding flat ocean floor. The mud may have become intermixed during slope debris flow. An important characteristic of this facies is its unconsolidated state. The strata have normal grading, or display massive bedding that parallels the present slope, and are unconformable with the underlying outcrops. The dips vary from almost zero to about 20 degrees. The muddy gravels distributed on top of landslide blocks and in submarine canyons have almost horizontal bedding, but those on piedmonts adjacent to landslide block scarps have very steep bedding. In video images, this facies appears similar to the volcaniclastic breccia, and when consolidated, they are identical.

The clasts are highly variable. Their shapes range from polymictic subangular to subrounded, and they include volcanic sand, sandy volcaniclastic rocks, and pieces of picritic and aphyric basalt. Some basaltic clasts are reddened by oxidation and are variably vesiculated. These clasts may have originated from subaerial eruptions. The clasts of sandy volcaniclastic facies indicate that the muddy gravel facies consists in part of reworked volcaniclastic sediments. The characteristics of this facies are similar to those of cohesive mudflows or debris flow deposits [e.g., *Blair and McPherson*, 1994; *Einsele*, 1992]. Such deposits are common on submarine deltas in submarine canyons and as a talus cones on the piedmonts of fault scarps. These sedimentary environments are consistent with the present localities, and it is likely that the same sedimentation processes continue today.

3.3.2. Talus. Talus is widely distributed at the base of each megablock, and consists mainly of broken pieces of the muddy gravel layer (Fig. 7E), followed by fragments of the sandy volcaniclastic facies, lava, and clasts of the volcaniclastic breccia. The talus was likely derived by mass movements of disrupted muddy gravel layers that draped the flanks of the megablocks. Some talus derived from muddy gravel layers resembles rubbly lava because of its manganese coating (Fig.7E). Much talus is not covered by thick mud, apparently having accumulated more rapidly than the mud. Dredge hauls from such landslide areas are likely to be dominated by talus deposits

3.3.3. Massive siltstone. Outcrops of massive silt were observed during dive S511 at 3800 m depth. In video images, these rocks resemble massive lava or sandstone because of manganese coatings (Fig. 7F). Thin (< 10 cm) silt layers are also intercalated with muddy gravel. These were likely deposited from the fine-grained debris elutriated from the upper parts of debris flows.

3.3.4. Mud. The ocean floor in the landslide area is covered by pale yellowish brown to pale brown mud. This material varies from silt to fine sand, and its thickness varies to over several tens cm. Mud was not seen on the lava terraces in dive S500 and S510, and seems unrelated to the local slope angle. Thick deposits of mud were found on top of the other megablocks and in the submarine canyons. Such mud layers were likely deposited repeatedly and may have been deposited from fine-grained suspensions associated with moving debris flows and turbidites (Fig. 7G).

4. DISCUSSION

4.1. Interpretive geologic map of the Nuuanu and Wailau Landslides

Based on available bathymetry but without direct observations of deep-sea features, *Moore et al.* [1994] inferred that slumps and debris-avalanche deposits are distributed widely offshore of northeast Oahu and northern Molokai. The present study examined the geology of parts of this area in more detail, combining manned submersible investigations, interpretation of dive videos, and analysis of newly obtained bathymetric maps. The landslide area consist mainly of five topographic types: megablocks (slump blocks), submarine canyons, volcanic-island aprons, hummocky terrain, and abyssal plain (Figs. 8 and 9).

4.1.1. Megablocks. The landslide megablocks are bounded by fault–line scarps. The observed megablocks are composed mainly of volcaniclastic breccia and sandy volcaniclastic facies, which include terrigenous clasts. Some are associated with thin massive lava flows. Most megablocks, including the large Tuscaloosa Seamount, consist chiefly of volcaniclastic deposits. Only two of the observed megablocks have lava flows at their top. The lavas have preserved their topographic characters, and the edges of the megablocks correspond to boundaries of the lava flows. Because these lavas and the associated autobrecciated lava are products of the same type of eruptions, they are mapped as single units.

Individually, the megablocks appear to be unfractured, intact blocks. Collectively, the megablocks can be graphically fitted together to form a thin, tabular slab with a low aspect ratio. This suggests that they did not move downslope as a turbulent flow. The distances from the landmass of Oahu to the present location of these megablocks is too far to be explained by a simple rota-

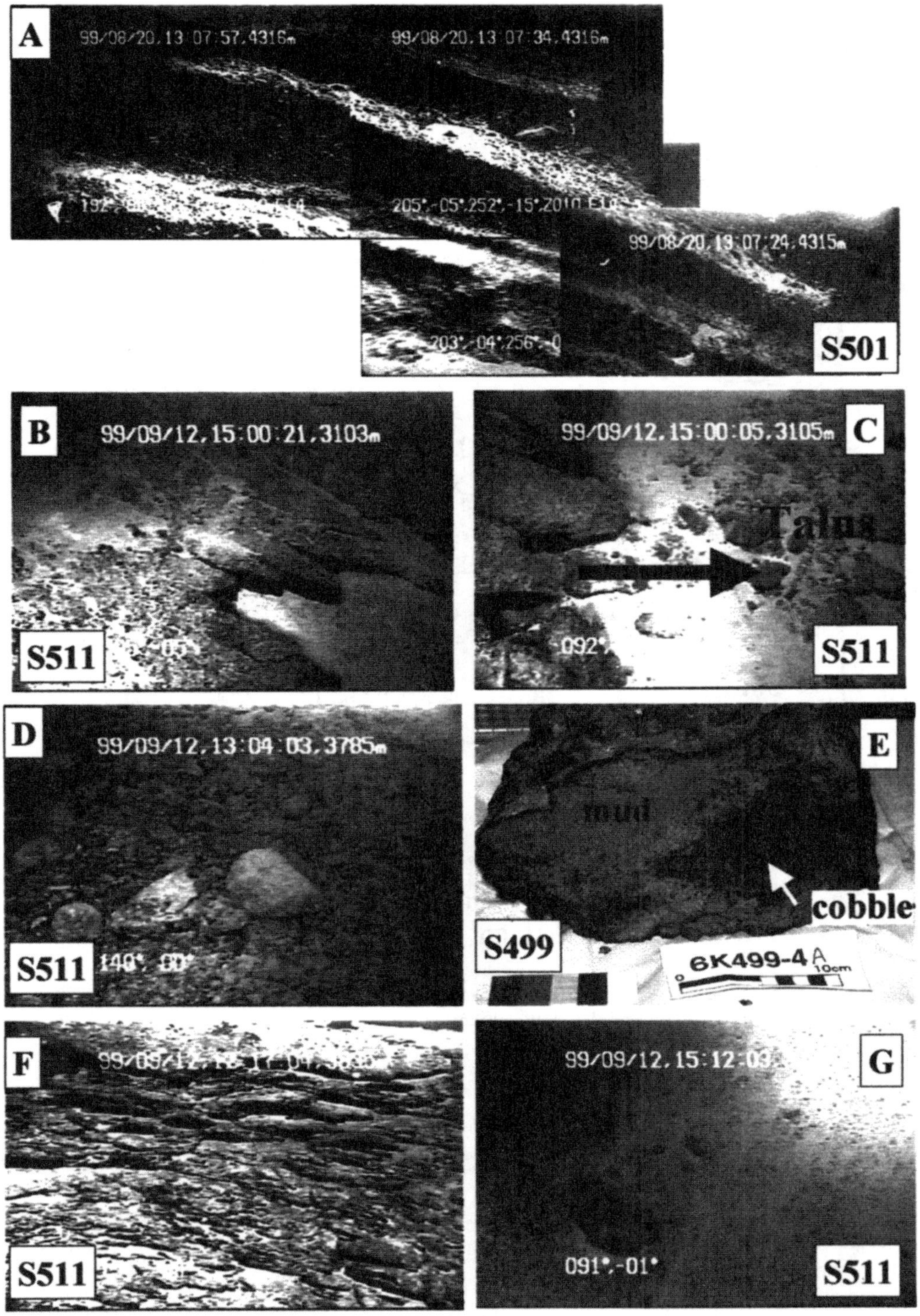

Figure 7. Unconsolidated epiclastic deposits. (A) Multi-layered muddy gravel facies showing slight imbrication and grading. (B-C) Talus breccia of muddy gravel clasts. (E) Sample of representative talus and its sawed surface. (F) Sheared siltstone. (G) Representative piedmont of a megablock shows reworking of surface rocks; field of view is about 2 m wide.

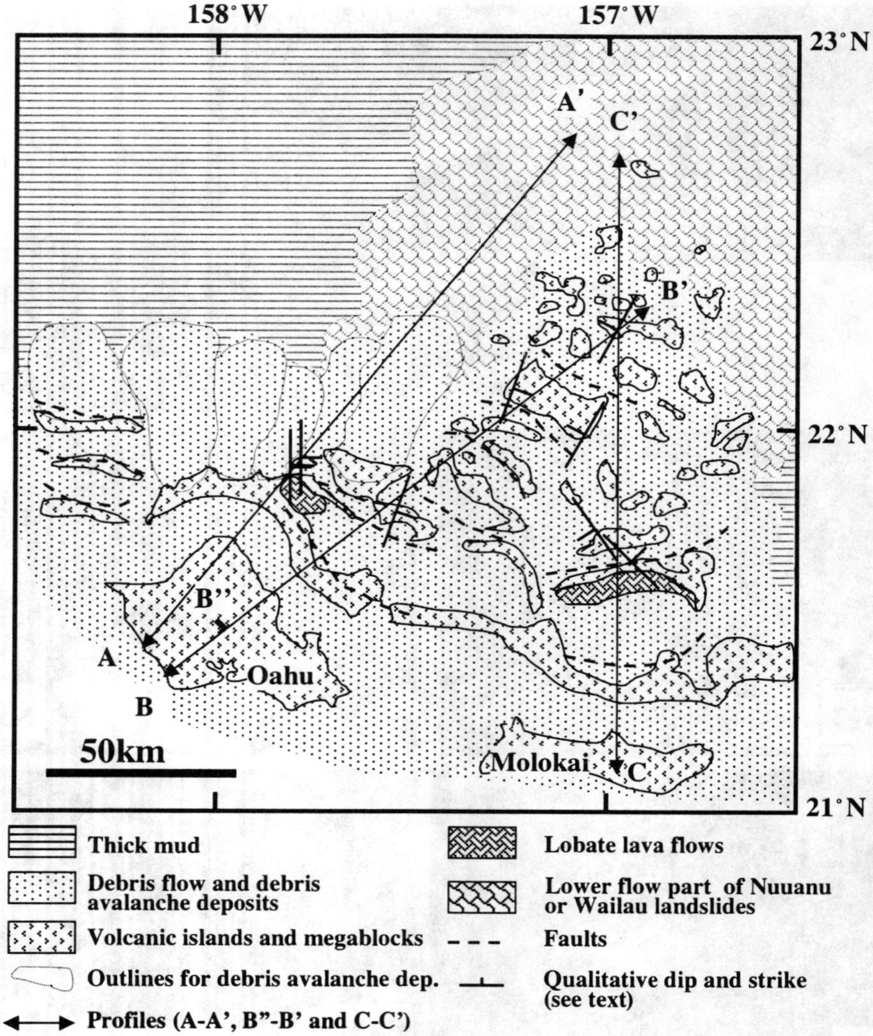

Figure 8. Geologic map of the Nuuanu and Wailau landslide areas, made by combining the data from route maps (see accompanying CD-ROM, this volume), in-situ samples, and bathymetric maps.

tional landslide. Consequently, explanations for the downslope movement of the megablocks need to include both slumping and downslope sliding.

4.1.2. *Flat and gentle-slope areas.* Gently dipping slopes and flats include submarine canyons, aprons of volcanic islands, hummocky terrains, and abyssal plains. Submarine canyons are common between megablocks, and are covered by muddy gravel facies, mud, and talus. Most of the submarine canyons can be traced back to the northeastern cliff of Oahu. These pathways have been used to transport terrigenous volcaniclastic materials from Oahu to deep-sea fans, and they are still active. This suggests that the muddy gravel facies distributed in relatively wide submarine canyons were deposited by turbidites or debris flows. Mud may have been deposited from suspended fine particles derived from the debris flows. In contrast, the muddy gravel facies in submarine canyons that head on megablocks formed by collapse of draped sediment from nearby steep cliffs. These collapsed sediments formed small submarine deltas or cones at the foot of the megablocks. Debris-avalanche deposits of the Nuuanu and Wailau landslides cannot be observed in the submarine canyons and the flat tops of megablocks because of the thick covering of muddy gravel facies.

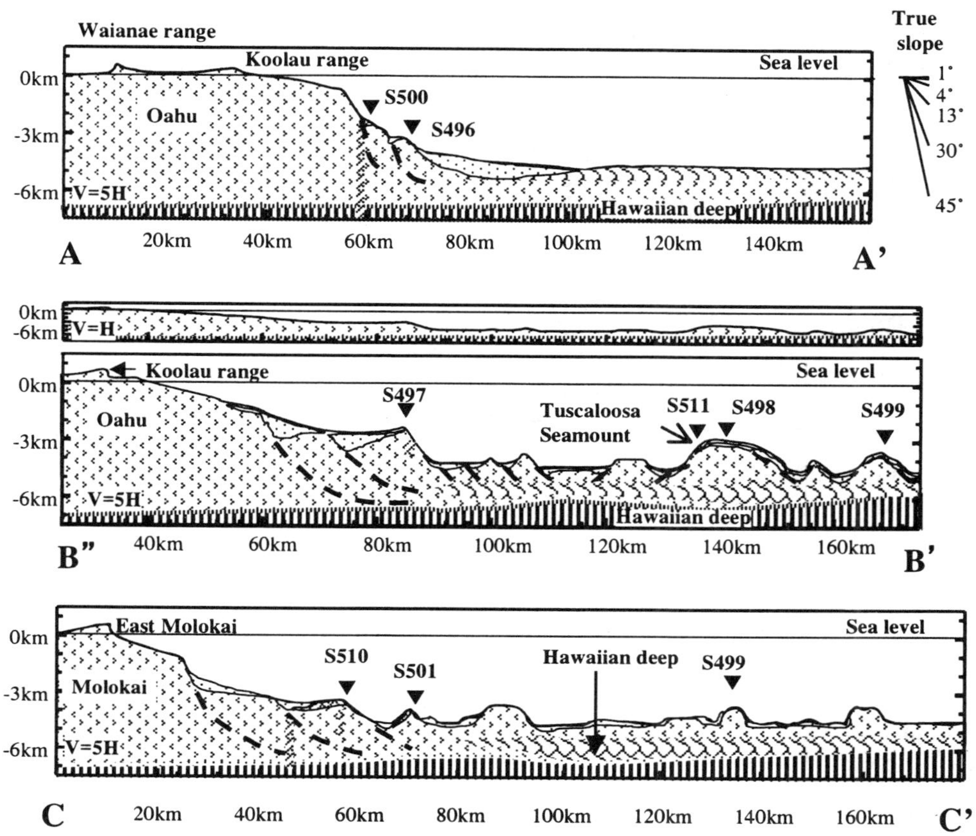

Figure 9. Geologic profiles of northeast Oahu and north Molokai. Profile locations and patterns shown in Figure. 8; vertical exaggeration x5. The middle profile shows B''-B' without vertical exaggeration.

Five minor debris-avalanche terranes offshore of Oahu are defined by well-developed amphitheaters and downslope hummocky regions. It is likely that the muddy gravel facies is distributed in these debris-avalanche terranes.

4.2. Reconstruction of Oahu

4.2.1. Topographic reconstruction. The newly obtained bathymetric map outlines the fine structure of the megablocks [Moore and Clague, this volume]. The megablocks have steep cliffs, and where these cliffs face each other across an intervening low area, they can be geometrically fitted together like a giant jigsaw puzzle [see also Moore and Clague, this volume]. It is suggested that these megablocks once formed a single huge slab. This slab slumped and slid downslope, separating during the landslide event into the large blocks we see today (Fig. 10). Tilting angles, sliding distances, and azimuths for each block can be interpreted from their present-day geometries. Inferred slumping directions are about the same as proposed by *Moore* [1964].

As seen in the bathymetric map, megablocks can be graphically fitted together by combining parallel translation and rotation. The reconstructed huge trapezoidal slab is approximately 50 km long, 40 km wide, and rises about 1.6 km above the surrounding sea floor. This huge mass is too large to have been derived from the eastern subaerial part of Oahu [*Macdonald et al.,* 1983]. However, the volume of this huge mass can be crudely balanced by the depression offshore of northeastern Oahu in the depth range of 1000 to 3000 m.

The balance between the huge mass and the depression in the reconstructed apron of northeast Oahu (Fig.10) is consistent, not only in plan but also in section, when a small amount of tilting is included. The submarine apron of northeastern Oahu has been reconstructed to accommodate the jigsaw-like fit of the megablocks. It is also note-

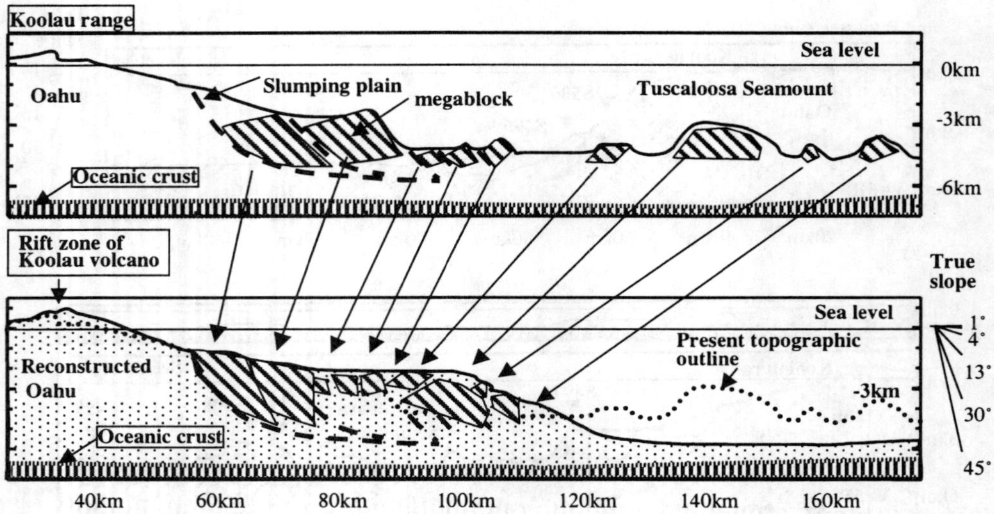

Figure 10. Profiles showing interpreted reconstruction of the pre-landslide apron on the northeastern side of Oahu. These profiles lie along B''–B' of Fig. 8; vertical exaggeration x5.

worthy that the toe of the pre-failure volcaniclastic apron is at the axis of the Hawaiian Deep.

4.2.2. The submarine-apron model. The geometrically reconstructed volcanic apron of northeast Oahu can be related to a submarine-apron model (Fig. 11). The volcaniclastic breccia is mainly observed in the upper part of the stratigraphic sequences—a relationship similar to alluvial-fan deposits and topset beds of submarine fans elsewhere. The sandy volcaniclastic facies stratigraphically beneath the volcaniclastic breccia can be interpreted to be grain-flow deposits and turbidites, of the type forming the foreset beds of a typical submarine delta. Many beds of the sandy volcaniclastic facies have dips of up to 30 degrees. These inclined strata are noteworthy , and their dips may be related to primary sedimentary structures, not to the slump-induced tilting of the megablocks.

In this submarine delta model, it is proposed that sand- and silt-size bottom-set beds are distributed beneath the foreset sandy volcaniclastic beds. Details of the lower part of the stratigraphic sequence cannot be observed because they are covered by the muddy gravel facies. The depth of the reconstructed delta surface, about 1000 to 2000 m below present sea level, is deeper than might be expected from the interpreted sedimentary environment. However, because Oahu has subsided about 1300 m [Mark and Moore, 1987], the topset beds may have been near sea level just before the collapse. In short, it is concluded that the volcaniclastic breccia was deposited in shallow water to form the upper parts of the large alluvial fan.

4.3. Why did the Huge Landslide Occur Northeast of Oahu?

4.3.1. Slope instability? Many factors can lead to slope instability, including overloading, internal structure, slope angle, composition of the geologic units, climate effects, and changing internal pore pressure. The frequency of landslides may be expected to increase with an increasing sedimentation rate [*Scheidegger,* 1982]. For example, the Mississippi delta has a sedimentation rate of more than 1000 m/m.y. [*Wetzel and Kohl,* 1986], and submarine mass wasting occurs frequently along the delta front [*Prior and Coleman,* 1982]. Oahu was built by volcanism that persisted for about 1Ma [*Clague and Dalrymple*, 1987]. As the volcanic island grew rapidly above sea level, subaerial lava flows entered the sea and shattered to form huge volumes of volcaniclastic debris, as hyaloclastite and its related rocks were supplied to the growing submarine delta by debris flows and turbidities. The island slopes have accumulated volcaniclastic materials about 4000 m thick in less than 1 million years. The sedimentation rate is up to four times higher than that of the Mississippi delta, suggesting that the huge volcaniclastic aprons around the volcanic island were gravitationally unstable. Furthermore, the volcaniclastic deposits, with distinct bedding, are probably distributed widely on the apron of the island. Bedding surfaces in the sandy volcaniclastic facies dip about 30 degrees seaward and could serve as potential slip planes for bedding-plane failure.

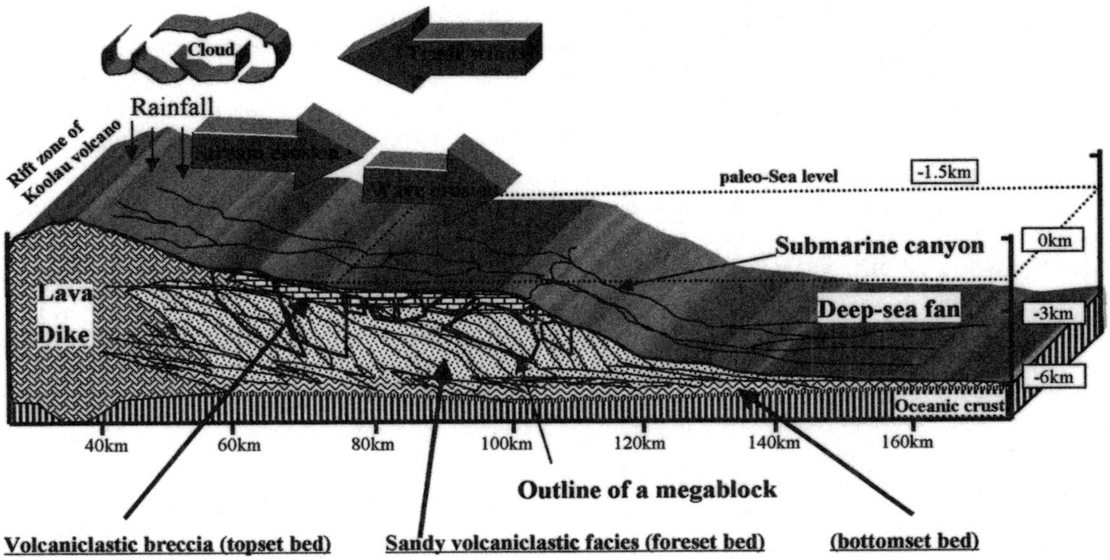

Figure 11. Schematic view of the paleo-Oahu Island before the giant landslide. Profile lies along B–B' of Fig. 8; vertical exaggeration x5.

In the study area, the transportation distance from the sediment provenance to the sedimentary basin is much shorter than that of the continental margin. Volcaniclastic rocks can accumulate to a great thickness in their sedimentary basin. The angle of repose of sedimentation depends on the size of its fragments [*Montgomery,* 1992], suggesting that coarse volcanic clasts can be deposited with a high angle of repose in such sedimentary environments. Indeed, the average slope angles on aprons of the Hawaii Islands are steeper than those of the continental margins. The slope angles of volcanic islands can be stabilized even at steep angles. This suggests that the frequency of slope failures decreases with increasing clast size, but a possible implication is that the size of a landslide, when is eventually occurs, is larger.

The most important prerequisite for a giant landslide is a large and unstable apron of volcaniclastic deposits on the volcano's submarine flanks. The stability of the slopes in volcanic aprons is controlled by the balance between the growth of the volcano and the rates at which volcaniclastic debris is supplied to its submarine apron.

Erosion is the main process transporting volcanic rocks to ocean basins when eruption takes place subaerially. In the Hawaii Islands, erosion systems are not equal in every direction, and are mainly of two types; wave and stream erosion systems. Both erosion systems are accelerated by trade winds [*Macdonald et al.,* 1983] since the largest waves and heaviest rainfall occur on the windward side of the islands. Growth of the pre-failure volcaniclastic apron has been enhanced by debris supplied by subaerial erosion from the windward, rainy sides of Oahu and Molokai. Thus, the exceptional size of the landslides described here may have been influenced by heavy trade-wind rainfall (Fig. 11).

4.3.2. Triggers for the giant landslides? The rapid increase of slope angle, strong ground motions associated with earthquakes, and the structural effects of magma intrusion have been proposed as triggers for structural failures in volcanic terranes [e.g., *McGuire,* 1996; *Elsworth and Voight*, 1996]. Magma intrusion is especially important, because it is closely related to the basic structure of volcanic islands. Whether the lavas erupt at the summit crater or not is critical for any discussion of landslides on a volcanic flank.

Vent locations for the lava flows in this area can be inferred from two key observations: (1) The lava morphology is lobate type, which results from high eruption rates (Umino et al., 2000). If lava has enough time to interact with air or seawater, its physical properties change due to cooling. Then, lava will solidify as strained pillows or hyaloclastite, not lobate morphology. (2) Based on the sulfur content in glass rims, the eruption depths at dives S500 and S510 were about 1000 m [*Shinozaki et al., this volume*]. Such a depth is consistent with the original position of the pre-failure volcanic apron. Therefore, these two types of evidence suggest that the vents should be located near the outcrops, not at a distant on–land rift zone of Koolau or east Molokai. The positions also correspond to near the heads of the Nuuanu and Wailau landslides, as seen in dives S500 and S510, respectively. These inferred magma extrusion sites are parallel to the Hawaiian Deep, on the northern extension of the Molokai – Kilauea trend in Jackson et al. (1972), where an incipient volcanic center may have been located. Thus, it is possible that magma was intruded, and erupted, near the head of the Nuuanu landslide. However, the megablocks and the Tuscaloosa seamount consist mainly of volcaniclastic debris, not lava flows or intrusive rocks. To better understand the relationship between magma intrusion and landslides in this area, additional studies are required in proximal shallow-water areas (< 2000 m).

5. CONCLUSIONS

The submarine geology of the Nuuanu and Wailau landslides was investigated during the 1999 cruise of the SHINKAI 6500 submersible and its mother ship the YOKOSUKA. Eight manned-submersible dives were made, and about 100 in-situ rock samples were collected at depths of 2500 to 4500 m. Based on video images and rock samples, the outcrops were divided into three lithologic types, each with several facies: products of volcanic eruptions, coarse volcaniclastic deposits, and unconsolidated and reworked volcaniclastic sediments.

The gently dipping slopes and flat parts of this area are covered with muddy gravel facies and mud, which are the recent reworked sediments of a submarine channel system and small-scale slope failure. The slumped megablocks, including the giant Tuscaloosa Seamount, consist mainly of volcaniclastic breccia and sandy volcaniclastic rocks. The volcaniclastic breccia is considered to have originated as debris-flow deposits in an alluvial fan system, based on their sedimentary characteristics.

The steep cliffs of the megablocks facing one-another across intervening low zones can be fitted together like a jigsaw puzzle. The resulting huge reconstructed trapezoidal slab can be matched with a depression offshore of eastern Oahu at depths of 1000 to 3000 m. This huge apron is thought to have been composed mainly of debris produced by ocean-entry processes and to a lesser extent by subaerial erosion. Instabilities in this apron, caused by some combination of sedimentary over steepening, the shaking action of large earthquakes, and the intrusion of basaltic dikes into near shore parts of the apron, triggered the giant landslides. Finally, the growth of the pre-failure delta may have been enhanced by debris supplied by subaerial erosion from the windward, rainy sides of Oahu and Molokai. Thus, the exceptional dimensions of the landslides described here may have been influenced by heavy trade-wind rainfall.

Acknowledgments. The author thanks dive observers, E. Takahashi (dive S496), M. Garcia (dive S497 and S500), J. Moore (dive S498 and S501), S. Takarada (dive S499), T. Kanamatsu (dive S510) and T. Ui (dive S511), for collecting samples and video images, the captain H. Tanaka and crews of the SHINKAI 6500 and YOKOSUKA operation teams, and the scientific parties of Legs 1 and 2 of K99 for collecting SeaBeam data. The author also thanks; (1) the scientific parties, especially chief scientist J. Naka and co-chief scientist E. Takahashi, and members of the Pheasant Memorial Laboratory for helpful discussions on the reconstruction model, (2) T. Matsuda and E. Honza for helpful suggestions on sedimentary basin models, and (3) W. Duffield, R. Fiske, P. Lipman, M. Garcia, and J. Moore for their reviews and suggestions that improved the original manuscript.

REFERENCES

Batiza, R. and J.D.L. White, Submarine lavas and hyaloclastite, *in* Sigurdsson, H. (ed.), *Encyclopedia of volcanoes,* Academic Press, London, 361-381, 2000.

Blair, T. C., and J. G. McPherson, Alluvial fans and their natural distinction from rivers based on morphology, hydraulic processes, sedimentary processes, and facies assemblages. *J. Sediment. Res., A64*, 450-489, 1994.

Blair, T. C. and J. G. McPherson, Grain-size and textural classification of coarse sedimentary particles. *J. Sedimentary Res*., 69, 6-19, 1999.

Carracedo, J.C., Growth, structure, instability and collapse of Canarian volcanoes and comparisons with Hawaiian volcanoes. *J. Volcan. Geothermal. Res.*, 94, 1-19, 1999.

Clague, D.A., and G.B. Dalrymple, The Hawaiian-Emperor volcanic chain part I. *U.S.Geol. Surv. Prof. Pap.,* 1350, 5-54, 1987.

Einesele, G., *Sedimentary basins.* Springer-Verlag, New York 628p., 1992.

Elsworth, D. and B. Voight, Evaluation of volcano flank instability triggered by dyke intrusion. *Geol. Soc. Spec. Pub.,* 110, 45-53, 1996.

Embley, R.W. and R.D. Jacobi, Distribution and morphology of large submarine sediment slides and slumps on Atlantic continental margins, *Marine Geology*, 2, 205-228, 1978.

Fisher, R.V., Rocks composed of volcanic fragments and their classification. *Earth-Sci. Rev.*, 1, 287-298, 1966.

Folk, R.L., P.B. Andrews, and D.W. Lewis, Detrital sedimentary rock classification and nomenclature for use in New Zealand. *New Zealand J. Geol. Geophys*. 13, 937-968, 1970.

Fornari, D.J., J. G. Moore, and L. Calk, A large submarine sand-rubble flow on Kilauea volcano, Hawaii. *J. Volcan. Geothermal. Res.,* 5, 239-256, 1979.

Griffith, R.W. and J.H. Fink, Solidification and morphology of submarine lava: A dependence on extrusion rate. *J. Geophys. Res.*, 97, 19729-19737, 1992.

Holcomb, R.T. and R.C. Searle, Large landslides from oceanic volcanoes. *Marine Geotechnology*, 10, 19-32, 1991.

Jackson, E.D., E. A. Silver and G. B. Dalrymple, Hawaiian-Emperor chain and its relation to Cenozoic circumpacific tectonics. *Geol. Soc. Amer. Bull.*, 83, 601-618, 1972.

Jones, J.G. and P.H.H.Nelson, The flow of basalt lava from air into water-its structural expression and stratigraphic significance. *Geol. Mag.*, 107, 13-19, 1970.

Labazuy, P., Recurrent landslides events on the submarine flank of Piton de la Fournaise volcano (Reunion Island) *Geol. Soc. Spec. Pub.* 110, 295-306, 1996.

Londsdale, P. and R. Batiza, Hyaloclastite and lava flows on young seamounts examined with a submersible. *Geol. Soc. Amer. Bull.*, 91, 545-554, 1980.

Lipman, P.W., W.R. Normark, J.G. Moore, J.B. Wilson, and C.E. Gutmacher, The giant submarine Alika debris slide, Mauna Loa, Hawaii. *J. Geophys. Res.*, 93, 4279-4299, 1988.

Lipman, P.W., T.W. Sisson, T. Ui, J. Naka, J. Morgan, and J. Smith, Ancestral submarine growth of Kilauea volcano and instability of its south flank, this volume.

Macdonald, G.A., A.T. Abbott, and F.L. Peterson, *Volcanoes in the Sea* (second edition) University of Hawaii Press, Honolulu 517 p., 1983.

Mark, R.K. and J.G. Moore, Slopes of the Hawaiian ridge. *US Geol. Surv. Prof. Pap.*, 1350, 101-107, 1987.

McGuire, W.J., Volcano instability: a review of contemporary themes. *Geol. Soc. Spec. Pub.*, 110, 1-23, 1996.

Montgomery, C.W., Mass movements. *Environmental geology* (third edition), Wm. C. Brown Publishers, 158-160, 1992.

Moore, J.G. , Giant submarine landslides on the Hawaiian ridge. *U.S.Geol. Surv.Prof. Pap.*, 501, D95-D98, 40 , 1964.

Moore, J.G. and D.A. Clague, Volcano growth and evolution of the island of Hawaii. *Geol. Soc. Amer. Bull.*, 104, 1471-1484, 1992.

Moore, J.G. and D.A. Clague, Mapping the Nuuanu and Wailau landslides in Hawaii, this volume.

Moore, J.G., D.A. Clague, R.T. Holcomb, P.W. Lipman, W.R. Normark, and M.E.Torresan, Prodigious submarine landslides on the Hawaiian ridge. *J. Geophys. Res.*, 94, 17465-17484, 1989.

Moore, J.G., W.R. Normark, and R.T. Holcomb, Giant Hawaiian landslides Annu. *Rev. Earth Planet. Sci.*, 22, 119-144 ,1994.

Prior, D.B. and J.M. Coleman, Active slides and flows in unrconsolidated marine sediments on the slopes of the Mississippi delta. In: S. Saxov and J. K. Nieuwenhuis (Editors), *Marine slides and other mass movements*, Plenum Press, New York, 21-49, 1982.

Saxov, S., Marine slides - some introductory remarks. In: S. Saxov and J. K. Nieuwenhuis (Editors), *Marine slides and other mass movements*, Plenum Press, New York, 1-7, 1982.

Scheidegger, A.E., On the tectonic setting of submarine slide. In: S. Saxov and J. K. Nieuwenhuis (Editors), *Marine slides and other mass movements*, Plenum Press, New York, 11-20, 1982.

Shinozaki, K., Z.-Y. Ren, and E. Takahashi, Geochemical and petrological characteristics of Nuuanu and Wailau landslide blocks, this volume.

Smith, M.S. and J.B. Shepherd, Tsunami waves generated by volcanic landslides:an assessment of the hazard associated with Kick'em Jenny. *Geol. Soc. Spec. Pub.*, 110, 115-123, 1996.

Sohn, Y. K., Depositional processes of submarine debris flows in the Miocene fan deltas, Pohang Basin, SE Korea, with special reference to flow transformation. *J. Sediment. Res.*, 200, 491-503, 2000.

Stillman, C.J., Giant Miocene landslides and the evolution of Fuerteventura, Canary Islands. *J. Volcan. Geothermal. Res.*, 94, 89-104, 1999.

Umino, S., P. W. Lipman, and S. Obata, Subaqueous lava flow lobes, observed on ROV KAIKO dives off Hawaii. *Geology*, 28, 503-506, 2000.

Walker, R.G., Generalized facies models for resedimented conglomerates of turbidite association. *Geol. Soc. Amer. Bull*, 86, 737-748, 1975.

Wetzel, A. and B. Kohl, Accumulation rates of Mississippi fan sediments core during Deep Sea Drilling Project Leg 96. In Bouma AH, Coleman JM, Meyer AW, et al., Init. Rept. DSDP 96: 595-600, US Govt. Print Office, Washington, 1986.

Yamagishi, H., Morphological and sedimentological characteristics of the Neogene submarine coherent lavas and hyaloclastites in Southwest Hokkaido, Japan. *Sedimentary Geology*, 74, 5-23, 1991.

H. Yokose, Department of Earth Sciences, Faculty of Science, Kumamoto University, 2-39-1 Kurokami, Kumamoto 860-8555, Japan.

Major Element Geochemistry of Glasses in Turbidites as Source Indicators: Implications for the Nu‘uanu and Wailau Giant Submarine Landslides

S.B. Sherman and M.O. Garcia

Dept. of Geology and Geophysics, University of Hawai'i, Honolulu, Hawaii

E. Takahashi

Earth and Planetary Sciences, Tokyo Institute of Technology, Tokyo, Japan

The giant Nu‘uanu and Wailau landslides resulted from the collapse of Ko‘olau and East Moloka‘i volcanoes. The turbidity currents associated with these landslides carried portions of these volcanoes great distances. Glass sands and silts were recovered from four piston cores that may have sampled these and other turbidites. Three cores were taken along a transect extending 100 to 350 km northeast of O‘ahu. The fourth core was taken 40 km north of Moloka‘i and contains ~20 turbidites. The major element compositions of >500 glasses from these cores are combined with paleontological and paleomagnetic data to determine the source volcano and relative ages for these turbidites, Sulfur was analyzed to infer the eruption water depth for the glasses. Our results show that the thick turbidite layer in each of the two proximal cores from the O‘ahu transect are probably related mostly to subaerial East Moloka‘i lavas transported by the Wailau landslide. The silt-size glasses from the distal transect core have Ko‘olau-like compositions. These glasses may be related to the Nu‘uanu landslide or a secondary collapse of Ko‘olau landslide blocks. The core sampled near Moloka‘i contains glasses from the deep submarine and subaerial eruptions of rejuvenated volcanism in its upper section and glasses with shield lava compositions in the middle of the core. The abundant turbidite intervals in this core indicates that Moloka‘i volcano has repeatedly collapsed.

1. INTRODUCTION

Hawaiian volcanoes are enormous edifices that frequently experience slope failure, generating landslides, debris flows and turbidity currents. The islands of O‘ahu and Moloka‘i are sources for two giant landslides, the Nu‘uanu and Wailau, respectively (Fig. 1). The Nu‘uanu landslide is related to the catastrophic slope failure of Ko‘olau volcano, on the northeastern side of O‘ahu, and is one of the largest landslides on earth. Massive blocks from this slide are found up to 230 km away from O‘ahu [e.g., *Moore et al.*, 1989; *Naka, et al*, 2000]. Nearby East Moloka‘i volcano also collapsed producing the Wailau landslide. These giant Hawaiian landslides are described as debris avalanches by Moore et al. [1994] and they undoubtedly generated turbidity currents that extend many hundreds of kilometers from the islands [*Rees et al.,* 1993]

Hawaiian Volcanoes: Deep Underwater Perspectives
Geophysical Monograph 128
Copyright 2002 by the American Geophysical Union

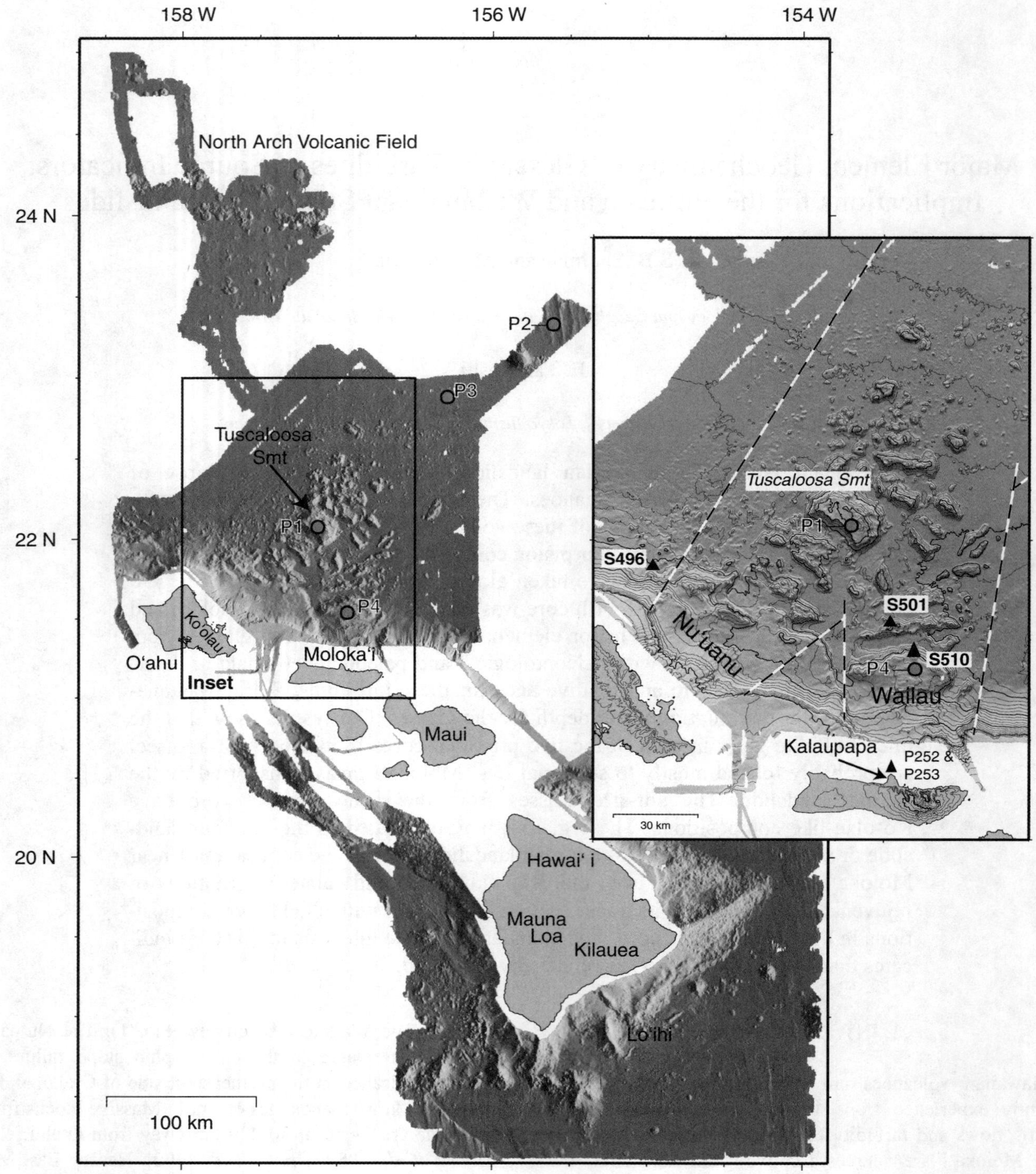

Fig. 1. Shaded relief maps of the ocean floor bathymetry around the Hawaiian islands based on surveys made during the 1998 and 1999 JAMSTEC cruises. The locations of the piston cores P1-4, and of the several Hawaiian islands and volcanoes discussed in the text are shown. The insert map shows the locations of SHINKAI 6500 (S) and PISCES V (P) dives, the postulated boundaries of the Nu'uanu and Wailau debris deposits, and the rejuvenated-stage volcano, Kalaupapa.

as observed for other Hawaiian landslides [*Moore et al.*, 1989; *Garcia and Hull*, 1994].

Turbidites are a common component of piston cores taken near the islands of Hawai'i [*Schreiber*, 1969] because of the frequent collapse of these islands [e.g. *Moore et al.*, 1994]. The abundant volcaniclastic material in these turbidites commonly includes fresh volcanic glass [*Edsall*, 1975; *Garcia and Hull*, 1994]. These fresh glasses have wide compositional ranges indicating that they sampled a major portion of Hawaiian volcanoes and not just an eruptive sequence. [*Garcia*, 1996a]. Thus, turbidites associated with giant landslides can provide insights into the history of these volcanoes that otherwise would be difficult to observe. Interpreting this history is complicated for the turbidites recovered off the northeast coast of O'ahu because they have two potential source volcanoes, Ko'olau and East Moloka'i and the deposits from the Nu'uanu and the Wailau landslides overlap one another (Fig. 1). Moore [1964] postulated that Wailau was the older deposit, although subsequent studies suggest that Wailau is younger than the Nu'uanu deposit [*Moore et al.*, 1989; *Kanamatsu et al.*, this volume; *Moore and Clague*, this volume].

One of the purposes of this study is to determine if it is possible to use the geochemistry of glasses from turbidites associated with these slides to distinguish the relative ages of the landslides. Glass is considered a reliable proxy for magma composition and is less susceptible to complications due to accumulation of phenocrysts or xenocrysts than whole rocks. In addition, glass sulfur concentration is thought to be a good indicator of the water depth of eruption for submarine basaltic lavas [e.g., *Moore and Fabbi*, 1971]. The major element composition of turbidite glasses have been used successfully to identify the source volcano for other Hawaiian landslides [e.g., *Garcia and Hull*, 1994]. Subaerial lavas from Ko'olau volcano are geochemically distinct from lavas of other Hawaiian volcanoes [*Frey et al.*, 1994], so it was anticipated that discriminating between the two potential source volcanoes for the turbidites north of O'ahu and Moloka'i islands would be feasible.

Glasses in turbidites from four piston cores collected north of O'ahu and Moloka'i were analyzed by electron microprobe for major elements and sulfur concentrations. These results are compared with the chemistry of glasses from other Hawaiian volcanoes and integrated with paleomagnetic and paleontological studies of these piston cores. We conclude that turbidites in three of the piston cores contained glasses that were derived predominantly from the island of Moloka'i. Two of these cores contain a thick turbidite that is probably related to the Wailua landslide. The core taken nearest to Moloka'i sampled several geochemically distinct alkalic submarine and subaerially erupted glasses that resemble East Moloka'i rejuvenated lavas. The fourth and most distal piston core contains older glasses that are geochemically similar to Ko'olau lavas. Thus, the Nu'uanu landslide is probably older than the Wailau landslide.

2. PISTON CORE LOCATION AND STRATIGRAPHY

Four, approximately seven-m-long piston cores were taken from deposits thought to be related to the Nu'uanu and Wailau landslides during the 1998 Japan Marine Science and Technology Center (JAMSTEC) cruise by the R/V *Kairei* [*Naka et al.*, 2000]. Piston cores 1 through 3 (P1 and P3), were taken along a transect extending 350 km NE from the island of O'ahu (Fig. 1). P1 and P3, taken closer to the island (100 and 250 km, respectively), have a single, well bedded, dark brown sand to silt sized volcaniclastic turbidite layer (Fig. 2). Piston core 2 (P2) has a thick interval at its base with silt-sized volcanic glass fragments but lacks a distinctive dark brown sandy layer. The presence of volcanic glass in this core was identified based on its high magnetic susceptibility [*Kanamatsu et al.*, this volume]. The total thickness of the turbidite layer in these three cores is unknown due to inflow of other sediment at the base of these cores (Fig. 2). The fourth piston core (P4) was taken 40 km north of the island of Moloka'i. It contains 22 bedded volcaniclastic turbidite layers with varying proportions of glass, mineral, fossils, and rock fragments (Fig. 2).

P1 was collected from a basin on top of Tuscaloosa seamount at a water depth of 3012 m, which is above the carbonate compensation depth (~ 4200 m; *Berger and Winterer*, 1974). Tuscaloosa is the largest block (~ 30 km long, 17 km wide and >1.8 km thick) associated with the Nu'uanu landslide [*Moore et al.*, 1989]. This core consists predominantly of yellowish–brown, nannofossil ooze and dark, yellowish-brown volcanic silt and sand with calcareous fossils. There is a dark brown volcaniclastic turbidite layer at ~ 4.5 meters below sea floor (mbsf) that is at least 26 cm thick [*Naka et al.*, 2000]. Within this thick turbidite layer there are four intervals of medium to coarse (0.2 to 2 mm) volcanic sand (Fig. 2). Fresh glass was sampled from each of these intervals for chemical analysis. Paleomagnetic studies place the top of the Jaramillo (0.99 Ma) at 3.97 mbsf for this core (Fig. 2). Fossils from just below this magnetic reversal give an age between 2.0 –1.77 Ma [*Kanamatsu et al*, this volume; *Herrero-Bervera et al.*, in press].

P2 and 3 are located between the Hawaiian Deep and the Hawaiian Arch in water depths of ~4200, and 4900 m, respectively. Both cores are composed mostly of heavily

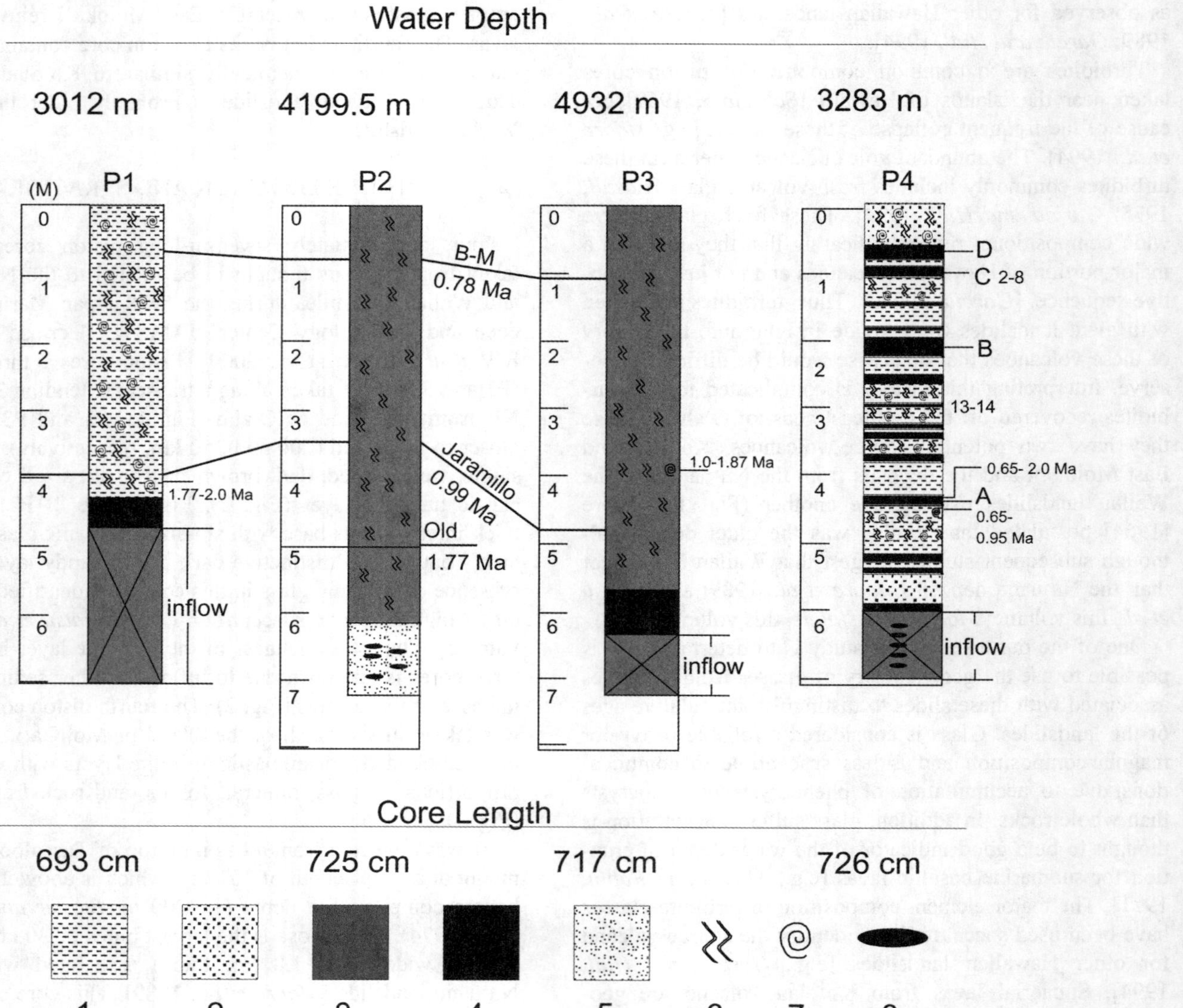

Fig. 2. General stratigraphy for piston cores P1-4 (after Naka et al., 2000 and Kanamatsu et al., this volume). The column patterns are: 1. yellowish-brown nannofossil ooze; 2. yellowish-brown nannofossil volcanic silt; 3. brown clay; 4. volcanic sand; 5.volcanic silt; 6. heavily bioturbated; 7. fossils present 8. lenses of brown clay clasts (inflow?). Water depths for core recovery and length of each core are given for reference. B-M is the abbreviation for Bruhnes-Matayama. See text for description of paleomagnetic and paleontological data..

bioturbated brown clay [*Kanamatsu et al.*, this volume]. P2 has an ~1 m thick very dark brown silt-size volcaniclastic turbidite layer with clay clast lens starting at 6.2 mbsf and continuing to the base of the core where inflow into the core occurred. Paleomagnetic studies indicate that these glasses are older than 1.77 Ma *[Herrero-Bervera, et al.*, in press]. There were no fossils identified in P2. The turbidite layer at the base of P3, 5.5 mbsf, is at least 60 cm thick with normal grading and parallel laminations [*Kanamatsu et al.*, this volume]. Four intervals of fine to medium grained (0.2-0.1 mm) volcanic sand and silt with fresh glass were sampled for chemical analyses. Fossils ~1.7 m above the turbidite layer have ages of 1.87-1.0 Ma [*Kanamatsu et al.*, this volume]. Piston core P3 has magnetostratigraphy similar to P1 (Fig. 2).

P4 was taken just north of the island of Moloka'i on the top of a large detached block in a water depth of 3283 m. This core contains 22 distinct black-colored volcaniclastic sand layers interlayered with nannofossil-rich calcareous ooze [*Kanamatsu et al*, this volume]. The seven thickest turbidites were sampled but glass was found in only four of them (Fig. 2; A-D). Fossils were examined from three horizons within this core. The plantonic formanifera *Globorotalia truncatulinoides* and *Globoratalia tosaensis* were found at 4.20-4.30 mbsf (turbidite interval P4A) and have an age between 2.0-0.65 Ma [*J. Resig, pers. comm.*]. Fossils from 4.41 have age ranges of 1.77-0.65 Ma and 0.95-0.65 Ma but those from 5.47 mbsf gave ages of 0.51 to 0.41 Ma [*Kanamatsu et al.*, this volume]. The older ages are probably for reworked fossils, which is a common problem in volcaniclastic turbidites deposited on archpelagicaprons [e.g., *Garcia and Hull*, 1994].

3. GLASS CHEMISTRY

3.1. Methods

Glass compositions were measured at the University of Hawai'i and at the Tokyo Institute of Technology (see [*Shinozaki et al*, this volume] for Tokyo analytical techniques). A five-spectrometer Cameca SX-50 electron microprobe with an accelerating voltage of 15 kV, 10 nA current, and a 15 micron wide beam was used for the University of Hawai'i analyses. Smithsonian VG2 and A99 glass standards were used to calibrate the major elements, and mineral standards were used to calibrate the minor elements (orthoclase for K_2O, apatite for P_2O_5, and troilite for S). Glass standards were analyzed before and after the samples. Na_2O was analyzed first in order to minimize Na loss during analyses. Counting times for the major elements were between 60 to 130 s, except for Na_2O (40 s). Counting times for minor elements were 50 s for K_2O, 110 s for P_2O_5 and 120 s for S. A PAP-ZAF matrix correction was applied to all analyses. The reported analyses represent averages of 5 to 15 points/sample. The glass standard, JB-2 was analyzed in order to evaluate interlaboratory differences between the two sets of microprobe data. There is good agreement for most of the major elements (see Table 1 for an average of ~ 20 points for each lab). Compared to the University of Hawai'i data, the Tokyo data have ~ 1% higher values for SiO_2, and ~ 1.0 % lower values for FeO, MgO, and CaO. No corrections were made to the data in the appendix tables for these minor differences.

3.2 Major Element and Sulfur Chemistry

Approximately 100 to 150 chips of glass were analyzed from each of the piston cores, except P4 where closer to 70 glass fragments were analyzed. The vast majority of the totals for the glass analyses are between 99 to 99.8 wt% with K_2O/P_2O_5 ratios of 1.5-2.0 (Tables 1, 2, and Appendix table on CD). Therefore, the glasses are relatively fresh and have experienced minimal post-eruptive alteration [e.g. *Garcia et al.*, 1989; *Frey et al*, 1994]. The compositional range in major element chemistry for the glasses from each of the piston cores is broad, as is the range within individual turbidite intervals. Alkalic and tholeiitic glasses were found in all of the piston core turbidite layers except in P2, which contains only tholeiitic glasses (Fig. 3). Only minor geochemical differences were found in the glasses from the multiple coarse horizons in the thick turbidites from P1 and P3 (Appendix table on CD). Thus, the glasses within each turbidite layer were combined to characterize these two cores. P4 has four turbidite layers with chemically distinct glass.

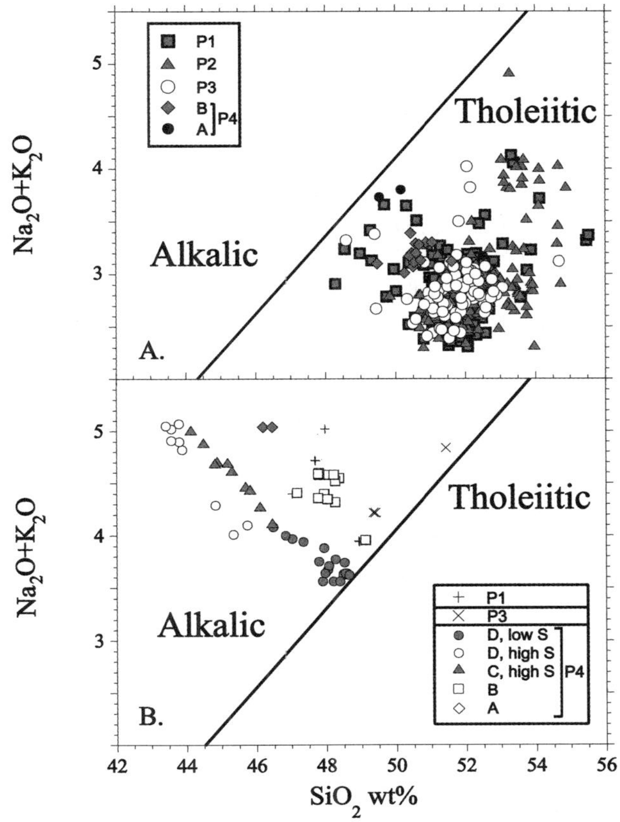

Fig. 3. Total alkalis (Na_2O + K_2O) verus SiO_2 for volcanic glasses from piston cores P1-4. A. Tholeiitic glasses; these glasses span a wide compositional range. B. Alkalic glasses; these glasses form distinct nearly linear trends. No alkalic glasses were found from P2. Groupings are based on the Macdonald and Katsura [1964] dividing line.

The tholeiitic glasses from P1 and P3 have similar major

TABLE 1. Representative Microprobe Analyses for Piston Core Tholeiitic Glasses

SiO_2	TiO_2	Al_2O_3	FeO*	MgO	CaO	Na_2O	K_2O	P_2O_5	S	Total
Piston Core 1										
48.27	2.85	14.97	12.00	7.24	11.70	2.51	0.40	0.21	0.020	99.47
49.27	2.94	14.60	12.62	6.41	10.30	2.83	0.58	0.26	0.020	99.21
49.67	3.39	13.26	12.08	6.08	10.68	2.93	0.73	0.42	0.020	99.26
50.81	2.85	13.56	11.66	6.53	10.92	2.58	0.52	0.26	0.006	99.68
51.31	2.84	13.18	11.99	6.03	10.47	2.55	0.62	0.33	0.001	99.31
52.25	2.29	13.61	10.43	7.13	10.80	2.33	0.38	0.24	0.005	99.42
53.37	2.91	14.97	10.24	5.70	8.20	3.35	0.72	0.32	0.004	99.78
53.75	2.48	13.94	10.14	6.27	10.14	2.62	0.42	0.23	0.008	99.99
55.46	1.71	14.77	9.73	6.02	9.19	2.99	0.33	0.16	0.001	99.65
Piston Core 2										
50.58	2.35	13.93	10.49	8.01	11.81	2.26	0.30	0.18	0.016	99.19
50.83	2.85	13.76	10.76	6.79	11.39	2.43	0.37	0.27	0.015	99.47
51.08	2.85	13.95	10.34	6.89	11.19	2.41	0.40	0.28	0.026	99.42
52.16	2.51	14.66	9.66	6.71	11.09	2.44	0.36	0.20	0.015	99.81
52.43	2.09	14.16	10.32	7.39	10.44	2.47	0.37	0.20	0.005	99.87
53.18	2.81	15.44	10.41	5.56	7.50	3.18	0.65	0.36	0.007	99.10
53.62	2.83	15.41	10.42	5.62	7.54	3.38	0.65	0.34	0.006	99.79
Piston Core 3										
48.59	2.84	14.82	11.48	7.82	10.22	2.85	0.47	0.25	0.001	99.34
49.40	2.99	14.40	11.46	6.80	10.99	2.86	0.52	0.27	0.004	99.69
50.82	2.66	13.53	11.14	6.70	11.33	2.36	0.35	0.24	0.020	99.14
51.36	2.92	13.62	11.29	6.43	10.44	2.48	0.37	0.24	0.060	99.20
52.75	2.50	13.44	11.05	6.18	10.46	2.41	0.54	0.26	0.002	99.58
53.06	2.48	14.11	10.14	6.25	10.44	2.43	0.44	0.21	0.005	99.56
54.67	2.27	13.78	10.59	5.63	9.54	2.61	0.51	0.25	0.001	99.87
Piston Core 4B										
49.47	3.18	13.42	11.75	6.52	11.44	2.58	0.52	0.35	0.001	99.23
50.25	3.00	13.36	11.92	6.33	10.96	2.45	0.56	0.36	0.020	99.21
50.57	3.21	12.84	12.90	5.63	10.39	2.66	0.63	0.38	0.040	99.25
50.72	2.92	13.20	12.66	5.93	10.42	2.55	0.58	0.35	0.030	99.35
51.61	3.36	12.81	12.67	5.71	10.19	2.58	0.54	0.33	0.020	99.81
Piston Core 4A										
50.15	3.78	14.59	12.53	5.80	8.36	3.46	0.35	0.42	0.015	99.45
52.50	2.84	13.48	11.34	6.08	10.40	2.56	0.61	0.26	0.004	100.1
49.54	3.84	12.60	13.28	5.39	10.34	2.87	0.86	0.46	0.017	99.20
JB-2 UH										
53.09	1.10	14.92	13.06	4.77	10.12	1.92	0.40	0.07	0.004	99.46
JB-2 TIT										
53.42	1.09	15.02	12.52	4.56	9.77	2.01	0.41	0.02	0.011	98.83

The Japanese basalt standard JB-2 was fused and run as an inter-laboratory comparison for analyses from the University of Hawaii (UH) and the Tokyo Institute of Technology (TIT). Values are an average of 20 analyses for each lab. All other data in this table are from UH and are an average of at least five spot analyses. See text for methods used. Total iron is listed as FeO*.

TABLE 2. Representative Microprobe Analyses for Piston Core Alkalic Glasses

SiO_2	TiO_2	Al_2O_3	FeO*	MgO	CaO	Na_2O	K_2O	P_2O_5	S	Total
Piston Core 1										
47.02	4.77	13.05	13.64	5.34	10.23	3.38	0.54	0.54	0.018	99.00
47.67	4.14	13.60	13.71	5.22	9.56	3.57	1.15	0.58	0.012	99.21
47.97	4.88	12.56	13.58	4.66	9.32	3.77	1.25	0.72	0.012	98.74
48.92	3.85	13.75	12.18	5.94	10.95	3.05	0.90	0.46	0.017	99.43
Piston Core 3										
49.34	2.98	15.24	11.01	6.43	9.92	3.49	0.73	0.37	0.019	99.52
51.42	3.55	14.82	11.17	5.98	7.37	3.74	1.06	0.39	0.000	99.49
Piston Core 4D (low S)										
46.47	2.40	14.64	12.05	6.24	12.61	3.44	0.64	0.27	0.034	98.82
46.83	2.10	14.71	12.05	6.35	12.68	3.41	0.59	0.24	0.034	99.01
47.34	2.33	14.72	12.00	6.44	12.66	3.35	0.59	0.23	0.023	99.68
47.93	2.09	14.42	12.21	6.35	12.51	3.34	0.54	0.21	0.032	99.64
48.39	1.81	14.92	11.54	6.80	12.38	3.16	0.40	0.19	0.026	99.62
48.48	1.80	15.05	11.37	6.79	12.46	3.21	0.41	0.20	0.027	99.79
48.64	1.81	14.81	11.46	6.63	12.36	3.19	0.43	0.21	0.027	99.57
Piston Core 4D (high S)										
43.57	3.17	13.97	12.32	6.35	13.41	3.82	1.21	0.41	0.098	98.32
43.79	3.00	13.95	12.20	6.32	13.58	3.87	1.18	0.46	0.091	98.42
44.83	2.69	14.61	11.98	6.46	13.10	3.37	0.92	0.31	0.118	98.38
45.74	2.45	14.84	11.86	6.26	12.77	3.33	0.77	0.29	0.080	98.38
Piston Core 4C										
44.13	3.05	14.30	12.36	6.07	13.30	3.81	1.16	0.47	0.084	98.73
44.49	3.24	14.36	12.35	5.98	13.40	3.79	1.11	0.36	0.086	99.21
45.30	2.82	14.63	12.02	6.26	12.97	3.62	1.00	0.35	0.086	99.07
45.69	2.62	14.73	12.03	6.25	12.75	3.59	0.88	0.30	0.089	98.92
46.10	2.44	14.75	11.85	6.52	12.54	3.49	0.79	0.26	0.105	98.84
46.45	2.37	14.86	11.68	6.54	12.35	3.39	0.73	0.24	0.102	98.70
Piston Core 4B										
47.16	4.19	13.95	13.08	5.70	10.12	3.41	1.04	0.56	0.009	99.22
47.79	4.10	13.91	13.05	5.46	9.87	3.58	1.01	0.49	0.014	99.27
48.06	5.11	12.22	14.94	4.54	9.46	3.28	1.33	0.64	0.033	99.60
48.36	4.10	14.04	12.92	5.50	9.79	3.60	0.95	0.51	0.014	99.80
49.11	3.28	14.02	11.72	6.12	11.72	3.12	0.84	0.49	0.012	99.72
Piston Core 4A										
46.19	4.49	13.56	13.46	5.03	10.48	3.72	1.32	0.67	0.010	98.94
46.45	4.57	13.53	13.44	4.92	10.43	3.71	1.33	0.61	0.012	99.01

See text for methods used.

element compositions but glasses in P1 have more variable FeO (Fig. 4) and slightly lower Al_2O_3 contents than glasses in P3. Most glasses in both cores have MgO contents between 6 and 8 wt% and a SiO_2 abundance that clusters around 52.0 wt% $\pm$ 1.0 wt %. Aluminum oxide and CaO both correlate positively with MgO for the glasses in P1 and P3. There are broad inverse correlations between FeO (Fig. 4), Na_2O, K_2O, TiO_2, and P_2O_5 with MgO for the majority of the tholeiitic glasses from P1 and P3. The sulfur content of most glasses from P1 and P3 are low (<0.04 wt%), but there are a few glasses from P1 with higher sulfur concentrations, > 0.8 wt% (Fig. 5).

Tholeiitic glasses from the more distal P2 core have higher SiO_2 (> 53 wt%) and Al_2O_3, (> 14.2 wt%) and lower FeO, (< 10.2 wt%; Fig. 4) TiO_2, and CaO from the P1 and P3 glasses. The glasses in P2 form broader ranges on many

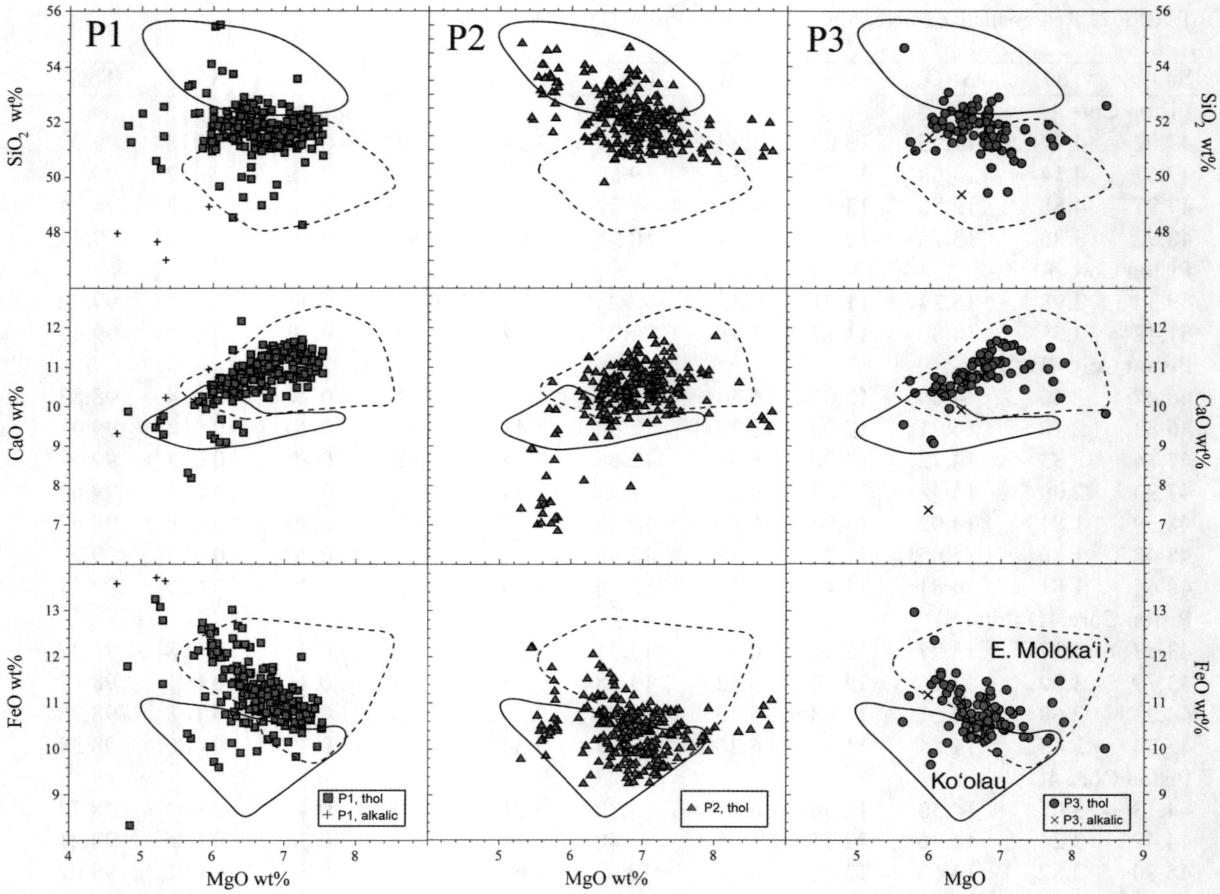

Fig. 4. MgO variation diagrams for SiO_2, CaO, and FeO in glasses from cores P1 –3. Most of the glasses from P2 fall within the Ko'olau field (solid field) [M. Garcia unpublished glass data from Ko'olau dikes and lava flows]. In contrast, the P1 and 3 glasses mostly fall outside of the Ko'olau field and in the East Moloka'i field (dashed field) [D. Clague unpublished, Garcia, unpublished, and E. Takahashi, unpublished, for East Moloka'i data from S501 and S510.].

of these plots at a given MgO value than the tholeiitic glasses from P1 and P3, indicating a wide spectrum of parental magma compositions for the lavas that produce the glasses in P2. Similar to glasses from cores P1 and P3, the majority of the glasses in P2 have S concentrations below 0.04 wt. % but some have higher S contents (Fig. 5).

The glasses in P4 have the largest range in SiO_2 of any of the cores (43-53.5 wt%; Fig. 6). Only four of the 22 turbidite layers in P4 contain fresh glass; they are labeled A through D up section. All four of these layers contain alkalic glasses, but only the two lower layers, P4A and P4B, also have tholeiitic glasses (Figs. 3 and 6). Most of the tholeiitic and alkalic glasses in P4 form relatively coherent negative trends on the FeO-MgO variation diagram but divergent trends or offset trends between layers for the other major elements (Fig. 6). The tholeiitic glasses in P4 have a lower and a more limited range in SiO_2 (~ 49– 52 wt%) and MgO (5.2-6.5 wt%) than tholeiitic glasses from the other piston cores (Figs. 4 and 6). All of the tholeiitic glasses have lower S concentrations (< 0.05 wt%; Fig. 5). The alkalic glasses form two broadly distinctive groups; the two lower turbidites have lower MgO (less than 6 wt%) and CaO (less than 11.5 wt%) concentrations than the two upper layers, P4C and P4D (Fig. 6). Interestingly, the higher CaO glasses have a negative correlation of CaO with MgO, while the P4A and P4B glasses form a positive correlation (Fig. 6). The CaO contents of the P4C and P4D glasses are among the highest ever reported for Hawaiian glasses. The alkalic glasses in the two lower layers have relatively constant SiO_2 concentrations over a range in MgO, whereas P4C and P4D have a positive SiO_2-MgO correlation.

Sulfur concentrations for glasses in P4 also have a bimodal distribution; only the glasses from the upper two layers have high S concentrations (Fig. 5). The uppermost turbidite interval, P4D, contains two distinct groups of alkalic glasses based on their S contents. The higher S group (> 0.075 wt%) has a limited range in MgO, (~6.0 to 6.5 wt%) and the lower S group (<0.04 wt%) has a slightly more variable MgO content (6 to 7 wt%; Fig. 6). The high S glasses in P4D have the highest CaO contents with the lowest SiO_2 concentrations for a given MgO (Fig. 6). The lower S glasses from P4D define coherent, positive trends for SiO_2 and Al_2O_3 with MgO and form negative correlations for FeO and CaO (Fig. 6). TiO_2, Na_2O, K_2O, and P_2O_5 also show negative correlations with MgO. All of the glasses from the turbidite interval P4C are alkalic and have relatively high S (0.08-0.10 wt %) concentrations, similar to the high S alkalic glasses from P4D. Both of these high S glasses have relatively low totals, ~ 98 wt%, which may be the result of higher volatile contents. Water contents up to 1.4 wt% were measured in alkalic glasses from the North Arch [*Dixon et al*, 1997]. The glasses in P4C and the low S glasses in P4D form coherent trends on the MgO variation diagrams (Fig. 6). Most of the high S glasses in P4D are chemically distinct from the glasses in P4C and the low S glasses in P4D glasses (Fig. 6).

4. SULFUR: AN INDICATOR OF DEPTH OF ERUPTION

Sulfur concentrations in volcanic glass are useful for determining water eruption depths (e.g., *Moore and Fabbi*, 1971). For example, subaerial Hawaiian glasses typically have low S abundances, (<0.04 wt%; [*Davis*, 2001]) indicative of substantial sulfur degassing, whereas most submarine erupted Hawaiian glasses have higher S values [e.g., *Moore and Fabbi*, 1971; *Dixon et al.*, 1991; *Garcia and Hull*, 1994]. Glasses with sulfur abundances greater than 0.09 wt% are considered undegassed and to have erupted in water depths > 1 km [e.g., *Moore and Clague*, 1992]. Sulfur concentrations between >0.04 to 0.09 wt% indicate eruption in shallow to moderate water depths (< 1 km), with partial degassing [e.g., *Moore and Clague*, 1992; *Garcia and Hull*, 1994].

Sulfur values for the piston core glasses range between <0.001 to 0.15 wt% (Figs. 5a and 5b). The majority, however, have S concentrations below 0.04 wt%, indicating that they erupted either subaerially or in very shallow marine conditions. This includes many of the tholeiitic glasses and all of the alkalic glasses in P1 and P3. There are, however, some tholeiitic glasses in both cores that have moderate S concentrations (0.04 to 0.09 wt %) indicating that they erupted in water depths <1 km. Higher S values (0.09 to 0.15 wt %) were found for a few glasses from P1 suggesting the eruptions were in water depths > 1 km. Piston core 2 also has some tholeiitic glasses with S >0.04 wt% (Fig. 5) indicating they were submarine erupted. The glasses from P4 fall into two distinct groups: one with

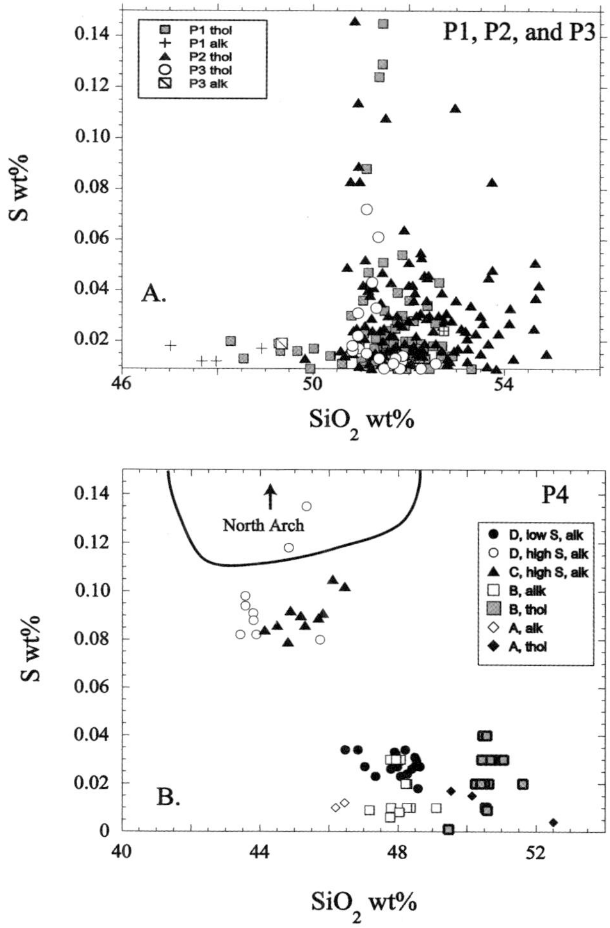

Fig. 5. SiO_2 versus S variation diagrams for glasses in turbidite layers. A. Tholeiitic and alkalic glasses from piston cores P1-3. Most of the glasses from these cores have low S contents (<0.04 wt%) indicating that they were erupted subaerially. Those with S contents between 0.04 to 0.09 were probably erupted in water depths <1 km. Those with >0.09 wt% S were probably erupted in water depths >1 km based on studies of S contents in Hawaiian tholeiitic basalts [e.g., Moore and Fabbi, 1971; Garcia and Hull, 1994]. B. Tholeiitic and alkalic glasses from P4. All of the glasses from layers P4A and P4B have low S contents indicating subaerial or very shallow submarine eruption conditions. Some of the glasses from layers C and D contain much higher S contents indicating eruption under deeper marine conditions (> 1 km). North Arch alkalic field from Dixon et al. (1997).

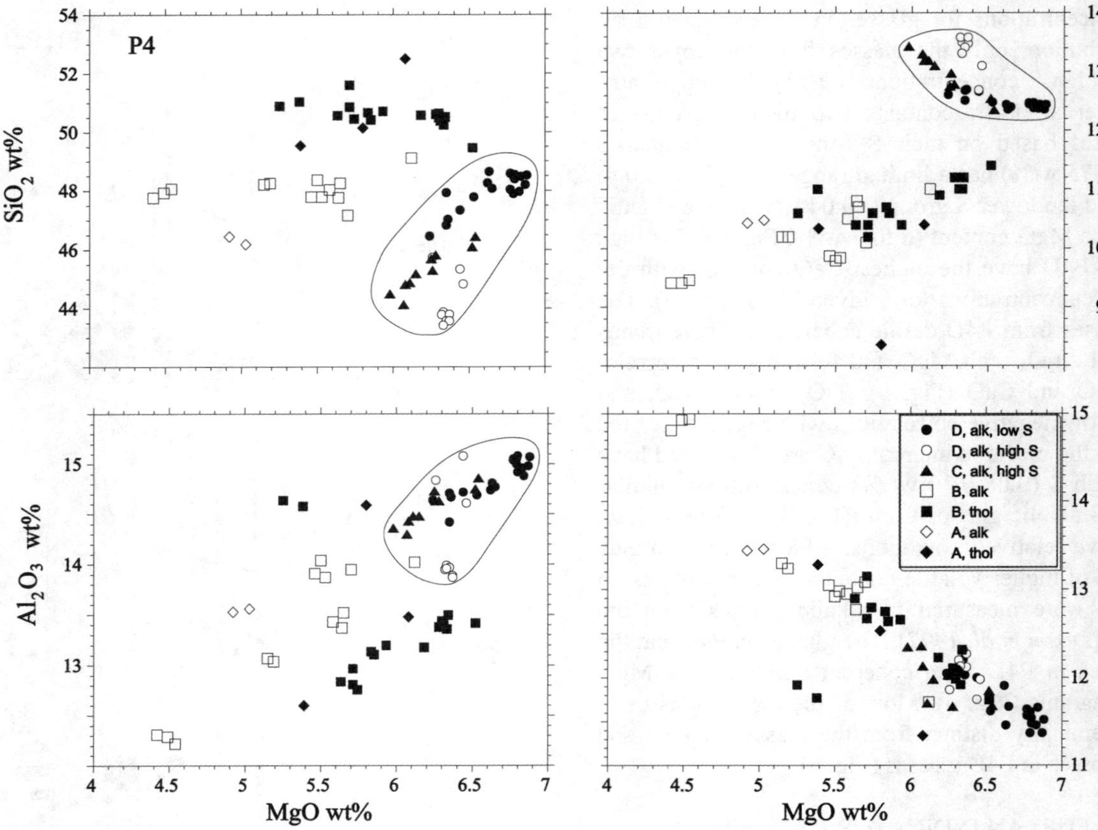

Fig. 6. MgO variation diagrams for alkalic and tholeiitic glasses from P4. Note the widely divergent trends for the glasses from the different layers for all oxides except FeO. The glasses from these four layers have distinct parental magmas. Glasses from P4C and P4D are enclosed in a field to distinguish from the glasses in layers P4B and P4A.

higher S concentrations (>0.06 wt%) that were erupted at moderate to deep submarine depths and one with lower S (<0.04) that were subaerially erupted (Fig. 5). The implications of these depths of eruption are discussed in the next section.

5. MAJOR ELEMENTS AS INDICATORS OF SOURCE VOLCANO

Major elements have proved useful in distinguishing lavas from different Hawaiian volcanoes [e.g., *Frey and Rhodes*, 1993]. For example, subaerial Ko'olau and submarine Lo'ihi lavas represent opposing tholeiitic end-member compositions for many major elements [e.g., *Frey et al.*, 1994; *Hauri*, 1996]. The utility of these major element differences is enhanced by the relatively consistent magma composition during the shield building stage for many Hawaiian volcanoes. For example, major element compositions for Mauna Loa lavas have not changed for at least 240,000 years [*Garcia et al.*, 1995b]. Similarly, Kilauea rocks show no change in composition beyond the range observed for historical lavas over the last 350 ka [*Quane et al.*, 2000].

In contrast, Ko'olau lava compositions are more variable [*Jackson et al.*, 1999]. New evidence from a 685 m deep drill hole and Nu'uanu landslide rocks indicate that deeper Ko'olau rocks have Mauna Loa chemical affinities [*Haskins and Garcia*, 2000; *Shinozaki et al.*, this volume]. Therefore, the diagnostic chemistry of the Ko'olau subaerial lavas is not representative of the entire shield building stage for this volcano. Since no other Hawaiian volcano to date has lavas with this distinctive high SiO_2 [e.g., *Hauri*, 1996], we assume that glasses with high SiO_2 (53.5 wt%) are from Ko'olau volcano.

The turbidite glasses from P1-4 each have a greater range in SiO_2 values (e.g., ~ 47 to ~ 55.5 wt % for P1; Fig. 4) than any single Hawaiian shield volcano. This is not related to wide variations in degree of differentiation because MgO, a key indicator of differentiation in Hawaiian basalts [e.g., *Wright*, 1971], has a limited range in the piston core glasses (Fig. 4). However, glasses from different

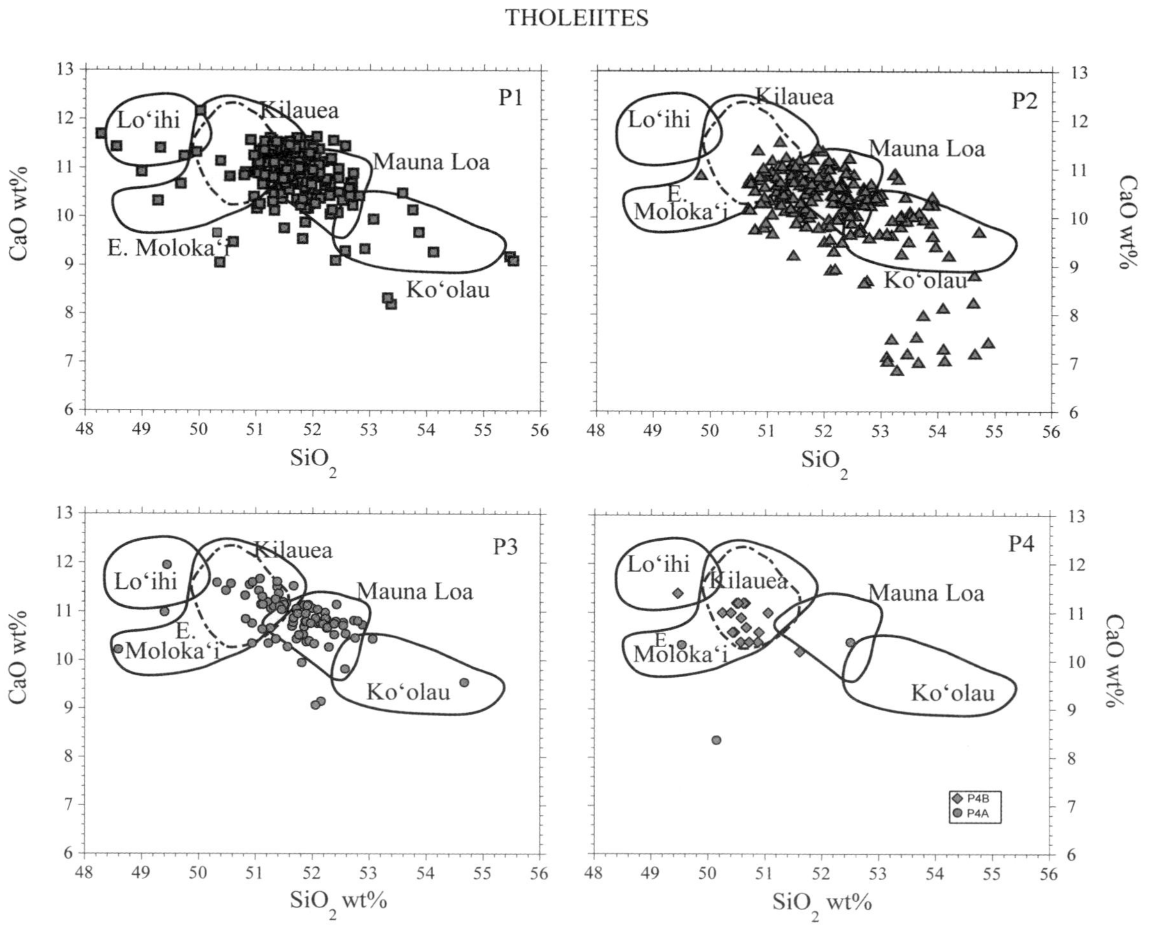

Fig. 7. CaO vs. SiO_2 variation diagrams for tholeiitic glasses from P1-4. The majority of the glasses from piston cores P1 and 3 fall in the Mauna Loa field with overlap into the East Moloka'i and Kilauea fields. Glasses in P2 range widely in composition but overlap and extend below the Ko'olau field. Nearly all of the P4 glasses plot within the East Moloka'i field. Data for fields are from: M. Garcia, unpublished for Ko'olau; Garcia et al. [1995b] and Garcia [1996b] for Mauna Loa; Garcia, unpublished for Kilauea; Garcia et al. [1993] and Garcia et al. [1995a] for Lo'ihi; Garcia, unpublished, E. Takahashai, unpublished, and D. Clague, unpublished for East Moloka'i.

stages or from several volcanoes may be present within a single turbidite layer.

The most likely source volcanoes for the turbidite glasses in P1-4 are Ko'olau and East Moloka'i. These glasses may be related to one of the giant landslide events, Nu'uanu or Wailau, or possibly smaller landslides on the flanks of these volcanoes [*Yokose*, this volume]. This last scenario is especially likely for P4, which has 22 thin turbidite intervals. In addition to higher SiO_2, subaerial Ko'olau lavas have higher Al_2O_3 and Na_2O and lower FeO and CaO (Fig. 7) than other Hawaiian tholeiitic lavas. These features are also seen in dike glasses (Fig. 4). The only glass chemistry data from East Moloka'i are from tholeiitic lavas and breccias collected during two *Shinkai* 6500 dives (S501 and S510; Fig. 1; tholeiitic East Moloka'i field in Fig. 4) and from eight alkalic and one tholeiitic lava taken during two *Pisces V* dives [P252 and P253; Clague, unpublished data]. Dive P252 sampled the submarine extension of the Kalaupapa peninsula, a rejuenivation stage feature. Dive P253 rocks were collected near the peninsula on the shelf break in slope. The alkalic glasses recovered from these two dive sites have MgO concentrations between 6 and 6.5 wt%. The compositions of these alkalic glasses along with those from Lo'ihi [*Garcia et al.*, 1993; *Garcia et al.*, 1995a] and the rejuvenation stage North Arch alkalic lavas [*Dixon et al.*, 1997] are

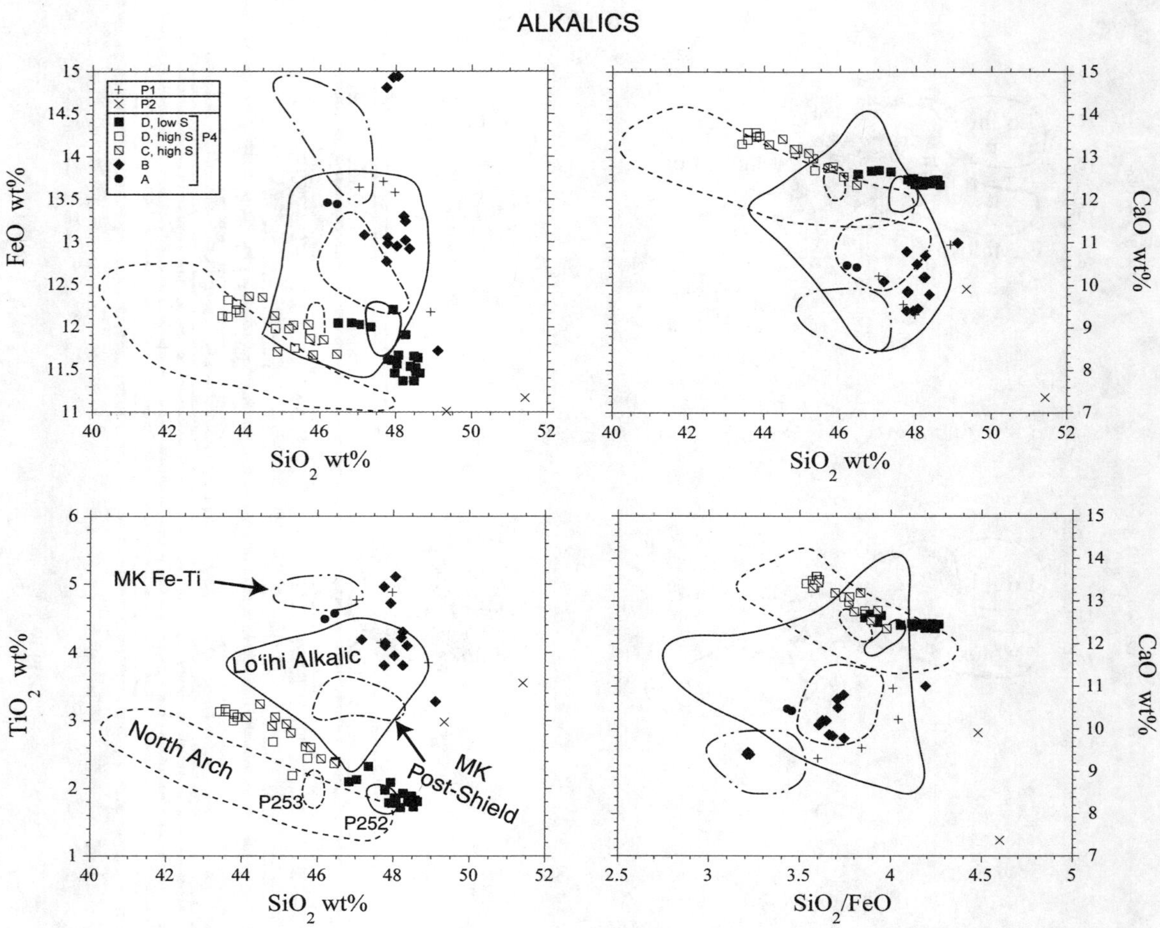

Fig. 8. **Plots of FeO, TiO_2, and CaO vs. SiO_2, and CaO vs SiO_2/FeO for alkalic glasses from piston cores P1, P3, and** P4. Glass data for fields from: Garcia et al. [1993] and Garcia et al. [1995a] for Lo'ihi; Frey et al., 1990 and Frey et al., 1991 for post-shield Mauna Kea, D. Clague, unpublished for PISCES V dives off (P252 and P253) East Moloka'i; and Dixon et al. [1997], for North Arch glasses.

plotted as fields for comparison with the alkalic glasses recovered from P1, P3, and P4 (Fig. 8).

The turbidite in P1 and P3 were deposited at approximately the same time period (Fig. 2) and have similar glass compositions (Figs. 4 and 7). Many of the tholeiitic glasses from these two piston cores have compositions similar to glasses from East Moloka'i, Mauna Loa, and Kilauea but both turbidites contain a few glasses with high silica, Ko'olau-like compositions (Fig. 7). It is difficult to definitively determine the source volcano for these glasses based on geochemistry alone. Geochemically the glasses are similar to both the Mauna Loa (Ko'olau?) and East Moloka'i volcanoes. However, the presence of alkalic glasses within these deposits is inconsistent with a Ko'olau origin because no alkalic lavas are known from this volcano prior to its rejuvenation stage (0.03-0.9 ma [*Clague and Dalrymple*, 1987]), which occurred after the turbidites in cores P1 and P3 were deposited (~ 1.0-1.5 Ma). Therefore, the rare Ko'olau composition glasses in these layers may have been picked up by the turbidites as they traveled across the ocean floor. Entrainment of ocean floor sediments and even scouring of Eocene basement rocks has been reported for Pleistocene turbidites from other avalanche deposits [e.g., *Garcia and Hull*, 1994].

P1 was taken from the top of the 1.7 km high Tuscaloosa block from the Nu'uanu landslide. Only a turbidity current from a giant debris avalanche would deposit sand on top of such a tall block [e.g., Garcia and Hull, 1994]. The only major debris avalanche northeast of O'ahu other than Nu'uanu was the Wailau landslide [Moore et al., 1994].

Thus our preferred interpretation is that the turbidite layers in cores P1 and P3 are related to the Wailau landslide. Based on the dominance of low sulfur content glasses in both turbidite layers, the Wailua landslide involved subaerially erupted lavas and therefore, must have occurred after the emergence of East Moloka'i volcano above sea level. One indication of the energy involved in this landslide event is that the current that carried the sands to the top of Tuscaloosa Seamount had to scale a ~1.7 km high wall after traveling at least 75 km from East Moloka'i.

Paleomagnetic evidence indicates that the silt-sized glass in P2 is older than the turbidite layers in cores P1 and P3 (Fig. 2). There is no evidence in core P2 of a turbidite layer deposited at the corresponding time as the turbidite in cores P1 and P3. The absence of this layer in P2 may be related to the location of the piston core site which is in the shadow of a large landslide block (Fig. 1). Some of the glasses in P2 have characteristic Ko'olau chemistry, lower FeO, CaO, TiO_2, and higher SiO_2/FeO, Al_2O_3, although others have Mauna Loa-like compositions (Fig. 7). Based on the new results of Haskins and Garcia [2000] and Shinozaki et al. [this volume], the glasses with typical Ko'olau composition may represent material from the younger, outer carapace of this volcano, while the glasses with Mauna Loa chemical affinities may be from an older section of the volcano. Strong support for a Ko'olau origin for the glass in P2 comes from the presence of glasses with distinctively low CaO abundances (~7-8 wt%) and high SiO_2 (53.1-54.9 wt%). Glasses with this composition have only been found in Hawai'i on the submarine flanks of Ko'olau volcano (dive S496; Fig. 1) [*Shinozaki et al.*, this volume]. The presence of these geochemically distinct glasses and the absence of alkalic glasses in P2 indicate that Ko'olau volcano is the most likely source. A wide range of S contents were found in glasses from P2 but most have <0.04 wt%, suggesting that most of the glasses were derived from lavas erupted subaerially or under shallow marine conditions (< 1 km).

Although it appears that the glasses in the silty interval at the base of P2 are related to the collapse of the shallow flanks of Ko'olau volcano, the relationship of this deposit to the Nu'uanu landslide is less certain. The thickness of the silty interval in P2 is unknown. It is possible that the layer was much thicker and overlies coarser grained material from the Nu'uanu landslide. However, what was recovered was a relatively thin layer of silt which is in striking contrast to the abundance nearby of one kilometer or bigger sized blocks from the Nu'uanu landslide. Therefore, an alternative possibility is the silty turbidite in P2 is derived from a later collapse of one of the large Nu'uanu blocks. Such collapses are now known to have been associated with some of these slide blocks [*Yokose*, this volume]. These results are consistent with the Nu'uanu landslide occurring before the Wailau landslide.

The strong geochemical similarity between the glasses in P4 with samples from East Moloka'i volcano (Figs. 4 and 7) and close proximity of where P4 was taken to the volcano (Fig. 1) provide compelling evidence that all of the turbidites in this core are from East Moloka'i. For example, all but two of the tholeiitic glasses from P4 fall in the East Moloka'i data field and none fall in the Ko'olau field (Fig. 7). Also, the alkalic glasses in the two uppermost turbidite intervals in this core (C and D) have similar chemistry to the rejuvenated alkalic glasses of Kalaupapa volcano on East Moloka'i. Thus, East Molokai volcano is most likely the source for the turbidites in P4.

All of the turbidite layers in P4 (including those that were studied here) are probably younger than the Wailau landslide based on the 0.4-0.5 Ma fossil age determined for near the base of the core (Fig. 2). The presence of multiple, thin, turbidite layers in P4 documents the frequent collapses of East Moloka'i volcano. Three of the turbidites produced from these collapses involved low S content glasses indicating that the lavas were subaerially erupted. These low S glasses could have been deposited as pillow lavas on the submarine flanks of the volcano after subaerial eruption [e.g., *Davis*, 2001] but pillow lava glasses from the block where P4 was taken and a nearby block have high S (> 0.10 wt%; Garcia, unpub. data). The upper two turbidite layers contain only alkalic glasses. The sulfur and major element compositions of these glasses indicate that they were derived from two eruptions. Both layers contain glass with lower silica contents (< 47 wt%) and high S contents of these glasses indicating they were erupted under relatively deep (>1 km) submarine conditions. The uppermost layer also has degassed glasses with higher silica contents, although they lie along the same variation trend as the undegassed glasses (Fig. 8). These results indicate that the rejuvenation stage eruptions that formed Kalaupapa peninsula included a submarine eruption. The Kalaupapa eruptions have been dated at 0.34-0.57 Ma [*Clague et al.*, 1982]. If the fossil age of 0.41-0.51 Ma near the base of P4 is correct, turbidites P4C and P4D post-date these eruptions.

The alkalic glasses from turbidite layers P4A and P4B are geochemically distinct from the rejuvenation stage East Moloka'i alkalic glasses (P252 and P253 fields) and plot in the Lo'ihi alkalic field (Fig. 8). However, such compositions are also produced during the post-shield stage [*Frey et al.*, 1991]. The low S content of the glasses in P4A and P4B indicates that they are from the subaerial post-shield stage rather than the deep submarine pre-shield stage.

6. CONCLUSIONS

Four new piston cores were collected north of O'ahu and Moloka'i islands with the hope of sampling glasses in turbidites related to the giant Nu'uanu and Wailau landslides. One of the purposes of this sampling was to determine the age relationship between these landslides. Three of the cores were collected off northeast O'ahu (P1, 94 km, P2, 300 km, and P3, 215 km away from the island). The fourth core was collected 40 km north of Moloka'i. Analyses of glasses in turbidites in these cores indicate the following history for the cores:

Cores P1 and P3. Magnetostratigraphy and glass chemistry indicates that the thick turbidite layer in these two cores are probably related to the same event. The glasses were dominantly erupted in the subaerial or shallow submarine environment and were derived from East Moloka'i volcano. They may have been mobilized by the collapse of the Wailau landslide. The rare Ko'olau glass fragments in these layers may have been picked up by the turbidites as they traversed the seafloor northeast of Ko'olau volcano.

Core P2. The low S content of the glasses from the silty turbidite in this core were dominantly erupted under subaerial and shallow submarine conditions. The distinctive but wide range in composition of the P2 glasses indicates that they were probably derived from different structural levels within Ko'olau volcano. The glasses in this turbidite layer may have been deposited by the Nu'uanu landslide or an alternative scenario is they were derived from a smaller collapse after the Nu'uanu event but before the turbidites in P1 and P3. There is no evidence for the Wailua landslide event in this core.

Core P4. The many thin turbidite layers in this core resulted from repeated small collapses of East Moloka'i volcano that post-date the Wailua landslide. Four of the layers contain abundant glass fragments. The lower two layers of the glass-bearing turbidites contain both alkalic and tholeiitic glasses that were erupted subaerially. The tholeiitic glasses from these two layers have compositions similar to East Moloka'i tholeiites and the alkalic glasses have chemical compositions similar to post-shield lavas. The upper two layers contain only alkalic glasses and they have a strong compositional similarity to rocks from the rejuvenation stage Kalaupapa volcano on East Moloka'i. The presence of high sulfur contents in many of these alkalic glasses documents the presence of deep water (>1 km), rejuvenation stage eruptions for East Moloka'i.

Acknowledgments. We would like to thank the captains and crew of the R/V Kairei and Yokosuka, and the dedicated onboard science team, especially chief scientist Jiro Naka and co-chief E. Takahashi for the opportunity to participate in the JAMSTEC Hawai'i cruises and for their collaboration during interpretation of cruise results. Many thanks to D. Clague for sharing his unpublished data for Moloka'i. Heidi Comer assisted with sample preparation, E. Herrero-Bervera gave valuable help in appreciating the paleomagnetic results for these cores, and John Smith provided the bathymetric maps for Figure 1. Mahalo nui loa to S.K. Rowland for his constructive comments on an early draft of this paper and to J. G. Moore, J. Naka, and an anonymous reviewer for their reviews. This work was supported by National Science Foundation grant OCE-9729894. This is SOEST contribution no. 5577

REFERENCES

Berger, W. H., and E. L. Winterer, Plate stratigraphy and the fluctuating carbonate line, *in Pelagic Sediments on Land and Under the Sea*, edited by K. J. Hsu and H. C. Jenkins, Spec. Publ. 1, 11-98, International Association of Sedimentologists, Blackwell, Oxford, 1974.

Clague, D.A, and G.B. Dalrymple, The Hawaiian-Emperor Volcanic Chain, part 1, Geologic Evolution, *U.S. Geol. Surv. Prof. Pap.* 1350, pp. 5-54, 1987.

Clague, D.A., C. Dao-Gong, R. Murnane, M.H. Beeson, M.A. Lanphere, and G.B. Dalrymple, W. Friesen, R.T. Holcolm, Age and petrology of Kalaupapa Basalt, Molokai Hawai'i, *Pac. Sci.*, 36, 411-420, 1982

Davis, M.G., Volatiles in Glasses from Mauna Loa Volcano, Hawai'i: Implications for degassing, contamination, and growth [Masters thesis]: Honolulu, Hawai'i, University of Hawai'i, 86 p. 2001.

Dixon, J.E., D.A. Clague, and E.M. Stolper, Degassing history of water, sulfur, and carbon in submarine lavas from Kilauea, Volcano, Hawai'i, *J. Geol.*, 99, 371-394, 1991.

Dixon, J. E., D.A. Clague, P. Wallace, and R. Poreda, Volatiles in alkalic basalts from the North Arch Volcanie field, Hawai'i: Extensive degassing of deep submarine-erupted alkalic series lavas, *J. Petrol.*, 38, 911-939, 1997.

Edsall, D.G., Submarine geology of volcanic ash deposits: Age and magmatic composition of Hawaiian and Aleutian tephra; Eocene to Recent [Ph.D. thesis]: Palisades, New York, Columbia University, 260 p, 1975.

Frey, F.A., M.O. Garcia, and M.F. Roden, Geochemical characteristics of Koolau Volcano: Implications of intershield geochemical differences among Hawaiian volcanoes, *Geochim. Cosmochim. Acta*, 58, 1441-1462, 1994.

Frey, F.A., M.O. Garcia, W.S. Wise, A. Kennedy, P. Gurriet, and F. Albarede, The evolution of Mauna Kea volcano, Hawaii: petrogenesis of tholeiitic and alkalic basalts, *J. Geophys. Res.*, 96, 14347-14375, 1991.

Frey, F.A., and J.M. Rhodes, Intershield geochemical differences among Hawaiian volcanoes: Implications for source compositions, melting process, and magma ascent paths, *Phil. Trans. R. Soc. Lond.*, 342, 121-136, 1993.

Frey, F.A., W.S. Wise, M.O. Garcia, H. West, S.T. Kwon, and A.

Kennedy, Evolution of Mauna Kea Volcano, Hawaii: Petrologic and geochemical constraints on postshield volcanism, *J. Geophys. Res*., 95, 1271-1300, 1990.

Garcia, M.O., Turbidites from slope failure on Hawaiian volcanoes, in *Volcano Instability on the Earth and Other Planets* vol 110, edited by W.J. MeGuire, A.P. Jones, and J. Neuberg, Geological Society Special Pub, 281-294, 1996a.

Garcia, M.O., Petrography and olivine and glass chemistry of lavas from the Hawai'i Scientific Drilling Project, *J. Geophys. Res*., 101, 11701-11713, 1996b.

Garcia, M.O., J.P. Foss, H.B. West, and J.J. Mahoney, Geochemical and Isotopic Evolution of Loihi Volcano, Hawaii, J. Petrol., 36, 1647-1674, 1995a.

Garcia, M.O., and D.M. Hull, Turbidites from giant Hawaiian landslides: Results from Ocean Drilling Program Site 842, *Geology*, 22, 159-162, 1994.

Garcia, M. O., B. A. Jorgenson, and J. J. Mahoney, An evaluation of temporal geochemical evolution of Loihi summit lavas: Results from *Alvin* Submersible dives, *J. Geophys. Res.*, 98, 537-550, 1993.

Garcia, M.O., T.P. Hulsebosch,. and J.M. Rhodes, Olivine-rich submarine basalts from the southwest rift zone of Mauna Loa Volcano: Implications for magmatic processes and geochemical evolution in *Mauna Loa Revealed: Structure, Composition, History, and Hazards, Geophys. Monogr. Ser.*, vol. 92, edited by J. M. Rhodes and J. P. Lockwod, pp. 219-239, AGU, Washingon, D.C., 1995b.

Garcia, M.O., D. Muenow, K. Aggrey, and J. O'Neil, Major element, volatile, and stable isotope geochemistry of Hawaiian submarine tholeiitic glasses, *J. Geophys. Res*., 94, 10525-10538, 1989.

Haskins, E.H., and M.O. Garcia, Scientific drilling of a Hawaiian volcano: Does Koolau have a skin? EOS Trans., 81, F1349, 2000.

Hauri, E. H., Major-element variability in the Hawaiian mantle plume, *Nature*, **382**, 415-419, 1996.

Herrero-Bervera, E., G.P.L. Walker and J. Guerrero-Garcia, The Nuuanu and Wailau Giant Landslides: Insights from paleomagnetic and Anisotropy of Magnetic Suceptibility (AMS) Studies, *Phys. Earth Planet. Int*., in press.

Jackson, M.C., F.A. Frey, M.O. Garcia, and R.A. Wilmoth, Geology and geochemistry of basaltic lava flows and dikes from the trans-Koolau tunnel, O'ahu, Hawai'i, *Bull. Volcanol.*, 60, 381-401, 1999.

Kanamatsu, T., E. Herrero-Bervera, and G. McMurtry, Magnetostratigraphy of deep-sea sediments from areas adjacent to the Hawaiian Islands: Implications for ages of turbidites derived from submarine landslides, *this volume*.

Macdonald, G., and T. Katsura, Chemical composition of Hawaiian lavas, *J. Petrol*., 5, 82-133, 1964.

Moore, J. G., Giant submarine landslides on the Hawaiian Ridge, *U. S. Geol. Surv. Prof. Pap*., 501-D, D95-98, 1964.

Moore, J.G., and D.A. Clague, Volcano growth and evolution of the island of Hawai'i, *Geol. Soc. Am. Bull.*, 104, 1471-1484, 1992.

Moore, J. G. and D. A. Clague, Mapping of the Nuuanu landslides in Hawaii, *this volume*

Moore, J.G., D.A. Clague, R. Holcomb, P.W. Lipman, W. Normark, and M.E.Torresan, Prodigious submarine landslides on the Hawaiian ridge, *J. Geophys. Res., 94*, 17465-17484, 1989.

Moore, J.G., and B.P. Fabbi, An estimate of the juvenile sulfur content of basalt, *Contrib. Mineral. Petrol.*, 33, 118-127, 1971.

Moore J. G., W. R. Normark and R.T. Holcomb, Giant Hawaiian landslides, *Annual Rev. Earth Planet. Sci*., 22, 119-144, 1994.

Naka, J. and others, Tectono-magmatic processes investigated at deep-water flanks of Hawaiian volcanoes, *Eos, Trans*, AGU, 81, 221,226-227, 2000.

Quane, S.L. M.O. Garcia, H. Guillou, and T.P. Hulsebosch, Magmatic history of the East Rift Zone of Kilauea Volcano, Hawai'i based on drill core from SOH 1, *J. Volcanol. Geotherm. Res*., 102, 319-338, 2000.

Rees, B.A., R.S. Detrick, and B.J. Coakley, Seismic stratigraphy of the Hawaiian flexural moat, *Geol. Soc. Amer.Bull.* 105, 189-205, 1993.

Schreiber, B.C., New evidence concerning the age of the Hawaiian Ridge, *Geol. Soc. Amer. Bull.*, 80, 2601-2601, 1969.

Shinozaki, K., R. Zhong-Yuan, and E. Takahashi, Chemical and petrological characteristics of Nuuanu and Wailau landslide blocks, *this volume*.

Wright, T.L., Chemistry of Kilauea and Mauna Loa lavas in space and time, *U.S. Geol. Surv. Prof. Paper*, 735, 1-49, 1971.

Yokose, H, Landslides of northeastern Oahu, Hawaii: SHINKAI 6500 submersible investigation, *this volume*.

Volcanic Breccia and Hyaloclastite in Blocks from the Nuuanu and Wailau Landslides, Hawaii

David A. Clague

Monterey Bay Aquarium Research Institute, Moss Landing, California

James G. Moore

United States Geological Survey, Menlo Park, California

Alicé S. Davis

Monterey Bay Aquarium Research Institute, Moss Landing, California

Steep slopes of giant landslide blocks in the Nuuanu and Wailau landslides expose fragmental volcanic rocks subdivided into monomict and polymict hyaloclastite and breccia. The various samples form as 1) secondary slope-mantling unlithified polymict breccia consisting of clasts set in a mud matrix; 2) monomict and polymict hyaloclastite and polymict breccia, with zeolite cement, that form downslope of the shoreline where lava flows enter the sea; and 3) monomict breccia formed by tectonic fragmentation of glassy submarine-erupted pillow basalt. Lavas erupted from single volcanoes are highly variable in major-element composition, even during their tholeiitic shield stage, making it difficult to identify which landslide block was derived from which volcano. Low-temperature fluids circulate through the fragmental deposits on the flanks of the volcanoes, partially altering the glass to palagonite and cementing the volcaniclastic rocks with Na- and K-rich zeolites. Volcano spreading along low-angle thrust faults laterally transports deep early submarine pillow lava out to the flank of the volcano where it crops out as tectonic monomict breccia. The faults underlying the landslide blocks are within this tectonized core of the volcano, not simply within the shallow slope deposits of hyaloclastite and breccia. The Nuuanu landslide predates the 1.5 Ma Wailau landslide.

1. INTRODUCTION

During the summer of 1999, a Japan-US cooperative research program mapped, explored, and sampled two giant landslides in the central Hawaiian region [*Naka et al.*,

Hawaiian Volcanoes: Deep Underwater Perspectives
Geophysical Monograph 128
Copyright 2002 by the American Geophysical Union

2000]. These Pleistocene features are among the largest landslides known. The Nuuanu landslide northeast of Oahu, and the Wailau landslide north of Molokai swept down the flanks of their respective volcanoes and deposited large blocks and finer debris out as far as 200 km from the island shores [*Moore*, 1964, *Moore et al.*, 1989]. During the research program the submersible *Shinkai 6500* operated by the Japan Marine Science and Technology Center (JAMSTEC) made eight deep dives primarily on the giant blocks of these landslides. The bulk of the material collected during these dives was volcaniclastic rock comprised of breccia and hyaloclastite. *Moore and Fiske* [1969] first proposed that the submarine flanks of Hawaiian volcanoes were covered by a carapace of fragmental volcanic debris and subsequent studies [*Clague et al.*, 1994; *Moore and Chadwick,* 1995] concluded that such fragmental debris constitutes a significant part of the volcanic edifice. More recently, *Garcia and Davis* [2001] proposed that volcaniclastic rocks are a minor component of the southern flank of Mauna Loa Volcano and by inference, all Hawaiian volcanoes.

Although several types of volcaniclastic rocks were collected—some as far as 118 km from the present shoreline and as deep as 4469 m—they have several features in common. Most contain so much basalt glass that they almost certainly were quenched by seawater. In addition, the low content of volcanic volatiles remaining in the quenched glass indicates that most lava was degassed subaerially. Downhill flow of this lava across the shoreline and into the sea caused it to quench and shatter producing the breccia and hyaloclastite. The enormous amount of this subaerially-derived volcanic rock that mantles the flanks of these volcanoes holds clues to the structure of the volcanoes as well as the variations in the composition of erupted lava since the time that the erupting vent grew above sea level. Moreover, the nature of this and other types of volcaniclastic material in the landslide can provide information on landslide timing and mechanisms. We studied the textures, physical properties, and compositions of the glasses in the volcaniclastic rocks recovered during eight *Shinkai 6500* dives north of Oahu and Molokai in 1999 to address these questions.

2. SAMPLING

Dives S496 and S500 collected 13 and 14 samples, respectively, from the north slope of Oahu (Figure 1). Dives S497, S498, S499 and S511, collected 9, 7, 16, and 11 samples, respectively, from Nuuanu landslide blocks northeast of Oahu, with dives S498 and S511 at opposite ends of Tuscaloosa Seamount, the largest block in the deposit. Dives S510 and S501 collected 14 and 9 samples, respectively, from the south-facing slopes of the two innermost large blocks in the Wailau landslide. Nearly all the outcrops observed on the landslide blocks in the Nuuanu and Wailau landslides consist of volcaniclastic rocks (Figure 2).

We studied 51 of the samples, including all of the volcaniclastic rocks. Most of the remaining samples are pillow lava fragments collected from breccia or talus deposits, which we did not analyze. We analyzed the only *in-situ* pillow lava samples from outcrops (S500-R6, and S510-R6A, -R6B, and –R6C) [*Yokose*, this volume], in both cases the shallowest samples collected on the dives. The volcaniclastic rocks can be subdivided into several types that have different origins and histories.

2.1. Unlithified Polymict Volcanic Breccia

Unlithified polymict breccia (Figure 3a) was observed mantling the slope during several of the dives and was collected during all four dives on blocks in the Nuuanu landslide (S497-R5; S498-R2, -7; S499-2A, -R4A, -R4B, -R6B; S511-R3B, -R3C, -R5, and –R6A) and during one dive on the Wailau landslide (S510-1A). The rocks generally are aggregations of sideromelane, lithic, and hyaloclastite fragments that are loosely bound in a muddy matrix. Their polymict character is demonstrated by multiple glass compositions in single samples. Most of the pillow lava fragments described elsewhere [*Shinozaki et al*, this volume; *Tanaka et al.*, this volume] are probably large clasts from these loosely consolidated deposits. The samples are aggregations of fragments from random outcrops farther upslope, not part of the outcropping volcaniclastic sequence in the landslide blocks. Such relations were difficult to identify from the submersible, so that some samples collected as outcrop turned out to be this type of secondary breccia. Some of the samples have thick Mn-oxide crusts.

These breccias are characterized by their friable character in hand sample and by the lack of zeolite cement between clasts in thin section. Some clasts are themselves hyaloclastite fragments, commonly with different alteration histories. Clasts with sideromelane and olivine completely altered to secondary phases can be adjacent to unaltered sideromelane clasts containing fresh olivine. These breccias, while providing a "channel sample" of the rocks upslope, are not stratigraphically significant.

2.2. Lithified Polymict Breccia

Lithified polymict breccias contain vitric and lithic fragments with varied textures, mostly of crystalline basalt from flow interiors; hyaloclastite fragments are absent. The

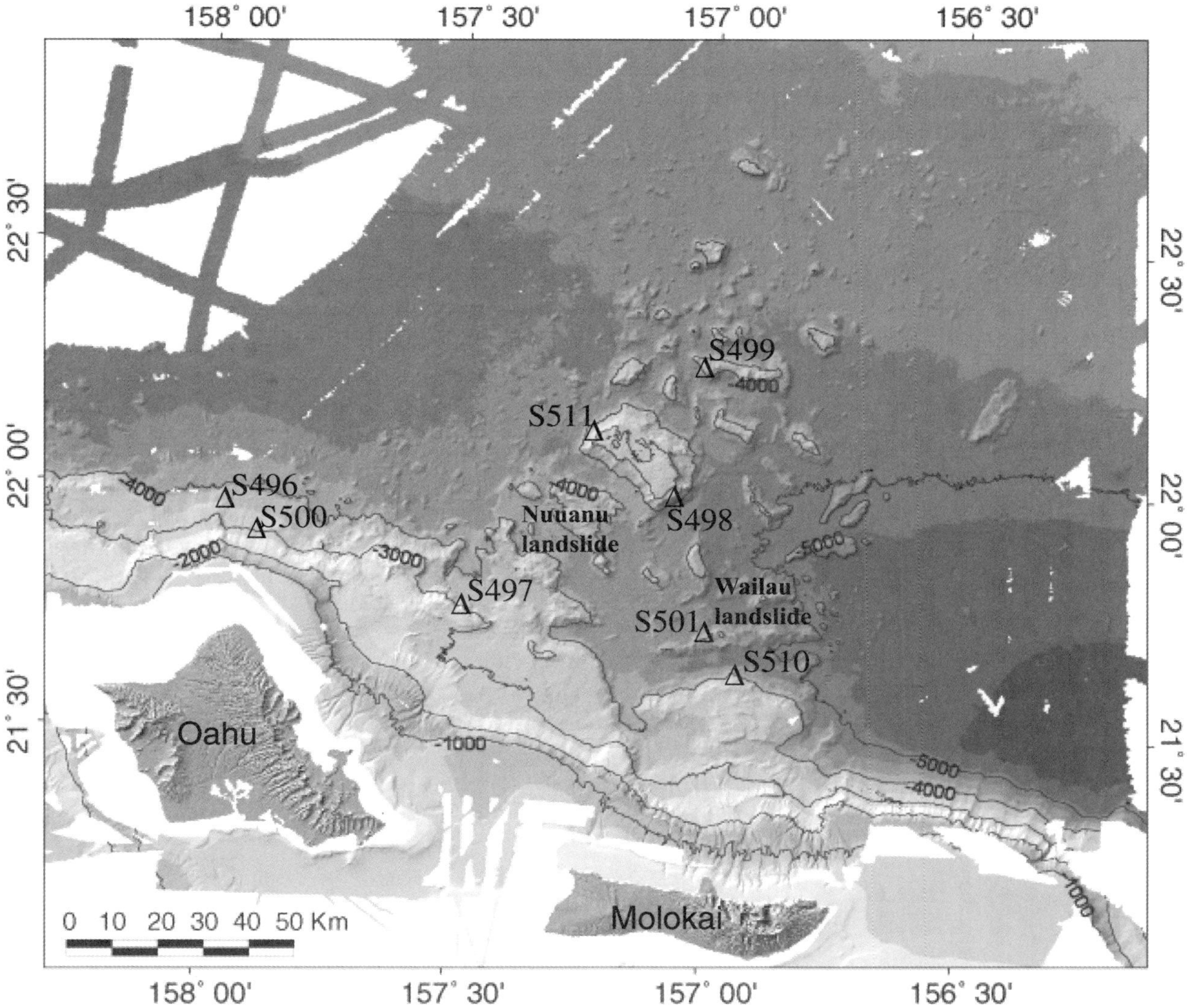

Figure 1. Bathymetric map of the Nuuanu and Wailau landslides showing the locations of the 8 *Shinkai 6500* dives in 1999. The bathymetry shown is a merged grid of SeaBeam 2100 data collected by the *R/V Kairei* in 1998 and the *R/V Yokosuka* in 1999 and Simrad EM300 data collected by C&C Technologies on the *M/V Ocean Alert* in 1998 under contracts from the U.S Geological Survey (nearshore Oahu, *Dartnell and Gardner* [1999]) and the Monterey Bay Aquarium Research Institute (nearshore Molokai, *MBARI Mapping Team* [2000]). Figure modified from *Moore and Clague* [this volume].

samples, cemented by Na- and K-rich zeolites (semiquantitative microprobe analyses indicate phillipsite, analcite, and an unidentified phase), are not friable in hand sample. Sideromelane fragments have rims of birefringent fibrous palagonite. Their polymict character is best demonstrated by the range of glass compositions found in single samples. The lithified polymict volcanic breccia samples are abundant rocks with three samples from the north slope of Oahu (S496-4, -10, -11), two from the Nuuanu landslide blocks (S499-R2B and S511-1) and three from the Wailau landslide blocks (S510-2A, -2B and S510-4B2).

2.3. Polymict Hyaloclastite

Polymict hyaloclastites are similar to lithified polymict breccias, except that the fragments are sand-sized. Sideromelane fragments have a wide range of compositions and lithic fragments are generally abundant (Figure 3b). The glass fragments are cemented by Na- and K-rich zeolites and rimmed by birefringent fibrous palagonite. In the most altered samples, sideromelane is replaced by yellow-to-orange isotropic gel palagonite. The rocks range from undeformed to highly sheared in narrow (mm) zones. In some

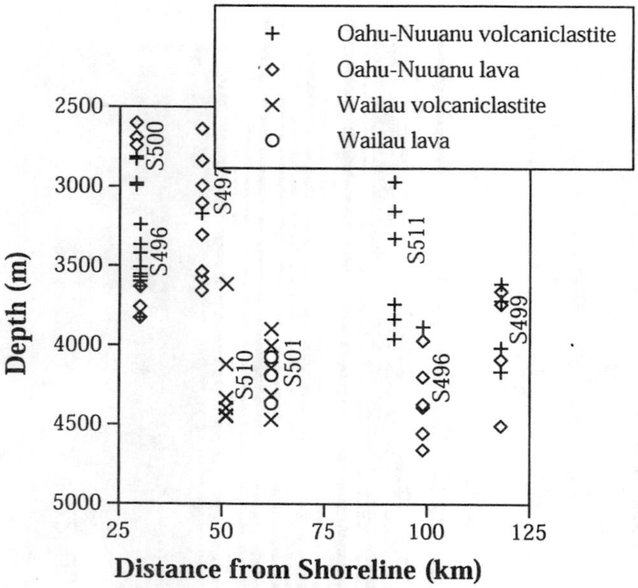

Figure 2. Plot showing the locations and depths of the volcaniclastic and lava samples collected during the *Shinkai 6500* dives.

samples (S496-R7 and S500-R3B), most glass fragments are identical but a few grains are compositionally distinct. This common lithology is represented by samples S496-R7, -R8, -R12; S500-R3B, -R4A, -R4B, and S511-R4A. Most samples are from the north slope of Oahu.

2.4. Monomict Volcanic Breccia

Monomict volcanic breccia was recovered during both dives on Wailau landslide blocks (dives S501 and S510), and the most distal block in the Nuuanu landslide (dive S499). There are two types of monomict volcanic breccia with the first consisting almost entirely of compositionally homogeneous sideromelane and tachylite (Figure 3c). Most volcaniclastic rocks from dive S501 (S501-R1, -R3, -R5B, -R7, -R8) and three from dive S510 (S510-R3A, -R4B, -R5B) are monomict volcanic breccia of this type. The second type, mostly from dive S499 on an elongate block in the Nuuanu slide (samples S499-R1, -R3B, R5A, and R5B) with one sample from the Wailau landslide blocks (S510-4B), consists of holocrystalline basalt. Both types are strongly deformed. In the first type, the sideromelane is fresh, with minimal palagonite rims on fragments and the rocks themselves have only thin Mn-oxide crusts. Zeolites are absent from these samples and the matrix appears to be finely comminuted sideromelane. The holocrystalline volcanic breccia samples have shear zones consisting of finely fragmented basalt.

For dive S510, the vitric samples (S510-R3A and S510-R5B) and the holocrystalline sample (S510-R4B) are stratigraphically interbedded with polymict volcanic breccia and capped by a pillow basalt flow (samples S510-R6A, R6B, and R6C). For dive S501, the monomict volcanic breccia samples are interbedded with polymict breccia from which pillow fragments were collected.

2.5. Monomict Hyaloclastite

Monomict hyaloclastite consists almost entirely of sideromelane fragments of uniform composition (Figure 3d). The glass fragments are subrounded to angular and altered on their outer margins to birefringent fibrous palagonite. The matrix is commonly cemented by Na- and K-rich zeolites. These samples are variably altered. The most altered (sample S496-R13) has no fresh sideromelane or olivine; in less altered samples most sideromelane and all olivine is fresh (S496-R2, -R5, -R6, -R13 with no fresh glass; S499-R8A, R8B, and R8C; and S511-R4B). In the more altered samples the sideromelane fragments are replaced by a yellow-to-orange isotropic gel palagonite. Lithic fragments are rare, although sample S496-R2 contains common lithic fragments. Most of these are relatively undeformed, except for S499-8A, -R8B, and –R8C. The monomict and polymict hyaloclastite are intergradational.

2.6. Pillow Lava Fragments

Pillow lava samples were only inadvertently included in this study of the volcanic fragmental rocks. All of the lava samples are submarine quenched pillow lava fragments. One of these samples, S500-R6, had an unusual texture in hand-sample that suggested it might be a tectonic breccia (not borne out in thin section). The remaining three pillow fragments (S510-R6A, R6B, and R6C) were the shallowest samples recovered on S510. *Shinozaki et al.* [this volume] describe a large number of the pillow basalt fragments and present microprobe compositions of the glass rims and whole-rock analyses. *Tanaka et al.* [this volume] present whole-rock trace element and isotopic analyses of many of the same samples.

3. CHEMICAL COMPOSITION OF GLASS FRAGMENTS

Electron microprobe analyses of glass fragments and glass rinds on pillow fragments were done at the U.S. Geological Survey in Menlo Park, California on a JEOL Superprobe using natural and synthetic standards. The analytical precision and accuracy are described by *Davis et al.*

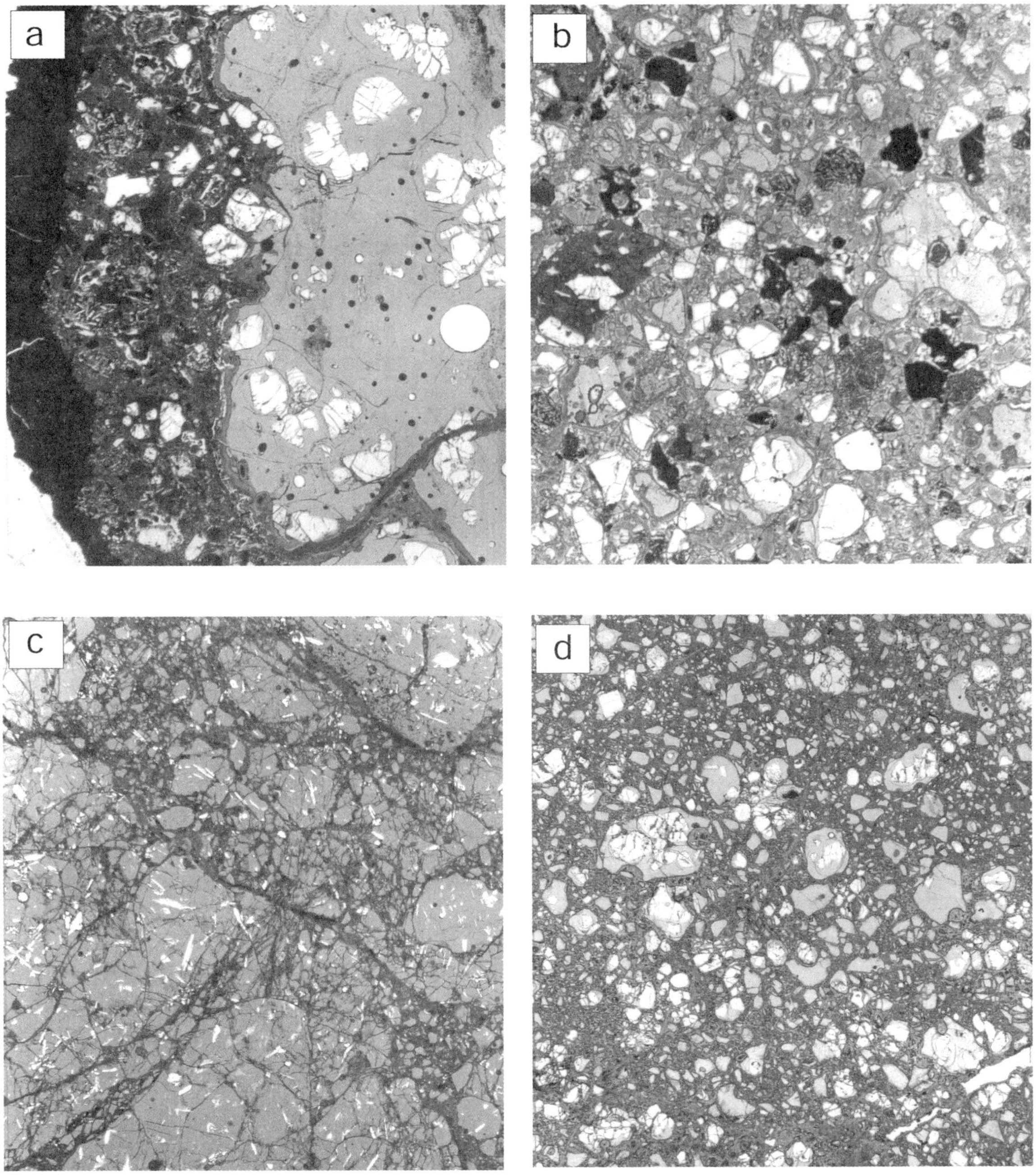

Figure 3. Photomicrographs showing the textures of four types of volcaniclastic rock recovered. a) Unlithified polymict breccia (S499-R6B) containing olivine phenocrysts. Note the thick Mn-oxide crust, the large glass clast with a thin palagonite rind, and the muddy matrix between the clast and the Mn-crust. b) Polymict hyaloclastite with abundant lithic fragments and olivine crystals (S510-R4B-2). The grains are cemented by Na-and K-rich zeolites and the glass grains are rimmed by palagonite. c) Monomict breccia composed of homogeneous submarine-erupted low-SiO_2 tholeiitic basalt glass (S501-R3) containing plagioclase microlites and some olivine microphenocrysts. The glass is fresh and has been fractured during shearing. d) Monomict hyaloclastite with intergranular cement of Na- and K-rich zeolites and olivine crystals (S511-R4B). The glass fragments are rimmed by palagonite. The images are each approximately 1.5 cm across.

[1994]. In samples where all the analyzed grains were statistically identical, the analyses were averaged, as were statistically identical analyses within a single sample. Analyses used here had totals between 98.4 and 100.6% and are presented and plotted on a dry-reduced normalized basis. Oahu north slope samples include 50 compositional groups based on 181 individual analyses on 14 samples, Wailau samples include 51 compositional groups based on 204 individual analyses on 19 samples, and Nuuanu samples include 47 compositional groups based on 186 individual analyses on 16 samples. We analyzed only a few pillow rinds from the Oahu slope and Wailau landslide (S500-R6 for Nuuanu and S510-R6A, -R6B, and –R6C for Wailau). *Shinozaki et al.* [this volume] present microprobe analyses of glasses for the same volcaniclastic rocks and for additional pillow rinds. Tables 1 to 3 present representative electron microprobe analyses for samples from the north slope of Oahu, the Nuuanu and the Wailau landslides, respectively.

In addition, we analyzed seven high-S monomict breccia and pillow rind glasses from the Wailau landslide for the volatile constituents H_2O, CO_2, F, S, and Cl, and five of these for trace elements (Table 4). Analytical techniques and precision and accuracy of the data are described by *Hauri et al.* [in press] for the volatile constituents and by *Shimizu and Hart* [1982] for the trace elements.

Comparisons are made to glass compositions from Loihi, Kilauea, and Mauna Loa Volcanoes, where extensive databases of submarine quenched glass data exist (references in caption to Figure 4). Unfortunately, data from Koolau, and East and West Molokai Volcanoes are almost entirely of whole-rock analyses and are more difficult to directly compare to the glass data from the *Shinkai 6500* samples.

3.1. North Slope of Oahu

The glasses from dives S496 and S500 on the north slope of Oahu have a broad range of compositions (Table 1; Figure 4a), both in terms of fractionation (9.2 to 5.6 wt.%MgO) and SiO_2 contents (50.6 to 54.5 wt.%). Most of the analyzed glasses in polymict breccia, polymict hyaloclastite, and monomict hyaloclastite have low S contents (<300 ppm, Figure 4b), indicating that they were subaerially erupted and degassed prior to quenching in seawater. However, the pillow lava S500-R6 and a group of chemically related glasses (high-MgO, moderate-SiO_2) from polymict breccia sample S496-R4 have high S contents (Figure 4b) and erupted submarine. *Shinozaki et al.* [this volume] attempted to correlate the compositions of the recovered samples to that of the Koolau, West Molokai and East Molokai Volcanoes. Only three of our samples from the north slope of Oahu resemble the distinctive high-SiO_2 (SiO_2>54 wt.%), low-FeO composition of subaerial Koolau Volcano [*Frey et al.*, 1994; *Roden et al.*, 1984, 1994]: pillow basalt S500-R6 and monomict hyaloclastite samples S496-R5 and -R6. The latter two samples are identical and distinct from most Koolau lavas in having CaO contents of only 8.0% and Al_2O_3/CaO ratios of 1.9 (Table 1 and *Shinozaki et al.*, [this volume]). These three samples are olivine-phyric and lack orthopyroxene phenocrysts, despite the high-SiO_2 content of their glass. Forty of 50 glass compositions have SiO_2<53 wt.% (Figure 5) and encompass the compositional range of Kilauea and Mauna Loa Volcanoes.

3.2. Nuuanu Landslide

Glass analyses from Nuuanu landslide blocks, like those from the slope of Oahu, have a large compositional range (Table 2, Figure 4c), both in degrees of fractionation (5.2-7.2 wt.% MgO) and SiO_2 content (49.3-54.5 wt.%). The highest-SiO_2 glass, from sample S499-6B, closely resembles lava from subaerial Koolau Volcano, but the remainder consist of lower SiO_2 glasses that again overlap the ranges of Kilauea and Mauna Loa lavas; 40 of 47 glass compositions have SiO_2<53 wt.% (Figure 5). The lowest-SiO_2 glass (S511-6A a, Table 2) is highly differentiated and is within the range of tholeiitic basalt compositions from Loihi Seamount (Figure 5). All the glasses from the Nuuanu landslide have low S contents (Figure 4d), indicating that they erupted subaerially and flowed into the sea.

3.3. Wailau Landslide

The glasses from the Wailau landslide blocks vary even more widely than those from the slope of Oahu and the Nuuanu landslide blocks: 9.2-4.1 wt.% MgO and 48.8-55.8 wt.% SiO_2 (Table 3, Figure 4e). The lowest-SiO_2 samples resemble transitional basalt from Loihi Seamount and plot close to the boundary separating Hawaiian tholeiitic and alkalic basalt on an alkali-silica diagram (not shown). The glasses from the Wailau landslide overlap with the compositions of glasses from the Nuuanu landslide and the north slope of Oahu; 44 of 51 glass compositions have SiO_2<53 wt.% (Figure 5).

Most glasses, including the two transitional basalts (S510-2B i and p, Table 3) have low S contents (Figure 4f), indicating subaerial eruption and quenching as the flows entered the sea. Other glasses, including all the monomict volcanic breccia samples, have high S contents

Table 1. Representative microprobe glass analyses for samples from the north-east slope of Oahu

Sample	S496-R2	S496-R4a	S496-R4f	S496-R5	S496-R6	S496-R7a	S496-R7c	S496-R8a
Lithology	mh	pb	pb	mh	mh	ph	ph	ph
Depth (m)	3826	3631	3631	3625	3571	3552	3552	3597
#of analyses	16	2	1	5	5	23	2	1
SiO_2	50.79	51.82	51.28	54.46	54.52	52.49	53.13	52.78
TiO_2	3.43	2.27	2.27	2.52	2.52	2.37	2.17	2.52
Al_2O_3	14.12	13.98	13.28	15.20	15.17	14.26	14.10	13.97
FeO	11.30	10.42	10.66	9.91	9.85	10.40	10.43	11.30
MnO	0.13	0.15	0.16	0.12	0.14	0.16	0.18	0.15
MgO	5.68	7.71	9.21	5.61	5.67	6.58	6.52	5.62
CaO	10.13	10.67	10.29	7.96	7.96	10.71	10.49	10.59
Na_2O	3.09	2.23	2.14	3.31	3.30	2.34	2.35	2.34
K_2O	0.79	0.35	0.34	0.57	0.57	0.42	0.35	0.44
P_2O_5	0.51	0.27	0.24	0.30	0.30	0.26	0.23	0.27
S	0.007	0.105	0.112	0.002	0.005	0.014	0.027	0.021
Cl	0.018	0.023	0.021	0.008	0.004	0.009	0.004	0.012

Sample	S496-R8g	S496-R10a	S496-R10c	S496-R11a	S496-R11d	S496-R11e	S496-R12a	S496-R12d
Lithology	pht	pb	pb	pb	pb	pb	ph	ph
Depth (m)	3507	3422	3422	3420	3420	3420	3368	3368
#of analyses	2	1	3	4	1	1	2	8
SiO_2	52.37	51.38	53.45	51.30	53.63	51.15	51.91	51.57
TiO_2	2.14	3.32	2.11	3.23	2.12	2.43	2.62	2.47
Al_2O_3	13.88	13.94	14.11	13.97	14.31	14.58	14.02	13.86
FeO	10.76	10.63	9.85	11.19	10.02	10.79	10.67	10.58
MnO	0.15	0.20	0.17	0.16	0.13	0.14	0.19	0.17
MgO	7.48	5.85	6.70	5.79	6.51	6.56	6.34	7.26
CaO	10.38	10.55	10.54	10.31	10.16	11.16	11.37	11.29
Na_2O	2.30	2.79	2.31	2.82	2.37	2.37	2.19	2.19
K_2O	0.32	0.81	0.48	0.72	0.48	0.50	0.36	0.35
P_2O_5	0.20	0.52	0.26	0.47	0.24	0.30	0.27	0.24
S	0.014	0.007	0.006	0.009	0.009	0.020	0.035	0.014
Cl	0.002	0.017	0.004	0.017	0.018	0.012	0.011	0.010

Sample	S500-R3Ba	S500-R3Bh	S500-R3Bi	S500-R4Aa	S500-R4Ab	S500-R4Ba	S500-R4Bb	S500-R6
Lithology	ph	ph	ph	mh	mh	mh	mh	pillow rim
Depth (m)	2994	2994	2994	2830	2830	2830	2830	2815
#of analyses	2	2	2	9	6	18	2	3
SiO_2	51.77	51.83	51.49	50.60	51.00	53.47	52.37	54.42
TiO_2	2.74	2.30	2.31	3.56	3.22	2.38	2.46	2.32
Al_2O_3	13.72	14.04	14.01	14.06	14.05	14.84	15.25	14.41
FeO	11.51	10.81	10.95	11.34	11.31	10.30	9.19	9.35
MnO	0.16	0.14	0.18	0.17	0.17	0.15	0.14	0.17
MgO	6.21	7.37	7.48	5.60	5.68	6.00	6.99	6.39
CaO	10.95	10.72	10.86	10.23	10.46	9.40	9.92	9.43
Na_2O	2.27	2.25	2.16	3.11	2.90	2.66	2.90	2.49
K_2O	0.39	0.31	0.33	0.81	0.71	0.52	0.50	0.59
P_2O_5	0.26	0.20	0.21	0.49	0.46	0.26	0.24	0.34
S	0.021	0.019	0.017	0.008	0.013	0.014	0.012	0.074
Cl	0.012	0.014	0.015	0.026	0.024	0.007	0.029	0.016

[a]m=monomict, p=polymict, b=breccia, h=hyaloclastite, u=unlithified

Table 2. Representative microprobe glass analyses for samples from the Nuuanu landslide

Sample	S497-R5a	S497-R5f	S498-R7	S499-R2Aa	S499-R2Ab	S499-R2Bb	S499-R2Bc	S499-R4A
Lithology[a]	upb	upb	upb	upb	upb	upb	upb	upb
Depth (m)	3172	3172	3884	4165	4165	4165	4165	4016
#of analyses	2	2	2	2	3	2	2	3
SiO_2	53.65	53.48	52.15	51.73	52.47	53.26	52.73	51.46
TiO_2	2.18	2.74	2.37	2.55	2.56	2.61	2.30	2.71
Al_2O_3	14.30	13.54	14.19	14.13	14.34	13.96	13.97	13.78
FeO	9.10	11.14	10.93	11.38	9.86	9.98	10.26	11.36
MnO	0.17	0.19	0.17	0.18	0.17	0.20	0.19	0.16
MgO	6.85	5.48	6.46	6.26	6.52	6.05	6.26	6.27
CaO	10.82	10.03	10.80	10.85	11.15	10.88	11.46	10.95
Na_2O	2.31	2.53	2.30	2.23	2.27	2.31	2.25	2.52
K_2O	0.38	0.55	0.35	0.37	0.37	0.48	0.26	0.38
P_2O_5	0.24	0.31	0.28	0.29	0.27	0.27	0.20	0.28
S	0.001	0.006	0.008	0.017	0.025	0.007	0.021	0.021
Cl	0.011	0.006	0.006	0.019	0.010	0.007	0.095	0.103

Sample	S499-R6Ba	S499-R6Bc	S499-R6Bd	S499-R8A	S499-R8B	S499-R8C	S511-R1c	S511-R1e
Lithology	upb	upb	upb	mh	mh	mh	pb	pb
Depth (m)	3718	3718	3718	3612	3612	3612	3959	3959
#of analyses	3	1	1	20	23	12	2	8
SiO_2	51.87	54.52	52.92	51.86	52.18	51.73	51.86	51.32
TiO_2	2.43	1.97	2.25	2.39	2.41	2.35	2.53	2.44
Al_2O_3	14.22	14.87	14.31	14.32	14.38	14.35	14.11	14.08
FeO	9.98	8.87	10.07	9.84	9.84	10.08	9.69	10.62
MnO	0.15	0.12	0.16	0.15	0.16	0.15	0.17	0.16
MgO	7.27	6.50	6.79	7.23	6.86	6.93	6.80	6.62
CaO	11.21	10.15	10.53	11.42	11.41	11.56	12.06	11.95
Na_2O	2.24	2.50	2.42	2.25	2.22	2.32	2.15	2.22
K_2O	0.33	0.27	0.32	0.31	0.29	0.31	0.36	0.33
P_2O_5	0.25	0.20	0.21	0.23	0.22	0.19	0.23	0.23
S	0.022	0.028	0.002	0.018	0.016	0.015	0.018	0.025
Cl	0.012	0.000	0.021	0.008	0.011	0.010	0.012	0.012

Sample	S511-R3B	S511-R3Ch	S511-R3Cm	S511-R4Ab	S511-R4B	S511-R5	S511-R6A a	S511-R6Ab
Lithology	upb	upb	upb	ph	mh	upb	upb	upb
Depth (m)	3740	3741	3741	3329	3329	3156	2972	2972
#of analyses	4	6	2	2	15	4	3	1
SiO_2	51.64	51.63	52.39	52.63	51.75	52.75	49.32	51.31
TiO_2	2.56	2.40	2.90	2.34	2.50	2.69	4.26	3.41
Al_2O_3	13.75	13.92	13.22	14.11	13.68	13.53	13.57	13.88
FeO	10.25	10.88	11.62	9.70	10.11	10.77	12.13	11.18
MnO	0.18	0.17	0.17	0.18	0.17	0.17	0.18	0.18
MgO	6.94	7.24	5.85	7.09	7.18	5.96	5.59	5.21
CaO	11.78	10.83	10.79	11.19	11.67	10.57	10.61	11.15
Na_2O	2.30	2.32	2.36	2.15	2.35	2.56	2.91	2.65
K_2O	0.35	0.36	0.42	0.34	0.34	0.69	0.85	0.60
P_2O_5	0.23	0.24	0.24	0.24	0.23	0.31	0.55	0.39
S	0.007	0.012	0.019	0.025	0.012	0.001	0.011	0.014
Cl	0.021	0.013	0.012	0.013	0.007	0.008	0.017	0.031

[a]m=monomict, p=polymict, b=breccia, h=hyaloclastite, u=unlithified

Table 3. Representative microprobe glass analyses for samples from the Wailau landslide

Sample	S501-R1	S501-R3	S501-R5B	S501-R7	S501-R8	S510-R1Ab	S510-R1Ae	S510-R2Ab
Lithology[a]	mb	mb	mb	mb	mb	upb	upb	pb
Depth (m)	4469	4314	4142	4005	3898	4448	4448	4443
# of analyses	14	8	11	4	16	4	2	2
SiO_2	49.04	49.32	51.40	51.38	51.63	51.99	51.4	53.96
TiO_2	2.72	3.09	2.53	2.15	2.77	3.30	2.36	1.94
Al_2O_3	15.03	14.86	14.33	13.29	13.86	12.52	13.30	14.88
FeO	11.80	12.54	11.06	10.63	12.20	13.10	10.83	9.68
MnO	0.17	0.18	0.17	0.16	0.18	0.21	0.13	0.16
MgO	6.57	6.33	6.39	8.09	5.84	5.39	7.76	7.01
CaO	11.53	10.26	11.17	11.73	10.43	10.36	11.31	8.90
Na_2O	2.38	2.55	2.19	1.94	2.27	2.39	2.24	2.77
K_2O	0.33	0.42	0.32	0.24	0.34	0.40	0.32	0.44
P_2O_5	0.27	0.32	0.27	0.21	0.31	0.32	0.21	0.24
S	0.114	0.101	0.122	0.130	0.129	0.020	0.093	0.007
Cl	0.019	0.041	0.058	0.038	0.030	0.014	0.031	0.001
Sample	S510-R2Af	S510-R2Ai	S510-R2Ak	S510-R2Bc	S510-R2Bf	S510-R2Bg	S510-R2Bh	S510-R2Bi
Lithology	pb	pb	pb	pb	pb	pb	pb	pb
Depth (m)	4443	4443	4443	4443	4443	4443	4443	4443
# of analyses	2	2	2	1.000	2	1	2	2
SiO_2	55.75	55.26	51.61	51.75	50.83	53.56	54.63	48.78
TiO_2	3.47	2.33	2.94	2.65	2.32	2.06	2.18	3.35
Al_2O_3	13.87	15.09	13.27	13.94	12.90	14.61	14.92	13.79
FeO	11.47	10.10	12.46	9.73	10.75	9.98	10.06	11.43
MnO	0.12	0.12	0.20	0.17	0.18	0.13	0.12	0.19
MgO	4.07	5.41	5.70	6.60	9.24	6.26	6.07	6.50
CaO	7.16	7.92	10.93	11.65	10.89	9.90	8.15	11.88
Na_2O	2.75	2.94	2.21	2.83	2.22	2.81	3.05	2.90
K_2O	0.85	0.52	0.38	0.38	0.33	0.44	0.52	0.73
P_2O_5	0.47	0.29	0.29	0.26	0.22	0.24	0.29	0.41
S	0.006	0.006	0.010	0.015	0.052	0.000	0.010	0.020
Cl	0.015	0.021	0.007	0.009	0.061	0.004	0.007	0.023
Sample	S510-R2Bj	S510-R2Bp	S510-R3A	S510-R4B2c	S510-R4B2d	S510-R4B	S510-R5B	S510-R6
Lithology	pb	pb	mb	ph	ph	m b	mb	pillow rims
Depth (m)	4443	4443	4396	4332	4332	4332	4121	3615
# of analyses	2	2	8	1	6	8	7	10
SiO_2	51.33	48.80	49.64	51.04	52.00	52.43	49.54	51.33
TiO_2	2.26	3.59	2.77	2.43	2.68	2.37	2.80	2.91
Al_2O_3	13.94	14.20	14.66	13.61	13.86	13.93	14.63	13.96
FeO	10.36	11.69	11.51	10.31	10.10	10.40	11.56	11.51
MnO	0.13	0.18	0.16	0.20	0.17	0.19	0.18	0.19
MgO	8.08	5.90	6.64	7.65	6.53	6.56	6.63	5.75
CaO	11.06	11.13	11.39	11.81	11.58	11.48	11.39	10.57
Na_2O	2.22	3.08	2.48	2.26	2.42	2.12	2.51	2.74
K_2O	0.33	0.87	0.33	0.43	0.37	0.28	0.33	0.54
P_2O_5	0.25	0.51	0.24	0.25	0.26	0.23	0.24	0.35
S	0.014	0.029	0.146	0.010	0.017	0.013	0.156	0.121
Cl	0.029	0.027	0.027	0.021	0.015	0.011	0.028	0.025

[a]m=monomict, p=polymict, b=breccia, h=hyaloclastite, u=unlithified

Table 4. Ion probe trace-element and volatile analyses of submarine-erupted glass from the Wailau landslide

Sample	501-1	501-3	501-5B	501-7	501-8	510-3A	510-6A
Sc	37.5	43.6	49.0	49.3	42.2	-	-
Cr	229	154	220	377	78.9	-	-
Rb	7.92	10.1	8.03	7.31	9.09	-	-
Sr	318	339	308	275	296	-	-
Y	31.5	35.4	29.0	25.6	31.7	-	-
Zr	146	174	135	116	153	-	-
Nb	13.6	16.0	12.7	10.6	14.8	-	-
Ba	55.8	71.2	52.3	44.1	58.3	-	-
La	11.4	13.6	10.5	9.04	12.0	-	-
Ce	27.1	32.2	25.4	21.6	28.2	-	-
Nd	19.1	22.2	17.6	15.3	19.3	-	-
Sm	5.69	6.80	5.47	4.77	6.00	-	-
Eu	1.76	2.05	1.63	1.49	1.78	-	-
Gd	5.88	7.63	6.01	5.09	6.56	-	-
Dy	6.22	6.77	5.61	4.86	6.06	-	-
Er	2.95	3.27	2.67	2.30	3.02	-	-
Yb	2.69	3.08	2.54	2.21	2.81	-	-
Hf	4.32	5.17	4.14	3.32	4.57	-	-
H_2O	1.00	0.55	0.42	0.45	0.36	0.26	1.15
CO_2	66	104	41	27	65	67	91
F	470	548	396	339	492	435	616
S	1220	880	1720	1520	1330	1740	1490
Cl	161	401	777	439	229	300	260

Analyses done on a Cameca IMS-6F ion probe at Carnegie Institute of Washington, Department of Terrestrial Magnetism, using instrument parameters and standardization described by *Hauri et al.* [2000]. Trace element values are averages of two separate analyses except 501-5B, which is a single analysis. Volatile analyses are averages of 3 separate analyses. All analyses in ppm except H_2O, which is in wt.%.

(>1000 ppm, Figure 4f) indicative of submarine eruption. Some fragments in these breccias are mildly vesicular, indicating volatile supersaturation at their eruption depths such as occurs at Loihi Seamount [*Dixon and Clague,* 2001]. These samples have glass compositions that correlate with their petrography. Samples S501-R1, S501-R5B, S510-R3A, S510-R3B, and S510-5B (6.39-6.64 wt.% MgO) are nearly aphyric. Samples S501-R3, and S501-R8 (6.33-5.84 wt.% MgO) are mildly plagioclase phyric, and sample S501-R7 (8.09 wt.% MgO) is picritic. Fractionation of the glasses does not vary systematically with stratigraphic position. The picritic sample includes some fist-sized basalt fragments of the same lava.

The monomict breccias contain most of the lower-SiO_2 glasses from the two Wailau dives. The glasses from the deeper samples from each dive have the lowest SiO_2 contents (49.0-49.6%) and the stratigraphically higher samples contain glasses with about 51.3-51.6% SiO_2. For dive S501, on the outer slope of the second large block from shore, the break from low-SiO_2 tholeiitic basalt to moderate-SiO_2 tholeiitic basalt occurs between 4314 and 4142 m and for dive S510, on the outer slope of the first large block from shore, it occurs between 4121 and 3615 m.

These monomict breccia glasses also contain high CO_2 contents (27-104 ppm, Table 4), high H_2O (0.26 to 1.15 wt.%), and high Cl (161 to 777 ppm). These volatile concentrations confirm that these lavas erupted under pressures equivalent to water depths of 640 to 2480 m using 42 ppm CO_2 solubility in Kilauea tholeitiic basalt melt per km water depth [*Dixon et al.*, 1991]. The collection depths of the samples do not correlate with the inferred eruption depth (Figure 6). The monomict breccia samples were collected 1840 to 3360 m deeper than their inferred eruption depths, whereas the pillow lava S510-R6 was collected about 1450 m deeper than its inferred eruption depth.

The trace element data for five of the Wailau landslide monomict breccia samples show that the low- and high-SiO_2 glasses have similar trace element characteristics (Table 4, Figure 7). These similarities extend to ratios of incompatible trace elements such as Zr/Nb=10.7±0.2, Zr/Y=4.71±0.18, Nb/Y=0.44±0.04, $[La/Sm]_C$=1.23±0.03, and Rb/Sr=0.028±0.003. These ratios are generally between those of Kilauea and Mauna Loa lavas [e.g., *Rhodes et al.*, 1989]. The Cr content of the glasses increases with increasing MgO content, but Sc shows no obvious trend. The abundances of the incompatible trace elements do not

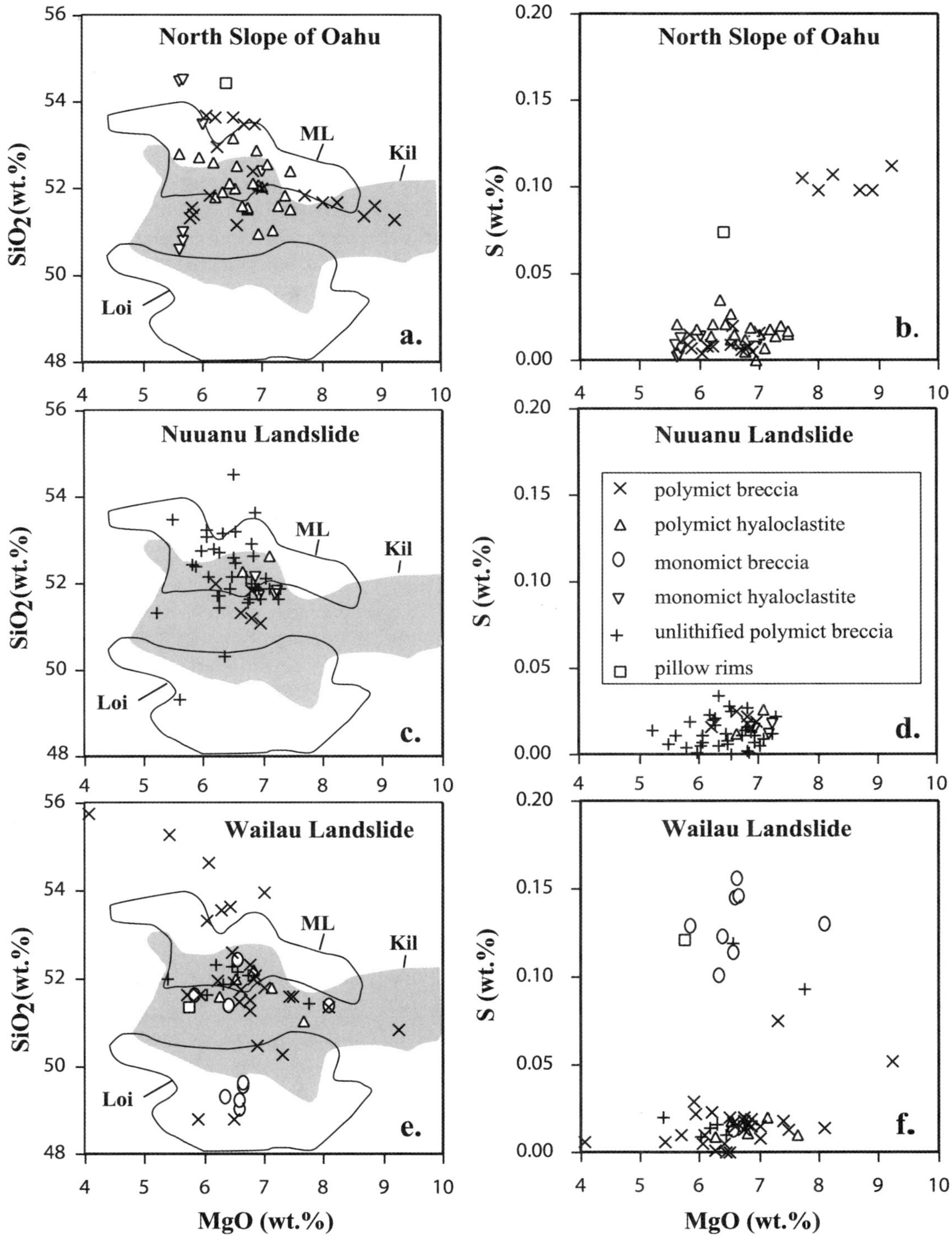

Figure 4. Dry-reduced, normalized compositions of glass fragments from the northeast slope of Oahu, the Nuuanu landslide, and the Wailau landslide. Fields for Loihi (labeled Loi; tholeiitic and transitional basalts only), Kilauea (shaded field labeled Kil), and Mauna Loa (labeled ML) are based on glass data from *Garcia et al.*[1989, 1995], *Moore and Clague* [1992], *Clague et al.* [1995, 2000], and *Dixon and Clague* [2001].

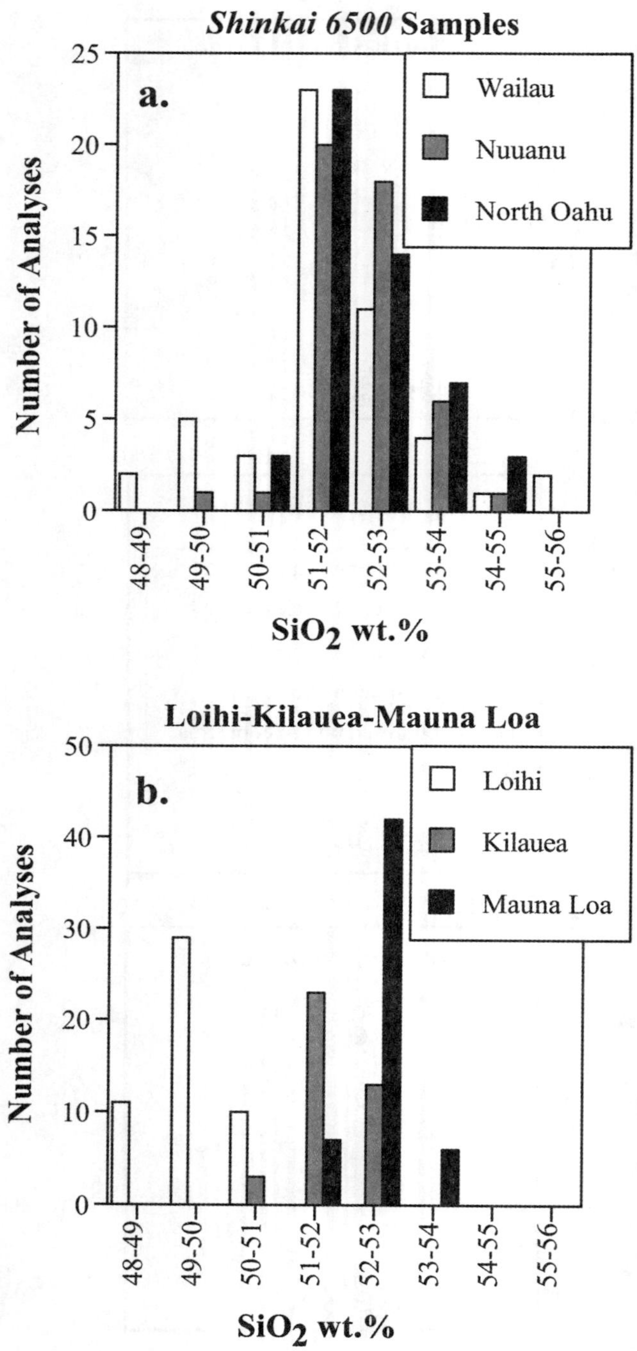

Figure 5. Histograms of the number of glass analyses as a function of dry-reduced, normalized SiO_2 content of glass for a) samples from the north slope of Oahu and the Wailau and Nuuanu landslides and b) pillow basalt rim from Loihi, Kilauea, and Mauna Loa Volcanoes. Fields based on data in *Garcia et al.* [1989, 1995], *Moore and Clague* [1992], *Clague et al.* [1995], and *Dixon and Clague* [2001].

increase systematically with increasing differentiation (decreasing MgO) or SiO_2 content.

4. PHYSICAL MEASUREMENTS ON THE SAMPLES

We measured the average thickness of palagonite on glass, the average thickness of Mn-oxide encrusting the rocks (both at sea and in the lab), the average size of the 5 largest clasts in the breccia samples, and the bulk water-saturated density of the samples (Figure 8). Palagonite (Figure 8a) is thicker on samples from the northeast slope of Oahu (average of 10 samples is 135 μ, range 50-300 μ) and the Nuuanu landslide (average of 9 samples is 185 μ, range 10-700 μ) and thinner on samples from the Wailau landslide (average of 7 samples is 55 μ, range 5-200 μ). Likewise, Mn-oxide encrustations (Figure 8b) are thicker on samples from the northeast slope of Oahu (average of 16 samples is 2.3 mm, range 0.1-5.65 mm) and the Nuuanu landslide (average of 30 samples is 1.1 mm, range 0.1-3.5 mm) than from the Wailau landslide (average of 18 samples is 0.3 mm, range 0.1-0.9 mm).

The average size of the 5 largest clasts in the breccia samples (Figure 8c) is about 1.5 cm (range 0.2-4.9 cm). Water-saturated densities are 2.1 to 3.04 g/cm^3 (Figure 8d). Dense samples are mostly lava containing abundant olivine whereas those with low density are fragmental rocks that contain little olivine and have abundant pore space. The average density of 2.4 g/cm^3 is consistent with moderately uncemented pore space, which means these rocks have low seismic velocities.

5. ORIGIN OF THE VOLCANICLASTIC ROCKS

Despite the varied textures, the volcaniclastic rocks have formed by only three major processes.

5.1. Sedimentary Landslide Block-mantling Debris

Unlithified polymict breccias formed by remobilization of rocks exposed on the steep slopes of the landslide blocks. These rocks were not found on the northeast slope of Oahu, but are common on the slopes of the landslide blocks where they mantle the steep slopes. They include sand, gravel, and cobble-sized fragments from outcrops farther upslope. The deposits observed from the submersible and some of the collected samples have moderately thick Mn-oxide crusts, suggesting that deposition occurred long ago, perhaps soon after the landslide blocks were emplaced. The Mn-crusts obscured the true character of these

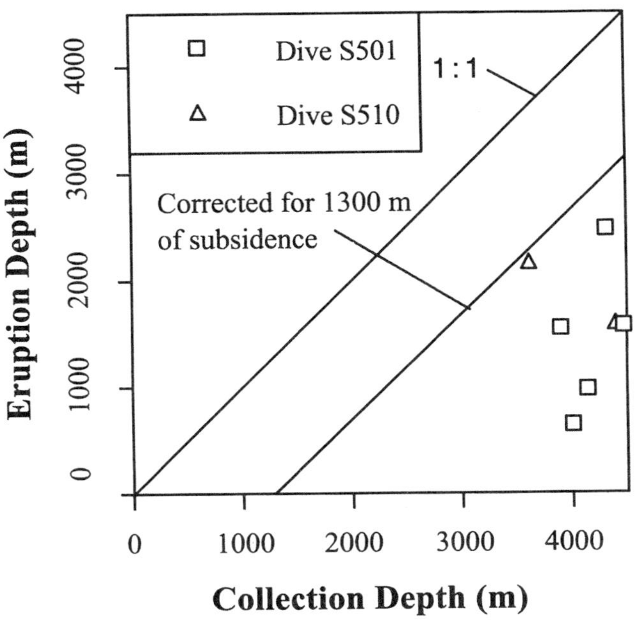

Figure 6. Plot showing eruption depth calculated from the CO_2 contents of the glasses as a function of their collection depth. Depth calculated using 42 ppm CO_2 solubility per km water depth from *Dixon e al.* [1991]. The estimated 1300 m of subsidence is from *Clague and Moore* [in press].

deposits and prompted collection of many of these sedimentary samples. *Yokose* [this volume] describes these and other sedimentary deposits in detail and subdivides them into several types.

5.2 Volcanic Island Slope-mantling Debris

The monomict hyaloclastite, polymict hyaloclastite, and lithified polymict breccia are the coarse (breccia) and fine (hyaloclastite) facies of volcanic fragmental debris formed by lava-seawater interactions when lava flows entered the sea, as has been occurring on Kilauea Volcano since 1986 [*Mattox and Mangan.*, 1997]. The low S contents of the glasses indicate that the flows erupted subaerially and degassed before being quenched and fragmented in the sea. Such debris creates a smooth offshore slope on Kilauea [Moore and Fiske, 1969; *Clague et al.*, 1994; *Moore and Chadwick*, 1995] and also makes up about 75% of the submarine section in the Hilo drill hole of the Hawaii Scientific Drilling Project [*Stolper et al.*, 2000]. The monomict samples are debris from a single eruption whereas the polymict ones are mixed debris from several eruptions or from eruptions that were not compositionally homogeneous, probably as the material is transported down slope. Most lithic fragments in these rocks are the crystalline interiors of lava flows emplaced at the shoreline. These rocks were deposited as volcanic debris on the flank of the islands, but were cemented by zeolites, lithified during burial, and sheared before or during emplacement of the landslide blocks. The Na- and K-rich zeolite cement requires introduction of both Na and K in excess of that released during alteration of the glass indicating that water flowed through these porous rocks. The presence of unaltered glass in most samples further suggests that the circulating water was low-temperature. Some thin zones may have had more vigorous fluid flow that resulted in complete alteration of glass and olivine in these horizons.

5.2. Tectonic Breccia

The monomict breccias from the Wailau landslide formed as pillow basalt flows that were submarine erupted, as demonstrated by their high volatile contents. The high H_2O and S contents are consistent with eruption before the summit of the volcano breached sea level. Most of the samples also have high-Cl contents, suggesting that they

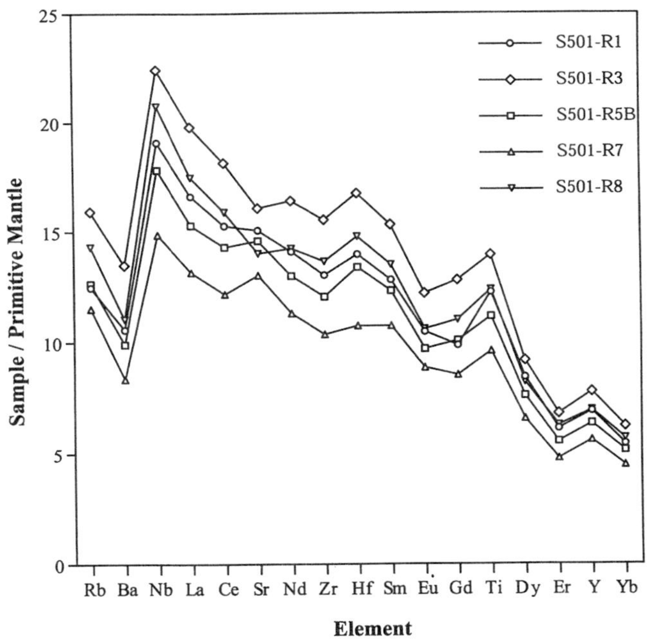

Figure 7. Trace element data normalized to primitive mantle values from *Sun and McDonough* [1989] for 5 monomict breccia samples of submarine erupted glass from the Wailau landslide. Despite a range of SiO_2 contents in these glasses, the trace element patterns are extremely similar to one another. Note that concentrations are plotted on a linear scale instead of the usual exponential scale.

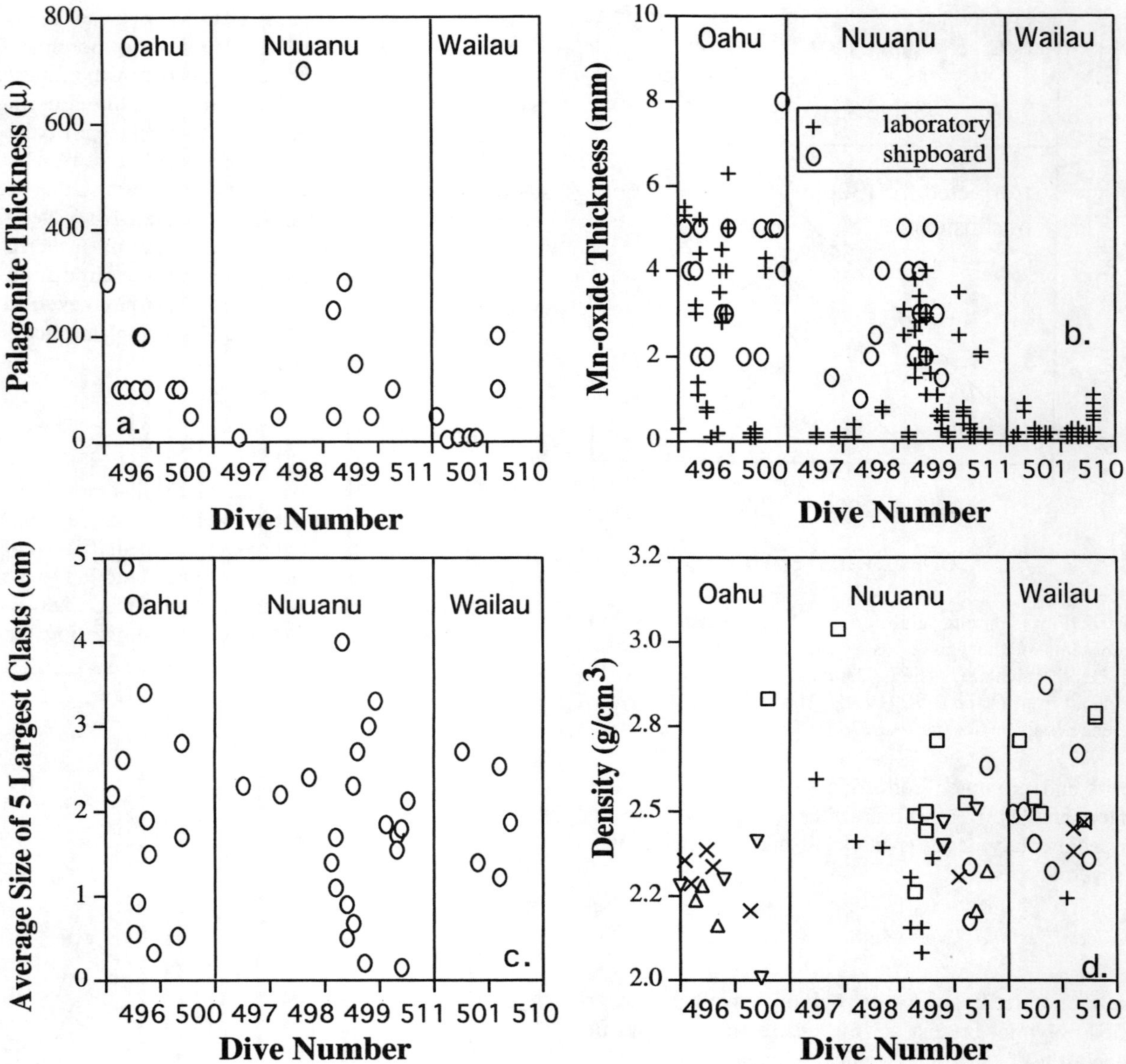

Figure 8. Plots showing palagonite thickness, Mn-oxide thickness, average size of the 5 largest clasts, and water-saturated densities of samples from the dives. The data are arranged by dive and region with sample numbers within a dive increasing from left to right.

erupted when the magma chamber within the growing volcano was below sea level and circulating hydrothermal fluids were seawater-derived [*Kent et al.*, 1999; *Clague and Dixon*, 2000; *Dixon and Clague*, 2001]. This places eruption of these lavas relatively early in the tholeiitic shield stage of one of the two volcanoes making up Molokai. The crystalline monomict breccias from the Nuuanu landslide may have formed at a similar stage in the evolution of Koolau Volcano, but the lack of glass prevents us from determining whether they were submarine or subaerially erupted.

The glassy lavas were tectonically sheared and granulated to form breccia during movement along faults thus producing fault gouge or tectonic breccia. The faults along which they formed do not appear to be those that bound the landslide blocks, suggesting they were low-angle thrust faults similar to those seismically imaged within the south flank of Kilauea [*Morgan et al.*, 2000; *Hills et al.*, this volume].

The eruption depths calculated from the CO_2 contents of the glasses indicate that these lavas erupted at depths1840-3660 m shallower than their sites of collection. Even accounting for island subsidence of about 1300 m since the end of shield building [*Clague and Moore*, submitted], these rocks are 540 to 2360 m deeper than their eruption depths. These analytical data demonstrates that the blocks, previously interpreted to be emplaced as landslide debris, have actually moved downslope from their point of origin.

6. DISTRIBUTION OF LITHOLOGIES

The distal ends of the Wailau and Nuuanu landslides ended up in the same region, making it difficult to assign specific landslide blocks to specific source volcanoes. Such a demarcation is important in understanding the mechanics of landslide emplacement and the tsunami hazards such landslides pose. One goal of the dives was to detemine which volcano was the source of which landslide blocks [*Shinozaki et al.*, this volume; *Tanaka et al.*, this volume]. Unexpectedly, the blocks from Wailau and Nuuanu landslides and the north slope of Oahu have a similar range of lava compositions (e.g., Figure 5), rather than consisting of lava samples with a narrow compositional range that could serve to "fingerprint" the source volcano. The abundance of volcaniclastic rocks, most polymict, further complicated establishing a stratigraphy within this wide range of basalts. The new geochemical data demonstrate that Hawaiian volcanoes erupt a wide range of lava compositions during their growth, even during their tholeiitic shield stage, contrary to previous studies that suggested that each volcano has a unique geochemical signature [*Frey and Rhodes*, 1993].

7. THE STRUCTURE OF SUBMARINE VOLCANIC SLOPES

The data also demonstrate that the submarine flanks of Koolau and East Molokai volcanoes, and presumably all Hawaiian volcanoes, consist largely of volcaniclastic rocks, as originally proposed by *Moore and Fiske* [1969] and contrary to the recent model of *Garcia and Davis* [2001]. The shallower part of this volcaniclastic section consists of debris formed when subaerial lava flows entered the sea, fragmented, and moved down the steep submarine slopes. These mostly polymict hyaloclastites and breccias comprise much of the stratigraphy in even the most distal blocks of the giant Nuuanu landslide and in the outer part of the south flank of Kilauea [*Lipman et al.*, 2000, this volume]. Nearer to shore, the deep parts of several blocks of the Wailau landslide (all >4000 m depth) consist of monomict tectonic breccia that formed as fault gouge in pillow lava sequences, probably along low-angle thrust faults that carried material outwards from deep within the early submarine part of the volcanic edifice. The base of the Wailau landslide is well within the submarine-erupted pillow basalt portion of the volcano and is not simply low-angle slumping of the fragmental rocks deposited on the slope of the island. Tectonic shuffling of the lithostratigraphy is seen in the section collected during dive S510, which appears to consist of interbedded tectonic and sedimentary breccias. That the deep interior of the volcano is apparently transported laterally along low-angle faults supports the general idea of volcano spreading originally proposed by *Borgia and Treves* [1992]. *Lipman et al.* [2000, this volume] describe abundant volcaniclastic rocks from the flank of Kilauea including early alkalic compositions, but none of their samples are texturally similar to, nor formed in the same manner as, the monomict tectonic breccias from the Wailau landslide blocks.

8. TIMING OF THE NUUANU AND WAILAU LANDSLIDES

Several new observations bear on the relative ages of the two landslides and the absolute age of the Wailau landslide. The main evidence about the relative timing of the two landslides comes from the amount of alteration and the thickness of Mn-oxides on rocks collected during the *Shinkai 6500* dives. Samples from the Nuuanu landslide and the northeast slope of Oahu have thicker palagonite on glass and Mn-oxide encrustations than on samples from the Wailau landslide (Figures 8a and b). These data suggest that the Nuuanu landslide is older than the Wailau landslide.

The general abundance of subaerially-erupted (degassed) glass in the hyaloclastites and breccias from the blocks argues that the giant landslides occur late in the history of the volcanoes, after they have built large islands and deposited vast amounts of volcanic debris on their submarine flanks.

Two geochemical observations constrain the absolute age of the Wailau landslide. First, the topmost undeformed pillow basalt flow from dive S510 (R6) is presently located only 1450 m deeper than its eruption depth, based on the CO_2 content of the glass, and within error of the inferred 1300 m of island subsidence [*Clague and Moore*, submitted]. This observation suggests that this flow erupted near its present position after the landslide blocks were emplaced (but prior to island subsidence) and, therefore, that the catastrophic landslide occurred while East Molokai Volcano was still active. The presence of these flows further indicates that volcanic activity accompanies the catastrophic emplacement of the landslide debris.

Second, the two transitional basalt samples (S510-2B analyses i and p in Table 3) were subaerially erupted, based on their low S contents. We therefore conclude that these lavas are from the postshield stage when Hawaiian volcanoes revert back to transitional and then alkalic lavas rather than the alkalic-transitional preshield or Loihi phase of activity, since such lava should be submarine-erupted. West Molokai does not have any known transitional basalt; the tholeiitic shield is directly overlain by postshield-stage hawaiite flows [*Clague*, 1987], so these grains apparently were not derived from West Molokai. On East Molokai, the tholeiitic shield lavas are overlain by a thick sequence of interbedded tholeiitic, transitional, and mildly alkalic basalt [*Beeson*, 1976; *Clague and Beeson*, 1980] and then by hawaiite, mugearite, and trachyte of the alkalic postshield stage. The incorporation of several transitional basalt fragments, but no alkalic basalt fragments, into the landslide blocks suggests that the catastrophic slide occurred after eruption of at least a few transitional lava flows at the end of shield building when tholeiitic and transitional basalts were both erupting.

The ages of these eruptive stages on East Molokai are well determined, so the age of the Wailau landslide can be estimated. The landslide occurred near the end of shield building when tholeiitic and transitional basalts were erupting at about 1.52±0.07 Ma [*McDougall*, 1964; ages corrected for modern decay constants by *Clague and Dalrymple*, 1987] and prior to 1.49 Ma, the beginning of eruption of postshield-stage hawaiite, mugearite and trachyte [*McDougall*, 1964], which are not represented in the landslide debris. This age is significantly older than the 1.0 Ma age estimated by *Kanamatsu et al.* [this volume] based on interpretation of paleomagnetic data on piston cores containing turbidite sand layers. It is possible that either the paleomagnetic correlations are incorrect or that the turbidite does not represent the one spawned by the Wailau landslide.

A similar argument for the timing of the Nuuanu landslide from Koolau Volcano is hampered by lack of any known transitional or alkalic postshield-stage lavas on the subaerial part of the volcano. The presence of a few Koolau-like high-SiO_2 lava fragments in the Nuuanu debris could also be used to infer that the landslide occurred at the end of the shield stage although it is not known if this unusual composition occurs only at the end of the Koolau shield stage, as implied by *Tanaka et al.* [this volume], or repeatedly during its growth.

9. CONCLUSIONS

1. The flanks of Hawaiian volcanoes, at least in regions away from rift zones, consist largely of hyaloclastite and volcanic breccia formed by fragmentation when subaerial flows entered the sea.
2. Lavas erupted from single volcanoes are highly variable in major-element composition, even during the tholeiitic shield stage, making it difficult to identify which landslide block was derived from which volcano.
3. Low-temperature fluids circulate through the fragmental deposits on the flanks of the volcanoes, partially altering the glass to palagonite and cementing the volcaniclastic rocks with Na- and K-rich zeolites.
4. Volcano spreading laterally transports deep early submarine pillow lava out to the flank of the volcano where it crops out as tectonic monomict breccia.
5. The faults underlying the landslide blocks are within the tectonized core of the volcano, not simply within the shallow slope deposits of hyaloclastite and breccia.
6. The Nuuanu landslide predates the Wailau landslide.
7. Subaerially erupted transitional basalt compositions found in the Wailau blocks correlate with lava compositions found only at the end of the shield stage on East Molokai, suggesting that the catastrophic Wailau landslide occurred about 1.5 Ma.

Acknowledgments. We thank JAMSTEC for inviting us to participate on their dive program around Hawaii and chief scientists Jiro Naka and Eiitchi Takahashi on the *R/V Yokosuka* for their skillful leadership during the cruise. The collection and descriptions of these samples is a credit to the entire shipboard crew and science party. The microprobe analyses were made at the U.S. Geological Survey in Menlo Park, California with the able assistance of Robert Oscarson and the ion probe analyses were made at the Department of Terrestrial Magnetism at the Carnegie Institution of Washington with considerable assistance from Erik Hauri and Jianhua Wang. Peter Lipman and Ryoji Tanaka provided helpful reviews that significantly improved the manuscript. DAC and ASD acknowledge the generous support of the David and Lucile Packard Foundation.

REFERENCES

Beeson, M.H., Petrology, mineralogy, and geochemistry of the East Molokai Volcanic Series, *U.S. Geol. Surv. Prof. Paper 961*, 53 p., 1976.

Borgia, A., and B. Treves, Volcanic plates override the ocean crust: Structure and dynamics of Hawaiian volcanoes, *Geol. Soc. Spec. Publ. London, 60*, 277-299, 1992.

Clague, D.A., Petrology of West Molokai Volcano. *Geol. Soc. Am. Abs., 19*, 366, 1987.

Clague, D.A., Beeson, M.H., Trace element geochemistry of the East Molokai Volcanic Series, Hawaii, *Am. J. Sci. 280-A*, 820-844, 1980.

Clague, D.A., and G.B. Dalrymple, The Hawaiian-Emperor Volcanic Chain: Part I. Geologic Evolution. *U.S. Geol. Surv. Prof. Paper 1350*, 5-54, 1987.

Clague, D.A., D.A. Davis, J.L. Bischoff, J.E. Dixon, and R. Geyer, Lava bubble-wall fragments formed by submarine hydrovolcanic explosions on Loihi Seamount and Kilauea Volcano, *Bull Volcanol., 61*, 437-449, 2000.

Clague, D.A., and J.E. Dixon, Extrinsic controls on the evolution of Hawaiian ocean island volcanoes, *Geochem. Geophysics, Geosystems, 1*, 1999GC000023, 12 p., 2000.

Clague, D.A., K.A. Hon, J.L. Anderson, W.W. Chadwick, Jr., and C.G. Fox, Bathymetry of Puna Ridge, Kilauea Volcano, Hawaii, *U.S. Geol. Surv. Misc. Field Studies Map, MF-2237*, 1994.

Clague, D.A., and J.G. Moore, The headwall of the giant Wailau landslide, Molokai, Hawaii, *Bull. Volcanol.*, in press.

Clague, D.A., J.G. Moore, J.E. Dixon, and W.B. Friesen, Petrology of submarine lavas from Kilauea's Puna Ridge, Hawaii. *Journal of Petrology*, *36*, 299-349, 1995.

Dartnell, P. and J.V. Gardner, Sea-floor images and data from multibeam surveys in San Francisco Bay, Southern California, Hawaii, the Gulf of Mexico, and Lake Tahoe, California-Nevada, *U.S. Geol. Surv. Digital Data Series, DDS-55*, 1999.

Davis, A.S., D.A. Clague, and W.B. Friesen, Petrology and mineral chemistry of basalt from the Escanaba Trough, southern Gorda Ridge, *U.S. Geol. Surv. Bull., 2022*, 153-170, 1994.

Dixon, J.E., and D.A. Clague, Volatiles in basaltic glasses from Loihi Seamount, Hawaii: Evidence for a relatively dry plume component, *J. Petrol., 42*, 627-654, 2001.

Dixon, J.E., D.A. Clague, and E.M. Stolper, Degassing history of water, sulfur, and carbon in submarine lavas from Kilauea Volcano, Hawaii, *J. Geol., 99*, 371-394, 1991.

Frey, F.A., and D.A. Clague, Geochemistry of diverse basalt types from Loihi Seamount, Hawaii: petrogenetic implications, *Earth Planet. Sci. Lett., 66*, 337-355, 1983.

Frey, F.A. M.O. Garcia, and M.F. Roden, Geochemical characteristics of Koolau Volcano: Implications of intershield geochemical differences among Hawaiia volcanoes, *Geochim. Cosmochim. Acta, 58*, 1441-1462, 1994.

Frey, F.A., and J.M. Rhodes, Intershield geochemical differences among Hawaiian volcanoes: Implications for source compositions, melting process, and magma ascent paths., *Phil Trans. Royal Soc London A, 342*, 121-136, 1993.

Garcia, M.O., and M.G. Davis, Submarine growth and internal structure of ocean island volcanoes based on submarine observations of Mauna Loa Volcano, Hawaii, *Geology, 29*, 163-166, 2001.

Garcia, M.O., D.J.P. Foss, H.B. West, and J. Mahoney, Geochemical and isotopic evolution of Loihi Volcano, Hawaii, *J. Petrol. 36*, 1647-1674, 1995a.

Garcia, M.O., Hulsebosch, T.P., and J.M. Rhodes, Olivine-rich submarine basalts from the Southwest Rift zone of Mauna Loa Volcano: Implications for magmatic processes and geochemical evolution, in *Mauna Loa Revealed: Structure, composition, history, and hazards*, ed. by J.M. Rhodes and J.P. Lockwood, *Am. Geophys. Union Monograph 92*, 219-239, 1995b.

Garcia, M.O., D. Muenow, K. Aggrey, and J. O'Neil, Major element, volatile, and stable isotopic geochemistry of Hawaiian submarine tholeiitic basalt, *J. Geophys. Res., 94*, 10525-10538, 1989.

Hauri, E.H., J. Wang, J.E. Dixon, P.L. King, C. Mandeville, S. Newman, SIMS analysis of volatiles in silicate glasses, 1: Calibration, matrix effects and comparisons with FTIR, *Chem. Geol.*, in press.

Hills, D.J., J.K. Morgan, G.F. Moore, and S.C. Leslie, Structural variability along the submarine south flank of Kilauea Volcano, Hawaii, from a multichannel seismic reflection survey, this volume.

Kanamatsu,T., E. Herrero-Bervera, and G.M. McMurtry, Magnetostratigraphy of deep-sea sediments from areas adjacent to the Hawaiian Islands: Implication for ages of turbidites derived from submarine landslides, this volume.

Kent, A.J., D.A. Clague, M. Honda, E.M. Stolper, I.D. Hutcheon, and M.D. Norman, Widespread assimilation of a seawater-derived component at Loihi Seamount, Hawaii, *Geochem. Cosmochim. Acta, 63*, 2749-2761, 1999.

Lipman, P.W., T.W. Sisson, T. Ui, J. Naka, J. Morgan, and J. Smith, Ancestral submarine growth of Kilauea Volcano and instability of its south flank, this volume.

Lipman, P.W., T.W. Sisson, T. Ui, and J. Naka, In search of ancestral Kilauea Volcano, *Geology, 28*, 1079-1082, 2000.

Mattox, T.N., and M.T. Mangan, Littoral hydrovolcanic explosions: a case study of lava-seawater interaction at Kilauea Volcano, Hawaii, *J. Volcanol. Geotherm. Res., 75*, 1-17, 1997.

MBARI Mapping Team, MBARI Hawaii multibeam survey, *Monterey Bay Aquarium Research Institute Digital Data Series No. 2*, 2000.

McDougall, I., Potassium-argon ages from lavas of the Hawaiian Islands, *Geol. Soc. Am. Bull. 75*, 107-128, 1964.

Moore, J.G., Giant submarine landslides on the the Hawaiian Ridge, *U.S. Geol. Surv. Prof. Paper 501-D*, D95-D98, 1964.

Moore, J.G., and W.W. Chadwick, Jr., Offshore geology of Mauna Loa and adjacent areas, Hawaii, in *Mauna Loa Revealed: Structure, composition, history, and hazards*, ed. by J.M. Rhodes and J.P. Lockwood, *Am. Geophys. Union Monograph 92*, 21-44, 1995.

Moore, J.G., and D.A. Clague, Volcano growth and evolution of the island of Hawaii, *Geol. Soc. Am. Bull., 104*, 1471-1484, 1992.

Moore, J.G., and D.A. Clague, Mapping the Nuuanu and Wailau landslides in Hawaii, this volume.

Moore, J.G., D.A. Clague, R.T. Holcomb, P.W. Lipman, R.W. Normark, and M.E. Torresan, Prodigious submarine landslides on the Hawaiian Ridge, *J. Geophys. Res., 94*, 17,465-17,484, 1989.

Moore, J.G., and R.S. Fiske, Volcanic substructure inferred from dredge samples and ocean bottom photographs, *Geol. Soc. Am. Bull., 80*, 1191-1201, 1969.

Morgan, J.K., G.F. Moore, D.J. Hills, and S.C. Leslie, Overthrusting and sediment accretion along Kilauea's mobile south flank, Hawaii: evidence for volcanic spreading from marine seismic reflection data, *Geology, 28*, 667-670, 2000.

Naka, J., Takahashi, E., and others, Tectono-magmatic processes on submarine flanks of Hawaiian Volcanoes: Japanese-USA cooperative studies. *EOS, 81*, 221-227, 2000.

Rhodes, J.M., K. Wenz, C.A. Neal, J.W. Sparks, and J.P. Lockwood, Geochemical evidence for the invasion of Kilauea's magmatic plumbing system by Mauna Loa magma, *Nature, 337*, 257-260, 1989.

Roden, M.F., F.A. Frey, and D.A. Clague, Geochemistry of tholeiitic and alkalic lava from the Koolau Range, Oahu, Hawaii: Implications for Hawaiian volcanism, *Earth Planet. Sci. Lett., 69*, 141-158, 1984.

Roden, M.F., T. Trull, S.R. Hart, and F.A. Frey, New He, Sr, Nd, and Pb isotopic constraints on the constitution of the Hawaiian plume: Results from Koolau Volcano, Oahu, Hawaii, *Geochim. Cosmochim. Acta, 58*, 1431-1440, 1994.

Shimizu, N., and S.R. Hart, Applications of the ion microprobe to geochemistry and cosmochemistry, *Ann. Rev. Earth Planet. Sci., 10*, 483-526, 1982.

Shinozaki, K., Z.-Y. Ren, and E. Takahashi, Geochemical and petrological characteristics of Nuuanu and Wailau landslide blocks, this volume.

Stolper, E., D. DePaolo, and D. Thomas, Hawaii Scientific Drilling Project: Overview of project goals and stratigraphy, *EOS, 81*, F1341, 2000.

Sun, S.S., and W.F. McDonough, Chemical and isotopic systematics of oceanic basalts: implications for mantle composition and processes, in *Magmatism in the Ocean Basins*, by Saunders, A.D. and M.J. Norry, eds., *Geol. Soc. Spec. Pub., 42*, 313-345, 1989.

Tanaka, R., E. Nakamura, and E. Takahashi, Geochemical characteristics of Koolau Volcano and the Nuuanu and Wailau landslide deposits, Hawaii, this volume.

Wright, T.L. , Chemistry of Kilauea and Mauna Loa in space and time, *U.S. Geol. Surv. Prof. Pap. 735*, 40 p., 1971.

Yokose, H., Landslides on the windward flanks of Oahu and Molokai, Hawaii: *Shinkai 6500* submersible investigations, this volume.

David A. Clague, Alicé S. Davis, Monterey Bay Aquarium Research Institute, 7700 Sandholdt Road, Moss Landing, CA 95039-9644

James G. Moore, U.S. Geological Survey, 345 Middlefield Road, Menlo Park, CA 94025

Geochemical and Petrological Characteristics of Nuuanu and Wailau Landslide Blocks

Ken Shinozaki, Zhong-Yuan Ren, and Eiichi Takahashi

Magma Factory, Earth and Planetary Sciences, Tokyo Institute of Technology, Japan

Major-element chemistry and petrology of ocean-floor rocks northeast of Oahu are compared with Koolau (Oahu), West Molokai, and East Molokai volcanoes in order to assess the source regions for the giant Nuuanu and Wailau landslides. In total, ca. 200 rocks were recovered during the 1998-99-JAMSTEC cruises; most are volcanic breccia and hyaloclastite, with subordinate pillow lava and subaerial lava. About 80 samples of lava and large clasts were analyzed by XRF for bulk chemistry; glass fragments in hyaloclastite (ca. 300) were analyzed by EPMA. Most rocks on the north submarine slope of Oahu, where undisturbed by the Nuuanu landslide, have petrologic and geochemical characteristics similar to subaerial Koolau volcano. Rocks within the Nuuanu landslide (derived from the north flank of Koolau) are typical Mauna Loa-like tholeiite; none has silica-rich Koolau-type chemistry. Rocks in the Wailau landslide (derived from East Molokai) are relatively silica-poor Kilauea-like tholeiite. These results indicate that the Nuuanu landslide records multiple growth stages at Koolau. "Koolau-type" tholeiite, which has been regarded as an end member of Hawaiian tholeiite magma, comprises only the late shield stage of this volcano.

1.INTRODUCTION

Joint US-Japan marine expeditions were undertaken in Hawaii during the summers of 1998 and 1999 using JAMSTEC ships, personnel, and deep submergence equipment. Two ROV (Remotely Operated Vehicle) KAIKO dives, 9 manned submersible SHINKAI dives, and 7 dredge hauls recovered nearly 200 samples from north flank of Koolau volcano and blocks in the Nuuanu and Wailau landslides (see *Yokose* [this volume], Figure 1). This paper reports petrologic descriptions, whole-rock XRF chemical analyses, and EPMA (electron microprobe) compositions of glasses in hyaloclastite and quenched margins of pillow lavas to assess source region of the landslide blocks. Trace elements and isotopes of selected rocks were studied by *Tanaka et al.* [this volume]. The lavas from Koolau volcano have been studied by many workers [e.g., *Frey et al.,* 1994; *Roden et al.,* 1984; 1994; *Jackson et al.,* 1999; *Takeguchi,* 1999]. Because only limited geochemical and petrologic data exist for the West and East Molokai volcanoes [*Beeson*, 1976; *Clague et al.*, 1982], additional samples were obtained to evaluate landslide sources.

Major element chemistry and phenocrysts of subaerial Koolau lavas (SiO_2 rich, ubiquitous orthopyroxene) are distinct from the volcanoes on Molokai (West Molokai volcano is similar to Mauna Loa, East Molokai to Kilauea). Rocks recovered from the northern flank of Oahu, undisturbed by the Nuuanu landslide, are petrologically indistinguishable from subaerial Koolau. Those from the Nuuanu landslide blocks differ from subaerial Koolau, having lower SiO_2 and lacking orthopyroxene phenocrysts. Rocks from the Wailau landslide are generally similar to subaerial East

Hawaiian Volcanoes: Deep Underwater Perspectives
Geophysical Monograph 128
Copyright 2002 by the American Geophysical Union

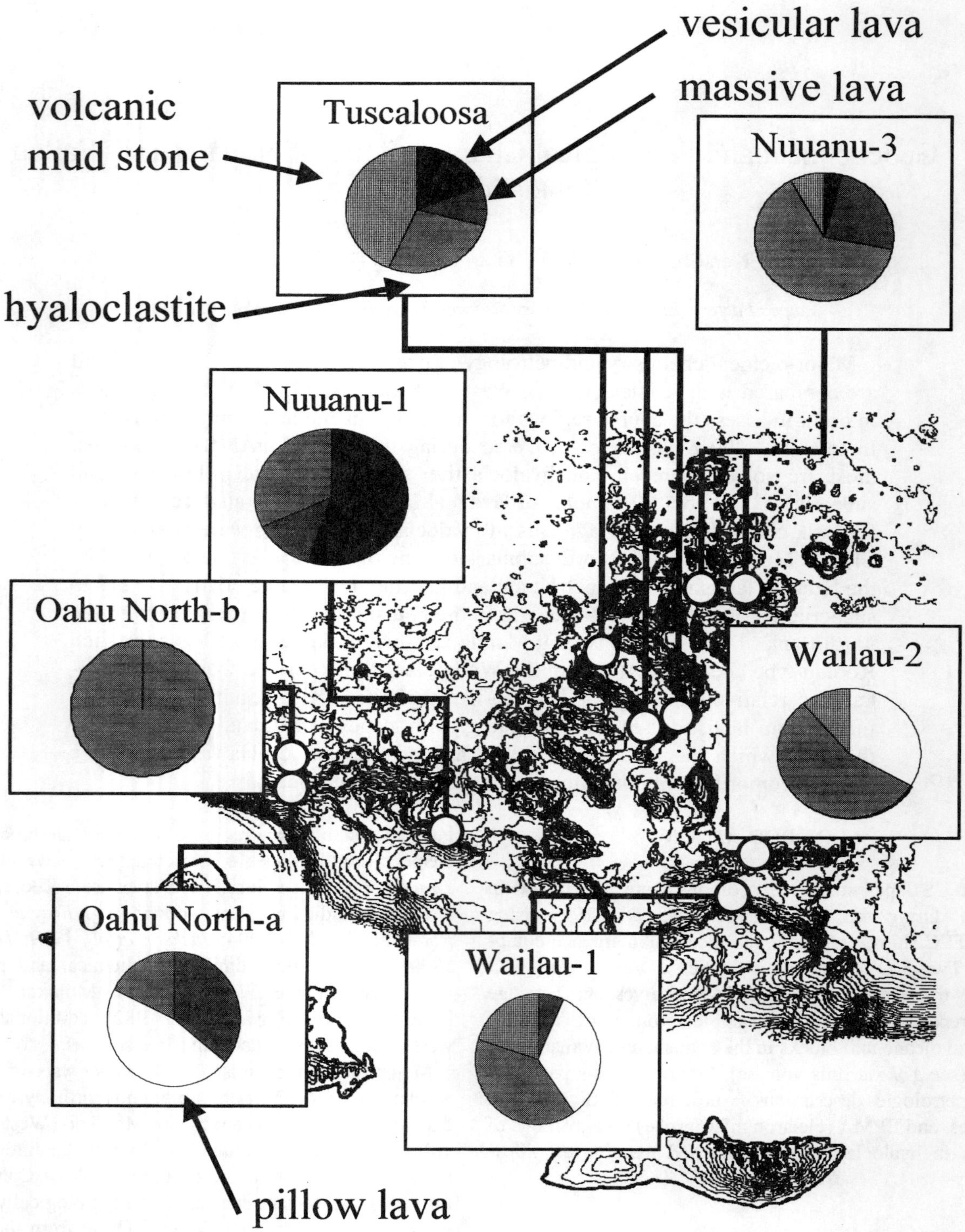

Figure 1. Frequency of the rock types in submarine samples, divided into volcanic mudstone, vesicular lava, massive lava, and hyaloclastite. Massive lava includes fragments of pillow lava, dike rocks, and some subaerial lava

Molokai volcano, from which they were derived. Our results, together with geologic reconstruction of the Nuuanu landslide [*Moore and Clague,* this volume; *Yokose,* this volume] and trace element and isotopes on these rocks [*Tanaka et al*., this volume], indicate that the Koolau volcano has experienced a multiple-stage growth history. Koolau's earlier stages, exposed by the giant landslide, have magma compositions similar to Kilauea and Mauna Loa.

2.ROCK TYPES

The submarine samples (Appendix, Table 1) are divided into four categories (Figure 1); vesicular lava of subaerial origin, massive lava (pillow fragments and dike rocks, maybe some subaerial lava), hyaloclastite, and volcanic mudstone-siltstone.

At the Oahu North sites on the submarine slope of Koolau volcano, hyaloclastites are dominant in deeper parts (Oahu North-b, 3000-4000 m), whereas pillow lava is abundant in shallower (Oahu North-a, 2000-3000 m). A similar transition from mostly volcaniclastic rocks upward to pillow lavas, at the Hilina slump on the south flank of Kilauea volcano [*Lipman et al.,* this volume], is interpreted as the flank portion of a typical Hawaii volcano that underwent a volcano-tectonic growth cycle [*Morgan et al.,* 2000].

Samples from the Nuuanu landslide vary in rock type. Rocks from the Nuuanu-3 site, on a landslide block 150 km from Oahu, are mainly hyaloclastite, whereas those from the Nuuanu-1 site (50 km from Oahu) are dominated by subaerially erupted vesicular lava (Figure 1). Volcanic mudstone is abundant in samples from Tuscaloosa Seamount (Nuuanu-2), the largest slide block in the Nuuanu slide. The abundant mudstone is mainly from dredges at the base of Tuscaloosa and may not characterize the interior of the block [*Yokose*, this volume]. The proportions of the rock types at Tuscaloosa, excluding the mudstone, is intermediate between those at the Nuuanu-1 and Nuuanu-3 sites (Figure 1).

How much of the Nuuanu-1 block represents the subaerial lava flank of Koolau volcano is controversial. Some rocks have a brecciated matrix; others are hyaloclastite (Appendix, Table 1). The abundant vesicular lava with coarse intergranular groundmass (Figure 2b), and absence of hyaloclastites in upper parts of the Nuuanu-1 site, suggests that part of this block comes from subaerial Koolau. The changes in slide-block rock types with distance from Oahu, from dominantly subaerial lava to abundant hyaloclastite (Figure 1), may record a progression in lithologies from shallower to deeper parts of the collapsed Koolau volcano. This is in harmony with geologic reconstruction of Nuuanu landslide by *Moore and Clague* [this volume] and *Yokose* [this volume]. Rocks from the Wailau slide blocks also show some differences, although hyaloclastite is dominant and pillow lava is next at both sites.

3. PHENOCRYST ASSEMBLAGES

All rocks (Appendix Table-1) were sectioned and examined with a petrographic microscope (Figure 3). The volume percentage (mode) of phenocrysts is dominated by olivine in most rocks (Figure 2a, b, d). The most distinctive petrologic feature is the high percentage of orthopyroxene (including microphenocrysts) in the pillow lavas at the Oahu North-a site (Figure 2c, d), in sharp contrast to the absence of orthopyroxene phenocryst at all other sites (except Tuscaloosa, Nuuanu-2). The abundance of orthopyroxene phenocrysts in Koolau lavas, noted by *Frey et al.* [1994], was confirmed by study of 250 samples from the upper 1000 m of Koolau [*Takeguchi*, 1999]. The mineralogical similarity suggests that the pillow lavas at the Oahu North-a site represent the submarine flank of Koolau volcano at the same growth stage as its subaerial part. Hyaloclastite glass at the deeper Oahu North-b site contains less orthopyroxene (Figure 3), but the glass chemistry is similar to that in the pillows. Paucity of orthopyroxene in the Oahu North-b hyaloclastite rocks may be due to repression of microphenocryst growth in the quenched glass (Figure 2a).

In contrast, most rock specimens from the Nuuanu blocks lack Koolau-type phenocryst assemblages. They are dominated by olivine and subordinate plagioclase and clinopyroxene (similar to most Hawaiian tholeiites [*BVSP,* 1981]). The marked difference in phenocryst assemblage must be associated with changes in magma chemistry between the Nuuanu landslide blocks and the subaerial Koolau volcano. Rocks from the Wailau landslide contain more plagioclase than those from the Nuuanu slide; this may be representative of East Molokai volcano.

4. WHOLE-ROCK ANALYSES

4.1 Analytical Procedures

Whole rock compositions were determined by XRF (X-ray fluorescence analyzer, Rigaku-3550 with Rh X-ray tube) at Tokyo Institute of Technology. After samples were sliced, least altered parts were crushed to several millimeter size, then carefully handpicked to avoid contamination or alteration. Submarine samples were rinsed ultrasonically in pure water, with repeat cycles for more than 2 hours; then the fragments were pulverized in an agate mortar. 400 mg of rock powder (dried at 110°C for 24 hours) was mixed

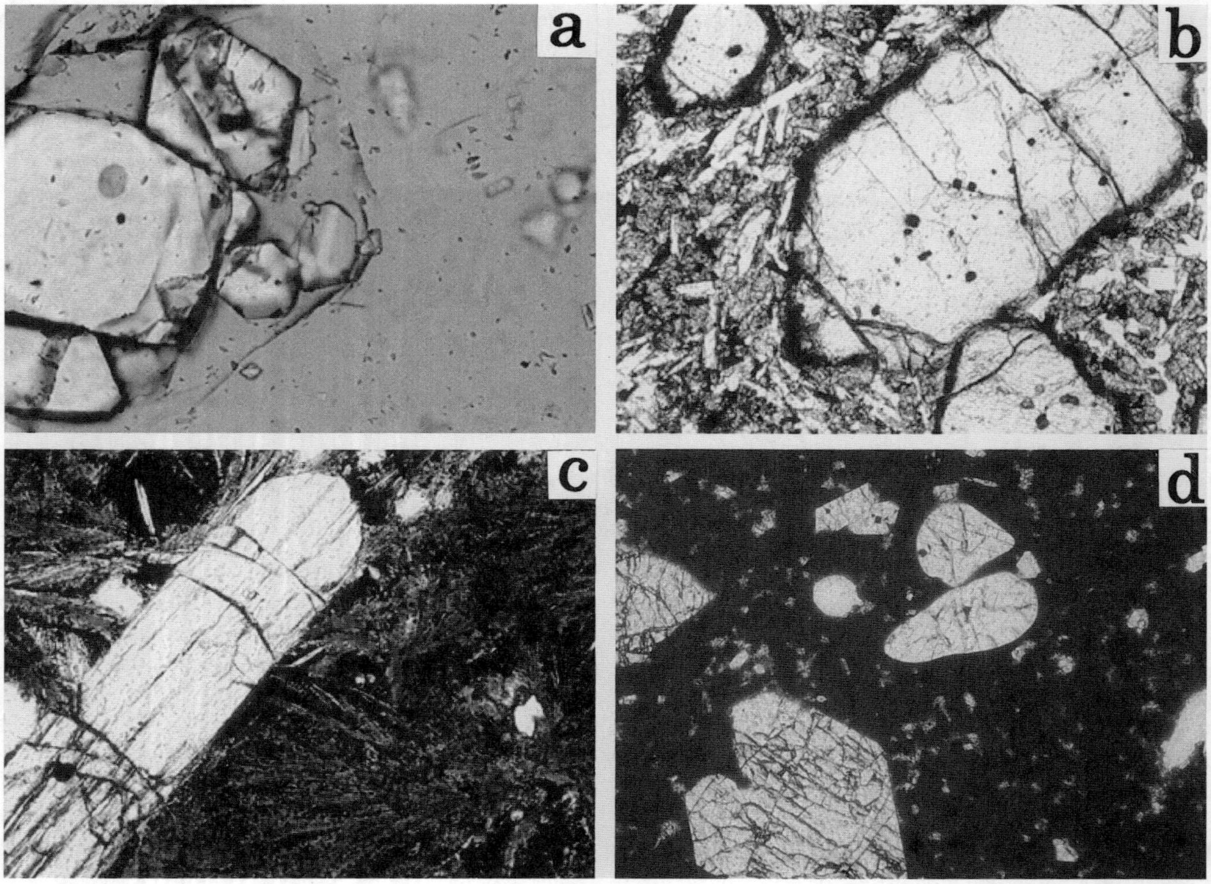

Figure 2. Photomicrographs of representative rock specimens. Width of the photos: 1.2 mm for b and d; 0.4 mm for a and c. a: Olivine phenocrysts in clear glass of hyaloclastite (S496-R6), which has a high Al_2O_3/CaO ratio (see Fig. 7e, and Appendix Table-3). b: Picrite with coarse intergranular groundmass (S497-R7) and 24% vesicles. c: Euhedral orthopyroxene phenocryst in a pillow lava (S500-R9A). d: Euhedral olivine phenocryst and rounded orthopyroxene (resorbed) in a pillow fragment (S500-R2A). Many lavas at the Oahu North sites contain either euhedral or rounded orthopyroxene.

with 4.000 g of $Li_2B_4O_7$ flux and fused at 1200°C for 7 min in a Pt crucible. Because the fusion is in air, all iron in the fused disk is assumed to be in the ferric oxidation state.

4.2 Subaerial Rocks

XRF data for subaerial rocks from Koolau, West Molokai, and East Molokai volcanoes are shown in Figure 4a-f. The Koolau data are from *Takeguchi* [1999]. The Molokai samples were collected in reconnaissance prior to the 1998 Hawaii cruise. Because a thick pile of transitional to alkalic lava covers East Molokai volcano [*Stearns and Macdonald*, 1947; *Beeson*, 1976], we tried to collect tholeiite from deep canyons. A section was sampled for the upper 700 m of lavas in Kamalo Gulch, on the south flank of East Molokai volcano, but much of the lava is transitional to alkalic basalt (Appendix, Table 2).

When tholeiitic basalts with 6-8 wt% MgO are compared, SiO_2 content changes systematically from Koolau (52-54 wt%), to West Molokai (49-51 wt%), and then to East Molokai (47-49 wt%) volcanoes (Figure 4a). The Koolau tholeiites also have lower CaO and TiO_2 and higher Al_2O_3/CaO than tholeiites from the two volcanoes on Molokai (Figure 4c-e). Tholeiites from East Molokai have higher TiO_2 than those from West Molokai volcano at given MgO (Figure 4d). Accordingly, there should be systematic changes between the rocks from Nuuanu (derived from Koolau volcano) and Wailau (from East Molokai volcano) landslide blocks. The relatively low SiO_2 content of tholeiite in East Molokai volcano suggests that it is a Kea-

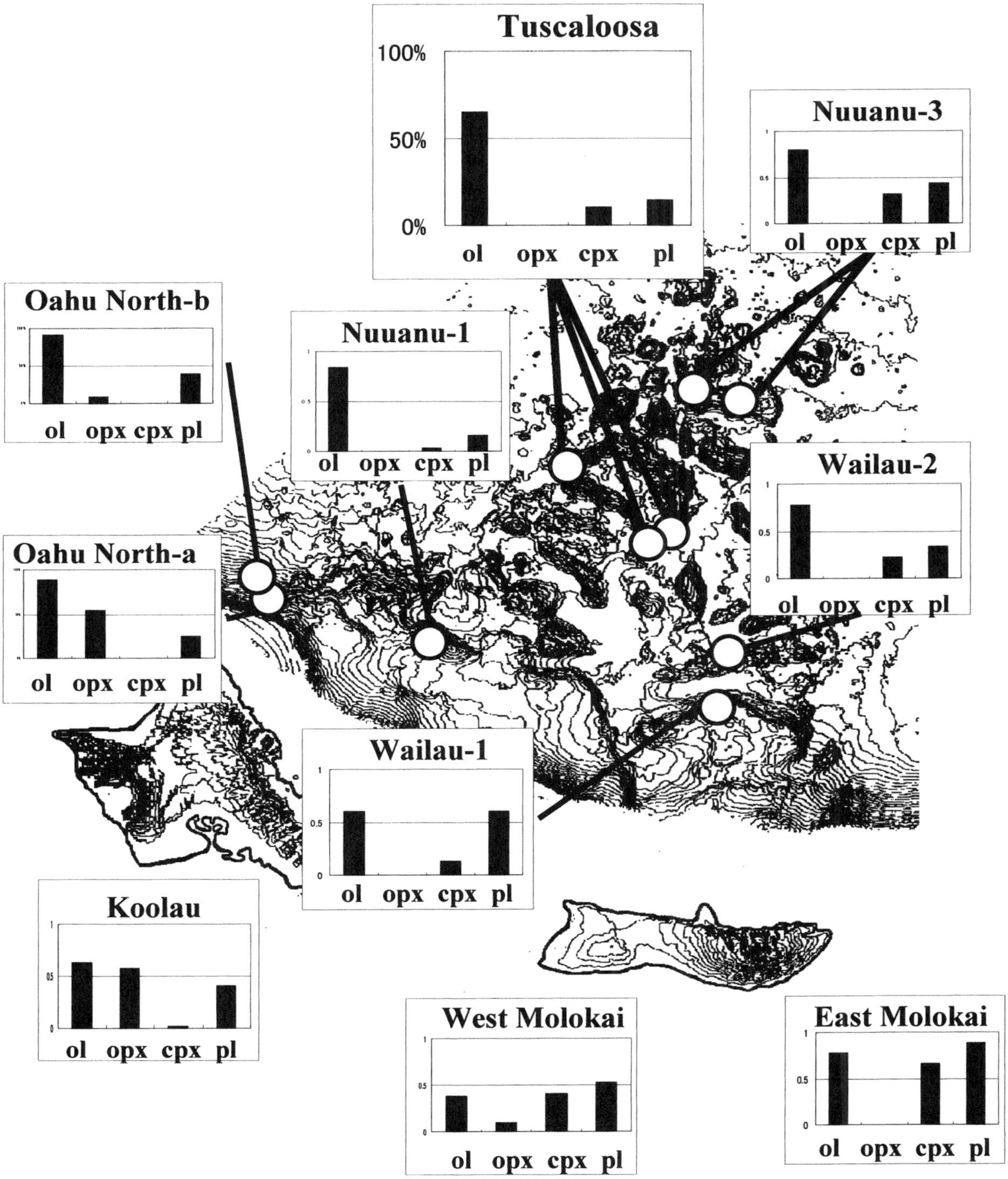

Figure 3. Frequency histogram of phenocrysts in thin sections. Each bar in the graph represents the number of thin sections in which the phenocryst occurs, relative to the total number of sections studied. Thus, the frequency of appearance does not represent modal abundance, and the bar graphs do not sum to 100%.

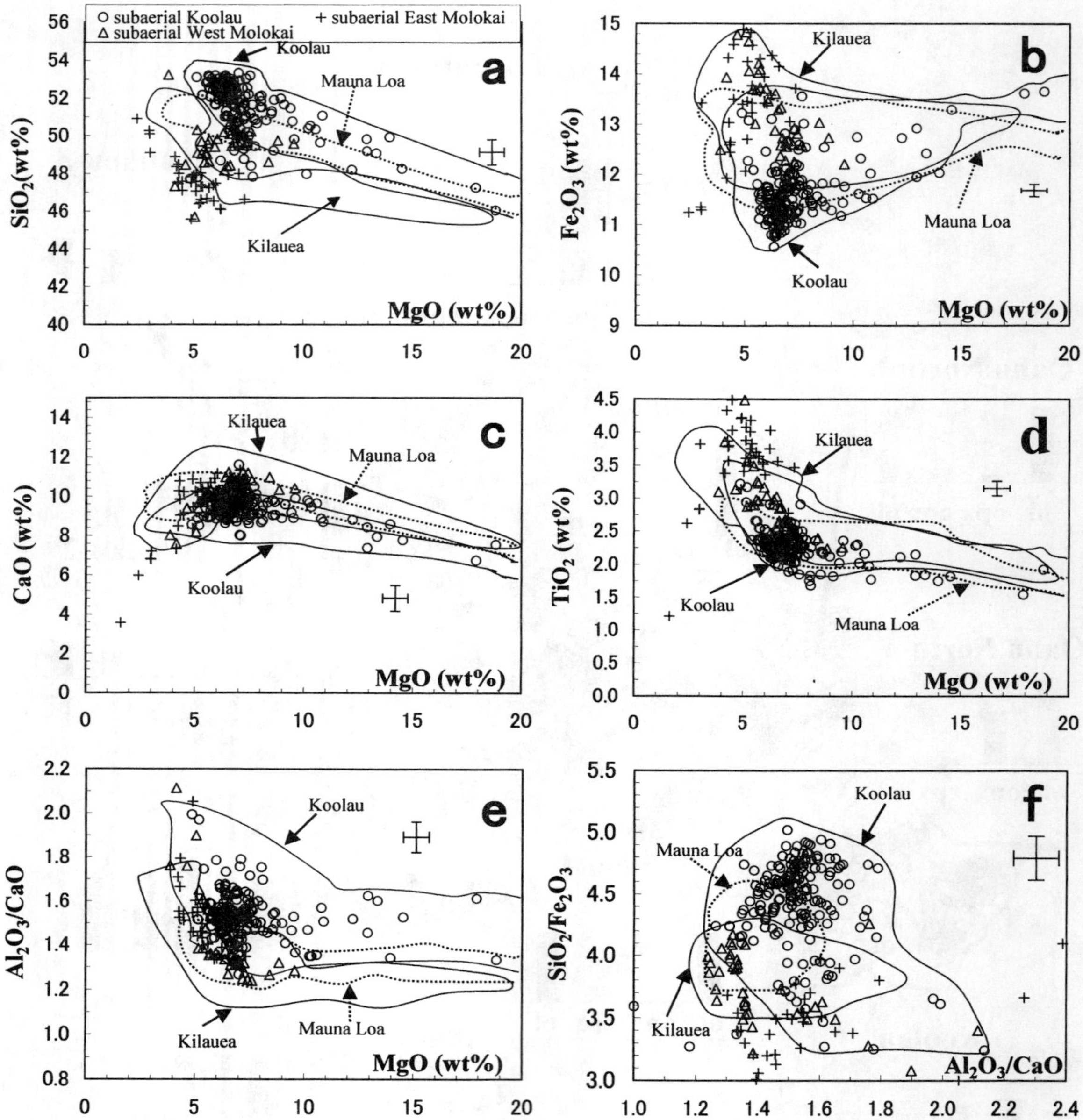

Figure 4. Whole rock compositions of subaerial Koolau volcano (green; unpublished data from *Takeguchi* [1999]), West Molokai volcano (purple) and East Molokai volcano (red). Data for the Molokai volcanoes are newly determined (Appendix, Table 1-2). Compositional field for Kilauea and Mauna Loa lavas (data set compiled by the Hawaii Volcano Observatory) are shown for comparison. Error bars are based on calculated 2σ deviations.

trend volcano like Kilauea, whereas West Molokai may resemble Loa-trend volcanoes. Because our samples from East and West Molokai volcanoes are limited, and fractionated rocks are dominant, the total compositional spectrum in these volcanoes is unconstrained (Figure 4). For comparison with the landslide rocks, therefore, we use the compositional ranges for Kilauea and Mauna Loa lavas.

4.3 Submarine Rocks

In MgO variation diagrams (Figure 5a-h), rocks recovered from both Oahu North sites are systematically higher in SiO_2, Na_2O, and K_2O and lower in FeO, CaO, and TiO_2 than rocks from the Nuuanu and Wailau slides blocks. Accordingly, they have distinctive Al_2O_3/CaO ratios (Figure

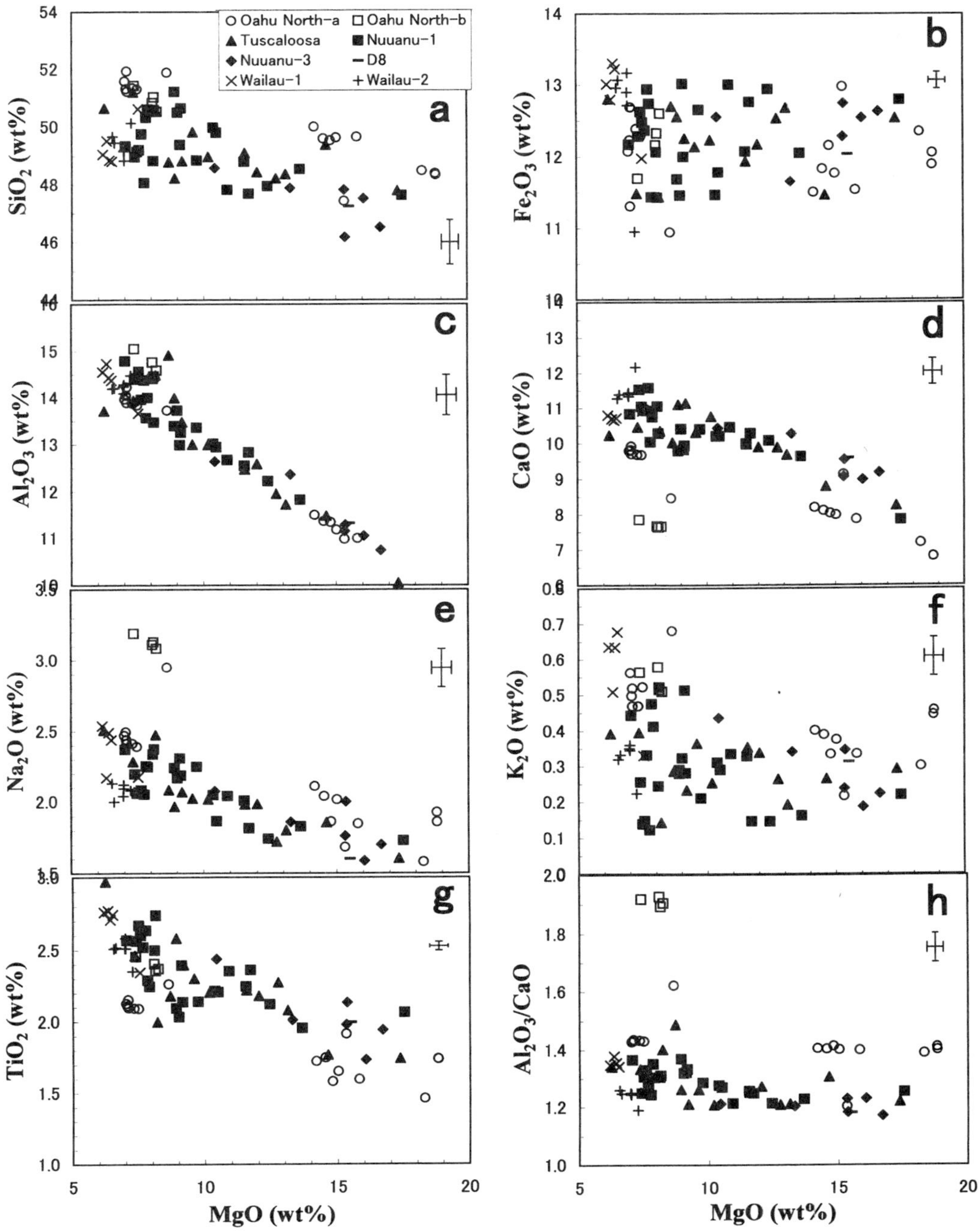

Figure 5. MgO variation diagrams of submarine samples (Appendix, Table 2). Oahu North rocks, open symbols; Nuuanu landslide blocks, solid symbols; Wailau landslide blocks, crosses. Error bars are as in Fig.4.

5h). Large clasts in a hyaloclastite (S496-R6, Figure 2a) have high Na_2O and Al_2O_3/CaO ratios, due to their unique glass chemistry (Figure 7). Trace element and isotope signatures of these clasts demonstrate that they belong to the Koolau-type [*Tanaka et al.*, this volume].

Compositions of rocks from the Nuuanu landslide (closed symbols in Figure 5) and those from the Wailau landslide (crosses) overlap in all the diagrams, but the latter tend to be slightly lower in SiO_2 and higher in FeOt, CaO, and K_2O. These chemical features also characterize Kilauea tholeiite in comparison with Mauna Loa [e.g., *Wright,* 1984; *Frey and Rhodes,* 1993]. The slight compositional difference between Nuuanu and Wailau slide blocks are consistent with the interpretation that East Molokai is a Kilauea-type volcano and that most blocks in the Nuuanu landslide were derived from a Mauna Loa-like volcano.

4.4 Rock Chemistry and Nature of Submarine Outcrops

Yokose [this volume] constructed route maps for each dive site based on videotape images from submersible cameras, reports of the dive observers, and features of the recovered samples. He also classified the outcrops: 1) outcrop exposing interior of the landslide block, 2) local talus deposit, 3) debris-avalanche deposit related to the landslide, 4) unclassified. Because the landslide traveled a great distance, rocks in the debris-avalanche deposits may have been mixed and are not necessarily related to category 1) and 2). In order to evaluate this possibility, rock compositions were plotted by the outcrop classifications of *Yokose* [this volume] (Figure 6). The compositions are similar to modern lavas in Kilauea and Mauna Loa (Figure 4), without systematic differences among rocks of different origin from the same site (Figure 6). Rocks from the Oahu North sites have higher SiO_2 and Al_2O_3/CaO and lower CaO and TiO_2. Compositions from the Nuuanu and Wailau landslide blocks overlap, and most plot in the compositional field of modern Mauna Loa and Kilauea lavas.

5. GLASS CHEMISTRY

5.1 Analytical Procedures

Hyaloclastite is the dominant rock type recovered from the Nuuanu and Wailau landslide blocks (Figure 1). Some large clasts (3-5 cm) in hyaloclastite could be analyzed by XRF; smaller glassy clasts and quenched margins of pillow lavas were characterized by EPMA (electron probe micro analyzer) analysis (Appendix Table 3, Figure 7), using the JEOL-8800 at Tokyo Institute of Technology. The beam current was 1.20×10^{-8} A; counting times were 20 sec for major elements (Si, Al, Ca, Mg and Fe) and 30-60 sec for minor elements (Mn, P, K, Ti, and S). Na was analyzed first in each analysis for 10 sec, to minimize its possible loss during analysis. An internal glass standard (JB-2) was analyzed before and after each batch analysis, and machine drift of the EPMA was regularly monitored (Appendix, Table 3, p. 2). Hydrated glass adjacent to the palagonite rinds that mantle glass fragments yielded low counting rates for all elements (analyses with totals <98 wt%, Appendix, Table 3). Only analyses with totals of 98.0-100.5 wt% were plotted in Figure 7 and used in the following discussion. Analytical uncertainty is 1-2 % for major elements, 5-10% for minor elements.

5.2 Chemical Distinction Between Oahu North Sites and Landslide Blocks

At constant MgO, glass from the Oahu North site (green in Figure 7) is higher in SiO_2 and Na_2O, lower in CaO, TiO_2 and FeO, and notably higher in Al_2O_3/CaO ratio than glass from Nuuanu (blue) and Wailau (red) landslide blocks (Figure 7a-e). The Oahu North site glasses are almost identical to whole-rock composition of subaerial Koolau lavas (marked fields in Figure 7) and are distinct from other Hawaiian shield volcanoes.

Clague et al. [this study] also studied hyaloclastite glass chemistry by EPMA. Although the studied rocks overlap and analytical techniques are similar, the conclusions differ. *Clague et al.* found no systematic difference in the glass SiO_2 wt%, recalculated volatile-free, between Oahu North sites and those in Nuuanu landslide (their Figures 4-5), and concluded that glass chemistry cannot identify source regions of the landslide blocks. In our analyses, however, differences are present for all oxides, not just for SiO_2 (Figure 7). Furthermore, our glass chemical trends are similar to the XRF bulk-rock compositions (cf. Figure 7 with Figures 5-6). Accordingly, rock chemistry is useful for identifying the source regions and reconstruction of the landslides.

5.3 Significance of Hyaloclastite S496-R6

Hyaloclastite sample S496-R6 at the Oahu North-b site has a distinctive glass composition. Matrix glass and clasts from this specimen have low CaO (7.5-8 wt%) and high Al_2O_3/CaO ratios (1.9-2.0) (Figure 7c,e) in both EMPA and XRF analyses (Appendix, Table 2-3) Trace element compositions of this sample plot in the Koolau field, verifying the existence of such melt compositions in Koolau volcano [*Tanaka et al.*, this volume].

The S496-R6 glass composition is important because no similar composition has been previously reported from Hawaiian volcanoes other than Koolau. Among glass fragments in the piston cores, only a turbidite layer at 6-7 m depth in the PC-2 core (250 km NE of Oahu) contains S496-R6 type glasses [*Sherman et al.,* this volume]. According to the magnetic stratigraphy of the cores [*Kanamatsu et al.,* this volume], the turbidite in PC-2 occurred at ca. >1.8 or >2.6 Ma and is most likely the distal facies of the Nuuanu landslide. The presence of S496-R6 type glass in the PC-2 core indicates that the northeastern flank of the Koolau volcano collapsed by the Nuuanu landslide contained hyaloclastite similar to that found at the Oahu North-b site (Figure 8).

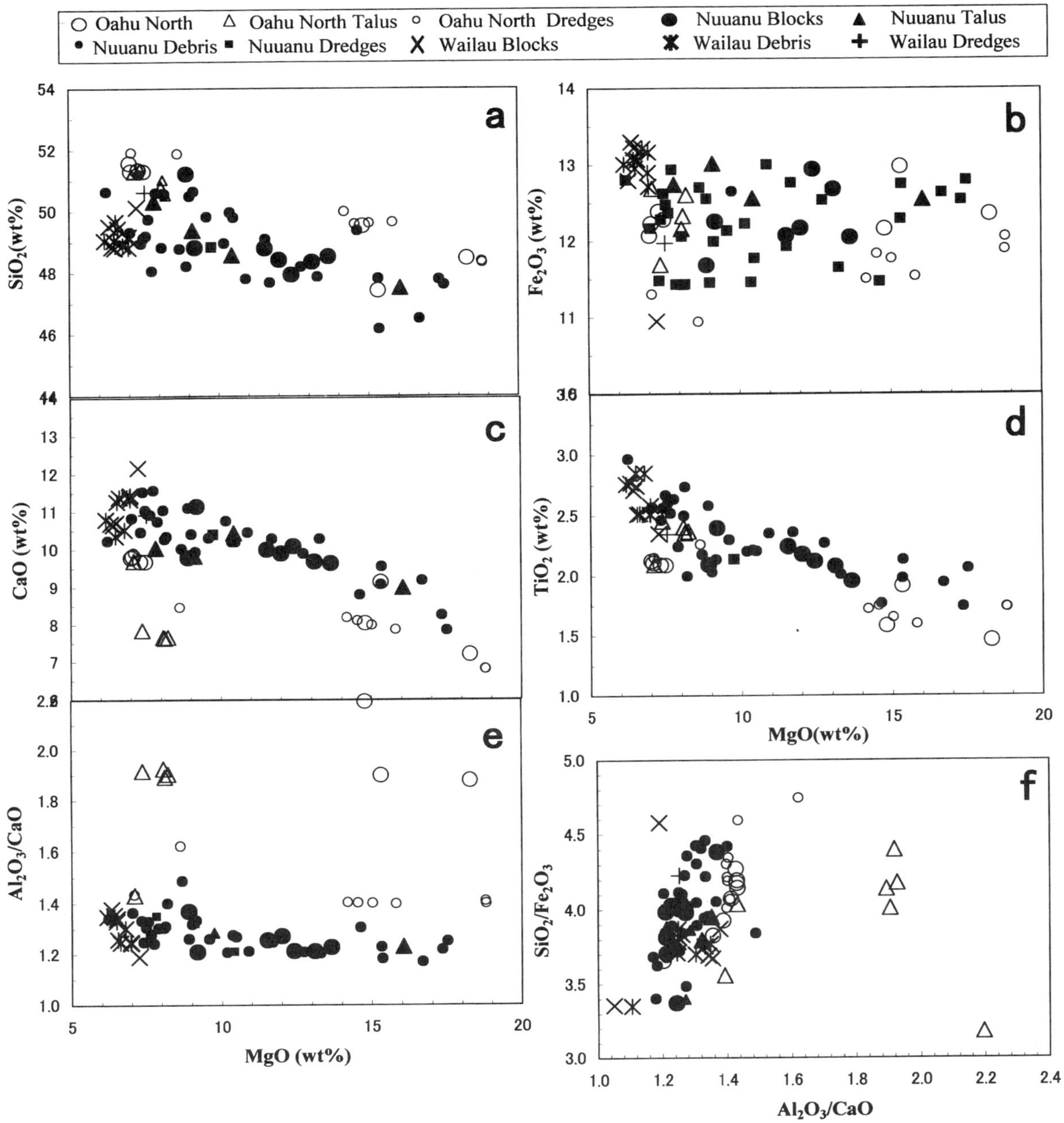

Figure 6. Compositions of submarine specimens (same data set as Fig.5), plotted by nature of outcrop [*Yokose*, this volume]. At any site, rocks from interiors of the landslide blocks do not differ from those in talus or in debris. Error bars are as in Fig.4.

5.4 Transitional Basalt in Wailau Landslide

Glass compositions from the Nuuanu and Wailau landslides are similar and cluster in MgO-variation plots with lava compositions similar to many Hawaiian shield volcanoes (Figure 7a-e). Although hyaloclastites from Nuuanu landslide were collected from a wide area (Nuuanu 1-3: Figure 2), glass compositions are relatively homogeneous compared to those from the Wailau landslide. Wailau glasses include a silica-poor composition (S501-R3), which plots close to the transition between tholeiite and alkalic basalts [*Macdonald and Katsura*, 1964]. The silica-poor glasses are also low in sulfur (Figure 7f), indicating their subaerial origin. Thus they are unlikely to be alkalic rocks of the pre-shield stage [e.g., *Sisson et al.,* this volume]. Wailau slide blocks, therefore, include the late alkalic cap of East Molokai volcano.

6. DISCUSSION

6.1 Hyaloclastite Abundance

Hyaloclastite is the dominant rock on the submarine slide blocks, except for the Nuuanu-1 site. The abundance of the hyaloclastite in both Nuuanu and Wailau blocks can be understood from the shield-volcano model of *Moore and Fisk* [1969]. In this model, the flanks of the Hawaiian volcanoes consist largely of glassy volcanic sandstone and breccia, derived from ocean entry of subaerially erupted lava. The on-going Puu Oo eruption, which is producing enormous amounts of volcanic glass at the shoreline, documents the formation of hyaloclastites on the flank of Kilauea volcano.

The dominance of hyaloclastite in the study area is consistent with the structure of collapsed volcanoes. Both Koolau and East Molokai volcano contain rift zones running parallel to the landslide detachment plane (NW-SE for Koolau [*Stearns and Vaksvik,* 1935] and W-E for East Molokai [*Stearns and Macdonald*, 1947]). Accordingly, the collapsed flank of each volcano is expected to consist of volcaniclastic rocks and subordinate pillow lavas [*Moore and Fisk,* 1969].

The abundant hyaloclastites may be a factor in the instability of the volcano and the gigantic landslides [*Moore and Clague,* this volume]. During transport of the slide blocks, their basal should have eroded by friction. Hyaloclastites may have increased mobility by providing slip planes. Transport of the largest blocks over long distances (to 150 km for Nuuanu) and up-slope [*Moore et al.*, 1989] must be related to a low frictional stress at bases of the sliding blocks.

6.2 Age of Nuuanu and Wailau Landslides

All the submarine samples have Mn coatings of variable thickness (Appendix, Table-1). Since the Mn coatings grow in the deep-water environment over time, their thickness can be related to the age of the outcrops. Thickness of Mn coating may be dependent on 1) duration of outcrop exposure (exposure age), 2) ocean depth, 3) water temperature, 4) depositional rate of abyssal clay, etc. The maximum thickness of the Mn coating on samples from the Oahu North site is greater than 6 mm, similar to the rocks on top of Tuscaloosa Seamount. In contrast, the maximum Mn thickness on rocks from the Wailau landslide is less than 3 mm in all specimens. These results are consistent with the estimated ages of the Nuuanu landslide (>1.8 Ma) and Wailau landslide (~1 Ma) by *Kanamatsu et al.* [this volume, their model 2].

Clague et al. [this volume] made similar observations of Mn thickness as well as EPMA glass analysis of hyaloclastite. The two studies, however, make different estimates on the timing of the Wailau landslide. *Clague et al.* estimated that Wailau landslide took place at the late shield stage of East Molokai volcano when both transitional and tholeiite basalts were active (1.5 Ma). This is based on their interpretation that tholeiite pillow lava at the top of Wailau-1 block (S510-R6) erupted after the landslide. According to their estimate judged from CO_2 content in the rim glass, the pillow lava now is 1450 m deeper than its eruption depth, due to later subsidence of the volcano. However, the Wailau-1 block is well beyond the East Molokai edifice (see Figure 1), and we think their explanation is difficult.

Alternatively, we interpret that the pillow lava on Wailau-1 block as erupted prior to the Wailau landslide and transported to its position by the landslide. The intact lava morphology might be due to slow slumping, rather than catastrophic slide motion. Presence of transitional basalt (e.g., S501-R3) in the Wailau landslide blocks supports post-shield formation of the landslide (~1 Ma).

6.3 Distribution of Nuuanu and Wailau Landslide Blocks

The primary goal of this study is to identify source regions for the landslide blocks. Although the available sample localities are limited relative to the numerous slide blocks, boundaries seem reasonable for three major zones (Oahu North, Nuuanu, Wailau) in the study area (Figure 8). Unexpectedly, rocks in the Nuuanu blocks differ from lavas exposed on Koolau volcano. Compositions of lava and hyaloclastite in the Nuuanu landslide are broadly similar to modern Mauna Loa. Because these compositions overlap with Wailau landslide rocks, discrimination of landslide sources from chemistry is difficult

In contrast, rock compositions obtained from areas undisturbed by the landslides (Oahu North and D-8 sites) differ from those in the landslide blocks. The rocks from these sites are directly comparable to subaerial parts of the volcanoes (Koolau for Oahu North, West Molokai for D-8).

6.4 Growth History Koolau Volcano

The difference between rocks of Nuuanu landslide and subaerial Koolau volcano permit two possibilities; 1) Nuuanu landslide was derived from a missing volcano with Mauna Loa-like chemistry located northeast of Koolau vol-

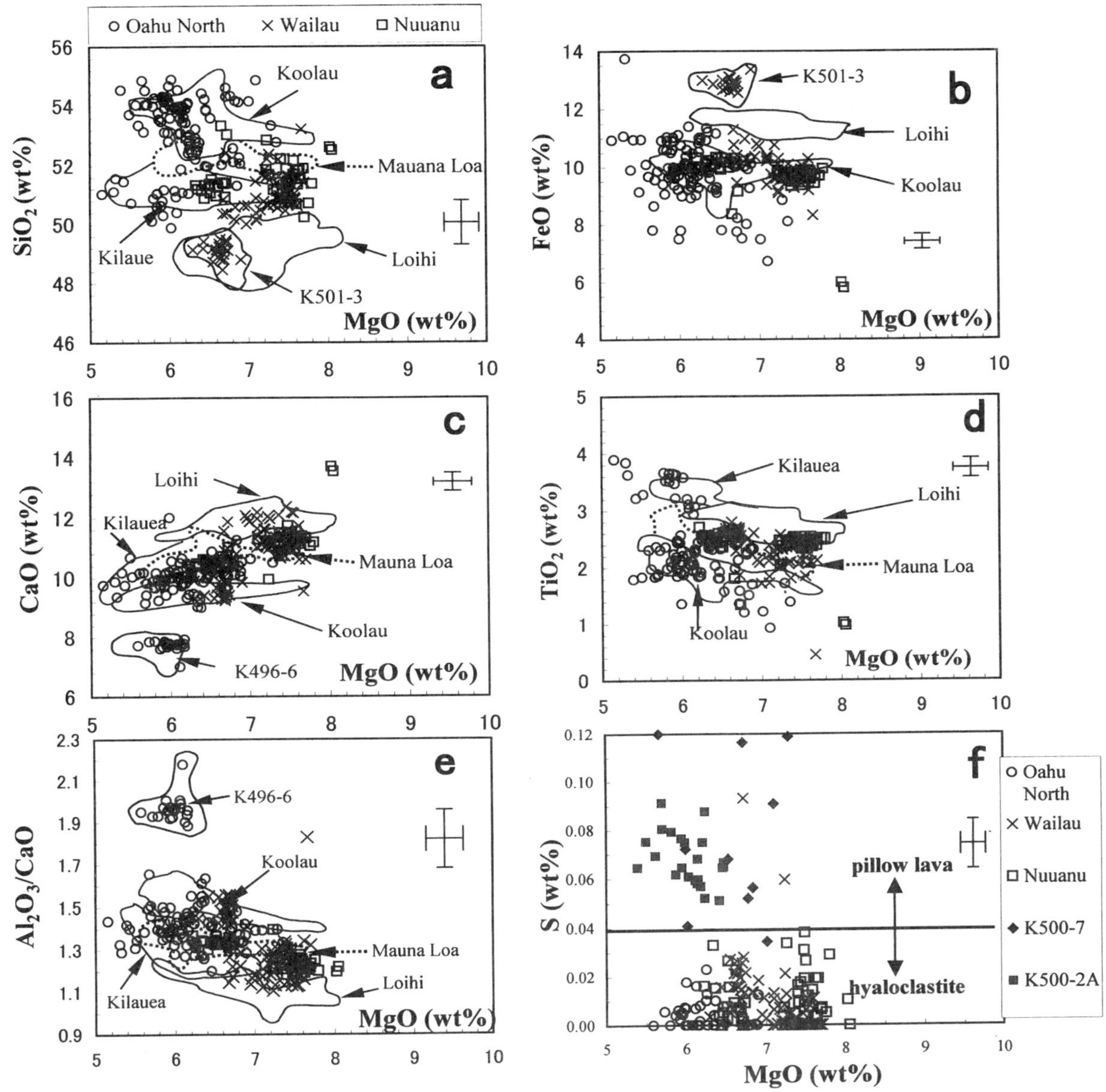

Figure 7. Composition of glass fragments in hyaloclastite and rims of pillows. Glasses in pillow lavas, filled symbols; those in hyaloclastites, open symbols. Compositional fields for Koolau, Mauna Loa, Kilauea, and Loihi glass fields are from *Frey et al.,* [1994]. Error bars are based on calculated of 2 σ deviations for JB-2

cano, or 2) Nuuanu blocks are from the interior of Koolau volcano that has multiple growth stages (at least Mauna Loa-type, followed by Koolau-type).

Topographic reconstructions show that volume of the Nuuanu landslide (2.8-3.8x10^3 km^3) can be accommodated by the amphitheater (missing mass) of Koolau volcano [*Satake et al.*, this volume, their Figure 3], so that missing volcano hypothesis is unnecessary. Furthermore, the presence of S496-R6 type glass in a distal facies of Nuuanu landslide (PC2) indicates that a relatively thin layer of Koolau-type material covered the north flank of Koolau before landsliding (Figure 8a).

The sequential growth history of Koolau volcano is further clarified by isotope data [*Tanaka et al.*, this volume]. Pb isotope compositions change progressively from Kilauea type to Mauna Loa type from Nuuanu-1, to Tuscaloosa, to

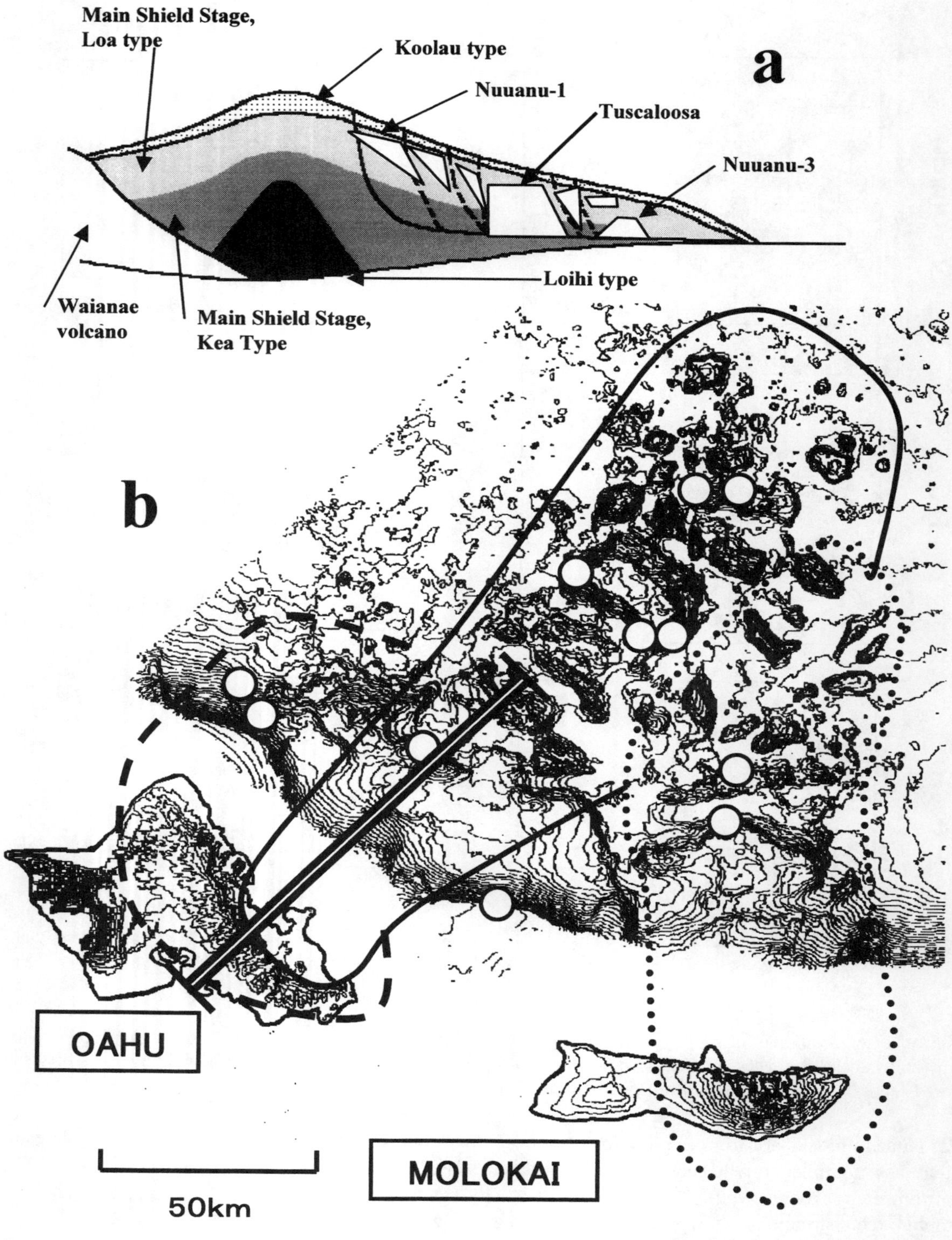

Figure 8. Cross section of Koolau volcano before Nuuanu landslide (a), and map distribution of the landslide blocks (b) showing boundaries between Nuuanu (solid line) and Wailau slides (dotted). Dashed line, inferred pre-slide shape of Koolau volcano

Nuuanu-3 blocks [*Tanaka et al.*, this volume, their Figure 2]. Furthermore, the progressive change is a function of stratigraphic level within given slide blocks (see Pb isotope data for Nuuanu-1, Tuscaloosa, and Nuuanu-3 blocks). These results indicate that Koolau lavas were stratified upward from Kilauea-type to Koolau-type (Figure 8a).

Additional fieldwork to study the change from Mauna-Loa-like and typical Koolau lavas on Oahu located this transition near the north base of Nuuanu Pali (at about 300 m elevation), confirming the results of *Jackson et al.* [1999] from the Trans-Koolau Tunnel. The transition was also identified at a lower horizon along the Puu Manamana hiking trail. No unconformity was observed in the lava succession, and the Nd, Sr, and Pb isotopes of lava in the transition zone shift from Mauna Loa type to Kilauea type in a narrow interval [~10m: *Tanaka et al.,* this volume]. Furthermore, *Haskins and Garcia* [2000] found a transition from Koolau-type to Mauna Loa-type lavas in drill core at Kalihi valley on the northern flank of Koolau volcano.

The strike and dip of lava in the Koolau shield [*Stearns and Maconald,* 1947] suggest that the Mauna Loa and the Koolau-type parts may be stratified as shown in Figure 8a. The transition is sharp, and chemical and isotopic compositions in the upper 1000 m of lava in the Koolau shield are relatively homogeneous (Figure 2 of *Tanaka et al.* [this volume]). Taking the well studied Makapuu Point at the eastern end of Oahu [*Frey et al.,* 1994] as a type locality for the late-stage products ("Koolau-type"), we name these as *Makapuu stage* and the lower Mauna-Loa-like and Kilauea-like lavas as *main shield stage* of Koolau volcano. Previously, only the *Makapuu stage* had been recognized, but the *main shield stage* is far larger in volume (Figure 8a). All the landslide blocks thus far studied in the Nuuanu landslide are main shield stage, whereas the rocks on the northern submarine slope (Oahu North-a, -b) formed during the Makapuu stage. The Makapuu stage lava is approximately 1000 m thick at the summit caldera (as exposed on Nuuanu Pali), and its volume would be of the order of 500-1000 km^3.

7. CONCLUSIONS

1.Hyaloclastite is the dominant rock type in submarine landslide blocks except at the Nuuanu-1 site.

2.The thickness of Mn coatings indicate that the Nuuanu landslide is much older than Wailau.

3.Most rocks from the submarine slope north of Oahu (Oahu North sites) show chemical compositions identical to the subaerial Koolau volcano.

4.Compositions of the rocks in the Nuuanu landslide blocks are different from those in the subaerial Koolau volcano but are similar to modern Mauna Loa.

5.Rocks in the Wailau landslide, derived from East Molokai volcano, are similar to those in Nuuanu but slightly poorer in SiO_2.

6.Sulfur contents of pillow lava indicate submarine eruption of the pillows at their present locations; the submarine slope of the Oahu North sites is undisturbed by landslide displacements.

7.Koolau volcano which supplied Nuuanu landslide has experienced a multi-stage compositional growth; from Kilauea-type, to Mauna Loa-type, and finally Koolau-type.

Acknowledgments. The 1998-99 Hawaii cruises were conducted by Jiro Naka (Chief Scientist) and a joint Japanese and US scientist team. We acknowledge collaborations of Drs. M. Garcia, J. Moore, D. Clague, H. Yokose, T. Kanamatsu, and J. Naka on the ship. Drs. R. Tanaka and E. Nakamura of ISEI worked with us on geochemical analyses, and Dr. H. Yokose on submarine geology. Drs. M. Garcia, S. Sherman, R. Tanaka, and H. Yokose reviewed early version of manuscript. The present manuscript benefited significantly by scientific reviews of Drs. T.L. Wright and J.M. Rhodes and thorough editorial works by Dr. P. Lipman.

REFERENCES

Beeson, M.H., Petrology, mineralogy, and geochemistry of the East Molokai Volcanic Series, *U.S. Geol. Surv Prof. Paper 961*, 53 p., 1976.

BVSP, *Basaltic Volcanism on the Terrestrial Planets*, 1286 pp., Pergamon Press, Oxford, 1981.

Clague, D.A., C. Dao-Gong, R. Murnane, M.H. Beeson, M. Lanphere, G.B. Friesen, and R.T. Holcomb, Age and petrology of the Kalaupapa Basalt, Molokai, Hawaii, *Pac. Sic., 36,* 411-420, 1982.

Frey, F.A., M.O. Garcia, and M.F. Roden, Geochemical characteristics of Koolau volcano: implications of intershield geochemical differences among Hawaiian volcanoes, *Geochim. Cosmocim. Acta., 58*, 1441-1462, 1994.

Frey, F.A., and J.M. Rhodes, Intershield geochemical differences among Hawaiian volcanoes: implications for source compositions, melting process and magma ascent paths, *Phil. Trans. R. Soc. Lond. A, 342*, 121-136, 1993.

Garcia, M.O., Pliocene-Pleistocene volcanic sands from site 842: products of giant landslide, in R.H. Wilkens, et al., eds., *Proceedings of the Ocean Drilling Program, Scientific Results, Vol. 136*, 53-63, 1993.

Garcia, M.O., T.P. Hulsebosch, and J.M. Rhodes, Olivine-rich submarine basalts from the Southwest rift zone of Mauna Loa volcano: Implications for magmatic processes and geochemical evolution, in *Mauna Loa Revealed: Structure, Composition, History and Hazards, Geophys. Monograph 92*, 219-239, 1995.

Haskins, E.H., and M.O. Garcia, Scientific drilling of a Hawaiian volcano: Dose Koolau have a skin? *EOS Trans. Am. Geophy. Union, 81*, No. 48, F1349, 2000.

Jackson, M.C., F.A. Frey, M.O. Garcia, and R.A. Wilmoth, Geology and petrology of basaltic lavas and dikes of the Koolau Volcano in the Trans-Koolau exploratory tunnels, *Bull. Volcan., 60*, 381-401, 1999.

Kanamatsu, T., E. Herrero-Bervera, and G.M. McMurtry, Magnetostratigraphy of deep-sea sediments from areas adjacent to the Hawaiian Islands: Implication for ages of turbidites derived from submarine landslides, *this volume*, 2001.

Lipman, P.W., T.W. Sisson, T. Ui, J. Naka, J. Morgan, J. Smith, Ancestral submarine growth of Kilauea Volcano and instability of its south flank, *this volume,* 2001.

Macdonald, G., and T. Katsura, Chemical composition of Hawaiian lavas, *J.Petrol., 5*, 82-133, 1964.

Moore, J.G., Giant submarine landslides on Hawaiian Ridge, *U.S.Geol. Surv. Prof. Pap., 501-D*, D95-D98, 1964.

Moore, J.G., and R.S. Fisk, Volcanic structure inferred from dredge samples and ocean-bottom photographs, Hawaii, *Geol. Soc. Amer., Bull., 80,* 1191-1202, 1969.

Moore, J.G., and D.A. Clague, Volcano growth and evolution of the island of Hawaii, *Geol. Soc. Amer., Bull., 104*, 1471-1484, 1992.

Moore, J.G., and D.A. Clague, Mapping the Nuuanu and Wailau landslides in Hawaii, *this volume,* 2001.

Moore, J.G., D.A. Clague, R.T. Holcomb, P.W. Lipman, W.R. Normak, and M.E. Torreson, Prodigious submarine landslide on the Hawaiian Ridge, *J. Geophys. Res., 94,* 17465-17484, 1989.

Morgan, J., G.F. Moore, D.J. Hills, and S. Leslie, Overthrusting and sediment accretion along Kilauea's mobile south flank, Hawaii: Evidence for volcanic spreading from seismic reflection data, *Geology, 28*, 667-670, 2000.

Roden, M.F., F.A. Frey, and D.A. Clague, Geochemistry of tholeiitic and alkalic lava from the Koolau range, Oahu, Hawaii: implications for Hawaiian volcanism, *Earth Planet. Sci. Lett., 69,* 141-158, 1984.

Roden, M.F., T. Trull, S.R. Hart, and F.A. Frey, New He, Sr, Nd and Pb isotopic constraints on the constitution of the Hawaiian plume: Results from Koolau Volcano, Oahu, Hawaii, *Geochim. Cosmochim. Acta, 58*, 1431-1440, 1994.

Satake, K., J.R. Smith, and K. Shinozaki, Volume estimate and tsunami modeling for the Nuuanu and Wailau landslides, *this volume,* 2001.

Sherman, S., M.O. Garcia, and E. Takahashi, Geochemistry of volcanic glasses from piston cores taken north of Oahu and Molokai Islands, *this volume,* 2001.

Sisson, T.W., P.W. Lipman, and J. Naka, Submarine alkalic through tholeiitic shield-stage development of Kilauea volcano, Hawaii, *this volume,* 2001.

Stearns, H.T., and K.N. Vaksvik, Geology and Ground-Water resources of the Island of Oahu, Hawaii, *Territory of Hawaii, Division of Hydrography, Bull 1*, 479pp., 1935.

Stearns, H.T., and G. Macdonald, Geologic map of Molokai island, Hawaii, *U.S. Geologicl Survey*, 1947.

Takahashi, E., and K. Nakajima, Melting processes in the Hawaiian mantle plume head: an experimental study, *this volume,* 2001.

Takeguchi, R., Primary magma composition of Koolau volcano, Hawaii, *MsC Thesis, Tokyo Institute of Technology,* 84pp. March, 1999.

Tanaka, R., E. Nakamura, and E. Takahashi, Geochemical characteristics of Koolau volcano and the Nuuanu and Wailau landslide deposits, Hawaii, *This volume,* 2001.

Wright, T.L., Origin of Hawaiian Tholeiite: A metasomatic model, *J. Geophys. Res., 89,* 3233-3252, 1984.

Yokose, H., Submarine geology of giant landslides in the Northeast Oahu: SHINKAI 6500 submersible investigation, *this volume,* 2001.

Ken Shinozaki, now at Schlumberger K.K., Data Management Center, Fuchinobe, Kanagawa Prefecture, Japan, sinozaki@tokyo.oilfield.slb.com

Zhong-Yuan Ren, Earth and Planetary Sciences, Tokyo Institute of Technology, 2-12-1 Ookayama, Meguro-ku, Tokyo 152-8551, Japan, ren@geo.titech.ac.jp

Eiichi Takahashi, Earth and Planetary Sciences, Tokyo Institute of Technology, 2-12-1 Ookayama, Meguro-ku, Tokyo 152-8551, Japan, etakahas@geo.titech.ac.jp (corresponding author)

Geochemical Evolution of Koolau Volcano, Hawaii

Ryoji Tanaka and Eizo Nakamura

The Pheasant Memorial Laboratory for Geochemistry and Cosmochemistry (PML), Institute for Study of the Earth's Interior, Okayama University at Misasa, Japan

Eiichi Takahashi

Earth and Planetary Sciences, Tokyo Institute of Technology, Tokyo, Japan

Major and trace-element and Sr, Nd, and Pb isotopic compositions have been determined for Koolau shield lavas and Nuuanu and Wailau landslide deposits, Hawaii. The Koolau shield is composed of older main shield-stage and younger Makapuu-stage lavas. Lavas of the main shield-stage are exposed low in the subaerial Nuuanu Pali and in Nuuanu landslide deposits. Main shield-stage lavas resemble Loa- or Kea-type compositions; they are characterized by lower Ba/Nb, La/Nb, Pb/Nd, Sr/Nb, Nd/Nb, Zr/Nb, and $^{87}Sr/^{86}Sr$, and by higher $^{143}Nd/^{144}Nd$ and $^{206}Pb/^{204}Pb$ than those of Makapuu-stage lavas (Koolau-type). Sr-Nd-Pb isotope systematics suggest that Koolau shield lavas were generated from at least three endmembers, such as: Enriched Makapuu (EMK), Depleted Makapuu (DMK), and Kea (KEA) components. The proportion of these endmembers successively changed from earlier KEA-dominated to later EMK-dominated. Koolau shield-stage lavas span nearly the entire range of the isotopic compositions reported previously for shield-stage lavas from all Hawaiian islands, and reveal secular changes in the amounts of plume components.

1. INTRODUCTION

Hawaiian volcanoes are the most comprehensively studied hot spot tracks on the earth. It is broadly accepted that the Hawaiian plume is composed of several components, such as depleted upper mantle and slab material recycled through subduction zones deep into the mantle [e.g. *Tatsumoto,* 1978; *Hoffman and White,* 1982; *Stille et al.,* 1986b; *Hauri,* 1996; *Eiler et al.,* 1996; *Lassiter et al.,* 1996; *Lassiter and Hauri,* 1998]. Among the Hawaiian shields, Koolau lavas are the most geochemically and isotopically enriched, and their compositions are referred to as the "Koolau component" or "Koolau endmember", of the Hawaiian plume [e.g. *Stille et al.,* 1986a, b] (in this paper, hereafter denoted as the Koolau-type). Lanai and Kahoolawe shield lavas also belong to this type [*West et al.,* 1987; *Basu and Faggart,* 1996]. The Koolau-type is characterized by the highest SiO_2, $^{87}Sr/^{86}Sr$, and $^{187}Os/^{188}Os$, and the lowest $^{143}Nd/^{144}Nd$ and $^{206}Pb/^{204}Pb$ in Hawaiian shields [*Tatsumoto,* 1978; *Hoffman and White,* 1982; *Stille et al.,* 1986a, b; *Hauri,* 1996; *Eiler et al.,* 1996; *Lassiter and Hauri,* 1998]. Such geochemical characteristics have led to the interpretation that Koolau-type reflects contributions from recycled oceanic crust in its source [*Tatsumoto,* 1978; *Stille et al.,* 1986b; *Hauri,* 1996; *Lassiter and Hauri,*

Hawaiian Volcanoes: Deep Underwater Perspectives
Geophysical Monograph 128
Copyright 2002 by the American Geophysical Union

1998]. Another endmember composition among the Hawaiian shields is represented by the Mauna Kea shield and Kilauea lavas, called Kea-type [*Stille et al.*, 1986b]. The Kea-type is characterized by the lowest $^{87}Sr/^{86}Sr$ and $^{187}Os/^{188}Os$, and the highest $^{143}Nd/^{144}Nd$ and $^{206}Pb/^{204}Pb$ in Hawaiian shields [*Tatsumoto,* 1978; *Stille et al.*, 1986b; *Hauri,* 1996; *Lassiter and Hauri,* 1998]. The Sr, Nd, and Pb isotopic compositions of the Kea-type are closest to the depleted upper mantle composition among Hawaiian shield lavas, suggesting enrichment of recycled lithospheric mantle [*Stille et al.*, 1986b] or entrained upper mantle [*Hauri,* 1996; *Lassiter et al.*, 1996] in the mantle plume. However, Os and O isotopes are distinct from the MORB-field and indicate the involvement of recycled, hydrothermally altered oceanic crust or lithospheric mantle components [*Eiler et al.*, 1996; *Lassiter and Hauri,* 1998]. Most other shield lavas, such as Mauna Loa, Haleakala, and West Molokai have compositions between these two endmembers [Stille et al., 1986a] (lavas with these intermediate compositions are hereafter referred to as the Loa-type).

Although many geochemical data have been reported from the Hawaiian shields, most are restricted to near-surface samples because only a few hundred meters of lavas out of a total thickness of 6-20 km have been exposed by faulting and erosion in the <5Ma Hawaiian shields. Thus, our knowledge regarding the geochemical characteristics of each shield could be biased by disproportionate sampling of the last-shield magmatism. Two methods allow access to the shield's interior: drilling such as the Hawaii Scientific Drilling Project (HSDP) in the Mauna Kea shield [*Stolper et al.*, 1996], and underwater samples from landslide deposits. Samples Koolau volcano previously have been collected only from a restricted stratigraphy in the shallow part of the shield [*Stille et al.*, 1986a, *Roden et al.*, 1984, 1994; *Frey et al.*, 1994]. Recently, *Jackson et al.* [1999] found one sample at the bottom of the Trans-Koolau tunnel section that shows Kea-type $^{87}Sr/^{86}Sr$ and $^{143}Nd/^{144}Nd$ compositions, and they suggested that the source composition of the Koolau shield varied during its growth. Therefore, the geochemical evolution of Koolau shield from its base to its top might have important information for the evolution of the Hawaiian plume.

During 1998-1999, JAMSTEC-Hawaii cruises constructed a new bathymetric map and collected samples from the widespread Nuuanu and Wailau landslide blocks offshore to the north of Oahu and Molokai [*Smith and Satake,* this volume] (Figure 1). Geological studies indicate that the Nuuanu blocks (Nuuanu-1, 2, and 3) and Wailau blocks (Wailau 1 and 2) are landslide deposits derived from Koolau and East Molokai shields, respectively [*Shinozaki et al.*, this volume; *Yokose,* this volume]. Samples include subaerial- and submarine-erupted lava fragments in volcanic breccia or conglomerate in Nuuanu and Wailau blocks, as well as autochthonous pillow lavas at the Oahu North site [*Satake et al.*, this volume; *Shinozaki et al.*, this volume; *Yokose,* this volume]. The phenocryst assemblages and major element compositions of lava fragments and hyaloclastite glasses from the Nuuanu landslide deposits, however, range more widely in composition than in the subaerial Koolau shield [*Shinozaki et al.*, this volume]. The Nuuanu landslide deposits therefore provide samples of the interior of the Koolau shield.

In this paper, we document the major and trace element abundances, and Sr, Nd, and Pb isotopic compositions of basaltic fragments in the Nuuanu landslide deposits, as well as autochthonous subaerial and submarine Koolau shield lavas, in order to interpret the geochemical evolution of the Koolau shield. In addition, we examine the geochemical characteristics of East and West Molokai shield from submarine-erupted lavas in Wailau landslide deposit and in the northwestern flank of the West Molokai shield.

2. GEOLOGICAL BACKGROUND AND SAMPLE DESCRIPTIONS

Koolau volcano consists of shield-stage tholeiitic lavas (1.8-2.7 Ma; *Doell and Dalrymple,* [1973]), overlain by rejuvenated-stage alkaline lavas of the Honolulu Volcanics (<0.6 Ma; *Lanphere and Dalrymple,* [1980]). Two rift zones extend to the northwest and east-southeast from the caldera. The shield-stage lavas are composed of at least two stages, from older to younger: main shield-stage and Makapuu-stage [*Shinozaki et al.*, this volume] (Figure 2, Table 1). Makapuu-stage is the highest shield lavas exposed in an approximately 800 m thick section from the top of the shield [*Takeguchi,* 1999; Figure 10 in *Takahashi and Nakajima,* this volume], and all subaerial samples of the shield previously studied [e.g. *Macdonald and Katsura,* 1964; *Roden et al.*, 1984; *Stille et al.*, 1986a; *Frey et al.*, 1994; *Takeguchi,* 1999] are from the Makapuu-stage. Koolau-type lava is, thus, equivalent to Makapuu-stage. Subaerial-erupted or shallow submarine-erupted pillow lavas and hyaloclastite of the Makapuu-stage occur in the northwestern flank of the shield (Oahu North site) [*Shinozaki et al.*, this volume]. Samples from the Oahu North site are autochthonous pillow lavas in the upper part (Oahu North-a), and lava fragments in volcanic breccia and hyaloclastite in the lower part (Oahu North-b). Olivine is the dominant phenocryst in Makapuu-stage lavas. Subordinate orthopyroxene phenocrysts are a unique petrographical characteristic of Koolau-type lavas among the

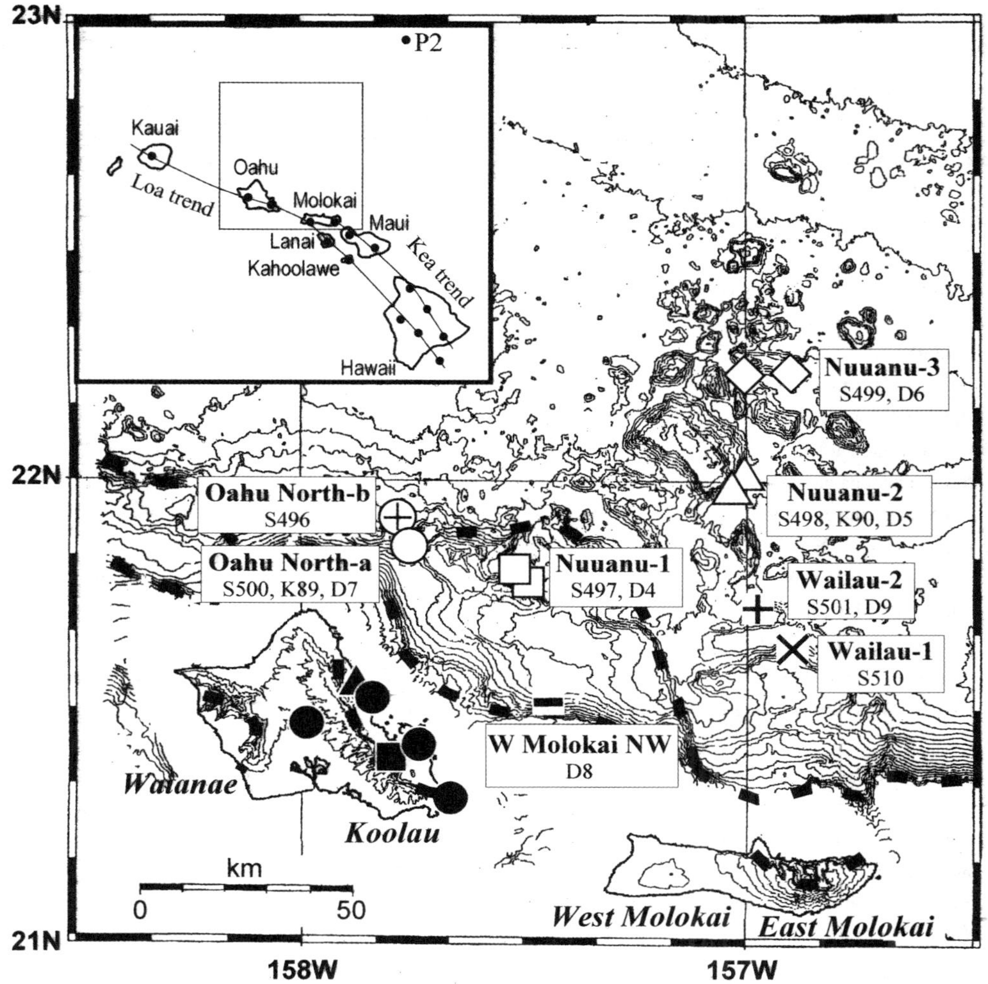

Figure 1. Location map showing analyzed Koolau subaerial lavas: SHINKAI dive sites (S-), ROV-KAIKO dive sites (K-), and dredge sites (D-) [modified after *Yokose,* this volume]. Piston core site PC2 is 350 km northeast of Oahu. The thick dashed line shows faults and scarps [*Yokose,* this volume]. Sample symbols are used in all subsequent plots. Lines of Kea and Loa volcanic trends (inset map) are after *Jackson et al.* [1972].

Hawaiian shields, as is well known [*Frey et al.,* 1994; *Takeguchi,* 1999].

Main shield-stage lavas have been found in Nuuanu landslide blocks [*Shinozaki et al.,* this volume], but not from the subaerial Koolau shield. Nuuanu blocks (Nuuanu 1-3) are landslide deposits derived from the main shield-stage of Koolau [*Shinozaki et al.,* this volume]. An interpreted stratigraphic section of the Nuuanu blocks indicates that the upper part of the main shield-stage is dominated by subaerial-erupted lava fragments (Nuuanu-1 and upper part of Nuuanu-2); the lower part of the stage is dominated by shallow submarine-erupted lava fragments (lower part of Nuuanu-2 and Nuuanu 3) [*Shinozaki et al.,* this volume; *Yokose,* this volume]. Phenocryst assemblages and major element compositions of main shield-stage lavas are similar to those of Mauna Kea or Mauna Loa [*Shinozaki et al.*, this volume]. The uppermost main shield-stage lavas near the base of Nuuanu Pali are Loa- or Kea-type, as is the stratigraphically lowermost lava in the Trans-Koolau (T-K) tunnel [*Jackson et al.* 1999]. These lavas contain variable amounts of olivine, clinopyroxene, and plagioclase phenocrysts (Table 2). Orthopyroxene phenocrysts only occur in LK4, with a modal abundance of 0.1 vol.%.

Based on the few geochemical data available for East and West Molokai shields, *Stille et al.* [1986b] inferred that East and West Molokai belong to Kea- and Loa-type,

Table 1. Occurrence and petrography of the analyzed samples

Sample No.	Occurrence and rock type	elevation (m)	vesicular (%)	phenocryst	alteration
Subaerial samples of the Koolau shield					
Subaerial Makapuu stage (●)					
WAF-2	Wheeler Air Force drilling core	(-54.6)	5*	oliv, opx	A
WAF-36	Wheeler Air Force drilling core	(-180.3)	4*	oliv	B2
WAF-39	Wheeler Air Force drilling core	(-194.3)	19*	oliv, opx, pl	B2
KOL43	Maakua Gulch dike	80	2*	-	A
KOL48	Maakua Gulch dike	80	<1*	oliv	A
KOL14	Makapuu Head lava flow	80	4*	oliv, opx	B1
MH-11	Makapuu Head lava flow	104	5*	oliv	B1
Lowermost Makapuu stage (▲)					
PM1	Puu Manamana elevation	61	14*	oliv, pl	B2
PM7	Puu Manamana elevation	255	6*	oliv, opx, pl	B2
Uppermost main shield stage (■)					
LK1	Nuuanu Pali lava flow	*260*	30*	oliv	B2
LK2	Nuuanu Pali lava flow	*266*	22*	oliv, pl	B2
LK3	Nuuanu Pali lava flow (bolder)	*262*	26*	oliv, cpx, pl	B2
LK4	Nuuanu Pali lava flow	*250*	13*	oliv, cpx, pl	B2
Submarine samples in landslide deposits as well as autochthonous pillow lavas					
Oahu North-a (○)					
K89-6	autochthonous picritic pillow lava	-2800	4*	oliv	A
D7-1	autochthonous aphyric pillow lava	-3000	4*	oliv, opx	A
S500-1	picritic pillow lava in volcanic breccia	-2980	1	oliv	A
S500-5B	picrite lava	-2696	1	oliv	A
S500-9A	fragment of pillow lava (aphyric basalt)	-2602	1-2	oliv	C1
Oahu North-b (⊕)					
S496-6C	aphyric basalt fragment in talus volcanic breccia	-3571	<1*	oliv	C2, D1
S496-6E	aphyric basalt fragment in talus volcanic breccia	-3571	<1*	oliv	C2, D1
Nuuanu-1 (□)					
D4-1	subaerial lava fragment in volcanic breccia	-3000	26*	oliv, pl	B2, C1
D4-2	subaerial lava fragment in volcanic breccia	-3000	27*	oliv	B2
S497-3	olivine basalt fragment in conglomerate	-3657	5	oliv	B2, D1
S497-6	picritic basalt lava	-3657	24	oliv	B2
S497-7	olivine basalt fragment in conglomerate	-3657	5-30	oliv	B2, D1
Nuuanu-2 (△)					
K90-1	subaerial lava	-3800	26*	oliv, pl	B2
K90-3	subaerial lava	-3000	23*	oliv, pl	B2
D5-18	dike	-4000	0*	pl	B2
S498-4	olivine rich basalt in autobrecciated lava	-4383	0	oliv, cpx	D2
Nuuanu-3 (◇)					
D6-1	pillow lava		<1*	oliv	B1
S499-6A	olivine basalt fragment in conglomerate	-3719	15	oliv	B2
S499-6B	olivine basalt fragment in conglomerate	-3719	7*	oliv	B2

Table 1. (continued)

Sample No.	Occurrence and rock type	elevation (m)	vesicular (%)	phenocryst	alteration
Wailau-1 (×)					
S501-4	basalt lava fragment in conglomerate	-4189	2	oliv, pl, cpx	D1
S501-7	autobrecciated lava	-4006	0	oliv	C1
Wailau-2 (+)					
D9-1			3*	oliv, pl	A
S510-6A	olivine basalt pillow lava	-3615	>2	ol, pl	B1
S510-6C	olivine basalt pillow lava	-3615	<1	ol, pl	C1
West Molokai NW rift zone (−)					
D8-1	pillow lava	-2800	1*	oliv, cpx, pl	B1, D2

Samples of subaerial Makapuu stage lavas are from *Takeguchi* [1999] and of subaerial main shield stage are collected by K. Shinozaki and R. Tanaka. Submarine samples denoted D is dredged samples in 1998, K is collected by Kaiko dive in 1998, and S is collected by Shinkai 6500 dive in 1999. Occurrence and sample descriptions of submarine samples are after *Shinozaki et al.* [this volume]. The vesicular % with asterisk is counted by modal analysis (2000-3000points) in this study. Symbols in parentheses for each location is same as those used in Fig. 1 to Fig. 7. Abbreviations for phenocrysts; oliv: olivine, cpx: clinopyroxene, opx: orthopyroxene, pl: plagioclase. Criteria for alteration; A: fresh, B1: fresh but some olivine has iddingsite rim, B2: all olivine have iddingsite rim, C1: secondary zeolite slightly crystalized along the vesicule; C2: secondary zeolite moderately crystallized along the vesicule; D1: some mafic mineral has alteration rim; D2: matrix glass is slightly replaced by palagonite. Symbols shown are corresponding to figures.

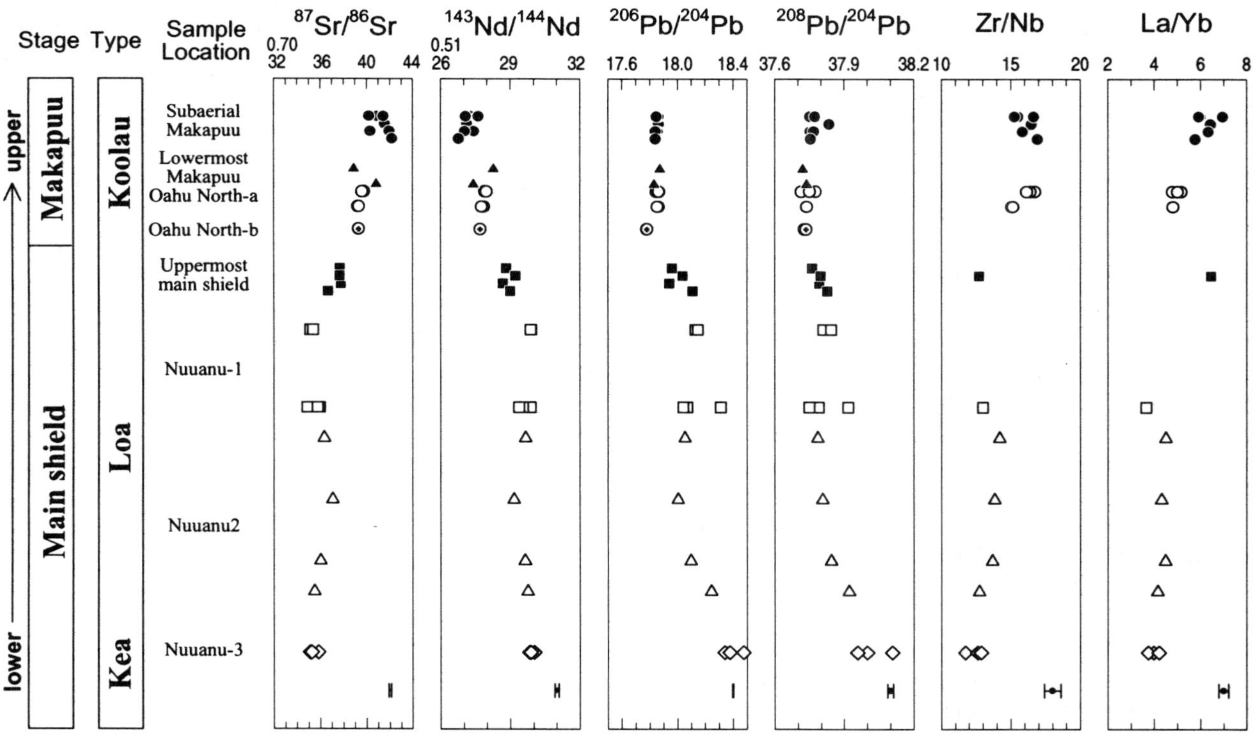

Figure 2. Compositions and estimated stratigraphic order of samples from subaerial Koolau (filled symbols), and submarine- collected samples (open and open with cross symbols), after *Takeguchi* [1999], *Shinozaki et al.* [this volume], *Yokose* [this volume], *Satake et al.* [this volume], and *Takahashi and Nakajima* [this volume]. Error bars for isotopic compositions are the mean of two standard deviations in analytical reproducibility, and for trace element ratios are one standard deviation.

Table 2. Modal proportions (vol.%) of phenocrysts in uppermost main shield stage samples (based on 3000 point counts)

	LK1	LK2	LK3	LK4
Phenocryst				
olivine	0.9	1.9	3.7	19.4
clinopyroxene	-	-	1.3	0.6
orthopyroxene	-	-	-	0.1
plagioclase	-	1.3	1.8	1.3
Microphenocryst				
olivine	0.7	1.7	0.9	2.5
clinopyroxene	0.0	0.4	2.5	40.9
plagioclase	0.0	4.6	7.2	29.7
oxide	0.0	0.0	0.0	5.5
Matrix	98.3	90.1	82.5	-

respectively. Wailau landslide blocks (Wailau 1 and 2), which contain deep submarine-erupted pillow lava and lava fragments, were derived from East Molokai shield [*Satake et al,* this volume; *Shinozaki et al.*, this volume; *Yokose,* this volume]. Autochthonous pillow lava was also dredged from the northwestern flank of the West Molokai shield [*Shinozaki et al.,* this volume].

3. ANALYTICAL PROCEDURE

All samples were analyzed for major and trace elements and Sr-Nd-Pb isotopic compositions at the Pheasant Memorial Laboratory (PML), Institute for Study of the Earth's Interior, Okayama University at Misasa (Table 3). Rock specimens were crushed by a jaw crusher or in an iron mortar to coarse chips 3-5 mm in diameter, and then fresh chips were carefully hand-picked. The chips from the submarine samples were washed with 0.1N HNO_3 for twenty minutes in an ultrasonic bath to remove surface contamination. Subsequently, the chips were rinsed with deionized water in an ultrasonic bath at least three times, and then dried at 80°C for 12 hours. The washed chips were ground using an alumina pack mill or in a silicon nitride mortar to obtain powder with grain size below 200 mesh.

Concentrations of major elements, Ni, and Cr were obtained for glass beads containing a lithium tetraborate flux (10 to 1 dilution of sample) using a Philips PW2400 X-ray fluorescence spectrometer (*Takei and Nakamura,* written comun.). H_2O^+ and H_2O^- were determined by gravimetrically, and FeO contents were determined by titration methods [*Yokoyama and Nakamura,* 2001]. Trace elements were analyzed using a Yokogawa PMS2000 inductively coupled plasma mass spectrometer fitted with a flow injection system using the methods of *Makishima and Nakamura* [1997], *Makishima et al.* [1997, 1999], and *Yokoyama et al.* [1999]. Boron, Li, Zr, and Hf were determined by isotope dilution and other trace elements by the calibration-curve method. All of the major and trace element analyses were duplicated, and replicate analyses were within 3-5% relative percent difference.

The analytical procedure for chemical separation and mass spectrometry followed *Yoshikawa and Nakamura* [1993] for Sr isotope measurements, *Makishima and Nakamura* [1991] for Nd isotope measurements, and *Koide and Nakamura* [1990] and *Kuritani and Nakamura* (written comun.) for Pb isotope measurements. All powdered samples for isotopic analyses were leached with 6N distilled HCl at 80°C for 12 hours [*Hanyu and Nakamura,* 2000] before sample decomposition. After the acid leaching, 20-50% of the Sr, 30-90% of the Nd, and 20-80% of the Pb were leached out of samples. The residue was then totally decomposed in HF-$HClO_4$ for Sr and Nd isotope measurements, and in HF-HBr for Pb isotope measurements. Mass spectrometry was carried out with modified Finnigan MAT261 [*Nakano and Nakamura,* 1998] and MAT262 thermal ionization mass spectrometers in static-multi collection mode. Normalizing factors to correct isotopic fractionation for Sr and Nd are $^{86}Sr/^{88}Sr$= 0.1194, $^{146}Nd/^{144}Nd$=0.7219. For Pb isotope analysis, 100ng Pb were measured and the data were normalized to NBS981 standard [*Todt et al.,* 1996], for which the 2σ standard deviations for $^{206}Pb/^{204}Pb$, $^{207}Pb/^{204}Pb$, and $^{208}Pb/^{204}Pb$ were 0.0015, 0.0022, and 0.0070, respectively (n=8). Measured ratios of the standard materials were $^{87}Sr/^{86}Sr$=0.71024±2 (least digit, 2σ) for 100ng of NBS987 (n=6), $^{143}Nd/^{144}Nd$=0.511839±9 (2 σ) for 100ng of La Jolla (n=4). Analytical reproducibilities for basaltic standard powders from Geological Survey of Japan were $^{87}Sr/^{86}Sr$= 0.703659±11 (2 σ) and $^{143}Nd/^{144}Nd$=0.513099±2 (2 σ) for JB2 (n=4), and $^{206}Pb/^{204}Pb$=18.293±3, $^{207}Pb/^{204}Pb$=15.533 ±4, and $^{208}Pb/^{204}Pb$=38.244±12 (2 σ) for JB3 (n=5).

4. RESULTS

4. 1. Major Element Composition

Analyzed samples are tholeiites, except for an alkaline basalt from the drill hole at the Wheeler Air Force Base (WAF36) (Figure 3a). The SiO_2 contents are higher in Makapuu-stage than in main shield-stage at a given MgO (Figure 3b). The SiO_2-Al_2O_3/CaO plot shows that Makapuu-stage samples have higher Al_2O_3/CaO than main shield-stage samples. The Al_2O_3/CaO ratios of lava fragments in the Oahu North-b are the highest among the tholeiitic lavas in Makapuu-stage, and are similar to those of matrix hyaloclastite glasses (Figure 3c).

Major element compositions for Makapuu-stage lavas are discussed in more detail by *Shinozaki et al.* [this volume] and *Takeguchi* [1999].

4. 2. Subaerial Weathering and Seawater Reaction

Post-magmatic, low-temperature weathering results in decreased K_2O/P_2O_5, SiO_2, and Na_2O and increased K/Rb contents in Hawaiian basalts [*Feigenson et al.*, 1983; *Lipman et al.*, 1990; *Chen et al.*, 1991; *Frey et al.*, 1990, 1994]. Samples from the lowermost Makapuu-stage (PM1, PM7) and most of the uppermost main shield-stage (LK1, LK2, LK4) have high K/Rb (>1000) than the other samples, which tends to negatively correlate with low K_2O/P_2O_5 ratios (<1) (Figure 4a). These samples also have low Cs, B, and Li abundances (Table 3). Samples WAF39, PM1, PM7 and LK1, are high in H_2O^+ (>1.5wt.%), suggesting the presence of secondary clay minerals associated with high vesicularity (Table 1). Thus, these samples appear to have been weathered, although they lack the REE-Y enrichment or negative Ce anomalies associated with a secondary phosphate phase [*Fodor et al.*, 1992].

Several samples from Nuuanu-1 and Oahu North-b have high Li and H_2O^+ (Figure 4b), suggesting low temperature alteration on the sea floor [*Muehlenbachs and Clayton*, 1972; *Staudigel et al.*, 1996]. For the Oahu North site, moderately altered volcanic breccias (S496-6C and S496-6E) show Li and H_2O enrichments; fresh pillow lavas do not (Table 1, Figure 4b). Subaerial-erupted lava fragments from Nuuanu-1 (D4-1, D4-2, S497-3, S497-7) are weakly to moderately altered (Table 1). The altered samples have relatively higher Li/Nb, and tend to have intermediate K/Rb (661-1133) and K_2O/P_2O_5 (0.95-1.24) ratios (Figure 4c). Thus, samples from Nuuanu-1 appear to have been weakly weathered on land, and subsequently altered in the ocean.

These samples are excluded from Figures 2 and 5 and discussion of the trace elements.

4. 3. Trace Element Compositions

Previous geochemical studies of Makapuu-stage tholeiite show high ratios of Ba/Nb, La/Nb, Sr/Nb, and Zr/Nb relative to Mauna Kea and the Mauna Loa shield lavas [*Roden et al.*, 1984, 1994; *Frey et al.*, 1994; *Jackson et al.*, 1999]. The variations of Ba/Nb, La/Nb, Pb/Nb, Sr/Nb, Nd/Nb, and Zr/Nb in Makapuu-stage lavas are relatively small (Ba/Nb=12±1, La/Nb=1.33±0.07, Pb/Nb=0.14±0.01,

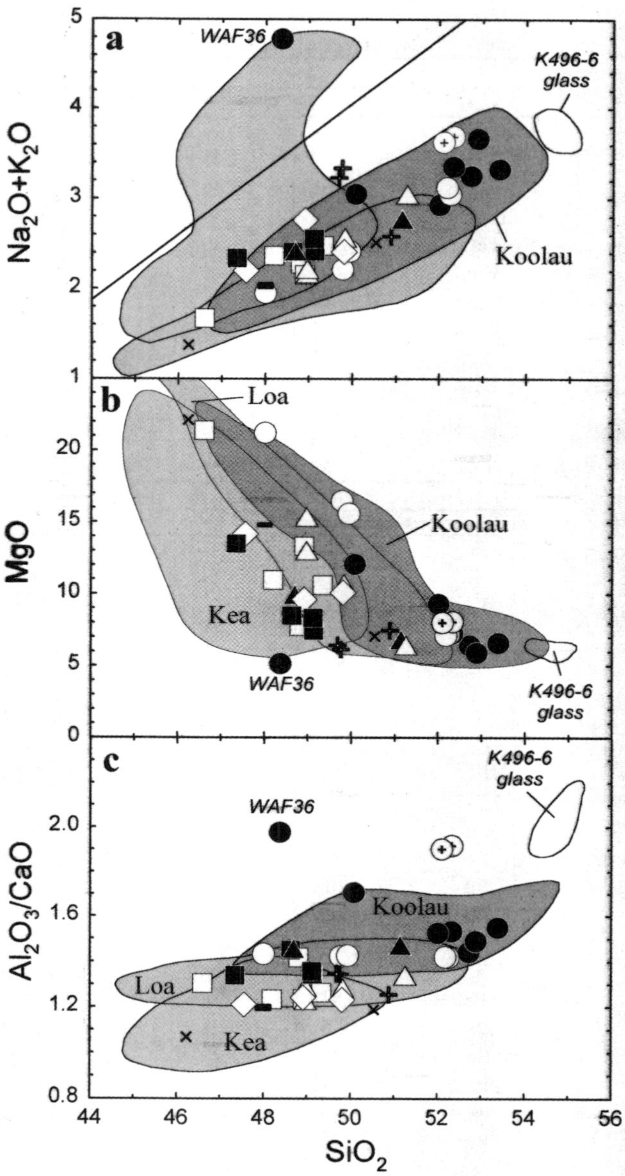

Figure 3. Silica-variation diagrams for the Koolau shield and the submarine-erupted East Molokai and West Molokai shield lavas. Abbreviations are the same in Figures 1-2. The high Al_2O_3/CaO hyaloclastite glass compositions from Oahu North-b (S469-6) are from *Shinozaki et al.* [this volume]. The compositional range for shield-stage tholeiitic lavas from Mauna Kea, Mauna Loa, and Koolau volcanoes from GEOROC database during 1980-2000 (http://georoc.mpch-mainz.gwdg.de) and *Takeguchi* [1999]. The Koolau-type is restricted to the Makapuu-stage. Solid line in Figure 3a is tholeiitic-alkalic dividing line after *Macdonald and Katsura* [1964].

Table 3. Major, trace, and Sr-Nd-Pb isotopic compositions of the studied samples

	Subaerial Makapuu stage							Lowermost Makapuu	
wt.%	WAF-2	WAF-36	WAF-39	KOL43	KOL48	KOL14	MH-11	PM1	PM7
SiO_2	53.24	48.31	49.55	52.83	52.97	52.46	52.13	50.35	48.37
TiO_2	2.09	3.11	2.09	2.09	2.41	2.26	2.00	2.18	2.22
Al_2O_3	14.02	16.46	12.03	13.97	13.91	13.96	13.36	14.50	14.08
Fe_2O_3	2.54	7.30	5.67	2.55	2.25	2.17	3.35	4.65	4.63
FeO	7.64	5.30	6.40	8.07	8.28	8.27	7.18	6.50	7.05
MnO	0.146	0.148	0.156	0.153	0.152	0.149	0.152	0.157	0.166
MgO	6.57	5.14	11.91	6.43	5.92	7.28	9.31	6.52	9.68
CaO	9.05	8.33	7.05	9.69	9.34	9.11	8.76	9.94	9.76
Na_2O	2.71	3.95	2.48	2.73	2.85	2.91	2.67	2.56	2.27
K_2O	0.609	0.826	0.536	0.529	0.822	0.457	0.266	0.137	0.101
P_2O_5	0.265	0.425	0.365	0.251	0.344	0.310	0.257	0.232	0.233
H_2O^+	1.25	0.85	1.96	0.90	0.94	0.92	0.85	2.25	1.65
H_2O^-	0.28	0.11	0.55	0.21	0.06	0.18	0.20	0.58	0.40
total	100.4	100.3	100.7	100.4	100.2	100.4	100.5	100.6	100.6
trace elements (ppm)									
Cr	239	61.1	654	185	159	259	506	264	470
Ni	130	173	641	111	98	178	285	123	307
Li	4.08	8.16	6.59	5.89	5.54	4.56	5.13	6.23	3.92
B	2.68	1.68	1.83	1.73	3.13	2.82	1.97	0.900	0.601
Rb	8.34	11.72	6.88	8.60	10.3	6.88	3.27	0.560	0.438
Sr	447	670	398	417	455	455	427	413	431
Y	26.3	34.3	30.3	24.7	29.2	26.2	26.3	31.4	25.2
Cs	0.0760	0.0461	0.0829	0.119	0.132	0.0676	0.0299	0.0072	0.0054
Ba	107	169	176	105	132	114	101	86.0	63.1
La	11.8	17.0	14.5	11.3	13.9	12.1	10.7	12.6	11.4
Ce	28.8	42.8	33.9	27.1	35.1	30.2	25.4	28.2	28.6
Pr	4.33	6.16	5.08	3.95	4.96	4.46	3.83	4.54	3.97
Nd	20.1	30.2	23.5	18.9	24.7	22.1	19.0	22.4	19.8
Sm	5.11	7.64	5.80	4.76	6.29	5.67	4.72	5.70	5.03
Eu	1.84	2.64	2.00	1.71	2.13	1.96	1.72	2.06	1.82
Gd	5.17	7.56	5.66	5.02	6.21	5.52	4.79	6.03	5.13
Tb	0.817	1.14	0.967	0.781	0.972	0.887	0.795	1.01	0.809
Dy	4.52	6.51	5.32	4.40	5.41	4.91	4.33	5.71	4.62
Ho	0.882	1.20	1.00	0.843	1.08	0.951	0.853	1.12	0.920
Er	2.12	2.81	2.33	2.13	2.52	2.23	1.99	2.66	2.13
Tm	0.290	0.373	0.343	0.306	0.346	0.308	0.285	0.371	0.311
Yb	1.84	2.44	2.03	1.88	2.35	1.91	1.85	2.54	1.98
Lu	0.253	0.324	0.284	0.268	0.294	0.263	0.240	0.325	0.263
Pb	1.21	1.70	1.48	1.21	1.44	1.24	1.04	1.14	1.31
Th	0.709	0.991	1.04	0.709	0.848	0.741	0.605	0.660	0.762
U	0.204	0.283	0.319	0.210	0.247	0.208	0.148	0.143	0.148
Zr	136	202	152	125	166	149	126	139	143
Hf	3.36	4.86	3.67	3.12	4.14	3.68	3.20	3.42	3.57
Nb	8.25	13.00	12.31	8.18	9.97	9.40	7.45	8.74	8.75
Ta	0.574	0.908	0.861	0.578	0.703	0.648	0.548	0.617	0.595
isotopes									
$^{87}Sr/^{86}Sr$	0.704156	0.704091	0.704031	0.704020	0.704146	0.704196	0.704219	0.703889	0.704083
$^{143}Nd/^{144}Nd$	0.512714	0.512729	0.512744	0.512763	0.512708	0.512704	0.512677	0.512828	0.512741
$^{206}Pb/^{204}Pb$	17.863	17.848	17.858	17.865	17.849	17.841	17.841	17.873	17.832
$^{207}Pb/^{204}Pb$	15.449	15.431	15.441	15.444	15.434	15.434	15.415	15.437	15.427
$^{208}Pb/^{204}Pb$	37.836	37.752	37.753	37.778	37.774	37.768	37.755	37.721	37.738

Table 3. (continued)

	Uppermost Main shield stage				Oahu North-a					Oahu North-b	
wt.%	LK1	LK2	LK3	LK4	K89-6	D7-1	S500-1	S500-5B	S500-9A	S496-6C	S496-6E
SiO_2	48.36	48.02	48.58	47.10	49.90	52.25	49.97	47.98	52.22	51.63	51.53
TiO_2	2.51	2.11	2.40	2.38	1.49	2.09	1.51	1.30	2.10	2.29	2.32
Al_2O_3	14.28	14.76	14.17	11.91	10.60	13.78	10.99	9.04	13.80	14.26	14.39
Fe_2O_3	3.63	3.99	3.46	2.86	2.79	3.03	2.29	2.13	3.50	2.51	3.02
FeO	8.00	7.67	7.79	9.13	8.04	7.63	8.30	8.74	7.26	7.69	7.35
MnO	0.172	0.170	0.167	0.174	0.161	0.158	0.160	0.157	0.159	0.124	0.126
MgO	7.34	8.39	8.24	13.42	16.51	7.20	15.61	21.22	7.12	7.95	7.92
CaO	10.62	10.18	10.43	8.89	7.45	9.71	7.71	6.31	9.72	7.44	7.57
Na_2O	2.23	2.25	2.28	2.12	1.86	2.54	2.04	1.64	2.57	3.08	3.07
K_2O	0.157	0.131	0.251	0.215	0.360	0.505	0.383	0.308	0.554	0.559	0.521
P_2O_5	0.251	0.233	0.272	0.285	0.189	0.263	0.187	0.164	0.254	0.275	0.272
H_2O^+	1.70	1.41	1.06	0.88	0.84	1.05	0.94	0.92	1.08	2.16	1.89
H_2O^-	0.37	0.19	0.16	0.02	0.11	0.18	0.17	0.15	0.06	0.57	0.68
total	99.6	99.5	99.3	99.4	100.3	100.4	100.3	100.1	100.4	100.5	100.7
trace elements (ppm)											
Cr	261	372	414	741	940	300	898	1150	293	253	265
Ni	102	147	166	540	793	127	725	1206	129	299	292
Li	3.39	5.02	4.53	4.90	6.99	6.00	5.38	5.08	5.95	14.72	19.59
B	0.638	0.447	0.873	0.613	3.47	3.38	2.80	2.90	5.57	7.04	8.06
Rb	1.13	0.33	2.56	1.33	5.32	7.06	4.84	4.20	7.57	6.85	8.01
Sr	421	400	449	418	277	389	291	239	372	414	412
Y	29.0	25.7	28.4	25.0	19.7	27.1	18.5	16.3	26.9	26.2	27.1
Cs	0.0125	0.0086	0.0297	0.0157	0.0561	0.0694	0.0467	0.0448	0.0945	0.156	0.203
Ba	66.8	67.8	88.2	114	64.8	90.2	63.7	53.2	89.4	74.3	77.4
La	15.7	10.5	13.8	13.0	7.21	10.4	7.22	6.19	9.95	10.4	10.6
Ce	37.1	25.4	33.4	31.0	18.5	26.6	18.4	15.9	25.6	26.5	27.5
Pr	5.14	3.79	4.87	4.60	2.69	3.92	2.65	2.30	3.70	3.88	3.93
Nd	24.0	18.5	23.8	22.0	13.4	18.6	13.0	11.5	18.3	19.3	19.6
Sm	5.69	4.86	5.85	5.56	3.41	4.90	3.53	3.02	4.84	5.17	5.29
Eu	1.98	1.73	2.09	1.96	1.26	1.73	1.30	1.11	1.77	1.89	1.96
Gd	5.82	5.10	6.01	5.71	3.79	5.18	3.72	3.31	5.21	5.47	5.66
Tb	0.941	0.821	0.966	0.867	0.621	0.860	0.604	0.527	0.861	0.888	0.905
Dy	5.36	4.62	5.35	4.74	3.51	4.90	3.46	3.05	4.78	4.86	5.00
Ho	1.03	0.885	1.02	0.906	0.668	0.902	0.678	0.601	0.934	0.903	0.935
Er	2.46	2.11	2.44	2.08	1.67	2.34	1.63	1.46	2.23	2.11	2.17
Tm	0.337	0.298	0.339	0.289	0.239	0.304	0.234	0.201	0.318	0.286	0.285
Yb	2.18	1.93	2.13	1.79	1.51	2.00	1.44	1.28	2.07	1.72	1.76
Lu	0.293	0.249	0.281	0.234	0.212	0.283	0.204	0.179	0.274	0.218	0.221
Pb	1.43	1.02	1.21	1.10	0.814	1.11	0.808	0.689	1.15	1.26	1.28
Th	1.080	0.626	0.909	0.948	0.454	0.613	0.452	0.389	0.638	0.705	0.715
U	0.244	0.101	0.245	0.180	0.149	0.200	0.144	0.130	0.216	0.241	0.238
Zr	160	131	153	149	89.5	129	88.7	75.2	124	135	136
Hf	3.84	3.16	3.80	3.78	2.27	3.27	2.20	1.94	3.30	3.65	3.62
Nb	13.8	8.56	12.0	12.3	5.35	7.87	5.51	4.97	8.18	8.87	8.77
Ta	0.997	0.605	0.843	0.856	0.391	0.555	0.389	0.352	0.567	0.629	0.622
isotopes											
$^{87}Sr/^{86}Sr$	0.703777	0.703768	0.703766	0.703667	0.703960	0.703976	0.703962	0.703925	0.703932	0.703930	0.703932
$^{143}Nd/^{144}Nd$	0.512868	0.512882	0.512922	0.512899	0.512793	0.512790	0.512797	0.512787	0.512777	0.512771	0.512771
$^{206}Pb/^{204}Pb$	17.942	17.960	18.038	18.111	17.866	17.849	17.864	17.864	17.856	17.779	17.781
$^{207}Pb/^{204}Pb$	15.423	15.443	15.441	15.453	15.447	15.430	15.440	15.437	15.437	15.420	15.422
$^{208}Pb/^{204}Pb$	37.793	37.761	37.799	37.828	37.777	37.716	37.750	37.738	37.739	37.729	37.736

Table 3. (continued)

	Nuuanu-1					Nuuanu-2			
wt.%	D4-1	D4-2	S497-3	S497-6	S497-7	K90-1	K90-3	D5-18	S498-4
SiO_2	47.81	48.26	48.38	46.95	48.12	50.16	50.10	51.36	48.52
TiO_2	2.22	2.70	1.90	1.41	2.22	2.35	2.35	2.89	2.06
Al_2O_3	12.41	14.11	11.67	8.87	12.35	12.52	12.38	13.33	11.61
Fe_2O_3	5.95	5.11	3.51	2.27	5.03	2.00	3.30	4.39	1.90
FeO	6.42	7.56	7.55	9.82	6.13	9.51	8.27	7.61	9.24
MnO	0.163	0.174	0.164	0.175	0.176	0.177	0.171	0.186	0.168
MgO	10.87	7.69	13.19	21.55	10.35	10.15	10.53	6.23	12.67
CaO	10.06	9.95	9.45	6.81	9.77	10.16	9.75	10.05	9.54
Na_2O	2.06	2.03	1.96	1.47	2.12	2.15	2.14	2.58	1.97
K_2O	0.293	0.224	0.175	0.211	0.299	0.283	0.410	0.439	0.200
P_2O_5	0.237	0.237	0.145	0.123	0.291	0.197	0.240	0.294	0.193
H_2O^+	1.68	2.28	2.00	0.48	2.91	0.63	0.68	1.05	1.62
H_2O^-	0.59	0.08	0.64	0.04	1.17	0.07	0.11	0.32	0.46
total	100.8	100.4	100.7	100.2	100.9	100.3	100.4	100.7	100.1
trace elements (ppm)									
Cr	571	317	960	1004	707	711	656	96.7	788
Ni	307	137	531	1168	367	679	339	78.9	463
Li	13.3	24.0	16.1	5.91	16.6	4.97	5.99	9.39	4.96
B	7.92	36.7	6.75	3.29	4.14	2.70	3.84	13.5	5.78
Rb	2.15	1.99	1.63	2.65	4.29	4.51	5.50	10.4	3.73
Sr	313	284	264	216	356	285	311	344	280
Y	27.0	32.1	23.9	16.9	27.1	20.2	26.7	33.6	25.2
Cs	0.0236	0.0509	0.0664	0.0277	0.0920	0.0449	0.0549	0.3802	0.0840
Ba	63.3	75.7	41.3	52.8	77.3	58.0	69.9	88.3	44.7
La	9.24	11.1	7.35	4.86	10.5	6.85	9.45	12.0	8.42
Ce	23.7	27.1	19.0	12.7	25.9	17.6	24.2	32.0	21.9
Pr	3.67	4.22	2.82	1.90	3.81	2.59	3.59	4.52	3.25
Nd	17.7	20.9	14.1	9.56	18.7	13.2	18.0	22.9	16.0
Sm	4.68	5.72	3.78	2.68	4.81	3.64	4.93	6.10	4.35
Eu	1.71	2.02	1.41	1.02	1.75	1.29	1.71	2.24	1.57
Gd	5.35	6.20	4.40	3.12	5.18	3.87	5.25	6.75	4.85
Tb	0.834	0.983	0.737	0.512	0.838	0.627	0.867	1.07	0.805
Dy	4.92	5.73	4.35	2.99	4.87	3.66	4.99	6.33	4.65
Ho	0.965	1.14	0.854	0.594	0.936	0.728	0.920	1.20	0.916
Er	2.33	2.57	2.08	1.49	2.24	1.75	2.29	2.95	2.25
Tm	0.322	0.372	0.298	0.211	0.313	0.254	0.325	0.437	0.320
Yb	2.10	2.51	1.82	1.33	1.97	1.59	2.10	2.67	2.03
Lu	0.307	0.321	0.249	0.185	0.274	0.225	0.280	0.350	0.273
Pb	0.868	1.01	0.532	0.599	1.31	0.809	0.895	1.13	0.70
Th	0.659	0.763	0.548	0.417	0.805	0.539	0.646	0.806	0.636
U	0.253	0.244	0.150	0.116	0.559	0.165	0.228	0.608	0.219
Zr	121	141	90.3	72.5	126	99.5	131	160	113
Hf	3.22	3.88	2.55	2.02	3.44	2.64	3.44	4.24	3.02
Nb	8.40	10.3	7.33	5.62	10.4	7.19	9.22	11.7	8.86
Ta	0.621	0.747	0.538	0.398	0.730	0.504	0.644	0.853	0.622
isotopes									
$^{87}Sr/^{86}Sr$	0.703513	0.703537	0.703487	0.703596	0.703576	0.703707	0.703634	0.703603	0.703550
$^{143}Nd/^{144}Nd$	0.512990	0.512986	0.512986	0.512954	0.512937	0.512916	0.512965	0.512963	0.512975
$^{206}Pb/^{204}Pb$	18.133	18.149	18.313	18.072	18.043	18.004	18.054	18.098	18.244
$^{207}Pb/^{204}Pb$	15.444	15.447	15.446	15.436	15.440	15.447	15.449	15.450	15.453
$^{208}Pb/^{204}Pb$	37.810	37.844	37.918	37.791	37.750	37.807	37.787	37.844	37.922

Table 3. (continued)

	Nuuanu-3			Wailau-1		Wailau-2			W. Molokai
wt.%	D6-1	S499-6A	S499-6B	S501-4	S501-7	D9-1	S510-6A	S510-6C	D8-1
SiO_2	47.64	50.16	48.87	50.21	45.95	51.21	49.58	49.62	48.01
TiO_2	2.04	2.35	2.45	2.37	1.41	2.32	2.75	2.73	2.01
Al_2O_3	11.61	12.52	12.91	14.06	8.02	13.53	14.34	14.27	11.34
Fe_2O_3	2.21	2.00	2.80	4.36	1.65	2.10	3.96	4.05	2.15
FeO	9.32	9.51	8.76	5.88	9.95	9.15	7.58	7.61	9.08
MnO	0.169	0.177	0.170	0.163	0.169	0.169	0.164	0.170	0.163
MgO	14.20	10.15	9.60	7.19	22.17	7.49	6.11	6.42	14.85
CaO	9.59	10.16	10.40	11.73	7.43	10.79	10.67	10.60	9.47
Na_2O	1.94	2.15	2.27	2.30	1.21	2.25	2.61	2.58	1.70
K_2O	0.265	0.283	0.492	0.230	0.192	0.348	0.713	0.645	0.343
P_2O_5	0.173	0.197	0.234	0.203	0.126	0.272	0.309	0.323	0.222
H_2O^+	1.34	0.63	0.82	1.49	1.49	1.03	1.26	1.16	0.95
H_2O^-	0.33	0.07	0.00	0.45	0.36	0.14	0.15	0.05	0.23
total	100.8	100.3	99.8	100.6	100.1	100.8	100.2	100.2	100.5
trace elements (ppm)									
Cr	917	634	689	259	1926	287	232	229	884
Ni	542	268	290	137	1012	109	77.7	85.8	603
Li	8.72	5.19	12.4	8.83	7.60	4.97	3.34	3.60	4.30
B	3.47	3.03	4.23	5.26	3.02	3.25	10.19	7.93	3.89
Rb	3.00	4.42	9.00	2.69	3.28	5.19	9.68	11.7	5.31
Sr	221	287	306	317	157	323	426	446	326
Y	22.0	29.1	30.3	30.3	18.2	31.3	32.6	34.3	24.1
Cs	0.0579	0.0476	0.0884	0.104	0.0501	0.0666	0.102	0.173	0.0547
Ba	28.4	50.4	64.9	67.2	30.7	53.4	93.6	97.8	68.2
La	7.59	8.45	9.89	9.73	5.60	8.34	12.3	13.0	8.75
Ce	19.5	21.6	25.5	25.3	14.8	22.3	32.0	33.9	23.6
Pr	2.98	3.26	3.73	3.73	2.24	3.29	4.67	4.89	3.25
Nd	14.1	16.5	18.6	18.1	11.2	17.2	23.2	24.3	16.4
Sm	3.91	4.64	5.10	4.96	2.99	4.98	6.02	6.34	4.42
Eu	1.42	1.73	1.86	1.79	1.11	1.80	2.13	2.25	1.63
Gd	4.50	5.34	5.91	5.54	3.45	5.69	6.52	6.79	4.98
Tb	0.722	0.899	0.979	0.932	0.578	0.947	1.05	1.09	0.777
Dy	4.21	5.18	5.60	5.47	3.45	5.34	5.94	6.32	4.55
Ho	0.863	1.03	1.08	1.06	0.667	1.04	1.17	1.23	0.843
Er	2.06	2.49	2.64	2.59	1.63	2.62	2.82	2.91	2.04
Tm	0.309	0.354	0.370	0.377	0.235	0.350	0.398	0.418	0.296
Yb	1.90	2.25	2.35	2.35	1.45	2.23	2.53	2.61	1.68
Lu	0.249	0.304	0.314	0.321	0.197	0.305	0.333	0.344	0.231
Pb	0.574	0.727	1.01	1.25	0.441	0.749	1.11	1.13	0.653
Th	0.615	0.657	0.758	0.811	0.461	0.510	0.899	0.919	0.660
U	0.144	0.209	0.289	0.810	0.163	0.365	0.402	0.311	0.226
Zr	105	120	133	134	72.3	126	167	170	118
Hf	2.78	3.31	3.49	3.39	1.93	3.32	4.21	4.13	3.10
Nb	8.95	9.51	10.3	11.2	6.68	7.30	11.5	11.8	9.02
Ta	0.618	0.693	0.751	0.788	0.475	0.536	0.824	0.824	0.610
isotopes									
$^{87}Sr/^{86}Sr$	0.703512	0.703582	0.703527	0.703530	0.703494	0.703553	0.703750	0.703746	0.703503
$^{143}Nd/^{144}Nd$	0.513003	0.512982	0.512987	0.512984	0.512991	0.513008	0.512888	0.512910	0.513012
$^{206}Pb/^{204}Pb$	18.476	18.341	18.378	18.620	18.602	18.234	18.185	18.200	18.523
$^{207}Pb/^{204}Pb$	15.464	15.459	15.449	15.474	15.465	15.469	15.455	15.472	15.508
$^{208}Pb/^{204}Pb$	38.108	38.000	37.958	38.179	38.143	37.892	37.904	37.963	38.094

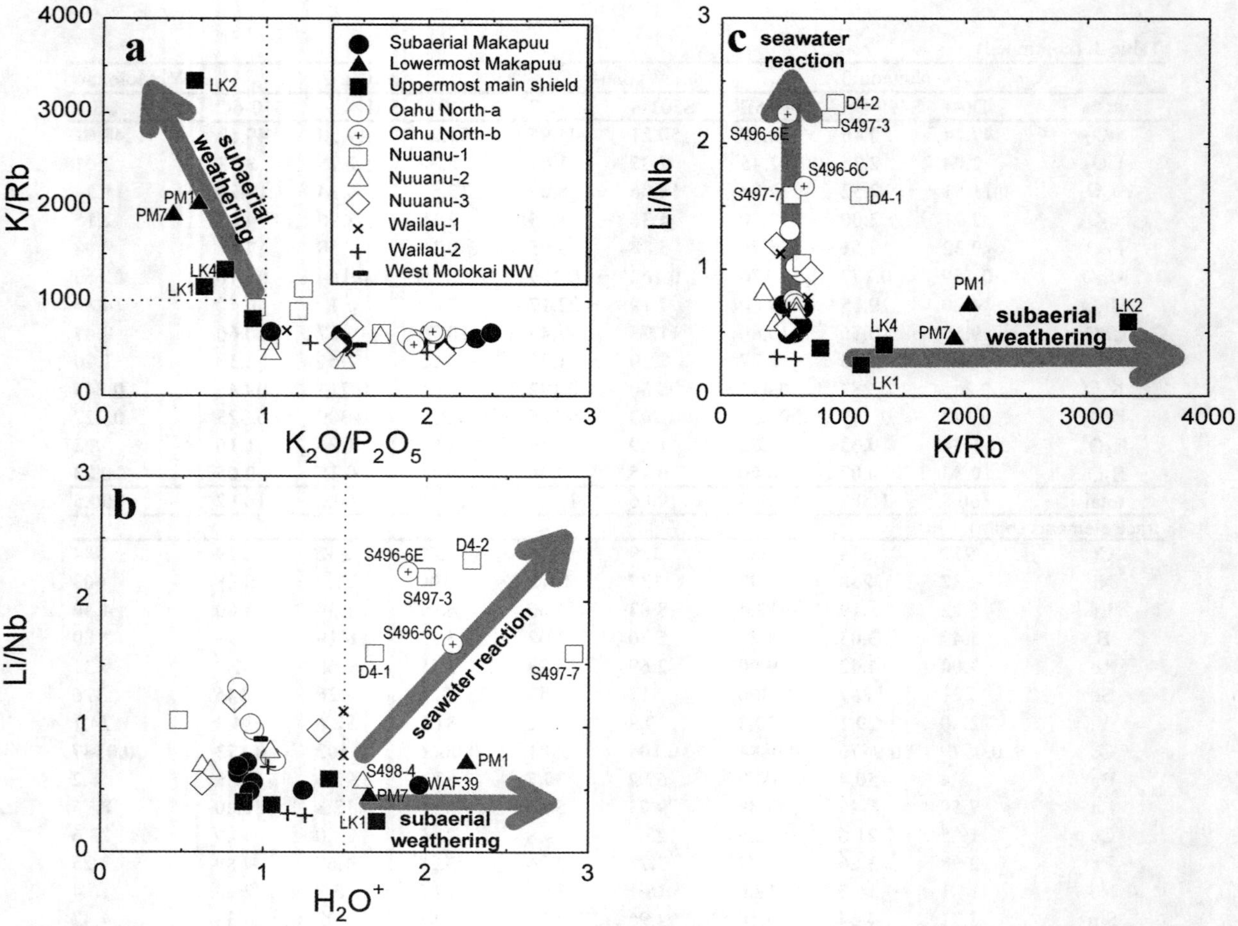

Figure 4. (a) Effects of post-magmatic subaerial low-temperature weathering in Koolau and submarine samples, i.e. increasing K/Rb with decreasing K_2O/P_2O_5. (b) Post-magmatic submarine low temperature alteration, i.e. increasing H_2O^+ with decreasing Li/Nb. The high H_2O^+ in LK1, PM1, PM7, and WAF39 without enrichment of Li/Nb would be caused by subaerial weathering. (c) Degree of subaerial and submarine low-temperature alteration. The labeled samples, which plot in the area designated by dotted lines in Figure 4a from uppermost main shield-stage and lowermost Makapuu-stage have relatively high K/Rb; those from Nuuanu-1 and Nuuanu-2 have relatively high Li/Nb. The altered samples in Figure 4a–c are omitted from the trace element plots in Figures 2, 5, and A1.

Sr/Nb=50±4, Nd/Nb=2.4±0.1, Zr/Nb=16.0±0.7), and higher than those of lavas from Mauna Kea and Mauna Loa (Figure 5). Subaerial Makapuu lavas have higher Ba, La, Sr, Nd, Zr, and La/Yb than those of the Oahu North-a samples (Figure 5). These incompatible trace element compositions correlate positively with SiO_2 and negatively with MgO, except for alkaline basalt WAF36 (figures not given).

Samples from Nuuanu blocks (Nuuanu-1, 2, and 3) have narrow ranges of incompatible elements (La/Nb=0.94 ±0.07, Sr/Nb=32±5, Nd/Nb=1.8±0.1, and Zr/Nb=13.1±0.8), similar to those from Mauna Kea and Mauna Loa (Figure 5). The La/Yb ratios of Nuuanu block samples are lower than those of Makapuu-stage (Figure 5). The La/Yb vs. Ba/Nb, La/Nb, Pb/Nb, Sr/Nb, Nd/Nb, and Zr/Nb of Nuuanu blocks essentially overlap the ranges from Mauna Loa (Figure 5).

Sample LK3 from the uppermost main shield-stage is intermediate in La/Nb between Makapuu-stage and Nuuanu block lavas (Figure 5). On the other hand, Ba/Nb, Pb/Nb, Sr/Nb, Nd/Nb, and Zr/Nb of LK3 are similar to those from the Nuuanu blocks, and La/Yb is within the range of the subaerial Makapuu-stage and lowermost Makapuu-stage lavas.

Incompatible trace elements abundances and ratios from the submarine-erupted samples of Wailau blocks and the

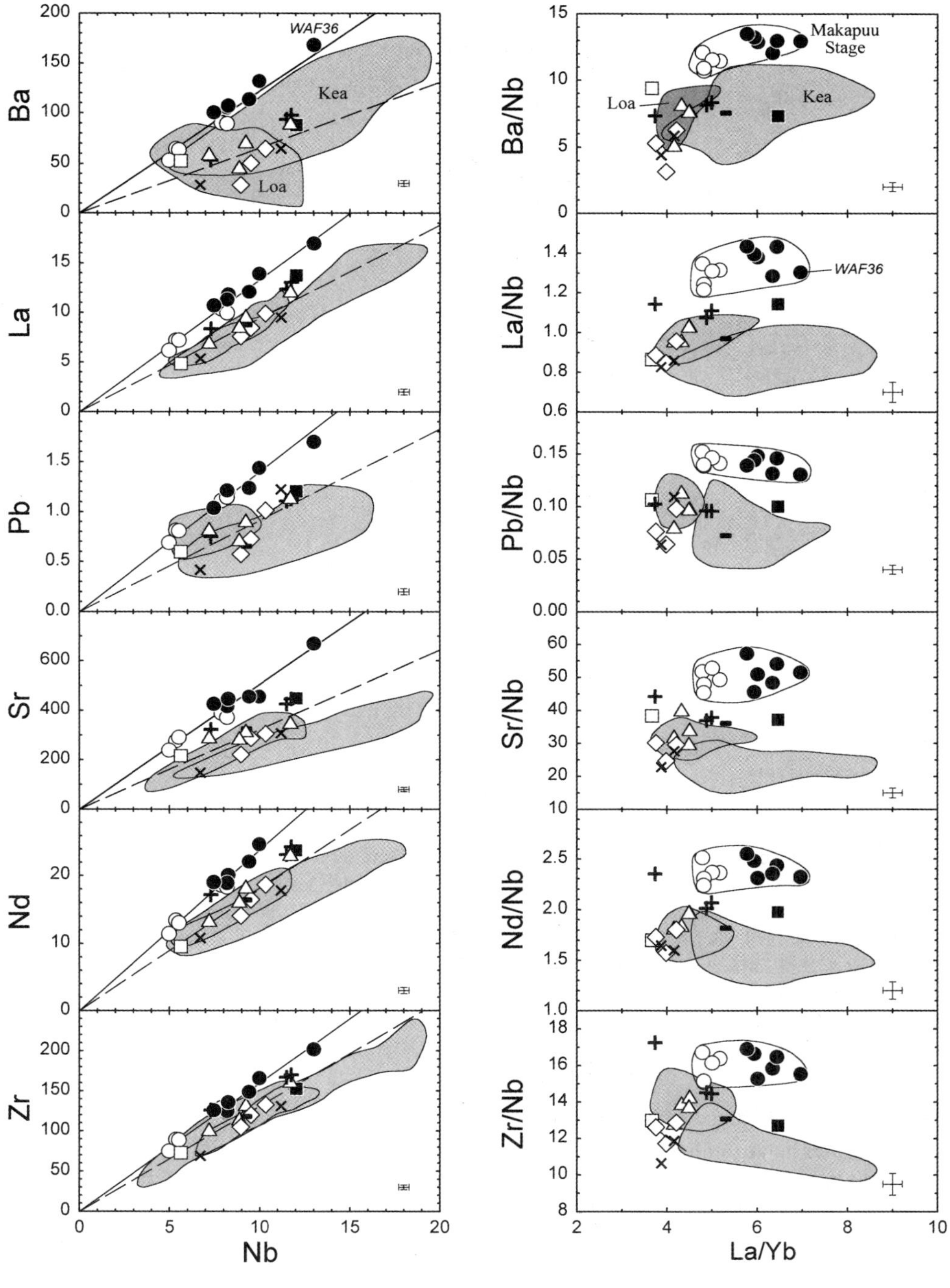

Figure 5. Trace element abundances (ppm) and ratios. Symbols are the same as in Figures 1-2. Solid and broken lines are regression lines for Makapuu-stage lavas and main shield-stage lavas, respectively. Compositional ranges for Mauna Kea and Mauna Loa shield tholeiite are from *Albarède* [1996], *Hauri and Kurz* [1997], *Hofmann and Jochum* [1996], *Kurz et al.* [1995], *Lipman et al.* [1990], *Norman and Garcia* [1999], *Rhodes and Hart* [1995], *Rhodes* [1988, 1996], and *Yang et al.* [1994, 1996]. Error bars represents one standard deviation for analytical reproducibility. Linear solid lines and broken lines are regression lines for Makapuu-stage lavas and for Main shield-stage lavas in the Koolau shield, respectively.

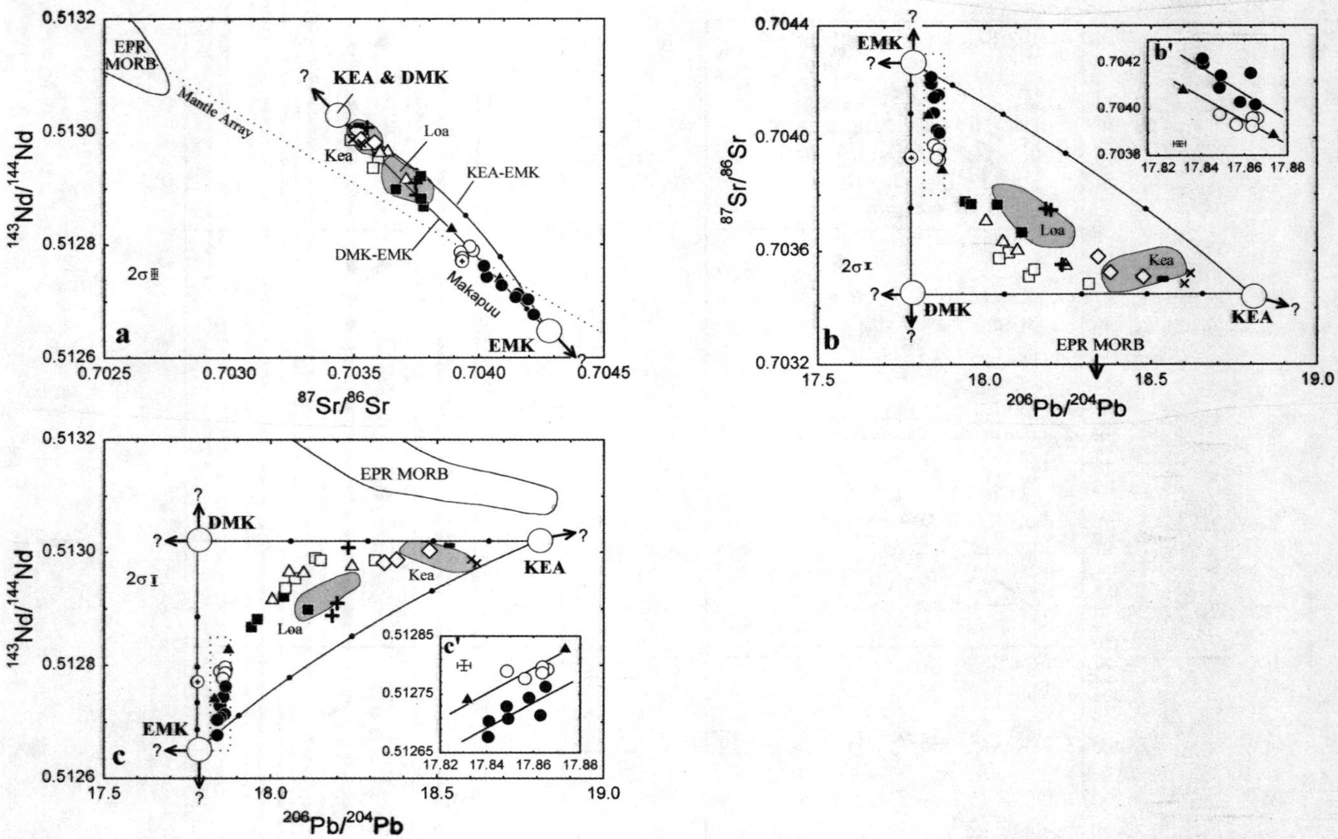

Figure 6. Sr-Nd-Pb isotopic-ratio diagrams. Abbreviations are the same as in Figure 1-2. Enlargements for Makapuu-stage lavas (shown by dotted square) are shown as insets on Figure 6b and Figure 6c. Solid lines in insets and are linear regression lines for subaerial upper part of Makapuu and for the lowermost Makapuu and Oahu North-a, respectively. Data sources for Mauna Kea and Mauna Loa, *Lassiter et al.* [1996], *Lassiter and Hauri* [1998], and *Abouchami et al.* [2000]; for EPR MORB, *Mahoney et al.* [1994], *Bach et al.* [1994], *Harpp* [1995], *Niu et al.* [1996], and *Regelous et al.* [1999]. Mantle Array in Figure 6a shows a tie line between Sr-Nd isotopic compositions of EPR MORB and Bulk Silicate Earth [*Zindler and Hart,* 1986]. KEA, DMK, and EMK are proposed endmember isotopic compositions for the Koolau shield lavas (Table 4). Dots on the mixing lines represent 20% mixing increments. Error bars are the same as in Figure 2. More detailed explanation is given in the text.

West Molokai NW mostly overlap those of Nuuanu blocks (Figure 5). Wailau-2 samples have higher Ba/Nb, La/Nb, Sr/Nb, Nd/Nb, and Zr/Nb than those of Wailau-1.

4. 4. Sr, Nd, and Pb Isotopes

The Sr, Nd, and Pb isotopic compositions of Koolau shield lavas successively change from Kea-, through Loa-, to Koolau-type; $^{87}Sr/^{86}Sr$ increases and $^{143}Nd/^{144}Nd$, $^{206}Pb/^{204}Pb$, and $^{208}Pb/^{204}Pb$ decrease with stratigraphic order from lower to upper (Figure 2). Furthermore, the ranges in these isotopic data encompass most of the Hawaiian shield lavas documented previously. On a $^{87}Sr/^{86}Sr$ - $^{143}Nd/^{144}Nd$ plot, a well defined negative linear trend (r^2=0.97) can be fitted to the entire array of samples.

The Sr, Nd, and Pb isotopic compositions of Makapuu-stage lavas are similar those of Koolau compositions published previously [*Tatsumoto,* 1978; *Stille et al.*, 1986a, *Roden et al.*, 1984, 1994; not plotted in Figures 6-7]. The isotopic compositions of the alkaline basalt (WAF-36) are also within the range of Makapuu-stage lavas, but differ from rejuvenated-stage Honolulu Volcanics ($^{87}Sr/^{86}Sr$= 0.7033-0.7034, $^{143}Nd/^{144}Nd$=0.51302-0.51306, $^{206}Pb/^{204}Pb$ =18.0-18.2) [*Stille et al.*, 1986a; *Lassiter et al.*, 2000]. The Sr, Nd, and Pb isotopic compositions of the subaerial Makapuu, lowermost Makapuu, and Oahu North-a samples show linear correlation; only $^{206}Pb/^{204}Pb$ of Oahu North-b lava fragments deviates slightly from the linear array to a lower value (Figures 6-7). The steep slope in the $^{206}Pb/^{204}Pb$ - $^{207}Pb/^{204}Pb$ plot represents a mixing correlation between at least two components (Figure 7).

Samples from the uppermost main shield-stage and Nuuanu blocks (Nuuanu 1-3) have lower $^{87}Sr/^{86}Sr$ and higher $^{143}Nd/^{144}Nd$ and $^{206}Pb/^{204}Pb$ than those of Makapuu-stage (Figures 6-7). The Sr and Nd isotopic compositions of the uppermost main shield-stage lavas resemble those from Mauna Loa. On the other hand, isotopic compositions of the Nuuanu blocks overlap those of Mauna Loa and Mauna Kea. On $^{206}Pb/^{204}Pb$ - $^{87}Sr/^{86}Sr$ and $^{206}Pb/^{204}Pb$ - $^{143}Nd/^{144}Nd$ diagrams (Figure 6b and c), Koolau shield lavas show a curved correlation with little variation of $^{206}Pb/^{204}Pb$ in Makapuu-stage lavas. Among the Nuuanu-2 samples, the Pb isotopic compositions tend to correlate with stratigraphic level (Figure 2).

Sr, Nd, and Pb isotopic compositions of submarine-erupted samples of Wailau-1 and –2 and West Molokai NW are essentially included in the variation of main shield-stage lavas of Koolau Volcano (Figures 6-7). In details, however, these compositions (except for $^{208}Pb/^{204}Pb$) deviate from trends of the main shield-stage (Figures 6-7, inserts).

5. ENDMEMBER COMPONENTS OF KOOLAU SHIELD

A linear trend on the $^{87}Sr/^{86}Sr$ - $^{143}Nd/^{144}Nd$ diagram indicates mixing of at least two endmember components for the Koolau shield lavas; one is close to the most enriched Makapuu-stage lavas (denoted as Enriched Makapuu, EMK) and the other to the most enriched Mauna Kea lavas (KEA) (Figure 6a). This interpretation is also supported by the fact that Ba/Nb, La/Nb, Pb/Nb, Sr/Nb, Nd/Nb, and Zr/Nb ratios of the Makapuu-stage lavas are higher than those of the main shield-stage lavas, irrespective of SiO_2 and MgO concentrations. These incompatible elements are more enriched in EMK source material than KEA (Figure 5). In contrast, the variation of Pb isotopic compositions cannot be explained by two component mixing (Figure 7). Although the Pb isotopic compositions of the main shield-stage vary between those of Makapuu-stage and Mauna Kea, the Makapuu-stage lavas show a nearly vertical trend in $^{206}Pb/^{204}Pb$ - $^{207}Pb/^{204}Pb$ with a relatively small variation of $^{206}Pb/^{204}Pb$ that is different from the trend of main shield-stage (Figure 7). Furthermore, the trend in $^{206}Pb/^{204}Pb$ - $^{208}Pb/^{204}Pb$ seems to be nearly vertical, though this is less clear compared with $^{206}Pb/^{204}Pb$ - $^{207}Pb/^{204}Pb$. These observations imply that a third endmember exists in addition to EMK and KEA. The three -endmember hypothesis can be evaluated from the Sr - Nd

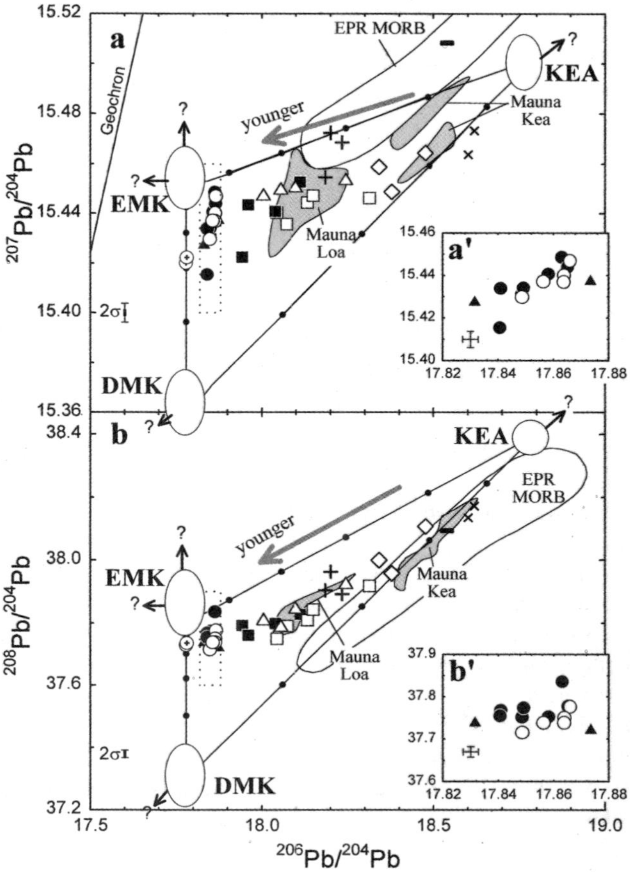

Figure 7. Lead isotope diagrams. Enlargements for Makapuu-stage lavas (shown by dotted square) are shown as insets on Figure 7a and Figure 7b. Abbreviations are the same as in Figures 1-2. Data sources for Mauna Kea and Mauna Loa, and East Pacific Rise MORB glasses (EPR MORB) are from *Abouchami et al.* [2000]. KEA, EMK, and DMK are same as in Figure 6. Dots on the mixing lines represent 20% mixing increments. Error bars are the same as in Figure 2.

- Pb isotopic systematics. If all the Koolau shield lavas formed by simple binary mixing of EMK and KEA, the fitted curves on $^{206}Pb/^{204}Pb$ - $^{87}Sr/^{86}Sr$ and $^{206}Pb/^{204}Pb$ - $^{143}Nd/^{144}Nd$ diagrams (Figure 6b and c) require the $(Pb/Sr)_{EMK}/(Pb/Sr)_{KEA}$ and $(Pb/Nd)_{EMK}/(Pb/Nd)_{KEA}$ ratios to be 5 and 11, respectively, where subscript EMK and KEA denotes EMK and KEA components. These values are approximately one-order magnitude higher than the observed values [$(Pb/Sr)_{Makapuu}/(Pb/Sr)_{Kea}=0.8\pm0.2$ (1σ) and $(Pb/Nd)_{Makapuu}/(Pb/Nd)_{Kea}=1.5\pm0.2$ (1σ)] obtained from the average Sr, Nd and Pb concentrations in Mauna Kea and Makapuu-stage lavas (Table 4). Thus, the Sr-Nd-Pb isotope systematics of Koolau shield lavas do not satisfy such a binary mixing and also require a third endmember

Table 4. Isotope compositions and element abundances of proposed endmembers

components	Enriched Makapuu (EMK)	Depleted Makapuu (DMK)	Kea (KEA)
$^{87}Sr/^{86}Sr$	>0.70427	<0.7035	<0.7035
$^{143}Nd/^{144}Nd$	<0.51265	>0.5130	>0.5130
$^{206}Pb/^{204}Pb$	<17.78	<17.78	>18.80
$^{207}Pb/^{206}Pb$	>15.45	<15.37	>15.50
$^{208}Pb/^{204}Pb$	>37.8	<37.2	>38.4
Sr (ppm)	470	*200**	205
Nd (ppm)	23	*10**	18
Pb (ppm)	1.4	*0.5**	0.76

Sr, Nd, and Pb concentrations of EMK is average value of the subaerial Makapuu lavas in this study and of KEA is average value of Mauna Kea lavas after the references in Fig. 5. *: Sr, Nd, and Pb concentrations of DMK is estimated by the assumption described in the text.

component, as envisaged based on the Pb isotope systematics.

Although Makapuu-stage lavas exhibit linear arrays in the $^{206}Pb/^{204}Pb$ - $^{87}Sr/^{86}Sr$ and $^{206}Pb/^{204}Pb$ - $^{143}Nd/^{144}Nd$ spaces as a whole, lowermost Makapuu-stage and Oahu North-a are more isotopically depleted in Sr and Nd than the subaerial Makapuu-stage lavas, forming parallel trends with negative (Figure 6b) and positive (Figure 6c) slopes in the fine-scale plots. The two parallel trends may reflect involvement of the third endmember component, referred to as Depleted Makapuu (DMK), which has lower $^{87}Sr/^{86}Sr$ and higher $^{143}Nd/^{144}Nd$ than those of EMK but nearly identical $^{206}Pb/^{204}Pb$ to EMK.

The major and trace element compositions also differ among the two parallel trends in the Sr-Nd Pb isotope diagrams (Figure 6); (1) subaerial Makapuu-stage is higher in incompatible trace elements (Ba, Nb, Sr, Pb, La, Nd, Zr), and in Al_2O_3/CaO at a given SiO_2 content, than the lowermost Makapuu-stage and Oahu North-a, (2) Al_2O_3/CaO of the former decreases with increasing SiO_2 while the latter shows nearly constant Al_2O_3/CaO ratios, (3) the La/Yb of the former is higher than the latter. These major and trace element signatures of two parallel isotopic trends in Makapuu-stage indicate that DMK is more depleted in incompatible trace elements than EMK, as well as in Sr and Nd isotopic compositions.

In order to evaluate the mixing ratio among three endmember components (KEA, EMK, and DMK), the Sr, Nd, and Pb contents and their isotopic composition will be estimated. Since the exact endmember isotopic compositions are difficult to determine, possible compositions are estimated in the following way: The $^{87}Sr/^{86}Sr$, $^{143}Nd/^{144}Nd$, and $^{206}Pb/^{204}Pb$ of KEA and EMK are assumed to be close to the ends of the isotopic trends as shown in Figure 6, and their Sr/Nd, Pb/Sr, and Pb/Nd ratios are determined based on the averages from Mauna Kea and subaerial Makapuu-stage lavas (Table 4). Using the $^{206}Pb/^{204}Pb$ in Figure 6, the $^{207}Pb/^{204}Pb$ and $^{208}Pb/^{204}Pb$ of KEA are determined by the intersections between the linear trends of Mauna Kea lavas [*Abouchami et al.*, 2000] on $^{206}Pb/^{204}Pb$ - $^{207}Pb/^{204}Pb$ and $^{206}Pb/^{204}Pb$ - $^{208}Pb/^{204}Pb$ diagrams (Figure 7). Constraints for the estimation of $^{87}Sr/^{86}Sr$, $^{143}Nd/^{144}Nd$, and $^{206}Pb/^{204}Pb$ of DMK are: (1) Sr, Nd, and Pb contents of the DMK are lower than those of EMK, (2) $(Sr/Nd)_{DMK}$, $(Pb/Sr)_{DMK}$, and $(Pb/Nd)_{DMK}$ are within the range of Hawaiian basalts, and (3) $^{87}Sr/^{86}Sr$ and $^{143}Nd/^{144}Nd$ of DMK are isotopically less enriched than EMK. If the $^{87}Sr/^{86}Sr$ and $^{143}Nd/^{144}Nd$ of DMK are between those of KEA and EMK (Figure 6), the isotopic compositions of Koolau shield lavas cannot be satisfied by a ternary mixing model. Hence, $^{87}Sr/^{86}Sr$ and $^{143}Nd/^{144}Nd$ of DMK must be more depleted than those of KEA. In this study, we fix the $^{87}Sr/^{86}Sr$ and $^{143}Nd/^{144}Nd$ of DMK to those of Kea. The $^{208}Pb/^{204}Pb$ and $^{207}Pb/^{204}Pb$ of DMK are fixed to those of the intersections of the linear trend of Mauna Kea [*Abouchami et al.*, 2000] and the vertical line with the lowermost $^{206}Pb/^{204}Pb$ of the Makapuu-stage lavas in Figure 7. Finally, the $^{207}Pb/^{204}Pb$ and $^{208}Pb/^{204}Pb$ of EMK are determined to satisfy all the Pb isotopic compositions of Koolau shield lavas by ternary mixing (Figure 7). The resulting endmember compositions are summarized in Table 4.

6. EVOLUTION OF KOOLAU MAGMATISM AND MANTLE PLUME

The Sr-Nd-Pb isotope systematics of the Koolau shield investigated in this study reveal that the source materials for the shield lavas have changed successively. At the beginning of the main shield-stage, the mantle source was dominated by KEA (>60%) with a small component of EMK (<10%), resulting in Kea-type isotopic compositions (Figures 6-7). Subsequently, the proportion of KEA component decreased from 60 to 20% as DMK increased from 40 to 70% and EMK increased slightly (<20%), forming Loa-type isotopic compositions. During the later main shield-stage, the proportion of EMK increased to approximately 40%, as KEA decreased to 20%. In the Makapuu-stage, equivalent to the latest shield stage of Koolau Volcano, the proportion of EMK component drastically increased relative to DMK from 40 to 90%, with only minor KEA (<10%). The $^{206}Pb/^{204}Pb$ - $^{207}Pb/^{204}Pb$ and $^{206}Pb/^{204}Pb$ - $^{208}Pb/^{204}Pb$ of Mauna Loa [*Abouchami et al.*, 2000] overlap those of the main shield-stage lavas of Koolau Volcano (Figure 7). This suggests that the Mauna Loa shield lavas may also have formed by mixing of EMK, DMK and KEA. Similar sequentially evolving compositions of broadly tholeiitic lavas have recently been documented for Kilauea volcano, based largely on newly acquired underwater samples [*Sisson et al.*, this volume].

KEA component in this study is close to that previously proposed as a "Kea component" elsewhere [e.g. *Stille et al.*, 1986b; *Lassiter and Hauri*, 1998; *Bichert-Toft, et al.*, 1999; *Abouchami et al.*, 2000]. In contrast, EMK and DMK components are newly determined in this study for Hawaiian volcanoes. The $^{87}Sr/^{86}Sr$, $^{143}Nd/^{144}Nd$, and $^{206}Pb/^{204}Pb$ of EMK are the same as those previously proposed as the Koolau component [e.g. *Stille et al.*, 1986b; *Lassiter and Hauri*, 1998; *Bichert-Toft et al.*, 1999] of two-component mixing with Kea-type. The main difference between EMK and the previously obtained Koolau component [e.g. *Stille et al.*, 1986b; *Lassiter and Hauri*, 1998; *Bichert-Toft et al.*, 1999] is the trend of $^{207}Pb/^{204}Pb$ and $^{208}Pb/^{204}Pb$ in EMK to higher values than in the Makapuu-stage lavas (Figure 7). On the other hand, the existence of a DMK-like endmember component that is depleted in Pb isotopic compositions was obtained by *Abouchami et al.* [2000] from the linear isotopic trends of Mauna Kea lavas on $^{206}Pb/^{204}Pb$ - $^{207}Pb/^{204}Pb$ and $^{206}Pb/^{204}Pb$ - $^{208}Pb/^{204}Pb$ diagrams (linear mixing lines between KEA and EMK in Figure 7). In addition to *Abouchami et al.* [2000], this study indicates that the DMK component is more depleted than KEA in Sr and Nd isotopic compositions (Figure 6).

Among the three endmembers, EMK is the most enriched in SiO_2, Al_2O_3/CaO, incompatible trace elements, and Sr and Nd isotope compositions, suggesting that EMK could be recycled crustal material as widely proposed for Koolau lavas [e.g. *Hofmann and White*, 1982; *Hauri*, 1996; *Lassiter and Hauri*, 1998]. Both DMK and KEA are depleted in incompatible trace elements and Sr and Nd isotopic compositions compared to EMK, indicating that these two endmembers are chemically and isotopically depleted mantle. Although DMK and KEA are not discriminated in terms of Sr and Nd isotopic compositions, they are distinctive in Pb isotope systematics (Figure 7). The Sr-Nd-Pb isotope systematics of KEA and DMK do not resemble East Pacific Rise MORB (Figs 6 and 7), suggesting that neither KEA nor DMK have been involved with Pacific MORB source mantle. Currently, KEA and DMK are too unconventional as plume components to be interpretable in schemes of crust/mantle recycling through geological time.

7. THE SOURCE OF WEST AND EAST MOLOKAI SHIELD LAVAS

The similarity of incompatible trace elements and Sr-Nd-Pb isotopic compositions between East Molokai lavas and the main shield-stage lavas in Koolau Volcano suggest that the East Molokai shield could have been generated from the same endmember components as Koolau shield lavas (Figures 5-7), even though the stratigraphic order of submarine-erupted East Molokai lavas is uncertain. In contrast, the Sr-Nd-Pb isotopic compositions of a submarine-erupted sample from West Molokai NW resembles that of main shield-stage lavas in Koolau Volcano, except for higher $^{207}Pb/^{204}Pb$ (Figures 6-7). The subaerial West Molokai shield lavas have Sr-Nd-Pb isotopic compositions within the range of Mauna Loa rocks (not plotted in Figures 6-7) [*Stille et al.*, 1986b, *West et al.*, 1987] with relatively higher $^{207}Pb/^{204}Pb$ in the variation of the Mauna Loa lavas ($^{206}Pb/^{204}Pb$=18.07-18.17, ^{207}Pb / ^{204}Pb = 15.44 - 15.46, ^{208}Pb / ^{204}Pb = 37.73 - 37.75). Therefore, West Molokai lavas have relatively higher $^{207}Pb/^{204}Pb$ than either Mauna Loa or Mauna Kea shield at a given $^{206}Pb/^{204}Pb$, suggesting a fourth endmember with higher $^{207}Pb/^{204}Pb$ than KEA.

East Molokai and West Molokai shields contain both Kea- and Loa-type compositions, just like the Koolau shield, even though the subaerial shield lavas in East Molokai and West Molokai have been previously interpreted as Kea type and Loa type, respectively [*Still et al.*, 1986b; *West et al.*, 1987]. This is inconsistent with the interpretation that the two volcanic loci of in the Hawaiian volcanic chain [*Jackson et al.*, 1972] (Figure 1) have distinct Loa- and Kea-type geochemistry, respectively [*Tatsumoto*, 1978; *Stille et al.*, 1986b; *Hauri*, 1996].

8. CONCLUSIONS

There are two stages in Koolau Shield; main shield-stage and Makapuu-stage. In both the subaerial shield and northwestern submarine flank, main shield-stage lavas are buried by Makapuu-stage lavas. So-called Koolau-type lavas, as recognized previously, are restricted

Table A1 Trace element concentrations (ppm) of glass fragments in the PC2 piston core

	High Al_2O_3/CaO type				Low Al_2O_3/CaO type								
Sample	G2	I6	J1	L7	A6	C3	D2	D7	F1	F9	H6	I5	J10
level	90m	90m	90m	90m	30m	60m	60m	60m	90m	90m	90m	90m	90m
SiO_2^*	53.28	53.46	53.10	53.11	51.01	51.22	52.10	50.78	51.46	50.74	51.93	51.48	51.33
Al_2O_3/CaO^*	2.32	2.20	2.24	2.20	1.50	1.39	1.47	1.33	1.51	1.44	1.32	1.34	1.36
Sr	449	461	449	440	299	375	276	287	345	307	302	344	-
Y	27.7	28.0	27.7	28.1	28.9	29.7	24.5	28.5	32.0	29.6	30.8	34.6	26.8
Zr	180	182	178	176	131	155	104	132	141	130	141	171	172
Nb	11.6	11.8	11.1	11.1	10.8	13.9	8.0	10.8	8.1	10.2	10.9	13.4	10.8
Ba	102.9	107.2	101.7	105.0	58.2	103.4	59.4	52.1	73.5	62.6	70.6	86.7	98.3
La	13.1	13.2	13.4	13.6	9.4	13.1	7.8	10.1	11.1	9.8	10.2	12.7	13.0
Ce	33.4	34.4	34.2	35.3	25.7	34.9	21.0	27.2	28.9	25.3	27.9	34.4	33.0
Pr	4.78	4.96	5.08	4.87	3.63	4.57	2.97	4.00	4.05	3.93	3.87	5.11	4.91
Nd	24.5	24.9	25.1	24.2	19.7	22.9	15.8	19.6	20.3	18.7	19.7	23.3	24.3
Sm	6.41	6.31	6.86	6.27	5.39	6.19	4.28	5.70	5.76	5.22	5.53	6.93	6.63
Eu	2.47	2.33	2.33	2.33	1.86	2.16	1.53	2.10	2.22	1.80	2.03	2.26	2.26
Gd	6.35	6.30	6.10	6.22	4.38	5.78	4.15	4.87	5.91	4.50	5.19	5.37	5.89
Dy	5.97	6.06	6.55	6.46	6.09	6.12	5.04	6.96	6.56	6.12	6.58	7.67	6.33
Er	2.16	2.14	2.08	1.92	2.44	2.29	1.63	2.55	2.18	2.17	2.32	2.75	2.01
Yb	1.78	1.85	1.96	1.93	2.14	2.22	1.84	2.39	2.27	2.01	2.32	2.93	1.89
Lu	0.28	0.34	0.37	0.33	0.30	0.36	0.30	0.32	0.39	0.30	0.36	0.42	0.36

*: SiO_2 (wt.%) and Al_2O_3 are after *Sherman et al.* [this volume]. Trace element compositions of the glasses were analyzed using the Cameca 5f Secondary Ion Mass Spectrometer at the PML. The analytical procedure is essentially same as those in *Nakamura et al.* [1998] and *Nakamura and Kushiro* [1998].

to the Makapuu-stage. The uppermost part of main shield-stage lavas is exposed on land only low in Nuuanu Pali, and most of the main shield-stage lavas were obtained from the Nuuanu landslide blocks offshore north of Oahu.

Makapuu-stage lavas are higher than the main shield-stage lavas in Ba/Nb, La/Nb, Pb/Nb, Sr/Nb, Nd/Nb, and Zr/Nb (with relatively small variations of these ratios), and are more enriched in Sr and Nd isotopic compositions. The Sr-Nd-Pb isotopic compositions of Koolau shield lavas successively change with stratigraphic order, from Kea-type to Loa-type in the main shield-stage, to the latest Koolau-type in Makapuu-stage. Samples of the Koolau shield form a negative linear trend in a $^{87}Sr/^{86}Sr$ - $^{143}Nd/^{144}Nd$ diagram, and curved trends in $^{206}Pb/^{204}Pb$ - $^{87}Sr/^{86}Sr$ and $^{206}Pb/^{204}Pb$ - $^{143}Nd/^{144}Nd$ diagrams in which the Makapuu-stage has a characteristically small variation in $^{206}Pb/^{204}Pb$. Although these individual isotopic trends might suggest mixing of two endmembers for Koolau shield lavas, the Pb isotopic compositions require an additional component. These isotope systematics reveal that the mantle source for Koolau shield lavas contained at least three-endmember, denoted as Enriched Makapuu (EMK), Depleted Makapuu (DMK), and Kea (KEA). Both DMK and KEA are similarly depleted in Sr and Nd isotopic compositions ($^{87}Sr/^{86}Sr < 0.7035$, $^{143}Nd/^{144}Nd > 0.5130$), but the $^{206}Pb/^{204}Pb$ of KEA (>18.80) is distinctively higher than DMK (<17.78). In contrast, EMK is isotopically enriched in $^{87}Sr/^{86}Sr$ (>0.70427) and $^{143}Nd/^{144}Nd$ (<0.51265) compared with DMK and KEA. The $^{207}Pb/^{204}Pb$ and $^{208}Pb/^{204}Pb$ of EMK (>15.45 and >37.8) are higher than those of DMK (<15.37 and <37.2, respectively). Higher SiO_2, Al_2O_3/CaO and incompatible trace elements than other endmembers also characterize EMK.

Involvement of the three-endmembers changes with stratigraphic order from lower to upper in the Koolau shield lavas. Early main shield formation was dominated by a KEA source, with only a small proportion of EMK, resulting in a Kea-type composition. Subsequently, the proportion of KEA decreased, corresponding with a large increase in DMK and a small increase in EMK, forming a Loa-type composition. High in the section, the proportion of EMK drastically increased versus DMK, forming the Koolau-type composition. Based on the geochemical characteristics of these endmember components, EMK is inferred to represent a basaltic component, and KEA and DMK to represent two distinctively different depleted mantle components in the Hawaiian plume

Submarine-erupted West Molokai shield lavas also encompass the range of Sr-Nd-Pb isotopic compositions from Kea- to Loa-type; East Molokai shield lava has similar isotopic compositions to Kea-type. These data, combined with results for subaerial erupted lavas in both shields [*Stille et al.*, 1986b; *West et al.*, 1987], indicate that the Molokai shields possess both the Kea- and Loa-type compositions, just as at Koolau shield.

Koolau Volcano has generally been thought to have distinctive geochemical characteristics among Hawaiian volcanoes, based on isotope and trace element studies of samples from subaerial outcrops. This study shows that the Koolau shield lavas newly sampled from subaerial sequences and from submarine landslide blocks have isotopic compositions which cover almost the entire range reported previously for shield stage lavas from all Hawaiian islands. The evolution and geochemical structure of Hawaiian mantle cannot be modeled based solely from study of surface lavas, as is typically done for oceanic island basalts.

APPENDIX. TIMING OF NUUANU LANDSLIDE: CONSTRAINTS FROM THE TRACE ELEMENT COMPOSITIONS OF PC2 GLASS

The absence of Makapuu-stage lava fragments in the Nuuanu blocks seems to indicate that the landslide occurred before Makapuu-stage magmatism. Due to the small proportion of Makapuu-stage lavas in the shield,

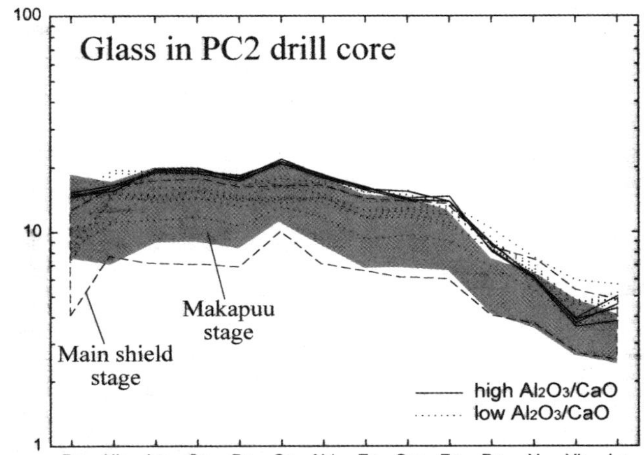

Figure A1. Primitive mantle-normalized trace elements compositions of glass fragments from PC2 piston core. The trace element data are normalized to the primitive mantle values of *Sun and McDonough* [1989]. The shaded area and the area outlined with a broken line are the ranges for Makapuu-stage and main shield-stage tholeiite compositions obtained in this study.

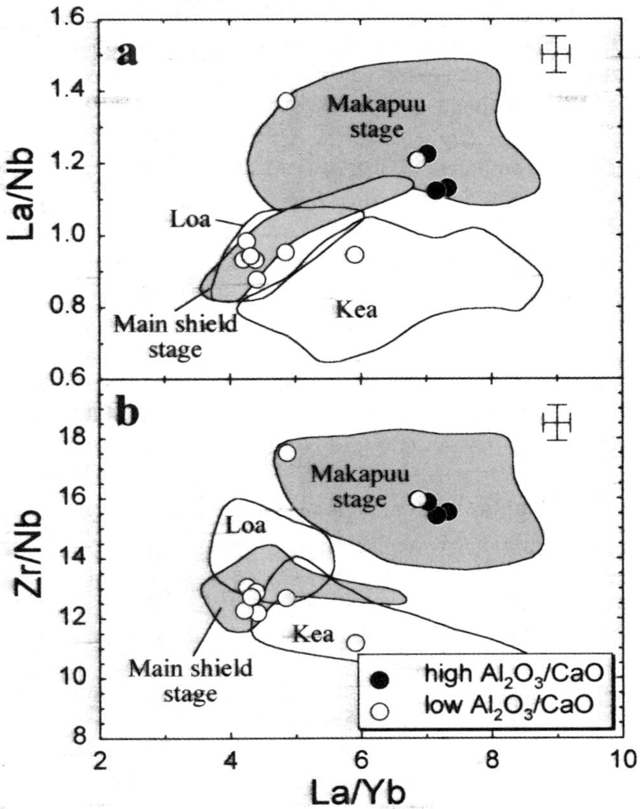

Figure A2. Trace element relationship for glass fragments recovered from PC2 piston core. Fields for Mauna Kea and Mauna Loa are the same as in Figure 6, and for Koolau are from *Frey et al.* [1991, 1994], *Jackson et al.* [1999], *Roden et al.* [1984, 1994], and this study. Error bars represent one standard deviation for analytical reproducibility.

Makapuu-stage lavas, occur in the PC2 piston core located 250 km northeast of Oahu [*Sherman et al.*, this volume] (Figure 1, Table A1). Magnetostaratigraphic study of the turbidite sand fragments from PC2 reveals that they were deposited at ~2 Ma, preceding the Wailau landslide at ~1 Ma [*Kanamatsu*, this volume]. *Sherman et al.* [this volume] concluded that the glasses in PC2 were erupted from the Koolau shield under subaerial or shallow marine conditions, because some glass fragments have high Al_2O_3/CaO (>2.0) compared with most other glasses (Al_2O_3/CaO <1.8) (Table A1). The Al_2O_3/CaO ratios of these glasses are higher than those of Koolau, East Molokai, and West Molokai lavas, except for K496 hyaloclastite in Oahu North-b [*Shinozaki et al.*, this volume] (Figure 3c). Therefore, trace element compositions of representative glass fragments in PC2 were analyzed in order to distinguish their geochemical characteristics (Table A1).

limited sampling might fail to collect Makapuu-stage lavas in the Nuuanu blocks. However, glass fragments, with major element compositions similar to those of the

Trace element compositions of the high Al_2O_3/CaO glasses have higher Ba, Sr, Zr and LREEs and lower HREEs compared to the low Al_2O_3/CaO glasses (Figure A1). Positive Sr anomalies appear in the high Al_2O_3/CaO glasses, but not in most low Al_2O_3/CaO glasses (Figure A1). The trace element compositions of high Al_2O_3/CaO glasses resemble those in Makapuu-stage lavas of Koolau Volcano (Figures A1-A2). Relationships among La/Yb vs. La/Nb and Zr/Nb allow clear discrimination between Makapuu-stage and main shield-stage lavas, or between Koolau and Loa and Kea types (Figure A2). The high Al_2O_3/CaO glasses and two of the low Al_2O_3/CaO glasses (F1 and J10 in Table 5) have higher La/Nb, Zr/Nb, and La/Yb than the others (Figure A2). These glasses also have trace element compositions characteristic of Makapuu-stage lavas. Most of the other low Al_2O_3/CaO glasses fall in the range of main shield-stage lavas. Most of the low Al_2O_3/CaO glasses seem to be Loa-type or Kea-type. Accordingly, the trace elements compositions of PC2 glasses confirm that the Nuuanu landslide occurred during or after Makapuu-stage magmatism.

Acknowledgements. Many thanks to K. Shinozaki for constructive discussions and field assistance, R. Takeguchi for denoting samples, T. Kuritani, H. Takei, T. Moriguti, A. Makishima, T. Yokoyama, N. Takeuchi, and all the other PML members for expert analytical assistance and discussions, and K. Ozawa for variable suggestions and encouragement. We thank to G. E. Bebout, M. J. Walter, and K. Koga for improving the manuscript, P. Lipman, and F. Frey and E. Hauri for constructive reviews. We greatly acknowledge P. Lipman for his helpful suggestions and editing. This research was supported in part by grants-in-aid for Scientific Research from Ministry of Education, Culture, Sports, Science and Technology of Japanese Government and Japan Society for the Promotion of Science to E. N. and by JAMSTEC to E. T.

REFERENCES

Abouchami, W., S. J. G. Galer, and A. W. Hofmann, High precision lead isotope systematics of lavas from the Hawaiian Scientific Drilling Project, *Chem. Geol.*, 169, 187-209, 2000.

Albarède, F., High-resolution geochemical stratigraphy of Mauna Kea flows from the Hawaii Scientific Drilling Project core, *J. Geophys. Res.*, B101, 11841-11853, 1996.

Bach, W., E. Hegner, J. Erzinger, and M. Satir, Chemical and isotopic variations along the superfast spreading East Pacific Rise from 6 to 30° S, *Contr. Min. Petrol*, 116, 365 - 380, 1994.

Basu, A. R. and B. E. Faggart, Jr., Temporal isotopic variations in the Hawaiian mantle plume: The Lanai anomaly, the Molokai Fracture Zone and seawater-altered lithospheric component in Hawaiian volcanism, in *Earth Processes: Reading the Isotopic Code, Geophys. Monogr. Ser.* vol. 95, edited by A. Basu and S. Hart, pp. 149-159, AGU, Washington, 1996.

Blichert-Toft, J., F. A. Frey, and F. Albarède, Hf isotope evidence for pelagic sediments in the source of Hawaiian basalts, *Science*, 285, 879-882, 1999.

Chen, C-Y, F. A. Frey, M. O. Garcia, and B. Dalrymple, The tholeiite to alkalic basalt transition at Haleakala Volcano, Maui, Hawaii: petrogenetic interpretations based on major and trace element and isotope geochemistry, *Contrib. Mineral. Petrol.*, 106, 183-200, 1991.

Doell, R. R. and G. B. Dalrymple, Potassium-argon ages and paleomagnetism of the Waianae and Koolau Volcanic Series, Oahu, Hawaii. *GSA bull.*, 84, 1217-1242, 1973.

Eiler, J. M., K. A. Farley, J. W. Valley, A. W. Hofmann, and E. M. Stolper, Oxygen isotope constraints on the sources of Hawaiian volcanism, *Earth Planet. Sci, Lett.*, 144, 453-468, 1996.

Feigenson, M. D., A. W. Hofmann, and F. J. Spera, Case studies on the origin of basalt. II. The transition from tholeiitic to alkalic volcanism on Kohala Volcano, Hawaii, *Contrib. Mineral. Petrol.*, 84, 309-405, 1983.

Fodor, R. V., F. A. Frey, G. R. Bauer, and D. A. Clague, Ages, rare-earth element enrichment, and petrogenesis of tholeiitic and alkalic basalts from Kahoolawe Island, Hawaii, *Contrib. Mineral. Petrol.*, 110, 442-462, 1992.

Frey F. A., M. O. Garcia, W. S. Wise, A. K. Kennedy, P. Gurriet, and F. Albarède, The evolution of Mauna Kea volcano, Hawaii: Petrogenesis of tholeiitic and alkalic basalts, *J. Geophys. Res.*, B96, 14347-14375,1991.

Frey, F. A., M. O. Garcia, and M. F. Roden, Geochemical characteristics of Koolau volcano: implications of intershield geochemical differences among Hawaiian volcanoes. *Geochim. Cosmocim. Acta*, 58, 1441-1462, 1994.

Frey, F. A., W. S. Wise, M. O. Garcia, H. West, S. -T. Kwon, and A. Kennedy, Evolution of Mauna Kea Volcano, Hawaii: petrologic and geochemical constraints on postshield volcanism, *J. Geophys. Res.*, B95, 1271-1300, 1990.

Hanyu, T. and E. Nakamura, Constraints on HIMU and EM by Sr and Nd isotopes re-examined. *Earth, Planets, and Space*, 52, 61-70, 2000.

Harpp, K. S., Geochemical variations in a single basaltic flow at 9°30'N on the East Pacific Rise, *Proc. ODP, Sci. Results*, 142, 9-22, 1995.

Hauri, E. H., Major-element variability in the Hawaiian mantle plume, *Nature*, 382, 415-419, 1996.

Hauri, E. H., and M. D. Kurz, Melt migration and mantle chromatography, 2: a time-series Os isotope study of Mauna Loa volcano, Hawaii, *Earth Planet. Sci, Lett.*, 153, 21-36, 1997.

Hofmann, A. W. and K. P. Jochum, Source characteristics derived from very incompatible trace elements in Mauna Loa and Mauna Kea basalts, Hawaii Scientific Drilling Project, *J. Geophys. Res.*, B101, 11831-11839, 1996.

Hofmann, A. W. and W. M. White, Mantle plumes from ancient oceanic crust. *Earth Planet. Sci, Lett.*, 57, 421-436, 1982.

Jackson, E. D., E. A. Silver, and G. R. Dalrymple, Hawaiian-Emperor chain and its relation to Cenozoic circumpacific tectonics, *GSA Bull.*, 83, 601-618, 1972.

Jackson, M. C., F. A. Frey, M. O. Garcia, and R. A. Wilmoth, Geology and petrology of basaltic lavas and dikes of the Koolau Volcano in the Trans-Koolau exploratory tunnels, *Bull. Volcanol.*, 60, 381-401, 1999.

Kanamatsu, T. and E. Herrero-Bervera, Magnetostratigraphy on marine sediments in an adjacent area of large Hawaiian submarine Landslides, *this volume.*

Koide, Y, and E. Nakamura, Lead isotope analyses of standard rock samples, *Mass Spectroscopy,* 38, 241-252, 1990.

Kurz, M. D., T. C. Kenna, D. P. Kammer, J. M. Rhodes, and M. O. Garcia, Isotopic evolution of Mauna Loa volcano: A view from the submarine southwest rift, in *Mauna Loa Revealed: Structure, Composition, History, and Hazards, Geophys. Monogr. Ser.,* vol. 92, edited by J. M. Rhodes and J. P. Lockwood, pp. 289-306, AGU, Washington, 1995.

Lanphere, M. A. and G. B. Dalrymple, Age and strontium isotopic composition of the Honolulu Volcanic Series, Oahu, Hawaii, *Am. J. Sci.,* 280-A, 736-751, 1980.

Lassiter, J. C. and E. H. Hauri, Osmium-isotope variations in Hawaiian lavas: evidnce for recycled oceanic lithosphere in the Hawaiian plume, *Earth Planet. Sci, Lett.,* 164, 483-496, 1998.

Lassiter, J. C., D. J. DePaolo, and M. Tatsumoto, Isotopic evolution of Mauna Kea volcano: Results from the initial phase of the Hawaii Scientific Drilling Project, *J. Geophys. Res.,* B101, 11769-11780, 1996.

Lassiter, J. C., E. H. Hauri, P. W. Reiners, and M. O. Garcia, Generation of Hawaiian post-erosional lavas by melting of a mixed lherzolite/pyroxenite source, *Earth Planet. Sci, Lett.,* 178 269-284, 2000

Lipman, P. W., J. M. Rhodes, and G. B. Dalrymple, The Ninole basalt - implications for the structural evolution of Mauna Loa Volcano, Hawaii, *Bull. Volcanol.,* 53, 1-19, 1990.

Macdonald, G. A. and T. Katsura, Chemical composition of Hawaiian lavas, *J. Petrol.,* 5, 82-133, 1964.

Mahoney, J. J., J. M. Sinton, M. D. Kurz, J. D. MacDougall, K. J. Spencer, and G. W. Lugmair, Isotope and trace element characteristics of a super-fast spreading ridge: East Pacific Rise 13-23°S, *Earth Planet. Sci. Lett.,* 121, 173 - 193, 1994.

Makishima, A, E. Nakamura and T. Nakano, Determination of zirconium, niobium, hafnium and tantalum at ng/g levels in geological materials by direct nebulization of samples HF solution into FI-ICP-MS, *Geostand. Newslett.,* 23, 7-20, 1999.

Makishima, A. and E, Nakamura, Suppression of matrix effect in ICP-MS by high power operation of ICP: Application to precise determination of Rb, Sr, Y, Cs, Ba, REE, Pb, Th and U at ng/g level in a few milligram silicate sample, *Geostand. Newslett.,* 21, 307-319, 1997.

Makishima, A. and E. Nakamura, Precise measurement of cerium isotope composition in rock samples, *Chem. Geol.,* 94, 1-11, 1991.

Makishima, A., E. Nakamura, and T. Nakano, Determination of boron in silicate samples by direct aspiration of sample HF solutions into ICPMS, *Anal. Chem.,* 69, 3754-3759,1997.

Muehlenbachs, K. and R. Clayton, Oxygen isotope studies of fresh and weathered submarine basalts, *Can, J. Earth Sci.*, 9, 172-184, 1972.

Nakamura, E. and I. Kushiro, Trace element diffusion in jadeite and diopside melts at high pressures and its geochemical implication, *Geochim. Cosmochim. Acta*, 62, 3151-3160, 1998.

Nakamura, E., T. Suzuki, K. Kobayashi, A. Makishima, and M. Yoshikawa, Trace element partitioning between majorite and silicate melt under high pressures, *paper presented at Ann. Meeting of the Japan Earth and Planet. Sci. Joint Meeting*, Tokyo, Japan, 67, 1998.

Nakano, T. and E. Nakamura, Static multicollection of $Cs_2BO_2^+$ ions for precise boron isotope analysis with positive thermal ionization mass spectrometry, *Int. J. Mass Spectrom. Ion Processes,* 176, 13-21, 1998.

Niu, Y, D. G. Waggoner, J. M. Sinton, and J. J. Mahoney, Mantle source heterogeneity and melting processes beneath seafloor spreading centers: The East Pacific Rise, 18°-19°S, *J. Geophys. Res.,* B101, 27711 - 27733, 1996.

Norman M. D. and M. O. Garcia, Primitive magmas and source characteristics of the Hawaiian plume: petrology and geochemistry of shield picrites, *Earth Planet. Sci, Lett.,* 168, 27-44, 1999.

Regelous, M, Y. Niu, J. I. Wendt, R. Batiza, A. Greig, and K. D. Collerson, Variations in the geochemistry of magmatism on the East Pacific Rise at 10° 30'N Since 800 Ka, *Earth Planet. Sci. Lett.,* 168, 45 - 63, 1999.

Rhodes, J. M. and S. R. Hart, Episodic trace element and isotopic variations in historical Mauna Loa lavas: implications for magma and plume dynamics, in *Mauna Loa Revealed: Structure, Composition, History, and Hazards, Geophys. Monogr. Ser.,* vol. 92, edited by J. M. Rhodes and J. P. Lockwood, pp. 263-288, AGU, Washington, 1995.

Rhodes, J. M., Geochemical stratigraphy of lava flows sampled by the Hawaii Scientific drilling Project, *J. Geophys. Res.,* B101, 11729-11746, 1996.

Rhodes, J. M., Geochemistry of the 1984 Mauna Loa eruption: Implications for magma storage and supply, *J. Geophys. Res.,* B93, 4453-4466, 1988.

Roden, M. F., F. A. Frey, and D. A. Clague, Geochemistry of tholeiitic and alkalic Lava from the Koolau range, Oahu, Hawaii: implications for Hawaiian volcanism, *Earth Planet. Sci. Lett.,* 69, 141-158, 1984.

Roden, M. F., T. Trull, S. R. Hart, and F. A. Frey, New He, Sr, Nd and Pb isotopic constraints on the constitution of the Hawaiian plume: Results from Koolau Volcano, Oahu, Hawaii, *Geochim. Cosmochim. Acta*, 58, 1431-1440, 1994.

Satake, K., J.R. Smith, and K. Shinozaki, Volume estimate and tsunami modeling for the Nuuanu and Wailau slides, *this volume.*

Sherman, S., M. O. Garcia and E, Takahashi, Geochemistry of volcanic glasses from piston cores taken north of Oahu and Molokai Islands, *this volume.*

Shinozaki, K., Z, Ren and E, Takahashi, Geochemical and Petrological Characteristics of Nuuanu and Wailau landslide blocks, *this volume.*

Sisson, T. W., P. W. Lipman, and J. Naka, Submarine alkalic through tholeiitic shield-stage development of Kilauea volcano, *this volume.*

Smith, J., and K. Satake, Collapse features revealed by recent multibeam sonar surveys on the Hawaiian Ridge, *this volume.*

Staudigel, H., T. Plank, B. White, and H-U. Schmincke, Geochemical fluxes during seafloor alteration of the basaltivc upper oceanic crust: DSDP sites 417 and 418, in *Subduction Top to Bottom, Geophys. Monogr. Ser.* vol. 96, edited by G.E. Bebout, D. W. Scholl, S. H. Kirby, and J. P. Platt, pp. 19-38, AGU, Washington, 1996.

Stille, P., D. M. Unruh, and M. Tatsumoto, Pb, Sr, Nd and Hf isotopic evidence of multiple sources for Oahu, Hawwaii basalts, *Nature*, 304, 25-29, 1986a.

Stille, P., D. M. Unruh, and M. Tatsumoto, Pb, Sr, Nd, and Hf isotopic constraints on the origin of Hawaiian basalts and evidence for a unique mantle source, *Geochim. Cosmochim. Acta.*, 50, 2303-2319, 1986b.

Stolper, E. M., D. J. DePaolo, and D. M. Thomas, Introduction to special section: Hawaii Scientific Drilling Project, *J. Geophys. Res.,* B101, 11593-11598, 1996.

Sun, S.-s., and W.F. McDonough, Chemical and isotopic systematics of oceanic basalts: implications for mantle composition and processes, in *Magmatism in the Ocean Basins*, edited by A.D. Saunders, and M.J. Nory, pp. 315-345, geological Society Special publication, 1989.

Takahashi, E. and K. Nakajima, Melting process in the Hawaiian mantle plume head: an experimental study, *this volume*.

Takeguchi, R. Primary magma composition of Koolau Volcano, Oahu, Hawaii, Master Thesis, 84pp, Tokyo Institute of Technology, 1999.

Tatsumoto, M., Isotopic composition of lead in oceanic basalt and its implication to mantle evolution, *Earth Planet. Sci, Lett.*, 38, 63-87, 1978.

Todt, T., R. A. Cliff, A. Hanser, and A. W. Hoffmann, Evaluation of a ^{202}Pb-^{205}Pb double spike for high-precision lead isotope analysis, in *Earth Processes: Reading the Isotopic Code*, Geophys. Monogr. Ser. 95, edited by A. Basu and S. Hart, pp. 429-437, AGU, Washington, 1996.

West, H. B., D. C. Gerlach, W. P. Leeman, and M. O. Garcia, Isotopic constraints on the origin of Hawaiian lavas from the Maui Volcanic Complex, Hawaii, *Nature,* 330, 216-220, 1987.

Yang, H.-J., F. A. Frey, M. O. Garcia, and D. A. Clague, Submarine lavas from Mauna Kea volcano, Hawaii: Implications for Hawaiian shield-stage processes, *J. Geophys. Res.*, B99, 15577-15594, 1994.

Yang, H.-J., F. A. Frey, J. M. Rhodes, and M. O. Garcia, Evolution of Mauna Kea volcano: Inferences from lava compositions recovered in the Hawaii Scientific Drilling Project, *J. Geophys. Res.*, B101, 11747-11767, 1996.

Yokose, H. Submarine geology of giant landslides in the Northeast Oahu: SHINKAI 6500 submersible investigation, *this volume.*

Yokoyama, T. and E. Nakamura, Precise determination of ferrous iron in silicate rocks, *Geochim. Cosmochim. Acta.,* in press.

Yokoyama, T., A. Makishima and E. Nakamura, Evaluation of the coprecipitation of incompatible trace elements with fluoride during silicate rock dissolution by acid digestion, *Chem. Geol.*, 157, 175-187, 1999.

Yoshikawa, M. and E. Nakamura, Precise isotope determination of trace amounts of Sr in magnesium-rich samples, *Japan. J. Min. Pet. Econ. Geol.*, 88, 548-561, 1993.

Zindler, A. and S. Hart, Chemical Geodynamics, *Ann. Rev. Earth Planet. Sci.*, 14, 493-571, 1986.

Ryoji Tanaka, Eizo Nakamura, PML, Institute for Study of the Earth's Interior, Okayama University, Misasa, 682-0193, Japan (ryoji@misasa.okayama-u.ac.jp, eizonak@misasa.okayama-u.ac.jp)

Eiichi Takahashi, Earth and Planetary Sciences, Tokyo Institute of Technology, Meguro, Tokyo, 152-8551, Japan (etakahas@geo.titech.ac.jp)

Three-Dimensional Reconstruction and Tsunami Model of the Nuuanu and Wailau Giant Landslides, Hawaii

Kenji Satake

Active Fault Research Center, AIST-GSJ, Tsukuba, Japan

John R. Smith

University of Hawaii, School of Ocean and Earth Science and Technology, Honolulu, Hawaii

Ken Shinozaki

Tokyo Institute of Technology, Department of Earth and Planetary Sciences, Tokyo, Japan

The Koolau and East Molokai volcanoes are reconstructed and the volume changes associated with the Nuuanu and Wailau landslides are estimated. An objective and universal method is developed and applied to the digital bathymetric data obtained during recent cruises. Because the location of the headwall immediately after the landslides is poorly known, we model both subaerial and submarine headwalls. For the Wailau landslide, the missing slide volume is 1.5x 10^3 km^3 for the subaerial headwall and 0.8 x 10^3 km^3 for the submarine headwall. The total volume of the excess debris deposit is estimated as 1.5 x 10^3 km^3. For the Nuuanu landslide, the slide volume is approximately 3 x 10^3 km^3 for the subaerial headwall and 2 x 10^3 km^3 for the submarine headwall. The debris volume is estimated as 3 x 10^3 km^3. Tsunami generation and propagation from these landslides are computed on the present-day bathymetry. The tsunami reaches most of the Hawaiian Islands within half an hour from slide onset. Tsunami heights from the Nuuanu landslide are greater than 100 m on the northern coasts of Oahu and Molokai islands, and 20 – 30 m on the southern coasts. The tsunami arrives on the West Coast of North America in about 5 hours and Japan in 8 hours. The Nuuanu tsunami is strongly directed toward southern California where the tsunami heights would be as large as 70 m. The tsunami arrival times from the Wailau landslide are similar, but the water heights are about 60 – 70 % of those from the Nuuanu landslide and are differently oriented.

1. INTRODUCTION

Numerous submarine landslide morphologies have been mapped around the Hawaiian islands, and they are classified into slumps and debris avalanches [*e.g., Moore et al.*, 1994]. Slumps are characterized by wide and thick blocks moving slowly and coherently, whereas debris avalanches are

Hawaiian Volcanoes: Deep Underwater Perspectives
Geophysical Monograph 128
Copyright 2002 by the American Geophysical Union

distinguished by many smaller masses of debris distributed over longer regions suggesting fast-moving landslides. The Nuuanu and Wailau debris avalanche deposits, extending north of Oahu and Molokai islands, respectively, were first described and proposed as resulting from landslides by *Moore* [1964]. Nuuanu is one of the largest submarine landslides in the world [*Moore et al.*, 1989]. Tuscaloosa seamount, located nearly 100 km from the northern shore of Oahu island, is approximately 25 km wide, 12 km long and rises nearly 2 km above the seafloor. It has a total volume of about 600 km^3, and is interpreted as one of the Nuuanu debris blocks.

Tsunamis are one of the hazards caused by submarine landslides [*e.g.*, *Keating and McGuire*, 2000]. The 1975 Kalapana earthquake on the southeastern coast of the island of Hawaii (Big Island) was caused by sudden motion of the Hilina slump, the mobile south flank of Kilauea volcano [*Lipman et al.*, 1985; *Ma et al.*, 1999]. The resulting tsunami killed two campers. Debris avalanches may be more efficient in generating tsunamis than slumps, because of the faster motion. In 1741, a much smaller debris avalanche (total volume of about 2.5 km^3) at Oshima-Oshima near Hokkaido, Japan, generated large tsunamis in the Japan Sea and caused damage as far as the Korean coast [*Satake and Kato*, 2001]. No historical tsunami from a debris avalanche has been documented in Hawaii, but the huge size of the landslides might result in catastrophic tsunamis. Coral deposits found at 300 m above present sea level have been interpreted as a tsunami deposit [*Moore and Moore*, 1984].

Multibeam swath bathymetry surveys were conducted during the 1998 (R/V *Kairei*) and 1999 (R/V *Yokosuka*) JAMSTEC cruises around the Hawaiian islands, and detailed digital bathymetric data were obtained [*Naka et al.*, 2000; *Smith and Satake*, this volume]. High-resolution bathymetric data, merged with topography, are used to reconstruct the Koolau and East Molokai volcanoes and estimate the volume changes associated with the Nuuanu and Wailau debris avalanches. The volume change is then used to simulate tsunami generation and propagation around the Hawaiian islands and the Pacific Ocean with an estimate of the local and far field tsunami heights as the final products.

2. RECONSTRUCTION AND VOLUME ESTIMATES OF DEBRIS AVALANCHES

2.1. Data and Reconstruction Method

Digital bathymetric data obtained during the JAMSTEC cruises [*Smith and Satake*, this volume] are merged with other bathymetry and topography data compiled and archived at the University of Hawaii [*Dunnebier et al.*, 1994: *Smith*, 1994]. The University of Hawaii bathymetry data for the Hawaiian islands were compiled from a combination of multibeam data (mainly around the island of Hawaii), NOS hydrographic nearshore data, random trackline data from ships of opportunity (offshore), and USGS digital topography for the islands. These data are merged and regridded with an interval of 30" (arc second) or about 1 km.

Plate 1 is the current bathymetry/topography around Oahu and Molokai, which clearly shows highly eroded remnants of shield volcanoes. Oahu consists of the Waianae range to the southwest and the younger Koolau range to the northeast, the latter of which is interpreted as the source of the Nuuanu debris avalanche. Molokai is made up of the West Molokai and East Molokai volcanoes, the latter is believed to be the source of the Wailau debris avalanche. The Nuuanu and Wailau debris deposits extend more than 200 km from the present shorelines. The location of the landslide headwalls is debatable; *Moore et al.* [1989] assumed the steep subaerial cliff (Nuuanu Pali) as the headwall, while *Moore and Clague* [this volume] attribute a submarine slope more than 10 km offshore to the headwall.

The original topography and bathymetry before the debris avalanche is reconstructed and the volumes of slide and debris deposits are estimated by using an objective and universal method suitable to digital data. A debris avalanche or landslide is assumed to consist of the slide region where the mass has been removed and the debris deposit region where the mass has been placed. In this paper, the term "slide region" refers to the only area where mass was removed and does not include the area where debris was deposited. For simplicity, no overlap of the two regions is assumed in this model; the effect of this assumption will be discussed later (in section 4.1). To perform the reconstruction of original topography and bathymetry, the slide region is first defined and masked. The topography/bathymetry inside the masked region is then estimated by fitting a curved surface to the surrounding data outside the masked slide region. Next, the regional extent of the debris deposits is defined and masked. The original bathymetry, before the deposition of debris, is estimated from the surrounding data in a similar way. The volume change can be computed from the original (reconstructed) and current bathymetry data. The method was applied to the Oshima-Oshima volcano and the volumes of the 1741 sector collapse and debris deposits were successfully estimated [*Satake and Kato*, 2001].

In this method, topography and bathymetry $z(x, y)$ are reconstructed by solving

$$(1-T)\nabla^2(\nabla^2 z)+T\nabla^2 z=0 \tag{1}$$

where T is the tension parameter [*Smith and Wessel,* 1990]. The program *surface* of the GMT (Generic Mapping Tools) system software package (version 3.3.6) is used [*Wessel and Smith*, 1998], where a value of T=0.35 is suggested for steep topography data. When the tension parameter is set as T=0, the reconstructed surface provides the minimum curvature solution. At the other extreme, T=1 gives a harmonic surface without local minima or maxima except at constrained points. *Smith and Wessel* [1990] demonstrated that the minimum curvature solution (T=0) causes erroneous reconstruction for bathymetry. The most appropriate choice of the tension parameter is not obvious and should be chosen to satisfy conditions such as a reasonable shape for the volcano edifice based on geological and geomorphological considerations on a case by case basis.

2.2. Reconstruction of the East Molokai Volcano

The East Molokai volcano is reconstructed to estimate the Wailau slide volume by first defining the slide boundary as shown in Figure 1a. As stated in the previous section, location of the upper (southern) edge, or headwall, of the slide is debatable. Two different headwall locations, subaerial and submarine, are modeled. The subaerial headwall is set along the summit of the current East Molokai volcano (the peak is about 1000 m above sea level). The submarine headwall is set along the -1000 m contour, probably a former coast line (dashed line in Figures 1a, b). The lateral, or the eastern and western, boundaries are set outside the submarine amphitheater mapped north of the East Molokai volcano so that the entire slide is included. The lower (northern) limit, or foot, of the slide is assumed along the -4000 m contour as shown in the figure. The grid points inside this boundary are masked (Figure 1b) and the topography is estimated from surrounding data using the gridding scheme with tension parameters described in the previous section (equation 1).

Several assumptions regarding the topography and bathymetry are made, and constraints posed at several locations (Figure 1b). A steep slope separating the Hawaiian Ridge and Hawaiian Deep, approximately between the -1500 m and -4000 m contours, extends in the EW direction north of Oahu and Molokai. The same slope is assumed to have existed on the northern slope of the original East Molokai volcano before it collapsed. To provide this pre-slide surface for the modeling, it is necessary to assume that eight points along the lower boundary are located at the -4000 m contour, and eight parallel points lie at -1500 m (Figure 1b). For the subaerial headwall model, the locations and heights of the summit and caldera rim are assumed to correspond with the present East Molokai volcano. The caldera rim is about 10 km in diameter [*Langenheim and Clague*, 1987], the height of the rim is set at +1000 m, and the summit height is initially assumed to be +2000 m.

The difference between the reconstructed volcano and the current topography is input into the slide volume calculation, which is computed by integrating the height differences. The tension parameter T and the peak height are varied to examine the effects on the volume estimate. The northern slope and the shape of the reconstructed volcano depend on the tension parameter. For the subaerial headwall model, the reconstructed edifice becomes largest when T=0 with an estimated slide volume of 1.7 x 10^3 km^3, and is smallest for T=1 with a volume of 1.2 x 10^3 km^3. The reconstructed volcano is nearly circular in shape when T=0.6 (Figure 1c), with an estimated slide volume of 1.5 x 10^3 km^3. The assumed summit height, on the other hand, has only a small effect on the estimated slide volume. A difference of 1 km in the summit height (either a flat surface at 1000 m high or a steep peak 2000 m high) makes less than 0.1 km^3 difference in the volume. For the submarine headwall model, the tension parameter has a very small effect on the shape and reconstructed volume. The estimated volume becomes 0.9, 0.8 and 0.7 x 10^3 km^3 for T=0.0, 0.6 and 1.0, respectively.

In summary, the estimated volume of the Wailau slide is (1.5 ± 0.3) x 10^3 km^3 for the subaerial headwall model, while it is (0.8 ± 0.1) x 10^3 km^3 for the submarine headwall model.

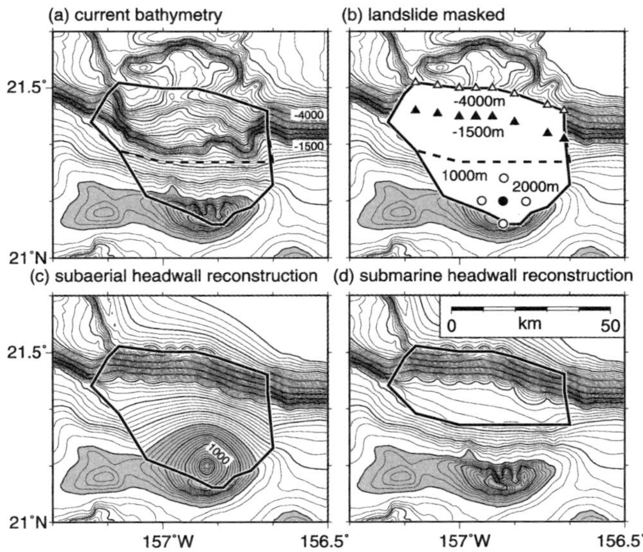

Figure 1. Reconstruction procedure for the Wailau slide from the East Molokai volcano. The contour interval is 100 m. (a) Current topography and bathymetry. The region for the Wailau slide is shown by thick lines. The dashed line shows the submarine headwall location. (b) The slide region is masked and reconstructed from the surrounding data outside the masked slide. Triangles represent the assumed pre-slide slope and circles the summit crater. (c) The East Molokai volcano reconstructed for the subaerial headwall model. (d) The East Molokai volcano reconstructed for the submarine headwall model.

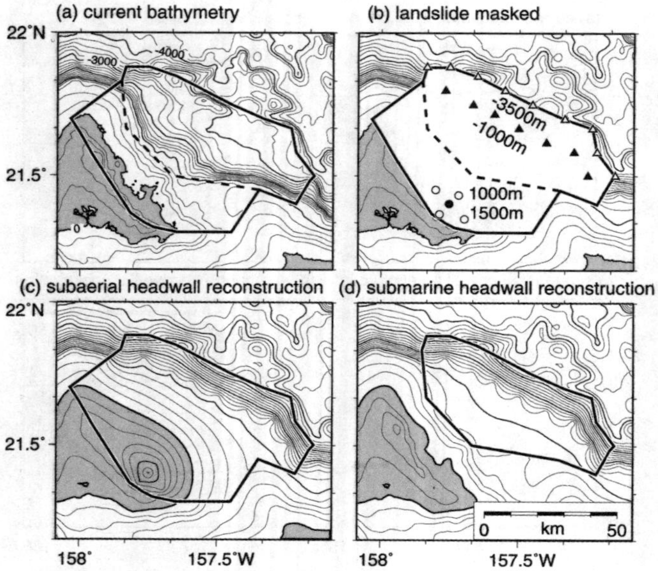

Figure 2. Reconstruction procedure for the Nuuanu slide from the Koolau volcano. The contour interval is 200 m. (a) Current topography and bathymetry. The region for the Nuuanu slide is shown by thick lines. (b) The slide region is masked and reconstructed from the surrounding data outside the masked slide. Triangles represent the assumed pre-slide slope and circles the summit crater. (c) The Koolau volcano reconstructed for the subaerial headwall model. Tension parameter T =0.7 and the reconstructed volcano has a symmetrical shape (reconstruction K1). (d) Reconstruction for the submarine headwall model.

2.3. Reconstruction of the Koolau Volcano

The Koolau volcano is reconstructed in a similar way to the East Molokai volcano. The slide boundary is less obvious for this case, thereby making the uncertainty larger. As in the East Molokai volcano reconstruction, both subaerial and submarine headwalls are modeled. The subaerial upper (southwestern) boundary is set along the summit of the current Koolau range (solid curve in Figure 2a). Presently, the highest point is about 800 m above sea level. For the submarine headwall model, the upper boundary is set along the -1000 m contour (dashed curve in Figure 2a). The lateral (northwestern and southeastern) boundaries are set outside the submarine amphitheater. The lower (northeastern) limit, or foot, of the slide is placed at about -3500 m contour as shown in the figure. This level is shallower than that assumed for the Wailau slide, because the -4000 m contour for the Nuuanu landslide includes some obvious debris blocks.

Like East Molokai volcano, the steep slope is assumed to continue from north of Oahu island to north of Molokai island. North of Oahu, the slope extends between the -1000 m and -3500 m levels. To incorporate this pre-slide slope into the modeling, nine points are positioned at -3500 m and seven points at -1000 m (Figure 2b). For the subaerial headwall model, we follow the caldera location estimated by *Langenheim and Clague* [1987], and the rim is set at +1000 m while the summit at +1500 m. The effect of the latter assumption on the estimated volume is again small, as noted in the East Molokai reconstruction.

For the subaerial headwall model, the effect of the tension parameter is much more significant than for the East Molokai case. When T=0, the reconstructed volcano becomes large and the coastline expands toward the northeast. At the other extreme, the reconstructed volcano becomes much narrower when T=1. The estimated slide volumes are also quite different; 4.2 and 2.3 x 10^3 km^3 for T=0 and 1, respectively. Visual inspection of the reconstructed volcano shape, however, would exclude such extreme cases. The reconstructed edifice is symmetric about the assumed ridge when T=0.7 with a slide volume of 2.9 x 10^3 km^3 (Figure 2c; named reconstruction K1). Volcano shape still appears natural for T=0.2, although the shape is asymmetric about the ridge. The northeastern flank is about twice as wide as the southwestern side. In this case, the steep submarine slope from the -3000 m level continues to the coast. The estimated slide volume is 3.8 x 10^3 km^3 (named reconstruction K2). These may be the two end members of the reconstruction, and we conclude that the slide volume ranges (2.9 –3.8) x 10^3 km^3.

For the submarine headwall model, the tension parameter has a smaller effect on the reconstructed slope. When T= 0, the reconstructed bathymetry has a 300 m deep hole just below the assumed headwall. When T= 1, the reconstructed topography still shows an arcuate amphitheater. The slide volumes for these extreme cases are 2.2 and 1.6 x 10^3 km^3, respectively. The reconstructed bathymetry before the slide shows a natural shape for T values between 0.7 and 0.2, with slide volumes ranging (1.9 – 2.1) x 10^3 km^3.

In summary, the estimated volume of the Nuuanu slide is (2.9 –3.8) x 10^3 km^3 for the subaerial headwall model, while it is (1.9 – 2.1) x 10^3 km^3 for the submarine headwall model.

2.4. Reconstruction of Debris Deposits

The original seafloor before the debris avalanche is next reconstructed and the total volume of debris estimated. The surface of the seafloor on which the avalanche deposits lie is at about -5000 m but not flat. To the north of Oahu and Molokai islands, the Hawaiian Deep extends in the ESE to WNW direction. The Hawaiian Deep is deeper to the east, at about -5700m, but shallower to the west at less than -5000 m (Plate 1). The seafloor becomes shallower beyond the Deep

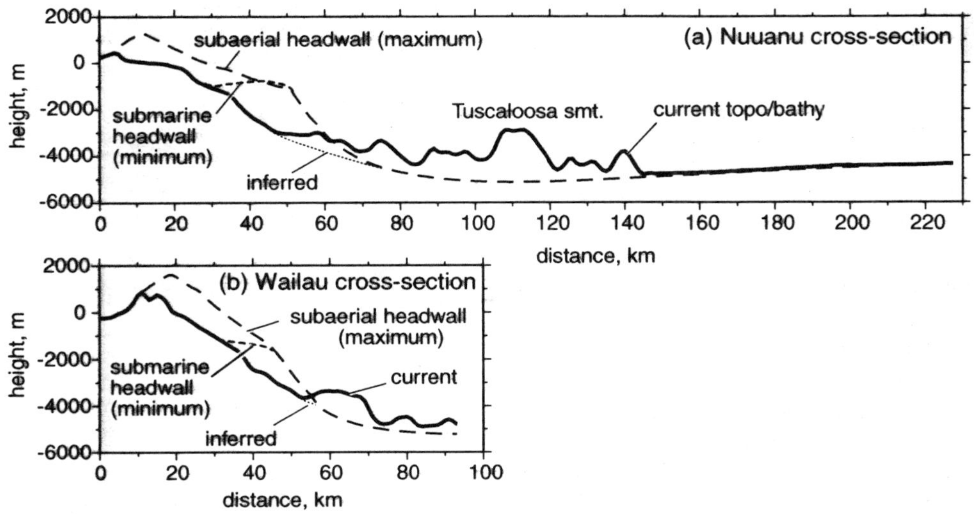

Figure 3. Cross-sections of the current and reconstructed topography/bathymetry for the (a) Nuuanu and (b) Wailau landslides. The thick curve is current profile. The dashed curves are reconstructed profiles for the subaerial and submarine headwalls. The dotted curve is inferred bottom profile (see text). The locations of the profiles are shown in Plates 2 and 3. Vertical exaggeration is 6.25.

toward the northeast, hence the debris avalanche must have climbed up this gentle slope unless the state of plate flexure was different than at present.

The regional extent of the avalanche deposit is estimated in Plate 2. Grid points within this boundary are masked, and the original bathymetry reconstructed from the surrounding data using the gridding scheme with varying tension parameter previously described. Again, the choice of the tension parameter makes a significant difference in the reconstructed surface. When T=0, the rebuilt Hawaiian Deep has a hole. At the other extreme, when T=1, the Koolau volcano unrealistically extends to the northeast and forms a saddle at about -4500 m, dividing the Hawaiian Deep. From visual inspections of reassembled bathymetry, we consider the best value for T is 0.35, with a possible range of $0.3 \leq T \leq 0.4$. Our criterion is that the reconstructed Deep should become deeper toward the east without creating local maxima (Plate 2). The reconstruction also depends on the surrounding topography used, and here the reconstructed volcano topography is utilized rather than the current bathymetry. When the K1 topography is used, the total volume of the debris is 5.4, 4.5, and 3.8 x 10^3 km^3 for T=0.3, 0.35, and 0.4, respectively. If the K2 topography is used, then the volumes increase by about 0.2 x 10^3 km^3 for each case.

From these results, we conclude that the total volume is (4.5 ± 1.1) x 10^3 km^3. The average thickness of the debris deposits, the volume divided by the area, is calculated as approximately 250 m.

2.5. Volume Balance between the Slides and Debris Deposits

Plate 3 shows the distribution of missing (slide) and excess (debris deposit) volumes. Total excess volume from the Wailau and Nuuanu slides has been estimated as (4.5 ± 1.1) x 10^3 km^3. Conversely, total missing volume of the both slides is calculated as (4.1 – 5.6) x 10^3 km^3 for the subaerial headwall model, and (2.6 – 3.0) x 10^3 km^3 for the submarine headwall model. Because the total debris volume is sandwiched by the two volume estimates with subaerial and submarine headwalls, these two may be considered as maximum and minimum estimates of the slide volumes.

Figure 3 shows representative topographic cross-sections through the Nuuanu and Wailau landslides. The reconstructed bathymetric profiles for the submarine and subaerial headwall models are almost identical below the assumed submarine headwalls, hence the difference in volume comes from the subaerial parts.

For the tsunami modeling, we adopt the volume estimate from the subaerial headwall for two reasons. First, it provides the maximum estimate of slide volume. Second, the estimated slide volume is closer the total debris volume. Considering the volume balance, *i.e.*, the missing volume matches the excess volume, the K1 reconstruction is preferred. Therefore, the lower slide volume for the subaerial headwall, 2.9 x 10^3 km^3, is used in the remainder of this paper.

The contributions from the Nuuanu and Wailau debris deposits are estimated. Although the reconstruction was performed for the entire debris deposit from both slides, they can be divided into two specified regions and the volumes compared. If the deposits are divided by their orientation (Plate 3), the volume of the Nuuanu and Wailau slides becomes 1.0 and 3.5 x 10^3 km^3, respectively. The Wailau deposit (1.0 x 10^3 km^3) is smaller than the maximum slide volume (1.5 x 10^3 km^3), but the Nuuanu deposit (3.5 x 10^3 km^3) is larger than the maximum slide volume (2.9 x 10^3 km^3). A trial and error analysis is performed so that the slide (missing) volume and the debris (excess) volume match for each slide. When the deposits are divided as shown in Plate 3, the volume for the Nuuanu landslide deposits is (3.0 ± 0.5) x 10^3 km^3 and the volume for the Wailau landslide deposits becomes (1.5 ± 0.5) x 10^3 km^3. In the above analysis, the error associated with the debris volume estimate is equally distributed between the two deposits. The average thickness of the debris deposits equals 270 m for the Nuuanu debris and 210 m for the Wailau deposit.

The volume changes associated with the Nuuanu and Wailau debris avalanches are summarized as follows. For the Nuuanu landslide, debris deposit volume is estimated as 3 x 10^3 km^3. The missing slide volume is approximately 3 x 10^3 km^3 for the subaerial headwall, while it is about 2 x 10^3 km^3 for the submarine headwall. For the Wailau landslide, the debris deposit volume is estimated as 1.5 x 10^3 km^3. The missing slide volume is 0.8 x 10^3 km^3 for the submarine headwall and 1.5 x 10^3 km^3 for the subaerial headwall. The larger volumes will be used for the tsunami modeling.

3. TSUNAMI MODELING

Using the estimated areas and volumes of the Nuuanu and Wailau landslides, numerical computations of tsunami generation and propagation are carried out. Present bathymetry and topography, rather than those at the time of landslides, are used for the computations of tsunami propagation and coastal tsunami heights, for two reasons. Firstly, there are many unknown factors and complications affecting morphologic reconstructions and integration of sea level changes in the Hawaiian islands. For example, at the time of the landslides (1 – 1.5 Ma) the southeastern islands such as Maui and Hawaii did not exist. Furthermore, shelf edges located at approximately 1200-1400 m depth around Oahu and Molokai islands may indicate former coastlines, suggesting that the volcanoes prior to landslides were higher above sea level than at present [*Moore and Clague*, this volume]. To accurately simulate the prehistoric tsunami, the bathymetry and sea level changes must be accurately reconstructed, though the detailed data do not exist. Secondly, there is no geologic evidence to confirm the

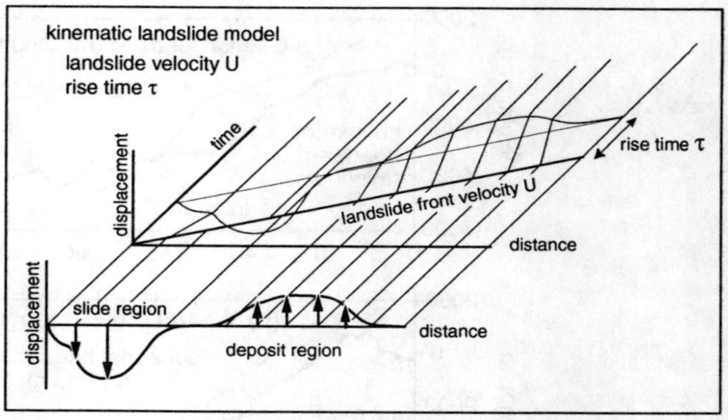

Figure 4. Schematic representation of the kinematic landslide model for tsunami generation. See text for explanation of values.

tsunami heights from the Nuuanu/Wailau landslides, even if one could recover the topography and assign the proper sea levels. Therefore, the purpose of the numerical tsunami computations in this paper is to roughly estimate tsunami heights from gigantic submarine landslides with the goals of future tsunami hazard assessment and providing clues of prehistoric tsunami characteristics.

3.1. Modeling Method

A landslide source of tsunamis, either subaerial or submarine, can be modeled in several ways. One method is to treat the landslide as a fluid flow and solve equations for multi-phase fluid flow [*Jiang and LeBlond*, 1993; *Imamura and Imteaz*, 1995; *Heinrich et al.*, 1998]. A second technique is to prescribe the water heights due to a landslide. *Harbitz* [1992] and *Ward* [2001] assumed a box-shaped slide while *Tinti et al.* [1997] introduced a block model of landslide. We have adopted the latter approach since the volume and bathymetry changes associated with the landslides have already been estimated.

A simple kinematic landslide model for the tsunami generation by submarine landsliding is introduced. The volume change (Plate 3) is used as the initial water height distribution. In other words, the change in topography/bathymetry from the reconstructed to the current morphology is used as the initial condition of tsunami propagation. The water height change at each point, negative in the slide region and positive in the debris deposit region, is performed during rise time τ. This change, or the landslide front, propagates outward from the peak of the reconstructed volcano with a horizontal velocity of U. Therefore, landslide motion is characterized by these two parameters, U and τ (Figure 4). It is similar to the kinematic source model of earthquakes, in which rupture velocity and rise time on faults are prescribed.

The rise time is fixed at $\tau = 5$ min (300 s). The landslide velocity is assumed to be $U = 50$ m/s, but is varied to lower (20 m/s) and higher (100 m/s) values to examine sensitivity to this factor. *Harbitz* [1992] assigned 20 m/s and 50 m/s for the lower and upper limits of submarine landslide velocity for the Storegga slides in the Norwegian Sea.

At the time of the 1980 Mount St. Helens eruption, the velocity of the first subaerial landslide was estimated at 50 – 70 m/s from analysis of a series of photographs [*Voight*, 1981]. The upper limit of horizontal velocity of the subaerial landslide could be approximately 100 m/s. The submarine landslide velocity may not exceed that of subaerial landslides because of friction between water and landslide material.

An observation made during the 1929 Grand Banks earthquake in the northwestern Atlantic Ocean may provide another constraint. This earthquake generated a tsunami on the east coast of the U.S. and Canada. The earthquake also caused damage to trans-Atlantic submarine cables. Six cables failed immediately after the earthquake and the other six failed successively up to 13 hours after the earthquake over a range of 350 nautical miles (600 km). *Heezen and Ewing* [1952] interpreted these cable failures as due to submarine slumping and turbidity currents with a velocity of up to 50 knots (25 m/s). This speed inferred for turbidity currents corresponds with the lowest speed (20 m/s) in our debris avalanche models.

The kinematic landslide model was applied to the 1741 Oshima-Oshima tsunami and the results suggest that a combination of vertical landslide velocity $Uz = 10$ m/s and rise time $\tau = 2$ min best explains the observed tsunami heights for that event [*Satake*, 2001]. The average slope of Oshima-Oshima volcano is about 0.13–0.17 [*Satake and Kato*, 2001], and thus the corresponding horizontal velocity U is about 60–80 m/s. A longer rise time is assumed for the Hawaiian landslides because of the larger maximum vertical change. The maximum vertical change is about 400 m for Oshima-Oshima, while it is nearly 2000 m for the Hawaiian submarine landslides. It would take 200 s (more than 3 min) to perform the maximum displacement change if the vertical landslide velocity is 10 m/s.

Tsunami propagation is calculated by the finite-difference method on actual bathymetry [*e.g.*, *Satake*, 1995], using nonlinear shallow-water equations. The governing equations are the equation of motion

$$\frac{\partial V}{\partial t} + (V \cdot \nabla) V = -g \nabla h \qquad (2)$$

where V is the horizontal velocity vector, h is the water (tsunami) height and g is the gravitational acceleration, and the equation of continuity

$$\frac{\partial (d+h)}{\partial t} = -\nabla \cdot \{(d+h) V\} \qquad (3)$$

where d is the water depth. The water depth changes as a result of the debris avalanche. Except for the slide source, the coastlines are fixed and no tsunami run-up on land is considered. The bottom friction is not included because of the relatively large grid size (see below).

Tsunami propagation is calculated on two different sets of grids: the Hawaiian grid system and Pacific grid system. The Hawaiian grid covers all the Hawaiian islands (161 – 154° W and 17 – 24° N) with a grid cell interval of 30" (arc second) or about 1 km. A total of 840 x 840 grid cells are included in the computational region (Plate 4), and the time step is set to 2 s to satisfy the stability condition of the finite-difference method. The tsunami generation and propagation for 2 hours in real time are computed so that full waveforms at all coastal points are obtained. The Pacific grid covers most of the Pacific Ocean (120°E – 100°W and 20°S – 60°N) with a grid cell interval of 5' (arc minute) or about 10 km. The number of grid cells is 1680 x 960, and the computational time step is 6 s to satisfy the stability condition. Computed tsunami heights are the average values on each cell, so the maximum tsunami height becomes greater if a smaller grid spacing is used. Comparison of the computed tsunami heights with recently observed heights from the 1993 Okushiri tsunami shows that the amplification factor (the ratio of the observed to the computed heights) is 1 – 3 for the 20" grid and 2 – 4 for the 1' grid [*Satake and Tanioka*, 1995]. This indicates that the actual tsunami may be locally higher than those computed here. The amplification factor may not be large for the Hawaiian (30") grid, but is at least 2 for the Pacific (5') grid.

3.2 Tsunamis from the Nuuanu Debris Avalanche

The Nuuanu slide is about 50 km long and the debris deposit extends about 150 km from the slide foot (Plate 3 and Figure 3). While the Nuuanu debris avalanche might have occurred as multiple events, the details are not known. Hence, we assume that a single landslide event created the current bathymetry and the tsunami, though we realize that this assumption may yield the upper bound of tsunami heights. With an assumed horizontal landslide velocity of 50 m/s and a rise time of 300 s, the landslide front reaches the slide foot at 1000 s (17 min) from the onset and the slide is completed at 1300 s (22 min). Debris deposition begins at 1000 s and is completed approximately 4300 s (72 min) from slide onset.

Plates A1 (on the accompanying CD-ROM) show snapshots of tsunami generation and propagation around the Hawaiian islands at 10 min intervals. The landslide front has not yet reached the slide bottom at 10 min, explaining why the water height disturbance is only negative. After 20 min, a positive peak following the negative front propagates outward from the slide region. The tsunami reaches the Molokai northshore in 20 min. At 30 min, the wavefront has surrounded Oahu and reached Kauai. By 40 min, the wavefront has reached Maui and passed the northern boundary of the computational area. At 50 min, the tsunami wavefront is reaching the northern shore of the island of Hawaii, and the front has surrounded the island in 60 min. Note that deposition of the debris deposit has not yet completed at this time.

The maximum tsunami heights in two hours from the slide onset are shown in Plate 4 and Figure 5. On Kauai, the tsunami heights are larger on the eastern coast where the maximum heights are about 40 m. The tsunami heights are lower on the southwestern coast, at about 30 m. The islands of Oahu and Molokai receive the maximum tsunami heights at about 100 m on the northern coasts. Tsunami heights are smaller on the southern coasts: about 30 m in Honolulu and Waikiki, and less than 20 m on the southern coast of Molokai. Similarly on Maui, the tsunami heights are found at about 50 m or more on the northern coast, but less than 20 m on the southern coast. The tsunami is larger, 50 – 80 m, on the northeastern coast of the island of Hawaii, but much smaller, generally 20 m or less, on the western and southern coasts.

Landslide velocities are varied and the tsunami heights compared in Figure 5. Except at a few places where the tsunami heights are very large, the computed tsunami heights are similar for U=100 m/s and 50 m/s. The computed heights for U= 20m/s are systematically lower, because it would take nearly an hour to complete the slide and about 3 hours to complete the debris deposition. The computed waveforms show that the maximum heights are registered by the first positive wave for the three cases.

The tsunami propagation in the Pacific Ocean is shown in Plates A2 (on CD-ROM). The tsunami propagates almost symmetrically from Hawaii. At about 5 hours, the wavefront reaches the west coast of North America and the Aleutian islands. By 6 hours, the tsunami has reached the Pacific coasts from California through the Aleutians. The tsunami front reaches northeastern Japan in 8 hours. At 12 hours, most of the Pacific coasts except for South America have been hit. Figure 6 shows the distribution of maximum tsunami height during the first 12 hours, showing strong azimuthal variation. The tsunami is largest in the northeastern direction, particularly toward southern California and the Pacific Northwest. The computed tsunami heights on the Pacific grid are 10 – 40 m in the Pacific Northwest (Washington and Oregon), 30 – 70 m in California and 5 –10 m on the Japanese coasts. Because these are the average

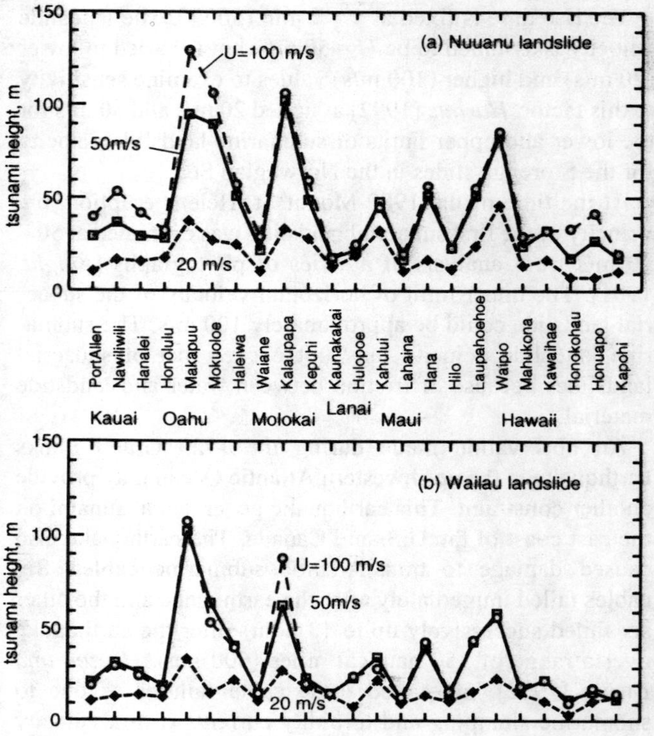

Figure 5. Comparison of computed tsunami heights for different landslide velocities (U=20, 50 and 100 m/s) for the (a) Nuuanu and (b) Wailau landslides.

values in the 5' (10 km) grid, the local tsunami height may be much higher, by more than a factor of two. The landslide velocity is varied to values of 20 m/s and 100m/s to examine the effect on the tsunami heights. The tsunami height distribution shows similar azimuthal patterns (*i.e.*, large amplitudes toward southern California) for different landslide velocities. The computed heights on the west coast of North America for U=100 m/s are similar to those resulting from 50 m/s, but those computed for U=20 m/s are systematically smaller. However, the computed heights on the Japanese coasts are similar for different landslide velocities.

3.3 Tsunamis from the Wailau Debris Avalanche

The Wailau slide is about 30 km long and the debris deposit extends at least 60 km from the slide foot (Plate 3 and Figure 3). The landslide front reaches the slide foot at 600 s (10 min) from the onset, and the slide is completed at 900 s (15 min). The debris runout distance is also less than the Nuuanu avalanche, so deposition of the debris completes in about 2100 s (35 min) from slide onset. This is much quicker than the Nuuanu landslide. Again, with lack of clear evidence of multiple events, we assume a single event for the tsunami computations.

Figure 6. Maximum tsunami heights in the northern Pacific Ocean from the Nuuanu landslide (U=50m/s). The gray scale indicates the maximum water height during 12 hours after the slide onset.

The tsunami propagation (Plates A3 on CD-ROM) is very similar to that of the Nuuanu avalanche. At 10 min, all the sea surface disturbance is negative. By 20 min, the first negative front reaches Oahu and Maui and the next peak has started to propagate. The wavefront is almost at the eastern coast of Kauai and the northern coast of Hawaii island in 30 min. At 40 min, the tsunami has reached Kauai, and the wavefront has reached all the islands by 50 min. The maximum tsunami height distribution is shown in Plate 5 and Figure 5. The largest tsunami heights of nearly 100 m are computed at the eastern edge of Oahu and northern coast of Molokai. The tsunami heights are also large, greater than 50 m, on the northern coasts of Maui and Hawaii. On the southern coasts of these islands the tsunami heights are less than 20 m. The effects of landslide front velocity on tsunami heights are the same as with the Nuuanu case. Computed tsunami heights are similar for U=50 and 100 m/s, but much smaller for U=20 m/s. When compared with the Nuuanu tsunami, the computed heights are 50 – 110% of the coastal points shown in Figure 5, with an average of 74%. The smaller volume change of the debris avalanche (~half) is reflected in the tsunami heights.

Tsunami propagation time for the Wailau event in the Pacific Ocean is similar to the Nuuanu tsunami. However, the maximum height distribution (Figure 7) is somewhat different from the Nuuanu case (Figure 6). First, the maximum water heights are much smaller than the Nuuanu tsunami. Second, on the west coasts of North America the tsunami heights are about 60% of those produced by the Nuuanu event, whereas the heights are about 70 % on the Japanese coasts. The azimuthal variation is also different. The directivity pattern shows that the amplitudes are larger in the north-to-northeast direction, and the maximum heights are located in the Queen Charlotte islands of Canada and the western Aleutians according to this model.

4. DISCUSSION

4.1. Landslide Volume

The subaerial and submarine headwall models may provide the maximum and minimum estimates of slide volume. The subaerial headwall has been long assumed since *Moore et al.* [1989], but field evidence such as non-disruption of the dike complex near the Nuuanu Pali or erosional features such as submarine canyons indicate that the subaerial cliffs are erosional remnants. *Moore and Clague* [this volume] argue that the headwall was about 10 – 20 km offshore along the shelf edge. A comparison of our Figure 3 and *Moore and Clague*'s similar figure indicates that their model is very similar to our submarine headwall model.

In terms of the volume balance alone, the subaerial headwall models provide volumes comparable to the total debris deposits. For the Nuuanu landslide, as noted before, the estimated missing slide volume for the submarine headwall is only about 2×10^3 km^3, while it is approximately 3×10^3 km^3 for the subaerial headwall, similar to the total debris volume (3×10^3 km^3). The situation is same for the Wailau landslide; the missing volume for the subaerial headwall, 1.5×10^3 km^3, matches the total debris volume, but the slide volume for the submarine headwall is only a half.

Figure 7. Maximum tsunami heights in the northern Pacific Ocean from the Wailau landslide (U=50m/s). The gray scale indicates the maximum water height during12 hours after the slide onset.

The above volumes estimated in this study are smaller than the previous estimates. *Moore et al.* [1989] estimated the Nuuanu landslide volume as 5 x 10^3 km^3 and the Wailau landslide as 1 x 10^3 km^3, assuming uniform thickness of 200 m and 75 m, respectively. Their "slide" area, however, includes both slide and debris deposit regions. Our average thickness in the debris deposit regions is 270 m for the Nuuanu and 210 m for the Wailau debris.

Some of the slide volume might have been scattered as turbidite sands. Turbidites are found in a much wider area than the debris deposit region assumed here. However, it is difficult to accurately estimate the turbidite volume from very limited piston core data, and the effects on tsunami generation may be minimal. We concentrate on the "solid" volume computable from bathymetry/topography changes.

In our volume computations we assumed that all the volume moved from the slide region to the debris region without overlap. In other words, the boundary separating the two regions is located on the original profile, and the position and elevation of the boundary is kept constant during the landslide. These assumptions may not be valid and the volume change might have occurred within the slide and/or debris regions. In such a case, the base of the slide may be located below the current bathymetry in the slide region and/or below the reconstructed bathymetry in the debris region, as shown in Figure 3. The actual slide bottom may be a smooth, concave-up curve (the so-called "Swedish circle") obtained by connecting the lowest curves in the figure. If this is the case, both slide and debris volumes would have been underestimated by the overlapping amounts. Because the amounts are common for both slide and debris, this correction will simply increase each volume estimate but not affect the volume balance.

Separation of the Nuuanu and Wailau debris deposits (Plate 3) was made in such a way that the debris (excess) volume matches the slide (missing) volume for each landslide. Careful inspection of Plate 3 shows that the Wailau debris blocks have two different orientations. Four large blocks that are nearly parallel with each other and resting east to southeast of Tuscaloosa seamount are parallel to other blocks in the Nuuanu debris field. Consequently, it is possible that these blocks belong to the Nuuanu deposit if a common failure direction is assumed. The boundary separating the two debris deposits may not be vertical; instead dipping in such a way that it lies deeper to the northwest. In this case, the boundary location on the seafloor would need to move east (or southeast) of that shown in Plate 3 to make the parallel blocks belong to the Nuuanu avalanche. The boundary at depth must be located farther west so that the volumes between the slide and debris for each volcano match. This means that the Nuuanu debris blocks lie atop the Wailau debris, requiring that the Wailau landslide be older. Alternatively, the Wailau avalanche could be younger than Nuuanu, and thinner deposits from the Wailau avalanche may blanket the older Nuuanu debris blocks. Paleomagnetic analysis of piston cores [*Kanamatsu et al.*, this volume] suggest the latter. In any case, the boundary between the two debris avalanche deposits is likely not vertical inferring they probably overlap.

In terms of volume change, the Storegga submarine

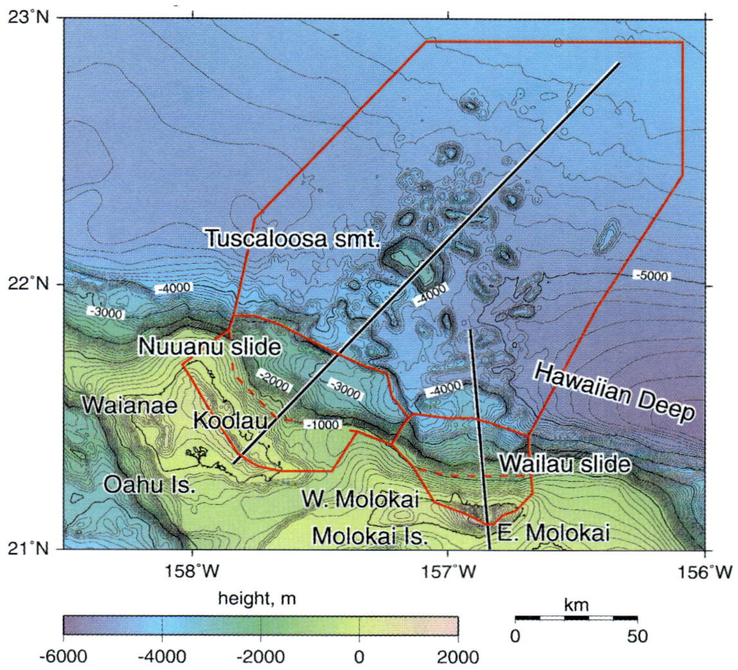

Plate 1. The current bathymetry compiled from both JAMSTEC and UH data. The grid cell size is 30 s (~1 km) and the contour interval is 100 m. The estimated regions of Nuuanu and Wailau slides and the debris avalanche deposits are shown by red lines. The red dashed lines show the submarine headwall locations.

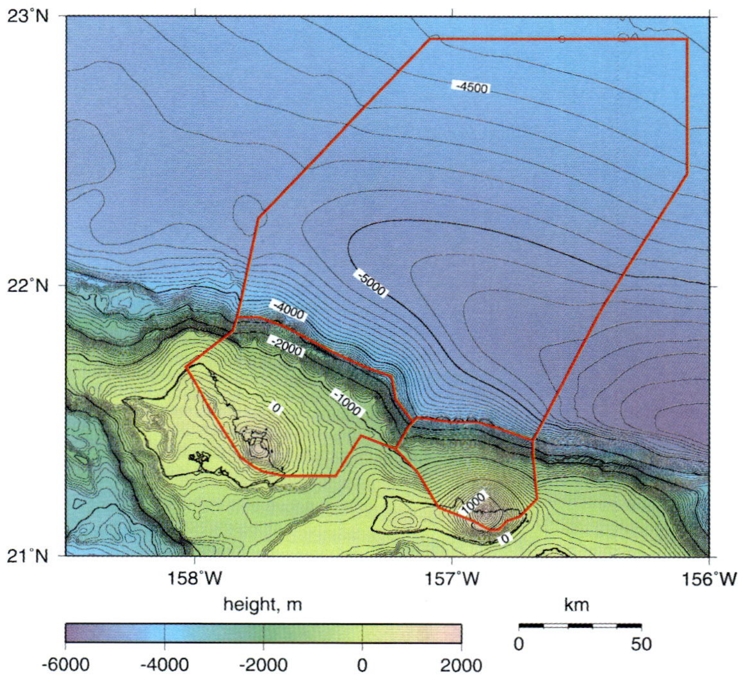

Plate 2. Bathymetry and topography before the Nuuanu and Wailau landslides. The slide and debris deposit regions are shown by red lines. The subaerial headwalls are assumed for the volcano reconstructions. The contour interval is 100 m. The thick black lines indicate the profiles shown in Figure 3. Present coastlines plotted for reference.

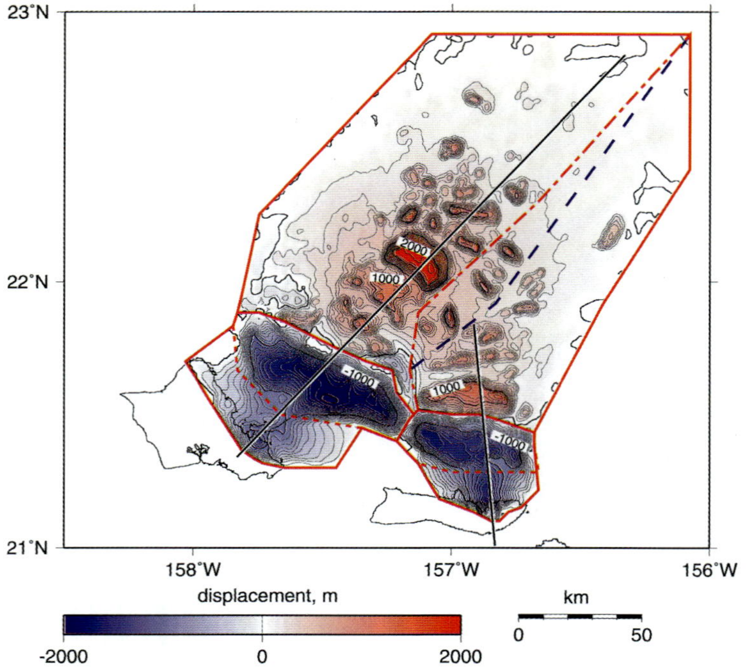

Plate 3. Difference between the original (reconstructed) and current topography/bathymetry, or difference between Plates 1 and 2. The blue color indicates missing volume from the slides and the red color indicates excess volume of debris deposits. The red solid lines show the boundaries for the slide and debris deposit regions. The red dotted lines are the submarine headwalls. The blue dashed line divides the debris deposits by their orientation, but resultant volumes do not match with the slide volumes. The red dashed line divides the debris deposits in such a way as to balance the volumes (see text). The contour interval is 100 m. The thick black lines indicate the profiles shown in Figure 3.

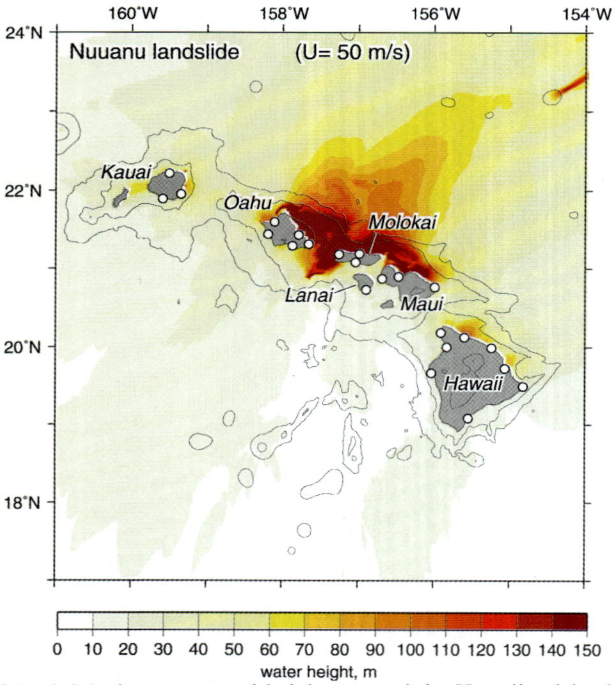

Plate 4. Maximum tsunami heights around the Hawaiian islands from the Nuuanu landslide. The color indicates the maximum water height within two hours of the slide onset using a landslide velocity U=50 m/s. Location of select points, for which the tsunami heights are compared in Figure 5, are shown.

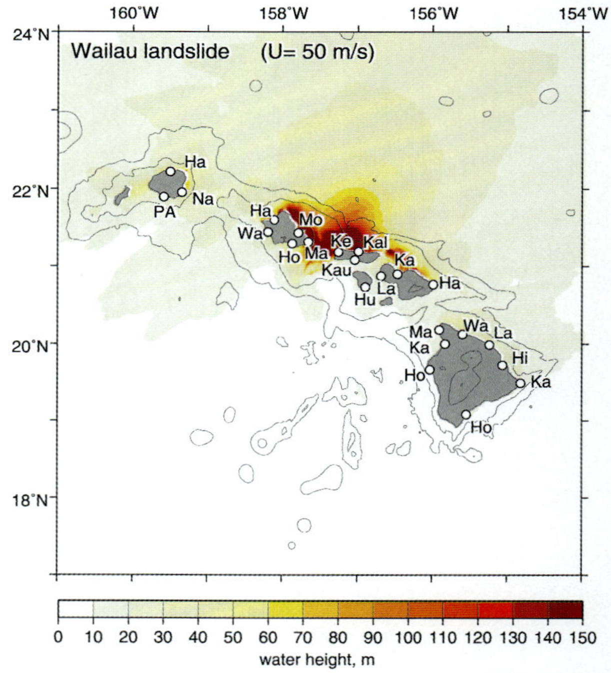

Plate 5. Maximum tsunami heights around the Hawaiian islands from the Wailau landslide. The color indicates the maximum water height within two hours of the slide onset using a landslide velocity of U=50 m/s. Open circles indicate the location of select points, for which the tsunami heights are compared in Figure 5.

landslide in the Norwegian Sea is reported to be larger than Nuuanu and Wailau combined. The total volume for Storegga is estimated as 5.6 x 10^3 km^3, with the maximum and average thicknesses of the landslide at 430 m and 160 m, respectively [*Bugge et al.*, 1988]. Geological studies indicate three landslide events. The first one probably occurred at 30 – 50 ka, with a volume of about 3.9 x 10^3 km^3, and maximum and average thicknesses of 280 m and 114 m, respectively. The second and third landslides occurred at 6 – 8 ka, with a total volume of 1.7 x 10^3 km^3 and maximum and average thicknesses of 330 m and 88 m, respectively. The first Storegga landslide is similar in size to the Nuuanu and Wailau landslides combined, and the total of the second and third Storegga landslides is similar to the Wailau landslide, but smaller than the Nuuanu landslide. *Harbitz* [1992] computed the tsunami runup heights of about 3 m on the east coast of Scotland assuming that the landslide velocity was 35 m/s, and stated that this height is consistent with those estimated from the tsunami deposits [*Dawson et al.*, 1988].

4.2. Tsunamis

When the velocity of a propagating landslide is similar to the local velocity of a tsunami, the tsunami amplitude can be increased because of energy coupling between the submarine landslide and resultant tsunami. However, in the cases of the Nuuanu and Wailau landslides, the typical water depth in the Hawaiian Deep is 5000 m and the tsunami phase velocity is about 220 m/s. This is much higher than any reasonable value for landslide velocity. Thus, the computed tsunami heights become smaller when the assumed landslide velocity is reduced.

The numerical modeling show that submarine landslides may cause devastating tsunamis. Distributions of maximum tsunami heights, in both the Hawaiian islands and the entire Pacific, show strong azimuthal variations. The Nuuanu landslide, in particular, sends waves with large heights toward southern California. The direction of submarine landslides has an important effect on the computed tsunami heights. Realistic landslide and tsunami modeling as shown in this paper may be an effective method for hazard mitigation.

5. CONCLUSIONS

The East Molokai and Koolau volcanoes were reconstructed and the volume changes associated with the Wailau and Nuuanu debris avalanches were estimated. For the Wailau landslide, the missing slide volume is (1.5 ± 0.3) x 10^3 km^3 for the subaerial headwall and (0.8 ± 0.1) x 10^3 km^3 for the submarine headwall. The total volume of the excess debris deposit is estimated as (1.5 ± 0.5) x 10^3 km^3. For the Nuuanu landslide, the slide volume is (2.9 – 3.8) x 10^3 km^3 for the subaerial headwall, where the lower value is preferable for the volume balance, and (1.9 – 2.1) x 10^3 km^3 for the submarine headwall. The debris volume is estimated as (3.0 ± 0.5) x 10^3 km^3.

Generation and propagation of the tsunamis were computed by using a kinematic landslide model. The tsunami heights on the northern coasts of Oahu, where the Nuuanu event originated, approach 100 m. The tsunami reaches most of the Hawaiian islands in 30 min. The maximum heights at major population points are: 37 m at Nawiliwili (Kauai), 28 m at Honolulu (Oahu), 48 m at Kahului and 17 m at Lahaina (both Maui), 23 m at Hilo, 13 m at Kailua-Kona (Hawaii). The tsunami arrives on the west coast of North America in 5 hours and Japan in 8 hours. Maximum tsunami heights are greater in the direction of landslide emplacement, which is toward southern California. Computed tsunami heights are 10 – 40 m in Washington and Oregon, 30 – 70 m in California and 5 –10 m on the Japanese coasts. The above numbers are average values of each grid cell, so the tsunami may locally be higher (by more than a factor of two) because the Pacific-wide tsunamis were computed on a 5' (10 km) grid.

Tsunamis from the Wailau landslide show similar arrival times, but the heights are smaller because of the smaller initial slide volume. Maximum tsunami heights are about 74% of those for the Nuuanu event on the Hawaiian islands, 60% on the west coast of North America, and 70 % in Japan. The Wailau tsunami produces the greatest wave heights in the Queen Charlotte islands of Canada and the western Aleutians because of the northward landslide direction.

Acknowledgments. We thank Eiichi Takahashi for continuous encouragement and discussion on the reconstruction. Jim Moore, Bill McGuire, Pete Lipman and Mike Garcia provided critical and constructive comments by which the paper was significantly improved.

REFERENCES

Bugge, T., R. H. Belderson, and N.H. Kenyon, The Storegga slide, *Phil. Trans. R. Soc. London, A 325*, 357-388, 1988.

Dawson, A.G., D. Long and N.H. Kenyon, The Storegga slides: Evidence from eastern Scotland for a possible tsunami. *Marine Geology, 82*, 271-276, 1988.

Duennebier, T., T. Reed, and HIGP Staff, *Northwestern Hawaiian islands: merged bathymetry and topography, [1:4,000,000]*, Sheet #2, Hawaii Seafloor Atlas, Hawaii Institute of Geophysics and Planetology, Honolulu, HI, 1994.

Harbitz, C.B., Model simulations of tsunamis generated by the Storegga slides, *Marine Geology, 105*, 1-21, 1992.

Heezen, B.C. and M. Ewing, Turbidity currents and submarine

slumps, and the 1929 Grand Banks earthquake, *Am. J. Sci., 250,* 849-873, 1952.

Heinrich, P., A. Mangeney, S. Guibourg, R. Roche, G. Boudon and J.-L. Cheminee, Simulation of water waves generated b a potential debris avalanche in Montserrat, Lesser Antilles, *Geophys. Res. Lett., 25*, 3697-3700, 1998.

Imamura, F. and M.M.A. Imteaz, Long waves in two-layers: governing equations and numerical model, *Sci. Tsunami Hazards, 13*, 3-24, 1995.

Jiang, L. and P. LeBlond, Numerical modeling of an underwater Bingham plasteic mudslide and the waves which it generates. *J. Geophys. Res., 98*, 10303-10317, 1993.

Kanamatsu, T., E. Herrero-Bervera and G. M. McMurtry, Magnetostratigraphy of deep-sea sediments from areas adjacent to the Hawaiian Islands: Implication for ages of turbidites derived from submarine landslides, this volume.

Keating, B. H., and W. J. McGuire, Island edifice failures and associated tsunami hazards, *Pure Appl. Geophys., 157*, 899-955, 2000.

Langenheim, A. M. and D. A. Clague, The Hawaiian-Emperor volcanic chain, Part II Stratigraphic framework of volcanic rocks of the Hawaiian Islands, *U.S.G.S. Professional Paper, 1350*, 55-84, 1987.

Lipman, P. W., J. P. Lockwood, R. T. Okamura, D. A. Swanson and K. M. Yamashita, Ground deformation associated with the 1975 Magnitude-7.2 earthquake and resulting changes in activity of Kilauea volcano, Hawaii, *U.S.G.S. Professional Paper, 1276*, 1-45, 1985.

Ma, K.-F., H. Kanamori and K. Satake, Mechanism of the 1975 Kalapana, Hawaii, earthquake inferred from tsunami data, *J. Geophys. Res., 104*, 13,153-13,167, 1999.

Moore, J., Giant submarine landslide on the Hawaiian Ridge. *U.S.G.S. Professional Paper, 501-D,* D95-98, 1964.

Moore, J. G. and G. W. Moore, Deposit from a giant wave on the island of Lanai, Hawaii, *Science*, 226, 1312-1315, 1984.

Moore, J. G. and D. A. Clague, Mapping the Nuuanu and Wailau landslides in Hawaii, this volume.

Moore, J. G., D. A. Clague, R. T. Holcomb, P. W. Lipman, W. R. Normark, and M. E. Torresan, Prodigious submarine landslide on the Hawaiian ridge, *J. Geophys. Res., 94*, 17465-17484, 1989.

Moore, J.G., W.R. Normark, and R.T. Holcomb, Giant Hawaiian landslides, *Ann. Rev. Earth Planet. Sci. 22*, 119-144, 1994.

Naka, J., E. Takahashi, D. Clague, M. Garcia, T. Hanyu, E. Herrero-Bervera, J. Ishibashi, O. Ishizuka, K. Johnson, T. Kanamatsu, I. Kaneoka, P. Lipman, A. Malahoff, G. McMurty, B. Midson, J. Moore, J. Morgan, T. Naganuma, K. Nakajima, T. Oomori, A. Pietruszka, K. Satake, D. Sherrod, T. Shibata, K. Shinozaki, T. Sisson, J. Smith, S. Takarada, C. Thornber, F. Trusdell, N. Tsuboyama, T. Ui, S. Umino, K. Uto and H. Yokose, Tectono-magmatic Processes investigated at deep-water flanks of Hawaiian volcanoes, *Eos, Trans. American Geophys. Union, 81*, 221, 226-227, 2000.

Satake, K., Linear and non-linear computations of the 1992 Nicaragua earthquake tsunami. *Pure and Appl. Geophys., 144*, 455-470, 1995.

Satake, K., Tsunami modeling from submarine landslides, *Proc. International Tsunami Symposium,* Seattle, 2001.

Satake, K. and Y. Kato, The 1741 Oshima-Oshima Eruption: Extent and Volume of Submarine Debris Avalanche, *Geophys. Res. Lett., 28*, 427-430, 2001.

Satake, K. and Y. Tanioka, Tsunami generation of the 1993 Hokkaido Nansei-oki earthquake. *Pure Appl. Geophys., 144*, 803-821, 1995.

Smith, J.R., *Island of Hawaii and Loihi submarine volcano, high resolution multibeam bathymetry around the Island of Hawaii, [1:75,000, 1:250,000, 1:500,000]*, Sheet #6, Hawaii Seafloor Atlas, Hawaii Institute of Geophysics and Planetology, Honolulu, HI, 1994.

Smith, J. R. and K. Satake, Deepwater multibeam surveys along the southeastern Hawaiian Ridge, explanation of CD-ROM, this volume.

Smith, W. H. F. and P. Wessel, Gridding with continuous curvature splines in tension, *Geophysics, 55*, 293-305, 1990.

Tinti, S., E. Bortolucci, and C. Vannini, A block-based theoretical model suited to gravitational sliding, *Natural Hazards, 16*, 1-28, 1997.

Voight, B., Time scale for the first moments of the May 18 eruption, *U.S.G.S. Professional Paper 1250*, 69-86, 1981.

Ward, S. N, Landslide tsunami, *J. Geophys. Res.,* in press, 2001.

Wessel, P. and W. H. F. Smith, New, improved version of Generic Mapping Tools released, *Eos, Trans. Am. Geophys. Union, 79*, 579, 1998.

K. Satake, Active Fault Research Center, AIST-GSJ, Tsukuba, Ibaraki 305-8567, Japan (kenji.satake@aist.go.jp)

K. Shinozaki, Schlumberger K.K., Data Management Center, Fuchinobe, Japan (sinozaki@tokyo.oilfield.slb.com)

J. R. Smith, SOEST/HURL, Univ. Hawaii, Honolulu, Hawaii (jrsmith@soest.hawaii.edu)

Introduction to Section 4
The Hawaiian Plume and Magma Genesis

Eiichi Takahashi

Earth and Planetary Sciences, Tokyo Institute of Technology, Tokyo Japan

Hawaii is currently the most magmatically productive hotspot on the earth. Its track across the Pacific plate has been recovered for more than 70 Ma. Although Hawaiian volcanoes and their magma genesis have been studied extensively compared with other hotspots, their deep-ocean features have been little known until recently. The final section of this volume contains papers on geophysical observations and magma genesis in the Hawaiian mantle plume.

Because of its central location in the Pacific Basin where the distribution of seismic stations is poor, seismic imaging of the mantle plume underneath Hawaiian hot spot is difficult. New seismic global tomography to look at the roots of major hot spots (e.g., Hawaii, Iceland, South Pacific, East Africa) have been carried out by *Zhao et al.* [this section] using body-wave first arrivals. Using advanced techniques (e.g., 3D-grid parameterization and incorporating the topography of Moho, 410 and 660 km discontinuities), they define structures under the hotspot with improved resolution. A low-velocity anomaly beneath Hawaii was found to extend westward down to at least the mid-lower mantle (ca. 1500 km depth). Slow velocity anomalies for other hotspots are found to extend to the core-mantle boundary but not as vertical pillars. Instead, *Zhao et al.* [this section] discovered that the plumes are deflected and they hypothesize that they are blown by mantle winds.

Heat flow anomalies around hot spots have been regarded as direct evidence of hot upwelling (mantle plume). Contrary to previous interpretations, *McNutt* [this section] shows that variations in ocean-bottom heat flow around the Hawaiian archipelago are best correlated with thermal conductivity of near-surface sediments. Observed heat flow patterns are largely determined by lateral flux in permeable aquifers in the oceanic crust, and thermal structure of the underlying mantle plume is therefore difficult to determine from heat-flow data.

Distinct anomalies in rare-gases (e.g., $^{3}He/^{4}He$ in lavas from Loihi Seamount is up to 35 times that in the atmosphere) are a key signature of mantle-plume magmas that originated from a deep undegassed reservoir in the mantle. *Kaneoka et al.* [this section] report new analyses of rare gas isotopes (He, Ne, Ar, Xe, and Kr) for deep-ocean samples from Loihi, Kilauea, and Koolau volcanoes. The new data and their previous studies document large variations in He isotopes among volcanoes in the Hawaiian hot spot. The highly variable He values are interpreted to represent different degrees of interaction between the rising plume magma and the uppermost asthenosphere. The variable values for the heavy rare gases document incorporation of seawater into the magma. *Kaneoka et al.* [this section] propose that the magnitude of these interactions with the upper mantle and crust is related to the evolutionary stage of the volcano.

Koolau volcano on Oahu island was known to be an unusual Hawaiian shield because its basaltic lavas are high in SiO_2, and have the highest $^{87}Sr/^{86}Sr$ and lowest $^{144}Nd/^{143}Nd$ isotopic ratios in Hawaii. Previous workers interpreted these features to recycling of old oceanic crust in the Hawaiian plume. *Garcia* [this section] examined olivine-rich lavas (picrite) recovered for the first time from the deep submarine flank of Koolau volcano. Although subaerial slopes of Koolau are covered by high-silica tholeiite, the sampled submarine parts contain abundant picrite. From detailed petrologic observations, Garcia concludes that the primary magma for Koolau volcano contains high MgO (14-15 wt%), as in other Hawaiian volcanoes.

Takahashi and Nakajima [this section] explore an alternative model for the Koolau magma genesis. In order to simulate melting processes in the Hawaiian plume consist-

Hawaiian Volcanoes: Deep Underwater Perspectives
Geophysical Monograph 128
Copyright 2002 by the American Geophysical Union

ing of peridotite and entrained oceanic crust (eclogite), they conducted series of high-pressure melting experiments. They showed that high-silica tholeiite characteristic of the subaerial Koolau volcano could be produced by direct partial melting of recycled eclogite at temperatures slightly below the peridotite dry solidus. At temperatures slightly above the peridotite dry solidus, the melt compositions produced from a basalt-peridotite hybrid source change drastically toward picritic. Based on the experiments, they propose a model of magma genesis to explain the abrupt change in magma composition during the growth history of Koolau, from Mauna Loa-type main shield stage to the silica-rich Makapuu stage [*Shinozaki et al.*, Section 3; *Tanaka et al.,* Section 3]. According to their model, temperatures within the Hawaiian plume are 100-150°C lower than previous estimates.

Seismic Images of Hotspots and Mantle Plumes

Dapeng Zhao and Fumiko Ochi

Geodynamics Research Center, Ehime University, Matsuyama, Japan

Eiichi Takahashi

Department of Earth and Planetary Sciences, Tokyo Institute of Technology, Tokyo

A novel global tomography approach has been used to study the deep structure and dynamics of hotspots and mantle plumes. Prominent plume-like slow anomalies are detected under hotspot regions from the crust down to the core-mantle boundary (CMB). Most of the hotspots in South Pacific and Africa are located above the slow anomalies in the lower mantle down to the CMB, suggesting that the mantle plumes under those hotspots may originate from the CMB. The Hawaiian and Iceland plumes are also found to originate from the CMB. However, there may be some small-scaled, weak plumes originating from the transition zone or middle mantle depths. The slow anomalies under hotspots usually do not show a vertical pillar shape, which suggests that plumes are not fixed in the mantle but can be deflected by the mantle winds. As a consequence, hotspots are not fixed but can wander on the Earth's surface, as evidenced by the recent geomagnetic and numeric modeling studies.

1. INTRODUCTION

Hotspots are regions of significant midplate volcanism, such as Hawaii, or very voluminous on/near ridge volcanism, such as Iceland. Island chains like the Hawaiian-Emperor chain could be viewed as the tracks left on the moving plate by a fixed source beneath it [*Wilson*, 1963]. The fixed source could in fact be partial melting of hot materials supplied by mantle plumes, which are thin, cylindrical, upwellings of hot, low-viscosity material [*Morgan*, 1971]. Plume material ascends buoyantly from a basal thermal boundary layer (which might be either at the base of the mantle or upper mantle), eventually interacting with the lithosphere at relatively shallow depths [*Sleep*, 1990, 1996].

Hotspots have an irregular but nonrandom distribution over the Earth's surface. They are preferentially located near divergent plate boundaries (mid-ocean ridges) and long-wavelength geoid highs, and are preferentially excluded from regions near the convergent plate boundaries [*Richards et al.*, 1988; *Weinstein and Olson*, 1989]. In addition, hotspots are generally located above low seismic velocity regions in the lower mantle [*Su et al.*, 1994] and ultra-slow velocity patches on the core-mantle-boundary (CMB) [*Williams et al.*, 1998].

Mantle plumes and hotspots have been extensively studied by using numerical modeling approaches with surface observables such as bathymetry, gravity and heat flow over hotspots as constraints [e.g., *Richards et al.*, 1988; *Sleep*, 1992, 1996; *Albers and Christensen*,

Hawaiian Volcanoes: Deep Underwater Perspectives
Geophysical Monograph 128
Copyright 2002 by the American Geophysical Union

1996; *Steinberger*, 2000]. Many of these models provide quantitative predictions of mantle flux; however, the plume diameter and temperature anomaly are not well constrained by these models.

Seismology provides an alternative approach of obtaining information on the structure of mantle plumes. Local and regional tomographic images have been made of plumes. These models provide high-resolution images of plumes in the upper mantle, but cannot determine the deep structure and provide little information on the depth extent and origin of plumes [e.g., *Tryggvasson et al.*, 1983; *Wolfe et al.*, 1997; *Foulger et al.*, 2000; *Ritter et al.*, 2001]. The global tomographic approach to studying mantle plumes provides lower resolution images of plumes in the entire mantle [e.g., *Bijwaard and Spakman*, 1999; *Ritsema et al.*, 1999]. Diffraction tomography was used to determine the Hawaiian plume structure at the base of mantle [*Ji and Nataf*, 1998]. Waveform modeling and other seismological methods are also used to study plume structures [*Shen et al.*, 1998; *Allen et al.*, 1999; *Ni et al.*, 1999].

In this work, we have adopted a new global tomography approach to study the seismic structure of mantle plumes. Our results shed new light on the structure and dynamics of plumes and hotspots.

2. METHOD AND DATA

Continuous efforts have been made during the last two decades to determine the three-dimensional (3-D) structure of the mantle [e.g., *Dziewonski*, 1984; *Inoue et al.*, 1990; *Su et al.*, 1994; *Bijwaard et al.*, 1998]. A new 3-D *P*-wave velocity model of the mantle was proposed recently, which has three distinct features compared to previous models [*Zhao*, 1999]. First, the topography of mantle discontinuities is taken into account in the tomographic inversions. The Moho depth ranges from about 10 km under oceans to 40-70 km under the continents [*Mooney et al.*, 1998]. The 410 and 660 km discontinuities exhibit depth variations of up to 36 km on a global scale [*Flanagan and Shearer*, 1998]. Such large depth changes of the discontinuities would greatly affect the travel times and ray paths of seismic waves, and hence their topography should be taken into account in the tomographic inversions. Local and regional tomographic studies have recognized the influences of the depth variations of seismic discontinuities, such as the Moho and the subducting slab boundary [*Zhao et al.*, 1992, 1994]. It is shown that when the discontinuity topography is considered, a more reasonable tomographic result can be obtained and the resulting 3-D velocity model fits the data better.

Secondly, a grid parameterization [*Zhao et al.*, 1992, 1994] was used to express the Earth structure. Previous global tomographic studies expressed the Earth structure using blocks [*Inoue et al.*, 1990; *Bijwaard et al.*, 1998] or spherical harmonic expansions [*Dziewonski*, 1984; *Su et al.*, 1994]. In the block models, the medium under study is divided into cubic blocks, seismic velocity in each block is assumed to be constant, and the velocity changes abruptly from block to block. Thus, artificial velocity boundaries are introduced into the model between blocks. The spherical harmonic approach is hard to image localized anomalies such as subducting slabs and magma chambers. With the grid approach, a 3-D grid net in the whole mantle is set up first. Then velocity perturbations at the grid nodes from the starting one-dimensional (1-D) velocity model [*Kennett and Engdahl*, 1991] are considered as unknown parameters. The velocity perturbation at any point in the model is computed by linearly interpolating the velocity perturbations at the eight grid nodes surrounding that point. Thus, in the grid model there are no artificial velocity boundaries as existing in the block models. The grid models can image both localized and large-scale anomalies [*Zhao*, 2001].

Thirdly, an efficient 3-D ray tracing technique [*Zhao et al.*, 1992] is used to compute travel times and ray paths. This technique is adaptable to a velocity model with complex velocity discontinuities and with 3-D velocity variations everywhere in the model. The principle of this ray tracing scheme is to use the pseudo-bending algorithm [*Um and Thurber*, 1987] and Snell's law iteratively to perturb an initial ray estimate until a converged solution (the ray with the shortest traveltime) is found.

Then a system of observation equations is constructed that relates the observed travel times to hypocentral and velocity unknown parameters. To conduct inversions of the observation equations we used the LSQR algorithm [*Paige and Saunders*, 1982] with damping and smoothing in a form similar to that of *Inoue et al.* [1990]. Hypocenters are relocated in the inversion process.

This new tomography approach was applied to the first and later arrival times compiled by the International Seismological Center (ISC) between 1964 to 1998, which were reprocessed by *Engdahl et al.* [1998]. The continental regions are generally well covered by the ISC stations, while the oceanic regions are poorly covered (Figure 1a). We selected a total of 7128 events that were recorded by over 50 stations with errors in hypocentral locations of < 10 km (Figure 1b). Their focal depths are well constrained by the use of depth phases (*pP*). In addition to first *P* arrivals, we also used three types of later phases in this study. They are the depth phases (*pP*), surface reflected waves (PP), and core reflected waves (PcP). Topography and bathymetry as well as the Moho depth variations are considered at the bounce points of pP and PP rays. In total, approximately one million arrival times were used in the tomographic inversions.

3. SYNTHETIC TESTS AND RESULTS

A checkerboard resolution test [*Inoue et al.*, 1990, *Zhao et al.*, 1992] was first conducted to evaluate the adequacy of the ray coverage and spatial resolution (Figure 2). To make a checkerboard, we assigned positive and negative velocity anomalies to the 3-D grid nodes. Synthetic arrival time data are calculated for the checkerboard model. Numbers of stations, events and ray paths in the synthetic data are the same as those in the real data. Then the synthetic data were inverted with the same algorithm as for the real data. The resolution is considered to be good for areas where the checkerboard image is recovered. In the preliminary analyses, we set up grid nodes in the whole mantle with a grid spacing of five degrees laterally and 15 to 330 km in depth (Figure 2). The iasp91 Earth model [*Kennett and Engdahl*, 1991] is used as the starting velocity model for the 3-D inversions. In the upper mantle the resolution is good under the continental regions but poor under oceans. For the lower mantle, the spatial resolution is generally good except for southern hemisphere.

Two more synthetic tests were conducted with the same procedure as the checkerboard test except for the input model (Figures 3 and 4). Plume-like slow anomalies in the entire mantle under Hawaii, Iceland and South Pacific are well recovered (Figure 3). Slow anomalies in the upper mantle are also reconstructed with little smearing to the lower mantle (Figure 4). We also conducted an inversion with the data from stations and earthquakes at Hawaii excluded. The results show that the Hawaiian slow anomaly in the lower mantle is well imaged though the upper mantle anomaly is not, indicating that the Hawaii anomaly is not a product of there being stations and earthquakes there.

Prominent slow anomalies are visible under the hotspot regions from the crust down to the CMB (Plates 1-4, Figure 5). The slowest anomalies in the lower mantle appear under South Pacific and Africa where many hotspots exist (Plates 1 and 4). Strong and wide high-velocity anomalies are visible in the transition zone and moderately fast anomalies exist in the lower mantle under the subduction zone regions (Plate 1). This result may suggest that most of the slab materials are stagnant for a long time in the transition zone before finally dropping down to the lower mantle. Some of the slabs may stay above the 660 km discontinuity and complete

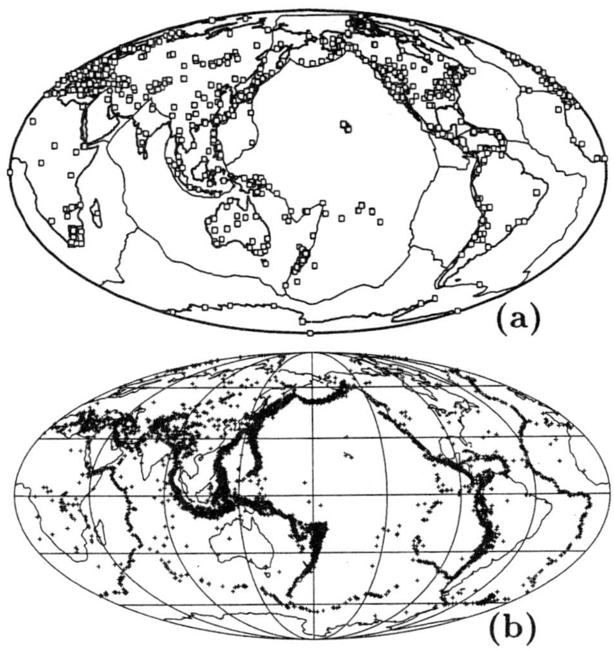

Figure 1. (a) Distribution of the ISC seismic stations used in this study. (b) Hypocentral distribution of 7128 earthquakes used in this study. These events generated approximately one million arrival times used in the tomographic inversions.

their convection within the upper mantle. The mid-ocean ridge slow anomalies are imaged at 420 and 550 km depths but are not well imaged at shallow depths (Plate 1) because of few ray samplings in the upper mantle under oceans (Figure 2).

A unique feature of the present study is that the depth variations of the Moho, and the 410 and 660 km discontinuities [*Mooney et al.*, 1998; *Flanagan and Shearer*, 1998] have been considered in our tomographic inversions. The effect of the discontinuity topography on the velocity tomography was evaluated by conducting several inversions with the same data set and method but assumed the Moho, 410 and 660 km discontinuities are flat. Only the major results of these inversions are presented here. Details of this analysis will be described in a separate paper.

The overall patterns of the global velocity images are stable, but there are considerable changes in the amplitudes and pattern of the velocity anomalies in those regions where the discontinuities have large depth variations. The Moho topography mainly affects the upper mantle images down to about 150 km depth. The 410 and 660 km discontinuity topography primarily influence the images in and around the mantle transition

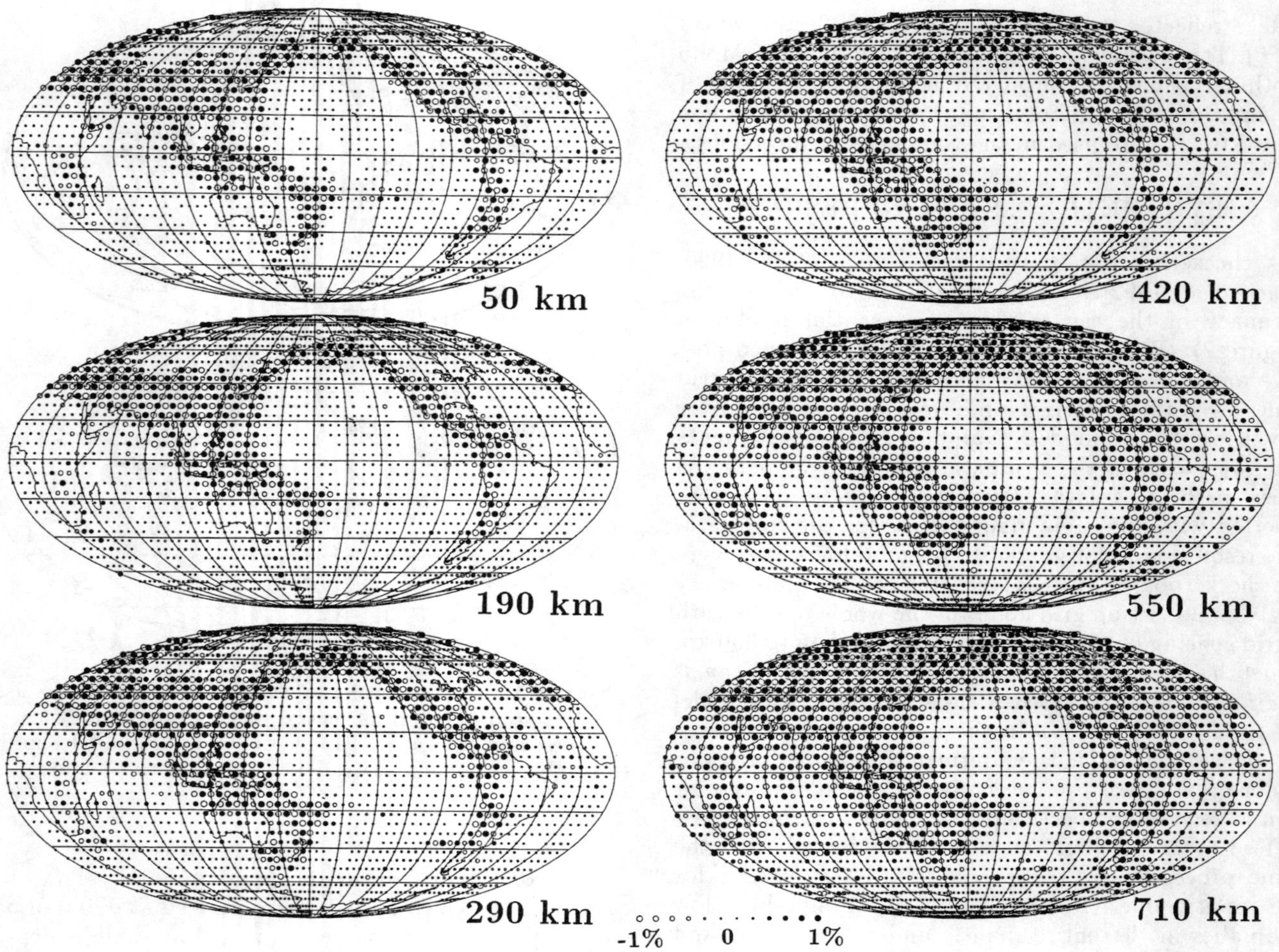

Figure 2. Results of a checkerboard resolution test for some representative depth slices in the mantle. Open and solid circles denote slow and fast velocities, respectively. The layer depth is noted below each map. The scale of velocity perturbations from the average velocity at each depth is shown at the bottom.

zone. For some areas like Australia, the amplitudes of velocity anomalies in the transition zone are changed by a factor of 2. We conducted a number of such inversions by changing the grid spacing, damping, smoothing and data sets. In general, the final travel time residual for the models including the discontinuity topography is reduced by 6-9% as compared to the models with flat discontinuities. This indicates that it is necessary to consider the discontinuity topography in the tomographic inversions in order to determine the mantle structure better.

4. DISCUSSION

4.1. Origin of Plumes

Among the 47 hotspots we adopted [*Richards et al.*, 1988], 19 are located above areas with the CMB velocity perturbation $(\Delta V/V) <$ -0.2%, 10 with $\Delta V/V > 0.2\%$, others with $-0.2\% < \Delta V/V <$ -0.2%. Most of the hotspots in South Pacific and Africa are located above the slow anomalies in the lower mantle down to the CMB (Plates 1, 4). *Williams et al.* [1998] showed that many hotspots are located above the ultra-slow velocity patches on the CMB. These results suggest that many mantle plumes under the hotspots may originate from the CMB. However, there are some hotspots that do not lie above the CMB low-velocity (low-V) areas (Plate 4). Those hotspots could be (a) associated with highly localized (and as yet undetected) low-V zones, (b) produced by lateral flow or deflection of plumes [*Loper*, 1991; *Sleep*, 1996; *Steinberger*, 2000], or (c) derived from a different depth or through a different process than the hotspots with CMB associated low-V areas [*Anderson*, 1975; *Williams et al.*, 1998].

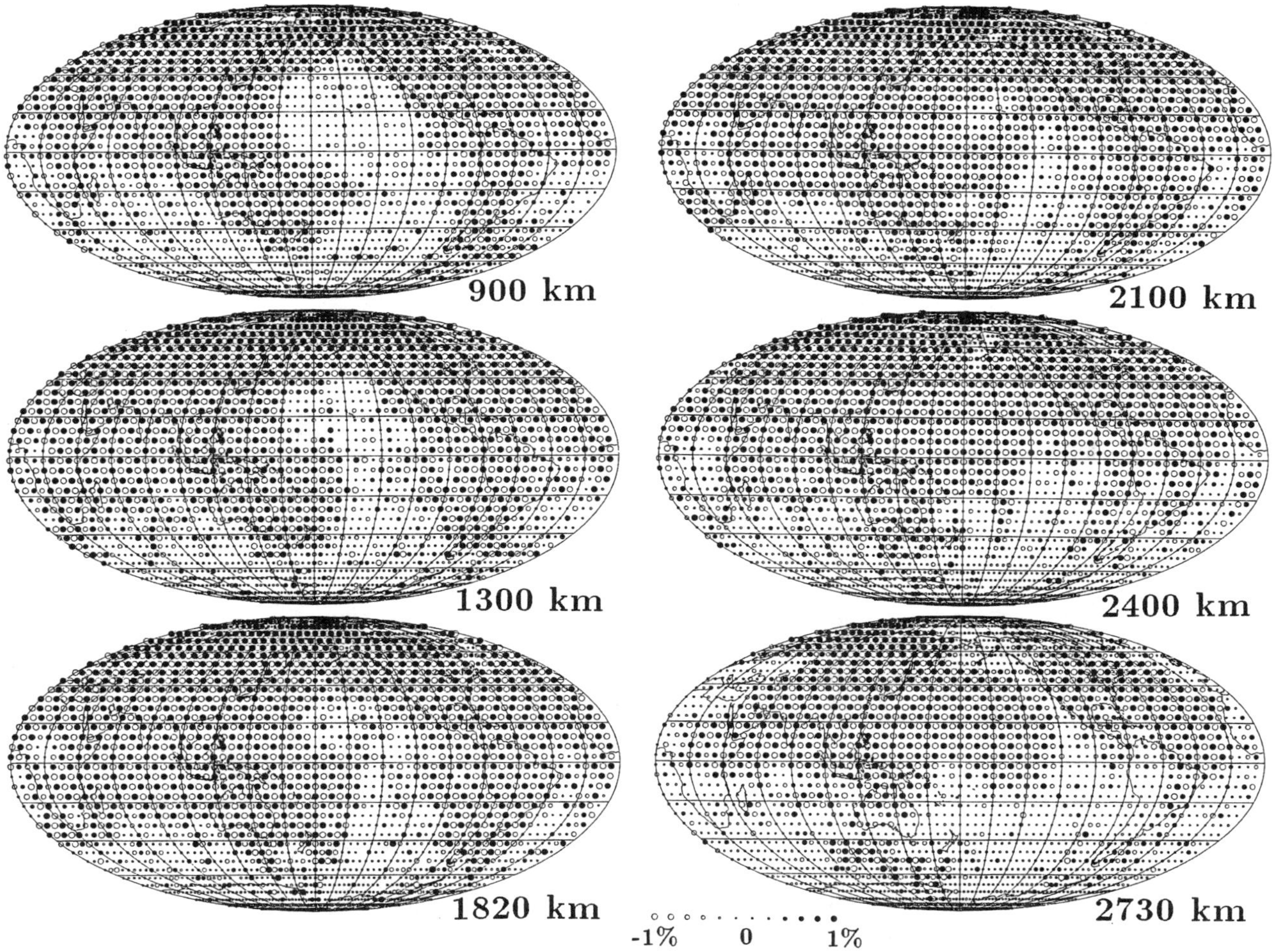

Strong plumes (e.g., Hawaii) could originate at the CMB but weak plumes may rise from the transition zone if there is a thermal boundary layer there [*Albers and Christensen*, 1996]. For example, some hotspots in South America and Africa may originate in the upper mantle through small-scale convection at the edges of continental cratons [*King and Ritsema*, 2000]. Thus, some mantle plumes may only extend to the upper mantle.

Prominent slow anomalies appear in some areas such as under the subducting Pacific slab beneath the Japan Islands (Plate 1) where the resolution of the tomography is very good (Figure 2). This feature may suggest that plumes may also exist in areas where there is no surface expression of a hotspot.

4.2. Deflection of Plume Conduits

The slow anomalies under hotspots usually do not show a vertical pillar shape (Plates 2, 3). This suggests that plumes are not fixed in the mantle but can be deflected due to the influences of mantle convection, or mantle winds [e.g., *Griffiths and Richards*, 1989; *Loper*, 1991]. Recent numerical experiments by *Steinberger* [2000] show that plume conduits can be tilted, with source regions at the D'' layer moving in the lowermost mantle flow, generally toward large-scale upwellings under southern Africa and the south central Pacific. Hotspot surface motion often represents the horizontal component of midmantle flow, which is frequently opposite to plate motion (i.e., toward ridges and away from subduction zones).

Under Hawaii, our tomography shows a tilted low-V zone in the entire mantle with the shallow end under the island of Hawaii and the CMB end located a few hundred kilometers north of Hawaii (Plate 2b). This result is consistent with a recent diffraction tomography [*Ji and Nataf*, 1998]. Our plume images show a whole mantle vertical low-V zone under Iceland (Plate 2, Figure 5), which are generally similar to the tomographic images determined recently by *Bijwaard and Spakman* [1999].

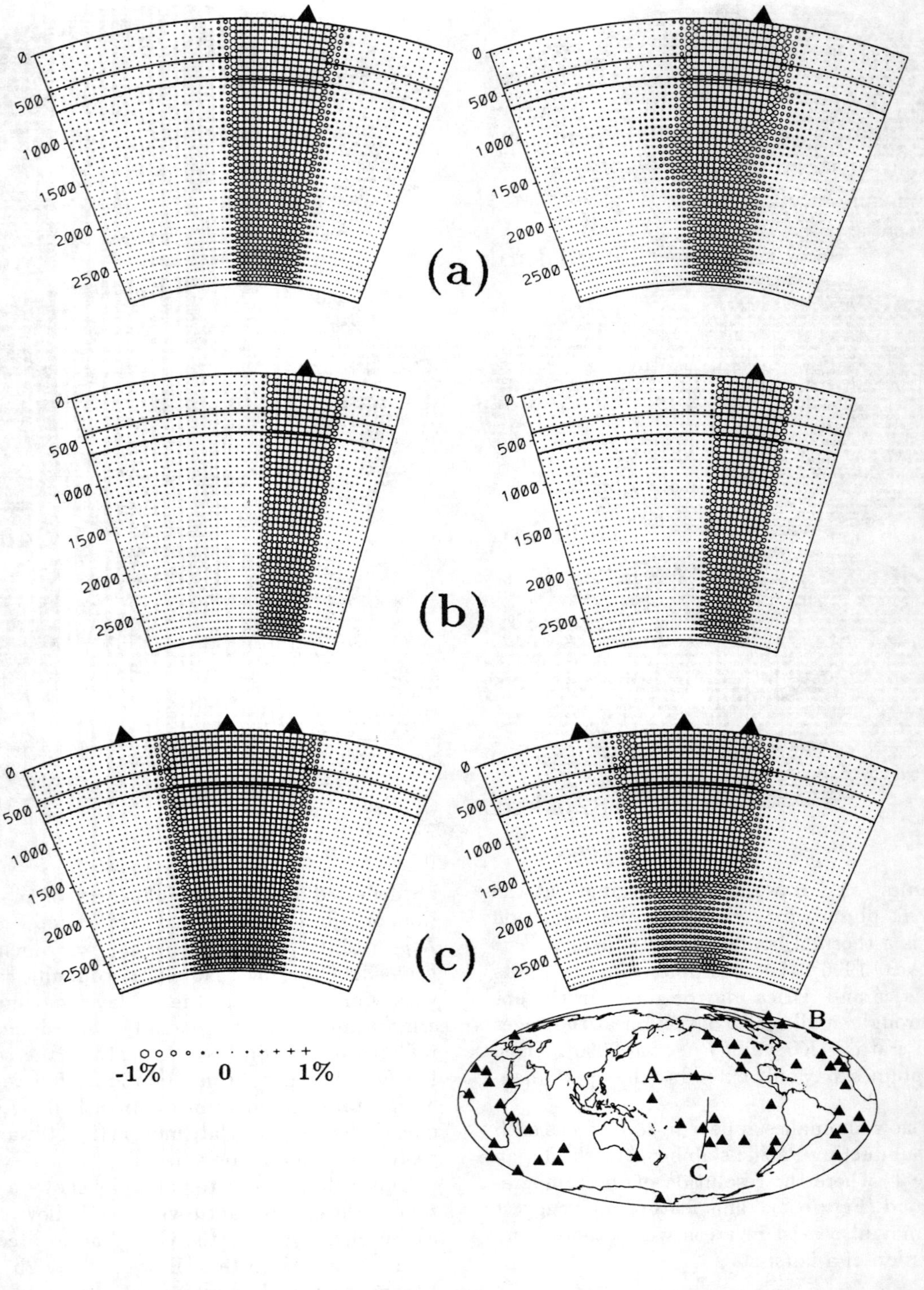

Figure 3. Vertical cross sections of *P*-wave velocity images under Hawaii (a), Iceland (b), and the South Pacific (c) along lines A, B and C in the insert map for a synthetic test. Left panels show the input model, while right panels show the inversion results. Circles and crosses denote slow and fast velocities, respectively. The velocity perturbation scale is shown at the bottom. Solid triangles denote the surface hotspots [*Richards et al.*, 1988; *Steinberger*, 2000].

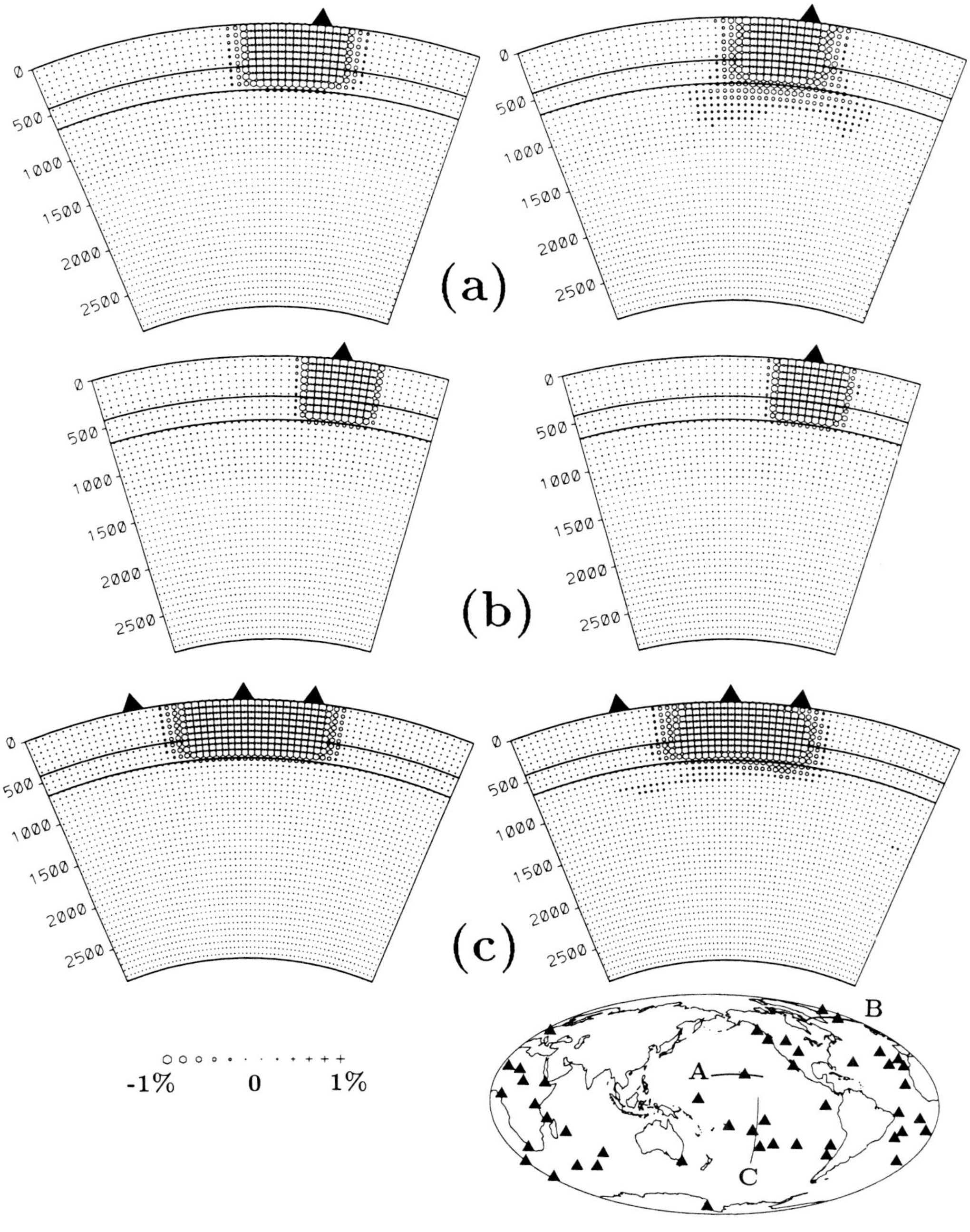

Figure 4. The same as Figure 3 but for another synthetic test.

The Iceland plume appears to be narrow in the east-west direction (Plate 2c) but is wider in the north-south direction (Plate 2d) which was also noted by *Foulger et al.* [2000]. The low-*V* zone extends toward the south at depths shallow than about 400 km depth (Plate 2d), which may be associated with the Iceland hotspot being located on the North Atlantic mid-ocean ridge. A part of the hot upwelling materials from the Iceland plume head may flow along the Reykjanes Ridge, south of Iceland [*Sleep*, 1996]. *Foulger et al.* [2000] proposed that the Iceland plume extends down to the mantle transition zone and no deeper. However, their tomographic images are based on the data recorded by a local seismic network and so have no resolution below the transition

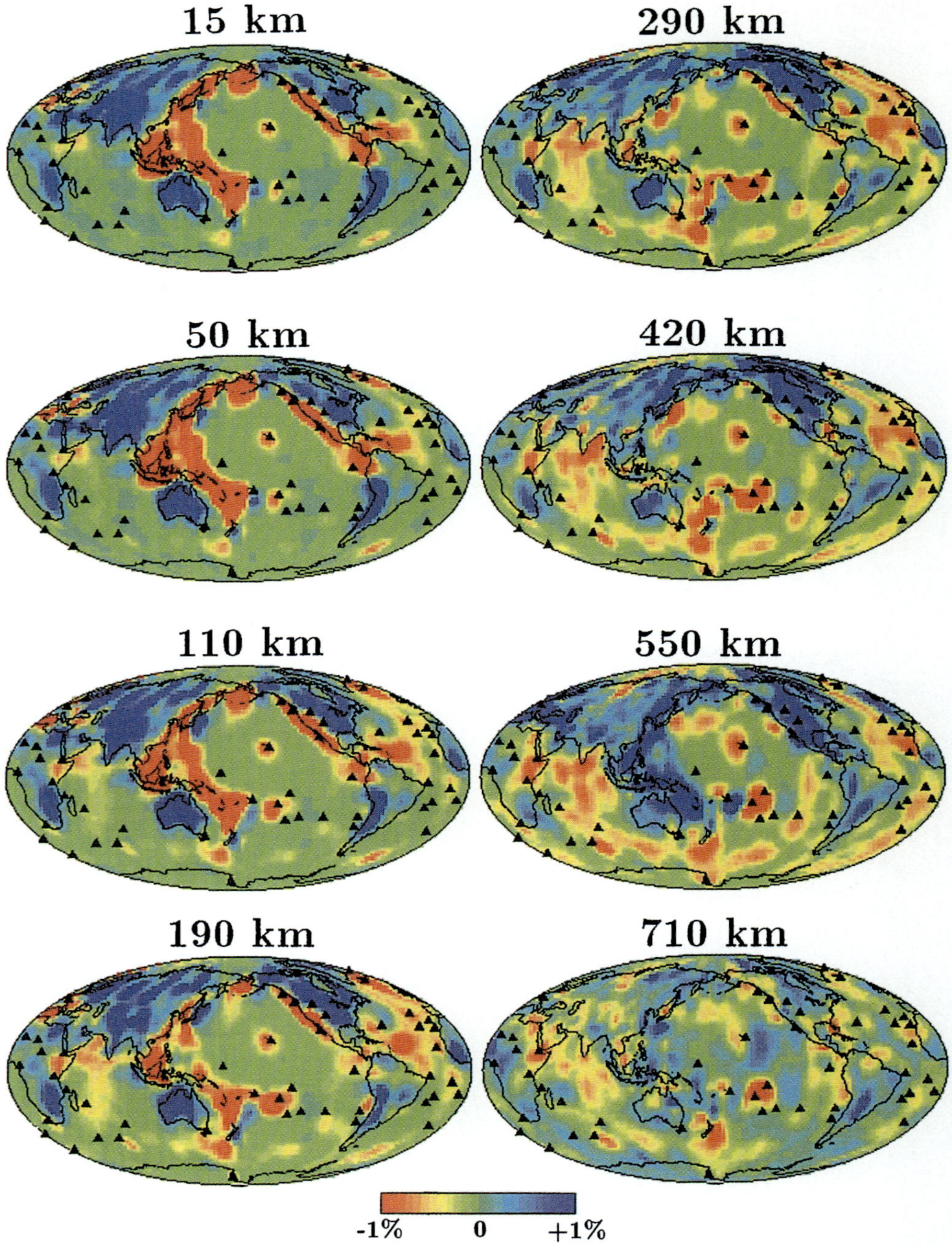

Plate 1. *P*-wave velocity perturbations from the average velocity for depth slices of the mantle. The depth of each layer is noted at the top of each map. Red and blue colors denote slow and fast velocities, respectively. The velocity perturbation scale is shown at the bottom. Solid triangles denote the surface hotspots [*Richards et al.*, 1988; *Steinberger*, 2000].

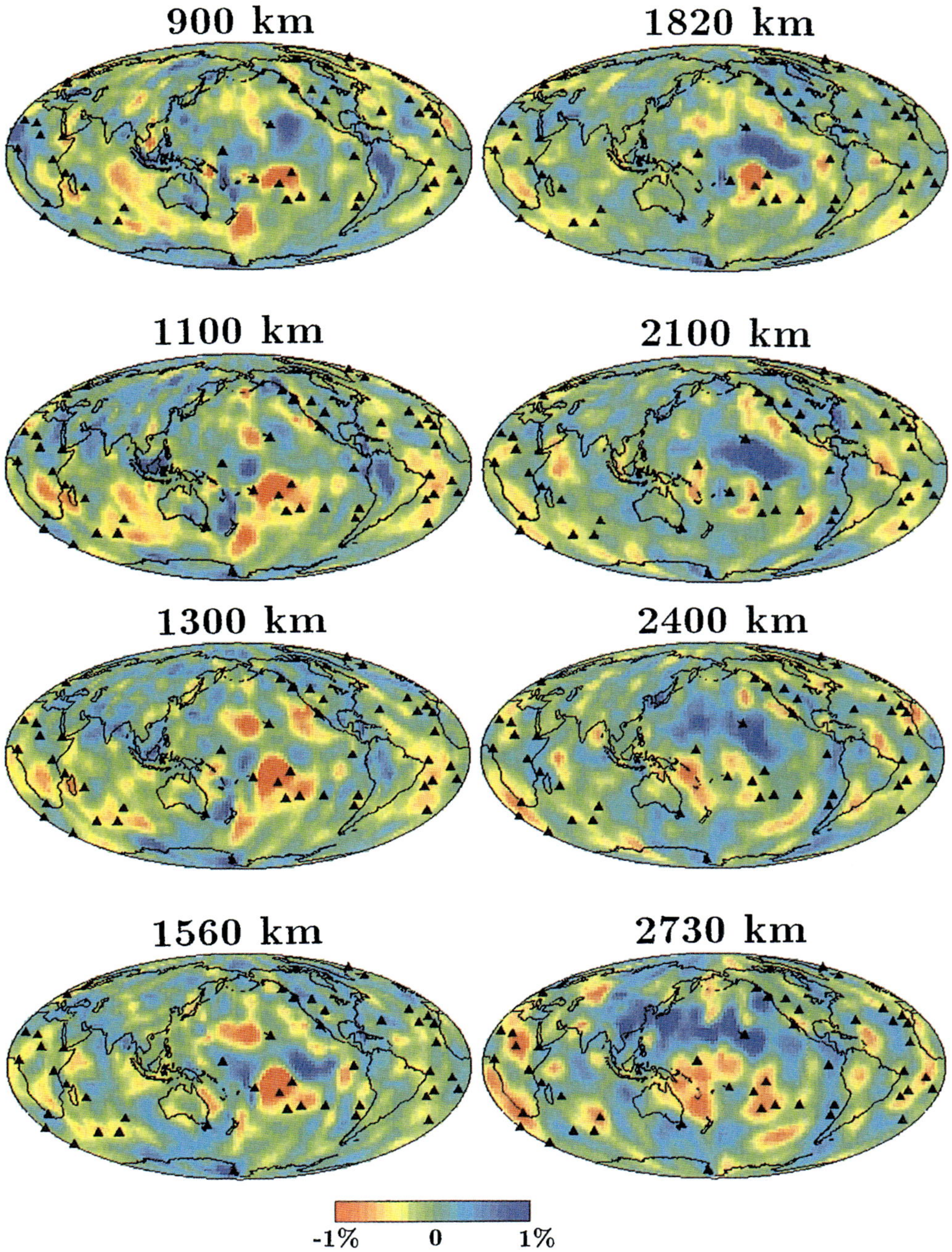

Plate 1 (continued)

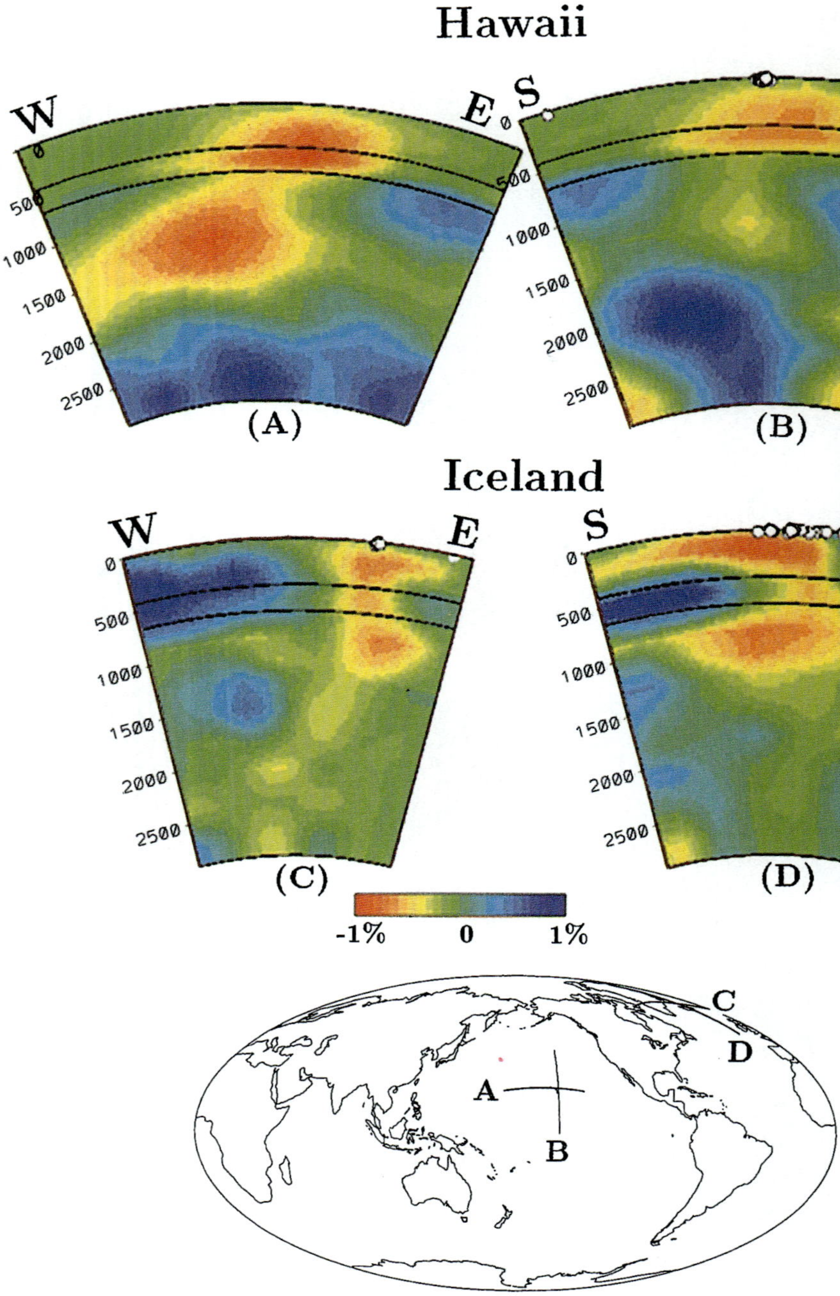

Plate 2. Vertical cross sections of *P*-wave velocity images under Hawaii (A) east-west and (B) north-south and Iceland (C) east-west and (D) north-south. Locations of the profiles are shown on the world map. Red and blue colors denote slow and fast velocities, respectively. The velocity perturbation scale is shown below the cross sections. Open circles denote earthquakes (compiled by the ISC) that occurred within a 40-km width from each of the profiles.

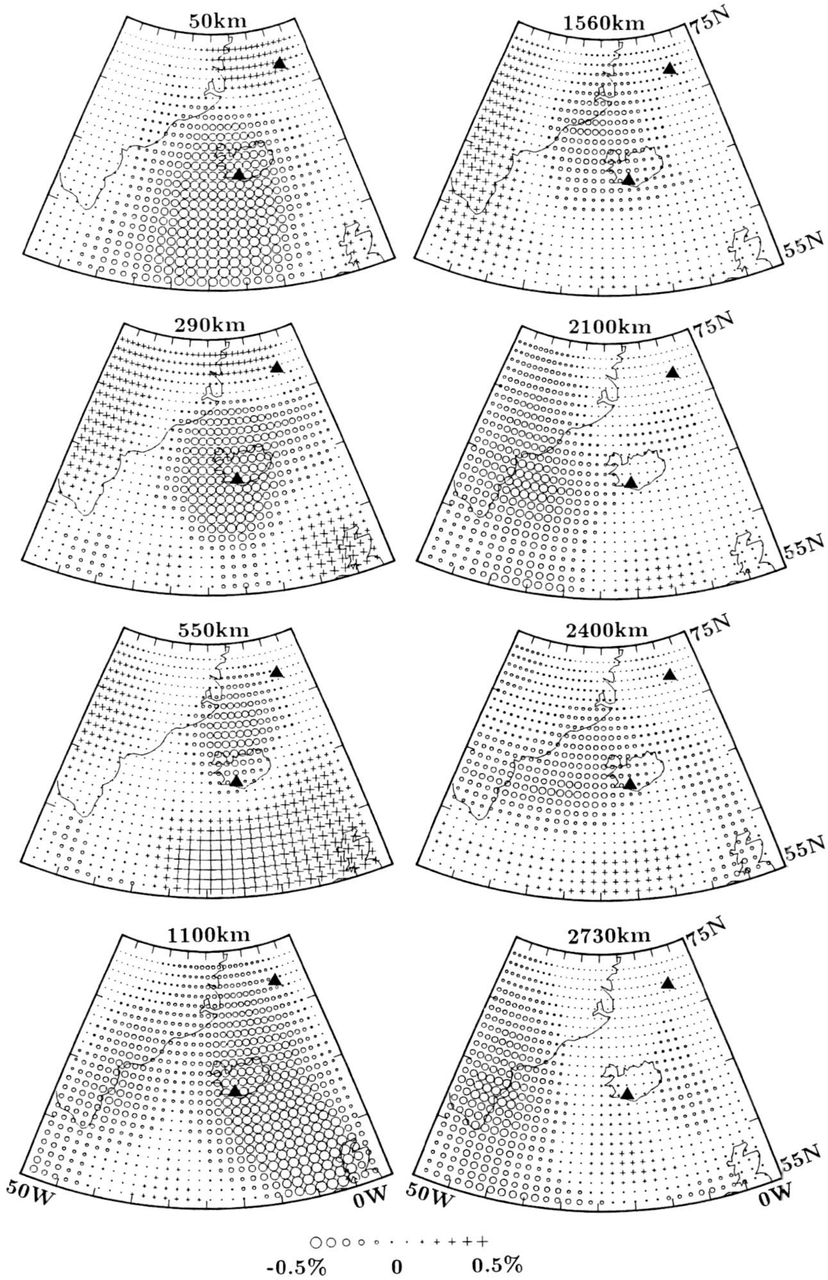

Figure 5. *P*-wave velocity images under the Iceland region. The depth of each layer is noted at the top of each map. Circles and crosses denote slow and fast velocities, respectively. The velocity perturbation scale is shown at the bottom. Solid triangles denote the Iceland and Jan Mayen hotspots.

zone. Our present global tomography and that of *Bijwaard and Spakman* [1999] show that the Iceland plume extends down to the CMB (Figure 5, Plate 2), which are confirmed with the synthetic tests (Figures 3b, 4b).

Under South Pacific and East Africa, huge low-V anomalies (> 1000 km wide) are visible in the mantle (Plate 3), which may represent superplumes [*McNutt and Fisher*, 1987; *Su et al.*, 1994; *Ritsema et al.*, 1999; *Niu et al.*, 2000]. *Steinberger* [2000] calculated the motion of hotspots in East Africa and the deformation of their underlying plume conduits with models of global mantle flow and suggested the presence of comparatively broad upwellings rather than localized plumes. This may explain the large and complex geometry of the low-V zones we imaged under East Africa (Plate 3c).

As a consequence of the deflection of mantle plumes, hotspots would not be fixed during the geological history but could wander on the Earth's surface, as suggested by the geomagnetic and numeric modeling studies, though their relative moving velocities are much smaller than that of the lithospheric plates [*Molnar and Stock*, 1987; *Griffiths and Richards*, 1989; *Steinberger*, 2000]. Mantle plumes and hotspots are a time dependent phenomenon [*Sleep*, 1992]. Plumes begin with an instability in the thermal boundary layer and decline when the hot material in the thermal boundary is tapped by the plume faster than it can be regenerated by the thermal conduction from below and recharged by influx from the sides. Alternatively, flow in the mantle locally may thicken the boundary layer faster than it is tapped by the plume.

4.3. Magma Production Rate of Hotspots

The size of the mantle plumes as seen in Plates 1-3 is not directly related to the extent of magma production at the hotspots. For example, the very large slow anomaly beneath the South Pacific (Plates 3a, 3b) may feed magma to several hotspots clustered there (Marquesas, Tahiti, and Tubuai). None of these hotspots or their total melt production rate are comparable to that of the Hawaiian plume [*Sleep*, 1990]. The velocity anomaly related to the Hawaiian hotspot, however, is smaller in size (Plates 2a, 2b).

In a standard plume model [e.g., *White and McKenzie*, 1989; *Watson and McKenzie*, 1991], the melt production rate of a given hotspot can be calculated from (a) the potential temperature of the upwelling flow, (b) the net flux of the upwelling, and (c) the thickness of the lithospheric lid. The relatively large magma production rate of the Iceland hotspot compared with the Africa and South Pacific hotspots may be due to the very thin lithosphere beneath the Iceland hotspot. Likewise, the large-scale upwelling beneath the South Africa (Plates 3c, 3d) may yield smaller melt production than that beneath the South Pacific (Plates 3a, 3b) because the African cratonic lithosphere is much thicker (~200 km thick) than that of the South Pacific (60-80 km thick) [*Zhang and Tanimoto*, 1993]. Given the very large magma production rate of the Hawaiian hotspot with a moderate lithosphere thickness (~80 km), its potential temperature should be much higher than other hotspots. The upwelling flow may be concentrated due to the low viscosity of this high temperature plume. A narrow conduit would be difficult to image using current tomographic techniques.

An alternative explanation for the difference in the magma production rate among the hotspots is the presence of variable amounts of eclogite (recycled old oceanic crust) entrained within a plume [*Cordery et al.*, 1997; *Takahashi et al.*, 1998]. A significant eclogite component in the Hawaiian plume has been postulated [e.g., *Hauri*, 1996]. Based on high-pressure experiments, *Takahashi and Nakajima* [this volume] proposed that the silica-rich tholeiite basalt magma at the final stage of Koolau volcano is almost pure partial melt of eclogite. They also proposed that picritic primary magmas in other Hawaiian volcanoes (e.g., Kilauea) are melting products of recycled eclogite together with surrounding peridotite. According to this interpretation, the very high magma production of the Hawaiian volcano is explained by a large amount of entrained eclogite in the plume and its potential temperature does not need to be higher than other less productive mantle plumes [*Takahashi and Nakajima*, this volume].

4.4. Excess Temperature of Plumes

Seismic velocity variations are affected by many factors, such as temperature, composition, partial melting, fluids, cracks, etc., but we still have no effective tools to discriminate between these contributions. If we assume that the slow velocities under the hotspot regions are entirely caused by high temperatures, we may use the method of *Karato* [1993] to estimate the excess temperatures of the plumes, which gives values in the range of 200-300 K. This is compatible with the estimates from numerical experiments [*Albers and Christensen*, 1996; *Steinberger*, 2000]. But the amplitudes of the velocity anomalies we obtained may be affected due to the damping and smoothing in the inversion process.

The temperature anomaly beneath the South Pacific has been estimated from the thinning of the mantle transition zone by *Niu et al.* [2000]. The excess temperature they estimated for the South Pacific plumes is only 100-200 K, which is similar to that for the Iceland plume estimated by *Shen et al.* [1998]. Note that these temperature anomalies are smaller than the excess potential temperature estimates (~300 K) based on the standard theory [*White and McKenzie*, 1989; *Watson and McKenzie*, 1991] but is consistent with the estimates from melting experiments of basalt/peridotite hybrid source [*Takahashi and Nakajima*, this volume]. The large slow anomalies beneath the South Pacific (Plates 3a, 3b) may consist of depleted peridotite yielding small amount of melts. They may be the former oceanic lithosphere subducted to the base of the man-

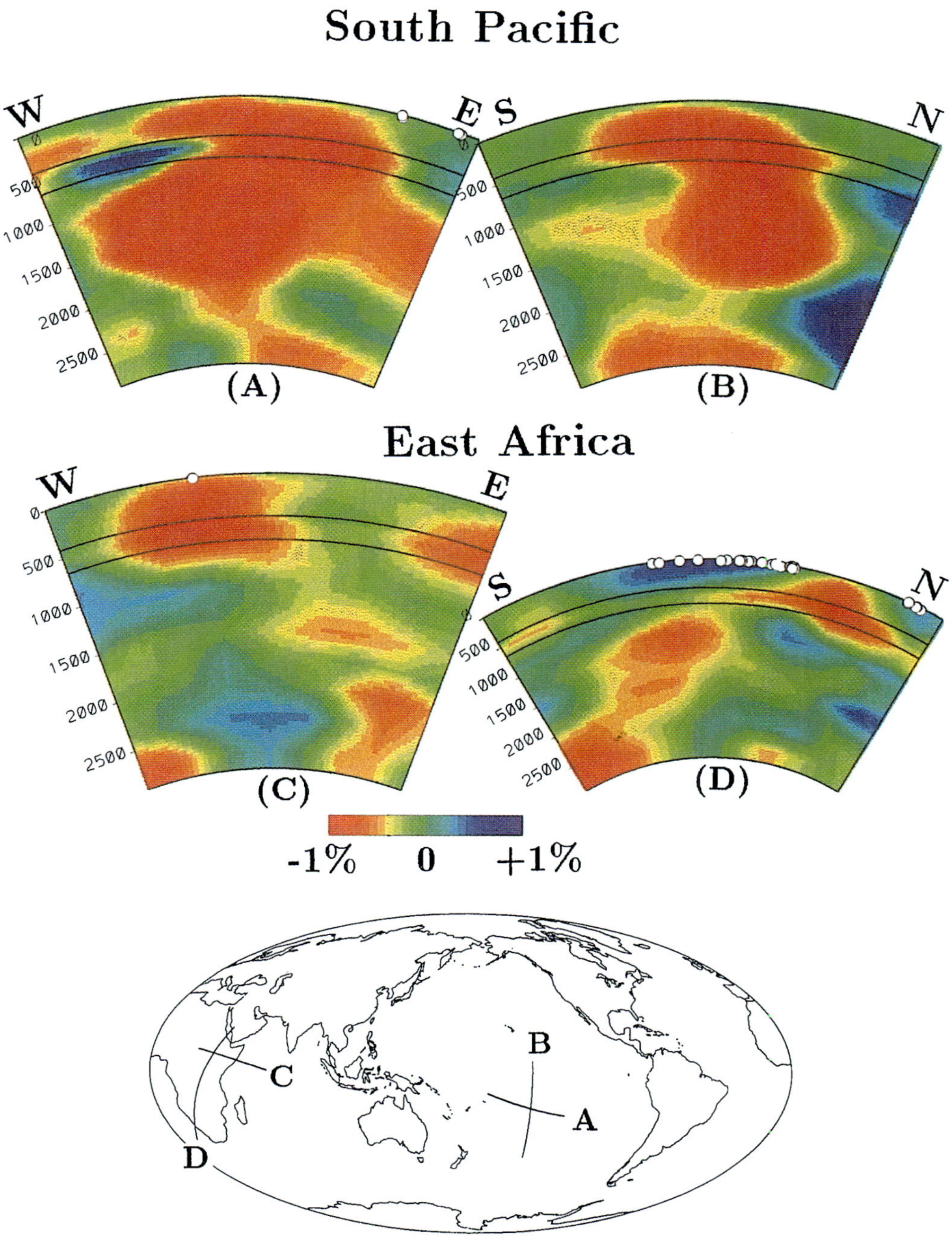

Plate 3. Vertical cross sections of *P*-wave velocity images under the South Pacific (A and B) and East Africa (C and D). Locations of the profiles are shown on the world map. Red and blue colors denote slow and fast velocities, respectively. The velocity perturbation scale is shown below the cross sections. Open circles denote earthquakes (compiled by the ISC) that occurred within a 40-km width from each of the profiles.

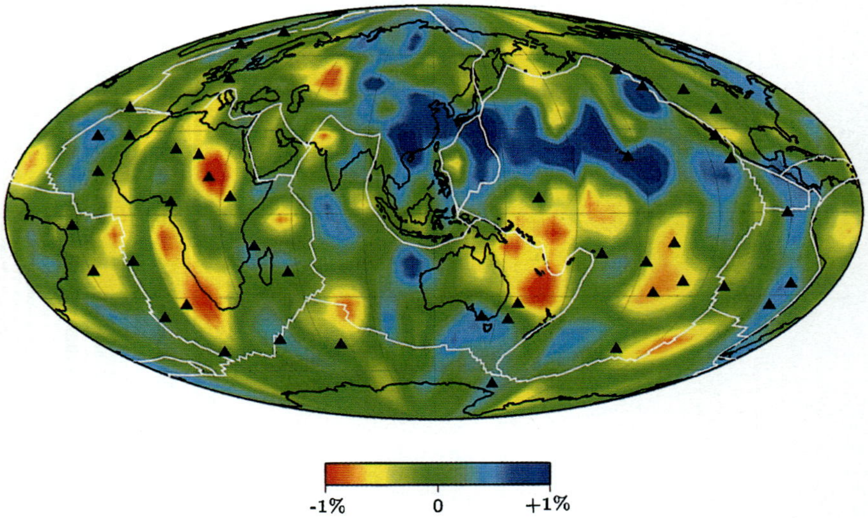

Plate 4. *P*-wave velocity image at the CMB. Red and blue colors denote slow and fast velocities, respectively. The velocity perturbation scale is shown at the bottom. Solid triangles denote the surface hotspots. White lines denote the plate boundaries.

tle, and separated from the oceanic crust component perhaps due to the melting differentiation at the CMB.

5. CONCLUSIONS

A new model of 3-D *P*-wave velocity structure of the mantle is determined by using a grid parameterization and a 3-D ray tracing technique and considering the depth variations of the Moho, 410 and 660 km discontinuities. Detailed resolution analyses and synthetic tests were conducted to confirm the major features of the obtained tomographic images. Plume-like slow anomalies are imaged clearly under hotspot regions from the crust down to the CMB. Hotspots in Hawaii, Iceland, South Pacific and Africa are located above the slow anomalies in the lower mantle down to the CMB, suggesting that mantle plumes under those hotspots may originate from the CMB. However, there may be some small-scaled, weak plumes originating from the transition zone or middle mantle depths. The slow anomalies under hotspots usually do not show a vertical pillar shape, which suggests that plumes are not fixed in the mantle but can be deflected by the mantle winds. Thus hotspots may not be fixed but can wander on the Earth's surface, as suggested by the recent geomagnetic and numeric modeling studies.

Acknowledgments. The authors are grateful to E. R. Engdahl for providing his reprocessed ISC data sets and to A. Gorbatov for his assistance at the data processing stage. P. Shearer helped to make use of his model for the topography of the 410 and 660 km discontinuities. M. Garcia, M. McNutt and an anonymous referee provided thoughtful reviews which improved the manuscript. This work was partially supported by a grant (Kiban-B No. 11440134) from the Japan Society for the Promotion of Science.

REFERENCES

Albers, M., and U. Christensen, The excess temperature of plumes rising from the core-mantle boundary, *Geophys. Res. Lett., 23*, 3567-3570, 1996.

Allen, R., G. Nolet, W. Morgan, K. Vogfjord, The thin hot plume beneath Iceland, *Geophys. J. Int., 137*, 51-63, 1999.

Anderson, D., Chemical plumes in the mantle, *Geol. Soc. Am. Bull., 86*, 1593-1600, 1975.

Bijwaard, H., and W. Spakman, Tomographic evidence for a narrow whole mantle plume below Iceland, *Earth Planet. Sci. Lett., 166*, 121-126, 1999.

Bijwaard, H., W. Spakman, and E. Engdahl, Closing the gap between regional and global travel time tomography, *J. Geophys. Res., 103*, 30,055-30,078, 1998.

Cordery, M., G. Davies, and I. Campbell, Genesis of flood basalts from Eclogite-bearing mantle plumes, *J. Geophys. Res., 102*, 20,179-20,197, 1997.

Dziewonski, A., Mapping the lower mantle: Determination of lateral heterogeneity in *P* velocity up to degree and order 6, *J. Geophys. Res., 89*, 5929-5952, 1984.

Engdahl, E.R., R. van der Hilst, and R.P. Buland, Global teleseismic earthquake relocation with improved travel times and procedures for depth determination, *Bull. Seismol. Soc. Am., 88*, 722-743, 1998.

Flanagan, M., and P. Shearer, Global mapping of topography on transition zone velocity discontinuities by stacking *SS* precursors, *J. Geophys. Res., 103*, 2673-2692, 1998.

Foulger, G., M. Pritchard, B. Julian, and J. Evans, The seismic anomaly beneath Iceland extends down to the mantle transition zone and no deeper, *Geophys. J. Int., 142*, F1-F5, 2000.

Griffiths, R., and M. Richards, The adjustment of mantle plumes to changes in plate motion, *Geophys. Res. Lett., 16*, 437-440, 1989.

Hauri, E., Major-element variability in the Hawaiian mantle plume, *Nature, 382*, 415-419, 1996.

Inoue, H., Y. Fukao, K. Tanabe, and Y. Ogata, Whole mantle *P* wave travel time tomography, *Phys. Earth Planet. Inter., 59*, 294-328, 1990.

Ji, Y., and H. Nataf, Detection of mantle plumes in the lower mantle by diffraction tomography: Hawaii, *Earth Planet. Sci. Lett., 159*, 99-115, 1998.

Karato, S., Importance of anelasticity in the interpretation of seismic tomography, *Geophys. Res. Lett., 20*, 1623-1626, 1993.

Kennett, B., and E. Engdahl, Traveltimes for global earthquake location and phase identification, *Geophys. J. Int., 105*, 429-465, 1991.

King, S., and J. Ritsema, African hot spot volcanism: Small-scale convection in the upper mantle beneath cratons, *Science, 290*, 1137-1140, 2000.

Loper, D., Mantle plumes, *Tectonophysics, 187*, 373-384, 1991.

McNutt, M., and K. Fischer, The South Pacific superswell, in *Seamounts, Islands, and Atolls, Geophys. Monogr. Ser., Vol. 43*, edited by B. Keating and R. Batiza, pp. 25-34, AGU, Washington, D.C., 1987.

Molnar, P. and J. Stock, Relative motions of hotspots in the Pacific, Atlantic, and India oceans since late Cretaceous time, *Nature, 327*, 587-591, 1987.

Mooney, W., G. Laske, and T. Master, CRUST 5.1: A global crustal model at $5° \times 5°$, *J. Geophys. Res., 103*, 727-747, 1998.

Morgan, W., Convection plumes in the lower mantle, *Nature, 230*, 42-43, 1971.

Ni, S., X. Ding, D. Helmberger, and M. Gurnis, Low-velocity structure beneath Africa from forward modeling, *Earth Planet. Sci. Lett., 170*, 497-507, 1999.

Niu, F., H. Inoue, D. Suetsugu, and K. Kanjo, Seismic evidence for a thinner mantle transition zone beneath the South Pacific Superswell, *Geophys. Res. Lett., 27*, 1981-1984, 2000.

Paige C., and M. Saunders, LSQR: An algorithm for sparse linear equations and sparse least squares, *Assoc. Comput. Mach. Trans. Math. Software, 8*, 43-71, 1982.

Richards, M., B. Hager, and N. Sleep, Dynamically supported geoid highs over hotspots: Observation and theory, *J. Geophys. Res., 93*, 7690-7708, 1988.

Ritsema, J., H. Jan der Heijst, and J. Woodhouse, Complex shear wave velocity structure imaged beneath Africa and Iceland, *Science, 286*, 1925-1928, 1999.

Ritter, J., M. Jordan, U. Christensen, and U. Achauer, A mantle plume below the Eifel volcanic fields, Germany, Earth Planet. Sci. Lett., 186, 7-14.

Shen, Y., S. Soloman, I. Bjarnason, C. Wolfe, Seismic evidence for a lower-mantle origin of the Iceland plume, *Nature 395*, 62-65, 1998.

Sleep, N., Hotspots and mantle plumes: some phenomenology, *J. Geophys. Res., 95*, 6715-6736, 1990.

Sleep, N., The dependence of mantle plumes: Some simple theory, *J. Geophys. Res., 97*, 20,007-20,019, 1992.

Sleep, N., Lateral flow of hot plume material ponded at sublithospherical depths, *J. Geophys. Res., 101*, 28065-28083, 1996.

Steinberger, B., Plumes in a convecting mantle: Models and observations for individual hotspots, *J. Geophys. Res., 105*, 11,127-11,152, 2000.

Su, W., R. Woodward, and A. Dziewonski, Degree 12 model of shear velocity heterogeneity in the mantle, *J. Geophys. Res., 99*, 6945-6980, 1994.

Takahashi, E., K. Nakajima, and T. Wright, Origin of the Columbia River basalts: melting model of a heterogeneous plume head, *Earth Planet. Sci. Lett., 162*, 63-80, 1998.

Takahashi, E., and K. Nakajima, Melting process in the Hawaiian plume: An experimental study, submitted to AGU Monograph *Evolution of Hawaiian Volcanoes* (this volume), 2001.

Tryggvasson, K., E. Husebye, and R. Stefansson, Seismic image of the hypothesized Icelandic hot spot, *Tectonophysics, 100*, 97-118, 1983.

Um, J., and C. Thurber, A fast algorithm for two-point seismic ray tracing, *Bull. Seism. Soc. Am. 77*, 972-986, 1987.

White, R., and D. McKenzie, Magmatism at rift zones: the generation of volcanic continental margins and flood basalts, *J. Geophys. Res., 94*, 7685-7729, 1989

Watson, S., and D. McKenzie, Melt generation by plumes: a study of Hawaiian volcanism, *J. Petro., 32*, 501-537, 1991.

Weinstein, S., and P. Olson, The proximity of hotspots to convergent and divergent plate boundaries, *Geophys. Res. Lett., 16*, 433-436, 1989.

Williams, Q., J. Revenaugh, and E. Garnero, A correlation between ultra-low basal velocities in the mantle and hot spots, *Science, 281*, 546-549, 1998.

Wilson, J., A possible origin of the Hawaiian islands, *Can. J. Phys., 41*, 863-870, 1963.

Wolfe, C., I. Bjarnason, J. VanDecar, and S. Soloman, Seismic structure of the Iceland mantle plume, *Nature 385*, 245-247, 1997.

Zhang, Y., and T. Tanimoto, High-resolution global upper mantle structure and plate tectonics, *J. Geophys. Res., 98*, 9793-9823.

Zhao, D., Whole mantle tomography with grid parameterization, 3-D ray tracing and topography of mantle discontinuities, *EOS Trans., AGU, 80*, F716, 1999.

Zhao, D., New advances of seismic tomography and its applications to subduction zones and earthquake fault zones: A review, *The Island Arc, 10*, 68-84, 2001.

Zhao, D., A. Hasegawa, and S. Horiuchi, Tomographic Imaging of *P* and *S* wave velocity structure beneath northeastern Japan, *J. Geophys. Res., 97*, 19,909-19,928, 1992.

Zhao, D., A. Hasegawa, and H. Kanamori, Deep structure of Japan subduction zone as derived from local, regional and teleseismic events, *J. Geophys. Res., 99*, 22313-22329, 1994.

D. Zhao and F. Ochi, Geodynamics Research Center, Ehime University, Matsuyama 790-8577, Japan. (e-mail: zhao@sci.ehime-u.ac.jp)

E. Takahashi, Department of Earth and Planetary Sciences, Tokyo Institute of Technology, Tokyo 152-8551, Japan. (e-mail: etakahas@geo.titech.ac.jp)

Heat Flow Variations Over Hawaiian Swell Controlled by Near-Surface Processes, not Plume Properties

Marcia K. McNutt

Monterey Bay Aquarium Research Institution, 7700 Sandholdt Road, Moss Landing, California

The commonly accepted explanation for the Hawaiian-Emperor chain of seamounts is that it is caused by drift of the Pacific plate over a relatively stationary plume of hot material upwelling from deep within the Earth. As the "classic" example of this phenomenon, geophysical data from the Hawaiian chain, particularly heat flow data, have been used to infer the physical properties of mantle plumes, such as their temperature and how they interact with the overlying plate. However, the heat flow data have defied simple explanation and are often at odds with other geophysical observations. Here I show that variations in heat flow are highly correlated with changes in the thermal conductivity of the surface sediments rather than with any expected variability in heat flux reaching the surface from a mantle plume. This correlation of heat flow with thermal conductivity is exactly what is predicted if the upper oceanic crust acts as a permeable aquifer in a flow system that laterally redistributes heat within the uppermost oceanic crust. The spatially- averaged mean of the heat flow values is well determined and indicates that the oceanic plate near Hawaii is hotter than expected for its age, consistent with the plume hypothesis. However, the effect of the fluid circulation is to obscure any heat flow pattern that could be used to infer the structure of the plume or its interaction with the overlying lithosphere.

1. INTRODUCTION

The voluminous midplate volcanic activity that has created the Hawaiian Islands and its extremely linear progression in age has been attributed to the drift of the Pacific plate over a mantle plume [*Morgan*, 1975]. The first heat flow experiment to systematically investigate the thermal influence of the Hawaiian plume sampled a transect located on the flexural arch just south of the chain (Figure 1) [*Detrick et al.*, 1981]. This survey (hereafter referred to as the South Arch or SA survey) indicated that at the current location of the hotspot ("zero age"), heat flow values are typical for average, 80- to 90-million year old oceanic lithosphere. Further to the northwest, on lithosphere that encountered the hotspot more than 6 million years ago, heat flow is anomalously high (Figure 2). This result was interpreted as firm evidence in favor of the lithospheric thinning model: the plume initially reheats the lower lithosphere to asthenospheric temperatures, forming a bathymetric swell by virtue of the lower density of the hot rock beneath the thinned plate [*Detrick and Crough*, 1978]. The slow rate of conduction of heat through the lithospheric lid prevents the heat flow signal from reaching the surface until millions of years later. The time delay between the uplift of the swell (zero age) and the detection

Hawaiian Volcanoes: Deep Underwater Perspectives
Geophysical Monograph 128
Copyright 2002 by the American Geophysical Union

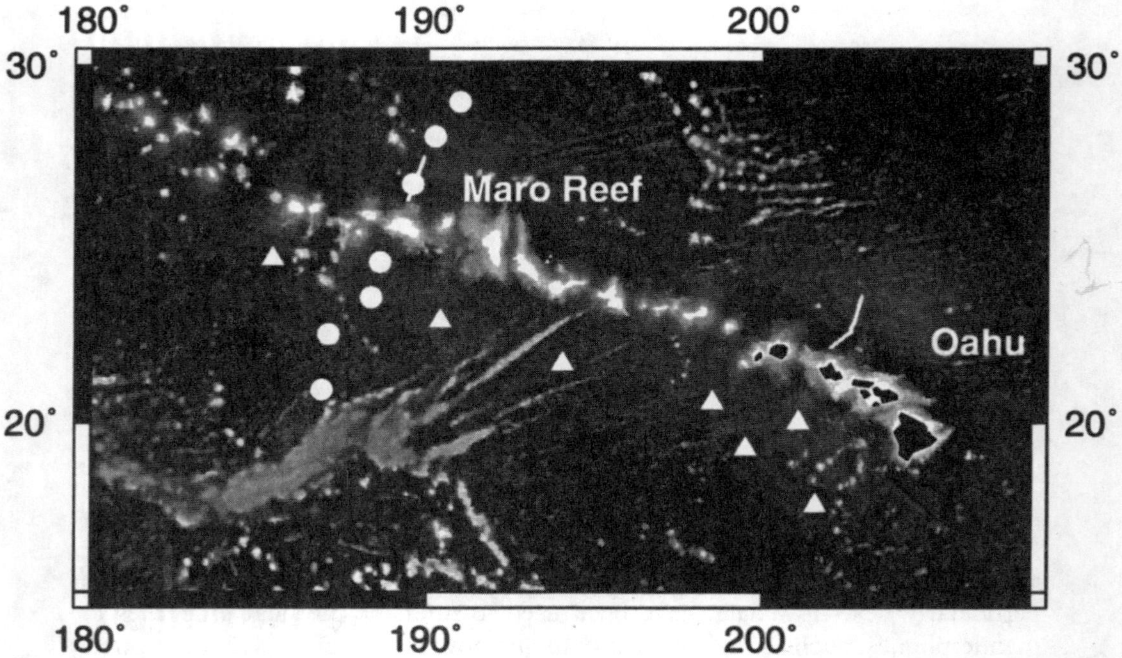

Figure 1. The Hawaiian Islands in the central Pacific and the location of the heat flow stations used in this analysis. The white triangles are the South Arch survey [*Detrick et al.*, 1981; *Von Herzen et al.*, 1982]. White dots are the Maro Reef survey [*Von Herzen et al.*, 1989]. The solid white lines north of Maro Reef and Oahu show the location of dense heat flow measurements used to constrain the hydrogeology of the oceanic crust near major volcanoes [*Harris et al.*, 200a,b].

of the heat flow anomaly downstream from the hotspot in the south arch data was used to determine the depth to which the Hawaiian plume thinned the lithosphere: about 40-50- km depth [*Detrick et al.*, 1981].

A subsequent heat flow expedition [*Von Herzen et al.*, 1989] sought to confirm the model by measuring the heat flow along a profile across the chain at the location of Maro Reef (Figure 1), where the heat flow signal from thinning the plate to only 40 km thickness [*Detrick et al.*, 1981] should be a maximum. This second survey (hereafter referred to as the Maro Reef or MR survey) failed to detect the expected Gaussian signature. Rather, the 8 heat flow values defined a "W" pattern (Figure 2), which is inconsistent with the lithospheric thinning model and other competing predictions of plume/lithosphere interaction. Theoreticians proposed that the saw-toothed structure in the heat flow data might indicate a very hot plume (450°C warmer than the surrounding mantle) [*Moore et al.*, 1998], but even that model did not provide a very good fit to the data and was at odds with petrological estimates of plume temperature [*McKenzie and Bickle,* 1988].

The most recent expedition to measure Hawaiian heat flow [*Harris et al.,* 2000a] investigated heat transfer on a finer scale than previous surveys. Using the "pogo" method for multiple penetrations [*Von Herzen*, 1987], heat flow was sampled over two finely spaced lines north of Oahu Island and Maro Reef (Figure 1). Each profile extended over 200 km with measurements taken every 1-2 km. The results of this detailed survey (hereafter referred to as the MO survey) cast yet more doubt on the validity of the lithospheric reheating hypothesis and on the possibility of recovering plume properties from Hawaiian heat flow data. To begin with, the mean heat flow (corrected for sedimentation) at Oahu (74.2 mW m^{-2} ± 14.6 mW m^{-2}) is greater than the global mean value for lithosphere of that age, 62 mW m^{-2} [*Stein and Stein*, 1992], although only marginally so given the large variability of the data about the mean. No heat flow anomaly at Oahu should be expected according to the lithospheric reheating hypothesis because Oahu is too close to zero age for the heat from a plume to have had sufficient time to conduct to the surface through the lithospheric lid. Furthermore, the heat flow at Oahu is slightly larger than that at Maro Reef (64.0 mW m^{-2} ± 4.9 mW m^{-2}), although the two means overlap given the uncertainties. The lithospheric reheating hypothesis would predict higher heat flow at Maro Reef since it is sufficiently far downstream from the hotspot for the heat to reach the surface. Finally, the variability along these ~200-km-long profiles (5 to 15 mW m^{-2}) is as large or larger than the heat flow change along or across the entire swell expected from a plume (e.g., Figure 2). At the very least, the first two heat flow surveys that only sampled sites several hundred kilometers apart are likely to have aliased the true variability in the heat flow around Hawaii.

Harris et al. [2000b] demonstrated that the magnitude and spatial wavelengths of the variability in heat flow cannot be fit by simple models that invoke either conductivity contrasts between sediments and volcanic basement or by diffuse flow models. Rather, they demonstrate that the amplitude and wavelength of the variation in heat flow along these two profiles is fit by a model in which pore fluid circulation is driven through a shallow, low-permeability aquifer by bathymetric relief associated with the volcanic edifice. Two sorts of flow are associated with this aquifer. The first is a moat-wide loop flow in which warm water is discharged through the volcano and cold seawater recharges the aquifer at basement outcrops on and beyond the flexural arch. The second flow is local cellular convection within the aquifer controlled by the pattern of topographic relief on the interface of the aquifer. The former produces heat flow variability at wavelengths of 30-100 km, while the latter produces variability at wavelengths from 5 to 40 km.

The purpose of this paper is to re-examine the heat flow data from the previous SA and MR surveys in light of the *Harris et al.* [2000a,b] result that explains heat flow anomalies at scales of a few 100 km within the context of the hydrogeology of the Hawaiian islands. The question I address is whether shallow fluid circulation is influencing the pattern of heat flow over dimensions of 1000 km, thus masking the signature from a mantle plume.

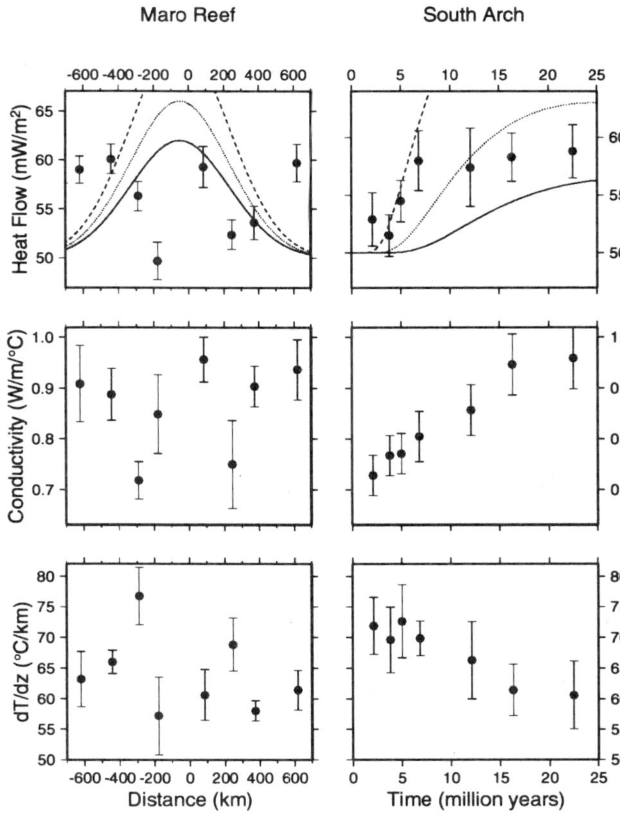

Figure 2. *Left.* Heat flow (*upper*), conductivity (*middle*) and vertical temperature gradient (*lower*) at Maro Reef plotted as a function of distance across the chain. *Right.* Heat flow (*upper*), conductivity (*middle*) and vertical temperature gradient (*lower*) along the south arch of the Hawaiian Islands plotted as a function of time since the plate drifted past the hotspot. Lines on the upper panels show the theoretical variations in heat flow expected if the hotspot reheated the lower lithosphere to asthenospheric values (~1350°C) as it passed over the thermal source. The amount of thinning assumed is 60 km (solid line), 50 km (dotted line), and 40 km (dashed line).

2. LOCAL PATTERNS IN REGIONAL SURVEYS

The first question that must be addressed is whether the magnitude and scales of variability in heat flow observed by *Harris et al.* [2000a,b] in the moats north of Maro Reef and Oahu are typical of other sites on the Hawaiian swell. Each of the heat flow values reported by *Detrick et al.* [1981] and *Von Herzen et al.* [1989] for the stations occupied along and across the Hawaiian swell is actually the mean of 10 to 20 individual heat flow values measured in a 10 km x 10 km patch of seafloor. Heat flow Q is the product of two quantities, thermal gradient, dT/dz, and conductivity, k: $Q=-k\ dT/dz$. In the case of the MR study, both *in situ* thermal conductivity and thermal gradient were measured independently. For the case of the older SA survey, *in situ* conductivity was not measured. One value of k was assigned to each study area based on shipboard analysis of conductivity from a piston core.

Figure 3 compares the individual values for k, dT/dz, and their product, Q, at site 4 of the MR study with the entire 200-km-long Maro Reef line from *Harris et al.* [2000a]. The latter data set passes through the same patch of seafloor occupied at site 4 in the MR study, and clearly reproduces the original result very well. Furthermore, the MR data from site 4 exhibits nearly the entire range of heat flow, conductivity, and thermal gradient observed over the longer MO line. What is worth noting in this figure is that on the local scale, heat flow is random with respect to k,

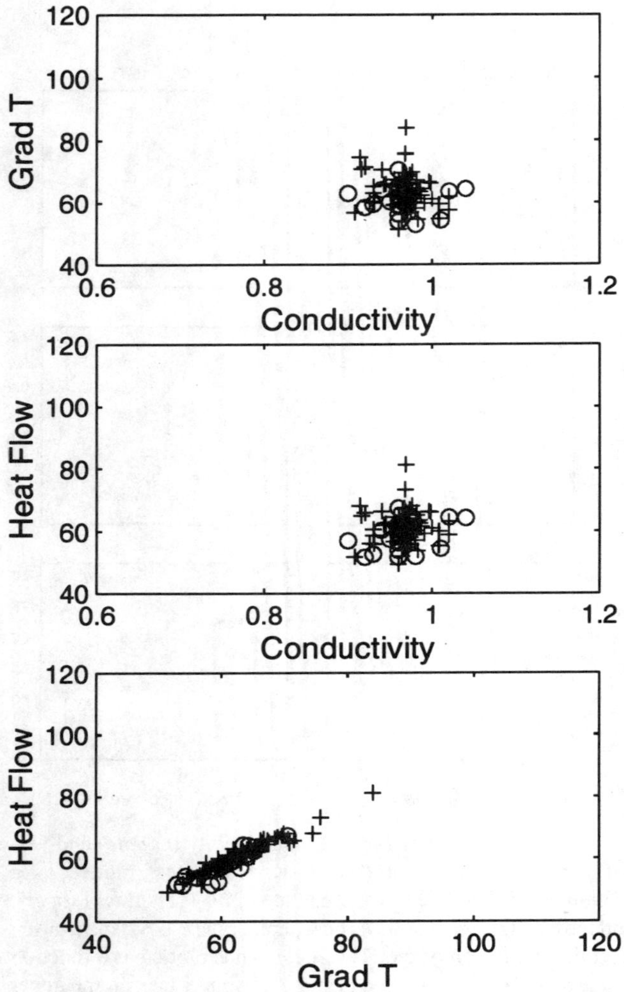

Figure 3. Comparison of individual penetrations of the heat flow probe in the flexural moat north of Maro Reef from the MR survey (open circles) [*Von Herzen et al.*, 1989] and the MO survey (plus signs) [*Harris et al.*, 2000a]. Temperature gradient is measured in °C/km, heat flow in mW/m^2, and conductivity in W/m/°C.

and k and dT/dz are uncorrelated. In fact, there is little variability is k whatsoever, and heat flow variations are completely determined by local variations in the near-surface thermal gradient. For example, in Figure 3 the variation in k is only 10% of its mean value at site 4 of the MR survey, whereas the variation in dT/dz is more than 50% of the mean.

The histograms of the distribution of individual heat flow penetrations from the SA and MR sites as plotted in *Von Herzen et al.* [1982; 1989] reveal that the scatter in heat flow values along the Hawaiian swell is similar to that seen in Figure 3. Small-scale variability in heat flow is 15 to 20 mW m^{-2} or more in each 10 km x 10 km patch regardless of location on the swell. Within each 10 km x 10 km site, the *in situ* conductivity values are relatively uniform, and variations in heat flow result from variations in thermal gradient. Although individual conductivity measurements were not available for the SA data, the variability in dT/dz at each of the SA sites is similar (~50% of the mean) to that for the more recent studies with *in situ* k measurements. Clearly the high variability in heat flow is caused by local fluctuations in thermal gradient in the near-surface sediments.

According to the modeling results of *Harris et al.* [2000b], small-scale convection within an aquifer of laterally varying thickness and shape produces a peak in the spectrum of heat flow at 10-km wavelength. In their simulations, relief on the basement aquifer of less than 1 km, as imaged in seismic reflection data, produces heat flow undulations of more than 30 mW m^{-2}. Therefore, it seems reasonable to conclude that if fluid circulation in a permeable aquifer reproduces the amplitude and scales of variability in the moats north of Oahu and Maro Reef [*Harris et al.*, 2000b], then similar processes are likely at work at other locations along the swell as well. Averaging the individual heat flow penetrations over scales of 10 km to several 10's of kilometers would effectively remove this small-scale variability.

3. THE CONDUCTIVITY FACTOR

At regional scales (> a few hundred kilometers), just the opposite pattern is apparent: variations in heat flow are controlled by variations in k, not dT/dz. On close inspection, it is clear on both the MR and SA surveys that heat flow is positively correlated with k and inversely correlated with dT/dz (Figure 2). This result is unexpected for the following reason. A plume should impose an abnormally high heat flux Q_o into the base of the plate. Because lateral temperature gradients are very small compared with the vertical gradient, conservation of energy requires that the anomalous heat flux be detected at the surface of the plate eventually if heat transfer through the lithosphere is via conduction only. In a region covered by low-conductivity surface sediment, the crust will warm up beneath the insulating blanket until the temperature gradient is high enough to offset the low k value. Thus two regions with the same basal flux Q_o should have the same surface heat flow because the near-surface temperature gradients will adjust to cancel out any effect from variations in the surface thermal conductivity.

In this scenario, the surface Q, if reflecting heat conducted to the surface from a plume, should be entirely uncorrelated with measured values of k. Furthermore, for a random distribution of sediment conductivity values with respect to the pattern of plume-generated heat flux, one would expect in general a positive correlation between temperature gradient and heat flow. The fact that the data display an inverse correlation of Q with dT/dz and a positive correlation between Q and k suggests that this model which assumes that surface heat flow anomalies reflect plume heat flux is fundamentally wrong.

Correlation between Q and k could occur if systematic errors in the measurement of k lead to high values being overestimated and low values being underestimated. However, conductivity measurements are very repeatable in the same location (e.g., Figure 3), and the values determined by different instruments and different methodologies are generally within the standard errors of the measurements (typically 0.02 to 0.10 $Wm^{-1}{}^{\circ}K^{-1}$) [*Harris et al.*, 2000a]. Furthermore, conductivity values tend to be consistent throughout a geologic region, with variations from region to region being well correlated with known variations in sediment properties (e.g., water content, carbonate content, etc.) [*Horai*, 1981; *Matsuda and Von Herzen*, 1986]. Therefore, there is no reason to believe that systematic errors exist.

4. THE CONSTANT-T MODEL

Suppose that instead of imposing a heat flux boundary condition at the base of the plate, as has been customary in the interpretation of heat flow over swells, it is assumed that the base of the conducting layer is an isotherm (Figure 4a). This physical model has been extremely popular in the interpretation of global heat flow to infer the thermal properties of lithospheric plates [e.g., *Parsons and Sclater*, 1977; *Stein and Stein*, 1992]. The model includes a surficial conducting layer in which the conductivity is allowed to vary laterally in accord with the observations from *in situ* conductivity measurements. Below that layer is another conducting layer of more uniform conductivity, corresponding to either older sediments or volcanic rock. This two-layer model is a sensible place to start for the following reasons. A one-layer model is clearly at odds with the observations, because in the steady state it would predict that all thermal gradients should be the same everywhere, with heat flow only varying on account of thermal conductivity. A three-layer model would introduce additional unknowns into the data inversion without adding any new flexibility in fitting the data. It is always possible to assign a thickness and thermal conductivity to a single lower layer such that it mimics the effect of two or more lower layers lying between the uppermost conducting layer and the isothermal base to the system. All that is necessary is that the equivalent single lower layer reproduce the same temperature and heat flux at the base of the uppermost layer as supplied by the more complex model.

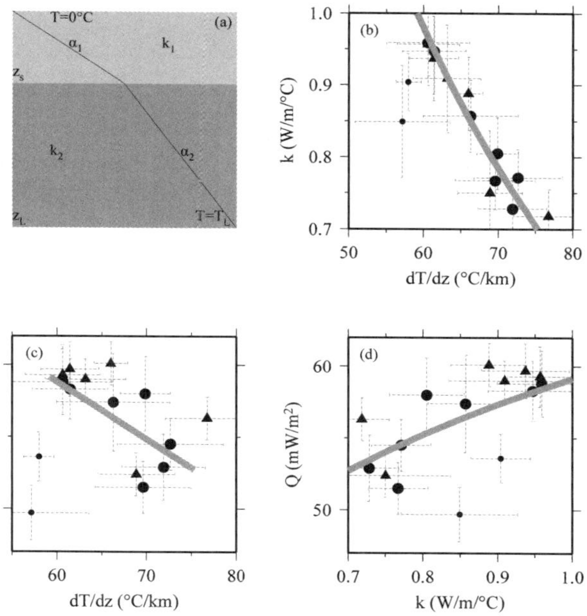

Figure 4. (*a*) Simple 2-layer model for heat transport near the Hawaiian islands. The upper layer of thickness z_s has thermal conductivity k_1 and temperature gradient α_1. The lower layer of thickness (z_L- z_s) has thermal conductivity k_2 and temperature gradient α_2. The lower boundary is maintained at temperature T_L by fluid circulation. The upper boundary is maintained at temperature T=0 °C by the ocean. (*b*) Plot of thermal conductivity k versus vertical temperature gradient dT/dz for the South Arch (triangles) and Maro Reef (dots). The thick line shows the least-squares best fit to the data as k_1 is varied in the model in (*a*), excluding the two Maro Reef outliers shown as small dots. These outliers correspond to the measurements at x=-177 km and x=375 km in Fig. 2. (*c*) Same as (*b*), but for heat flow Q as a function of dT/dz. (*d*) Same as (*b*), but for Q as a function of k.

4.1 Formulating the Model

In this simple scenario (Figure 4a), the surface sediment layer has thickness z_s and thermal conductivity k_1. The surface layer is underlain by a more conductive layer with conductivity k_2 and thickness (z_L- z_s). The thermal gradients in the surface and lower layers are α_1 and α_2, respectively. These conducting layers overlie a well-mixed region of low viscosity maintained at temperature T_L. The requirement that the heat flux be continuous at z_s is

$$Q = k_1\,\alpha_1 = k_2\,\alpha_2$$

The temperature must rise from T_0 to T_L over the depth z_L:

$$T_L - T_0 = \Delta T = \alpha_1 z_s + \alpha_2 (z_L - z_s)$$

These equations can be converted to a simple linear regression that can be solved for two unknown quantities, β and γ:

$$\alpha_1^{-1} = \beta + k_1 \gamma$$

$$\beta = z_s/\Delta T$$

$$\gamma = (z_L - z_s)/(k_2 \Delta T).$$

4.2 Fitting the Data

With this simple change in the boundary condition assumed at the base of the conducting layer, it is possible to produce the observed correlation between *Q* and *k* and the inverse correlation between *Q* and *dT/dz* in the MR and SA data sets (Figure 4). Of the 15 data points, all but 2 in the *k* versus *dT/dz* plot are fit to within one standard error. Eight of the 15 in the *Q* versus *dT/dz* plot are fit to within one standard error, and all but two are matched at the 2-sigma level. For *Q* versus *k*, 5 of 15 points are fit at the one-sigma level and all but 1 is fit at the two-sigma level. Furthermore, the theoretical curves reproduces the inverse correlation between *Q* and *dT/dz* and the positive correlation between *Q* and *k*, which would not be the case with a plume flux boundary condition at the base of the plate. Although there is still some scatter in the data from what is predicted by this very simple model, the fits are clearly superior to those in Figure 2 for the same data sets interpreted within the context of lithospheric thinning.

This least-squares best fit to the 15 data points from the Maro Reef and the South Arch surveys is obtained for β=0.0044 km/°C and γ=0.0128 m^2/mW. Removing the two points from the Maro survey that are extreme outliers to the trend defined by the other 13 points changes the fit only slightly: β=0.0048 km/°C and γ=0.0121 m^2/mW. Therefore, I believe that the parameters determined are robust and not influenced by outliers in this admittedly noisy data set.

4.3 Interpreting the Parameters

Additional information is required to convert the parameters β and γ to physical quantities. A reasonable assumption is that the surface layer of low conductivity z_s corresponds to an acoustically semi-transparent layer of thickness approximately 100 m seen pervasively on 3.5 kHz echo sounders in surveys in the central Pacific [*Dietz et al.,* 1954; *Filmer et al.*, 1994]. With this assumption, the computed value for ΔT is 21°C. The value for the conductivity in the lower layer might reasonably be in the range 1.5-3 W/m/°C, yielding values for z_L between 500 and 900 m.

None of these values is consistent with the assumption that the isothermal layer is at the base of the entire lithosphere and that the temperature of that layer corresponds to the temperature of the plume. In order to get plume-like temperatures for T_L and lithospheric-type thicknesses for z_L I would have to assume that the surficial layer of varying thermal conductivity is 6 km thick. Clearly there is no justification for arguing that variations in water content and lithification of surface sediments extends through the entire oceanic crust.

All of these numbers are, however, quite consistent with the assumption that the region of constant temperature at the base of the conducting layers is an aquifer within the uppermost oceanic crust, as invoked by *Harris et al.* [2000b] to explain moat-wide patterns of heat flow in the MO data. Variation in depth to the top of the oceanic crust within the region probably explains at least the worst fitting of the outliers in Figure 4, a heat flow station in the south moat of Maro Reef. In this scenario, lower conductivity sediment patches act as insulators, inhibiting heat release in those areas, while regions of higher conductivity sediments act as radiators for the heat carried by the circulating waters. Surface heat flow anomalies reflect the distribution of sediment insulators and conductors in the upper few hundred meters and thus would not reflect properties of mantle plumes. Such channelized flow is apparently required [*Fisher and Becker*, 2000] to explain the fact that the observed decrease in mean value of heat flow with age is only $1/10^{th}$ the drop expected based on conductive heat flow models [*Stein and Stein*, 1994].

5. DISCUSSION

Based on these results, the following picture emerges for the pattern of heat flow around the Hawaiian Islands (Figure 5). Fluid circulation through a shallow, permeable aquifer, tentatively identified as lying within the upper ocean crust, is widespread. On regional scales, this fluid circulation is driven by the topographic relief of the large Hawaiian volcanoes and serves to redistribute laterally the heat from the cooling magmas in the volcanoes and any anomalous flux from the mantle plume. Water circulates in this aquifer beneath a blanket of less permeable sediments that vary in their thermal conductivity along and across the chain on scales of 50 to several hundred kilometers. The variations in thermal conductivity arise from the regional geology (e.g., depth to the seafloor which controls carbonate content, proximity to sediment sources such as eroding volcanoes) and age (diagenesis). Local irregularities in the geometry or permeability structure of the aquifer lead to small-scale convection that produces relatively large (~10 mW m^{-2}) heat flow anomalies on 10-20 km scales corresponding to rising and sinking limbs of the convection. However, averaging heat flow values over 10 km to several 10's of kilometers smooths out the effect of this small-scale circulation. On regional scales, the heat conducted from the aquifer through the sediments to the surface of the seafloor is controlled by thermal conductivity of the overlying sediments. High-conductivity sediments act as radiators, whereas low-conductivity sediments act as insulators. Therefore, the pattern of heat flow anomalies on regional scales reflects the properties of the sediments overlying the aquifer.

There are several over-simplifications in this mathematical and conceptual model that can easily lead to misfit between the theoretical curves and data in Figure 4. For example, the aquifer will not lie everywhere at uniform depth, particularly in the flexural moat where the upper oceanic crust is flexed downward by the weight of the surface volcanoes. However, the fact that the coarse volcaniclastic sediments in the moat are likely to have low permeability could result in shallow fluid flow despite depression of the upper oceanic basement. Additionally, fluid flowing in the aquifer is not expected to be exactly of uniform temperature. It should be colder near seawater recharge zones than near discharge zones. However, the experimental protocol used in the SA and MR surveys were specifically designed to be as far as possible from basement outcrops where fluid venting might occur [*Von Herzen et al.,* 1982; 1989]. Therefore, the fact that the *k*-dependent pattern emerges in Figure 4 is likely the result of careful choice of survey sites on a regional basis such that they overlie the best mixed portions of the aquifer. Finally, the mathematical formulation assumes steady state. Ignoring time-dependent terms is justified because the heat flow data were already corrected for time-dependent sediment effects (i.e., the cooling effect of quickly depositing sediment in the flexural moat) [*Von Herzen et al.*, 1982; 1989; *Harris et al.,* 2000a].

If variations in thermal conductivity are the primary control on the pattern of heat flow around the Hawaiian Islands, as is suggested by Figure 4, then it is difficult to use such data to constrain the thermal effect of the hotspot.

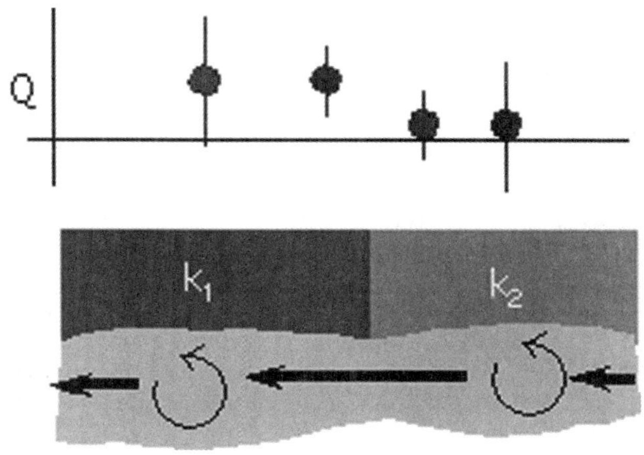

Figure 5. Cartoon showing conceptual model for fluid flow on the Hawaiian swell and its effect on surface heat flow. The basal aquifer contains two types of flow: a net transport from recharge to discharge zones and a superimposed small-scale circulation controlled by irregularities in the shape or permeability properties. The aquifer underlies sediments that vary in their thermal conductivity, *k*. In this case, $k_1 > k_2$, so that the heat flow is higher over the k_1 sediments. In regions of vigorous small-scale convection, the variability in the heat flow on scales of 1-10 km is much large than in areas unaffected by small-scale convection, as represented by the size of the error bars on the average heat flow in the region.

Although the effect modeled here is likely to be largest near major volcanoes on account of the large buoyancy-driven updraft they produce [*Harris et al.*, 2000b], the absence of any major variation in the mean heat flow for oceanic lithosphere as a function of age [*Stein and Stein*, 1992] suggests that that this is a more pervasive problem [*Fisher and Becker*, 2000]. In this regard, it is unfortunate that so much of the heat flow data base consists of single penetrations (that therefore are badly aliased by the local effects of fluid circulation) and/or measurements with no nearby determination of thermal conductivity [*Stein and Abbott*, 1991]. Future oceanic heat flow surveys should be designed with distances of less than a few kilometers between penetrations and local determination of thermal conductivity. While such surveys should reveal much about the hydrogeology of the upper oceanic crust, they may contain little information about lateral variations in mantle heat flux.

Acknowledgments. T. Seno and P. Wessel kindly reviewed the manuscript. This research was funded by the National Science Foundation (NSF-OCE9996275) and by the David and Lucile Packard Foundation.

REFERENCES

Detrick, R. S. and S. T. Crough, Island subsidence, hot spots, and lithospheric thinning, *J. Geophys. Res., 83*, 1236-1244, 1978.

Detrick, R. S., R. P. Von Herzen, S. T. Crough, D. Epp, and U. Fehn, Heat flow on the Hawaiian swell and lithospheric reheating, *Nature 292*, 142-143, 1981.

Dietz, R. S., H. W. Menard, and E. L. Hamilton, Echograms of the Mid-Pacific Expedition, *Deep-Sea Res., 1,* 258-272, 1954.

Filmer, P. E., M. K. McNutt, H. F. Webb, and D. J. Dixon, Volcanisn and archipelagic aprons in the Marquesas and Hawaiian Islands, *Marine Geophys. Res., 16*, 385-406, 1994.

Fisher, A. T., and K. Becker, Channelized fluid flow in oceanic crust reconciles heat-flow and permeability data, *Nature 403*, 71-74, 2000.

Harris, R. N., G. Garven, J. Georgen, M. K. McNutt, L. Christiansen, and R. H. Von Herzen, Submarine hydrology of the Hawaiian archipelagic apron, part 2: numerical simulation of coupled heat transport and fluid flow, *J. Geophys. Res., 105*, 21,371-21,385, 2000b.

Harris, R. N., R. P. Von Herzen, M. K. McNutt, G. Garven, and K. Jordahl, Submarine hydrology of the Hawaiian archipelagic apron, Part 1: Heat flow patterns north of Oahu and Maro Reef, *J. Geophys. Res., 105*, 21,353-21,370, 2000a.

Horai, K., Thermal conductivity of sediments and igneous rocks recovered during deep sea drilling project Leg 60, *Init. Rep. DSDP 26*, 451-463, 1981.

Matsuda, J., and R. P. Von Herzen, Thermal conductivity variation in a deep-sea sediment core its relation to H_2O, Ca and Si content, *Deep-Sea Res., 33,* 165-176, 1986.

McKenzie, D., and M. J. Bickle, The volume and composition of melt generated by entension of the lithosphere, *J. Petrol., 29*, 625-679, 1988.

Moore, G., G. Schubert, and P. Tackley, Three-dimensional simulations of plume-lithosphere interaction at the Hawaiian swell, *Science, 279*, 1998.

Morgan, W. J. Convection plumes in the lower mantle, *Nature, 253*, 708-710, 1975.

Parsons, B. E. and J. G. Sclater, An analysis of the variation of ocean floor bathymetry and heat flow with age, *J. Geophys. Res., 82*, 803-827, 1977.

Stein, C. A., and D. H. Abbott, Heat flow constraints on the South Pacific Superswell, *J. Geophys. Res., 96*, 16,083-16,099, 1991.

Stein, C., and S. Stein, A model for the global variation in oceanic depth and heat flow with lithospheric age, *Nature, 359*, 123-129, 1992.

Stein, C., and S. Stein, Constraints on hydrothermal heat flux through he oceanic lithosphere from global heat flow, *J. Geophys. Res., 99*, 3081-3095, 1994.

Von Herzen, R. P., Measurement of oceanic heat flow, *Methods Exp. Phys., 24*, 227-263, 1987.

Von Herzen, R. P., M. J. Cordery, R. S. Detrick, and C. Fang, Heat Flow and thermal origin of hot spot swells: The Hawaiian Swell revisted. *J. Geophys. Res. 94*, 13,783-13,799, 1989.

Von Herzen, R. P., R. S. Detrick, S. T. Crough, D. Epp, and U. Fehn, Thermal origin f the Hawaiian swell: Heat flow eidence and thermal models, *J. Geophys. Res., 87*, 6711-6723, 1982.

Marcia K. McNutt, Monterey Bay Aquarium Research Institute, 7700 Sandholdt Road, Moss Landing, CA 95039; (831) 775-1814; mcnutt@mbari.org

Noble Gas Systematics of the Hawaiian Volcanoes Based on the Analysis of Loihi, Kilauea and Koolau Submarine Rocks

Ichiro Kaneoka, Takeshi Hanyu, Junji Yamamoto and Yayoi N. Miura

Earthquake Research Institute, University of Tokyo, Tokyo, Japan

Noble gas isotopes were analyzed for glass and olivine phenocrysts of Hawaiian submarine volcanic rocks collected at depths of 2300-5700 m using submersibles "Shinkai 6500" and "Kaiko". Both crushing and heating methods were used for extracting noble gases. Loihi samples had $^{3}He/^{4}He$ of 21-32 Ra, $^{20}Ne/^{22}Ne$ of 10-12 and $^{40}Ar/^{36}Ar$ of 300-2700, while Kilauea samples had $^{3}He/^{4}He$ of 13-15 Ra, $^{20}Ne/^{22}Ne$ of 10.6-11.1, and $^{40}Ar/^{36}Ar$ of 800-3600. For all samples analyzed, the $^{129}Xe/^{130}Xe$ cannot be distinguished from the atmospheric value within analytical uncertainty. Olivine separates from two Koolau basaltic rocks collected at depths of about 3600 m and 3100 m have $^{3}He/^{4}He$ of about 19 Ra, whereas other Koolau samples collected at depths of less than 3000 m have $^{3}He/^{4}He$ of 14-15 Ra like subaerial Koolau rocks. The samples collected from Loihi, in the pre-shield stage, contain both higher and more variable values of $^{3}He/^{4}He$ compared to the shield-building stage of other volcanoes in the Hawaiian chain. The post-shield and rejuvenated stages show the lowest $^{3}He/^{4}He$, overlapping values obtained for MORB. We interpret this age trend to reflect an increasing degree of interaction between the rising plume magma and the uppermost part of asthenosphere. The extreme He isotope variation shown at Loihi may represent additional interaction of the plume magma with the old, altered oceanic crust. The variation of heavier noble gases may indicate incorporation of seawater into the magma before eruption.

1. INTRODUCTION

The Hawaiian Islands have been conjectured to represent typical hot-spot volcanism [*Wilson,* 1963; *Morgan,* 1971] and therefore, the geochemical characteristics of their volcanic materials should reflect deep source materials. Noble gas isotopes are chemically inert and include both radiogenic and stable isotopes for He, Ne, Ar, Kr and Xe. Hence, they are considered to be quite effective for identifying the characteristics of magma sources and the interaction of magmas with host rocks [*e.g., Kaneoka,* 1993]. Since it is thought that mantle plumes are derived from the Earth's deep interior, samples from the Hawaiian Islands may give us information on noble gas signatures of the deep mantle.

In earlier studies, however, only "excess" radiogenic components could be identified to have derived from the Earth's deep interior. In 1960s, excess ^{4}He and ^{40}Ar were identified in ultramafic nodules from Hualalai Volcano, Hawaii Island, and from Salt Lake Crater, Oahu Island [*Lovering and Richards,* 1964; *Funkhouser and Naughton,* 1968; *Gramlich and Naughton,* 1972]. Furthermore, large

Hawaiian Volcanoes: Deep Underwater Perspectives
Geophysical Monograph 128
Copyright 2002 by the American Geophysical Union

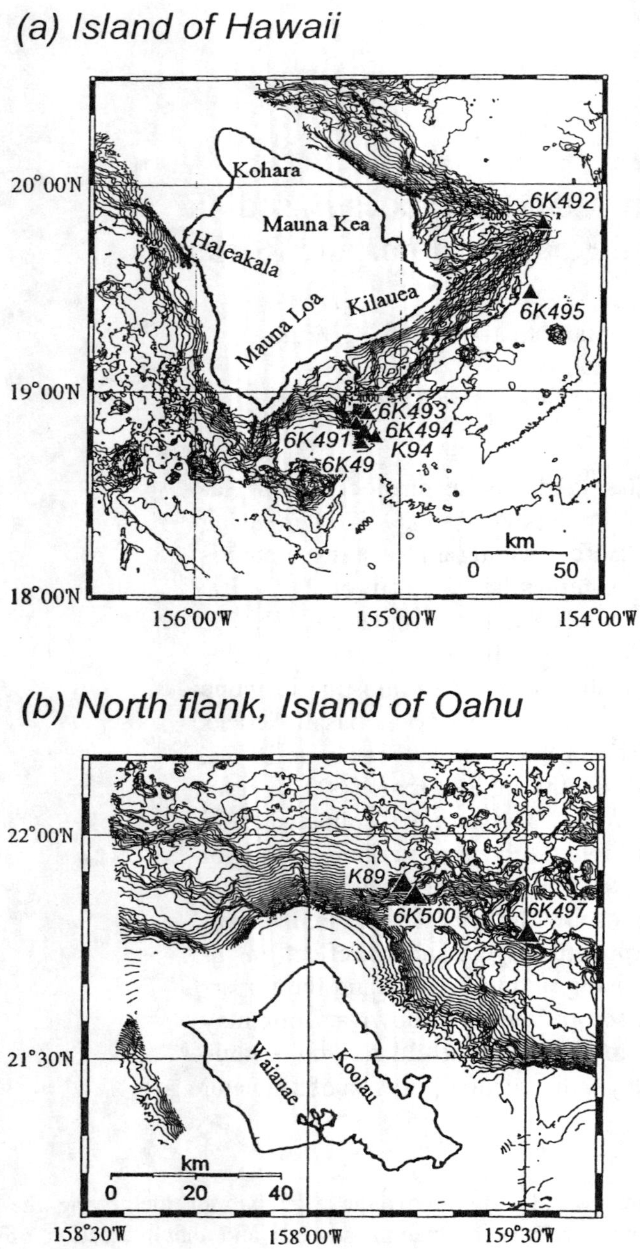

Figure 1. Maps for the island of Hawaii (a) and the north flank of Oahu (b) showing locations of submersible collected samples used in this study. Sampling sites are indicated by triangles.

plume [*e.g., Allègre et al.,* 1983; *Staudacher et al.,* 1986; *Sarda et al.,* 1988]. *Patterson et al.* [1990] argued that these values might have been caused by the atmospheric contamination due to the interaction of seawater with magma. Additional studies have revealed that higher $^{40}Ar/^{36}Ar$ (up to about 8000) are observed with high $^{3}He/^{4}He$ of about 25 Ra amounts of excess ^{40}Ar were found in submarine pillow basalts dredged from the east rift zone of Kilauea Volcano, Hawaii Island [*Dalrymple and Moore,* 1968; *Noble and Naughton,* 1968].

The occurrence of excess ^{129}Xe in Hawaiian samples was first reported in ultramafic nodules [*Hennecke and Manuel,* 1975]. $^{3}He/^{4}He$ higher than MORB (mid-oceanic ridge basalt) and $^{20}Ne/^{22}Ne$ higher than that of the atmosphere were reported for Kilauea fumaroles and pillow basalts [*Craig and Lupton,* 1976]. Such findings characterized the uniqueness of noble gas signatures for Hawaiian magmas. The occurrence of both high $^{3}He/^{4}He$ and excess ^{129}Xe in the same Hawaiian samples was also shown in lava flows and ultramafic nodules [*Kaneoka and Takaoka,* 1978; *Kaneoka et al.,* 1978]. *Kaneoka and Takaoka* [1980] demonstrated a systematic difference in the observed $^{3}He/^{4}He$ and the $^{40}Ar/^{36}Ar$ between olivine phenocrysts of lava flows and olivine in ultramafic nodules from Kilauea, Hualalai and Salt Lake Crater volcanoes. These results indicate that isotope signatures of olivine phenocrysts represent those of the plume magma and those of ultramafic nodule represent the signature of the MORB source. Similar results have been reported together with those for the other mantle sources [*Kyser and Rison,* 1982]. Variable $^{3}He/^{4}He$ (7.7-9.4 Ra) and $^{40}Ar/^{36}Ar$ (4100-9700) have been observed within a single garnet pyroxenite nodule from Salt Lake Crater, which has been interpreted as interaction of the mantle plume with the MORB-type lithospheric mantle [*Rocholl et al.,* 1996].

Loihi Seamount has been thought to represent the initial, pre-shield stage of hot spot volcanic activity [*Moor et al.,* 1982]. High $^{3}He/^{4}He$ (R) up to about 35 Ra (Ra is R for the air, 1.4×10^{-6}) has been found for Loihi samples, but values vary widely [*Kaneoka et al.,* 1983; *Kurz et al.,* 1983; *Rison and Craig,* 1983; *Allègre et al.,* 1986; *Staudacher et al.,* 1986]. Previous results have shown $^{40}Ar/^{36}Ar$ of less than 2000 and no clear evidence for the occurrence of excess ^{129}Xe [*e.g., Allègre et al.,* 1983; *Kaneoka et al.,* 1983; *Allègre et al.,* 1986; *Staudacher et al.,* 1986]. Based on such observations, some investigators have concluded that the atmosphere-like heavy noble gas isotopic compositions are representative of the mantle source for the Hawaiian mantle for Loihi dunite samples [*Trieloff et al.,* 2000]. Glassy Loihi samples have $^{40}Ar/^{36}Ar$ of up to about 3000 [*e.g., Hiyagon et al.,* 1992]. The occurrence of solar-type Ne has been demonstrated for Loihi and Kilauea samples [*Honda et al.,* 1991, 1993; *Valbracht et al.,* 1996, 1997]. Thus, the heavier noble gases reveal that Loihi samples differ from atmospheric values. However, the cause of near-atmospheric Xe isotope values for Loihi samples remains unknown.

The systematic variations in $^{3}He/^{4}He$ within and among Hawaiian volcanoes must be significant [*e.g., Kurz et al.,*

Table 1. Submarine samples used for noble gas analysis

Sample No.	Locality	Depth(m)	Material used	Rock type	Vesicularity(%)*
<u>Loihi</u>					
6K490-2	18°44.6'N, 155°11.4'W	4657	glass	picrite	S
6K490-3	18°44.7'N, 155°11.4'W	4598	olivine	picrite	S
6K490-8	18°45.5'N, 155°11.4'W	4383	glass	pillow basalt	S
6K491-5	18°46.1'N, 155°10.4'W	4137	glass, olivine	picrite	S
6K493-2	18°53.2'N, 155°10.0'W	4411	olivine	picrite	S
6K494-3	18°50.1'N, 155°13.5'W	2491	glass	picrite	S
6K494-6	18°50.5'N, 155°13.8'W	2323	glass	pillow basalt	S
6K94-7	18°47.0'N, 155°11.1'W	3900	glass	basalt	M
<u>Kilauea</u>					
6K492-4	19°48.0'N, 154°19.1'W	3958	glass, olivine	pillow basalt	S
6K495-7	19°29.6'N, 154°21.2'W	5568	glass	aphyric basalt	S
<u>Koolau</u>					
6K497-2	21°46.3'N, 157°29.9'W	3580	olivine	picritic basalt	M
6K497-6	21°46.3'N, 157°29.9'W	3103	olivine	picritic basalt	L
6K497-7	21°46.3'N, 157°29.9'W	2999	olivine	olivine basalt	L
6K500-1	21°51.4'N, 157°45.4'W	2980	olivine	picritic pillow lava	S
6K500-5B	21°51.0'N, 157°45.6'W	2696	olivine	picrite	S
6K500-6	21°50.8'N, 157°46.2'W	2815	olivine	picrite	S
K89-6	21°51.0'N, 157°46.0'W	2603	olivine	picritic basalt	—

N.B. All data are based on the reports for YOKOSUKA/SHINKAI 6500 (1999) and for KAIREI/KAIKO (1998) cruises by JAMSTEC.

* Vesicularity indicates only a rough estimate on board. S: less than 5%, M: 5-20%, L: more than 20%.

1983, 1987, 1995; *Kaneoka,* 1987; *Kurz and Kammer,* 1991]. The $^3He/^4He$ ratio seems to be correlated with variations in non-volatile isotopes such as $^{87}Sr/^{86}Sr$ and $^{143}Nd/^{144}Nd$ [*e.g., Kurz et al.,* 1982; *Staudigel at al.,* 1984; *Valbracht et al.,* 1996]. The other noble gas isotopes of Hawaiian samples are also variable. Thus, it is important to identify the cause of variation in observed noble gas values and infer their values in pristine components.

In the summer of 1998-1999, several Hawaiian submarine volcanoes were sampled by the submersibles "Shinkai 6500" and "Kaiko" of the JAMSTEC (Japan Marine Science and Technology Center) during a Japan-USA cooperative research program on Hawaiian Volcanism. In this study, noble gas isotopes for olivine phenocrysts and pillow glasses from basalts collected at depths of 2300-5700 m from Loihi, Kilauea and Koolau volcanoes (Figure 1) were analyzed in order to clarify the noble gas systematics for Hawaiian volcanism. Our results have revealed that several processes including the incorporation of noble gases into the plume magma from seawater and the lithosphere (mostly the oceanic crust) have affected the magmas before eruption.

2. SAMPLES

For noble gas analyses, only basaltic rocks with fresh glassy rims and/or those bearing fresh olivine phenocrysts were used. Eight Loihi samples from five sites (K94, 6K490, 6K491, 6K493, 6K494) were selected based on freshness and thickness of glassy rinds and/or freshness of olivine phenocrysts. However, the thickness of glassy rind is a maximum of about 4 mm and only a few mm or less in most cases, which is much thinner than that of typical MORB. Most Loihi samples were collected from the depth of more than 4000 m, but two samples (6K494-3, 6K494-6) collected from the depth of 2300-2500 m were selected for comparison with the deeper samples. From the Kilauea area, two samples from two sites (6K492, 6K495) at the depth of around 4000-4600 m were selected. In the case of Koolau area, seven samples from three sites (K89, 6K497, 6K500) were selected for noble gas analyses. These samples are mostly picritic or olivine basalts. They vary strongly in their extent of vesicularity (Table 1). Samples from the Loihi area are grouped mostly as tholeiitic (6K490-2, 6K490-3, 6K490-8, 6K491-5, 6K94-7), but 6K493-2 and 6K494-6 are transitional and 6K494-3 is alkalic [*T. Shibata,* 2001, *personal communication*]. Their chemical compositions will be reported elsewhere [*Shibata et al.,* 2001].

3. EXPERIMENTAL

Glassy rims were carefully chipped from rock samples. Only fresh and homogeneous glasses with no visible inclusions were handpicked with tweezers using a binocular microscope. Chip sizes ranged from 1mm to 5 mm. Our studies revealed that glassy chips are least affected by secondary contamination [*Kumagai,* 1999]. To obtain olivine phenocryst samples, rocks were crushed to 1-2 mm size. Olivine crystals were carefully handpicked using a binocular scope to remove any altered grains. Loihi and Kilauea samples were leached for 30 minutes with 2N-HNO3 at 60°C. However, Koolau samples were not leached because only crushing experiments on olivines were planned for these samples and we thought that the leaching effect would be minimal. This was followed by washing in acetone, ethanol and distilled water. After drying, samples were weighed and loaded into the degassing apparatus. However, no clear difference was observed in results obtained from acid-leached and non-leached samples.

Two kinds of degassing methods (stepwise heating and crushing in vacuum) were applied to get information on the noble gas components in vesicles and inclusions in olivine crystals. Stepwise heating was conducted by using a furnace with Ta resistance heaters. Two to four temperature steps were applied ranging from 600 to 1900°C. Heating temperature was controlled with an applied current to generate heat for heaters and calibrated by using an optical pyrometer. The estimated temperature uncertainty is about 20°C.

For vacuum crushing, a press-type crusher was operated at 120°C using a charcoal trap cooled by liquid nitrogen to avoid the adsorption of heavy noble gases on fresh surfaces created during crushing. Based on the previous experiments, the degassing efficiency of our crushing method is about 20-30% for the type of samples analyzed here. Noble gases were purified with two or three titanium-zirconium traps heated at 700-800°C. Five noble gases (He, Ne, Ar, Kr, Xe) were separated from each other using cold traps with activated charcoal and a cryogenic pump. Noble gas isotopes were measured on a sector-type mass spectrometer (VG5400) with three types of collectors (Farady cup, Daly multiplier and ion counting system). The resolving power of this instrument is about 600, which is sufficient to separate $^3He^+$ from HD^+ and $^3H^+$. Helium isotopic ratio was calibrated by repeated measurements of the standard gases ($^3He/^4He$ = 6.04 ± 0.29 Ra) from Kaminoyama well, Yamagata, Japan. For Ne, Ar, Kr and Xe analyses, isotopic ratios were calibrated using diluted air. For Ne analyses, $^{20}Ne^+$ and $^{22}Ne^+$ were corrected for interference of doubly charged $^{40}Ar^{++}$ and $^{44}CO_2^{++}$, respectively. $^{40}Ar^{++}/^{40}Ar^+$ was 0.25, and $^{44}CO_2^{++}/^{44}CO_2^+$ was 0.008. To estimate noble gas abundance, the absolute sensitivity of Ar was determined based on the amount of radiogenic Ar in the standard sample (MMhb-1 and JG-1 biotite). The relative sensitivities for He, Ne, Kr and Xe to Ar were determined by using an air standard. An uncertainty of about 5% was assigned to the abundance of each noble gas based on the repeated analyses of the air standard. More details on experimental procedures at Earthquake Research Institute, University of Tokyo were reported by *Hanyu et al.* [1999].

4. RESULTS

4.1 Noble Gas Abundance

The amounts of ^4He and ^{40}Ar degassed by crushing are generally less than those degassed by heating (Table 2, Figures 2 and 3). In the case of the sample 6K494-3, however, the glass shows large amounts for all noble gases even by the crushing method.

For the sample 6K491-5, glass and olivine separates were analyzed for comparison. Glass shows higher amounts of noble gases, especially for He and Ne, compared to the olivine separates (*cf.,* Table 2). Since glass represents melt phase and olivine crystals correspond to an early formed mineral phase from a melt, this trend is expected [*e.g., Kaneoka et al.,* 1983; *Kaneoka,* 1993]. For glass samples with the $^{20}Ne/^{36}Ar$ of less than one from total heating or crushing (*i.e.,* 6K494-3, 6K494-6, K94-7, 6K495-7), all noble gases except He seem to be affected by secondary atmospheric components. Although Ne is more soluble than Ar in a basaltic magma, these samples are depleted in Ne compared to Ar and their $^{40}Ar/^{36}Ar$ ratios are much smaller than 1000. These results imply the incorporation of the atmospheric component into the magma.

To evaluate a water pressure effect on the retention of noble gases in the glassy phase, the total amounts of ^4He and ^{40}Ar in glass samples degassed by heating were compared for samples collected at different water depths. If this effect holds, a sample collected at a deeper site will contain more gases than a sample collected at a shallower site. Sample 6K494-6, collected at the depth of about 2300 m, contains smaller amounts of ^4He and ^{40}Ar compared to samples collected at more than 4000 m (Figures 2 and 3). However, sample 6K494-3, collected at the depth of about 2500 m and degassed by crushing, has amounts of ^{40}Ar similar to those of samples collected at deeper sites. The noble gas isotopic ratios for this sample (Table 2) suggest the incorporation of atmospheric components into the glass without He degassing. Contamination is reflected in both abundance patterns and in the isotopic ratios of noble gases except for He. Hence, although the number of samples is limited, the results reported here are consistent with a pressure effect of seawater inhibiting gas loss, especially for He.

4.2 Noble Gas Isotopic Ratio

4.2.1 Helium. No difference was found in the $^3He/^4He$ between glass and olivine in either this study, or a previous

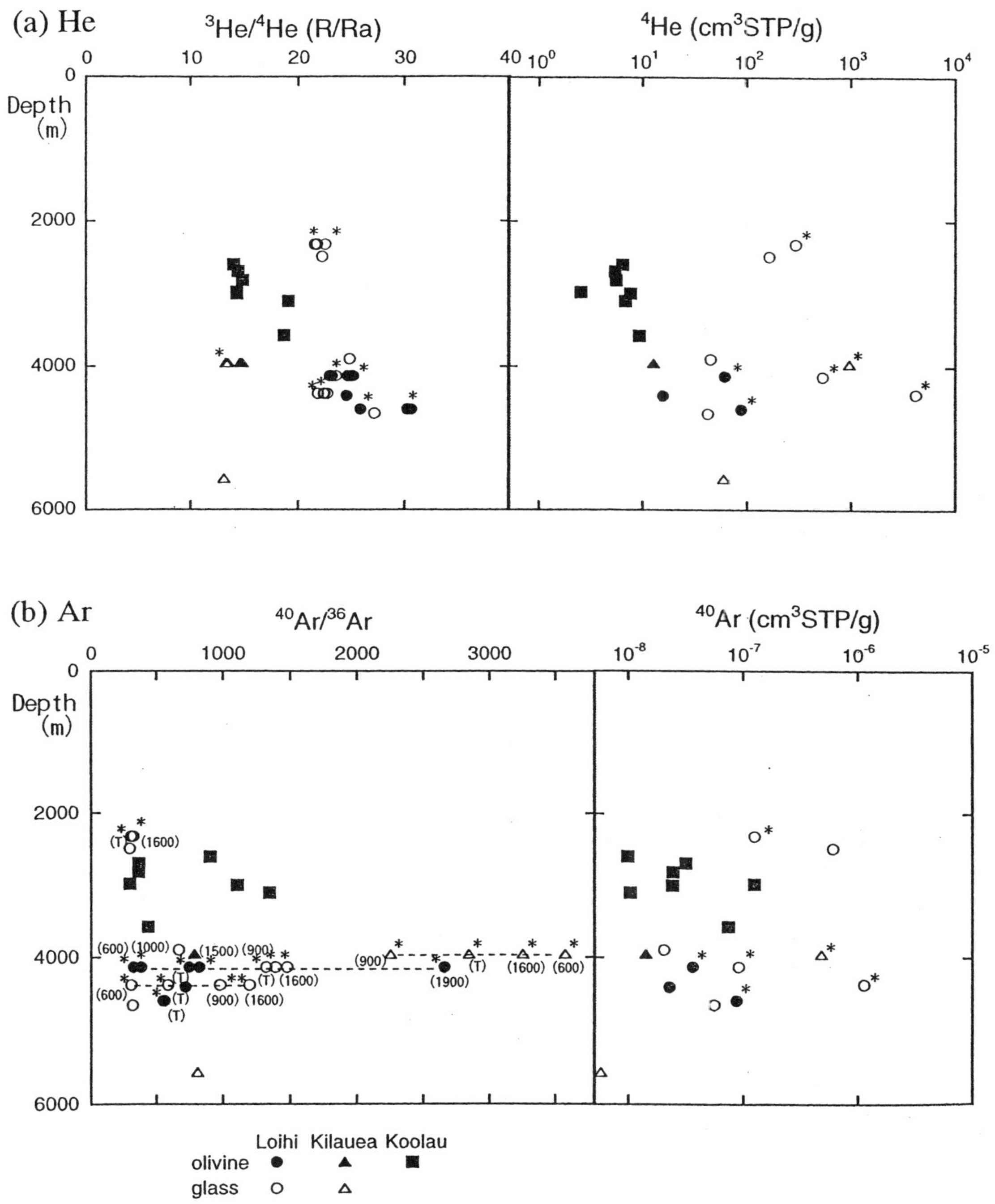

Figure 2. The isotopic ratio and the abundance plotted against the collection depth for samples from Loihi, Kilauea and Koolau; (a) He, and (b) Ar. An asterisk indicates the result for a sample degassed by the heating method, while a datum without an asterisk indicates a result by the crushing method. Values in parentheses indicate the degassing temperatures and (T) shows the value as a total for each temperature fraction. For the abundance, only the value as a total is plotted for the heating method.

Table 2. Noble gas abundances in Hawaiian submarine samples

Sample		Weight (g)	T (℃)	^{4}He (10^{-9})	^{20}Ne (10^{-12})	^{36}Ar (10^{-12})	^{84}Kr (10^{-12})	^{132}Xe (10^{-12})
Loihi								
6K490-2	glass	1.853	C	42.8	213	178	6.74	0.349
6K490-3	olivine	1.335	700	7.76	10.6	26.1	0.992	0.166
			1900	80.6	90.4	131	9.41	1.36
			total	88.4	101	157	10.4	1.53
6K490-8	glass	1.948	600	598	925	1030	28.2	1.12
			900	3320	1110	734	27.5	1.73
			1600	323	181	180	8.64	1.10
			total	4240	2220	1940	64.4	3.94
6K491-5	glass	1.115	600	155	6.44	0.676	0.0622	0.0341
			900	357	81.7	31.3	1.22	0.124
			1600	21.2	51.6	33.7	1.55	0.224
			total	534	140	65.8	2.83	0.382
	olivine	1.589	600	0.304	2.13	7.15	0.254	0.0249
			1000	3.09	6.82	28.2	1.06	0.0899
			1500	15.4	3.11	6.47	0.481	0.0948
			1900	43.2	12.0	7.40	0.353	0.0387
			total	62.0	24.1	49.2	2.15	0.248
6K493-2	olivine	1.577	C	15.8	70.6	31.9	0.903	0.0643
6K494-3	glass	1.852	C	164	1600	2050	61.6	3.05
6K494-6	glass	2.189	600	81.3	5.64	2.11	0.119	0.0356
			900	216	123	205	7.69	0.647
			1600	0.252	27.3	190	9.38	0.620
			total	297	156	398	17.2	1.30
K94-7	glass	1.219	C	45.4	23.7	30.9	0.991	0.0668
Kilauea								
6K492-4	glass	1.965	600	329	42.1	7.03	0.266	0.0297
			900	627	336	126	2.16	0.141
			1600	14.1	34.2	35.1	0.804	0.0255
			total	970	412	168	3.23	0.197
	olivine	1.211	C	12.8	147	18.1	0.571	0.0426
6K495-7	glass	1.932	C	60.4	6.19	7.09	0.227	0.0205
Koolau								
6K497-2	olivine	1.529	C	9.28	70.1	171	3.74	0.144
6K497-6	olivine	1.531	C	6.87	13.5	7.74	0.250	0.0170
6K497-7	olivine	0.397	C	7.69	27.3	22.1	0.785	0.0960
6K500-1	olivine	0.604	C	2.55	229	421	8.81	0.324
6K500-5B	olivine	1.502	C	5.46	54.0	87.4	1.88	0.0810
6K500-6	olivine	1.548	C	5.62	66.4	68.4	1.58	0.104
K89-6	olivine	1.056	C	6.45	28.5	10.9	0.348	0.0400

N.B., C : Degassed by crushing. Unit in abundance is cm^3STP/g. All tabulated data were corrected for blanks. Unit in blanks is cm^3STP.

Blanks for crushing were as follows (in cm^3STP): ^{3}He, (1-2)x10^{-15}; ^{4}He, (0.6-3)x10^{-11}; ^{20}Ne, (2-3)x10^{-12}; ^{36}Ar, (0.8-1.3)x10^{-12}; ^{84}Kr, (2-4)x10^{-14}; ^{132}Xe, (0.7-8)x10^{-15}.

Blanks for typical heating temperatures were as follows (in cm^3STP): at 600°C, ^{3}He, (1-3)x10^{-15}; ^{4}He, (1-1.3)x10^{-10}; ^{20}Ne, (1-3)x10^{-12}; ^{36}Ar, (0.8-2)x10^{-12}; ^{84}Kr, (1-6)x10^{-14}; ^{132}Xe, (1-6)x10^{-15}; at 1600°C, ^{3}He, (2-3)x10^{-15}, ^{4}He, 1x10^{-10}; ^{20}Ne, (2-3)x10^{-12}; ^{36}Ar, (2-4)x10^{-12}; ^{84}Kr, (0.8-3)x10^{-13}; ^{132}Xe, (1-8)x10^{-14}.

one on Hawaiian basalts [*Kaneoka et al., 1983*]. Thus, the ^{3}He/^{4}He observed in either phase can be regarded as representative of erupted lava (Table 3). The ^{3}He/^{4}He for Loihi samples vary from 21 to 32 Ra, which is similar to the range reported previously [*e.g., Kurz et al.*, 1983; *Rison and Craig,* 1983; *Kaneoka,* 1987; *Valbracht et al.*, 1997]. Even samples collected from the depths of more than 4000 m show ^{3}He/^{4}He with a similar range of variation. Therefore, such variation in the ^{3}He/^{4}He can be regarded as a characteristic feature for Loihi samples independent of the erupted depth.

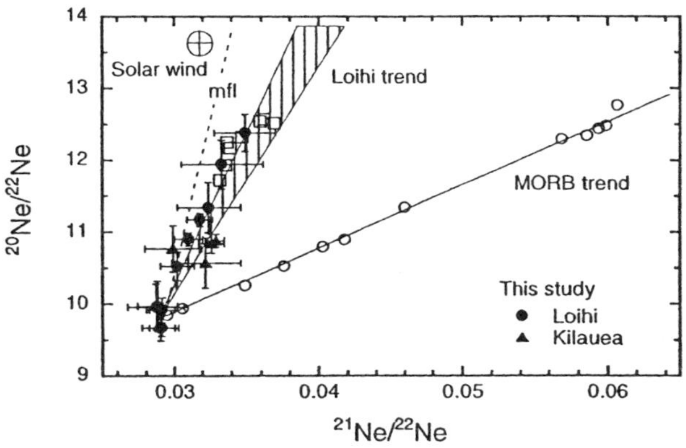

Figure 3. Plot of Ne three-isotopes for samples from Loihi and Kilauea. Data for $^{20}Ne/^{22}Ne$ ratio with the uncertainties of less than 5% are plotted. For comparison, data for Loihi dunite (open squares) [*Trieloff et al.,* 2000] and popping rock (open circles) [*Moreira et al,* 1998] are plotted. mfl: mass fractionation line.

Although the samples collected at the depth of 2300-2500 m seem to have similar $^3He/^4He$ (22 Ra), we cannot regard the value as a representative because previous results show larger variation for samples collected at similar shallow depths [*Kaneoka et al., 1983; Kurz et al., 1983; Rison and Craig, 1983; Allègre et al., 1986; Staudacher et al., 1986*].

The two Kilauea samples collected at depths of about 4000 m (6K492-4) and 5600 m (6k495-7) show similar $^3He/^4He$ to those previously collected from Kilauea at depths of 740 to 5000 m (12-17 Ra) [*Kurz et al.,* 1982; *Honda et al.,* 1993; *Valbracht et al.,* 1996].

The Koolau Series samples collected diving dives 6K500 and K89 at depths of less than 3000 m show similar $^3He/^4He$ (14-15 Ra), which corresponds to the highest $^3He/^4He$ reported for subaerial rocks of Koolau Volcano [*Roden et al.,* 1994]. However, two samples (6K497-2, 6K497-6) collected from depths of more than 3100 m show distinctly higher $^3He/^4He$ (19 Ra). Such a high $^3He/^4He$ was not observed previously among subaerial Koolau samples.

4.2.2 Neon. Many Loihi samples show $^{20}Ne/^{22}Ne$ and $^{21}Ne/^{22}Ne$ that are higher than those of the atmospheric values and they lie near the line of Loihi trend (Figure 4). The highest $^{20}Ne/^{22}Ne$ (about 12.4) was observed in the 900°C fraction of glass phase for sample 6K491-5, while no such value was observed for olivine separate from the same rock due to the relatively large analytical uncertainties.

Two Kilauea glass samples also show the higher $^{20}Ne/^{22}Ne$ and $^{21}Ne/^{22}Ne$ than those of the atmospheric values (Table 2). These samples deviated slightly to the right of the Loihi trend (Figure 4), suggesting slightly more effect of nucleogenic ^{21}Ne to $^{21}Ne/^{22}Ne$ of the Kilauea magma than Loihi. .

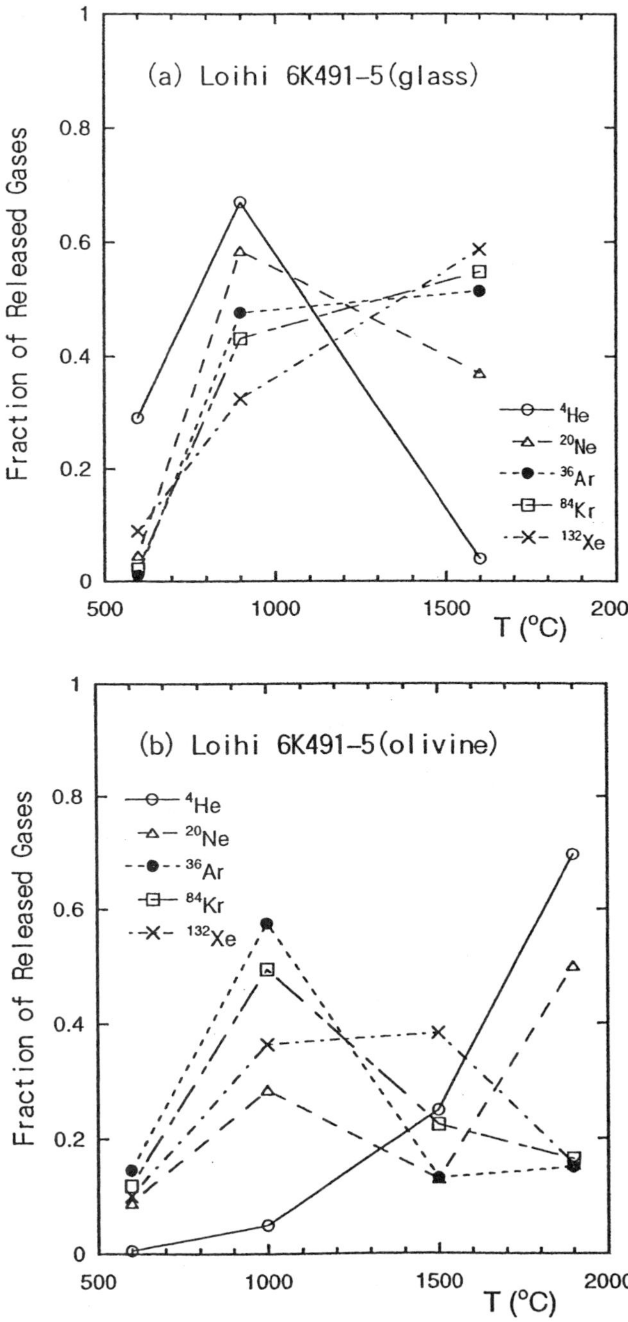

Figure 4. Noble gas release patterns plotted against the degassing temperature for the Loihi sample 6K491-5; (a) glass, and (b) olivine.

The Ne isotopic ratios of the Koolau olivines cannot be discerned from the atmospheric values ($^{20}Ne/^{22}Ne$ = 9.80, $^{21}Ne/^{22}Ne$ = 0.0290) within their analytical uncertainties, possibly due to the presence of altered olivine in these

Table 3. Isotopic ratios for Hawaiian submarine samples

Sample		T (°C)	^{3}He/^{4}He (Ra)	^{20}Ne/^{22}Ne	^{21}Ne/^{22}Ne	^{38}Ar/^{36}Ar	^{40}Ar/^{36}Ar
Loihi							
6K490-2	glass	C	27.2 ± 1.4	9.92 ± 0.10	0.0290 ± 0.0010	0.1883 ± 0.0013	317.8 ± 3.4
6K490-3	olivine	700	25.9 ± 1.2	*	*	0.1833 ± 0.0060	552.1 ± 17.2
		1900	30.7 ± 1.2	*	*	0.1890 ± 0.0015	565.0 ± 7.1
		total	30.3 ± 1.2	*	*	0.1882 ± 0.0017	563.2 ± 7.4
6K490-8	glass	600	22.8 ± 0.3	9.97 ± 0.35	0.0288 ± 0.0014	0.1877 ± 0.0026	306.5 ± 3.7
		900	22.5 ± 0.3	10.90 ± 0.09	0.0310 ± 0.0010	0.1885 ± 0.0032	979.6 ± 14.0
		1600	21.9 ± 0.4	11.17 ± 0.09	0.0318 ± 0.0009	0.1891 ± 0.0008	1197 ± 3
		total	22.4 ± 0.3	10.53 ± 0.20	0.0302 ± 0.0012	0.1881 ± 0.0018	590.0 ± 4.8
6K491-5	glass	600	25.1 ± 1.3	11.11 ± 1.22	*	0.1935 ± 0.0116	387.9 ± 9.0
		900	23.1 ± 1.2	12.38 ± 0.26	0.0349 ± 0.0021	0.1887 ± 0.0011	1322 ± 12
		1600	23.2 ± 1.9	11.34 ± 0.35	0.0324 ± 0.0022	0.1898 ± 0.0012	1478 ± 14
		total	23.6 ± 1.2	11.94 ± 0.34	0.0333 ± 0.0028	0.1893 ± 0.0008	1393 ± 13
	olivine	600	25.1 ± 11.4	9.78 ± 2.25	0.0403 ± 0.0162	0.1884 ± 0.0011	326.1 ± 2.9
		1000	23.0 ± 2.4	10.17 ± 0.93	0.0306 ± 0.0067	0.1882 ± 0.0006	381.7 ± 3.4
		1500	24.7 ± 1.5	9.91 ± 2.05	0.0296 ± 0.0119	0.1879 ± 0.0020	826.0 ± 8.2
		1900	25.2 ± 2.4	11.05 ± 0.73	0.0389 ± 0.0077	0.1886 ± 0.0039	2662 ± 36
		total	25.0 ± 1.9	10.54 ± 1.09	0.0355 ± 0.0087	0.1883 ± 0.0007	750.8 ± 6.7
6K493-2	olivine	C	24.6 ± 1.0	*	*	0.1885 ± 0.0015	723.9 ± 4.1
6K494-3	glass	C	22.3 ± 1.2	9.92 ± 0.17	0.0291 ± 0.0009	0.1880 ± 0.0006	296.4 ± 2.7
6K494-6	glass	600	22.6 ± 0.5	*	*	0.1875 ± 0.0027	305.1 ± 1.4
		900	21.6 ± 0.6	9.67 ± 0.12	0.0291 ± 0.0009	0.1887 ± 0.0006	310.9 ± 0.8
		1600	21.7 ± 2.8	9.96 ± 0.32	0.0287 ± 0.0020	0.1890 ± 0.0006	324.1 ± 0.7
		total	21.8 ± 0.5	9.67 ± 0.18	0.0290 ± 0.0013	0.1888 ± 0.0005	317.0 ± 0.6
K94-7	glass	C	24.9 ± 1.0	*	*	0.1898 ± 0.0033	670.4 ± 9.9
Kilauea							
6K492-4	glass	600	13.5 ± 0.3	10.57 ± 0.35	0.0322 ± 0.0024	0.1873 ± 0.0023	3577 ± 21
		900	13.3 ± 0.3	10.88 ± 0.09	0.0329 ± 0.0006	0.1889 ± 0.0016	2252 ± 11
		1600	14.8 ± 2.2	10.77 ± 0.32	0.0299 ± 0.0020	0.1885 ± 0.0020	3257 ± 24
		total	13.4 ± 0.3	10.84 ± 0.13	0.0326 ± 0.0009	0.1886 ± 0.0012	2845 ± 12
	olivine	C	14.6 ± 0.7	*	*	0.1875 ± 0.0336	790.5 ± 110.0
6K495-7	glass	C	13.1 ± 0.7	11.05 ± 1.18	0.0338 ± 0.0098	0.1892 ± 0.0109	813.0 ± 39.8
Koolau							
6K497-2	olivine	C	18.7 ± 0.4	9.92 ± 0.16	0.0296 ± 0.0008	0.1870 ± 0.0004	437.6 ± 0.8
6K497-6	olivine	C	19.1 ± 0.4	10.03 ± 0.22	0.0296 ± 0.0017	0.1871 ± 0.0015	1346 ± 6
6K497-7	olivine	C	14.3 ± 0.5	9.50 ± 0.31	0.0270 ± 0.0026	0.1883 ± 0.0017	1107 ± 4
6K500-1	olivine	C	14.3 ± 0.6	9.85 ± 0.17	0.0290 ± 0.0013	0.1866 ± 0.0007	296.9 ± 1.1
6K500-5B	olivine	C	14.4 ± 0.5	9.85 ± 0.16	0.0297 ± 0.0017	0.1870 ± 0.0012	366.3 ± 1.6
6K500-6	olivine	C	14.8 ± 0.4	9.80 ± 0.15	0.0298 ± 0.0014	0.1871 ± 0.0006	362.6 ± 0.9
K89-6	olivine	C	14.0 ± 0.4	9.97 ± 0.20	0.0294 ± 0.0025	0.1888 ± 0.0011	905.9 ± 10.2

N.B., Analytical uncertainty indicates one standard deviation in the measured value. The uncertainty for the standard sample is not included.

slightly older samples (2-3 Ma) analyzed in this study [*K. Uto*, 2000, personal communication] compared to those from Loihi and Kilauea.

4.2.3 Argon. ^{40}Ar/^{36}Ar values vary greatly in our sample suite (296 to 2845) because they are quite sensitive to the secondary atmospheric contamination. For example, Loihi samples range from 296 to 2662 for ^{40}Ar/^{36}Ar (Table 3, Figure 2). However, most of the samples have values of less than 1500. In particular, two glass samples (6K494-3, 6K494-6) collected at the depth of 2300-2500 m have ^{40}Ar/^{36}Ar close to the atmospheric value, though one sample was degassed by crushing and the other by stepwise heating. Since we tried to select the freshest parts among samples collected from each site, the ^{40}Ar/^{36}Ar probably reflects the conditions of Loihi's magma extrusion system at those depths.

Samples from Kilauea show relatively high ^{40}Ar/^{36}Ar, up to 3600 in the 600°C fraction and about 2800 as the total for the glass of the sample 6K492-4 (Table 3, Figure 2). Olivine in sample 6K492-4 and glass in sample 6K492-7 yielded values of $^{40}Ar/^{36}Ar$ of about 800 in samples where gas was extracted by crushing. This implies a complicated relationship between the olivine phase and the glass for the sample 6K492-4. Olivine separates from the Koolau have $^{40}Ar/^{36}Ar$ of 300-1300 for gases extracted by crushing. These variable ratios probably reflect the secondary incorporation of the atmospheric components.

4.2.4 Krypton and xenon. Only atmospheric isotopic compositions have been observed for Kr both in MORB and oceanic island basalt (OIB) samples. All present samples show atmospheric compositions of Kr within the analytical uncertainties of 3%. For Xe isotopes, a recent investigation of Loihi dunites, using multiple crushing to degas samples, revealed the occurrence of excess ^{129}Xe and ^{136}Xe and other related Xe isotopes [*Trieloff et al.,* 2000]. In the present study, we tried to find such excess components by applying both the stepwise heating and crushing methods, but could not distinguish the compositions from the atmospheric composition within the analytical uncertainties of 3%. Similar results were obtained for Loihi other samples collected at depths of 4000-5000 m [*e.g., Valbracht et al.,* 1997].

5. DISCUSSION

5.1 $^3He/^4He$ Ratios of the Koolau Series

The Koolau samples analyzed in this study can be divided into two groups based on their $^3He/^4He$. The first group consists of five shallow samples (depth 2600-3000 m) collected during dives K89 and 6K500 and one sample collected during 6K497. These samples have constant $^3He/^4He$ of 14-15 Ra, which is equivalent to the highest $^3He/^4He$ of subaerial rocks of the Koolau Series [*Roden et al.,* 1994]. The second group, consisting of two deeper samples from dive 6K497 (depth more than 3100 m), is characterized by significantly higher $^3He/^4He$ (19 Ra) than the first group. These different helium isotopic ratios cannot be explained by atmospheric contamination, because there are no systematic differences in ^4He concentration between the two groups and $^3He/^4He$ of the first group (14-15 Ra) is constant regardless of ^4He concentration (Figure 2). Since both higher and lower $^3He/^4He$ ratios were observed among the samples from the 6K497 site, the different helium isotopic characteristics may be related to the depth rather than the locations where samples were collected. However, a section of submersible collected Mauna Loa lavas had a greater degree of He isotope variability than the Koolau Series and show no correlation with collection depth [*Kurz et al.,* 1995]. Hence, this point should be checked for the Koolau samples in a future work.

The distinct isotopic variation for the submarine part of the Koolau Series indicates that the source component of the volcano changed during the shield stage. It is generally conjectured that Loihi magmas are representative of the most primitive source in the asthenosphere, whereas the source of shield stage magmas has been affected by interaction with asthenospheric/lithospheric materials of similar $^3He/^4He$ to MORB [*e.g., Kaneoka,* 1987; *Kurz and Kammer,* 1991]. Hence, the shield stage magmas have moderately primitive helium isotopic characters and their $^3He/^4He$ ratio is variable between 10 and 20 Ra. A recent Os isotope study gave evidence for incorporation of recycled materials in the Koolau magmas [*Lassiter and Hauri,* 1998]. Since recycled materials should have low $^3He/^4He$ [*e.g., Graham et al.,* 1992; *Hanyu and Kaneoka,* 1997], the decrease in $^3He/^4He$ for the Koolau lavas upsection (19 Ra to 14-15 Ra) may indicate incorporation of recycled materials into the plume magma However, we cannot excluded the possibility of the addition of the asthenospheric component of noble gases into the plume magma. This point is discussed in section

5.5. *Noble Gas Abundance Controlled by the Environmental Conditions of Samples*

MORB glasses normally contain ^4He on the order of 10^{-6} cm^3STP/g [*e.g., Ozima and Podosek,* 1983], while Hawaiian glasses typically show the ^4He of less than 1 x 10^{-6} cm^3STP/g. Even most glass samples from Loihi have ^4He contents of the order of 10^{-7} cm^3STP/g [*e.g., Kurz et al.,* 1983; *Rison and Craig,* 1983; *Valbracht et al.,* 1997]. Since Loihi samples show $^3He/^4He$ higher than MORB, the source material of the Loihi magma is considered to be more enriched in ^3He than that of MORB, indicating that Loihi's source should also contain more ^4He than MORB [*e.g., Kaneoka and Takaoka,* 1985]. However, the apparent relationship for the noble gas abundance does not support this hypothesis. This phenomenon is sometimes called the "He paradox" [*Anderson,* 1998].

In this section we look at factors that could control noble gas abundance. Although solubility of noble gases is controlled by the chemical composition of magmas [*e.g., Jambon et al.,* 1986; *Lux,* 1987], the difference in composition between MORB and OIB is too small to explain their noble gas abundance differences. Glassy material represents the melt phase of a magma, while phenocrysts contain noble gases controlled by the distribution coefficient between a crystal and a melt [*e.g., Kaneoka,* 1993]. The distribution coefficient of He between olivine and a basaltic magma is around 0.01 or less [*Hiyagon and Ozima,* 1986]. Noble gas

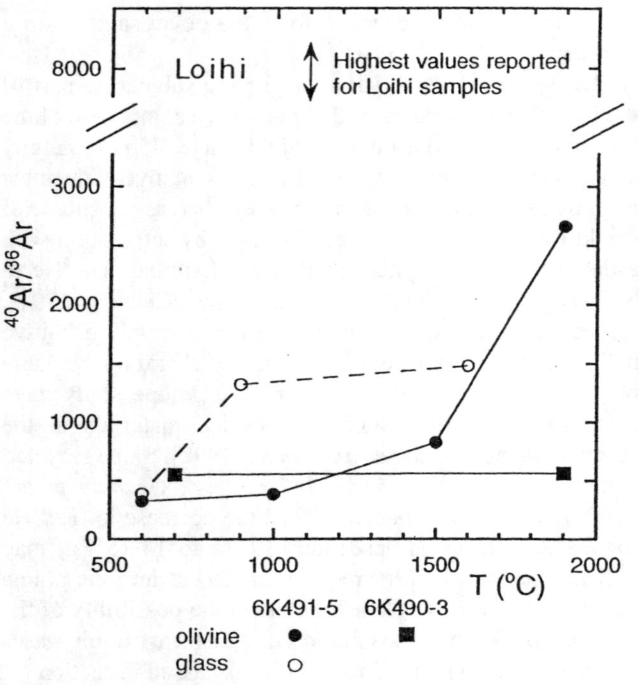

Figure 5. The $^{40}Ar/^{36}Ar$ ratio vs. the degassing temperature for Loihi samples (6K490-3, 6K491-5). The highest values reported for Loihi samples are about 8000 [*Trieloff et al.*, 2000] and shown here as the assumed pristine component of the plume source for comparison.

data of MORB have been taken for glass phases. For OIB, it is easier to use phenocrysts from subaerial rocks or nodules than to collect glass samples around an island using a submersible or by dredging. As long as olivine and/or clinopyroxene crystals are equilibrated with the magma, isotopic ratios should be the same. Thus, one can use the phenocryst data to characterize the regional difference in the isotopic ratios, but not for discussing noble gas abundance.

We also consider the possible effects of the thickness of glassy rinds as they might affect noble gas retention. In the present study, the sample 6K490-8 glass contains ^4He with the total amount of 4.2 x 10^{-6} cm^3STP/g, which is one of the highest values so far observed as OIB glass samples, approaching values characteristic of MORB. This sample was collected at the depth of 4383 m and the glass rind has a thickness of about 4 mm, which is the maximum among our samples, but still thinner than the 10 mm thickness characteristic of MORB samples. Other glass samples with thinner rinds collected from similar depths have smaller amounts of ^4He. Therefore, we consider the possibility that the thickness of glass might control the abundance of ^4He. Detailed studies made on a single MORB pillow show that the outermost layer and the boundary layer with a partly crystallized phase have been affected by secondary effects together with loss of magmatic components [*Kumagai and Kaneoka,* 1998]. Hence, when fresh glass is not sufficiently thick, it may lose magmatic gases and/or incorporate atmospheric components from seawater. For Hawaiian rocks, we have never seen glass rinds as thick as observed in MORB pillows. Vesicularity of glass samples seems also to control the retentivity of gases in samples [*Kurz and Jenkins,* 1981]. Furthermore, He in MORB has been reported to be enriched with respect to Ne due to elemental fractionation [*Honda and Patterson,* 1999]. Thus, the observed He abundance in MORB and OIB glasses may be largely controlled by secondary effects rather than primary characteristics of their source materials. Although He loss is controlled by CO_2 degassing, C/He ratios are still variable with more than one order of the magnitude for Loihi samples [*Trull et al.,* 1993]. This may also reflect the selective loss of He compared to CO_2 reflecting higher diffusivity of He (more than one order of the magnitude in a silicate melt) [*Watson,* 1994].

5.3 Trapping Sites of Noble Gases in Hawaiian Rocks and Minerals

Since noble gases have no electric charge and relatively large atomic sizes, they are thought not to occupy any regular crystal site, but occur in crystal imperfections, vesicles (e.g., Kurz and Jenkins, 1981), fluid inclusions, grain boundaries (e.g., Kurz et. al., 1996), or are dissolved in glasses. It is generally believed that noble gases trapped in vesicles and fluid inclusions might be released by crushing, but fusion of the rock would be required to release noble gases trapped in the remaining sites [*e.g., Rison and Craig,* 1983]. In the present study, either heating or crushing methods were applied noble gases from each sample. Previous experiments have revealed that the noble gas components degassed by crushing are also degassed up to about 1000°C by the heating method [*e.g., Hanyu et al.*, 1999].

Glass and olivine phases of lighter noble gases such as He and Ne were more easily degassed at lower temperatures and heavier gases such as Ar, Kr and Xe were degassed at higher temperatures for sample 6K491-5 (Figure 4). Up to the intermediate temperature of 900°C, more than 90% of ^4He was degassed from glass, while for Ar, Kr and Xe more than 50% of gases were degassed at the highest temperature (1600°C) (Figure 4a). This trend of degassing characteristics may reflect the difference in the diffusion coefficient of each noble gas in glass. From laboratory experiments, diffusion coefficients of noble gases in a basaltic glass were determined [*Hiyagon,* 1994] and the followings are values at 1000°C in the unit of cm^2/sec: He- 5.5x10^{-6}; Ne- 6.6x10^{-7}; Ar- 1.8x10^{-11}; Kr- 5.4x10^{-13}, Xe- 4.6x10^{-14}.

The release patterns of olivine in sample 6K491-5 are quite different from those of the glass phase (Figure 4b). The amount of ^{4}He increased with increasing temperature. More than 70% of ^{4}He was released in the highest temperature fraction (1900°C). However, the Ne and Ar trends are more variable with the 1500°C fraction having less gas than the1000°C fraction. In the highest temperature fraction (1900°C), the released fraction of Ne and Ar was greatest (about 50%). Such release patterns indicate the occurrence in olivine of at least two trapping sites for noble gases. Similar results were obtained for olivine and pyroxene crystals of Siberian lherzolite mantle xenoliths [*Yamamoto,* 2001].

Variable amounts of shallow contamination are supported by the data for ^{40}Ar/^{36}Ar ratios (Figure 5). For the glass of sample 6K491-5, the fraction at the highest temperature indicates a slightly higher ^{40}Ar/^{36}Ar ratio. However, the olivine of sample 6K491-5 shows ^{40}Ar/^{36}Ar of less than 400 in the 600°C and 1000°C fractions, while the ratio increases with temperature to a value of about 2700 in the 1900°C fraction. Since large amounts of ^{40}Ar were degassed in the 1000°C fraction, we conclude that this component was probably trapped in bubbles and/or fluid phases, and is of secondary origin. This component was probably incorporated into olivine crystals during upwelling of magma. The component degassed at the highest temperature corresponds to a primary component trapped in the early stage of crystallization to the surface.

Atmospheric contamination via interaction with a seawater component is another factor to consider. In the case of the ^{3}He/^{4}He, both glass and olivine crystals in sample 6K491-5 show no difference within analytical uncertainty in all temperature fractions sampled (Figure 6), thus reflecting no observable effect of atmospheric contamination. The isotopic compositions of Ne and Ar are quite variable and a mixing trend is observed for Loihi samples between a component in the magma and the atmospheric component from seawater. There may have been more than one fluid involved, as has been suggested for a Samoan harzburgite [*Burnard et al.,* 1998]. We conclude that most noble gases trapped in fluid phases of Loihi samples reflect a variable addition of atmospheric components, probably from seawater-altered oceanic crust.

5.4 Possible Cause of Variation and the Source of Pristine Noble Gas Isotope Components in the Hawaiian Plume

The preceding sections show that noble gas data can help to identify different source components that affect Hawaiian plume magmas. The proportion of these components apparently changes during the evolution of a Hawaiian volcano. The following components appear to be important:

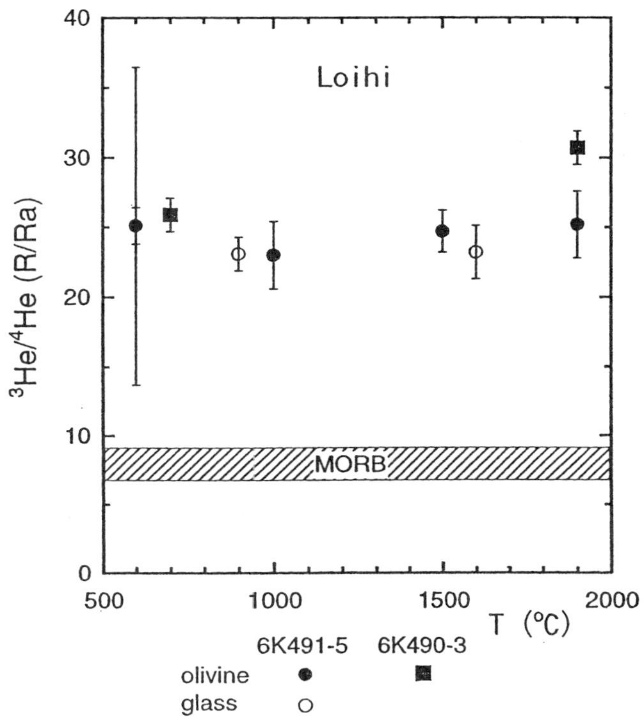

Figure 6. The ^{3}He/^{4}He ratio vs. the degassing temperature for Loihi samples (6K490-3, 6K491-5). The data for the MORB range are taken from the literature [*Ozima and Podosek,* 1983].

1. Pristine plume component as identified in dunite data obtained by multiple crushing [*Trieloff et al.,* 2000].

2. Crustal component derived from the upper lithosphere (probably oceanic crust). This is best identified by the variable ^{3}He/^{4}He observed in Loihi samples.

3. An atmospheric component that is most likely derived from interaction of magma with seawater. The most likely source of such contamination is altered oceanic crust. Noble gases heavier than He are diagnostic elements.

4. An asthenospheric component has been inferred from MORB-like noble gas isotope signatures observed in ultramafic nodules from Hawaiian Islands other than those of Loihi [*e.g., Kaneoka,* 1987]. The decrease of ^{3}He/^{4}He to the MORB value with the evolution of Hawaiian Volcanoes has been ascribed to the incorporation of this component as discussed in the next section.

5. Recycled component derived from old subducted slab in the mantle has been argued based on ^{187}Os/^{188}Os ratios [*e.g., Lassiter and Hauri,* 1998]. Some Hawaiian xenoliths have ^{40}Ar/^{36}Ar ratios of more than 5000 accompanied with almost atmospheric Xe, which were interpreted to indicate the incorporation of heavier noble gases from recycled

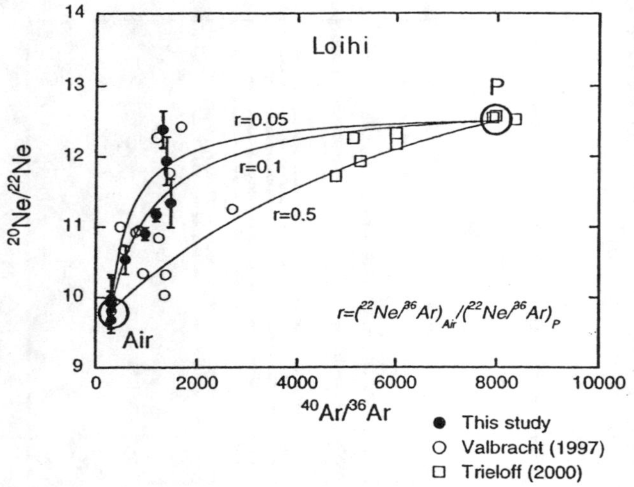

Figure 7. Plot of $^{20}Ne/^{22}Ne$ vs. $^{40}Ar/^{36}Ar$ for selected Loihi and Kilauea samples. "P" indicates the assumed end member of the Loihi magma source ($^{20}Ne/^{22}Ne = 12.5$, $^{40}A/^{36}Ar=8000$). "r" indicates a parameter for determining a mixing line between the air and the "P" component. Data for the $^{20}Ne/^{22}Ne$ with analytical uncertainties of less than 5% are plotted for Loihi samples (glass, olivine) together with reported values for Loihi basalts (glass, olivine) (open circles) [*Valbracht et al.,* 1997] and dunites (open squares) [*Trieloff et al.,* 2000].

components [*Kaneoka,* 1998]. This is the only supporting evidence for the occurrence of heavier noble gases from recycled materials. Based on laboratory experiments, it has been shown that He and Ne in sediments are easily removed during subduction process [*e.g., Hiyagon,* 1994]. Hence, recycling of He and Ne is thought to be not likely. For heavier elements (Ar, Kr, Xe), however, no consensus has been obtained about their behavior during recycling. It has been argued that the occurrence of the lower $^{40}Ar/^{36}Ar$ accompanied with higher $^{206}Pb/^{204}Pb$ for MORB glass might reflect atmospheric contamination from recycled materials [*Sarda et al.,* 1999]. However, if old, fine-grained sediments were incorporated into MORB magma at the ocean bottom, it will also explain the trend. Old, recycled materials accumulate radiogenic and nucleogenic noble gas isotopes such as ^{4}He, ^{21}Ne and ^{40}Ar over time. Once they are incorporated into magma, they are easily homogenized with the magmatic components. Unless samples show a uniform $^{3}He/^{4}He$ lower than MORB, it is difficult to identify the contribution of recycled materials from other mantle components. Thus, it is not always easy to separate the recycled component from the crustal component based on noble gas data alone.

Many studies of Loihi samples have found high $^{3}He/^{4}He$ ratios of up to 35 Ra [*e.g., Kurz et al.,* 1983] and solar-like noble gases in the plume source [*e.g., Honda et al.,* 1991]. Loihi samples had generally low $^{40}Ar/^{36}Ar$ ratios of less than 1000, but sometimes had higher ratios up to about 6000 [*e.g., Valbracht et al.,* 1997], and only atmospheric components were reported for Kr and Xe in most cases. Our results agree. However, recent studies based on the multiple crushing method have revealed a plateau $^{40}Ar/^{36}Ar$ ratio of about 8000 in dunites together with the occurrence of excess ^{129}Xe and other related Xe isotopes [*Trieloff et al.,* 2000]. Although the genetic relations of these xenoliths to plume magmas are not always clear, isotope signatures of He and Ne for all Loihi samples show the same trends. Hence, we assume here that these data approximate the pristine component (P) of the Loihi plume source.

If this assumption is correct, we can explain the relationship between Ne and Ar isotopes for Loihi samples. Loihi samples show a trend for Ne and Ar, which is clearly different from that of Loihi dunites (Figure 7). Assuming that "P" is an end member, the mixing curve between the "P" component and the atmospheric component indicates that the present samples roughly follow the mixing line with the "r" values of 0.05 to 0.1. In contrast, Loihi dunites seem to lie on the mixing line of r = 0.5. This indicates that the $^{22}Ne/^{36}Ar$ of the "P" component is roughly two times larger than the atmospheric value for Loihi dunites. This might mean that the dunite samples contain noble gases in residual crystalline phases, while glasses and olivine phenocrysts are melt phases.

Why do only the dunites among the Loihi samples show the pristine components of heavier noble gases? As indicated in the data for Loihi samples, glass and olivine phenocrysts in most Loihi basalts seem to have been affected by the atmospheric components via seawater before eruption. However, if the dunites are formed from the plume magma at a depth below which such contamination cannot affect the magma, it will retain the pristine component. As discussed in the previous section, however, even a single olivine might contain different components trapped in different sites of olivines. Only multiple crushing can separate them, because stepwise heating might mix such components by diffusion during heating.

How do we explain the high variation in $^{3}He/^{4}He$ ratios at Loihi? Figure 3 shows a unique end member for Ne with different degree of mixing with the atmospheric component. This requires the homogenization of Ne isotopes in the source material. Diffusivity of He is higher than that of Ne [*e.g., Ozima and Podosek,* 1983]. Therefore, it is quite difficult to attribute the variation in the $^{3}He/^{4}He$ to the addition of radiogenic ^{4}He to the plume source. This strongly suggests that variation in the $^{3}He/^{4}He$ must be due to the addition of radiogenic ^{4}He during the ascent of magma at relatively shallow depths where there was insufficient time for

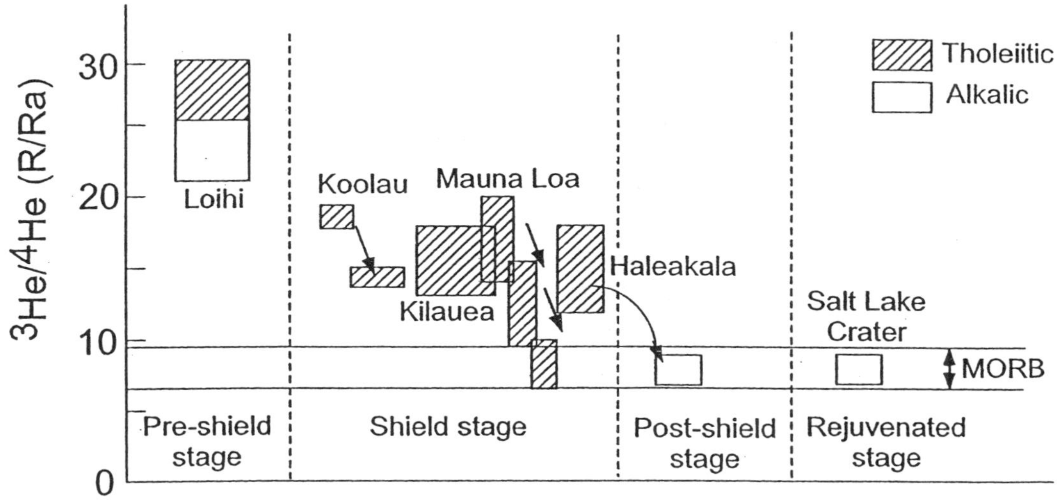

Figure 8. Schematic diagram to show the general trend of variation for the $^3He/^4He$ ratio with the evolution of Hawaiian volcanoes. In the shield-stage, only the general trend is shown for each volcano. No specific distinction has been made among each volcano in this stage. This figure is modified from *Kaneoka* [1987]. Salt Lake Crater samples represent those of Honolulu series. Additional sources of data include *Sarda et al.* [1988], *Kurz and Kammer* [1991], *Hiyagon et al.* [1992], *Honda et al.* [1993], and *Valbracht et al.* [1996, 1997].

homogenization to occur. The old lithosphere, especially the old oceanic crust, has accumulated the radiogenic ^4He. The Hawaiian oceanic crust has the formation age of about 100 million years [*Clague and Dalrymple,* 1987]. Assuming that the basaltic layer of the oceanic crust has the U content of 100 ppb with the Th/U ratio of 2.2 [*Taylor and McLennan,* 1985], it produces 1.8 x 10^{-6} cm^3STP/g of ^4He during 100 million years. The U content of the average oceanic lithosphere is not well known, but it is thought to be similar or less than that of the MORB source (about 10 ppb) [*Taylor and McLennan,* 1985]. Therefore, production of the radiogenic ^4He is much less than the oceanic crust. Because of lower diffusivity of Ne and Ar than He [*Ozima and Podosek,* 1983; *Lux,* 1987], addition of nucleogenic ^{21}Ne and radiogenic ^{40}Ar from the oceanic crust to a magma would be less compared to that of He. Furthermore, incorporation of the atmospheric components would largely dilute its effect, which is commonly observed for noble gas isotopes of heavier elements. This reasoning suggests that noble gas signatures in Loihi samples have been largely modified in the oceanic crust.

5.5 Systematic Variation of the $^3He/^4He$ During the Evolution of Hawaiian Volcanoes

The variation of $^3He/^4He$ in Hawaiian lavas has been interpreted to systematically decrease during evolution of a Hawaiian volcano (Figure 8) [*e.g., Kaneoka,* 1987; *Kurz and Kammer,* 1991; *Valbracht et al.,* 1996]. At the pre-shield stage represented by Loihi Volcano, the $^3He/^4He$ is high with the range of 20-35 Ra. At the shield stage, the values range from 10-20 Ra. At the post-shield stage, the $^3He/^4He$ ratio decreases to around 8 Ra, the average MORB value. In the rejuvenated stage, the lavas have $^3He/^4He$ ratios identical to MORB. This strongly suggests that progressive incorporation of lithospheric and/or asthenospheric material add MORB-type helium to the plume magma.

Several models have been proposed to explain the Hawaiian He isotope trend [*e.g., Kaneoka,* 1987; *Kurz and Kammer,* 1991; *Valbracht et al.,* 1996]. Incorporation of radiogenic ^4He into a magma from the oceanic crust is large due to its high diffusivity [*Ozima and Podosek,* 1983], resulting in a large decrease in $^3He/^4He$, but contamination would be less for other isotopes. Thus, the effect of the lithospheric component would be mostly reflected in short-term variation of the $^3He/^4He$ ratio. However, the general $^3He/^4He$ trend would be largely controlled by interaction with the asthenosphere. Other isotopes also show systematic variation with the stages of Hawaiian volcanism. Some Hawaiian volcanoes such as Koolau and Haleakala show lower $^{143}Nd/^{144}Nd$ ratios in tholeiites erupted during the shield stage compared to post-shield and rejuvenated alkaline basalts from the same volcano [*e.g., Chen and Frey,* 1983; *Stille et al.,* 1983]. This has been interpreted as the result of an increasing involvement of depleted mantle and/or lithosphere during the late stage of Hawaiian volcanism [*Chen and Frey,* 1983; *Kaneoka,* 1987; *Kurz et al.,* 1987].

The contribution of recycled materials originated from an old subducted slab in the mantle is another candidate to explain the variation in $^3He/^4He$. In this case, the decrease in the $^3He/^4He$ should correlate with the increase of the $^{87}Sr/^{86}Sr$ or the decrease of the $^{143}Nd/^{144}Nd$. As discussed in section 5.1, one cannot deny the possibility that incorporation of recycled materials have decreased the $^3He/^4He$. Such an effect is probably obscured by the contribution from an asthenospheric component. The $^3He/^4He$ of Mauna Loa samples decrease with the increase of the $^{87}Sr/^{86}Sr$ with time, but it approaches almost the MORB value in spite of the decrease in the $^{87}Sr/^{86}Sr$ for samples younger than 7 ka [*Kurz et al.,* 1996]. This may also reflect the contribution of a recycled component, but to a lesser degree. If recycled materials are retained in a relatively thin layer, however, its $^3He/^4He$ would be homogenized with the surrounding MORB-type helium due to much higher diffusivity of helium in silicates. Helium can move more than 5 km during 1 billion years by diffusion at depths of the upper-lower mantle boundary condition [*Hanyu and Kaneoka,* 1998], but much less for solid elements. Therefore, even if the effect of recycled materials is distinct for solid isotopes such as Os or Sr in the magma source, for helium it is not always easy to distinguish it from the asthenospheric component. Although Os isotopes suggest the incorporation of recycled materials into plume magmas [*e.g., Lassiter and Hauri,* 1998], the contribution of recycled materials is less for the other Hawaiian volcanoes compared to the Koolau Series. It should be emphasized that no Hawaiian samples have $^3He/^4He$ below MORB values except samples showing He loss.] Hence, the major trend for the decrease in the $^3He/^4He$ with the evolution of volcano is mainly controlled by the different degree of interaction of the plume magma with an asthenospheric component. This probably occurred within the uppermost part of the asthenosphere.

This general trend that the $^3He/^4He$ ratio decreases from the early to later stages of Hawaiian volcanism is roughly correlated with the variation of the solid element isotopes like $^{87}Sr/^{86}Sr$, although a slight time lag is observed [*e.g., Valbracht et al.,* 1996]. The contribution of the asthenospheric component to the plume magma should be also reflected in Ne isotopes. Thus far, precise measurements of Ne isotopes for Hawaiian volcanoes have been reported for only a few Hawaiian volcanoes. If such inference holds for samples from Salt Lake Crater, they are expected to form a trend close to that of MORB.

6. CONCLUSIONS

The noble gas studies for Loihi, Kilauea and Koolau samples collected by the submersibles "Shinkai 6500" and "Kaiko" of JAMSTEC, lead to the following conclusions.

(1) Loihi glass and olivine samples collected at the depth of 2300-4600 m show $^3He/^4He$ of 21-32 Ra, $^{20}Ne/^{22}Ne$ of 10-12 and $^{40}Ar/^{36}Ar$ of 300-2700. These values are similar to those reported previously. Stepwise heating and crushing show variable values, indicating different trapping sites and trapping temperatures.

(2) Kilauea glass and olivine samples collected at depths of 4000-5600 m have $^3He/^4He$ of 13-15 Ra, $^{20}Ne/^{22}Ne$ of 10.6-11.1 and $^{40}Ar/^{36}Ar$ of 800-3600. These values are similar to those reported previously.

(3) For the Koolau Series, olivine separates prepared from basaltic rocks collected at depths greater than 3100 m show $^3He/^4He$ of about 19 Ra. In contrast, those collected at depths of less than 3000 m have $^3He/^4He$ of 14-15 Ra, similar to the highest values for subaerial rocks of the Koolau Volcano.

(4) All Xe isotopic ratios are atmospheric within their analytical uncertainties of 3%. These values probably are the result of atmospheric contamination with seawater in the oceanic crust. .

(5) Noble gases trapped in olivine crystals consist of different components released at different temperatures. Secondary incorporation of noble gases has occurred in some olivine crystals, indicating a rather complicated history of olivine crystals in Loihi samples.

(6) The thickness of Loihi and Kilauea glassy rinds is thinner than that of MORB and may reduce their ability to retain magmatic noble gases. The glass thickness may control the apparent concentration of noble gases and may be one of the main reasons for the "He paradox".

(7) The large variation in the $^3He/^4He$ ratio for Loihi samples suggests the incorporation of the radiogenic ^4He from oceanic crust during magma transport at shallow depths. Evidence for atmospheric contamination from seawater, probably derived from interaction of rising magmas with oceanic crust, is also present for heavier noble gases.

(8) The occurrence of deeper recycled components in the mantle may also decrease the $^3He/^4He$ for Koolau samples, but its effect is limited.

(9) The general decrease in the $^3He/^4He$ of Hawaiian volcanoes, as they pass through different evolutionary stages, is well established. The $^3He/^4He$ approaches MORB-type values of around 8 Ra, typically observed in the post-shield stage and rejuvenated stages. The long-term decrease in $^3He/^4He$ is probably controlled by the different degree of interaction of the plume-derived magma with the

depleted component around the uppermost part of the asthenosphere. For Loihi, local variation in the $^3He/^4He$ ratio in limited time and space is probably controlled by the interaction of the plume magma with the oceanic crust.

Acknowledgements. We are grateful to Dr. J. Naka of JAMSTEC and Prof. E. Takahashi of Tokyo Institute of Technology for organizing the Japan-USA Cooperative Research on Hawaiian Volcanism and all colleagues on board who collected samples using submersibles "Shinkai 6500" and "Kaiko". We thank Prof. T. Shibata of Okayama University for giving us information on samples used in this study. We greatly appreciate Drs. M. Garcia, M. D. Kurz, J. Lupton and T. Wright for their careful and constructive reviews together with their endeavors to edit and improve English for the manuscript. This study has been financially supported in part by the Grant-in-Aid-for Scientific Research by the Ministry of Education, Science, Culture and Sports, Japan, to I.K. (Nos. 09440182, 09954023).

REFERENCES

Allègre, C. J., T. Staudacher, P. Sarda, and M. Kurz, Constraints on evolution of earth's mantle from rare gas systematics, *Nature,* 303, 762-766, 1983.

Allègre, C. J., T. Staudacher, and P. Sarda, Rare gas systematics: formation of the atmosphere, evolution and structure of the earth's mantle, *Earth Planet. Sci. Lett.,* 81, 127-150, 1986.

Anderson, D., The helium paradoxes, *Proc. Natl. Acad. Sci. USA,* 95, 4822-4827, 1998.

Burnard, P. G., K. A. Farley, and G. Turner, Multiple fluid pulses in a Samoan harzburgite, *Chem. Geology,* 147, 99-114, 1998.

Chen, C. Y., and F. A. Frey, Origin of Hawaiian tholeiites and alkalic basalt, *Nature,* 302, 785-789, 1983.

Clague, D. A., and G. B. Dalrymple, The Hawaiian-Emperor volcanic chain Part I, Geologic evolution, in*Volcanism in Hawaii* (eds., R. W. Decker, T. L. Wright, and P. H. Stauffer), *U.S.G.S. Prof. Paper* 1350, pp.5-54, 1987.

Craig, H., and J. E. Lupton, Primordial neon, helium and hydrogen in oceanic basalts, *Earth Planet. Sci. Lett.,* 31, 369-385, 1976

Dalrymple, G. B., and J. G. Moore, Argon-40: excess in submarine pillow basalts from Kilauea Volcano, *Science,* 161, 1132-1135, 1968.

Funkhouser, J. G., and J. J. Naughton, Radiogenic helium and argon in ultramafic inclusions from Hawaii, *J. Geophys. Res.,* 73, 4601-4607, 1968.

Graham, D.W., S.E.Humphris, W.J.Jenkins, and M.D.Kurz, Helium isotope geochemistry of some volcanic rocks from Saint Helena, *Earth Planet. Sci. Lett.,* 110, 121-131, 1992.

Gramlich, J. W., and J. J. Naughton, Nature of source material for ultramafic minerals from Salt Lake Crater, Hawaii, from measurement of helium and argon diffusion, *J. Geophys. Res.,* 77, 3032-3042, 1972.

Hanyu, T., and I. Kaneoka, The uniform and low $^3He/^4He$ ratios of HIMU basalts as evidence for their origin as recycled materials, *Nature,* 390, 273-276, 1997.

Hanyu, T., and I. Kaneoka, Open system behavior of helium in case of the HIMU source area, *Geophys. Res. Lett.* 25, 687-690, 1998.

Hanyu, T., I. Kaneoka, and K. Nagao, Noble gas study of HIMU and EM ocean island basalts in the Polynesian region, *Geochim. Cosmochim. Acta,* 63, 181-1201, 1999.

Hennecke, E. W., and O. K. Manuel, Noble gases in Hawaiian xenoliths, *Nature,* 257, 778-780, 1975.

Hiyagon H., Experimental studies on noble gas diffusion in basalt glass and partition between basalt melt and olivine, *Ph. D. Thesis, Univ. Tokyo,* 138pp., 1984.

Hiyaon, H., Retention of solar helium and argon in IDPs in deep sea sediments, *Science,* 1257-1259, 1994.

Hiyagon, H., and M. Ozima, Partition of noble gases between olivine and basalt melt, *Geochim. Cosmochim. Acta,* 50, 2045-2057, 1986.

Hiyagon, H., M. Ozima, B. Marty, S. Zashu, and H. Sakai, Noble gases in submarine glasses from mid-oceanic ridges and Loihi seamount: Constraints on the early history of the Earth, *Geochim. Cosmochim. Acta,* 56, 1301-1316, 1992.

Honda, M., and D. B. Patterson, Systematic elemental fractination of mantle-derived helium, neon, and argon in mid-oceanic ridge glasses, *Geochim. Cosmochim. Acta,* 63, 2863-2874, 1999.

Honda, M., I. McDougall, D. B. Patterson, A. Doulgeris, and D.A. Clague, Possible solar noble-gas components in Hawaiian basalts, *Nature,* 349, 149-151, 1991.

Honda, M., I. McDougall, D. B. Patterson, A. Doulgeris, and D. A. Clague, Noble gases in submarine pillow basalt glasses from Loihi and Kilauea, Hawaii: a solar component in the Earth, *Geochim. Cosmochim. Acta,* 57, 859-874, 1993.

Jambon,A.,H.Weber, and O.Braun, Solubility of He,Ne,Ar,Kr and Xe in a basalt melt in the range 1250-1600°C. Geochemical implications. *Geochim. Cosmochim. Acta, 50,* 401-408, 1986.

Kaneoka, I., Constraints on the characteristics of magma sources for Hawaiian Volcanoes based on noble-gas systematics, in *Volcanism in Hawaii* (eds. R. W. Decker, T. L. Wright, and P. H. Stauffer), *U.S.G.S. Prof. Paper* 1350, pp.745-757, 1987.

Kaneoka, I., Noble gas signatures of magmatic sources and processes, *Geochem. J., 27,* 201-211, 1993.

Kaneoka, I., Noble gas signatures in the Earth's interior-coupled or decoupled behaviour among each isotope systematics and problems related to their implication, *Chem. Geology,* 147, 61-76, 1998.

Kaneoka, I., and N. Takaoka, Excess ^{129}Xe and high $^3He/^4He$ ratios in olivine phenocrysts of Kapuho lava and xenolithic dunites from Hawaii, *Earth Planet. Sci. Lett.,* 39, 383-386, 1978.

Kaneoka, I., and N. Takaoka, Rare gas isotopes in Hawaiian ultramafic nodules and volcanic rocks: constraints on genetic relationships, *Science,* 208, 1366-1368, 1980.

Kaneoka, I., and N. Takaoka, Noble gas state in the Earth's interior - some constraints on the present state, *Chem. Geology, 52,* 75-95, 1985.

Kaneoka, I., N. Takaoka, and K. Aoki, Rare gases in mantle-derived rocks and minerals, in *Terrestrial Rare Gases*, edited by E.C. Alexander, Jr., and M. Ozima, Center for Academic Publications, pp.71-84, 1978

Kaneoka, I., N. Takaoka, and D. A. Clague, Noble gas systematics for coexisting glass and olivine crystals in basalts and dunite xenoliths from Loihi Seamount, *Earth Planet. Sci. Lett.*, 66, 427-437, 1983.

Kumagai, H., Variation of noble gas signatures controlled by tectonic conditions and magmatic processes: a case study for an area around the Rodriguez Triple junction in the Indian Ocean, *Ph. D. Thesis, Univ. Tokyo,* 127pp. 1999.

Kumagai, H., and I. Kaneoka, Variation of noble gas abundances and isotope ratios in a single MORB pillow, *Geophys. Res. Lett.*, *25*, 3891-3894, 1998.

Kurz, M. D., and W. J. Jenkins, The distribution of helium in oceanic basalt glasses. *Earth Planet. Sci. Lett.*, 53, 41-54, 1981.

Kurz, M. D., and D. P. Kammer, Isotopic evolution of Mauna Loa Volcano, *Earth Planet. Sci. Lett.*, 103, 257-269, 1991.

Kurz, M. D., W. J. Jenkins, and S. R. Hart, Helium isotopic systematics of oceanic islands: implications for mantle heterogeneity, *Nature,* 297, 43-47, 1982.

Kurz, M. D., M. O. Garcia, F. A. Frey, and P. A. O'Brien, Temporal helium isotopic variations within Hawaiian volcanoes: basalts from Mauna Loa and Haleakala, *Geochim.. Cosmochim. Acta,* 51, 2905-2914, 1987.

Kurz, M. D., W. J. Jenkins, S. R. Hart, and D. Clague, Helium isotopic variations in volcanic rocks from Loihi Seamount and the Island of Hawaii, *Earth Planet. Sci. Lett.*, 66, 388-406, 1983.

Kurz, M. D., T. Kenna, D. Kammer, J. M. Rhodes, and M. O. Garcia, Isotopic evolution of Mauna Loa volcano; a view from the submarine southwest rift, in *Mauna Loa Revealed: Structure, Composition, History, and Hazards*, edited by J.P. Lockwood, and J.M.Rhodes, AGU monograph #92, pp.289-306, 1995.

Kurz, M, D., T.C. Kenna, J. C. Lassiter, and D. J. DePaolo, Helium isotopic evolution of Mauna Kea Volcano: First results from the 1 Km drill core. *J. Geophys. Res.*, 101, 11781-11791, 1996.

Kyser, T. K., and W. Rison, Systematics of rare gas isotopes in basic lavas and ultramafic xenoliths, *J. Geophys. Res.*, 87, 5611-5630, 1982.

Lassiter, J.C., and E. H. Hauri, Osmium-isotope variations in Hawaiian lavas: evidence for recycled oceanic lithosphere in the Hawaiian plume, *Earth Planet. Sci. Lett.*, 164, 483-496, 1998.

Lovering, J. F., and J. R. Richards, Potassium-argon age study of possible lower-crust and upper-mantle inclusions, *J. Geophys. Res.*, 69, 4895-4901,1964.

Lux, G., The behavior of noble gases in silicate liquids: solution, diffusion, bubbles and surface effects, with applications to natural samples. *Geochim. Cosmochim. Acta,* 51, 1549-1565, 1987.

Moore, J.G., D.A.Clague, and W.R.Normark, Diverse basalt types from Loihi seamount, Hawaii, *Geology,* 10, 88-92, 1982.

Moreira, M., J. Kunz, and C. J. Allègre, Rare gas systematics in popping rock: Isotopic and elemental compositions in the upper mantle, *Science,* 279, 1178-1181, 1998.

Morgan, W. J., Convection plumes in the lower mantle, *Nature,* 230, 42-43, 1971.

Noble, C. S., and J. J. Naughton, Deep-ocean basalts; Inert gas content and uncertainties in age dating, *Science,* 162, 265-269, 1968.

Ozima, M., and F. A. Podosek, *Noble Gas Geochemistry,* Cambridge University Press, 367pp., 1983.

Patterson, D. B., M. Honda, and I. McDougall, Atmospheric contamination: a possible source for heavy noble gases in basalts from Loihi Seamount, Hawaii, *Geophys. Res. Lett.*, 17, 705-708, 1990.

Rison, W., and H. Craig, Helium isotopes and mantle volatiles in Loihi Seamount and Hawaiian Island basalts and xenoliths, *Earth Planet. Sci. Lett.*, 66, 407-426, 1983.

Rocholl, A., E. Heusser, T. Kirsten, J. Oehm, and H. Richter, A noble gas profile across a Hawaiian mantle xenolith: Coexisting accidental and cognate noble gases derived from the lithospheric and asthenospheric mantle beneath Oahu, *Geochim. Cosmochim. Acta,* 60, 4773-4783, 1996.

Roden, M. F., T. Trull, S. R. Hart, and F. A. Frey, New He, Nd, Pb, and Sr isotopic constraints on the constitution of the Hawaiian plume: Results from Koolau Volcano, Oahu, Hawaii, USA, *Geochim. Cosmochim. Acta,* 58, 1431-1440, 1994.

Sarda, P., T. Staudacher, and C. J. Allègre, Neon isotopes in submarine basalts, *Earth Planet. Sci. Lett.*, 91, 73-88, 1988.

Sarda, P., M. Morerira, and T. Staudacher, Argon-lead isotope correlation in Mid-Atlantic Ridge basalts, *Science,* 283, 666-668, 1999.

Shibata, T., H. Kagami, J. Naka, S. Umino, and K. Nogami, Petrology of picritic basalts from the basal parts of Loihi submarine volcano, Hawaii, 2001, (*to be submitted to J. Petrology*).

Staudacher, T., M. D. Kurz, and C. J. Allègre, New noble gas data on glass samples from Loihi seamount and Hualalai and on dunite samples from Loihi and Reunion Island, *Chem. Geology*, 56, 193-205, 1986.

Staudigel, H., A. Zindler, S. R. Hart, T. Leslie, C.–Y. Chen, and D. Clague, The isotope systematics of a juvenile intraplate volcanism: Pb, Nd, and Sr isotope ratio of basalts from Loihi Seamount, Hawaii, *Earth Planet. Sci. Lett.*, 69, 13-29, 1984.

Stille, P., D. M. Unruh, and M. Tatsumoto, Pb, Sr, Nd and Hf isotopic evidence of multiple sources for Oahu, Hawaii basalts, *Nature,* 304, 25-29, 1983.

Taylor, S. R., and S. M. McLennan, *The Continental Crust: its Composition and Evolution,* Blackwell Sci. Pub., 312pp., 1985.

Trieloff, M., J. Kunz, D. A. Clague, D. Harrison, and C. J. Allègre, The nature of pristine noble gases in mantle plumes, *Science,* 288, 1036-1038, 2000.

Trull, T., S. Nadeau, F. Pineau, M. Polvé, and M. Javoy, C-He systematics in hotspot xenoliths: Implications for mantle carbon contents and carbon recycling, *Earth Planet. Sci. Lett.*, 118, 43-64, 1993.

Valbracht, P. J., H. Staudigel, M. Honda, I. McDougall, and G. R. Davies, Isotopic tracing of volcanic source regions from Hawaii: decoupling of gaseous from lithophile magma components, *Earth Planet. Sci. Lett.,* 144, 185-198, 1996. .

Valbracht, P. J., T. Staudacher, A. Malahoff, and C. J. Allègre, Noble gas systematics of deep rift zone glasses from Loihi seamount, Hawaii, *Earth Planet. Sci. Lett.,* 150, 399-411, 1997.

Watson, E. B., Diffusion in volatile-bearing magmas, in *Volatiles in Magmas,* edited by M. R. Carroll, and J. R. Holloway, *Rev. Mineralogy,* 30, pp.371-411, 1994.

Wilson, J. T., A possible origin of the Hawaiian Islands, *Canadian J. Phys.,* 41, 863-870, 1963.

Yamamoto, J., Investigation of the subcontinental mantle based on noble gas isotopes, petrological and spectroscopic studies of Siberian mantle xenoliths. *Ph. D. Thesis, Univ. Tokyo,* 131pp., 2001.

I. Kaneoka, T. Hanyu, J. Yamamoto and Yayoi N. Miura, Earthquake Research Institute, University of Tokyo, Yayoi 1-1-1, Bunkyo-ku, Tokyo 113-0032, Japan. (kaneoka@eri.u-tokyo.ac.jp)

Submarine Picritic Basalts from Ko`olau Volcano, Hawai`i: Implications for Parental Magma Compositions and Mantle Source

Michael O. Garcia

Hawai'i Center for Volcanology, Department of Geology & Geophysics, University of Hawai'i, Honolulu, Hawai'i

Previous studies of Ko'olau Volcano subaerial basalts have invoked melting of recycled oceanic crust (garnet pyroxenite) as a mantle source to explain the unusually high SiO_2 (53-55 wt.%), moderate MgO (6-8 wt.%) and low olivine contents (<5 vol.%) of these lavas. To evaluate whether such usual lavas form the bulk of Ko'olau Volcano, we sampled for the first time the dissected flanks of this volcano and the landslide blocks derived from Ko'olau and the volcanoes on the neighboring island of Molokai using the SHINKAI 6500 submersible, the remotely operated vehicle (ROV) KAIKO and dredging. Olivine-rich basalts (>10 vol.%) with high MgO contents (14 to 31.5 wt.%) are abundant among submarine Ko'olau lavas. Many of these picritic lavas have accumulated olivine, including xenocrysts with one or more subgrain boundaries. However, euhedral, undeformed olivine with forsterite contents of 88-90% and normal zoning is the dominant crystal type in these Ko'olau submarine basalts. Olivine CaO contents are moderate (0.16-0.27 wt%) and NiO contents are generally high (up to 0.60 wt%) compared to olivines from other Hawaiian volcanoes. These results indicate that subaerial Ko'olau exposures give a biased sampling of the volcano and that the olivines in these submarine lavas grew at crustal depths in parental magmas with MgO contents of at least 14-15 wt.%. Thus, they were primarily derived from melting peridotite, like other Hawaiian tholeiitic magmas. The wide range in olivine CaO contents indicates that there is a much greater range in Ko'olau parental magma compositions than observed for subaerial lavas.

1. INTRODUCTION

Subaerial lavas from Ko'olau volcano are compositionally distinct among Hawaiian lavas in all aspects of their geochemistry [major element, trace elements and isotopes; e.g., Roden et al., 1984; 1994; Frey et al., 1994; Hauri, 1996; Takeguchi and Takahashi, in review]. They typically have lower olivine abundances (<5 vol.%) and MgO concentrations (6-8 wt%) and higher silica (53-55 wt%) than tholeiitic lavas from other Hawaiian shield volcanoes [e.g., Jackson et al., 1999]. The high silica concentration in Ko'olau basalts has been difficult to explain by crustal processes such as higher degrees of fractionation because Ko'olau lavas typically have lower concentrations of incompatible elements [Norman and Garcia, 1999]. Alternatively, it has been proposed that the mantle plume source for Ko'olau magmas contains a mafic component derived from recycled oceanic crust [eclogite; Hauri, 1996]. Estimates for the proportion of this component in the source for Ko'olau lavas have been as high as 100% [Takeguchi and Takahashi, in review].

Our understanding of Ko'olau volcano has been based entirely on collections of subaerial exposures. At many

Hawaiian Volcanoes: Deep Underwater Perspectives
Geophysical Monograph 128
Copyright 2002 by the American Geophysical Union

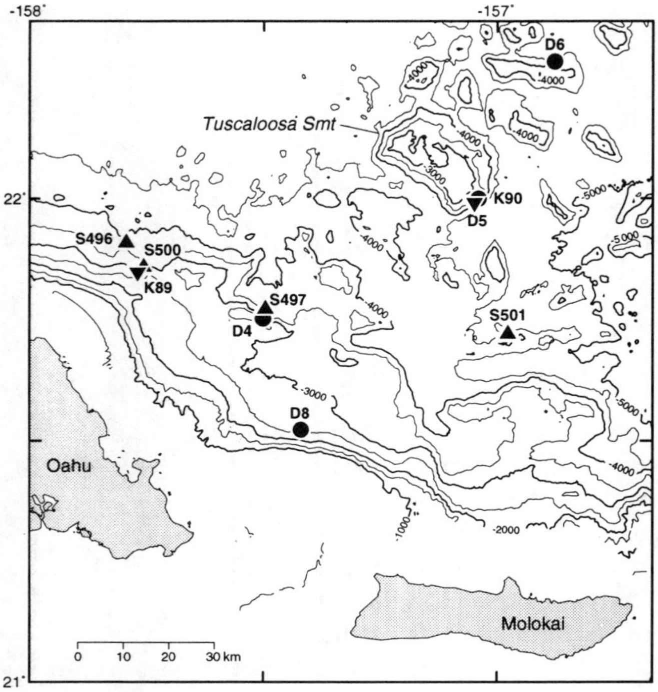

Figure 1. Bathymetric map of the area northward of the islands of Oahu and Molokai showing Nuuanu and Wailua landslide blocks and the locations where samples for this study were collected. Solid triangles show Shinkai 6500 submersible dive locations, and solid stars indicate ROV Kaiko dives and dredgehaul locations. Bathymetry for this map was obtained during the JAMSTEC 1998 and 1999 cruises in Hawai'i [Smith and Satake, this volume]. Contour interval is 500 m.

other Hawaiian volcanoes it has been shown that the subaerial lavas are less MgO-rich than those on the submarine flanks of these volcanoes [e.g, Garcia et al., 1989]. It has been proposed that these volcanoes have an internal density filter that allows lower density, differentiated magmas to erupted subaerially from a shallow, summit magma chamber and higher density, olivine-rich magmas to be intruded into the rift zones and erupted primarily offshore [Ryan, 1987]. One of the objectives of this study was to determine whether a density filter has operated within Ko'olau volcano. If so, previous studies presented a biased perspective of this volcano. In addition, it was anticipated that cold water temperature (~0°C) would allow the submarine rocks to avoid the intense tropical weathering that characterizes Ko'olau subaerial outcrops [e.g., Frey et al. 1994]. Marine expeditions were undertaken in 1998 and 1999 to sample for the first time the submarine flanks of Ko'olau and the landslide blocks derived from this volcano and those from the neighboring island of Molokai using the JAMSTEC deep submergence vehicles KAIKO and SHINKAI 6500, and dredging methods (Fig. 1). Whole-rock geochemical analyses (major and trace elements, and Sr, Nd and Pb isotope analyses) are given for many of these sample in the papers in this volume by Shinozaki et al. [2001] and Tanaka et al. [2001].

Our results show that olivine-rich basalts (>10 vol.%) are abundant among the submarine lavas recovered from Ko'olau Volcano in contrast to their rarity in subaerial exposures. The olivine crystals in these submarine rocks vary from euhedral, undeformed phenocrysts to strongly deformed xenocrysts. The forsterite content of both types of olivines range from 78-90% forsterite but most are 88-90%. There is no correlation of olivine composition with crystal morphology. CaO contents of all olivines are moderate (~0.16-0.27 wt%) indicating they grew at crustal depths. NiO contents of these olivines range widely (0.20-0.60 wt%) and they correlate with forsterite content. These results suggest there is a wide range in composition for Ko'olau parental magmas and that many of these magmas had MgO contents >14 wt.%, as have been found for other Hawaiian shield volcanoes. Thus, Ko'olau magmas were probably derived at least in large part from melting peridotite.

2. SAMPLES AND PETROGRAPHY

The olivine-rich lavas (>10 vol.% olivine) used in this study were obtained from the submarine flanks of Ko'olau volcano during SHINKAI 6500 submersible dives 496, 497 and 500 (SHINKAI samples are labeled with a 'S' prefix), KAIKO ROV dive 89 (KAIKO samples are labeled with a 'K' prefix), and dredgehaul 4 (dredged samples are labeled with a 'D' prefix). Olivine-rich lavas were obtained from landslide blocks adjacent to Molokai volcanoes during SHINKAI dive 501 and dredgehaul 8. The giant landslide block Tuscaloosa (~40 km long; Fig. 1) yielded olivine-rich lavas during KAIKO dive 90 and dredgehaul 5. An unnamed distal landslide block yielded olivine-rich lavas in dredgehaul 6 (Fig. 1). This block and Tuscaloosa Seamount were probably derived from Ko`olau volcano [Yokose, this volume].

All of the submarine rocks have a coating of Mn oxide on at least one surface. The Ko`olau samples typically have a thicker Mn-coating than those from Molokai volcanoes (1-5 mm vs. <1 mm). Beneath this coating, many of the submarine lavas appear to be remarkably unaltered. They contain fresh olivine and glass (Table 1). Others have variable degrees of olivine iddingization. Samples labeled as weakly altered in Table 1 have only thin rims (<0.005 mm thick) of iddingsite on olivines; those labeled as moderately altered have thicker rims (0.005-0.025 mm) and iddingsite

Table 1. Petrography of picritic lavas from Koolau and Molokai volcanoes

Sample	Rock MgO wt%	Max. pheno Fo%	Olivine pheno	Olivine xeno	Olivine mph	Cpx mph	Opx mph	Plag mph	Opaque mph	Gmass	Vesicles	Alteration
Koolau												
S496-3	-	89.2	9.2	10.4	7.2	-	-	-	-	73.2	0.8	weak
S497-2	21.2	89.7	12.6	15.2	4.2	<0.1	<0.1	-	-	68.0	9.6	moderate
S497-6	23.5	89.7	26.8	5.2	6.0	-	-	-	-	62.0	13.6	moderate
S497-9	-	85.4	25.0	4.4	2.6	-	0.6	-	1.2	66.2	1.6	moderate
S500-1	14.8	90.1	17.2	3.4	0.6	0.4^{X}	-	-	<0.1	78.8	<0.1	weak
S500-2B	30.6	89.3	35.8	7.8	5.8	3.2	<0.1	1.8	0.4	45.0	<0.1	moderate
S500-5B	22.5	90.3	22.8	6.2	2.6	-	-	-	<0.1	68.4	2.0	weak
S500-6	23.2	90.0	20.4	8.8	1.6	<0.1^{X}	-	-	<0.1	69.2	2.0	weak
K89-2	14.9	90.3	13.0	4.2	1.8	-	0.6	-	0.2	80.2	5.6	none
K89-4	18.6	89.5	13.4	21.0	3.0	-	-	-	<0.1	62.6	2.6	none
K89-5	18.6	90.1	13.6	11.2	4.0	0.4^{X}	-	-	<0.1	70.8	5.0	none
K89-6	15.9	90.3	14.0	11.6	1.4	<0.1^{X}	<0.1^{X}	<0.1^{X}	<0.1	72.8	3.2	none
D4-7	25.4	89.1	17.8	23.0	6.6	-	-	-	<0.1	52.6	6.0	none
D4-8	10.4	88.6	9.0	<0.1	3.0	<0.1	-	<0.1	-	88.0	22.8	none
Tuscaloosa												
K90-1	17.4	84.4	10.6	3.4	6.8	-	-	-	-	79.2	12.6	weak
D5-13	-	89.2	13.6	6.8	8.4	-	-	-	<0.1	71.2	33.6	weak
D5-16	14.7	90.0	14.2	5.0	9.8	-	-	<0.1	<0.1	71.0	2.4	none
Molokai												
S501-2	28.0	89.5	26.6	19.2	2.2	-	-	-	0.4	51.6	5.0	weak
S501-7	22.6	89.5	9.4	20.6	5.2	-	-	-	<0.1	64.8	0.4	weak
D8-1^{P}	15.5	88.5	6.8	11.2	0.6	8.8	-	5.6	<0.1	67.0	0.8	weak
D8-4	-	89.0	17.6	15.4	6.4	-	-	-	<0.1	60.6	1.6	none
Other block												
D6-1	16.6	88.8	8.6	12.0	2.2	0.2	-	<0.1	-	77.0	1.4	weak

Mineralogy and vesicularity (vesicles) values are in volume percent based on 500 point counts per sample with the mineralogy adjusted to be vesicle-free. Phenocrysts (pheno) are >1.0 mm; microphenocrysts (mph) <1.0 mm; xenocrysts (xeno) have one or more dislocation planes or are strongly resorbed. Cpx- clinopyroxene, Opx-orthopyroxene; Plag- plagioclase; Gmass-groundmass. Alteration level is based on petrographic observations; none - no signs of a alteration; weak - thin iddingsite rims on olivine; moderate - thicker iddingsite rims, alteration along fractures and matrix may be stained. ^{P} Mode includes 3.4 vol.% cpx phenocrysts and 1.4 vol.% plagioclase phenocrysts. ^{X} xenocrysts of pyroxene or plagioclase. Whole-rock MgO data from Shinozaki et al. (this volume).

along fractures. The vast bulk of the olivine in all samples is unaltered.

Olivine is the dominant mineral in the samples examined ranging in abundance from 12 to 49 vol.% (Table 1). The population of olivine crystals was subdivided into phenocrysts (>1 mm wide with euhedral to subhedral shape), xenocrysts (>1 mm wide with at least one subgrain boundary or resorbed margins), and euhedral microphenocrysts (0.1-1.0 mm wide). Many of the phenocrysts and microphenocrysts contain glass inclusions indicating that they grew in a magma. Most of the olivine xenocrysts are only weakly deformed (1-3 subgrain boundaries with only slight shifts in their extinction angles; 1-5°). However, some olivine crystals are strongly deformed with multiple kink bands with extinction offsets >10°, iron-oxide decorated dislocations forming cellular textures, abundant minute gas inclusions and anhedral shape. Olivine xenocrysts are generally more abundant in the Molokai lavas than those from Ko'olau and Tuscaloosa.

Microphenocrysts or xenocrysts (large crystals with resorbed margins) of plagioclase, clinopyroxene and/or orthopyroxene are present in some of the submarine Ko'olau and Molokai rocks but these minerals are usually rare (<1 vol.%; Table 1). Rare xenoliths of dunite and gabbro were observed in some of the lavas. Vesicularity is generally quite low in all of the samples. Only 4 of 22 samples have >10 vol.% and most have <5 vol.% vesicles (Table 1). None of the Molokai samples examined have >5 vol.%

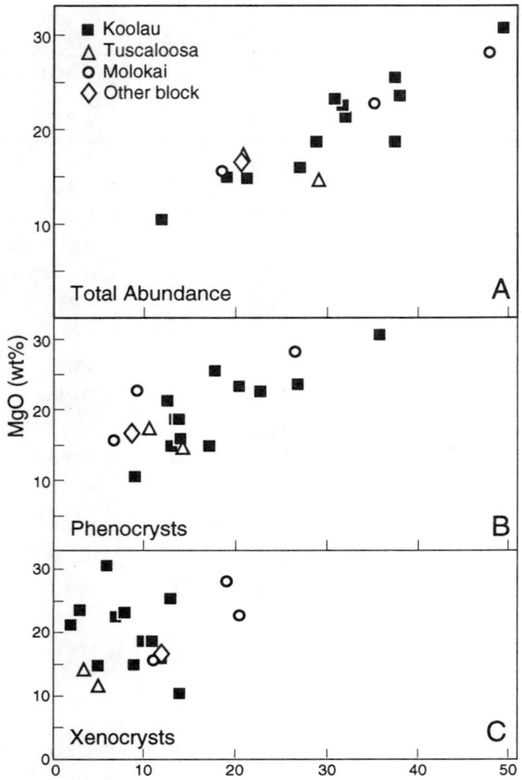

Figure 2. Plots of olivine abundance in basalts from Ko'olau, Molokai, Tuscaloosa and an unnamed landslide block (data from Table 1) versus whole-rock MgO content (in weight percent; data from Shinozaki et al., this volume). A. Total olivine abundance (phenocrysts, microphenocrysts, and xenocrysts); B. phenocryst abundance; and C. xenocryst abundance. Note the good correlation of whole-rock MgO content with total olivine. abundance. The two sigma errors for MgO are smaller than the size of the symbols; for the modes, the errors are about twice the size of the symbols for abundances <10 vol. % and 3-4 times the symbol sized for mineral abundances of 20-40 vol.% based on counting statistics.

vesicles. Subaerial Hawaiian lavas typically have higher vesicularity [>10 vol.%; Hawai'i Scientific Drilling Project, 2000] and submarine lavas generally have lower vesicularity (<10 vol.% for lavas erupted in 1 km or more of water; Moore, 1965]. Two of the three lavas from Tuscaloosa Seamount have vesicularities >10 vol.% (Table 1) indicating a possible subaerial or shallow submarine origin for these rocks.

The MgO contents of the Ko'olau and Molokai submarine olivine-rich rocks used in this study range from 14.7 to 30.6 wt% (except sample D4-8 with 10.4 wt%, which also has the lowest olivine content; Table 1). All of the other lavas qualify as picritic basalts [Le Bas, 2000]. There is a good positive correlation for whole-rock MgO content and total olivine content (phenocrysts, xenocrysts and microphenocrysts) for these lavas, and with olivine phenocryst abundance but not with xenocryst abundance (Fig. 2).

3. MICROPROBE METHODS

Olivine and spinel compositions were determined using the University of Hawai'i, five-spectrometer, Cameca SX-50 electron microprobe with SAMx automation. Operating conditions were a minimum spot size (~1 micron), 15 kV and 20 nA beam current. Counting time on the peak for each element was 60 s; background counting time was 30 s. Mineral standards from the Smithsonian collection (San Carlos olivine for Si and Mg, Natural Bridge Diopside for Ca, Minas Gerais Magnetite for Fe [Jareosewich et al., 1979]) and a synthetic NiO were used for the calibration and San Carlos olivine was used as an internal control to check instrument drift and reproducibility for olivine. For spinel, Smithsonian standards (chromite for Cr, San Carlos Olivine for Mg, ilmenite for Fe and Ti) and synthetic jadeite for Al were used; the chromite standard was used as an internal control. Two-sigma precision based on counting statistics is <1% for the major elements and 5-10% for minor elements (CaO and NiO in olivine and TiO_2 in spinel). A PAP-ZAF matrix correction was applied to all analyses. The analyses reported here are all an average of at least 3 spots in the core of each crystal. Detailed zoning profiles (5-micron steps near the margin, 20-40 micron steps in the core) were made for selected euhedral olivine phenocrysts.

4. MINERAL COMPOSITIONS

4.1 Olivine

Compositions were determined for 715 olivine crystals from 22 submarine Ko'olau and Molokai picritic basalts to evaluate their magmatic history. About 470 of these analyses are for Ko'olau lavas making this the most comprehensive study of olivine from any Hawaiian volcano. For each rock at least 30 crystals were analyzed including representatives of each crystal type. Representative olivine core analyses are given in Table 2. A complete data set is provided on the CD with this volume. The weight percent totals for these analyses vary from 99.1 to 100.5 wt%, with the vast majority between 99.3-100.1 wt%. The cation totals range from 2.989 to 3.013 based on 4 oxygens; most of the totals are between 2.994-3.004. These parameters are good indicators of the overall high quality of these data.

Olivine core compositions of Ko'olau picritic basalts range from 78.4-90.4% forsterite (Fo); olivines from the

Table 2. Representative analyses of olivine cores from Koolau submarine olivine-rich basalts

xtal	SiO_2	FeO	NiO	MgO	CaO	Sum	Fo%	xtal	SiO_2	FeO	NiO	MgO	CaO	Sum	Fo%
	S496-3								S500-1						
p	41.00	10.43	0.44	48.26	0.22	100.35	89.2	p	40.65	9.62	0.58	49.05	0.16	100.06	90.1
p	40.36	10.66	0.44	48.31	0.23	100.00	89.0	p	39.93	12.61	0.42	46.74	0.20	99.89	86.9
p	40.27	13.12	0.38	46.06	0.22	100.05	86.2	p	39.78	14.91	0.26	44.79	0.19	99.93	84.3
d	40.84	11.00	0.44	47.96	0.24	100.47	88.6	d	40.63	9.91	0.52	48.89	0.18	100.13	89.8
r	40.60	10.86	0.40	48.02	0.22	100.09	88.7	d	40.05	12.48	0.42	46.72	0.19	99.86	87.0
m	40.67	10.64	0.43	48.22	0.22	100.19	89.0	m	40.09	12.34	0.36	47.10	0.17	100.06	87.2
	S497-2								S500-2B						
p	40.79	10.00	0.49	48.65	0.23	100.17	89.7	p	40.61	10.33	0.40	48.41	0.27	100.03	89.3
p	39.65	15.12	0.35	44.39	0.22	99.72	83.9	p	40.56	10.63	0.42	48.13	0.26	100.00	89.0
d	40.45	9.91	0.44	48.76	0.23	99.78	89.8	p	40.03	12.28	0.41	47.09	0.26	100.06	87.2
d	39.87	13.75	0.41	45.60	0.24	99.87	85.5	p	39.55	14.62	0.30	44.77	0.26	99.49	84.5
r	40.65	10.98	0.45	47.94	0.22	100.23	88.6	d	40.39	10.42	0.41	48.32	0.23	99.77	89.2
m	40.23	12.56	0.36	46.57	0.22	99.96	86.9	d	39.68	14.92	0.35	44.50	0.23	99.67	84.2
	S497-6								S500-5B						
p	40.40	9.94	0.50	48.46	0.18	99.48	89.7	p-1	40.65	9.36	0.58	49.00	0.18	99.76	90.3
p	40.54	10.26	0.47	48.18	0.19	99.64	89.3	p-2	40.51	9.73	0.53	48.78	0.18	99.73	89.9
p	40.54	10.51	0.47	47.99	0.21	99.72	89.1	p-3	40.55	10.13	0.52	48.65	0.18	100.02	89.5
p	39.76	14.26	0.35	45.04	0.26	99.68	84.9	p	40.29	11.87	0.32	47.19	0.19	99.86	87.6
p	39.64	15.21	0.41	44.24	0.18	99.69	83.8	d	40.48	9.92	0.51	48.62	0.19	99.71	89.7
d	40.12	13.71	0.36	45.53	0.20	99.91	85.5	m	40.41	10.80	0.49	47.94	0.16	99.81	88.8
	S497-9								S500-6						
p	39.74	13.83	0.43	45.50	0.20	99.70	85.4	p-1	40.74	9.72	0.56	48.90	0.17	100.10	90.0
p	39.61	15.18	0.41	44.44	0.21	99.85	83.9	p-2	40.72	9.94	0.48	48.74	0.19	100.08	89.7
p	39.08	17.66	0.37	42.25	0.19	99.55	81.0	p-3	40.63	10.58	0.44	48.34	0.19	100.19	89.1
p	38.67	19.70	0.38	40.61	0.21	99.57	78.6	d	40.56	9.69	0.54	48.85	0.20	99.84	90.0
d	39.26	16.14	0.45	43.49	0.19	99.54	82.8	d	38.90	19.04	0.34	41.18	0.19	99.65	79.4
d	39.31	17.71	0.40	41.99	0.20	99.61	80.9	m	40.37	10.93	0.51	47.44	0.20	99.44	88.6
	D4-7								K89-02						
p-1	40.37	10.61	0.40	48.43	0.20	100.01	89.1	p	41.01	9.34	0.60	48.72	0.17	99.84	90.3
p-2	40.25	11.34	0.37	48.04	0.22	100.22	88.3	p	41.03	9.42	0.56	48.64	0.19	99.84	90.2
p-3	40.05	12.01	0.34	47.55	0.23	100.18	87.6	p	40.43	11.82	0.50	46.82	0.17	99.74	87.6
d	40.29	10.57	0.37	48.41	0.21	99.85	89.1	d	40.54	10.77	0.40	47.94	0.17	99.83	88.8
m	40.39	10.70	0.37	48.59	0.23	100.28	89.0	r	39.46	16.16	0.33	43.40	0.22	99.57	82.7
	D4-8								K89-04						
p	40.75	10.95	0.46	47.60	0.22	99.99	88.6	p-1	40.40	10.60	0.43	48.53	0.21	100.17	89.1
p	40.64	11.20	0.47	47.16	0.18	99.65	88.2	p-2	40.31	11.31	0.44	48.08	0.20	100.34	88.3
p	40.25	11.90	0.42	47.16	0.20	99.94	87.6	p-3	39.86	13.31	0.31	46.22	0.22	99.93	86.1
p	39.69	15.27	0.33	44.21	0.22	99.73	83.8	d	39.16	17.63	0.21	42.90	0.24	100.14	81.3
m	39.95	15.17	0.31	43.21	0.22	98.86	83.5	r	39.27	16.39	0.30	43.90	0.24	100.10	82.7
	K89-5								K89-06						
p	40.72	9.55	0.51	48.72	0.19	99.70	90.1	p	40.96	9.42	0.58	49.04	0.19	100.20	90.3
p	40.56	9.86	0.53	48.68	0.18	99.81	89.8	p	40.96	9.54	0.53	49.08	0.18	100.29	90.2
p	39.72	15.31	0.33	44.12	0.20	99.68	83.7	p	40.17	11.92	0.48	47.12	0.19	99.88	87.6
d	40.81	10.07	0.47	48.44	0.19	99.97	89.6	p	39.81	14.48	0.29	44.81	0.20	99.59	84.6
d	40.34	10.32	0.47	48.29	0.18	99.59	89.3	d	40.70	10.43	0.41	48.18	0.18	99.90	89.2
d	40.02	15.00	0.29	44.61	0.22	100.13	84.1	r	40.12	11.67	0.36	47.17	0.19	99.51	87.8

xtal-crystal; p-phenocrysts; d-deformed xenocryst; r-resorbed xenocryst; m-microphenocryst; Fo%-forsterite percent

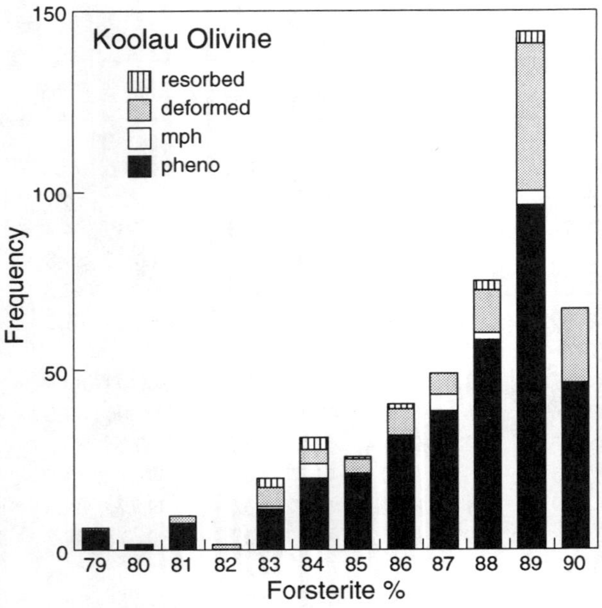

Figure 3. Histogram of forsterite percent in the cores of ~470 olivine phenocrysts, microphenocrysts, deformed xenocrysts, and resorbed but undeformed xenocrysts from a suite of 14 Ko'olau submarine olivine-rich basalts. Most of these olivines have forsterite contents of 88-90%. Xenocrysts have nearly the same distribution as phenocrysts. Resorbed crystals and microphenocrysts have bimodal distributions, which may be a function of too few analyzed crystals. Some of the microphenocrysts have relatively high forsterite contents (87-89%) indicating that they grew in moderately high MgO content magmas (12-14 wt%).

Molokai lavas range from 79.9 to 89.9% Fo. For Tuscaloosa Seamount, olivine varies in composition from 82.7-90.0% Fo and those from the other landslide block range from 82.4 to 89.2% Fo. These ranges for Ko'olau and Molokai olivine composition are typical of tholeiitic lavas from Hawaiian volcanoes [e.g., Clague et al., 1995; Garcia et al., 1995; Garcia, 1996]. Most of the Ko'olau olivines are Fo 88-90% (Fig. 3), which is similar to but slightly less forsteritic than some of the olivines from Mauna Loa submarine basalts (up to 91.3%, the highest reported for Hawaiian basalts; Garcia et al., 1995) but somewhat more forsteritic than most olivines from the Mauna Loa and Mauna Kea picritic lavas obtained by the Hawai'i Scientific Drilling Project [HSDP, 86-88%; Garcia 1996] and those from the east rift zone of Kilauea [Clague and Denlinger, 1994]. Some of the low Fo values may be an artifact of not cutting the probe sections through the core of the analyzed crystals [Pearce, 1984], which is also indicated by the skewed forsterite distribution (Fig. 3).

There is no systematic difference in forsterite content with crystal type in the submarine Ko'olau lavas (Fig. 3). The abundance of phenocrysts and deformed xenocrysts are greatest at 88-90% Fo and decrease markedly at lower forsterite contents, especially for phenocrysts. Similar results were reported for submarine picritic lavas from Kilauea and Mauna Loa [Clague and Denlinger, 1994; Garcia et al., 1995] and for subaerial HSDP picritic lavas [Garcia, 1996]. A bimodal distribution was observed for resorbed

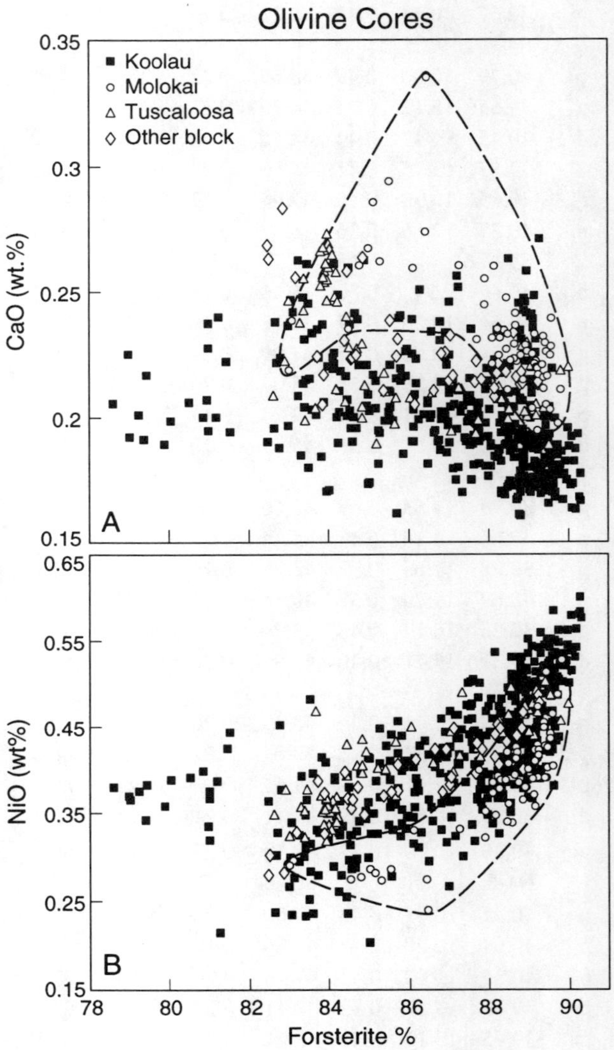

Figure 4. CaO and NiO concentrations versus forsterite percent in the core of olivines from Ko'olau, Molokai, Tuscaloosa and an unnamed landslide block basalts. The olivines from Molokai basalts (enclosed in field) tend to have higher CaO and lower Ni concentrations at a given forsterite value than olivines from Ko'olau basalts. Two sigma errors are about the size of the symbol for forsterite and 0.02 to 0.03 wt% for NiO and CaO.

but undeformed xenocrysts and microphenocrysts in the Ko'olau submarine lavas (83-84, 87-89%; Fig. 3) but this may be an artifact of too few data. It is important to note that four Ko'olau submarine lavas have euhedral microphenocrysts with high Fo contents (89%), indicating that these crystallized in relatively MgO-rich magmas (>13 wt%). For individual lavas, phenocrysts and deformed xenocrysts have similar large ranges in Fo content. The Fo content of undeformed xenocrysts is clearly lower than most phenocrysts in some lavas but not in others. Thus, the source of the resorbed xenocrysts and the composition of the magmas they entered were varied.

CaO contents of all of the olivines are moderate (0.16-27 wt%) and show a broad negative correlation with forsterite content (Fig. 4). The olivines from Ko'olau lavas generally have lower CaO contents than those from Molokai lavas. The olivines from the slide blocks overlap with both fields in CaO content. Previous studies of olivine in Hawaiian picritic basalts found a good correlation between CaO content in whole rock and olivine [Norman and Garcia, 1999], which agrees with experimental work [Libourel, 1999]. Thus, the wide range in Ko'olau olivine CaO content at a given forsterite value (Fig. 4) probably indicates a wide spectrum of Ko'olau parental magma CaO contents in contrast to observations for subaerial Ko'olau lavas [e.g., Frey et al., 1994]. The moderate level of CaO in the Ko'olau olivines indicates that they grew at crustal depths [e.g., Stormer, 1973; Larsen and Pedersen, 2000]. The broad negative trend of CaO with forsterite content in Ko'olau, Molokai and Tuscaloosa olivines agrees with experimental crystallization results [Jurewicz and Watson, 1988] suggesting that these olivines are magmatic and not mantle derived.

NiO contents in Ko'olau olivines range widely (0.20-0.60 wt%) compared to those from Molokai lavas (0.24-0.53 wt%; Fig. 4) and are higher than those reported for olivines from other Hawaiian shield volcanoes. Also, Ko'olau olivines tend to have higher NiO contents at a given forsterite content then those from Molokai (Fig. 4). The broad curved trend of Ni vs. forsterite content for Ko'olau olivine is thought to be indicative of crystal fractionation [e.g., Nakamura, 1995].

Detailed zoning profiles were made of selected euhedral, undeformed high forsterite (88-90%) phenocrysts from Ko'olau submarine picrites. All of these crystals are normally zoned. Some have classic crystal fractionation-produced profiles [c.f., Pearce, 1984]; other crystals have flat profiles from the core to within 50 to 200 microns from the rim. The crystals with flat core profiles re-equilibrated with the host magma prior to eruption. Thus, they probably had even higher forsterite contents prior to re-equilibration.

Table 3. Representative spinel analyses in Koolau submarine olivine-rich basalts

sample	TiO_2	Al_2O_3	Cr_2O_3	FeO*	MgO*	sum
K89-04						
p-1	0.96	13.00	48.51	24.30	11.32	98.09
p-2	1.15	13.61	46.31	25.76	10.97	97.80
p-3	1.71	15.46	41.21	28.12	11.13	97.63
matrix	1.15	13.14	47.43	26.06	10.52	98.30
S500-5B						
p-1	0.86	16.15	47.51	19.45	14.21	98.18
p-3	0.77	13.99	50.15	19.08	14.15	98.15
p-4	1.02	14.65	47.60	21.90	13.08	98.25
matrix	0.82	14.10	50.41	18.96	14.46	98.75
S500-6						
p-2	0.95	15.56	48.70	19.18	14.20	98.59
p-3	0.92	14.24	48.41	21.17	13.21	97.96
p-4	3.49	13.18	29.86	37.94	11.08	95.54
matrix	1.03	12.82	47.86	25.27	10.70	97.67
D4-7						
p-1	1.61	15.73	44.60	22.00	13.72	97.66
p-2	1.08	14.45	46.93	22.41	13.14	98.01
p-3	1.32	16.45	43.85	22.91	13.32	97.85
matrix	1.19	14.83	45.15	24.19	12.54	97.90

Spinels p-1 to p-4 are inclusions in olivines; see Table 2 for host mineral compositions.

4.2 Spinel Compositions

Core compositions were determined for 96 spinel crystals that occurred as either inclusions in olivine phenocrysts or in the matrix of four submarine Ko'olau picritic basalts to evaluate their magmatic history. Representative analyses are given in Table 3 and a complete data set is provided on the CD with this volume. The recalculated weight percent totals (adjusted for ferric iron) for these analyses vary from 97.4 to 99.5 wt%, with the vast majority between 98.0-99.2 wt%. These totals indicate that there are minor amounts of other elements that were not analyzed.

The compositions of matrix spinels are remarkably similar to those for inclusions in olivine phenocrysts for the Ko'olau submarine basalts (Table 3). Both have relatively have Cr numbers [100(Cr/Cr + Al)], ranging from 72.0-83.6 for matrix crystals and 71.3-85.3 for inclusions. These are among the highest reported for Hawaiian tholeiites [e.g., Maaløe, S., and B. Hansen, 1982; Wilkinson and Hensel, 1988; Clague et al., 1995], which indicate that these spinels grew in relatively primitive melts. The TiO_2

contents of most of the Koʻolau spinels are rather low (<2 wt%, with many <1 wt%). Mg#s [(Mg/Mg + Fe^{2+})100] of the spinels range from 33.9 to 57.9 but most range from 41 to 53. The systematic compositional trends observed for the spinel data indicate rapid crystallization avoiding the effects of post-eruption re-equilibration.

5. DISCUSSION

5.1 Rock Type Variations: Subaerial vs. Submarine

Olivine-rich lavas are relatively rare among subaerial Koʻolau lavas. For example, only two of the ~100 lavas studied in the Makapuu section of southeast Koʻolau volcano [Frey et al., 1994] have olivine abundances >10 vol.% [Garcia, unpublished data]. In both cases, the high olivine concentrations are due to settling of olivine within moderately olivine-phyric lavas (5-10 vol.% olivine). Another stratigraphic study of Koʻolau lavas and dikes in a tunnel found only 1 of 24 samples with olivine abundances >10 vol.% [11 vol.%; Jackson et al., 1999]. In contrast, about half of the submarine Koʻolau lavas we collected have >10 vol.% olivine [Shinozaki et al., this volume). The high abundance of picritic basalts on the flanks of Hawaiian volcanoes is in striking contrast to their paucity in flood basalts sequences both on land [e.g., Deccan; Krishnamurthy et al., 2000] and at sea [Ontong Java; Neal et al., 1997]. Although the magma from flood basalts and Hawaiian lavas are both thought to be derived from mantle plumes, the magmatic plumbing systems for these volcanic systems must be quite different to explain why picritic basalts are so abundant on Hawaiian volcanoes.

Subaerial Koʻolau lavas typically have relatively low MgO contents (6-8 wt%) and only ~8% have MgO concentrations >10 wt% [Frey et al., 1994; Jackson et al., 1999]. Among these subaerial lavas, only ~2% have MgO >13 wt% and qualify as picrites [Le Bas, 2000]. For the suite of recovered submarine Koʻolau lavas, 19 of 42 lavas have MgO >13 wt% [Shinozaki et al., this volume] and all but one of the olivine-rich sample have MgO contents >13 wt% (Table 1). Thus, subaerial exposures provide a biased sampling of Koʻolau volcano rock types and a density filter must have operated within this volcano, as was noted for other Hawaiian volcanoes [e.g., Garcia et al., 1989]. The density of the olivine-rich Koʻolau magmas would have been high (2.80 to 2.95 g cm^{-3}) compared to magma densities for subaerially erupted, weakly olivine-phyric or aphyric basalts (~2.70 g cm^{-3}). These results support the model of Ryan [1987] that magma is preferentially erupted on Hawaiian volcanoes based on its density. According to this model, lower density magmas rise within the ediface and erupt from the summit to build the subaerial portion of the volcano, while denser, olivine-rich magmas are intruded into rift zones and erupted on the submarine flanks of the volcano.

5.2 Origin of Olivine-rich Koʻolau Lavas and Implications for Parental Magma Composition

Studies of picritic basalts from Hawaiʻi and elsewhere have successfully utilized olivine compositions to infer the compositions of parental magmas [e.g., Francis, 1985; Wilkinson and Hensel, 1988; Hansteen, 1991; Rhodes, 1995; Garcia, 1996; Fram and Lesher, 1997; Krishnamurthy et al., 2000]. Many of the Koʻolau submarine basalts contain euhedral, undeformed olivines with forsterite contents of 90% and all but one of the lavas in this study (S497-9) contain at least Fo 88% olivines (Table 1). These high forsterite olivines (88-90%), the most prevalent composition within these lavas (Fig. 3), probably grew in magmas with Mg#s of 69-73 (Fig. 5). Five of the 12 Koʻolau samples used in this study fall within this Mg# range. If we assume that 10% of the total iron is ferric (based on the oxidation state of rapidly quenched subaerial and submarine Hawaiian basalts; Moore and Ault, 1965; Byers et al., 1985], a Fe/Mg K_d of 0.30±0.03 (which is consistent with the crustal origin of the olivines; Ulmer, 1989) and equilibrium crystallization, then the high forsterite olivines (88-90.3%) grew in magmas with 14-15 wt.% MgO.

A minimum estimate of 14-15 wt% MgO content for some Koʻolau parental magma is consistent with the range of parental magma compositions that have been proposed for other Hawaiian tholeiitic volcanoes based on studies of picritic lavas (13-17 wt% MgO; e.g., Clague et al., 1995; Rhodes, 1995; Garcia et al., 1995] and with the high MgO glass sands found at the foot of Kilauea Volcano (up to 15 wt%; Clague et al., 1991]. Such high MgO magmas would have been in equilibrium with and were probably derived from a peridotitic mantle source [Maaløe and Hansen, 1982].

The proposed Koʻolau parental magma MgO minimum estimate is substantially lower than observed for many of the submarine Koʻolau lavas (21.2-30.6 wt.%; Table 2). Furthermore, the forsterite contents of the olivines in these lavas are too low to have been in equilibrium with the whole rock composition (Fig. 5). Therefore, these high MgO Koʻolau lavas must have accumulated olivine. This interpretation is supported by the good correlation of whole-rock MgO content for these lavas with total olivine abundance (Fig. 2). The good correlation of MgO with total phenocryst abundance in these lavas (but not with their xenocryst abundance) suggests that accumulation of phenocrystic olivine is the major cause of the high MgO content in some Ko'olau submarine lavas.

What is the origin of the xenocrystic olivines? There is a wide compositional range for xenocrysts in individual lavas (e.g., Fo 82-90% in sample K89-6; Fig. 5). Some of the low Fo values may be an artifact of the thin section not cutting the center of the crystal [e.g., Pearce, 1984] but it is unlikely that all of the low values are related to sectioning problems. The strongly deformed olivines have anhedral margins and rare, small dunitic xenoliths are present in some Ko'olau submarine lavas. Thus, it seems likely that the strongly deformed olivines in these submarine basalts were probably derived by disaggregating dunitic xenoliths. Partially digested dunitic xenoliths are also present in many of the HSDP tholeiitic lavas, which contain abundant strongly deformed olivine xenocrysts [Garcia, 1996]. The moderate CaO contents and forsterite compositions of the Ko'olau deformed olivines indicate that they were probably derived from deformed cumulates. This reasoning has also been applied to explain the presence of deformed olivines in the lavas from the 1959 Kilauea eruption [Helz, 1987]. The survival of dislocations in the deformed olivines gives an indication of how long these features might persist in a magma. Dislocations in olivines are reported to have survived for at least 22 years in the molten Kilauea Iki lava lake [Helz, 1987]. Thus, some of the deformed Ko'olau olivines may have been picked up by ascending magmas decades before they were erupted.

5.3. Origin of Tuscaloosa and the Other Landslide Block Based on Olivine Composition

The debris from the Nuuanu (Ko'olau) and Wailau (Molokai) landslides overlap each other [Moore et al., 1994], which has stymied efforts to reconstruct the size and history of these two landslides. Even the relative age of the two slides has been an open questioned [Moore et al., 1994]. The source volcano for some of the blocks from the two landslides remains ambiguous. For example, although Tuscaloosa Seamount is thought to have been derived from Ko'olau Volcano [e.g., Yokose, this volume], the origin of the block to the north of it (which was sampled by dredge-haul 6) is unknown. The orientation of this unnamed block is parallel to the island of Molokai but there are no other blocks of similar size and orientation near it (Fig. 1).

Olivine composition may be useful in determining the source of the landslide blocks. Norman and Garcia [1999] found that olivines from subaerial Ko'olau lavas have distinctly lower CaO contents than olivines from other Hawaiian shield volcanoes and they related this difference to the distinct composition of Ko'olau lavas. This result was confirmed here but the range of Ko'olau olivine CaO content is much greater than previously reported, which implies a greater range of Ko'olau magma composition than was observed in subaerial lavas. Thus, low CaO content remains a distinctive feature of some but not all Ko'olau olivines (Fig. 5). NiO content may also be of use in determining the source of these blocks. Ko'olau olivines appear to have somewhat higher NiO contents at a given forsterite content than olivines from other Hawaiian shield volcanoes, especially at Fo contents of 84-87% [Norman and Garcia, 1999; Fig. 4]. One problem with this analysis is that we have only a limited amount of data for Molokai volcanoes. Nonetheless, there are no reported analyses of Molokai lavas that are similar to the distinctive chemistry of subaerial Ko'olau lavas [Takeguchi and Takahashi, in review].

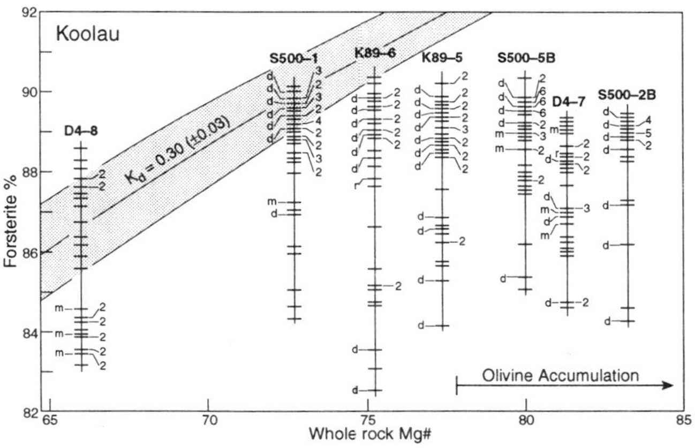

Figure 5. Plot of whole rock Mg# [(Mg/Mg + Fe^{2+}) 100], assuming 10% Fe^{3+}, versus forsterite percent in olivine cores from a suite of representative Ko'olau submarine basalts. The vertical lines represent individual flows and the horizontal lines represent olivine core compositions. Deformed xenocrysts are designated with a "d" at the end of the horizontal line, microphenocrysts by a "m" and resorbed but undeformed crystals by an "r". All other horizontal lines are for euhedral, undeformed phenocrysts. The numbers next to the lines refer to the number of analyzed crystals with that composition, with only one deformed, resorbed or microphenocryst shown for each line. There are wide variations in phenocryst and xenocryst compositions for most lavas but high forsterite content olivine crystals (88-90%) are dominate. Some of the olivines plot within the equilibrium field for their whole rock compositions for lavas with lower Mg# (<75). Olivines in rocks with higher Mg#s plot below the equilibrium field indicating that these rocks have probably accumulated olivine. The equilibrium field is based on a Fe/Mg partition coefficient for olivine/whole rock of 0.30 ±0.03 [Roeder and Emslie, 1970]. The arrow in the lower right corner shows the effect of olivine accumulation. The error bar is in the upper left corner is for 2 sigma.

Olivines in lavas recovered from the two landslide blocks

have higher NiO contents for a given forsterite in the range of Fo 83-86 than those from Molokai but overlap the Molokai field at higher Fo values (Fig. 4). CaO contents in olivines from the two landslide blocks show the same relationships with generally lower values than those from Molokai lavas but with considerable overlap at high and low forsterite values (Fig. 4). These results are consistent with a Ko'olau source for these blocks, which has been proposed based on reconstruction of the blocks [Yokose et al., this volume], but the olivine data are not sufficient to prove a Ko'olau origin.

6. SUMMARY

Submarine Ko'olau lavas are petrographically and geochemically diverse. Unlike subaerial Ko'olau lavas, many are remarkably olivine-rich (up to 49 vol.%) and relatively weakly altered. Euhedral, undeformed olivines with forsterite contents of 88-90% are the most common composition within a suite of 14 submarine picritic Ko'olau basalts. These olivines probably grew in magmas with 14-15 wt% MgO. This minimum estimate for Ko'olau parental magma MgO content is consistent with studies of other Hawaiian shield volcanoes and indicates a peridotite source for at least some Ko'olau magmas. The wide range in olivine CaO contents indicates that there is a much greater range in Ko'olau parental magma compositions than has been observed for subaerial lavas from this volcano.

The high MgO content of some Ko'olau lavas (22-30.6 wt%) is a consequence of olivine phenocryst accumulation based on the good correlation of olivine phenocryst abundance with whole-rock MgO content. Thus, the high MgO contents of the Ko'olau submarine lavas are not an indication of primary magma compositions. The deformed olivine xenocrysts in the Ko'olau picritic basalts were probably derived from disaggregated deformed dunitic cumulates from Ko'olau magmas. This interpretation is supported by the presence of dunitic inclusions in a few of the olivine-rich lavas.

Acknowledgments. I would like to thank the captains and crew of the R/V Kairei and Yokosuka, and the dedicated shipboard science team, especially chief scientist Jiro Naka and co-chief E. Takahashi, for the opportunity to participate in the JAMSTEC Hawai'i cruises and for their collaboration during interpretation of cruise results. Sarah Sherman, Kelly Kolysko-Rose and Heidi Comer assisted with sample preparation and in creating the olivine database for this study, and John Smith provided the bathymetric map for Figure 1. This work was supported by National Science Foundation grant OCE-9729894. This is SOEST contribution no. 5576.

REFERENCES

Byers, C., M.O. Garcia, and D.W. Muenow, Volatiles in pillow rim glasses from Loihi and Kilauea volcanoes, Hawai'i, *Geochim. Cosmochim. Acta, 49,* 1887-1896, 1985.

Clague, D.A. and R.P. Denlinger, Role of olivine cumulates in destabilizing the flanks of Hawaiian volcanoes, *Bull. Volcanol., 56,* 425-434, 1994.

Clague, D.A., W.S. Weber, and J.E. Dixon, Picritic glasses from Hawaii, *Nature, 353,* 553-556, 1991.

Clague, D.A., J.G. Moore, J.E. Dixon, and W.B. Friesen, Petrology of submarine lavas from Kilauea's Puna Ridge, Hawaii, *J. Petrol., 36,* 299-349, 1995.

Francis, D., 1985, The Baffin Bay lavas and the value of picrites as analogues of primary magmas. Contrib. Mineral. Petro., 89, 144-154

Fram, M.S. and C.E. Lesher, Generation and Polybaric Differentiation of East Greenland Early Tertiary Flood Basalts, *J. Petrol., 38,* 231-275, 1997.

Frey, F.A., M.O. Garcia, and M. Roden, Geochemical characteristics of Ko'olau Volcano: Implication of intershield geochemical differences among Hawaiian volcanoes, *Geochim. Cosmochim. Acta, 58,* 1441-1462, 1994.

Garcia, M.O., Petrography, olivine and glass chemistry of lavas from the Hawaii Scientific Drilling Project, *J. Geophys. Res., 101,* 11701-11713, 1996.

Garcia, M.O., T. Hulsebosch, and J.M. Rhodes, Olivine-rich basalts from the southwest rift zone of Mauna Loa volcano: Implications for magmatic processes and geochemical evolution, *AGU Monograph-- Mauna Loa Decade Volcano, 92,* 219-239, 1995.

Garcia, M.,O., D.W. Muenow, K. Aggrey, and J. O'Neil, Major element, volatile and stable isotope geochemistry of Hawaiian submarine tholeiitic glasses, *J. Geophys. Res., 94,* 10,525-10,538, 1989.

Hansteen, T.H., Multi-stage evolution of the picritic Maelifell rocks, SW Iceland: constraints from mineralogy and inclusions of glass and fluid in olivine. *Contrib. Min. Petro.*, 109, 225-239, 1991.

Hauri, E.H., Major-element variability in the Hawaiian mantle plume, *Nature, 382,* 415-419, 1996.

Hawaii Scientific Drilling Project Core Logs and Sumarizing Data, Vol. 1-4, 1999.

Helz, R.T., Diverse olivine types in lava of the 1959 eruption of Kilauea volcano and their bearing on eruption dynamics, *U. S. Geol. Surv. Prof. Paper 1350,* 691-722, 1987.

Jackson, M.C., F.A. Frey, M.O. Garcia, and R.A. Wilmoth, Geology and geochemistry of basaltic lava flows and dikes from the Trans-Ko'olau tunnel, Oahu, Hawaii, *Bull. Volcanol., 60,* 381-401, 1999.

Jarosewich, E., J.A. Nelen, and J.A. Norberg, Electron Microprobe Reference Samples for Mineral Analyses, *Mineral Sciences Investigations 1976-1977, 22,* 68-69, 1979.

Jurewicz, A., and E.B. Watson, Cations in olivine, Part 1, Calcium, *Contrib. Mineral. Petrol., 99,* 176-185, 1988.

Krishnamurthy, P., K. Gopalan, and J.D. Macdougall, Olivine Compositions in Picrite Basalts and the Deccan Volcanic Cycle, *J. Petrol.*, *41*, 1057-1069, 2000.

Larson, L.M., and A.K. Pederson, Processes in High-Mg, High-*T* Magmas: Evidence from Olivine, Chromite and Glass in Palaeogene Picrites from West Greenland, *J. Petrol.*, *41*, 1071-1098, 2000.

Le Bas, M.J., IUGS Reclassification of the High-Mg and Picritic Volcanic Rocks, *J. Petrol.*, *41*, 1467-1470, 2000.

Libourel, G., Systematics of calcium partitioning between olivine and silicate melt: implications for melt structure and calcium content of magmatic olivines, *Contrib. Mineral Petrol.*, *136*, 63-80, 1999.

Maaløe, S., and B. Hansen, Olivine phenocrysts of Hawaiian olivine: tholeiite and oceanite, *Contrib. Mineral. Petrol., 81,* 203-211, 1982.

Moore, J.G., Petrology of deep-sea basalt near Hawaii, *Am. J. Sci.*, *263*, 40-52, 1965.

Moore, J.G., and W.U. Ault, Historic littoral cones of Hawaii, *Pacific Science*, *19*, 3-11, 1965.

Moore, J.G., W.R. Normark, R.T. Holcomb, Giant Hawaiian landslides, *Ann. Rev. Earth Planet. Sci., 22*, 119-144, 1994.

Nakamura, M., Residence time and crystallization history of nickeliferous olivine phenocrysts from the northern Yatsugatake volcanoes, Central Japan, *J. Volcanol. Geotherm. Res.*, *66*, 81-100, 1995.

Neal, C.R., J.J. Mahoney, L.W. Kroenke, R.A. Duncan, M.G. Petterson, The Ontong Java Plateau, *AGU Monograph- Large Igneous Provinces*, 100, 183-216, 1997

Norman, M.D. and M.O. Garcia, Primitive magmas and source characteristics of the Hawaiian plume: petrology and geochemistry of shield picrites, *Earth Planet. Sci. Lett.*, *168*, 27-44, 1999.

Pearce, T.H., The analysis of zoning in magmatic crystals with emphasis on olivine, *Contrib. Mineral. Petrol.,* 86, 149-154, 1984

Roden, M.F., F.A. Frey, and D.A. Clague, Geochemistry of tholeiitic and alkalic lavas from the Ko'olau Range, Hawaii: Implications for Hawaiian volcanism, *Earth Planet. Sci. Lett.*, *69*, 141-158, 1984.

Roden, M.F., T. Trull, S.R. Hart, and F.A. Frey, New He, Nd, Pb, and Sr isotopic constraints on the constitution of Hawaiian Plume: Results from Ko'olau Volcano, Oahu, Hawaii, *Geochim. Cosmochim. Acta*, *58*, 1497-1446, 1994.

Rhodes, J.M., The 1852 and 1868 Mauna Loa Picrite eruptions: Clues to parental magma compositions and the magmatic plumbing system, *AGU Monograph: Mauna Loa Volcano*, *92*, 241-262, 1995.

Roeder, P.L., and R.F. Emslie, Olivine-liquid equilibrium, *Contrib. Mineral. Petrol., 29,* 275-289, 1970.

Ryan, M. P., Neutral buoyancy and the mechanical evolution of magmatic systems, *Geochem. Soc. Spec. Pub. 1,* 259-287, 1987.

Shinozaki, K., Z. Ren, and E. Takahashi, Geochemical and Petrological Characteristics of Nuuanu and Wailau landslide blocks, this volume, 2001.

Smith, J.R. and K. Satake, Multibeam bathymetry and sidescan data of the Hawaiian islands, 1998-1999, (explanation of accompanying CD-ROM), this volume, 2001.

Stormer, J., Calcium zoning in olivine and its relationship to silica activity and pressure, *Geochim. Cosmochim. Acta, 27,* 1815-21, 1973.

Takeguchi, R., and E. Takahashi, Primary Magma Composition in Ko'olau Volcano, Hawaii, *Earth Planet. Sci. Lett.*, in review, 2001.

Tanaka, R., E. Nakamura, and E. Takahashi, Geochemical evolution of Ko`olau volcano, Hawaii, this volume, 2001.

Ulmer, P., The dependence of the Fe^{2+}-Mg cation-partitioning between olivine and basaltic liquid on pressure, temperature and composition, *Contrib. Mineral. Petrol., 101,* 261-73, 1989.

Wilkinson, J. F. G. and H. D. Hensel, The petrology of some picrites from Mauna Loa and Kilauea volcanoes, Hawaii, *Contrib. Mineral. Petrol., 98,* 326-345, 1988.

Yokose, H., Submarine Geology of Giant Landslides in the Northeast Oahu: SHINKAI 6500 Submersible Investigation, this volume, 2001.

Michael O. Garcia, Department of Geology & Geophysics, University of Hawaii, Honolulu, HI, 96822, USA. (email: garcia@soest.hawaii.edu)

Melting Process in the Hawaiian Plume: An Experimental Study

Eiichi Takahashi and Katsuji Nakajima

Magma Factory, Earth and Planetary Sciences, Tokyo Institute of Technology, Tokyo, Japan

In order to understand the melting processes in the Hawaiian plume, which may involve a mixture of peridotite and entrained former oceanic crust (eclogite), we carried out basalt/peridotite sandwich melting experiments with NMORB-like CRB72-31, Archaean Fe-rich basalt CLG46 and a fertile peridotite KLB-1. Melts similar to a relatively unfractionated Koolau tholeiite (SiO_2=53, MgO=7 wt%) can be produced directly by partial melting of CLG46 at 2.5-3.0 GPa and 1350-1400°C. The conditions are slightly below the dry solidus of mantle peridotite. At or above the peridotite solidus, melts become saturated with olivine and are similar to picritic magma in Kilauea (SiO_2=48, MgO=16). In the basalt/peridotite sandwich melting experiments, melt proportion within the basalt domain decreases as temperature increases near the peridotite solidus. Due to this effect, the degree of partial melting in the basalt layer is low at a temperature just above the peridotite solidus. Accordingly, the picritic melts produced at temperatures slightly above the peridotite solidus are more enriched in incompatible elements (K, Ti) than those produced at somewhat lower temperatures in the basalt layer. Two fundamental melting modes were found in the basalt/peridotite hybrid magma source: partial melting of eclogite domains, and reactive melting of both eclogite and peridotite domains. These results help to explain the origin of the two magma types in Hawaiian volcanoes (Koolau-type tholeiite and Kilauea-type picrite). Based on the melting experiments and the growth history of Koolau volcano, a model of melting process in the Hawaiian plume is proposed which advocates lower plume temperatures and larger block size of entrained eclogite than previous studies.

1. INTRODUCTION

The Hawaii hot spot is currently the most magma productive hot spot on the Earth [*BVSP,* 1981]. Geophysical features of the mantle plume and magma genesis have been modeled on the basis of the Hawaii hot spot [e.g., *Watson and McKenzie* 1991; *Ribe and Christensen,* 1999]. According to the standard model of mantle plumes [e.g., *Davies,* 1990; *White and McKenzie,* 1995; *Campbell,* 1998], it is generally assumed that plumes originate either from the bottom of the mantle (CMB; core-mantle boundary) or the base of the upper mantle (600-700 km depth). Most previous workers believe that potential temperatures of plumes are significantly higher than shallow asthenosphere (>300°C). Accordingly, the upwelling flow exceeds the dry mantle solidus at some depth and partial melting of peridotite produces hot spot magma.

Recently, the significance of recycled oceanic crust (eclogite) component in generation of MORB and hot spot magmas has received much attention [e.g., *Hauri,* 1996;

Hawaiian Volcanoes: Deep Underwater Perspectives
Geophysical Monograph 128
Copyright 2002 by the American Geophysical Union

Table 1. Composition of the starting materials and some modern basalts

	CLG46	CRB72-31	NMORB	Iceland
SiO_2	50.77	50.76	50.55	47.74
TiO_2	1.11	1.48	1.31	1.59
Al_2O_3	15.15	16.02	16.38	15.12
FeO*	12.25	9.52	8.90	11.82
MnO	0.22	0.14	0.16	0.2
MgO	8.07	8.23	7.80	8.99
CaO	9.59	10.77	11.62	11.61
Na_2O	2.74	2.29	2.79	2.04
K_2O	0.01	0.39	0.09	0.19
Total	99.90	99.60	99.60	99.30

CLG46: Archean MORB collected at Pilbara Craton, Australia [Ohta et al., 1996].
CRB72-31: Columbia River Basalt, Asotin member [Takahashi et al., 1998].
NMORB: Normal MORB from the Mid Altantic Ridge (49-52N).[Schilling et al., 1983]
Iceland: [Schilling et al, 1983].

Hirschmann and Stolper, 1996; *Cordery et al.,* 1997; *Yasuda et al.,* 1998; *Campbell,* 1998; *Takahashi et al.,* 1998; *Lassiter and Hauri,* 1999; *Sobolev et al.,* 2000]. The mechanism of magma production in the hybrid partial melt zone (consisting of eclogite and peridotite), however, is not well understood. The role of the interface that separates these lithologically different domains in the source region is an essential but unknown factor in plume magma genesis.

The goal of this study is to model magma generation in the Hawaiian plume based on new experiments and the growth history of Koolau volcano [*Shinozaki et al,* this volume, Figure 8a]. Two sets of experiments were carried out: 1) melting of basalt/peridotite sandwich and 2) melting of Fe-rich Archaean tholeiite with or without peridotite matrix. The first sets of experiments were conducted to study the reaction kinetics of the basalt/peridotite interface and this was followed by a set of experiments to study the chemistry of melts produced by melting layered basalt and peridotite (sandwich melting experiment). Abrupt changes were observed in the experimental melt compositions as well as the degree of partial melting of the system when solidus of the host peridotite is reached. The abrupt changes found in our experiments give clues to understand the abrupt changes in magma compositions during the growth history of Koolau volcano [*Shinozaki et al.*, this volume; *Tanaka et al.*, this volume]. A model is proposed on melting process in the Hawaiian plume consisting of peridotite and entrained eclogite. Compared with previous models [e.g., *Watson and McKenzie,* 1991; *Frey and Rhodes,* 1993; *Hauri,* 1996; *Ribe and Christensen,* 1999], our model advocates lower plume temperatures and large blocks of entrained eclogite.

2. EXPERIMENTAL PROCEDURES

2.1 Starting Material

In the previous experiments, we used a modern NMORB (Normal Mid Ocean Ridge Basalt)-like starting material (CRB72-31) to study the petrogenesis of the CRBs (Columbia River Basalts) [*Takahashi et al.,* 1998]. Considering the time scale for global mantle circulation (1.7-2.5 Ga, see Figure 7 of *Silver et al.* [1988]), the composition of the recycled old oceanic crust in the Hawaiian plume may be closer to Archaean greenstones than modern MORB. The Archaean basalts are notably enriched in FeO and depleted in Al_2O_3 relative to modern MORBs [*BVSP,* 1981]. We, therefore, selected Archaean basalt CLG46, collected from Pilbara craton, western Australia (3.1-3.3 Ga; *Ohta et al.* [1996]), as a starting material. The CLG46 is higher in FeO and lower in Al_2O_3 and CaO than modern NMORB. Both the NMORB like CRB72-31 and the Fe-rich CLG46 were embedded in a peridotite KLB-1, whose phase relations are well known [e.g., *Takahashi,* 1986; *Zhang and Herzberg,* 1994;]. The compositions of the two basalts and the peridotite are listed in Table 1.

The rock starting materials were powdered and homogenized in an agate mortar for one hour. Then the rock powders were dried in a one-atmosphere furnace at 1000°C un-

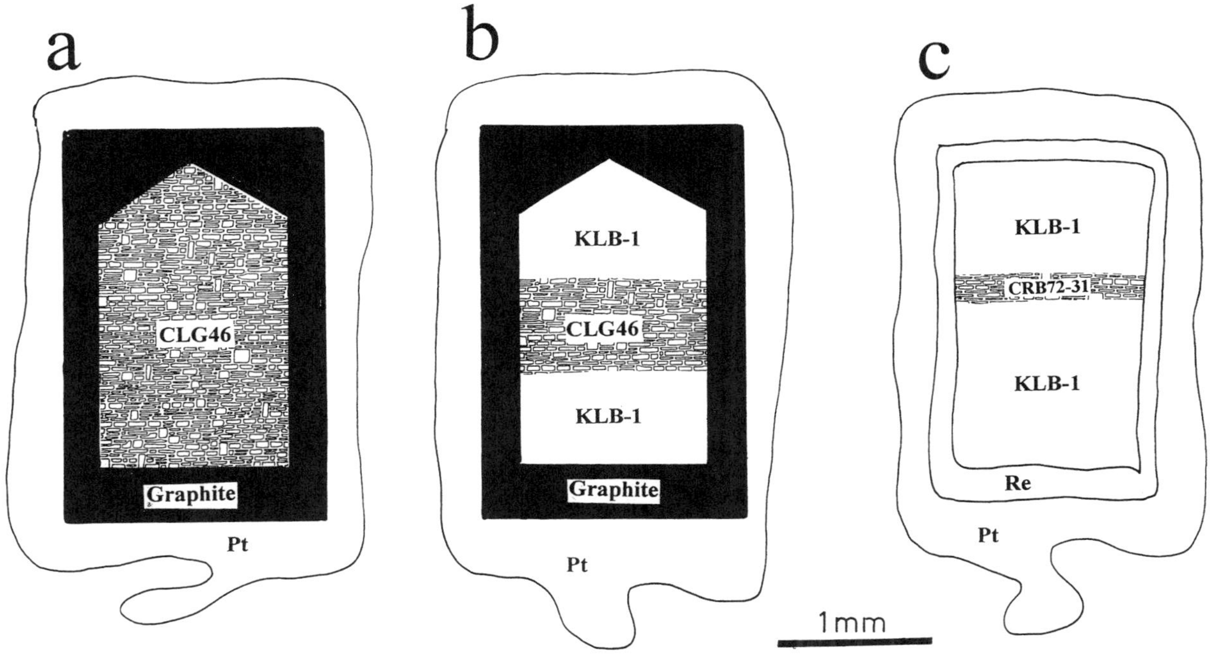

Figure 1. Experimental capsule configurations for basalt and basalt/peridotite starting materials. **a**: graphite/Pt double capsules were used for experiments for batch melting of Archaean basalt CLG46. **b**: sandwich melting of CLG46 with peridotite KLB-1. **c**: Re/Pt double capsules were used for sandwich partial melting of NMORB-like basalt CRB72-31 with peridotite KLB-1 with and without water.

der oxygen fugacity equivalent to NNO buffer by CO_2-H_2 gas mixture for several hours. Glass starting material with (1.0 wt% H_2O) or without water was prepared from the dry powder of CRB72-31 in a sealed Pt tube (3.0 mm outer diameter) at 1.5 GPa and 1200°C using a piston-cylinder apparatus. Hydrous peridotite (ca. 0.2 wt% H_2O) was prepared also by the piston-cylinder at 1.5GPa and 900°C in the stability field of amphibole.

2.2 *Experimental and Analytical Methods*

High-pressure experiments were carried out with a Kennedy-type (2.0-3.0 GPa) and a modified Boyd-England-type (1.0-1.5 GPa) piston-cylinder apparatuses. Experi-mental pressures were monitored with a digital pressure gauge and controlled to within 3% error by hydraulic valves. Temperatures were measured with W5Re-W26Re thermocouples and no correction was applied for pressure effect on thermocouple emfs (see Figure 3 of *Takahashi et al.,* [1993]).

Melting experiments were carried out using a talc-Pyrex cell with graphite heater and ALSIMAG spacer (see Figure 1 of *Takahashi* [1986]). Friction correction (-15%) was applied based on a calibration of quartz-coesite transition at 1000-1200°C and melting point of Au at 1-3 GPa and 1100-1200°C. Dry melting experiments were carried out with sealed graphite/Pt double capsules to avoid Fe-loss and inhibit access of moisture from the pressure medium [*Takahashi and Kushiro,* 1983]. Hydrous melting experiments were carried out using double capsules made of sealed Pt-tube and Re-inner capsule whose oxygen fugacity is estimated to be very close to the QFM buffer [*Matsukage et al.,* 2000]. Other experimental techniques are similar to those described in a previous study [*Takahashi et al,* 1998]. The crystalline phases in the run products were determined with an electron microprobe (JXA-8800 at Tokyo Inst. of Technology), using an accelerating voltage of 15 kV, specimen current of 12 nA and focused beam. The quenched melts were analyzed by EPMA with relatively large beam size (10μm) and more than 5 points were averaged. Counting times were 10-20 seconds for major elements (Si, Mg, Al, Ca, Fe and Na) and 30 seconds for minor elements (Ti, K, Mn, Cr and Ni).

Batch partial melting experiments were carried out for the Archaean basalt CLG46 using the graphite/Pt double capsule (Figure 1a). The basalt/peridotite experiments for CLG46 were carried out with the layered graphite/Pt capsule (Figure 1b). In the experiments using the MORB-like basalt CRB72-31, a thin wafer of basalt was embedded in the peridotite KLB-1 (Figure 1c). Both dry and wet starting materials were used in the CRB72-31/KLB-1 sandwich experiments at 1.5 and 3.0 GPa (see Table 2).

3.EXPERIMENTAL RESULTS

3.1 Phase Relations

The pressures and temperatures of the sandwich experiments (1200 and 1300°C at 1.5 GPa, and 1300 and 1400°C at 3 GPa) were chosen to study the chemical reactions between the eclogite and peridotite domains with and without the presence of partial melts (see open squares Figure 2a). Because our experiments were above the stability limit of amphiboles, both peridotite and basalt layers should have included very small amount of partial melts in the hydrous experimental charges.

For the CLG46 starting composition, clinopyroxene (cpx) appears in broad pressure and temperature range and is the only liquidus phase from 1.5 to 3 GPa [Figure 2b]. Garnet crystallizes after cpx at high-pressures (>1.5 GPa). Although liquidus phase relationships for cpx and garnet are very similar between CLG46 and CRB72-31, the stability field of plagioclase of the former is much narrower than the latter due to its lower CaO and Al_2O_3 content.

3.2 Textures of run Products

The basalt/peridotite boundaries [Figure 3 a,b,c] are coated with a reaction zone of orthopyroxene (opx), which was also reported by *Yaxley and Green* [1998]. The opx bands (30 to 50 μm in Figure 3c) were formed by the chemical reaction between silica-rich partial melts of eclogite and olivine in the peridotite matrix. At 3 GPa, opx reaction band was formed in dry experiments both at 1300 and 1400°C (SW-19, SW-21) and in wet experiments at 1300°C (SW-20) but was absent in a wet experiment at 1400°C (SW-22). The opx reaction band may have formed at the beginning of experiments near the peridotite solidus (cf. Figure 3b with Figure 3a,c). Due to the enhanced chemical reaction rate of partially molten peridotite matrix, melt in the basalt layer changed its composition and the opx reaction band diminished.

In the experiments at subsolidus temperature of peridotite, the thickness of the opx reaction band is positively related with the run temperatures and durations. The basalt peridotite boundary consists of a nearly monomineralic opx band, and their constituent minerals are more refractory (having higher Mg#) than those in the basalt layer.

3.3 Composition of Solid Phase

Significant compositional differences were observed between cpx and garnet in the basalt layer and in the peridotite layer in the dry melting experiments after 24 hours (SW-19 and SW-21 in Figure 4, see also solid compositions listed in Appendix Table 1 in CD-ROM). On the other hand, solid phase compositions are almost identical between the layers in the hydrous melting experiment at 3 GPa and 1400°C (SW-22 in Figure 4).

In order to examine the chemical reaction rate between the mafic layer and the peridotite layer, clinopyroxene crystals (cpx) were chosen both from peridotite and the basalt layer (CRB72-31, Figure 1c) and their Mg# = 100Mg/(Mg+Fe) and Al_2O_3 wt% were determined (Figure 4). At 3 GPa and 1300°C under dry condition, both basalt and peridotite remains subsolidus (see Figure 2a) and reactions proceed only by diffusions in solids. In the basalt domain where cpx coexists with garnet, cpx Al_2O_3 content is high but variable (up to 15 wt% see Figure 4a). Rapid crystallization from the starting material (basalt fused glass) enhances the heterogeneity of cpx alumina content in the basalt layer (e.g. SW-19 in Figure 4). On the other hand, Al_2O_3 content of peridotite matrix cpx is low (5-6 wt%) due to the presence of orthopyroxene and garnet [*Boyd,* 1970]. At 3 GPa and 1300°C, the chemical equilibrium between the basalt and peridotite layers was not achieve after 24 hrs even in the presence of a small amount of water (0.2 wt% in peridotite and 1 wt% in basalt in the starting material, see SW-20 in Figure 4).

At 3 GPa and 1400°C under dry condition, the basalt layer is extensively molten whereas peridotite layer remains under subsolidus (Figure 2a). In the dry starting material, molten basalt layer is separated from the peridotite by thin reaction bands of orthopyroxene (similar to Figure 3c) and therefore the chemical reactions between the basalt and peridotite layer proceeded only by solid diffusion across the boundary. Due to the sluggishness of solid diffusion, clinopyroxenes in basalt layer is compositionally isolated from the peridotite matrix during this experiment (SW-21 in Figure 4). Only at 3 GPa and 1400°C under hydrous condition, does substantial melting occur both in the basalt and peridotite layers (SW-22 in Figure 4). Extensive chemical mass transport has occurred for Fe, Mg, Si, Al, Ti and K by chemical diffusion and liquid percolation allowing clinopyroxene compositions in basalt and peridotite layers to reach chemical equilibrium.

3.4 Degree of Partial Melting in CLG46

The most striking observation in our experiments is the complex behavior in degree of partial melting (DPM) of the

Table 2. Experimental run conditions.

Run No.	Starting Material	Capsule	P. (GPa)	Temp (deg C)	time (hrs.)	Phase present
BM-1	CLG46	Gr/Pt	1.00	1100	26	cpx,opx, pl, melt
BM-2	CLG46	Gr/Pt	1.00	1150	26	cpx, opx, melt
BM-3	CLG46	Gr/Pt	1.00	1200	22	cpx, opx, melt
BM-4	CLG46	Gr/Pt	1.00	1250	20	opx, melt
BM-5	CLG46	Gr/Pt	1.50	1325	20	melt
BM-6	CLG46	Gr/Pt	2.00	1275	24	cpx, gar, melt
BM-7	CLG46	Gr/Pt	2.00	1300	20	cpx, gar, melt
BM-8	CLG46	Gr/Pt	2.00	1325	20	cpx, melt
BM-9	CLG46	Gr/Pt	2.00	1350	24	melt
BM-10	CLG46	Gr/Pt	2.10	1300	24	cpx, gar, melt
BM-11	CLG46	Gr/Pt	2.25	1325	22	cpx, gar, melt
BM-12	CLG46	Gr/Pt	2.25	1350	20	cpx, gar, melt
BM-13	CLG46	Gr/Pt	2.25	1375	22	cpx, melt
BM-14	CLG46	Gr/Pt	2.25	1425	20	melt
BM-15	CLG46	Gr/Pt	2.70	1400	22	cpx, gar, melt
BM-16	CLG46	Gr/Pt	2.70	1425	20	cpx, melt
BM-17	CLG46	Gr/Pt	2.70	1450	24	cpx, melt
BM-18	CLG46	Gr/Pt	2.70	1475	20	melt
BM-19	CLG46	Gr/Pt	3.00	1425	10	cpx,gar,melt
BM-20	CLG46	Gr/Pt	3.00	1450	10	cpx,gar,melt
BM-21	CLG46	Gr/Pt	3.00	1475	5	cpx,melt
BM-22	CLG46	Gr/Pt	3.00	1500	2	melt
SW-1	CLG46/KLB-1	Gr/Pt	2.00	1300	4	
SW-2	CLG46/KLB-1	Gr/Pt	2.00	1325	20	
SW-3	CLG46/KLB-1	Gr/Pt	2.00	1350	24	
SW-4	CLG46/KLB-1	Gr/Pt	2.70	1400	4	
SW-5	CLG46/KLB-1	Gr/Pt	2.70	1400	24	
SW-6	CLG46/KLB-1	Gr/Pt	2.70	1400	100	
SW-7	CLG46/KLB-1	Gr/Pt	2.70	1425	20	
SW-8	CLG46/KLB-1	Gr/Pt	2.70	1450	20	
SW-9	CLG46/KLB-1	Gr/Pt	2.70	1475	16	
SW-10	CLG46/KLB-1	Gr/Pt	3.00	1425	6.5	
SW-11	CLG46/KLB-1	Gr/Pt	3.00	1450	5.5	
SW-12	CLG46/KLB-1	Gr/Pt	3.00	1475	5	
SW-13	CLG46/KLB-1	Gr/Pt	3.00	1500	5	
SW-14	CLG46/KLB-1	Gr/Pt	3.00	1525	4	
SW-15	CRB72-31+KLB-1	Re/Pt	1.50	1200	24	
SW-16	CRB72-31*/KLB-1*	Re/Pt	1.50	1200	24	
SW-17	CRB72-31+KLB-1	Re/Pt	1.50	1300	23	
SW-18	CRB72-31*/KLB-1*	Re/Pt	1.50	1300	23	
SW-19	CRB72-31+KLB-1	Re/Pt	3.00	1300	23	
SW-20	CRB72-31*/KLB-1*	Re/Pt	3.00	1300	23	
SW-21	CRB72-31+KLB-1	Re/Pt	3.00	1400	25	
SW-22	CRB72-31*/KLB-1*	Re/Pt	3.00	1400	25	

CRB72-31*: contains 1.0 wt% H_2O, KLB-1*: contains 0.2 wt% H_2O

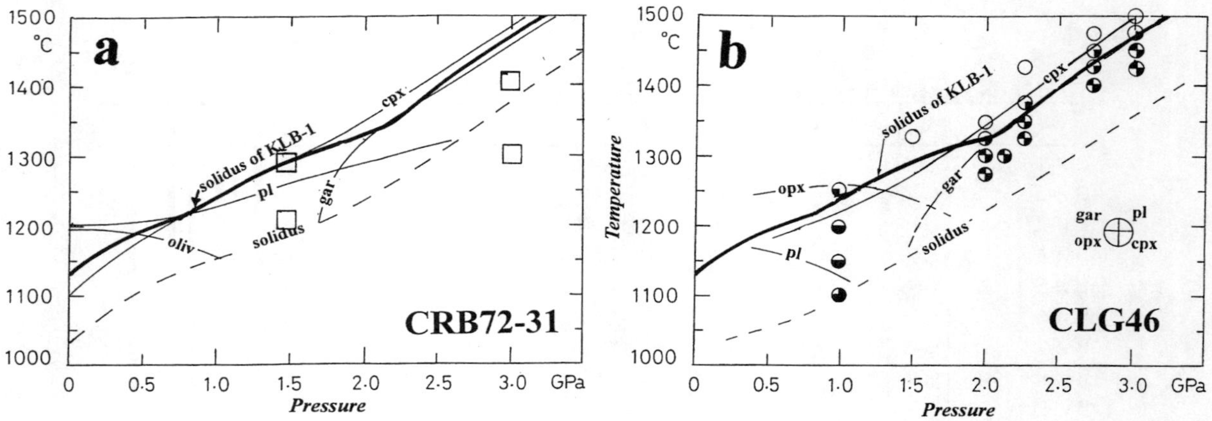

Figure 2. Melting phase relation for two basalts. **a**: NMORB-like basalt CRB-72-31 (experimental data for CRB72-31 from *Takahashi et al.* [1998]), **b**: Fe-rich Archaean basalt (see Fig.1a and Table 2). The dry solidus of peridotite KLB-1 (solid line in both **a** and **b**) was taken from *Takahashi* [1986]. Open squares in **a** indicate experimental conditions where basalt/peridotite experiments for CRB72-31/KLB-1 pairs (Fig. 1c, Fig. 4) were conducted. Filled sectors in b denote presence of crystal phases in the run products.

basalt starting material in the basalt/peridotite sandwich experiments. In a strict sense, DPM cannot be defined for the hybrid starting material with heterogeneous chemical composition between the layers. We have determined the volume fraction of the melt only within the basalt layer of CLG46 (Figure 1b) using BEIs (back-scattered electron images) and plotted them against temperature in Figure 5.

As expected, DPM increases both at 2 and 2.7 GPa for batch melting of CLG46. The DPM in the basalt/peridotite sandwich experiments also increases up to the dry solidus of the ambient peridotite but decreases dramatically when peridotite solidus temperature is reached. This huge drop in DPM must be due to the following factors; 1) migration of the partial melt from the basalt domain to the peridotite grain boundary, 2) change in Mg/Fe ratio of constituent phases in the basalt layer (such as SW-22 in Figure 4), 3) upward shift of the melting temperature of the basalt layer. As shown in the next section, melt composition in the basalt domain changes drastically across the narrow temperature zone from high-silica tholeiite (below peridotite solidus) to picritic at temperatures above the peridotite solidus (see Figure 6).

In order to test the reaction kinetics that controls DPM in basalt layer, additional experiments were carried out at 3.0 GPa with shorter run durations (4-6 hrs, SW-10 to SW-14) with the graphite capsule and thick basalt layer (Figure 1b). Similar to the experiment conducted at lower temperatures with thin basalt wafer (SW-19, SW-21), the basalt melt is isolated from the peridotite layer by the orthopyroxene reaction band at 1425°C (Figure 3c).

The run product formed at temperatures >50°C above the dry solidus of peridotite (SW-13, SW-14) consists of fully molten former basalt layer and partially molten peridotite (e.g. Figure 3a). As found previously [e.g., *Takahashi and Kushiro*, 1983], the picritic melt in the former basalt domain established chemical equilibrium with ambient peridotite even in the short runs. The melt composition at 1525°C (see Table 3) is very similar to the partial melt of the peridotite KLB-1 determined by the diamond aggregate technique [*Hirose and Kushiro*, 1993].

At the temperature just on the solidus of peridotite (1475°C), the orthopyroxene reaction band is developed and basalt peridotite interface is covered by very coarse clinopyroxene crystals growing inward from the interface (Figure 3b). This clinopyroxene is significantly higher in Mg# compared with clinopyroxene crystallized from the basalt at the same pressure and temperature (Figure 3, also see Appendix Table 1). This run product records the isothermal freezing kinetics of the basalt layer that takes place only near the peridotite solidus (see dashed lines with arrows in Figure 5).

3.5 Composition of Partial Melts

Melts formed at pressures higher than 2.0 GPa contain variable amount of quench crystals. MgO-rich melts formed at 3.0 GPa in the basalt layer of the sandwich experiments contain aggregates of dendrites when quenched (Fig. 3a, b). A broad electron beam (10-20 micron meter beam size) was employed to analyze these dendrites. At

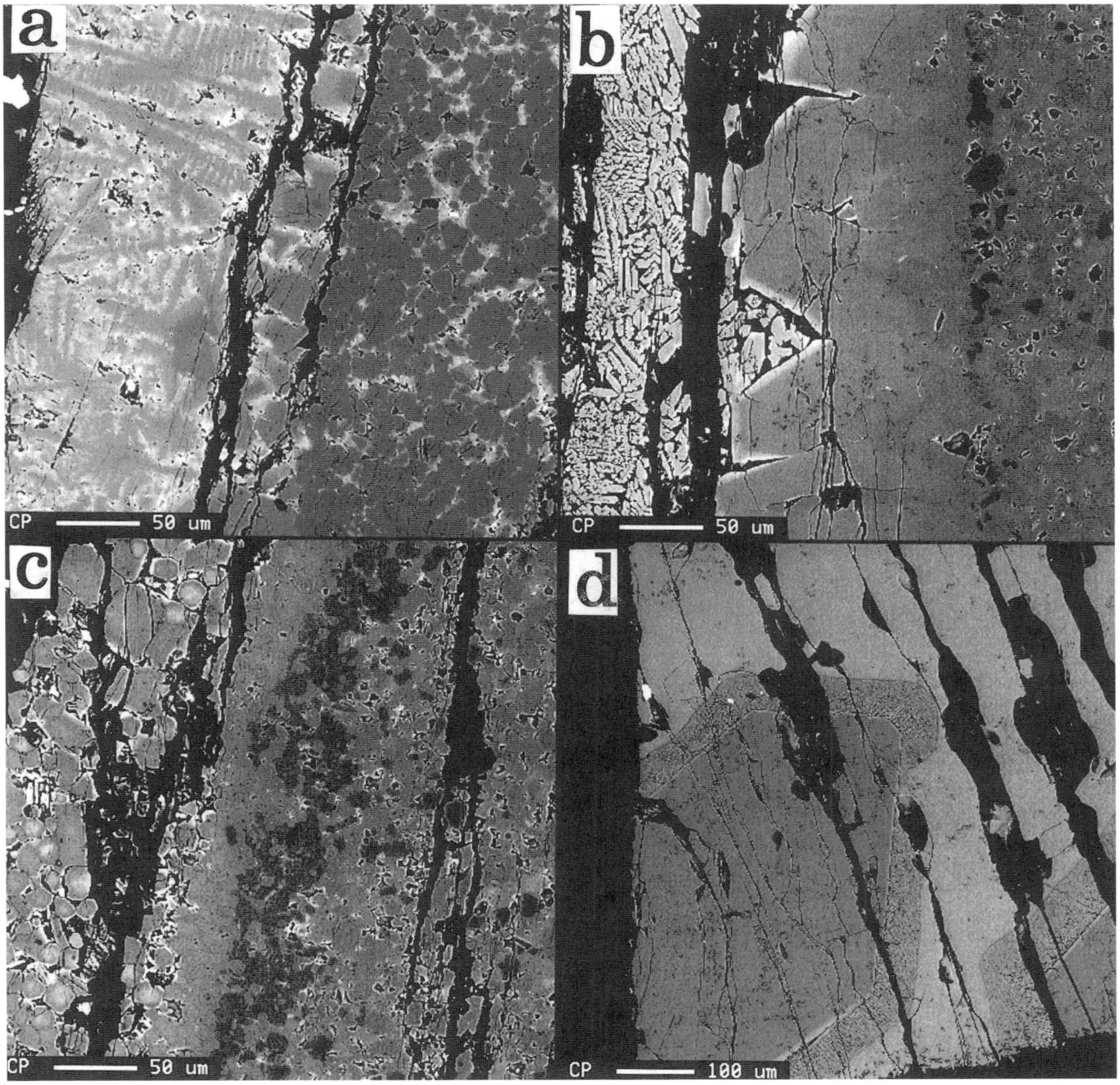

Figure 3. Back-scattered electron images of experimental run products. **a**: basalt-peridotite interface of SW-14 (3 GPa, 1525°C), **b**: basalt-peridotite interface of SW-12 (3GPa, 1475°C), **c**: basalt-peridotite interface of SW-10 (3GPa, 1525°C), **d:** large clinopyroxene mantled with quench crystals formed in a basalt melting experiment (BM-21, 3GPa, 1475°C). The variations in reaction kinetics at the basalt/peridotite interface at different temperatures are shown. **a**: basalt layer was fully molten and crystallized as dendritic quench crystals. **b**: large clinopyroxene crystallized inward from the original interface (thin orthopyroxene reaction band) to basalt layer. **c**: basalt melt is isolated from subsolidus peridotite by orthopyroxene reaction band (dark layer in **c**). **d:** large euhedral clinopyroxenes up to 0.5 mm were formed near the bottom of graphite capsule. Note **b** and **d** are run products obtained at the same pressure and temperature but with different starting materials (**b**: basalt/peridotite sandwich, **d**: basalt). The crystallization of the basalt layer shown in **b** is due to the enhanced chemical reaction of peridotite layer on the dry solidus. The basalt layer crystallizes nearly completely after long run durations (24-100 hrs, see Figure 5).

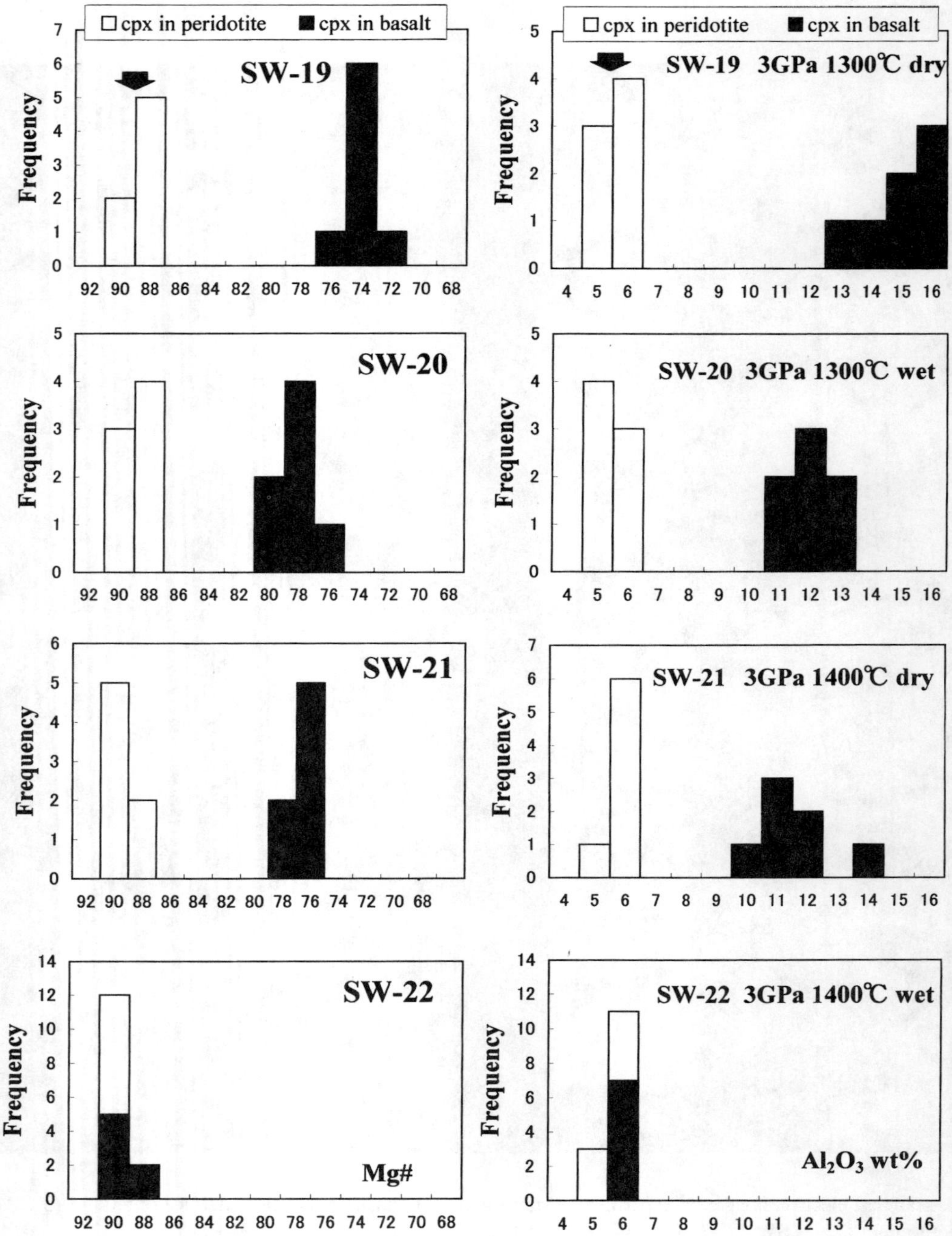

Figure 4. Compositional histogram of Mg# (=100Mg/(Mg+Fe) in mole fraction) and Al_2O_3 content of cpx in the CRB72-31/KLB sandwich experiments (Fig. 1c). Seven cpx crystals were chosen from the peridotite (open squares) and from the embedded basalt layer (solid circles). In most of the charges, chemical reaction between the peridotite matrix and the embedded basalt is limited and the pyroxene compositions are different between the two layers. Starting compositions for peridotite clinopyroxene are shown with arrows but those for basalt are undefined because fused basalt powder was used for these experiments. Only in the hydrous experiment near the peridotite dry solidus (SW-22), has chemical equilibrium between the basalt and peridotite (involving solid solutions in cpx) been established throughout the run product (2 mm in length, see Figure 1c) during the 25 hr experiment.

least three analyses were averaged to obtain melt compositions (Table 3).

In the basalt melting experiments, melts scattered in solid matrix were easily modified by rapid crystallization during quenching. Only the melts which satisfy the following criteria are shown in Table 3; $K_d(Fe/Mg)^{(oliv/melt)}$=0.30-0.36 or $K_d(Fe/Mg)^{(cpx/melt)}$=0.27-0.33 [*Takahashi and Kushiro,* 1983]. Glass compositions for CLG46 listed in Table 3 may be still different from those of the equilibrium partial melts due to the quench crystallization. Melt pockets formed in the basalt/peridotite thin wafer experiments (SW-15 to SW-22 in Table 2) were all rejected by this criteria. Solid phase compositions are listed in Appendix Table 1 in the CD-ROM.

Compositions of melts formed by melting of CLG46 (open squares in Figure 6) are not significantly different from those in basalt/peridotite reaction melting experiments at temperatures below the solidus of host peridotite. However, melt compositions are markedly different between the two sets of experiments at temperatures above the peridotite solidus (see tie lines in Figure 6). Melts formed at basalt/peridotite sandwich experiments are systematically higher in MgO and FeO but lower in SiO_2 and Al_2O_3 compared with the batch melt of CLG46 at the same pressure and temperatures. These differences can be understood from phase equilibrium observations. The sandwich experiment melts are saturated with olivine and opx as well as cpx and garnet whereas the melts in the basalt-only experiments coexist with only cpx and garnet (see Fig. 3 of *Yaxley and Green* [1998]).

4. DISCUSSION

4.1 Generation of Two Types of Primary Magmas in the Hawaiian Plume

The most important discovery from our melting study is that we demonstrated formation of two contrasting primary magmas in the melting of basalt/peridotite hybrid source. The first melt is a silica-rich tholeiite magma (SiO_2=53-54 wt%, MgO=6-7 wt%) similar to those in the late stage of Koolau volcano [*Frey et al.,* 1994] and the second is the picritic magma (SiO_2= 47-49, MgO=15-20 wt%), which has been considered to be primary in most Hawaiian shield volcanoes [e.g., *Norman and Garcia,* 1999; *Garcia,* this volume]. Based on petrologic study of 250 rock samples from Koolau volcano, *Takahashi and Takeguchi* [1998] concluded that nearly all olivines in studied picrite specimens are xenocrysts and they proposed that the silica-rich tholeiite is the primary magma in Koolau volcano.

The silica-rich tholeiite magma can be produced only in the relatively cool (or deeper) region of the plume where ambient peridotite remains subsolidus yet entrained eclogite blocks are substantially molten. Because the solidus of eclogite is significantly lower than that of peridotite (Figure 5) and the magma produced inside the eclogite pocket is effectively isolated with subsolidus peridotite by the opx reaction band (Figure 4c), the DPM in the eclogite pocket could become very high (theoretically up to 100%) before peridotite starts partial melting. *Yasuda et al.* [1998] and *Campbell* [1998] also predicted the generation of high-degree partial melts in eclogite pockets in the ascending plume.

In the peridotite-melting zone of the plume, presence of mafic material has only limited effect on major element chemistry of magma (such as Mg#), because its eutectitic relationship with the residual solids (olivine + orthopyroxene + clinopyroxene ± garnet) controls the melt chemistry. Addition of up to 50-wt% of MORB into peridotite does not increase SiO_2 of partial melt in the hybrid source [*Kogiso et al.,* 1998]. The systematic change of SiO_2 content in tholeiitic magmas between the Hawaiian shield volcanoes [*Frey and Rhodes,* 1993], therefore, is difficult to explain by simple addition of an eclogite component to the source. Moreover, the eclogite domain in the plume must be chemically isolated to produce primary high-silica magma

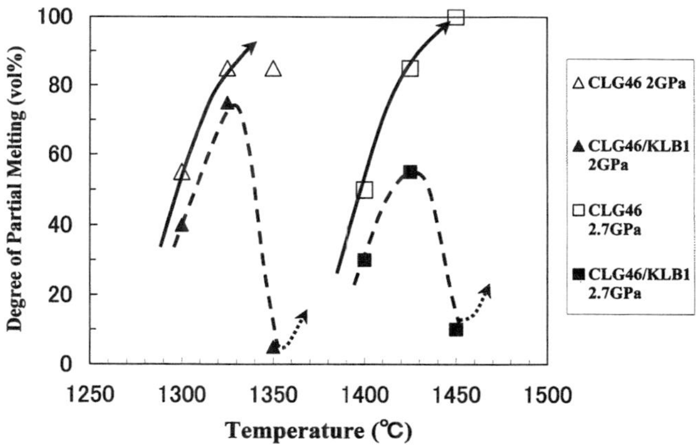

Figure 5. Degree of partial melting (DPM) of Archaean basalt CLG46 as a function temperature in experiments at 2.0 and 2.7 GPa. Open symbols indicate DPM in basalt melting experiments (Fig.1a) and solid symbols those of basalt layer in the basalt/peridotite sandwich melting experiments (Fig.1b). The DPM was determined from BEIs and X-ray maps for some elements similar to those in Figure 3. Because the DPM was determined only at a cross section of the charge, it has a large uncertainty (reproducibility is less than 10% but true uncertainty is difficult to determine). It is very important to note, however, that DPM in the sandwich experiment shows a sharp drop as the temperature approaches the peridotite dry solidus (shown as arrows). See text for explanation.

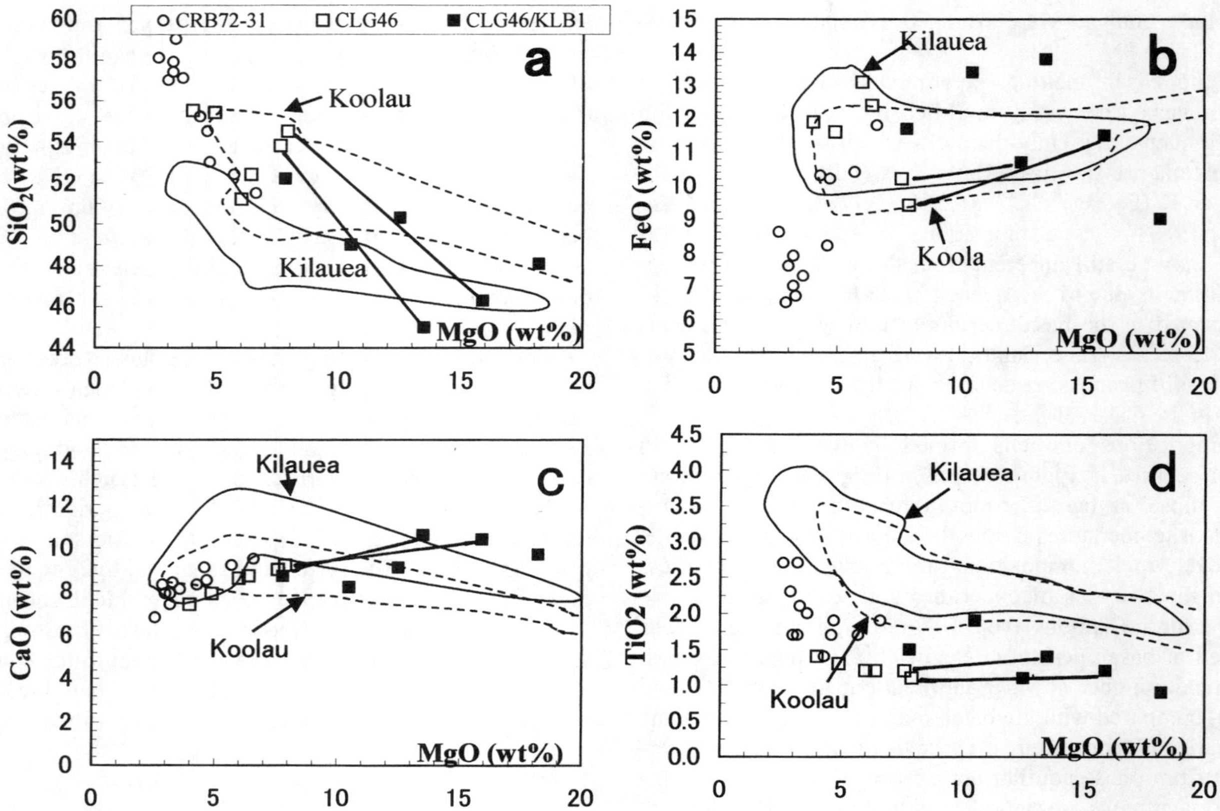

Figure 6. Composition of partial melts formed by partial melting of Fe-rich Archaean basalt CLG46 (open squares), those of NMORB-like CRB72-31 (open circles), and those in the basalt/peridotite sandwich experiments (closed squares). Data for CRB72-31 were taken from *Takahashi et al* [1998]. Compositional fields for tholeiite lavas of Koolau and Kilauea volcanoes [*Frey and Rhodes,* 1993; *Wolfe and Morris,* 1996] are also shown. Melts formed after 50% partial melting of CLG46 at 2.7 GPa are very close to the estimated primary magma for Koolau volcano (closed stars) by *Takahashi and Takeguchi* [1998] except TiO_2 and Al_2O_3 that are sensitive to phase equilibria and the composition of the starting material. See text for explanation.

for Koolau volcano. Mechanisms to allow compositional variation of tholeiite magma in Hawaiian shield volcanoes are discussed below.

4.2 Sharp Transition in the Growth History of Koolau Volcano and its Possible Origin

In Koolau volcano, the transition from main shield stage (with Mauna Loa like chemistry) to the final Makapuu stage (Koolau type silica-rich tholeiite) was found at least in three localities: North exit of H3 tunnel; *Jackson et al.* [1999]; west of Nuuanu Pali, and Puu Manamana hiking trail; *Shinozaki et al.* [this volume] and *Tanaka et al.,* [this volume]. On the west side of Koolau volcano, *Haskins and Garcia* [2000] found this transition at Kalihi valley, Honolulu by scientific drilling. According to *Shinozaki et al.* [this volume] and *Tanaka et al.* [this volume], the west of Nuuanu Pali transition is very sharp and abrupt. Over less than 10 m, the Sr, Nd and Pb isotopes changes significantly compared to the nearly constant isotopes for the entire Makapuu stage (see Figure 2 of *Tanaka et al.* [this volume]). Sharp change in both major element and isotopes between the main shield stage and the Makapuu stage and the relatively homogeneous chemistry of the lavas throughout the Makapuu stage indicate that some irreversible change had occurred in the melting region of the Hawaiian plume (>2.7 Ma ago, according to *Doell and Dalrymple* [1973]). Once the change occurred, the Koolau volcano produced another ~1000 km^3 of magma before it ceased.

Although the mechanism that caused the transition is unknown, we propose that the sharp transition corresponds to the onset of a rapid melt extraction from an eclogite block in the melting region of the Hawaiian plume. This is be-

Table 3 Melt compositions determined by EPMA.. Average of more than 3 point analysis for the basalt/peridotie sandwich experiment.

P. (GPa)	2.00	2.00	2.00	2.70	2.70	2.70	2.70	2.70	2.70	3.00	3.00	3.00
T. (□ C)	1300	1325	1350	1400	1425	1450	1425	1450	1400	1475	1500	1525
DPM vol%	55%	85	85	50	85	100	55	10	30	40	80	100
Starting Material	CLG46	CLG46	CLG46	CLG46	CLG46	CLG46	CLG46 KLB-1	CLG46 KLB-1	CLG46 KLB-1	CLG46 KLB-1	CLG46 KLB-1	CLG46 KLB-1
SiO_2	55.45	52.37	51.24	55.37	53.78	54.49	45.00	46.30	49.00	52.19	50.25	48.09
TiO_2	1.43	1.15	1.17	1.30	1.16	1.12	1.40	1.20	1.85	1.50	1.11	0.90
Al_2O_3	16.05	16.15	16.79	15.81	15.67	15.39	13.70	12.70	14.30	14.87	14.06	12.53
FeO*	11.92	12.44	13.10	11.61	10.22	9.42	13.80	11.50	13.40	11.68	10.71	9.02
MgO	4.02	6.42	5.95	4.88	7.62	7.85	13.50	15.90	10.50	7.81	12.47	18.15
CaO	7.37	8.68	8.58	7.89	8.98	9.17	10.58	10.38	8.18	8.66	9.11	9.72
Na_2O	3.74	2.76	3.15	3.13	2.55	2.55	2.00	2.00	2.75	3.15	2.24	1.57
K_2O	0.02	0.02	0.02	0.02	0.02	0.01	0.02	0.02	0.02	0.14	0.05	0.02
Total	100.00	100.00	100.00	100.00	100.00	100.00	100.00	100.00	100.00	100.00	100.00	100.00

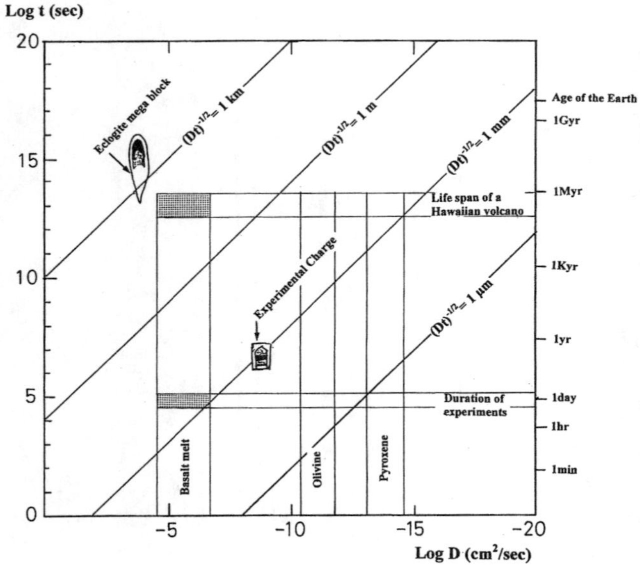

Figure 7. Comparison of diffusion distances that can be achieved in experimental charges (Figure 1) and in the melting zone of the Hawaiian plume. Diffusion coefficients for basalt melt, olivine and pyroxene are from *Takahashi and Kushiro* [1983]. Chemical isolation of molten basalt layer from peridotite (Figure 4) can be understood by sluggish chemical diffusions thorough pyroxene reaction bands (Figure 3d). Achievement of chemical equilibria in the 2-mm long, hydrous charge (SW-22 in Figure 4) can be explained by fast chemical diffusions in basalt melt (lower shaded box). During the lifetime of a Hawaiian shield volcano (0.5-1.0Ma, *Langenheim and Clague* [1987] or 1.0-1.4 Ma by *Guillow et al.* [1997]), compositional heterogeneity larger than several 100 m in thickness cannot be homogenized by chemical diffusion process (upper shaded box). Thus, large eclogite blocks entrained in the plume may hold as lithologically distinct domain and yield silica-rich (olivine under saturated) magmas.

cause the silica-rich tholeiite magma dominant at the late stage of Koolau volcano (SiO_2=53, MgO=7 wt%) could be produced by almost pure batch partial melting of recycled oceanic crust (Figure 6). Secondly, the isotope signatures of the Makapuu stage Koolau lavas indicate large contribution of eclogite derived melt component in the source [e.g., *Hauri,* 1996; *Lassiter and Hauri,* 1998; *Halliday,* 1999].

The Makapuu stage Koolau lavas show coherent chemistry both in major and minor elements and the straight mixing lines for all measured isotopes (see solid circles in Figures 3 to 7 of *Tanaka et al.* [this volume]). This is best explained if they were derived from a single eclogite block in the plume. The eclogite partial melt reaches the high-silica tholeiite composition at DPM>50 % (see Figure 6). In order to yield the total volume of Makapuu stage magma (~10^3 km^3 see Figure 8a of *Shinozaki et al.* [this volume]) from a single eclogite block at DPM=50%, its volume should be of the order of 10^3 km^3. Accordingly we propose that entrained eclogite blocks in the Hawaiian plume [*Hauri,* 1996] would be variable but may reach up to 10 km in size or larger. Characterization of the Makapuu stage primary magma in major, minor and isotopic compositions and their genesis by partial melting of recycled oceanic crust will be discussed in a forthcoming paper [*Takeguchi and Takahashi,* in preparation].

4.3 Melting Process in the Hawaiian Plume

Melting in the basalt/peridotite hybrid source is a kinetic process and chemical equilibrium may not be established on a large scale. The reaction rate between the basalt and peridotite layers depends on many factors besides temperature and pressure; i.e., presence or absence of partial melt in both layers, thickness of the basalt layer and its spacing. Accordingly, the scaling of the experimental results is critically important (Figure 7).

Size of our experimental charges (~1 mm) and that of large eclogite blocks in plume may differ by a factor of 10^6. Diffusion distances of some elements are shown against diffusion coefficients and time (Figure 7). In experimental runs at temperatures above the peridotite solidus, chemical equilibration between the basalt and the peridotite domain was established within one day (e.g., SW-22 in Figure 4). Because expected diffusion distance in solids is of the order of 1 μm for major elements, this chemical reaction rate is explained only by chemical diffusion in basalt melt (shaded area in Figure 7).

In a large-scale plume head (Figure 8), it is expected that melts in the eclogite domain may be difficult to homogenize with ambient peridotite partial melt even in a lifetime of a volcano if the size of the eclogite block is larger than 1 km (see Figure 7). Physical movement of the melts (percolation and liquid advection) may enhance the reaction rate and the necessary size for eclogite blocks to and the necessary size for eclogite blocks to keep chemical isolation might be even larger. On the other hand, if the melts in the eclogite blocks are isolated from the peridotite melts by reaction bands of orthopyroxene (which took place in our experiments below peridotite solidus; see Figure 3c), the reaction rate becomes much slower because the rate will be limited by solid diffusion through the reaction bands. In this case, melts in the eclogite domains may be distinct from the peridotite partial melt even for small isolated eclogite patches.

Magma produced near the axis of the plume head (Figure 8) may be mixtures of two types of melts 1) basaltic andesite melt formed by melting of eclogite (open symbols in Figure 6) and 2) picritic melts formed by the reactive melting of eclogite and peridotite (solid squares in Figure 6). If the eclogite blocks are large enough, or if the peri-

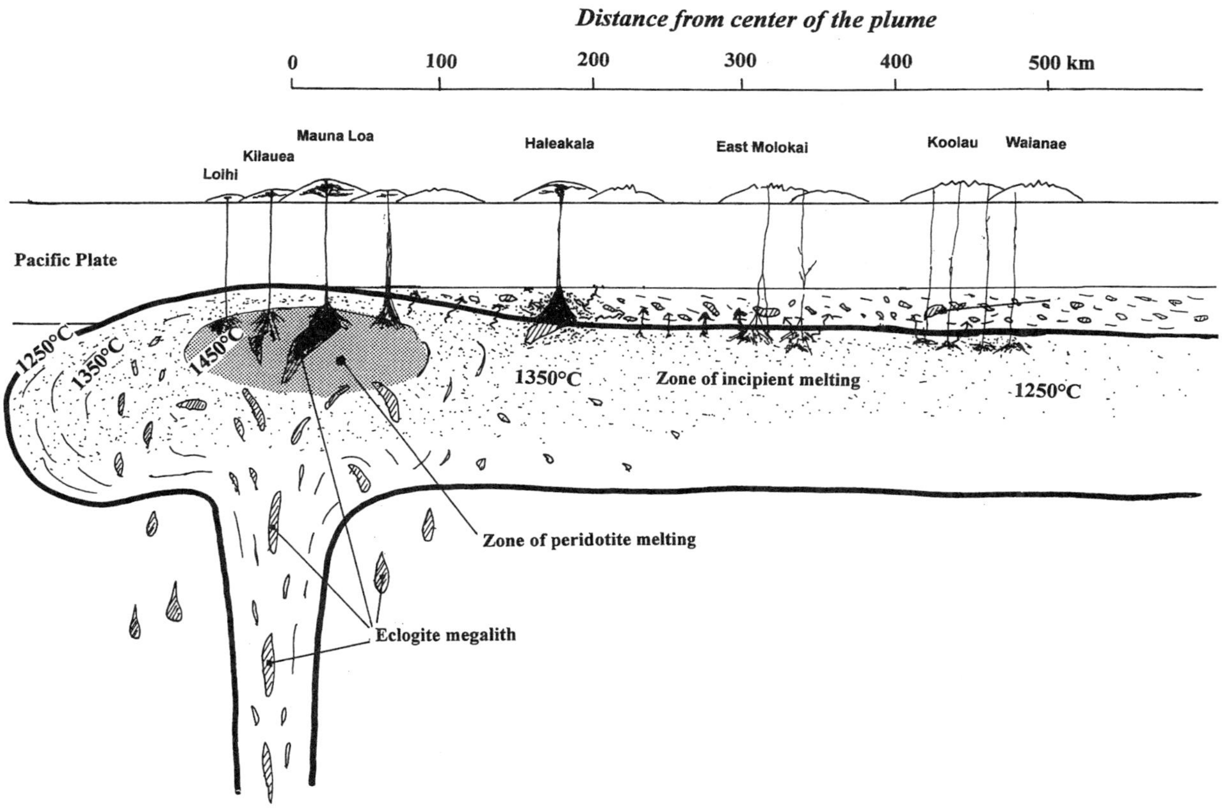

Figure 8. Schematic model of melting processes in the Hawaiian plume. The plume outline is drawn after *Ribe and Christensen* [1999] but internal plume temperature is assumed to be ca.150°C lower. Size of the entrained eclogite blocks may be up to 10^3km^3 (see text) and that of eclogite mega-blocks in the plume might determine the spacing of the shield volcano. Due to the low plume temperature, zone of peridotite partial melting is confined to the topmost part of the plume and much of the magma is supplied from the extensively molten eclogite mega-blocks.

dotite remained under subsolidus conditions, silica-rich partial melt from the eclogite blocks will be dominant in the plume head such as in the Makapuu stage of Koolau volcano.

If two types of magmas (picrite and silica-rich basaltic andesite) were formed separately and mixed subsequently, the blended magma may become readily olivine saturated in the peridotite partial melt zone. To explain the systematic difference in silica content of the Hawaiian tholeiite magma between the shield volcanoes [*Frey and Rhodes,* 1993; *Hauri,* 1996] and the occurrence of melt inclusions with diverse source rock signatures [*Sobolev et al,* 2000], the liquid blending process must take place in the region where the magmas have been isolated from their sources (conduit zone or magma chambers). In order to maintain chemical distinction of the magmas and to enable their blending processes, the presence of large (and perhaps vertically elongated streaks) eclogite blocks is needed (see Figure 8).

Recent high precision Pb-isotope studies on Hawaiian shield lavas show linear arrays within each shield volcano but not overlapping between the shields [*Abouchami et al, 1999*]. Moreover, *Abouchami et al.* [2000] reported that Pb-isotopes observed in HSDP cores (cross section of Mauna Loa and Mauna Kea) are segmented and they show linear arrays within each segment. This behavior of Pb-isotopes can be best explained if the eclogite blocks in the Hawaiian plume are large and the isotopic character of the tholeiite magma at given stage of a volcano is dominated by the melt component from the largest eclogite block. If a single eclogite block yielded the Makapuu stage lavas of Koolau volcano, its volume must be of the order of 10^3km^3.

4.4 Potential Temperature of the Hawaiian Plume

It is very important to determine the potential temperature of the Hawaiian plume because 1) Hawaii is the most active hot spot on the present earth; 2) many researchers believe that the Hawaiian plume may originate from the core-mantle boundary and therefore is very hot (e.g., *Williams et al.* [1998]; see also *Zhao et al.* [this volume]); 3) volcanic activity over the Hawaiian hot spot is among the best studied; 4) quantitative models have been proposed for the Hawaiian plume [e.g., *Watson and McKenzie*, 1991; *Ribe and Christensen*, 1999].

Previously, temperature of the Hawaiian plume was estimated to be as much as 1600°C in its melting zone [e.g., *Watson & McKenzie,* 1991; *Li et al.,* 2000]. In light of the new experiments, we propose that temperatures to yield Kilauea-type melting regimes at 2.7 GPa could be more than 150°C lower than previous estimates (1400-1450°C see Figures 5 and 6). We also showed that the temperature difference between the Koolau-type melting regime and the Kilauea-type melting regime is very small (of the order of 10°C see Figure 5). Although the growth history of Koolau volcano records melting episode in Hawaiian plume at 2.7-1.8 Ma [*Doell and Dalrymple,* 1973], we believe that our plume temperature is applicable to its present condition because there can be little fluctuation in temperature in a large plume over a short time period.

In the mantle melting due to adiabatic decompression [*McKenzie,* 1984], the isentropic P-T path of upwelling flow can be calculated by heat balance between the adiabat and the latent heat of melting. The peridotite must have a large excess temperature to yield magma in this model. For the Hawaiian plume with PMT=1560°C, melting would start at ~1700°C at about 5 GPa before arriving at 3 GPa and 1550°C, (path-A in Figure 9).

If the plume had experienced this high-temperature path, however, the eclogite blocks inside the plume must have undergone total melting at much elevated pressures (7-5 GPa) and should not have survived to low pressures (see path-A in Figure 9). Furthermore, silica-rich melts derived from the deep eclogite blocks may migrate upwards but will be blended with the picritic partial melts in the overlying peridotite partial melt zone. The silica-content of the peridotite melt, however, will not increase because coexisting solids buffer it [*Kogiso et al.,* 1998]. Thus, the silica-rich Koolau-type tholeiite magma cannot be produced under such high PMT. Distinct source rock chemical signatures found in melt inclusions in some Mauna Loa lavas [e.g. *Sobolev et al,* 2000] also may be difficult to produce in the high temperature plume. Therefore, we propose lower temperatures for the Hawaiian plume (path B in Figure 9, PMT=1400°C). Partial melting of peridotite takes place only at the uppermost part of the plume in our model (Figure 8).

The lower PMT for Hawaiian plume appears to be inconsistent with the common hypothesis that Hawaii must be the hottest because it is the most magma productive hot spot on the Earth. An excess temperature of ~300°C has been estimated for the Hawaiian plume from a seismic observation by *Li et al.* [2000]. However, the excess temperature in our model is not inconsistent with seismically observed excess temperatures for other plumes using extensive seismic arrays (i.e., 150°C for Iceland plume, *Shen et al.* [1998] and 100-200°C for the South Pacific Super Swell region, *Niu et al.* [2000]). Alternative to the high PMT, we propose that existence of large amount of entrained oceanic crust in the Hawaiian plume that explains its high magma productivity.

5. CONCLUSIONS

Silica-rich tholeiite magma (SiO_2=53, MgO=7 wt%) representing the final growth stage of Koolau (end member of the Loa trend volcano) can be produced by melting of recycled Fe-rich Archaean basalt (eclogite) at 2.5-3 GPa and 1400-1450°C. Kilauea-type picritic magma (Kea trend, SiO_2=48, MgO~15 wt%) with relatively high incompatible element concentrations (K, Ti, LREE) can be produced by interactive melting of eclogite and ambient peridotite at slightly above the solidus of the peridotite at 2.5-3 GPa and 1450-1500°C. Abrupt change in magma composition (ma-

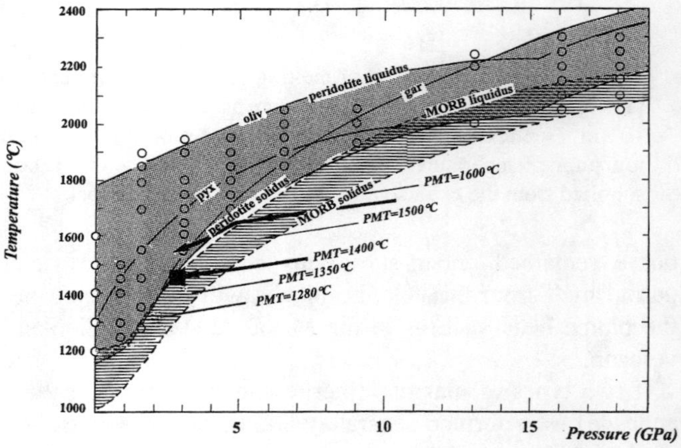

Figure 9. Melting phase diagram showing potential P-T paths for entrained eclogite in the Hawaiian plume. Liquidus and solidus for MORB eclogite are taken from *Yasuda et al*. [1994] and those for peridotite from *Takahashi et al.* [1993]. If the PMT for Hawaiian plume is high (e.g., 1560°C, *Watson and McKenzie* [1991]) melting of eclogite should starts as deep as 200 km. Chemical isolation of the eclogite and peridotite domain is difficult to maintain in this high temperature path. If silica-rich initial melts are supplied from the very deep eclogite pockets, magmas will be blended and become olivine saturated in the overlying extensive melting zone of peridotite. In order to preserve the unmixed magmatic components [*Halliday,* 1999; *Sobolev et al,* 2000], PMT should be 1400-1450°C. See text for explanations.

jor and trace elements including Sr, Nd, and Pd isotopes) observed at the final growth stage of Koolau volcano may correspond with the onset of melt extraction from a large (~10^3 km^3) eclogite block in the plume. The plume model with hybrid source may explain diversity in tholeiite magma types in Hawaiian volcanoes, provided that lithological distinction between the peridotite and entrained eclogite block is preserved until the final magma generation at the uppermost part of the plume. This requires potential temperature of the Hawaiian plume significantly lower than previous estimates (PMT=1400-1450°C).

Acknowledgements. This manuscript was written as a synthesis summarizing many thoughts on the melting processes in the Hawaiian plume emerged during the field study in Koolau volcano on Oahu (1996-1998) and the Nuuanu landslide in the north east of Oahu Island using the JAMSTEC research ships and submersibles (1998-1999). Special thanks are due to Mike Garcia who very kindly guided Hawaiian geology and petrology. Thoughtful reviews by Mark Hirschmann, Kevin Johnson and Mike Garcia improved the manuscript considerably. This work was supported by the grants 09304052 and 12002006 from the Ministry of Education, Japan.

REFERENCES

Abouchami, W., A. W. Hofmann, and S. J. G. Galer, Lead isotope anomaly of the Hawaiian plume, *EOS Transaction of American Geophysical Union, 80,No.46,* F1183, 1999.

Abouchami, W., S. J. G. Galer, and A. W. Hofmann, High precision lead isotope systematics of lavas from the Hawaii Scientific Drilling Project core, *Chem. Geol., 169,* 187-209, 2000.

Boyd, F. R., Garnet peridotite and the system $CaSiO_3$-$MgSiO_3$-Al_2O_3, *Mineral. Soc. Amer. Spec. Pap., 3,* 63-75, 1970.

BVSP, *Basaltic Volcanism on the Terrestrial Planets*, 1286 pp., Pergamon Press, 1981.

Campbell, I. H., The mantle's chemical structure: insights from the melting products of mantle plumes, *The Earth's Mantle*, edited by I. Jackson, Cambridge Univ. Press, 259-310, 1998.

Cordery, M.J., G.F. Davies, and I.H. Campbell, Genesis of flood basalts from eclogite-bearing mantle plumes, *J. Geophys. Res., 102,* 20179-20197, 1997.

Davies, G.F., Mantle plumes, mantle stirring and hot spot chemistry, *Earth Planet. Sci., Lett., 99,* 94-109, 1990.

Doell, R.P., and G. B. Dalrymple, Potassium-argon ages and paleomagnetism of the Waianae and Koolau Volcanic Series, Oahu, Hawaii. *Geol. Soc. Amer., Bull, 84,* 1217-1242, 1973.

Frey, F.A., and J. M. Rhodes, Intershield geochemical differences among Hawaiian volcanoes: implications for source compositions, melting process and magma ascent paths, *Phil. Trans. R. Soc. Lond. A, 342,* 121-136, 1993.

Frey, F.A., M. O. Garcia, and M. Roden, Geochemical characteristics of Koolau Volcano: Implication of intershield geochemical differences among Hawaiian volcanoes, *Geochim. Cosmochim. Acta, 58,* 1441-1462, 1994.

Garcia, M. O., Submarine picritic basalts from Koolau volcano, Hawaii: implications for parental magma compositions and mantle source, *this volume,* 2001.

Guillou, H., M. O. Garcia, and L. Turpin, Unspiked K-Ar dating of young volcanic rocks from the Loihi and Pitcairn seamounts, *Journal of Volcanology and Geothermal Research, 78,* 239-249, 1997.

Halliday, A. N., Unmixing Hawaiian cocktails, *Nature, 399,* 733-734, 1999.

Haskins, E.H., and M. O. Garcia, Scientific drilling of a Hawaiian volcano: Does Koolau have a skin? *EOS Transaction of American Geophysical Union, 81, No.48,* F.1349, 2000.

Hauri, E. H., Major-element variability in the Hawaiian mantle plume, *Nature*, *382*, 415-419, 1996.

Hirose, K., and I. Kushiro, Partial melting of dry peridotites at high pressures: determination of compositions of melts segregated from peridotite using aggregates of diamond, *Earth Planet Sci., Lett.,* 114, 477-489, 1993.

Hirschmann, M. M., and E. M. Stolper, A possible role for garnet pyroxenite in the origin of the "garnet signature" in MORB, *Contrib. Min. Petrol., 124,* 185-205, 1996.

Jackson, M. C., F. A. Frey, M. O. Garcia, and R. A. Wilmoth, Geology and geochemistry of basaltic lava flows and dikes from the Trans-Koolau tunnel, Oahu, Hawaii, *Bull. Volcanol., 60,* 381-401, 1999.

Kogiso, T., K. Hirose, and E. Takahashi, Melting experiments on homogeneous mixtures of peridotite and basalt: application to the genesis of oceanic island basalts, *Earth Planet. Sci. Lett. 162,* 45-61, 1998.

Langenheim, V. A. M., and D. A., Clague, The Hawaiian-Emperor volcanic chain, Part II. Stratigraphic framework of volcanic rocks of the Hawaiian islands, *Volcanism in Hawaii, U.S.G.S. Professional Paper 1350*, 1987.

Lassiter, J. C., and E. H. Hauri, Osmium-isotope variations in Hawaiian lavas: evidence for recycled oceanic lithosphere in the Hawaiian plume, *Earth Planet. Sci. Lett., 164,* 438-496, 1998.

Li, X., R. Kind, K. Priestley, S. V. Sobolev, F. Tilmann, X. Yuan, and M. Weber, Mapping the Hawaiian plume conduit with converted seismic waves, *Nature, 405,* 938-941, 2000.

Matsukage, K., K. Kubo, and E. Takahashi, Melting of dry peridotite KLB-1 revisited: Precise determination of phase relations and composition of solids. *EOS, Transaction of American Geophysical Union, 81,* No.48, F1383, 2000.

McKenzie, D. P., The generation and compaction of partially molten rock, *J. Petrol., 25,* 713-765, 1984.

Niu, F., H. Inoue, D. Suetsugu, and K. Kanjo, Seismic evidence for a thinner mantle transition zone beneath the South Pacific Superswell, *Geophys. Res. Lett., 27,* 1981-1984, 2000

Norman, M. D., and M. O. Garcia, Primitive magmas and source characteristics of the Hawaiian plume: petrology and geochemistry of shield picrites, *Earth Planet. Sci., Lett., 168,* 27-44, 1999.

Ohta, H., S. Maruyama, E. Takahashi, Y. Watanabe, and Y. Kato, Field occurrence, geochemistry and petrogenesis of the Archean Mid-Oceanic Ridge Basalts (AMORBs) of the Cleaverville area, Pilbara Craton, Western Australia, *Lithos, 37,* 199-221, 1996.

Ribe, N.M., and U.R. Christensen, The dynamical origin of Hawaiian volcanism, *Earth Planet. Sci., Lett., 171,* 517-531, 1999.

Shen, Y., S. Solomon, I. Bjarnason, and C. Wolfe, Seismic evidence for a lower-mantle origin of the Iceland plume, *Nature, 395,* 62-65, 1998.

Shinozaki, K., Z.-Y. Ren, and E. Takahashi, Geochemical and Petrological characteristics of Nuuanu and Wailau landslide blocks, *this volume,* 2001.

Silver, P.G., R.W. Carlson, and P. Olson, Deep slabs, geochemical heterogeneity, and the large-scale structure of mantle convection: Investigation of an enduring paradox, *Ann. Rev., Earth Planet. Sci., 16,* 477-541, 1988.

Sobolev, A.V., A.W. Hofmann, and I. Nikogosian, Recycled oceanic crust observed in 'ghost plagioclase' within the source of Mauna Loa lavas, *Nature, 404,* 986-990, 2000.

Takahashi, E., melting study of a dry peridotite KLB-1 up to 14 GPa: implications on the origin of peridotitic upper mantle. *J. Geophys. Res., 91,* 9367-9382, 1986.

Takahashi, E., and I. Kushiro, Melting of a dry peridotite at high pressures and basalt magma genesis, *Amer. Mineral., 68,* 859-879, 1983.

Takahashi, E., and R. Takeguchi, How hot are the mantle plume beneath Hawaii? *EOS, Transaction of American Geophysical Union, 79, No.45.* F587, 1998.

Takahashi, E., K. Nakajima, and T. L. Wright, Origin of the Columbia River basalts: melting model of a heterogeneous plume head, *Earth Planet. Sci. Lett., 162,* 63-80, 1998.

Takahashi, E., T. Shimazaki, Y. Tsuzaki, and H. Yoshida, Melting study of a peridotite KLB-1 to 6.5 GPa and the origin of basaltic magmas, *Phil. Trans. R. Soc. London, A342,* 105-120, 1993.

Tanaka, R., E. Nakamura, and E. Takahashi, Geochemical characteristics of Koolau volcano and Nuuanu and Wailau landslide deposits, Hawaii: Implication for the evolution of Hawaiian plume, *this volume,* 2001.

Watson, S., and D.P. McKenzie, Melt Generation by plumes: a study of Hawaiian volcanism. *J. Petrol., 32,* 501-537, 1991.

White, R. S., and D. P. McKenzie, Mantle plumes and flood basalts, *J. Geophys. Res., 100,* 17543-17585, 1995.

Williams, Q., J. Revenaugh, and E.Garnero, A correlation between ultra-low basal velocities in the mantle and hot spots, *Science, 281,* 546-549, 1998.

Wolfe, E. W., and J. Morris, Sample data for the geologic map of the island of Hawaii, *Miscellaneous Investigation Series, U.S. Geol. Survey,* pp.51, 1996.

Yasuda, A., T. Fujii, and K. Kurita, Melting phase relations of an anhydrous mid-ocean ridge basalt from 3 to 20 GPa: implications for the behavior of subducted oceanic crust in the mantle, *J. Geophys. Res., 99,* 9401-9414, 1994

Yasuda, A., T. Fujii, and K. Kurita, Ascending subducted oceanic crust entrained within mantle plumes. *Geophys. Res. Lett., 25,* 1561-1564, 1998

Yaxley, G. M., and D. H. Green, Reactions between eclogite and peridotite: mantle refertilisation by subduction of oceanic crust, *Schweiz. Mineral. Petrogr. Mitt., 78,* 243-255, 1998.

Zhao, D., F. Ochi, and E. Takahashi, Seismic images of mantle plumes under hotspot regions, *this volume,* 2001.

Zhang, J., and C. Herzberg, Melting experiments on an anhydrous peridotite KLB-1 from 5.0 to 22.5 GPa, *J.Geophys. Res., 99,* 17729-17742, 1994.

Eiichi Takahashi, Earth and Planet. Sci., Tokyo Institute of Tech, 2-12-1, Ookayama, Meguroku, Tokyo, 152-8551, Japan (email: etakahas@geo.titech.ac.jp)
(http: //www.geo.titech.ac.jp/takahashilab/takahashilab.e.html)

Katsuji Nakajima, now at Seiko Precision Co, Narashino City, Chiba, Japan.